Analysis of geological structures

Analysis of geological structures

N.J. PRICE

University College, London

J.W. COSGROVE

Imperial College, London

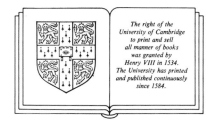

The right of the
University of Cambridge
to print and sell
all manner of books
was granted by
Henry VIII in 1534.
The University has printed
and published continuously
since 1584.

CAMBRIDGE UNIVERSITY PRESS

CAMBRIDGE

NEW YORK   PORT CHESTER   MELBOURNE   SYDNEY

Published by the Press Syndicate of the University of Cambridge
The Pitt Building, Trumpington Street, Cambridge CB2 1RP
40 West 20th Street, New York, NY 10011, USA
10 Stamford Road, Oakleigh, Melbourne 3166, Australia

First published 1990

Printed in Great Britain at the University Press, Cambridge

British Library cataloguing in publication data

Price, N. J. (Neville James), *1926–*
Analysis of geological structures
1. Geological structure
I. Title   II. Cosgrove, J. W. (John W.)
551.8

Library of Congress cataloguing in publication data

Price, Neville J.
Analysis of geological structures.
Includes index.
1. Geology, Structural. I. Cosgrove, J. W. (John W.)
II. Title.
QE601.P694 1989 551.8 88-34089

ISBN 0 521 26581 9 hard covers
ISBN 0 521 31958 7 paperback

Dedication: To JOAN

Contents

Preface

A knowledge of structural geology is fundamental to the understanding of the processes by which the Earth's crust has evolved. It is a subject which is not only of crucial importance to many academic researchers but has great practical value to petroleum and mining engineers and geologists.

Although we attempt to present a reasonably comprehensive review of geological structures, the majority of structures dealt with in this book are in the meso- and macroscopic range (i.e. structures which can be seen in hand specimen and in outcrop). However, the extremes of the size-scale are not completely neglected, for brief mention is also made of micro-mechanics as well as plate-tectonics.

The literature relating to geological structures is so extensive that we would certainly not claim to be encyclopaedic. We have therefore been selective in our approach and this must automatically cause concern to some expert readers who would wish a more detailed treatment of specific aspects. For example 'flower structures' are not specifically mentioned, because they relate only to the geometry of fault complexes usually revealed (or thought to be revealed) in seismic sections above larger faults. They are, however, implicit in our treatment of strike-slip and normal faulting.

In addition, because this book has taken a considerable time to prepare, those chapters which were completed first must, of necessity, omit reference to some of the more recently published work.

Unlike many other texts dealing with structural geology this book does not dwell upon the geometrical description of geological structures. Instead, emphasis is placed on mechanical principles and the way in which they can be used to interpret and understand how and why a wide variety of geological structures develop.

To this end, Chapter 1 presents a relatively elementary treatment of the fundamental principles, most of which are used and sometimes augmented in subsequent chapters of the book.

The arrangement of the book, following Chapter 1, is designed to introduce the reader to the use of these concepts in the gentlest possible manner. Knowing that many geologists are antipathetic to mathematics, and because the interested reader can refer to the original (cited) publications, we do not generally derive the equations we have used in our attempt to apply these principles to the many and varied problems of structural geology. Instead, we have concentrated on establishing the assumptions built into the various analyses and discuss the implications of the many pertinent equations. Such equations relating the interplay of physical parameters are of themselves of very considerable interest. However, it has long been our personal philosophy that their importance in casting light upon geological processes is only ultimately realised when they are quantified in terms of known data relating to such parameters as the angle of friction, elastic moduli, coefficients of viscosity etc. This quantification is rarely a simple task and is one which tends to fill geologists with acute apprehension. However, when such an exercise is possible, we are convinced that it enhances the geologist's understanding of natural processes.

We have made every attempt to present a unified approach to the mechanistic analysis of geological structures that, as far as we are aware, does not contain any mutually contradictory element. We trust that this approach will be welcomed by a wide range of geologists, earth scientists and engineers.

N.J.P. University College, London
J.W.C. Imperial College, London

Acknowledgements

It is difficult to acknowledge fully those who directly or indirectly have influenced us either by personal contact, in lectures, in the field, or by their written work. Indeed the numbers, who include our friends, colleagues and students, are so large that it would be an onerous task to name them. It will be apparent from the cited works within the text that we are indebted to many such people. We must, however, express our indebtedness to Ralph Hancock for his very considerable help in preparing the vast majority of the illustrations.

We are obliged to Shell and B.P. for their permission to include in this text the results of work carried out on their behalf and for their financial support in these tasks. Also we wish to thank the London Brick Company for their permission to refer to certain structures which occur in one of their clay-pits.

In addition we wish to thank various bodies, societies and publishers for their permission to reproduce various diagrams. These are as follows: Allen & Unwin (Fig. 18.25), American Assoc. Petroleum Geology (Figs. 2.18, 4.2, 4.7, 4.10, 4.13, 4.15, 4.17, 4.22, 4.25, 7.6, 7.7, 7.11, 8.8, 8.33(a), 11.3, 11.13, 12.3, 12.8), American Jour. Sci. (Figs. 2.7, 6.13(b), 11.4–5, 11.24, 11.27–29, 17.38), Amer. Geophysical Union (Figs. 4.11, 11.36, 14.9, 14.24), Amer. Inst. Min. Metall. Pet. Eng. (Fig. 4.39), Belgian Geol. Soc. (Figs. 10.2, 18.6, 18.29, 18.38, 18.42), British Association (Fig. 12.34), Butterworths (Fig. 9.16–17), Can. Journ Earth Sci. (Figs. 6.4, 6.19, 13.47, 18.14–15), Can. Petrol. Geol. Bull. (Fig. 7.5), Deutsch Geol. Ges. (Fig. 14.19), Earth & Plan. Sci. (Fig. 8.19), Earth Sci. Review (Fig. 18.40), Economic Geol. (Fig. 10.46), Geol. Assoc. London (Figs. 3.11, 9.7(b), 10.11, 17.10), Geol. Foren Stockholm Fohr. (Fig. 16.31), Geol. Mag. (Figs. 2.17–19, 3.9, 6.9, 9.18–19), Geol. Mijn. (Figs. 8.20, 8.32, 12.37, 17.17), Geol. Mitt. (Fig. 18.5(b), Geol. soc. Amer. (Figs. 3.27, 5.10, 5.11, 5.12, 6.6, 6.12, 7.16, 7.18, 8.17, 8.27–28, 8.30–31, 9.2, 9.4, 10.63, 11.19, 11.22, 11.25–26, 11.38–39, 12.4, 12.12–14, 12.21–23, 12.35, 12.35(c)–(d), 13.31–32, 14.4, 14.23, 16.19, 16.28–29, 17.3, 17.36–37, 17.42), Geol. Soc. China (Fig. 7.4(c), Geol. Soc. France (Figs. 12.33, 13.21, 13.29–30), Geol. Soc. London (Figs. 3.8, 3.18–22, 4.28, 4.48, 6.2, 7.12, 7.17, 7.20–24, 7.27–33, 8.33(b)–(e), 8.36, 9.47–48, 10.66, 11.45, 13.13, 13.15–17, 13.20, 13.37–39, 14.5, 16.54(c), 17.9, 17.17–18, 17.23, 18.20(a), 18.49–58), Geol. Rundsch. (Figs. 12.2, 12.7, 12.9, 16.19), Geol. Soc. S. Africa (Fig. 16.38), Geol. Surv. Canada (Figs. 6.25, 6.27, 6.28, 6.34, 13.33, 17.17), Geomechanics (Fig. 10.20), Geophys, J. R. astr. Soc. (Fig. 5.9), Geotechnique (Fig. 9.30), H.M.S.O. (Figs. 3.36, 4.42), Int. Ass. Sed. (Fig. 6.10), Int. Geol. Congress (Figs. 7.3, 7.4, 15.24, 17.11, 17.30, 17.32), Int. Journ. Rock Mech. Min. Sci. Abs. (Figs. 4.38, 4.40, 9.29), Int. Soc. Rock Mech. (Figs. 9.7–13, 9.25), Journ. Geol. (Figs. 1.64, 4.35, 16.10, 16.27), Journ. Geophys. Res. (Figs. 6.3, 18.16–17, 18.20), Journ. Mech. Phys. Solids (Fig. 16.24), Journ. Structural Geol. (Figs. 2.21–2, 8.40, 10.33–34, 11.43, 11.44, 16.22–23, 16.54(b), 16.55), McGraw–Hill (Figs. 1.18, 1.25, 10.14, 10.24, 10.28, 10.29–32, 11.14, 11.16, 15.26, 16.17–18, 18.39), Nature (Figs. 4.20, 4.56, 6.35, 8.35), Norsk Geol. Tidsskr. (Fig. 16.45), Oliver & Boyd (Figs. 6.1(b), 6.13(a)), Pergamon Press (Figs. 4.49, 4.50), Physics Earth Planet Inter. (Fig. 16.21), Rand McNally (Fig. 18.41), Roy. Soc. London (Figs. 1.46, 1.64, 1.72, 7.8–9, 10.61, 13.14, 15.20, 18.23), Science Progress (Figs. 14.1, 14.3), Tectonophysics (Figs. 2.10, 4.1, 4.44, 5.7, 6.11, 6.14(d), 6.16, 7.35–41, 8.14–16, 8.21–26, 8.29, 10.15, 10.26, 10.64–65, 11.17–18, 11.33–34, 11.37, 11.41–42, 12.5, 12.15–18, 12.20, 12.35(a), 12.38, 12.39–40, 13.12, 13.18–19, 13.34, 15.12, 15.18–19, 16.32, 16.44, 17.39), U.S. Geol. Survey (Figs. 3.30, 3.38), U.S. & Mexico Publ. Co. (Fig. 4.8), Wiley (figs. 8.9, 8.34, 10.49).

1 Fundamental principles

Introduction

In this opening chapter we introduce a variety of topics and basic principles, related to the mechanics of development of geological structures, which will be used throughout the book. As all the various aspects introduced cannot be treated in detail in an introductory chapter, some will be considered in greater detail later in appropriate chapters.

The elements of force and stress, which play such important roles in the mechanics of the development of geological structures, are considered at some length in the first section of this chapter. The derivation of the various stress equations is presented in some detail, so that the reader should quickly become familiar with the concepts used.

The second section is given to a relatively brief exposition of strain which, in the third section, leads to a treatment of stress–strain relationships. In this third section, Linear Elastic theory is dealt with at some length, but Viscosity and Plasticity theory receive relatively brief mention. The topic of fluids in the Earth's crust is then introduced and the importance of the concept of effective stress is then applied to criteria of brittle failure in compressive and also in tensile stress conditions. This is followed by an introduction to certain pertinent aspects of rock mechanics and the way in which various environmental factors and parameters influence the differential stress which can be supported by rock. The final section of the chapter is given to a very brief description of the terms and concepts relating to the various mechanisms of crystal deformation involved in the deformation of rock masses.

Forces and stress

Thermally and gravitationally activated movements within the mantle and crust are the prime causes of the force and stress fields which result in the development of folds, faults and minor structures of various kinds. In order to understand the mechanical processes which give rise to these events, the student of structural geology must have some grasp of the concepts of force and stress. The treatment given in this chapter is extremely elementary and most of the necessary steps in the mathematical arguments are presented in some detail. By such treatment it is hoped that those readers whose natural inclination is to 'skip the maths' may be

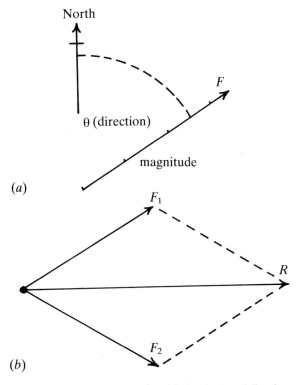

Fig. 1.1. (a) Force: a vector quantity with magnitude and direction. (b) Resolving by the parallelogram of forces.

induced to overcome their aversion and discover how simple these concepts are.

Force

Force is usually defined as any action which alters, or tends to alter, a body's state of rest or uniform motion in a straight line. When a force acts on a body it can be specified completely if one knows (i) its direction of action in space and (ii) its magnitude (Fig. 1.1(a)). Consequently, it is a *vector quantity*.

The magnitude of a force is measured by its effect. Thus, one can measure a force by the weight it will support. In dynamics, which is that aspect of mechanics which deals with motion, the magnitude of a force is measured by the motion it will induce in a given time. This is best seen from Newton's Second Law of Motion which states that 'rate of change of momentum is proportional to the force applied and takes place in the direction in which the force acts'. Thus, if a force F acts on a mass M there is a proportional change in momentum (where momentum is Mv and v is the velocity). As the mass does not change when a force is applied, the applied force produces a change in velocity (V), i.e. it produces an acceleration (a) so that the Second Law can be written as:

$$F = \text{const } Ma.$$

It is convenient to choose units of force such that the constant in this equation equals unity; i.e. $F = 1.0$ when $M = 1.0$ and $a = 1.0$. This fundamental equation then becomes:

$$F = Ma. \tag{1.1}$$

When the unit of mass is the gram, and the units of space and time are the centimetre and the second respectively, the force unit is known as the dyne, which is termed an absolute unit because its value is not dependent upon gravitational attraction. Alternatively, if the acceleration owing to Earth's gravity is defined as $g = 9.81 \text{ m s}^{-2}$, the force exerted by a mass of 1 kg at rest on Earth is 9.81 Newtons.

If two forces act at a single point then, because they are vector quantities, they may be combined graphically (Fig. 1.1(b)) by the parallelogram of forces. Similarly, a single force may be resolved into two or more components. There are, of course, an infinite number of ways in which this may be done, but in most analyses it is necessary, or convenient, to resolve the force into two directions at right angles to one another. Such an example is indicated in Fig. 1.2, in which a rectangular particle, of mass M, is resting on a planar surface inclined at an angle ϕ to the horizontal. The force F generated by this mass (M) in the gravitational field acts vertically as indicated. The component of tractive force (F_t) which acts parallel to the inclined surface is clearly given by:

$$F_t = F\sin\phi. \tag{1.2}$$

Similarly, the component (F_n) acting normal, or perpendicular, to the slope is:

$$F_n = F\cos\phi. \tag{1.3}$$

If the angle ϕ is small, the component of traction is also small and the particle will not slide down the slope, because of the resistance to movement provided by a frictional force. If, however, the angle ϕ is gradually increased, F_t also increases, while F_n decreases. When the angle ϕ reaches the critical value (ϕ_s), the frictional resistance to movement is overcome and the particle slides down the slope. This critical angle is a characteristic of the material, or materials, of which the particle and slope are made. It has been found by experiment, that when one body is in contact with another along a planar surface and these are moved laterally relative to one another, the frictional force tending to prevent motion is proportional to the normal reaction, or force, which acts on the surface along which sliding takes place. This constant ratio is termed the coefficient of dynamic or sliding friction (μ) and for the given example, shown in Fig. 1.2, is obtained from Eqs. (1.2) and (1.3), so that:

$$\mu = F_t/F_n = \frac{F\sin\phi_s}{F\cos\phi_s} = \tan\phi_s. \tag{1.4}$$

As we shall see later, this concept of sliding friction plays an important role in the mechanics of fracture and fault movement.

Stress

If a cube of granite with sides of 25 cm is submitted to an evenly distributed, compressive force of 10 tonnes (i.e. 10 000 kg) only infinitesimal deformation (strain) would be observed in the cube. If, however, the same load were applied to a cube of the same material with sides of one twentieth the length of the larger cube, the smaller granite cube would be pulverised by the action of the force. The magnitude of the applied force is the same in both instances and the difference in the behaviour of the two cubes is the result of the different stresses induced by the force, where the term stress (S) is defined as the force (F) per unit area (A), or:

$$S = F/A. \tag{1.5}$$

With reference to the example given above, the stress in the larger cube was 16 kg cm², i.e. about 16 atmospheres of pressure (16 bar or 1.6 megapascals (MPa)), while in the smaller cube it was 400 times larger at 6400 kg cm², or 6.4 kbar, which greatly exceeded the uniaxial crushing strength of granite.

In the example cited, it was tacitly assumed that the direction of application of force was perpendicular to the surface of the cube, so that there was no component of force acting parallel to the loading surfaces. The corresponding stress which acts perpendicular to a surface is defined as a *principal stress* when the shear stress acting on that surface is zero.

If only one principal stress acts on a body, as indicated in Fig. 1.3(*a*), this is termed *uniaxial compression*. When two or three principal stresses act, as shown in Fig. 1.3(*b*) and (*c*), the conditions are called *biaxial* and *triaxial compression* respectively. It will be

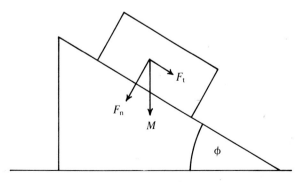

Fig. 1.2. Vertical, layer normal and layer parallel components of force of a mass, M, on a rigid, inclined plane at an angle ϕ.

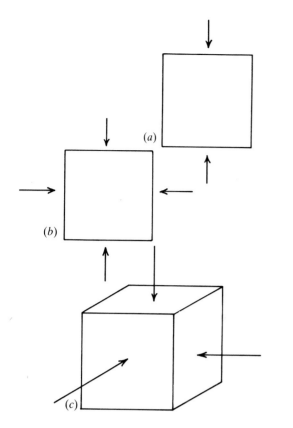

Fig. 1.3. Uniaxial, biaxial and triaxial compression; (*a*), (*b*) and (*c*) respectively.

shown later that the axes of the principal stresses, i.e. the directions in which the principal stresses act, are always at right angles to one another.

If one considers a plane within a body which is inclined to the direction of the applied force, or principal stress (Fig. 1.4), it is apparent that the force F has a component acting normal to, and also one parallel to, the internal plane. There are corresponding normal and shear stresses which are designated S_n and τ respectively.

It is often convenient to refer stresses to a coordinate system in which one of those coordinates is vertical and the other two are in a horizontal plane. Convention has it that the suffix z is used to indicate the vertically-acting stress S_z. The other stresses, oriented parallel to the x and y directions are then S_x and S_y. It is emphasised that, when using this terminology, S_x, S_y and S_z need not be, and in general are not, principal stresses. Shear stresses may also be referred to the coordinate system. The nomenclature used in a biaxial field is indicated in Fig. 1.5. Taken together, the two subscripts indicate the plane in which the shear stresses act; while the last of the subscripts indicates their direction of action.

Some of the earliest tests of materials were conducted on specimens (usually metal) which were subjected to a tensile force, thereby causing an elongation of the test specimen. The extension, and hence the strain, was considered to be a positive quantity. It therefore became the convention among engineers and physicists to regard the tensile stresses associated with this 'positive' strain to be positive also. Conversely, compressive stresses were considered to be negative quantities. Geologists, however, tend to use the opposite convention for, they argue, stresses in the

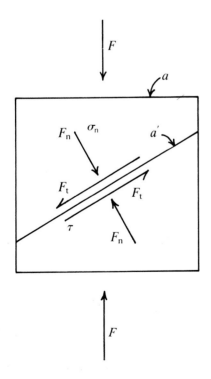

Fig. 1.4. Normal and shear forces on an internal plane within a cube subjected to uniaxial compression by force, F. a = area of cube face and a = area of internal plane.

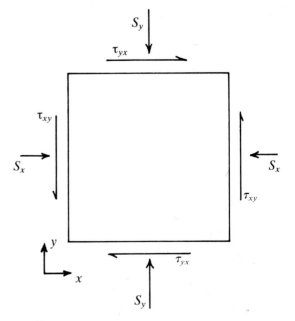

Fig. 1.5. Nomenclature for normal and shear stress in biaxial compression in the x and y directions.

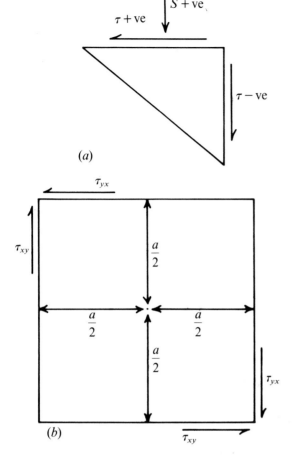

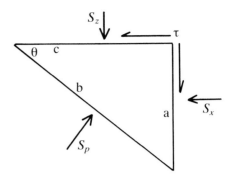

Fig. 1.7. Elemental prism acted on by normal and shear stresses.

(*a*)

(*b*)

Fig. 1.6. (*a*) Sign convention for shear stresses when compressive stresses are positive. (*b*) Moments induced in an element by shear stresses.

Moment = Force × Distance

Clockwise moment about centre O

$$M_c = 2\tau_{yx}aa/2 = \tau a^2$$

Anticlockwise moment about centre O

$$M_a = 2\tau_{xy}aa/2 = \tau a^2$$

element in a clockwise direction. However, the same argument may be applied to the pair of shear stresses τ_{yx}, which provide a couple $\tau_{yx}a^2$ which tends to rotate the element in an anticlockwise direction, but, as it is assumed that the element is in equilibrium, $\tau_{xy} = -\tau_{yx}$. Thus, in two-dimensional problems, only three stresses S_x, S_y and τ are required to define completely the stress system acting on any two-dimensional element.

Earth's crust are usually compressive and it is more convenient to deal with positive quantities. These sign conventions refer specifically to normal stresses. There are also corresponding sign conventions regarding shear stresses. The signs for shear stresses, when compressive normal stresses are taken as positive (the convention used throughout this book) are indicated in Fig. 1.6(*a*).

In analysing mechanical processes pertaining to the development of geological structures, it is usual to assume, when dealing with stresses, that strain is irrotational or that an element is rotating so slowly that one may consider it to be irrotational. If the sides of the element (Fig. 1.6(*b*)) have a length *a* and there is a shear stress (τ) acting on all sides, the shear force on sides AB and CD has a magnitude $\tau_{xy}a$ and because these forces act at a distance *a* from each other they form a couple of $\tau_{xy}a^2$, which tends to rotate the

It can be readily shown that if S_z, S_y and τ acting on two surfaces are known, then it is possible to calculate the orientation and magnitude of the two principal stresses. Consider a small, triangular prism of unit thickness and lengths of sides '*a*', '*b*' and '*c*', as shown in Fig. 1.7, with S_x and S_z acting perpendicular to sides '*a*' and '*c*' respectively. Suppose that side '*c*' makes an angle θ with '*b*' and that the stress S_p acting on '*a*' is wholly normal, then S_p is a principal stress. Because the thickness of the prism is unity, the areas of the sides are respectively *a*, *b*, and *c*. Moreover, it will be evident from Eq. (1.5) that if one multiplies a stress by an area, the resulting quantity is a force (which one can consider to act at the centre of the area). Hence, the force acting on face '*b*' is $S_p b$. Now, both the horizontal and vertical components of the force action on the prism must be in equilibrium. Equating the horizontal components one obtains:

$$S_p b \sin \theta = S_x a + \tau c$$

or

$$S_p \sin \theta = S_x a/b + \tau c/b.$$

But $a/b = \sin \theta$ and $c/b = \cos \theta$, so:

$$S_p \sin \theta = S_x \sin \theta + \tau \cos \theta.$$

Therefore,

$$S_p - S_x = \tau \cot \theta. \tag{1.6}$$

Equating the vertical components one obtains:

$$S_p b \cos \theta = S_z c + \tau a$$

or

$$S_p \cos \theta = S_z c/b + \tau a/b$$

so that

$$S_p \cos \theta = S_z \cos \theta + \tau \sin \theta.$$

Therefore:

$$S_p - S_z = \tau \tan \theta. \qquad (1.7)$$

If one multiplies Eq. (1.6) by Eq. (1.7) (remembering that $\tan \theta \cot \theta = 1.0$), one obtains:

$$\tau^2 = (S_p - S_x)(S_p - S_z)$$

or, by multiplying out and rearranging the terms:

$$S_p^2 - (S_x + S_z) S_p + (S_z S_x - \tau^2) = 0. \qquad (1.8)$$

Eq. (1.8) is a quadratic in S_p and the two roots may be obtained in the usual manner, so that:

$$S_p = \tfrac{1}{2}\{[S_x + S_z] \pm [(S_x - S_z)^2 + 4\tau^2]^{\frac{1}{2}}\}. \qquad (1.9)$$

Thus, provided the roots are real, the values of the two principal stresses can be calculated.

In order to ascertain the orientation of these principal stresses, it is necessary to determine the two corresponding values of θ. To obtain the required relationship between θ and the stresses, subtract Eq. (1.6) from Eq. (1.7), so that:

$$S_z - S_x = \tau(\cot \theta - \tan \theta).$$

It can be shown by simple trigonometrical manipulation that:

$$\cot \theta - \tan \theta = 2 \cot 2\theta$$

therefore

$$S_z - S_x = 2\tau \cot 2\theta$$

or

$$\cot 2\theta = \frac{(S_z - S_x)}{2\tau}. \qquad (1.10)$$

Also, from trigonometrical relationships it is known that $\cot 2\theta = \cot(2\theta - 180°)$ and from Eq. (1.10) it follows that:

$$\cot(2\theta - 180°) = \frac{(S_z - S_x)}{2\tau}. \qquad (1.10a)$$

Therefore, the stress data give rise to two angles θ and $\theta - 90°$ which define the orientation of the principal stresses.

Thus, knowing S_z, S_x and τ, the values and orientations of the corresponding principal stresses can be calculated. Alternatively, if the orientation and magnitude of the principal stresses are known, one can easily calculate the values of the normal and shear stresses acting on any plane which makes an angle θ with the axis of principal stress.

Consider first the uniaxial compression of a rectangular prism, represented in Fig. 1.8. A force F acts perpendicular to the end surfaces of the prism,

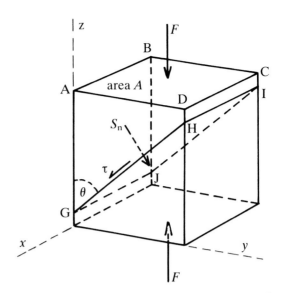

Fig. 1.8. Normal and shear stresses induced on an internal surface in a prism subjected to uniaxial compression.

which has a cross-sectional area A, so that the principal stress (S_z) is given by:

$$S_z = F/A.$$

Now consider an internal surface, GHIJ, which is oriented so that it makes an angle θ with the axis of principal stress. It will be seen that the force F has a component F_n which is normal to the internal plane, so that:

$$F_n = F \sin \theta.$$

Similarly, the force F has a component of traction F_t parallel to the surface, where:

$$F_t = F \cos \theta.$$

It is clear from Fig. 1.8 that the area A' of the internal plane is greater than the area A of the end sections of the prism and that:

$$A' = A/\sin \theta.$$

Using these equations, it is clear that the normal stress (S_n) acting on the internal plane is given by:

$$S_n = F_n/A' = F \sin^2 \theta / A = S_z \sin^2 \theta$$

therefore

$$S_n = S_z \sin^2 \theta. \qquad (1.11)$$

Similarly, the shear stress (τ) on this same plane is given by:

$$\tau = F_t/A' = F \sin \theta \cos \theta / A = S_z \sin \theta \cos \theta$$

therefore,

$$\tau = S_z \sin \theta \cos \theta. \qquad (1.12)$$

These *uniaxial stress equations* (Eqs. (1.11) and (1.12)) give the normal and shear stress on any plane inclined at an angle θ to the principal stress.

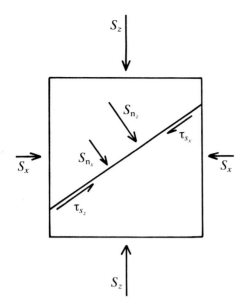

Fig. 1.9. Normal and shear stress components induced on an internal surface of a prism subjected to biaxial compression.

Consider now the condition of biaxial compression (Fig. 1.9). The normal stress S_n acting on the internal surface is compounded from two components of normal stress as the result of S_z and S_x. From Eq. (1.11), the component of normal stress that results from S_z (designated as S_{n_z}) is:

$$S_{n_z} = S_z \sin^2 \theta.$$

Similarly, the component of normal stress (S_{n_x}) that results from S_x is:

$$S_{n_x} = S_x \sin^2(90° - \theta) = S_x \cos^2 \theta.$$

Hence,

$$S_n = S_{n_z} + S_{n_x} = S_x \cos^2 \theta + S_z \sin^2 \theta. \qquad (1.13)$$

If the principal stresses are both compressive and S_z is greater than S_x, then it will be seen that the components of shear stress on the internal plane that result from these two principal stresses will have a different sense. The total shear stress, therefore, will be:

$$\tau = \tau_{S_z} - \tau_{S_x}.$$

Therefore, from Eq. (1.12),

$$\tau = (S_z - S_x)\sin \theta \cos \theta. \qquad (1.14)$$

These two equations (Eqs. (1.13) and (1.14)) are the *biaxial stress equations*.

As we shall see, it is often convenient to express Eqs. (1.13) and (1.14) in terms of the double angle 2θ. It can be shown from elementary trigonometry that:

$$\sin \theta \cos \theta = \frac{(\sin 2\theta)}{2}.$$

Consequently, Eq. (1.14) may be written as:

$$\tau = \frac{(S_z - S_x)}{2} \sin 2\theta. \qquad (1.15)$$

It can also be shown that:

$$\cos^2 \theta - \sin^2 \theta = \cos 2\theta$$

and

$$\cos^2 \theta + \sin^2 \theta = 1.$$

Therefore,

$$\cos^2 \theta = \frac{1 + \cos 2\theta}{2}$$

and

$$\sin^2 \theta = \frac{1 - \cos 2\theta}{2}.$$

Substituting these last two expressions in Eq. (1.13), gives:

$$S_n = \tfrac{1}{2}[(S_z - S_z \cos 2\theta) + (S_x + S_x \cos 2\theta)]$$

or

$$S_n = \frac{(S_z + S_x)}{2} - \frac{(S_z - S_x)}{2} \cos 2\theta. \qquad (1.16)$$

Eqs. (1.15) and (1.16) are particularly important because they lend themselves to graphical solutions of stress problems by the use of a technique developed by Otto Mohr.

Mohr expressed the stress equations (Eqs. (1.15) and (1.16)) graphically by plotting shear stress against normal stress. Knowing the magnitude of the principal stresses, the normal and shear stresses on any plane, with values of θ between 0° and 180°, can be determined using these equations. If the normal and shear stresses for all values of θ are plotted, they form a circle, known as the Mohr's stress circle, Fig. 1.10(a). Because there are no shear stresses acting on a surface perpendicular to an axis of principal stress, the stresses S_1 and S_3 plot on the S_n axis, at points B and A respectively. Clearly, the distance from the origin O to the mid-point C between A and B is $(S_1 + S_3)/2$ and the radius of the Mohr's circle BC $= (S_1 - S_3)/2$. The magnitude of τ_x is represented by XP. It will be seen that XP/XC $= \sin 2\theta$. It will also be noted that XC $= (S_1 - S_3)/2$, therefore the value of τ is as given by Eq. (1.15). Similarly, the magnitude of S_n equals CO minus CP, where:

$$CO = \frac{(S_1 + S_3)}{2}$$

and

$$CP = \frac{(S_1 - S_3)}{2} \cos 2\theta.$$

Thus the value of S_n is that given by Eq. (1.16). Hence, knowing the two principal stresses, points A and B on the diagram enables the Mohr's circle to be con-

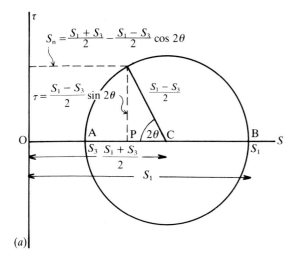

(a)

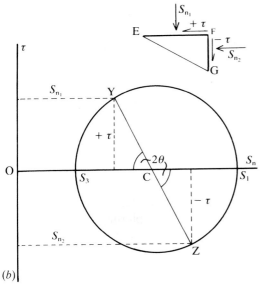

(b)

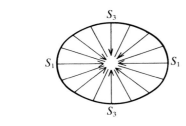

(c)

Fig. 1.10. (a) The Mohr's circle: a graphical representation of Eqs. (1.15) and (1.16). See text. (b) Construction of the Mohr's circle from values of normal and shear stress acting on two surfaces of given orientation, see text. (c) Diagrammatic representation of Eq. (1.13), known as the stress ellipse.

the data for surface FG plot at point Z. (It should be noted that the shear stresses on EF and FG are of opposite sign.) Connect points Y and Z, and where this line cuts the S_n axis, at C, one has the point equivalent to $\frac{1}{2}(S_1 + S_3)$. Also $CY = \frac{1}{2}(S_1 - S_3)$. Hence, if a circle of radius CY is drawn, the values of S_1 and S_3 can be read off. The orientation of the axis of maximum principal stress relative to surface EF can be determined by measuring angle OCY (2θ) and taking half the measured angle.

The properties of stress, and the implication of the stress equations, can be further understood by determining the values of normal stress of various values of θ, using Eq. (1.13) and representing this quantity by an arrow, pointing in the appropriate direction, whose length is proportional to its magnitude. The two principal stresses are at right angles and correspond to the maximum and minimum values of normal stress. As can be seen from Fig. 1.10(c), the normal stress vectors define an ellipse, the *stress ellipse* (ellipsoid in 3-D) whose axes coincide with the principal stresses.

If the principal stresses are equal the resulting situation is sometimes termed a *hydrostatic* stress, because it is the stress experienced by a fluid at rest (however, the term is also widely used to describe the corresponding stresses in a solid). It is often convenient to consider a non-hydrostatic stress as being made up of two parts. These are the *mean stress* ($\bar{S}$) and the *deviatoric stress* (S'). The mean stress is defined as:

$$\bar{S} = \frac{(S_1 + S_2 + S_3)}{3}$$

and, because it is only responsible for volumetric changes, can be considered to be the hydrostatic part of the stress system. The deviatoric stress is defined as:

$$S' = S_n - \bar{S}$$

and is a measure of how much the normal stress in any direction deviates from the mean, or hydrostatic, stress. The deviatoric stress is responsible for distortional deformation. A useful indication of a stress field's ability to cause deformation is the difference in magnitudes between the maximum and minimum principal stresses ($S_1 - S_3$), which is known as the *differential stress*.

Strain

There are a number of reasons why it is useful to determine the state of strain in rocks. For example, strain determination can be used to assess the original stratigraphic thickness of a deformed sedimentary sequence, or to calculate the amount of displacement across a shear zone. Moreover, by studying deformed

structed, and the normal and shear stresses acting on any plane making an angle θ with the maximum principal stress can be determined simply and quickly.

Conversely, if the values of the normal and shear stresses acting on two surfaces of given orientations are known, it is possible to determine the values and orientations of the related principal stresses. Thus, the data for surface EF of the inset diagram shown in Fig. 1.10(b) are plotted (let this define point Y) and objects of known original shape, strain determination may enable one to establish the amount of deformation necessary for the formation of various rock fabrics such as slaty cleavage. In addition, if the strain distribution within and around a geological structure, such as a fold, is known, it can be compared with the strain pattern predicted from, or assumed to exist in, a particular theoretical treatment of folding. The relevance of the theory to the formation of that fold can thereby be assessed.

When a rock is subjected to stress, the particles of the rock are displaced. The types of displacement fall into four categories; two rigid body displacements and two non-rigid body displacements. These are:

(i)   rigid body translation,
(ii)   rigid body rotation,
(iii)   distortion (shape change), and
(iv)   volume changes (Fig. 1.11).

Rigid body translation is the movement of a body through space without any change of shape of that body and in such a way that any line drawn on the body maintains the same orientation throughout the displacement (Fig. 1.11(a)). The displacements parallel to the x and y axes are termed u and v respectively and the resultant displacement vector (R) is the same for all parts of the body.

Rigid body rotation also involves the movement of the body through space without any change in shape of the body, but the displacement vectors of all points on the body are not the same and there is a single 'stationary' point about which the body rotates (Fig. 1.11(b)).

Distortion involves the movement of the particles with respect to each other and causes a change in shape of the body (Fig. 1.11(c)). The displacements of the particles are not the same and *displacement gradients* are set up, the magnitude of which gives a measure of the distortion or *strain*. For example, the strain e_x (extension) in the x direction is defined as a displacement gradient:

$$e_x = \frac{du}{dx}. \qquad (1.17)$$

Volume changes cause no change in shape, and although often termed *dilation*, can be either positive (Fig. 1.11(d)) or negative.

Displacement in rocks generally involves all four types but frequently one type of behaviour is dominant. For example, the movement on an overthrust may be predominantly rigid body translation but locally, at or near the decollement horizon, considerable distortion of the rock mass can occur. Taking another example, volume changes dominate during compaction and the dewatering of sediments associated with burial and diagenesis.

All the types of displacements shown in

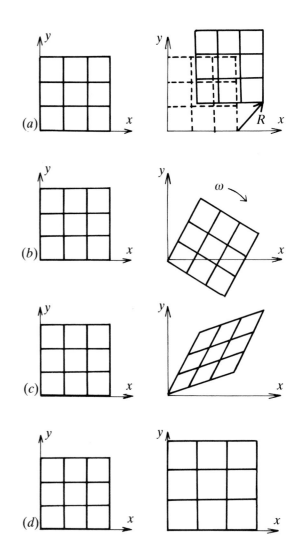

Fig. 1.11. Four types of displacement. (a) Rigid body translation. (b) Rigid body rotation. (c) Distortion (shape change). (d) Volume change.

Fig. 1.11, can be represented mathematically, singly or in combination, by displacement or transformation equations which relate the undisplaced body to the displaced body by giving new coordinates for a point (x', y') in terms of the old coordinates (x, y). For example, the transformation equations for rigid body translation are:

$$x' = x + u \qquad (1.18)$$

$$y' = y + v \qquad (1.19)$$

when u and v are constants.

Generally, it is not possible to determine the exact amount of rigid body displacement that a rock has experienced since its formation. However, if it contains objects of known, original shape or size, then the strain in the rock can be measured precisely. A considerable amount of effort has been directed towards determining the state of strain in rocks, and for a more detailed discussion of strain and techniques of measuring it, the reader is referred to books by

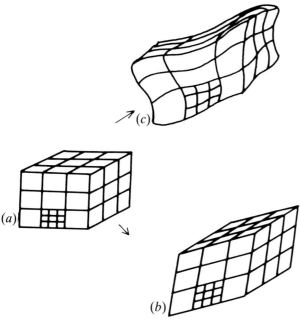

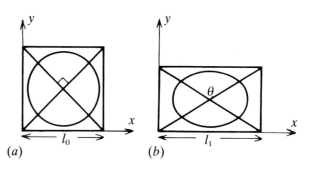

Fig. 1.13. Diagram showing the properties of homogeneous strain. See text for discussion. (*a*) Undeformed state. (*b*) Deformed state.

Fig. 1.12. (*a*) Undeformed body. (*b*) Homogeneously deformed body. (*c*) Heterogeneously deformed body. (One square is divided into small elements over which the strain can be assumed to be approximately homogeneous.)

Ramsay (1967), Means (1976) and Ramsay & Huber (1983). The following brief discussion is presented to cover aspects of the subject commented on and used in this book.

Strain is homogeneous where it is constant throughout a body and heterogeneous where the displacement-gradient varies across a body (Fig. 1.12). Although strain in rocks is generally non-uniform, it is possible to consider a small domain over which the strain is approximately homogeneous. By determining the states of strain in these domains, the more complex, heterogeneous strain state of a region can be inferred. The size of the domains over which the strain can be considered to be homogeneous varies enormously. For example, in a slate belt the domain may extend over many kilometres, whereas in a folded region, the domains may be only a few metres, or even less, in extent.

The sub-division of a heterogeneously strained region into smaller domains across which the strain is approximately homogeneous considerably simplifies the problem of strain determination. This problem can be simplified even further by considering strain in only two dimensions, i.e. on a rock face. The state of three-dimensional strain can be subsequently determined by combining two-dimensional strain data from three non-parallel faces.

During two-dimensional, homogeneous strain, straight lines remain straight, parallel lines remain parallel and a circle is deformed into an ellipse, the strain ellipse (Fig. 1.13). However, the length of a line and the angle between two lines generally change. Changes in angle indicate the amount of shear defor-

mation in a particular direction and changes in length give a measure of the elongation in that direction. A variety of parameters have been used to measure these changes.

Changes in the length of a line

There are several parameters used to measure changes in the length of a line and each has its own advantage. The most commonly used is the extension (e), which is defined as:

$$e = \frac{(l_1 - l_0)}{l_0} = \frac{dl}{l_0} \tag{1.20}$$

where l_0 and l_1 are the original and final length of a line. A more complex parameter used to measure length changes is λ the quadratic elongation:

$$\lambda = (1 + e)^2. \tag{1.21}$$

As we shall see, this parameter is used because it enables the strain equations (Eqs. (1.31) and (1.32)) to be expressed more simply.

Natural or logarithmic strain ϵ defined as:

$$\epsilon = \log_e(1 + e) \tag{1.22}$$

also gives a measure of the change in length of a line. Ramsay (1967) points out that Eq. (1.22) expresses the changes in length more realistically than e. For example, in a comparison of the two parameters along two lines, one of which is contracted to half its original length and the other expanded to twice its original length, the extensions, e, of the two lines are: $e = -0.5$, and 1.0 respectively, whereas the natural strains are $\epsilon = -\log_e 2$ and $\epsilon = +\log_e 2$. With very great contractions, e approaches -1, whereas ϵ approaches $-\infty$.

Changes in the angle between lines

The change in angle between two lines that originally intersected at right angles is known as the *angular shear* (ψ) (Fig. 1.15) and the *shear strain* γ is defined as:

$$\gamma = \tan \psi. \tag{1.23}$$

The angular shear and shear strain along a line may be either positive or negative depending upon whether

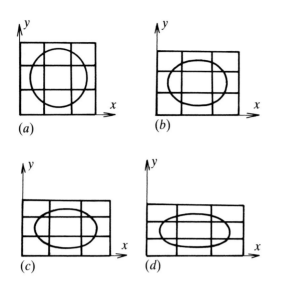

Fig. 1.14. (*a*) Undeformed body. (*b*)–(*d*) Various stages in pure shear deformation. Such deformations involve no area change.

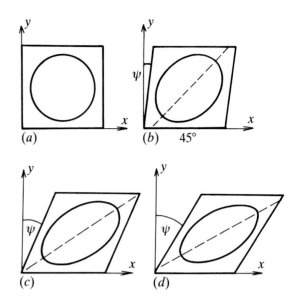

Fig. 1.15. (*a*) Undeformed body. (*b*)–(*d*) Stages in simple shear deformation, involving no area change.

the original normal to the line is deflected to the right (Fig. 1.15) or left respectively.

Two types of homogeneous strain that have been studied extensively by structural geologists are *pure shear* and *simple shear*. Extension parallel to the *x* axis involving no area change (Fig. 1.14) is an example of pure shear and the transformation equations representing this deformation are:

$$x' = (1 + e_x)x. \tag{1.24}$$

$$y' = (1 + e_y)y. \tag{1.25}$$

where e_x and e_y are related by $(1 + e_x) = 1/(1 + e_y)$.

The deformed state (Fig. 1.14(*d*)) is a state of finite strain which can be considered as being made up of a large number of incremental or *infinitesimal strains* (Fig. 1.14). The circle representing the undeformed state is deformed into an ellipse, the *strain ellipse*, whose major and minor axes indicate the magnitude and orientation of the maximum and minimum principal strains. During pure shear deformation, the axes of the strain ellipse do not rotate and the incremental and finite strain ellipses are coaxial. Such a deformation is known as an *irrotational* deformation. Simple shear results when a body is subjected to a uniform shear, parallel to some direction, involving no area change (Fig. 1.15) and the transformation equations that describe this deformation are:

$$x' = x + y \tan \psi \tag{1.26}$$

$$y' = y. \tag{1.27}$$

It can be seen from Fig. 1.15 that the axes of the strain ellipse rotate during simple shear. Such a deformation is known as a *rotational* deformation and except during the first increment of strain, the incremental and finite strain ellipses are non-coaxial.

The two sets of transformation equations (Eqs. 1.24 to 1.27) represent two specific examples of two-dimensional, homogeneous strain. They are both examples of the more general transformation equations:

$$x' = ax + by \tag{1.28}$$

$$y' = cx + dy \tag{1.29}$$

which represent any two-dimensional homogeneous strain.

Although it can be argued that geological deformations generally involve area changes (volume in 3-D) and that therefore the models of pure and simple shear deformations are geologically unrealistic, field observations show that deformation closely related to these two types of deformation (i.e. homogeneous flattening involving contraction in one direction and extension at right angles to this direction and localised shear deformation) are extremely common. For example, the deformation within large areas of some slate belts approximates very closely to homogeneous flattening. Alternatively, the deformation in other areas is found to be localised along bands of high shear strain. The development of 'shear' instabilities in rocks is discussed extensively in this book in the chapter on faulting and folding and it is instructive to consider the properties of simple shear in a little more detail.

Although we are interested in the effects of simple shear on homogeneous, isotropic material, the properties of this deformation can be studied conveniently with the aid of card deck models (Fig. 1.16). After the first increment of shear parallel to the cards, the original, undeformed circle is transformed into an ellipse whose axes are inclined at 45° to the shearing direction (Fig. 1.16(*b*)). The amount of shear strain

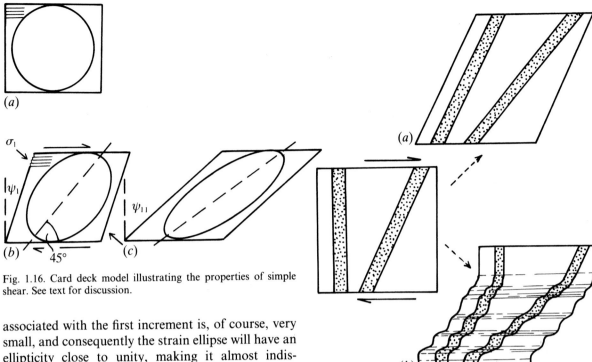

Fig. 1.16. Card deck model illustrating the properties of simple shear. See text for discussion.

Fig. 1.17. Deformation of two passive markers by (*a*) homogeneous, simple shear and (*b*) heterogeneous, simple shear.

associated with the first increment is, of course, very small, and consequently the strain ellipse will have an ellipticity close to unity, making it almost indistinguishable from the original circle. In Fig. 1.16(*b*), the ellipticity has been exaggerated for clarity.

In isotropic materials, the incremental strain ellipse and the stress ellipse are coaxial and it follows, therefore, that during simple shear, the principal compression direction is inclined at 45° to the shear zone margin (Fig. 1.16(*b*)). The minor structures, such as tension gashes and rock fabrics, that form within a shear zone in response to this stress are discussed in Chaps. 2, 6 and 18.

The shear strain across a shear zone formed by homogeneous simple shear is constant (Fig. 1.17(*a*)), whereas in most geological examples the amount of shear across a shear zone varies considerably. Such deformation is known as heterogeneous simple shear and the effects of homogeneous and heterogeneous simple shear on passive marker bands are shown in Fig. 1.17.

Graphical representation of two-dimensional strain

A convenient way of graphically representing a state of two-dimensional strain is to plot the two principal strains (i.e. the two axes of the strain ellipse) or λ_1 and λ_2 against each other (Fig. 1.18(*a*)). By definition $\lambda_1 > \lambda_2$ and therefore all states of strain fall to the right of a line inclined at 45° and which passes through the origin. The undeformed state, which is represented by a circle of unit diameter plots on this line at coordinates (1, 1) and the finite strain in Fig. 1.14(*d*), which resulted from pure shear deformation, plots at point X. If the intermediate states of strain between Fig. 1.14(*a*) and (*d*) are plotted on the graph, a trail, or path, of points between the undeformed state and final

strain state results. This is known as a *deformation path*.

The strain field shown in Fig. 1.18(*a*) can be divided into three sub-fields, 1, 2 and 3, (Fig. 1.18(*b*)) on the basis of the relationship between the deformed and undeformed state. Let us consider these relationships in turn and examine the types of structures which might develop in layers deformed in these three fields (which we assume are oriented so that the two principal strains lie in the plane of the layer).

In Field 3, the finite strain ellipse fits inside the undeformed circle, indicating that there has been a contraction in all directions. The structures that form in a layer subjected to such a strain field will be folds with axes in various directions (Fig. 1.18(*c*)). In Field 1, the undeformed circle lies inside the strain ellipse, showing that extension has occurred in all directions. Boudins or pinch-and-swell structures with necks which trend in various directions will be formed in a layer subjected to this strain field (Fig. 1.18(*c*)). In Field 2, the superposition of the undeformed and deformed states shows that extension occurs parallel to one principal strain and contraction parallel to the other. Thus, a layer subject to this strain field might develop fold axes in one direction with boudin necks at right angles to them (Fig. 1.18(*c*) and Chap. 16). Because the undeformed circle and strain ellipse 'intersect' in this strain field, there are two lines (planes in 3-D) which have the same length before and after deformation. These lines, which are termed

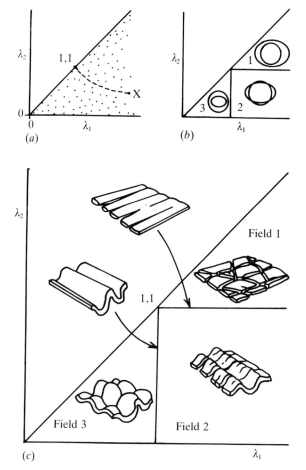

Fig. 1.18. (*a*) Graphical representation of two-dimensional strain coordinate (1, 1) in the undeformed state and X the deformed state. The dotted line represents the deformation path. (*b*) Subdivision of the strain field of (*a*) into three subfields. (*c*) Various structures associated with the 3 fields are shown. (After Ramsay, 1967.)

Lines of No Finite Longitudinal Strain (L.N.F.L.S.), separate areas of finite extension from areas of finite contraction (Fig. 1.19). It is interesting to note that the formation of simple boudins and folds is restricted to the strain states that straddle the boundaries between Fields 1 and 2 and 2 and 3 respectively.

For a state of two-dimensional strain to be completely represented, it is necessary to know the amount of rotation, ω, that has occurred (Fig. 1.11(*b*)). This can be represented on a three-dimensional graph with two axes, as shown in Fig. 1.18, and a third (at right angles to this) along which rotation is plotted (Fig. 1.20). The plane defined by the deformation path of pure shear, AX, and a line drawn from the undeformed state parallel to the rotation axis, is the plane on which all strains involving no area change (i.e. plane strains) will plot. The deformation path for simple shear (Fig. 1.15) therefore falls on this plane and is shown as a dotted line on Fig. 1.20.

The graphical representation of three-dimensional strain can be achieved by using the technique first

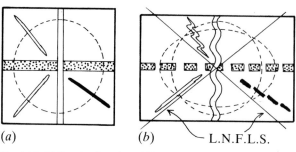

Fig. 1.19. (*a*) Randomly oriented lines in the undeformed state. (*b*) Lines of no finite, longitudinal strain separating region of extension from region of contraction

used by Zing (1935) and introduced into structural geology by Flinn (1962). It involves plotting the ratio, *a*, of the maximum and intermediate principal strains $(1 + e_1)/(1 + e_2)$ against the ratio, *b*, of the intermediate and minimum principal strain $(1 + e_2)/(1 + e_3)$ (Fig. 1.21(*a*)). On this graphical representation, which is known as a Flinn diagram, plane strain deformations, such as pure and simple shear, plot along a line inclined at 45° passing through the point (1, 1). This line separates *constrictional* deformation in which contraction parallel to at least two principal strain axes occurs and *flattening* deformation in which extension occurs parallel to two principal strain axes and contraction parallel to the third.

Ellipsoids associated with constrictional and flattening deformations are shown in Fig. 1.21(*b*), together with the structures that might form in a layer oriented to contain any two of the principal strains. The shape of an ellipsoid representing a particular state of strain can be conveniently indicated by giving the slope of the line joining the origin to the point on the graph representing the ellipsoid. The slope is given the symbol *k*, where:

$$k = \frac{(a - 1)}{(b - 1)}. \tag{1.30}$$

For pure constrictional deformation $k = \infty$ and the ellipsoid is described as being *prolate* or cigar-shaped.

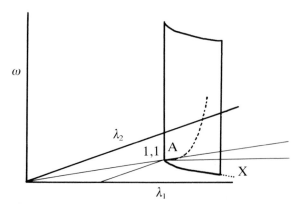

Fig. 1.20. Modification of the two-dimensional graph of Fig. 1.18 to include rotation (ω). See text for details.

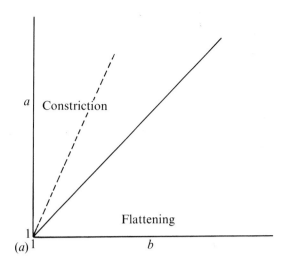

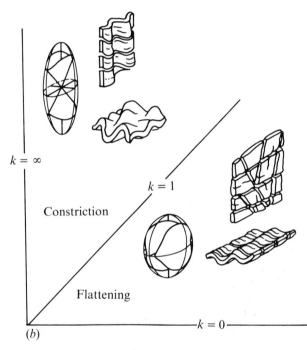

Fig. 1.21. (*a*) Graphical representation of three-dimensional strain, '*a*' is the ratio of major to intermediate axes of the strain ellipsoid and '*b*' the ratio of intermediate to minor axes. (*b*) Shape of the strain ellipsoids associated with constrictional and flattening deformation together with the shape of the surface of no finite, longitudinal strain and the structures that may form in the layer oriented to contain any two of the principal strains. (From Nicolas, 1984 after Brun, 1981.)

For pure flattening deformation, $k = 0$ and the ellipsoid is described as being *oblate* or pancake-shaped. Plane strain deformations which involve no volume change plot along the line $k = 1$.

Because it is the ratios of the principal strains that are plotted against each other on a Flinn diagram, it is not possible to record volume change that may accompany a deformation. However, by plotting $\log_e a$ against $\log_e b$, where a and b are as defined in Fig. 1.21(*a*), the Flinn diagram can be modified to enable volume changes to be represented (Ramsay & Wood, 1973).

The close relationship between stress and strain will already be apparent. Both are tensor quantities and can be represented in two dimensions by an ellipse and three dimensions by an ellipsoid. In addition, the strain equations (Eqs. (1.31) and (1.32)), which give the state of extensive and shear strain along, and normal to, a line at an angle θ to the maximum principal extension, are identical in form to the stress equations (Eqs. (1.13) and (1.14)) which give the state of normal and shear stress on a line at θ to the maximum principal compression.

$$\lambda' = \lambda'_1\cos^2\theta + \lambda'_2\sin^2\theta \tag{1.31}$$

$$\gamma' = (\lambda'_2 - \lambda'_1)\sin\theta\cos\theta \tag{1.32}$$

where λ' is the reciprocal of λ and is therefore known as the *reciprocal quadratic elongation* and γ' is γ.

Strain, like stress, can be represented graphically on a Mohr diagram (Fig. 1.22), in which γ' is plotted against λ'. Using Eqs. (1.31) and (1.32), the values of γ' and λ' for values of θ between 0° and 180° can be determined. When these data are plotted onto the graph, they describe a circle, the Mohr strain circle, which can be directly compared to the Mohr stress circle (Fig. 1.10). Any point A on the circle represents the state of strain (γ'_A, λ'_A) along a line inclined at θ to the principal extension direction λ_1. Moreover, the slope of the line joining point A to the origin gives the value of angular shear (ψ_A) along the line of interest.

Determination of strain

Techniques for determining the state of strain in rocks generally rely on the rocks possessing some object of known original size or shape. These objects are termed strain markers and the majority fall into one of four categories.

(i) Objects that were originally circular (spherical in 3-D).
(ii) Objects that were originally elliptical (ellipsoidal in 3-D).
(iii) Objects of known original size (length).

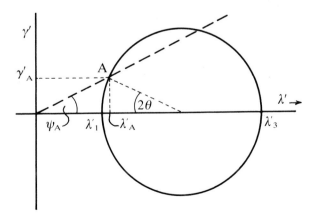

Fig. 1.22. Representation of strain on a Mohr diagram. See text for discussion.

(iv) Objects that originally possessed bilateral symmetry.

Only a brief outline is given here of how strain may be determined from these types of markers. The interested reader is referred to the excellent text by Ramsay & Huber (1983), where strain and strain determination are discussed in detail.

Strain determination from originally spherical markers

Spherical strain markers will be deformed into ellipsoids in a homogeneous strain field and these will appear as ellipses on any surface cutting the deformed rock (Fig. 1.23(a)). The long axes of the ellipses will all be parallel even for quite low values of finite strain. The ellipses represent the state of strain on the specific surface of interest and the ratio of the maximum and minimum extensions can be obtained by measuring and comparing the axes of any one of the ellipses. It is usual to measure a number of ellipses and plot the length of the axes against each other (Fig. 1.23(b)). The slope of the best-fit line passing through these points, tan θ, is the ratio of the long to short axis and therefore of the maximum and minimum principal strains. If it is known that the deformation involved no change in area (plane strain), then the actual values of the principal strains can be determined by measuring the area of an ellipse and calculating the diameter of the circle with the same area. The diameter of the circle is the original length and the lengths of the axes of the ellipse are the final lengths.

Strain determination from originally ellipsoidal markers

We noted above that ellipses resulting from the deformation of circular markers all have their long axes parallel to the maximum principal extension direction (Fig. 1.23(a)). This would not be so if the markers were originally elliptical, unless the long axis of the undeformed ellipse coincided with the principal extension direction (Fig. 1.24(b) (i)). Generally, there will be some angle (ϕ) between the long axis of the deformed elliptical particle and the principal exten-

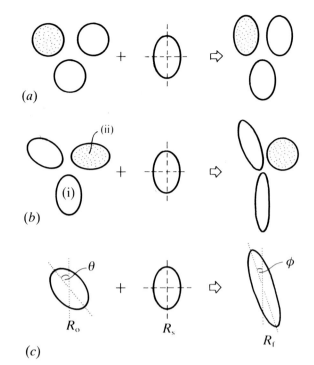

(a)

(b)

(c)

Fig. 1.24. (a) Effect of strain on circular markers. (b) Effect of strain on elliptical markers. (c) Effect of strain, R_s, on an elliptical marker, R_o, inclined at θ to the principal extension direction. The deformed marker (R_f) is inclined at ϕ to the extension direction where ϕ is the fluctuation.

sion direction. This angle, known as the *fluctuation*, decreases with increase in strain.

The problem of determining strain from originally ellipsoidal particles is summarised in Fig. 1.24(c), which shows an original elliptical marker with an axial ratio R_o inclined at an angle θ to the principal extension direction of an applied strain, axial ratio R_s. The result of the deformation is an elliptical particle, axial ratio R_f, inclined at an angle ϕ to the principal extension direction. The problem is to determine R_o and R_s from the deformed state, R_f.

It can be seen from Fig. 1.24(b) and (c), that the final shape, R_f, depends on the relative orientation of the original particle and the applied strain. When they constructively interfere (Fig. 1.24(b) (i)), the resulting ellipse has the greatest eccentricity and when they destructively interfere (Fig. 1.24(b) (ii)), the resulting ellipse has the least eccentricity. The axial ratio of these two extreme shapes are related to R_o and R_s by:

$$R_{f_{min}} = R_s/R_o \qquad (1.33)$$

$$R_{f_{max}} = R_s R_o. \qquad (1.34)$$

To determine the strain, R_s, and original shape, R_o, it is simply necessary to establish $R_{f_{min}}$ and $R_{f_{max}}$ and thus solve the equations. This can be achieved by using the following procedure. On a rock face (or photograph) containing deformed, originally elliptical, strain markers (Fig. 1.25(a)), measure the axial ratio R_f of each ellipse (long/short axis) and the orientation θ of the long axis with respect to some arbitrary datum line, and plot each ellipse on a graph of R_f against θ.

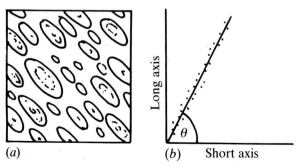

(a)   (b)   Short axis

Fig. 1.23. Determination of strain from markers that were originally circular (spherical in 3-D). See text.

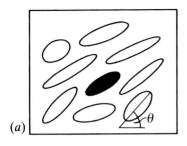

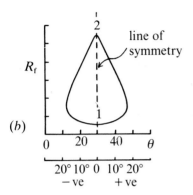

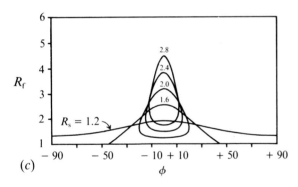

Fig. 1.25. (*a*) Homogeneous deformation of variously oriented ellipses. The black ellipse represents the strain ellipse. (*b*) Graph of the axial ratio, R_f, of the deformed ellipse against θ the orientation of the major axis with respect to an arbitrary reference line. (*c*) R_f/ϕ curves for a particular initial shape, R_o, for a range of applied strains R_s (1.2–2.8). (After Ramsay, 1967.)

If the original particles all had the same shape and did not possess a preferred orientation, the points on the graph would define a closed curve, symmetric about a particular value of θ(Fig. 1.25(*b*)) which is the angle between the datum line and the principal extension direction. The fluctuation of any ellipse can be measured from the graph, and is the angle between the line of symmetry of the curve and the point on the curve representing the deformed ellipse. The maximum and minimum values of R_f (points 2 and 1, Fig. 1.25(*b*)) can be read off from the graph and, using Eqs. (1.33) and (1.34), R_o and R_s can be determined.

If the magnitude of the original shape, R_o, is greater than the applied strain, R_s, then the data plot does not form a closed curve. Instead, a bell-shaped curve results (Fig. 1.25(*c*)). The maximum fluctuation in this example is 180° and the

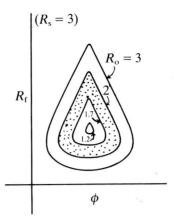

Fig. 1.26. R_f/ϕ curves for various values of original shape R_0.

maximum and minimum values of R_f can be read off as before.

In the example discussed above, it has been assumed that the original particles had the same initial shape, R_o. In practice, however, there is often a range of initial shapes. For any particular value of strain, each original shape (R_o) will have its own curve (Fig. 1.26) and, therefore, if there is a range of initial shapes, the data will plot in an area between two of the curves, the outer curve being followed by the ellipses with the largest R_o values and the inner curve by the ellipses with the smallest R_o values. Ellipses with R_o values between these two fall into the area between the two curves (dotted region, Fig. 1.26). It is therefore possible to use the R_f/ϕ plot to determine the range of original shapes of the deformed particles.

In addition, the existence of a preferred orientation of the undeformed particles can be detected from the R_s/ϕ plot because a preferred orientation causes the symmetric plots of Figs. 1.25 and 1.26 to become asymmetric. More details of this technique can be found in Ramsay (1967), Dunnet (1969), Lisle (1977 and 1985) and Ramsay & Huber (1983).

Strain markers with bilateral symmetry

Strain markers that originally possessed bilateral symmetry (usually fossils) have this symmetry destroyed during deformation unless the axis of symmetry coincides with one of the principal strain directions. The original, right-angular relationship (Fig. 1.27(*a*)) is lost and the change in angle gives a measure of the angular shear (ψ) and, therefore, shear strain (γ) (Eq. (1.23)) along the two lines A and B that were normal to each other. The values of ψ in these two directions have the same magnitude but are of opposite sign.

Strain determination from a marker of known original length

Unlike the strain markers discussed so far, the belemnites shown in Fig. 1.28 do not have the same prop-

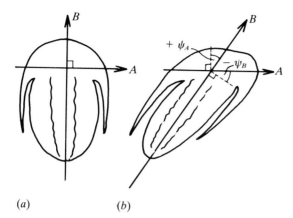

(a)   (b)

Fig. 1.27. Deformation of a strain marker with bilateral symmetry.

erties as the matrix rock. It is assumed that they deform by brittle deformation and that the segments of the belemnites undergo no change in length. If these assumptions are valid, then the length of the segments of belemnite in the deformed rock can be added together to give the original length, l_o, and the final length, l_1, measured directly from the deformed rock. From these data, the extension in the directions (A, B and C) of the belemnite shells can be determined (Eq. (1.20)).

Summary and comments on the use of strain

From the preceding sections, we see that in order to determine the orientation and magnitude of strain on a rock face it is necessary to establish the orientation and magnitude of the strain ellipse on that face. Sometimes this can be done directly by measuring the ellipses produced from originally circular (spherical in 3-D) strain markers, or by the measurements of initially elliptical (ellipsoid in 3-D) strain markers and the application of the R_f/ϕ technique. Alternatively, strain markers can be such that they do not yield the total strain on a particular face, but only the strain (extension and/or shear) in a particular direction. In order to construct the strain ellipse from such data, it is necessary to know the state of strain in more than one direction. However, provided that sufficient strain data are available from deformed objects of known original length or bilateral symmetry, the total strain state can be determined either geometrically

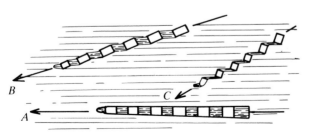

Fig. 1.28. Brittle deformation of belemnite shells. The original and final lengths can be measured and the extension in direction parallel to the shells calculated.

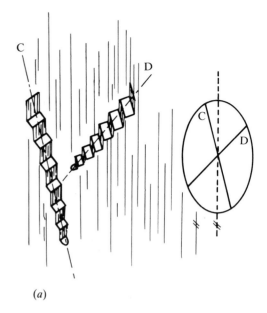

(a)

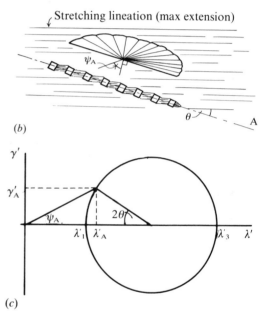

(b)

(c)

Fig. 1.29. (a) Geometric construction of the strain ellipse when the extension in two directions (C and D) is known. (b) Determination of the state of strain on a surface using a combination of extension and shear strain values determined from strain markers and the Mohr diagram, (c).

(Fig. 1.29(a) and (b)), graphically using a Mohr's circle construction (Fig. 1.29(c)) (see Ramsay, 1967; p. 75) or from the strain equations (1.31) and (1.32).

The three-dimensional state of strain can be determined by combining two-dimensional strain data from three non-parallel faces. These three strain ellipses uniquely define an ellipsoid which represents the state of strain in the rock completely.

When determining strain in rocks, it is often implicitly assumed that the strain markers and rock matrix have the same competence (resistance to deformation) and deform in the same way. If this is so, then the strain values determined from the markers give the

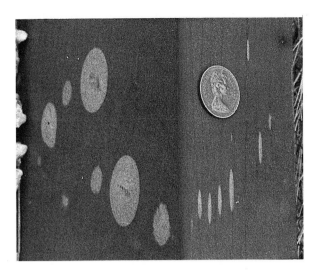

Fig. 1.30. Deformed reduction spots in Cambrian slate, N. Wales. These ellipsoidal spots were originally spherical and represent the 3-D state of strain in the rock. Ellipses on the left hand side are on the cleavage plane, coin is on a joint, normal to the cleavage.

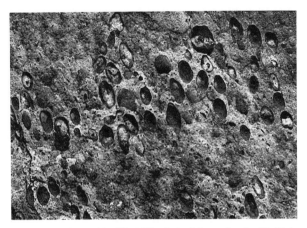

Fig. 1.31. Deformed lapilli tuff in Ordovician volcanics N. Wales. These originally spherical particles are now deformed into ellipsoids which represent the 3-D state of strain in the rock.

bulk strain that the rock has experienced. In practice, of course, differences in mechanical behaviour between a strain marker and its matrix may exist and all gradations between strain markers with identical properties to matrix and markers with very different properties occur. Examples of the former include calcium carbonate oolites in a limestone matrix, reduction spots in slate (Fig. 1.30) and lapilli tuffs (Fig. 1.31).

Reduction spots form around small, organic nuclei which oxidise on decay and reduce the purple (ferric) iron in the surrounding rock to its green (ferrous) state. If the rock is homogeneous and isotropic, then the effect of this reduction will extend the same distance from the nucleus in all directions, producing a spherical region of reduction with the nucleus at its centre. During deformation, this sphere is deformed into an ellipsoid. Lapilli tuffs are another example where strain markers and matrix have identi-

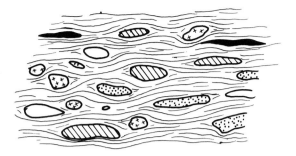

Fig. 1.32. Matrix-supported polymict conglomerate. Irregularity of the cleavage planes indicates large, local variation in strain. See text.

cal mechanical properties. The lapilli were originally spherical-shaped masses of tuffaceous material in a matrix of identical tuff and are thought to form around water droplets as rain falls through an erupting cloud of tuff.

An example of strain markers and matrix having very different mechanical properties and where the strain measured by the markers would not accurately reflect the bulk strain of the rock, would be a matrix-supported polymict conglomerate with a weak matrix. Each pebble type (granite, quartzite, shale etc.) will deform by different amounts and the matrix will deform more than any of the pebbles. The concentration of cleavage in the matrix (Fig. 1.32) and the deflection of the cleavage around the relatively competent pebbles is often observed in deformed conglomerates and is visual evidence of the uneven distribution of strains in the rock.

It is important to remember that the strain values determined for rocks indicate only the strain state of the markers. It is necessary to determine how closely the matrix resembles the marker in its deformation behaviour before commenting on how close the strain value obtained is to the bulk strain of the rock.

The lack of suitable strain markers is a constant source of frustration to those geologists who attempt to determine strain in rocks accurately. Even when a stratigraphic horizon contains markers and the strain state can be determined, it is not generally reasonable to assume that the same state of strain exists in the adjacent layers that do not contain the markers. Nevertheless, it is sometimes possible to extend the strain values measured in a particular horizon a considerable distance on either side of the marker layer. For example, if the strain had been determined in a lapilli tuff set in a thick sequence of identical tuff, the strain value could, with some confidence, be extended beyond the marker layer. Confirmation that the lapilli tuff and the surrounding tuff had identical mechanical properties would be provided if the rock were cleaved and the cleavage passed from the lapilli tuff into the surrounding tuff, and beyond, without refraction (Fig. 1.33).

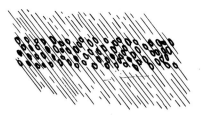

Fig. 1.33. Cleavage passing from lapilli tuff into the adjacent tuff without refraction, indicating that the strain in the lapilli tuff and adjacent tuff are the same.

In general, however, it is necessary to establish whether a determined strain value has regional significance or is related to some local structure. For example, considerable variation in the magnitude and orientation of strain may occur around a fold (Figs. 1.34, 10.23 and 17.38). Clearly, the magnitude and orientation of strain determined from some point in the area shown in Fig. 1.34 will be related to the fold and are, therefore, unlikely to have any regional significance.

Before committing themselves to the considerable amount of work generally involved in strain determination, geologists are advised to think carefully about how they intend to use the data. If their interest lies in the regional distribution of strain there would be little point in determining the strain at a locality where regional strain values are swamped by those associated with local structures.

Whilst acknowledging the use which strain determination can have in geology, it is impossible not to notice the large number of strain determinations described in the literature where the strain values obtained are not put to any use.

Stress–strain relationships and elements of rheology

Having dealt with stress and strain we shall now consider how these parameters are related. The structural geologist is, of course, basically interested in the stress–strain relationships of rock, but, as will become apparent, rock exhibits exceptionally complex behaviour. It is apposite, therefore, that we examine the concepts relating to the deformation and flow of ideal bodies (i.e. rheology). In this section of the chapter the elements of elastic theory are discussed first, followed by a brief introduction to the principles

of viscous and plastic behaviour. Diagrammatic model analogues of the various ideal bodies are then combined to show how more complicated rheological stress–strain relationships may be satisfactorily explained. This leads to a discussion of the theories relating to the brittle failure of rocks, the understanding of which underpins the geologist's ability to interpret many geological fractures.

Elasticity

When a material is subjected to stress it deforms. The material is said to be perfectly elastic if, when the stress is removed, the deformation completely and instantly disappears. The theory of elasticity most commonly used is based upon four simplifying assumptions, namely that the elastic material is (i) homogeneous (ii) isotropic (iii) the elastic strains are infinitesimal (extremely small) and (iv) the material has a linear stress–strain relationship. Materials are said to be homogeneous when they are everywhere composed of the same substance and are isotropic when their physical properties are the same in every direction. An example of a homogeneous and isotropic material is glass. Most crystalline materials depart, to some degree, from this ideal specification but, from the practical point of view, many rock types can be considered to be homogeneous and isotropic provided the scale is appropriately chosen. For example, a porphyritic granite in hand specimen will appear to be far from homogeneous and isotropic, but if a sufficiently large volume of the material is examined, variations in homogeneity and isotropy of this and similar rock types become negligible.

If the strains are infinitesimal, then the strain equations can be significantly simplified, because the second order terms (e_x^2, or $e_x e_y$) can be ignored. The value which one sets as the limit of infinitesimal strain is, of course, arbitrary and will depend upon the required accuracy of the calculations. For many practical purposes, it may be assumed that infinitesimal strains may be as large as 3 per cent without incurring significant error.

The fourth assumption, namely that the elastic stress–strain relationship is linear is, in part, related to mathematical convenience, for, prior to the availability of the computer, the analytical problems which resulted from non-linear relationships were grave. However, in addition, there are sound theoretical considerations which indicate that at the atomic level, provided displacement of atoms and hence strains in materials are small, the force-displacement (or the stress–strain relationship) will, in fact, be linear. Moreover, as long ago as 1660, Hooke noted that extension in a thin elastic rod was proportional to the applied tension, thus formulating Hooke's Law. This relationship was found to hold for a wide range of materials and later Young

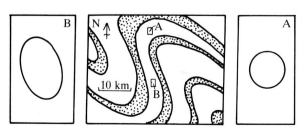

Fig. 1.34. Local strain variations around a major multilayer fold.

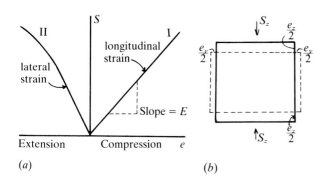

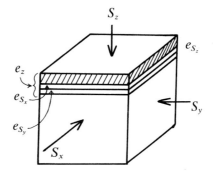

Fig. 1.35. (a) Linear stress-strain relationship in the z direction. (b) Cross-elastic, stress-strain relationship.

Fig. 1.36. Components of strain in the z direction attributable to the three orthogonal stresses σ_z, σ_y and σ_x.

suggested that the relationship between stress and strain could be expressed by:

$$S/e = E \qquad (1.35)$$

where E, a material constant, is known as Young's Modulus. It will be noted that strain (e) is dimensionless, so that, for dimensional homogeneity of Eq. (1.35), E must have the same dimensions as S (i.e. bars or MPa).

In practice, it is found that rocks do not always exhibit a completely linear stress–strain relationship (Fig. 1.35(a) (I)). However, the deviation from linearity is often quite small, so that one can conveniently quote a single value for the elastic modulus (E) of many rocks.

The stress S_z (Fig. 1.35(b)) gives rise not only to the strain e_z in the vertical direction, but also causes horizontal dilational strains, such that $e_x = e_y$. This effect was noted by Poisson early in the nineteenth century and he expressed the ratio of the horizontal to vertical strains resulting from the uniaxial stress as a second material constant. Thus:

$$e_x/e_z = e_y/e_z = v \qquad (1.36(a))$$

where v is *Poisson's ratio*. The reciprocal relationship is:

$$e_z/e_x = 1/v = m \qquad (1.36(b))$$

where m is *Poisson's number*.

Poisson's ratio, or Poisson's number, is also held by elastic theory to be absolutely constant. Indeed, from a consideration of the displacement of atoms in an idealised crystal structure, in response to applied stress, it can be shown that $m = 3.3$. However, aggregates of particles, such as many rocks are, do not show a perfectly linear stress–lateral strain relationship (Fig. 1.35(a) (II)). Consequently, in practice, the value of Poisson's number is not a constant but will vary slightly throughout the stress range. However, in the literature, it is not unusual to find that Poisson's number is assumed to be a simple material constant.

We can now proceed to show how the stresses and the corresponding elastic strains in three dimen-

sions can be related. Clearly, from Eqs. (1.35) and (1.36), it follows that:

$$e_x = e_y = e_z/m = S_z/(mE). \qquad (1.37)$$

Consider now the strains in the z direction which will result from the three principal stresses S_x, S_y and S_z, represented in Fig. 1.36. Each of the principal stresses will contribute to the total strain in the vertical direction. Thus, the stress in the z direction will give rise to a component of strain $e_{z'}$, where

$$e_{z'} = S_z/E. \qquad (1.35(a))$$

There will also be a tendency towards an extensile strain in the vertical direction caused by the horizontal principal stresses S_x and S_y which will give rise to two components of strain in the z direction (Fig. 1.35(b)). These may be designated $e_{z''}$ and $e_{z'''}$ respectively, where:

$$e_{z''} = S_x/(mE) \qquad (1.37(a))$$

and

$$e_{z'''} = S_y/(mE). \qquad (1.37(b))$$

Because all the stresses are compressive, the dilational components of strain $e_{z''}$ and $e_{z'''}$ will have the opposite sign to that of $e_{z'}$. Thus, the total strain in the z direction is:

$$e_z = e_{z'} - e_{z''} - e_{z'''}. \qquad (1.38)$$

By substituting Eqs. (1.35(a)) and (1.37(a, b)) in Eq. (1.38) the following relationship is obtained:

$$e_z = (S_z/E) - (S_x/mE) - (S_y/mE)$$

$$= \frac{1}{E}\left[S_z - \frac{1}{m}(S_x + S_y)\right]. \qquad (1.39)$$

By a similar argument, it follows that the three stress–strain equations in the x, y and z directions are:

$$e_z = \frac{1}{E}\left[S_z - \frac{1}{m}(S_x + S_y)\right] \qquad (1.40(a))$$

$$e_x = \frac{1}{E}\left[S_x - \frac{1}{m}(S_z + S_y)\right] \qquad (1.40(b))$$

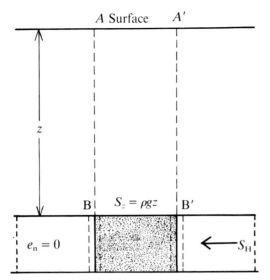

Fig. 1.37. Elemental 'cube' (stippled) in the Earth's crust in zero horizontal strain conditions. The dashed lines outside the cube show the position of the cube sides if it were not confined by the surrounding rock.

$$e_y = \frac{1}{E}\left[S_y - \frac{1}{m}(S_z + S_x)\right]. \qquad (1.40(c))$$

This set of equations forms the very foundation of elasticity theory and the reader should become familiar with their manipulation. For example, by making certain reasonable assumptions these equations may be used to establish the relationship between the vertical and horizontal stresses in the crust when subjected only to the force of gravity.

Consider a stable section of the Earth's crust of negligible relief which is subjected to gravitational loading and let it be assumed that there is absolutely zero lateral strain. Then a small, elemental cube within the crust (Fig. 1.37) will be subjected to a vertical stress (S_z) as a result of the weight of the superincumbent strata contained in the vertical column AA'B'B. If, for convenience, it is assumed that the material in the column has uniform properties, the vertical stress acting on the cube will be.

$$S_z = \rho_r gz$$

where ρ_r is the bulk density of the material, g is the gravitational constant and z is the depth from the surface in the appropriate units.

Let it be further assumed that the rock mass is completely elastic. Then, because of the elastic properties of materials, the vertical stress (S_z) will attempt to cause a lateral strain. However, we have assumed that lateral strain is completely inhibited in all directions by the material surrounding the cube, for each of the hypothetical, elemental cubes will attempt to give rise to a lateral strain and each cube cancels the effects of its immediate neighbour (Eq. (1.40)). Because the lateral strain is inhibited, it follows that there exists a lateral stress ($S_H = S_x = S_y$) which has a magnitude

sufficient to counterbalance the tendency to lateral strain induced by the vertical, gravitational load. The relationship between the vertical and the horizontal stresses may be established by using Eqs. (1.40(b) or (c)).

It has been specified that $e_x = e_y = 0$, also E is not equal to zero. Therefore one may write:

$$\left[S_x - \frac{1}{m}(S_z + S_y)\right] = 0.$$

But $S_z = S_V$ (the vertical stress) and $S_x = S_y = S_H$ (the horizontal stress), so the above relationship may be written:

$$\left[S_H - \frac{1}{m}(S_V + S_H)\right] = 0$$

Rearranging this, one arrives at:

$$S_H = \frac{S_V}{m - 1}. \qquad (1.41)$$

It will be realised that if $m = 2$, the horizontal and vertical stresses are identical, i.e. the system is analogous to the 'hydrostatic state' which develops in liquids at rest. Clearly, for the situation envisaged, the load caused by gravity cannot give rise to a horizontal stress greater than the vertical stress. Therefore, the lowest possible value one can have for Poisson's number (m) is 2.0.

Bulk, shear and other elastic moduli
Young's modulus and Poisson's number (or ratio) are the elastic parameters which are most easily determined experimentally. However, elastic behaviour may also be related to volume change or to angular distortion. Thus, the *Bulk modulus* (K) is defined as the relationship between a change in hydrostatic pressure (P) and the corresponding change in volume (V) of the elastic body. So that:

$$K = \frac{dP}{dV}. \qquad (1.42)$$

The Shear modulus (G) relates the small angular shear (ψ) to the shear stress (τ) (Fig. 1.38) by:

$$G = \frac{\tau}{\psi}. \qquad (1.43)$$

Although the shear and bulk moduli cannot readily be determined by direct measurement, it can be shown that both G and K can be expressed in terms of E and m, so that:

$$G = \frac{mE}{2(m + 1)} \qquad (1.44)$$

and

$$K = \frac{mE}{3(m - 2)}. \qquad (1.45)$$

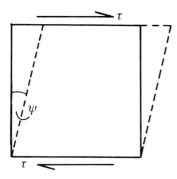

Fig. 1.38. Shear stress-shear strain relationship.

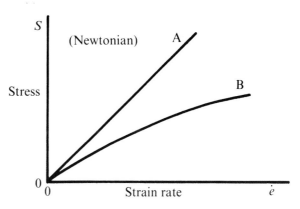

Fig. 1.39. Shear strain-rate relationships for Newtonian and non-Newtonian liquids.

For the sake of completeness, it may be noted that two other elastic parameters, defined by Lamé, are encountered in advanced treatises on elasticity theory. Lamé's parameters are G, (already defined above in terms of m and E), and λ, where:

$$\lambda = \frac{mE}{(m+1)(m-2)}.\qquad(1.46)$$

In some problems, Lamé's parameters are a convenient means of separating the components of volumetric and distortional strains which result from a given stress system. They will not, however, be used in this book.

Viscosity

When a liquid is poured into a rigid container of irregular shape, such as a jug, it completely conforms to the internal shape of the container in a relatively short time. This behaviour pattern reveals that a perfect liquid possesses no inherent shear strength, so that it will flow (albeit relatively slowly in the case of some liquids) under the action of an infinitesimally small shear stress. A higher rate of flow is associated with an increase in magnitude of the shear stress.

One can, for example, refer to the behaviour of water in lakes and streams to illustrate this relationship between shear stress and the rate of flow. Provided that the effects of wind, thermal action and the ingress and egress of streams are negligible, water in a lake is at rest and the pressures acting at any point within the body of water are the same in every direction, i.e. hydrostatic pressure is exerted. A stream flowing down a gentle slope progresses smoothly and slowly, but, on a somewhat steeper slope, the rate of flow is correspondingly increased. This rate of flow is related directly to the magnitude of the shear stress acting parallel to the slope; the greater the angle of slope, the larger is the shear stress and, consequently, the rate of flow of the stream.

For many liquids the relationship between the shear stress (τ) and the shear strain-rate ($d\gamma/dt$) is a linear one and can be expressed as:

$$\tau\bigg/\frac{d\gamma}{dt} = \eta\qquad(1.47)$$

where η is a material constant known as the coefficient of viscosity. If the shear stress is measured in dyne-s/cm^2 (1 dyne/cm^2 being approximately 10^{-3} grams/cm^2) and the rate of strain is measured in seconds, then the unit of viscosity is the poise. Thus, if a shear stress of 1 dyne/cm^2 acts on a liquid and gives rise to a strain rate of 1/s, the liquid has a viscosity of 1 poise (in SI units viscosity is expressed in Pa s, where 10 poise = 1 Pa s).

Liquids which obey Eq. (1.47) and hence give rise to the linear relationship illustrated by Curve A in Fig. 1.39 are known as Newtonian liquids, while those which exhibit a non-linear relationship (Curve B) are non-Newtonian. Comparison of Eqs. (1.43) and (1.47) show that they are very similar in form. Indeed, viscous behaviour is a mathematical analogue of elastic behaviour. The coefficient of viscosity (η) is analogous to the elastic shear modulus (G). The latter relates the shear strain to the shear stress, while the former relates the shear strain-rate to the shear stress.

Plasticity

Whereas viscosity theory relates to the behaviour pattern of a *liquid*, plasticity theory deals with the behaviour of a *solid*. The relationship between stress and strain for a rigid, 'perfectly' plastic body is illustrated in Fig. 1.40(a). At low loads a rigid, plastic body exhibits absolutely no distortion. Plastic flow begins only when the load reaches a critical yield strength when, at this critical stress it is theoretically possible for unlimited plastic deformation to develop. One would not expect perfectly rigid behaviour from a real substance even at low levels of stress, but would rather expect an initial elastic response to applied loads, (Fig. 1.40(b)) which represents an 'elastic-plastic body'. Furthermore, it is commonly found that once the initial yield stress has been reached and the first increment of plastic strain occurs, it is necessary to increase the stress slightly before further plastic deformation can take place. This is termed work- or strain-hardening and results in a stress–strain relationship of the type shown in Fig. 1.40(c).

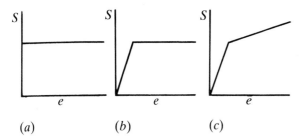

Fig. 1.40. Stress-strain relationships of (*a*) a rigid, plastic body, (*b*) an elastic, plastic body and (*c*) an elastic, plastic body with strain hardening.

Because of their high, elastic modulus (E), many metals exhibit stress–strain relationships which are extremely close to that of a rigid plastic and in 1864 Tresca established an empirical yield condition for such metals of the form:

$$(S_1 - S_3) = K_0 \qquad (1.48)$$

where K_0 is the yield strength in either simple extension or in uniaxial compression.

In 1913, von Mises proposed, on theoretical grounds, that plastic deformation would be related to the distortional strain energy of a body (discussed in Chap. 5) and in two-dimensional stress systems would obey the relationship:

$$S_1^2 + S_3^2 - S_1 S_3 = K_0^2. \qquad (1.49)$$

The difference in the yield strength (K_0) between Eqs. (1.48) and (1.49) does not differ by more than 15 per cent. However, experiments conducted on metals to distinguish between the two criteria have shown that the one proposed by von Mises better satisfies experimental data.

Examples of rheological behaviour

The reader may, at first, have some difficulty in discerning the essential difference between plastic and viscous behaviour, for both result in permanent deformation. It is recommended, therefore, that certain simple experiments be conducted using modelling clay (Plasticine) and silicone putty (which is sometimes called 'bouncing' or 'potty' putty). Cylinders of these two materials may be prepared and set side by side as indicated in Fig. 1.41(*a*). After a period of some 15 minutes it will be noted that the modelling clay is unaltered in shape while the silicone putty forms a smooth, relatively flat disc (Fig. 1.41(*b*)). The modelling clay would maintain its form indefinitely and is, in fact, a solid, while the silicone putty has flowed under its own weight and is a liquid.

Now take these same materials and shape them into spheres. Throw the sphere of modelling clay at some flat, rigid surface with considerable force. It will be noted that there is barely any bounce (or elastic behaviour) on impact but the sphere shows considerable and instantaneous, permanent deformation

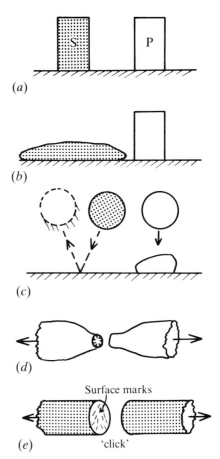

Fig. 1.41. (*a*) Cylinders of Plasticine (P) and 'bouncing putty' (S, stippled) at time $t = 0$. (*b*) Shape of bodies after time elapse of 10–15 minutes. (*c*) Response of spheres of Plasticine and bouncing putty to being dropped on a flat surface. (*d*) Necking and fibrous fracture surface of Plasticine subjected to extension. (*e*) Cylinder of bouncing putty extended to failure, showing surface features on newly developed fracture.

(Fig. 1.41(*c*)). If the silicone putty is now taken and dropped onto a flat surface it rebounds elastically in a manner comparable with a rubber ball. (The reader is warned against throwing the silicone putty, for the result could, literally, be shattering.) For very short periods, although subjected to relatively high stresses, the silicone putty which, it will be recalled, is a liquid, has behaved in a wholly elastic manner. Indeed, if the advice just given is ignored and the putty is thrown at some surface it is possible to induce completely brittle failure and 'shattering' of the 'liquid' sphere.

Thus, the significant difference between plastic and viscous behaviour is that, whereas *permanent plastic deformation can develop instantaneously, permanent viscous deformation is only possible if the liquid material is subjected to deforming stresses for a relatively long period*. The actual length of time will, of necessity, be related to the viscosity of the material and will range from milliseconds for water to many thousands or even millions of years for highly viscous materials.

It is also instructive to roll the two materials into

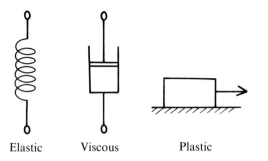

Fig. 1.42. Conceptual models 'spring', 'dashpot' and 'weight' respectively representing elastic, viscous and plastic behaviour.

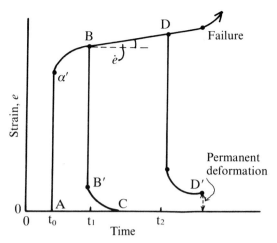

Fig. 1.43. Time–strain behaviour of an ideal body subjected to a constant load, $\dot{e}$ is the steady-state strain-rate of secondary creep.

long cylinders and then break them by rapid extension of the ends. The modelling clay will 'neck' before it fractures (Fig. 1.41(*d*)) and the fracture itself will exhibit a fibrous surface so typical of plastic fractures. The silicone putty, however, shows little or no necking before failure. Moreover, it emits a faint, but clearly audible, 'click' or sharp sound as the tensile fracture develops and usually exhibits characteristic marks on the fracture surface (Fig. 1.41(*e*)) which, as will be seen later (Chap. 2), are features associated with brittle or semi-brittle modes of fracture.

Diagrammatic analogues
It will be realised that the various equations presented in this section are mathematical models which have been proposed to explain the behaviour of elastic, viscous and plastic materials. These mathematical models may be represented in diagrammatic form where the elastic, viscous and plastic bodies are simulated by a 'spring', 'dashpot' and 'weight' respectively (Fig. 1.42). The analogy of a diagrammatic spring to represent the elastic behaviour of materials is apparent. Similarly, the frictional restraint to movement of a heavy block resting on a rough surface will clearly represent a reasonable analogue of the yield strength of a plastic solid. The dashpot, which consists of a piston which may slide within a cylinder, simulates viscous behaviour. This model may be thought of as a mechanical 'shock absorber' type of action. The piston may move easily at low velocities, or strain rates, but considerable stress must be applied to the model to drive the piston at high velocities, corresponding to high strain rates.

The behaviour pattern of natural material does not necessarily correspond closely to one of the ideal models. Indeed, the stress–strain relationships exhibited by natural materials frequently include components of all three kinds of behaviour.

One of the ways of testing materials such as rock, metals, ceramics etc., is to submit a specimen of regular shape (usually cylindrical) to a constant load and note the ensuing change of strain in the specimen with the passing of time. Such experiments are known as *creep tests*. Initially, as soon as the specimen is

subjected to a load (Fig. 1.43), it exhibits a corresponding elastic strain AA', which is followed by a period of decreasing rate of strain A'B, which is known as decelerating, or primary, creep. This phase is also termed time-elastic creep, for if at time (*t*) the load is removed from the specimen there will be an initial, instantaneous elastic recovery (BB') followed by a so-called time-elastic recovery B'C which is completed when the strain returns to zero. However, if the specimen is kept under load, it will enter into a phase of secondary, or steady-state, creep (BD). The strain associated with the secondary creep phase is not recoverable, for if the specimen is unloaded at time t_2 it once more exhibits only instantaneous and time-elastic recovery (DD'). The strain, however, is never fully recovered and there is a residual, irrecoverable strain as indicated in Fig. 1.43. But if the specimen is not off-loaded at time t_2, it will be observed that the phase of secondary creep passes into the stage of accelerating, or tertiary, creep which eventually leads to delayed failure of the specimen.

No simple rheological model exists which can represent tertiary creep. However, the instantaneous elastic, time-elastic and steady-state creep can certainly be represented by models. The instantaneous elastic strain can be simulated by a simple spring. Similarly, the constant strain rate, or secondary creep phase, and the primary creep curve can actually be represented by a simple dashpot and spring rigidly coupled in parallel respectively (Fig. 1.44), in the form known as a Kelvin (or Voight) unit. If a load is suddenly applied to such a unit, the response is a compromise between elastic and viscous behaviour. The natural response of the spring would be to take up a certain instantaneous, elastic strain (e_0 in Fig. 1.44(*b*)) but, because it is rigidly coupled to the dashpot (η_k), this instantaneous response is inhibited. The applied stress must cause viscous movement of the dashpot before the elastic strains can develop. Initially, the strain-rate of the dashpot is high but, as

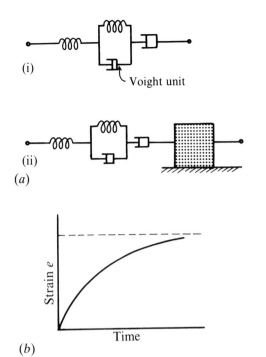

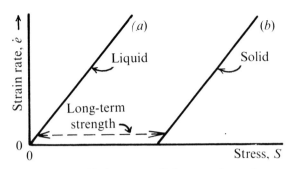

Fig. 1.45. Relationship between applied stress and secondary creep rates: (*a*) for complex non-Newtonian liquids and (*b*) for solids with viscous elements of behaviour when subjected to a stress in excess of some limiting value σ_0 representing a 'long-term' strength.

Fig. 1.44. (*a*) Linking of spring and dashpot elements in series and parallel to form (i) an elastico-viscous and (ii) a Bingham Body. (*b*) Time strain behaviour of a Voight model.

elastic strain develops in the unit, the differential stress acting upon the dashpot decreases and this, in turn, results in a decrease in the strain-rate of the dashpot. It can be shown that if a stress is suddenly applied to such a Kelvin unit, the subsequent strain–time relationship is exponential. If the unit is first subjected to a load and after reaching its equilibrium strain conditions is then suddenly off-loaded, the elastic energy stored in the spring will bring about a 'time-elastic recovery', as indicated by curve BC in Fig. 1.43 which will also be exponential in form.

It can be inferred from Fig. 1.43 that the material properties are complex. Nevertheless, during the period of secondary creep, if the relationship between the strain–rate ($\dot{e}$) and stress (S) is linear and passes through the origin (Fig. 1.45(*a*)), the material behaves as a viscous fluid; for no matter how small a differential stress is applied to the material, a finite strain would eventually ensue. However, if the relationship between $\dot{e}$ and S shows a positive intercept on the stress axis, it may be inferred that the intercept represents the long term yield strength of the material and that it is a solid. These concepts will later be considered in the light of data obtained from tests on rock.

To this point we have tacitly assumed that the materials considered have been dry. However, in the Earth's crust, rocks below the water table usually contain fluids, so before we continue with aspects of real rock failure, it is pertinent to indicate the importance of these fluids, especially as regards the concept of effective stress.

Fluids in rocks and the importance of effective stresses
In general, fluids in the crust will be water or brines, while hydrocarbons can occasionally be significant. The influence of fluids on the strength of rocks, which can be attributed to both chemical and mechanical effects, are considerable. Rehbinder & Lichtman (1957) noted that water caused the strength of a single crystal to be reduced to a tenth of the value it possessed when completely dry. Similarly, Price (1960) and Colback & Wiid (1965), in tests which included the chemical and mechanical effects, concluded that the uniaxial strength of completely saturated rock was only about 45 per cent of its oven-dry strength.

The chemical, or Rehbinder, effect is related to the fact that absorbed water reduces the surface energy of the walls of a pore, flaw or microfracture and so enhances the ease of propagation of that surface, provided suitable stress conditions exist. The adsorbed (or 'hard') water also has the effects, at elevated temperatures, of promoting pressure solution and the migration of inorganic species (an aspect which we shall discuss briefly later in this chapter). It can be argued that fluid is always adsorbed on surfaces in rock existing in natural environments, so the Rehbinder effect always operates. However, this chemical effect is often small compared with the mechanical influence of fluid pressures. These mechanical effects are twofold, namely:

(i) fluid pressure reduces the ability of rock to withstand the effects of differential stress (i.e. high fluid pressure reduces the strength of rock), and

(ii) fluid pressure influences the mode of deformation (i.e. at modest confining pressures and low fluid pressures rocks may deform as ductile material, but when the fluid pressure approaches the magnitude of the confining pressure the rock tends to behave in a brittle manner).

Both these situations are illustrated in Fig. 1.46, where fluid pressure (*p*) is expressed as a proportion of the total confining pressure (S_3),

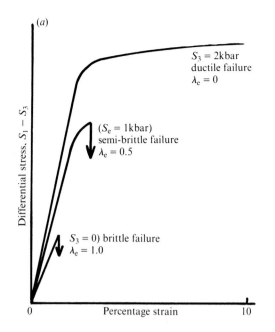

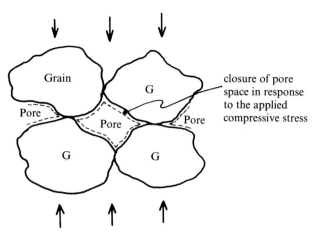

Fig. 1.47. Representation of fluid pressure acting in the pore spaces of a granular aggregate.

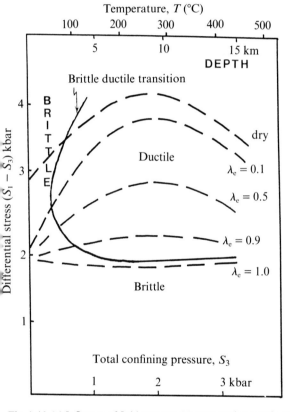

Fig. 1.46. (a) Influence of fluid pressure (p) expressed as a ratio (λ_e) of p to confining pressure (S_3), upon the mechanical behaviour of rock. (b) The brittle–ductile transition for Solnhofen limestone. (After Rutter, 1974.)

so that $\lambda_e = p/S_3$. As we shall see later, the symbol λ is used in the hydrocarbon industry to represent the ratio of fluid pressure (p) to the total vertical stress (S_z), sometimes termed the *geostatic pressure*, so that $\lambda = p/S_z$.

Rutter (1974) conducted experiments where, except for strain-rate, realistic crustal environments likely to be encountered at different depths were experimentally reproduced; that is, temperatures and confining pressures were varied in combination to represent depths in the crust between 0 and 14 km. To do this, it was assumed that the vertical stress gradient was about 250 bar/km and the temperature gradient was 30 °C/km. It will be seen (Fig. 1.46(*b*)) that brittle failure of Solnhofen limestone, or a similar rock type, would only be possible at depths of 7–14 km when the value of λ_e approaches unity. To understand these mechanical effects of fluid pressure, it is necessary to discuss the concept of *effective stress*.

Effective stresses

Consider a porous granular aggregate of the type represented in Fig. 1.47, which is subjected to a bulk compression (S), such that stresses (σ) occur in the granular aggregate and a fluid pressure (p) develops in the interstitial, or pore fluid. This problem was first addressed by Terzarghi (1923) and more recently brought to the attention of geologists by Hubbert & Rubey (1959). In its simplest from the relationship between the stresses are such that:

$$S - p = \sigma \tag{1.50}$$

where, as already noted, σ is the *effective stress* responsible for deformation of the granular mass.

The relationship between the total and effective stresses can be demonstrated by tests in apparatus of the type shown in Fig. 1.48, in which the total stresses (S_1 and S_3) are applied externally and a constant fluid pressure (p) is maintained (by an external pump). The effective principal stresses are:

$$\sigma_1 = S_1 - p \tag{1.51}$$

and

$$\sigma_3 = S_3 - p. \tag{1.52}$$

(We shall consider a more general effective stress relationship later in this chapter.)

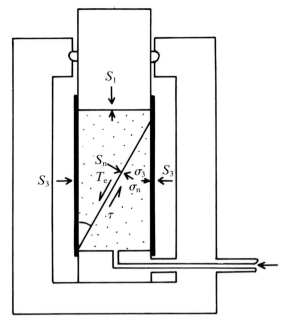

Fig. 1.48. Total and effective stresses acting on and within a cylinder of rock.

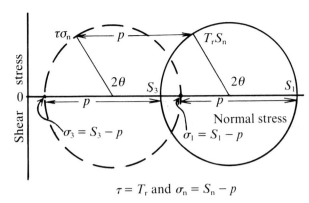

$$\tau = T_r \text{ and } \sigma_n = S_n - p$$

Fig. 1.49. Influence of fluid pressure (p) on total stresses, represented by Mohr's stress circles.

The normal and shear total stresses (S_n and T_r respectively) on an internal surface, which makes an angle θ with the axis of maximum compression (Fig. 1.48), previously given as Eqs. (1.15) and (1.16), are:

$$S_n = \frac{(S_1 + S_3)}{2} - \frac{(S_1 - S_3)}{2} \cos 2\theta \qquad (1.53)$$

and

$$T_r = \frac{(S_1 - S_3)}{2} \sin 2\theta. \qquad (1.54)$$

Rewriting Eqs. (1.53) and (1.54) in terms of effective stresses used in Eqs. (1.51) and (1.52) gives:

$$\sigma_n = \frac{[(S_1 - p) + (S_3 - p)]}{2}$$
$$- \frac{[(S_1 - p) - (S_3 - p)]}{2} \cos 2\theta$$
$$= \left[\frac{(S_1 + S_3)}{2} - \frac{(S_1 - S_3)}{2} \cos 2\theta \right] - p$$
$$= S_n - p \qquad (1.55)$$

and

$$\tau = \frac{(S_1 - p)}{2} - \frac{(S_3 - p)}{2} \sin 2\theta$$
$$= \frac{(S_1 - S_3)}{2} \sin 2\theta \qquad (1.56)$$
$$= T_r. \qquad (1.56(a))$$

Hence the pore fluid pressure reduces the normal stress acting on any arbitrary plane, but has no effect upon the shear stress (τ). This relationship can clearly be seen if the total and effective stresses are represented graphically by Mohr's circles (Fig. 1.49). It will be seen that the effective stress circle is transposed to the left by the amount p. The differential stress $(S_1 - S_3) = (\sigma_1 - \sigma_3)$ and the shear stress (τ) acting on any surface, which makes an angle θ to the axis of greatest principal stress, remains unchanged. Rock deformation in general, and especially the initiation and development of fractures, is related to effective stresses. It is now pertinent to deal with the various criteria of brittle failure, expressed in terms of effective stresses.

Relationship between principal stresses for brittle failure

When tested to failure under conditions of triaxial compression, many competent rocks exhibit a relationship between principal stresses at failure which is represented by the empirically derived linear equation:

$$\sigma_1 = \sigma_0 + k\sigma_3 \qquad (1.57)$$

where σ_0 is the *uniaxial compressive strength* and k is a constant. Such a linear relationship is shown in Fig. 1.50 (line a). However, other rock types exhibit a non-linear relationship between principal stresses at failure (Fig. 1.50, line b).

In these type of tests the specimens fail in shear. Usually the specimen fails along only one shear plane. However, occasionally specimens exhibit two *conjugate* shear planes, with opposite shear sense. The acute angle (2θ) between them, which is often considerably less that 90°, is bisected by the axis of maximum principal stress (σ_1). When the rock exhibits a linear relationship between principal stresses at failure, the angle (2θ) is constant for all values of confining pressure. However, if the rock exhibits a curved relationship between principal stresses at

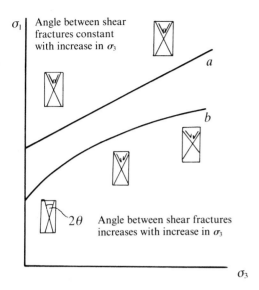

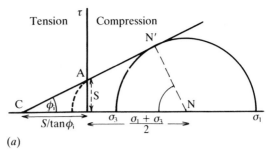

(a)

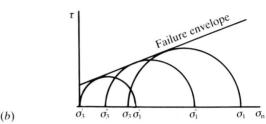

(b)

Fig. 1.50. (a) Linear relationship between principal stresses at failure. (b) Non-linear relationship between principal stresses at failure.

Fig. 1.51. Navier–Coulomb failure envelope (a) Failure envelope CAN' is the graphical expression of the Navier–Coulomb failure criteria (Eq. (1.59)) in which it is assumed to be valid in both tension and compression. (b) Experimentally determined failure envelope in the compressive region.

failure, this angle (2θ) is not constant, but increases as the confining pressure (σ_3) is increased.

Two main concepts have been proposed to explain these different experimental relationships, namely the Navier–Coulomb and the Griffith criteria: these we shall now briefly consider, in turn.

Navier–Coulomb criterion of brittle failure

This criterion of brittle shear failure is based upon Amonton's Law of frictional sliding which is given by:

$$\tau = \sigma_n \tan \phi \tag{1.58}$$

where τ and σ_n are the shear and normal stresses respectively acting on the potential fracture plane, and ϕ is the *angle of sliding friction* on that plane. However, it was recognised that, prior to the development of the fracture plane, the cohesion (C_0) of the rock must also be overcome. Hence the complete criterion can be expressed in the form:

$$\tau = C_0 + \sigma_n \tan \phi. \tag{1.59}$$

The relationship between the angle of sliding friction and the coefficient of friction (μ) is given by:

$$\mu = \tan \phi.$$

So that the complete shear failure criterion may be written as:

$$\tau = C_0 + \sigma_n \mu. \tag{1.59(a)}$$

In the simple, two-dimensional case considered here, the intermediate principal stress (σ_2) (which is assumed to act parallel to the shear plane and at right angles to the direction of shear) has, in theory, no influence upon the failure.

It has been indicated that stresses may be represented by a Mohr's circle. Also, it has been noted that

some rocks exhibit a linear relationship between principal stresses at failure. It is, therefore, a simple step to draw a series of Mohr's circles representing each particular pairing of σ_1 and σ_3 which gives rise to failure. The curve which is tangent to a series of such circles (Fig. 1.51(b)), represents the failure conditions for the material under test. Hence, a linear envelope (Fig. 1.51(b)) has the equation of the Navier–Coulomb shear failure criterion. From Fig. 1.51(a), it follows that:

$$\frac{(\sigma_1 - \sigma_3)}{2} = \left[\frac{(\sigma_1 + \sigma_3)}{2} + C_0 \cot \phi_i\right] \sin \phi_i$$

which simplifies to:

$$\sigma_1 = \frac{2C_0 \cos \phi_i}{(1 - \sin \phi_i)} + \sigma_3 \left[\frac{1 + \sin \phi_i}{1 - \sin \phi_i}\right]. \tag{1.60}$$

Hence, the Navier–Coulomb criterion satisfies those data for rock types with a linear relationship between principal stresses at failure. Furthermore, from Eqs. (1.58) and (1.60), it follows that:

$$k = \frac{1 + \sin \phi_i}{1 - \sin \phi_i} \tag{1.60(a)}$$

and that the uniaxial strength $\sigma_0 = 2C_0 k^{\frac{1}{2}}$; i.e. the uniaxial compressive strength is related to the cohesive strength and the angle of internal friction.

An important feature of the Navier–Coulomb criterion of failure is that the angle (θ), which a shear plane will make with the axis of maximum principal stress, can be predicted. To do this it is necessary to express the failure criterion (Eq. 1.59) in terms of the principal stresses (obtained from Eqs. (1.55) and (1.56)). This relationship, so derived, is then differentiated with respect to θ and the minimum and optimum conditions for shear failure are obtained. It can be shown for these optimum shear conditions, that:

$$\pm \theta = (45° - \phi_i/2)$$

or

$$2\theta = 90° - \phi_i \qquad (1.61)$$

where 2θ is the acute angle between conjugate shears. This angle (2θ) is represented graphically by the angle a tangent line to the envelope makes with the normal stress axis (Fig. 1.51).

An important shortcoming of the Navier–Coulomb criterion arises from the fact that the signs of the stresses are not taken into account. Consequently, the tensile strength (T_p) predicted by this criterion (Fig. 1.51) is:

$$T_p = C_0 \cot \phi_i.$$

For angles of ϕ_i less than 45° (and this applies to most sedimentary rocks) it follows that the predicted tensile strength is larger than the cohesive strength. However this prediction is at variance with experimental data, for the measured tensile strength is always smaller than, and is frequently approximately half, the cohesive strength.

Griffith criterion of brittle failure

The approach used by Griffith (1925) based on the interatomic bond-strength is a more fundamental appraisal of strength. It is possible to argue from atomic bonding theory that the theoretical tensile strength of ideal brittle solids is $T \simeq E/10$, where E is Young's modulus. For many strong rocks $E \simeq 10^6$ bar, so that $T \simeq 10^5$ bar. Griffith suggested that the vast discrepancy between the theoretical and the observed values of the tensile strengths of materials was a result of the intense local stress concentrations which developed at the tips of microscopic flaws. He assumed, for mathematical convenience, that the flaws were elliptical, with large eccentricity, and calculated the stress concentration around such elliptical cracks in a two-dimensional thin plate subjected to a tensile stress (σ_T) (Fig. 1.52). He demonstrated that the maximum tensile 'tip' stress becomes sufficiently large so that it equals the atomic bonding strength and so permits the flaw to propagate when:

$$\sigma_T = \sqrt{\frac{2AE}{\pi c}} \qquad (1.62)$$

where σ_T is the applied tensile stress, A is the surface energy of the flaw and $2c$ is the length of the major axis of the elliptical flaw.

Using reasonable values for the physical con-

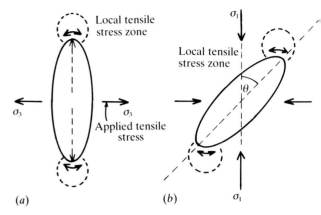

Fig. 1.52. Line drawing showing stress concentrations near the end of an elliptical flaw in a sheet subjected to compression at an acute angle (θ) to the long axis of the ellipse.

stants in Eq. (1.62), Griffith obtained values for the critical stress (σ_T) which were in good agreement with the measured values for the tensile strength of large specimens of glass. He presented even more striking evidence to support the theory by preparing 'fibres' of glass, free from flaws and surface imperfections, and found that the tensile strengths of such specimens were as much as 45 kbar which, of course, approximates reasonably closely to the theoretical atomic bonding strength.

Griffith then considered the two-dimensional problem of propagation of flaws in a sheet subjected to biaxial compression. He assumed that randomly orientated, elliptical micro-flaws, in such a sheet, were so spaced that the stress field associated with each flaw did not interfere with that of its neighbour. He showed that, even when the applied stresses are compressive, the 'tip' stresses associated with each flaw would be tensile and that these tip stresses would be a maximum when:

$$\cos 2\theta = \frac{\sigma_1 - \sigma_3}{2(\sigma_1 + \sigma_3)} \qquad (1.63)$$

where θ is the angle the long axis of the flaw makes with the axis of maximum principal stress (Fig. 1.52(b)). He also pointed out that, provided $\sigma_1 \neq \sigma_3$ and $(3\sigma_1 + \sigma_3) > 0$, the tensile stresses around the flaw reach the critical stress resulting in the spreading of the crack (which it was subsequently suggested eventually caused failure) when:

$$(\sigma_1 - \sigma_3)^2 + 8T(\sigma_1 + \sigma_3) = 0. \qquad (1.64)$$

Murrell (1958) expressed this non-linear relationship between principal stresses at failure as a Mohr's envelope (Fig. 1.53), with the equation:

$$\tau_2 + 4T\sigma_n - 4T^2 = 0. \qquad (1.65)$$

It may be noted that the 'cohesive strength' of the Navier–Coulomb shear failure criterion is twice the tensile strength, which is in good agreement with

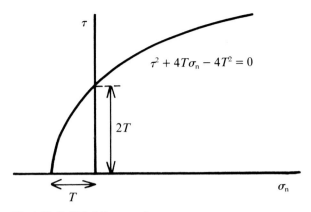

Fig. 1.53. Griffith failure envelope.

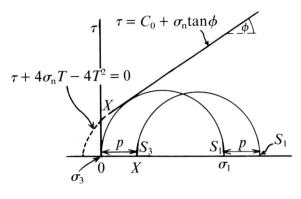

Fig. 1.54. Combined Griffith/Navier–Coulomb failure envelope. The junction is made at point X where the slope of the two criteria are the same.

values obtained from experimental data. This follows because $\tau = C_0$, when $\sigma_n = 0$, in Eq. (1.65).

The debate regarding which of the two criteria best fitted experimental data was resolved by the analysis by McClintock & Walsh (1962) who demonstrated that the two criteria were end members of a continuum of failure conditions. They noted that 'elliptical' micro-flaws, likely to give rise to failure, will have great eccentricity. Consequently, in a compressive stress field, these flaws will tend to close, or partially close. McClintock & Walsh used this concept to modify the Griffith theory and assumed that flaws closed when the normal stress across the flaw reached a certain value (σ_c) to give rise to an effective normal stress across the crack, given by:

$$\sigma_e = (\sigma_n - \sigma_c).$$

When the crack surfaces are in contact, shear stresses may be generated along the closed portion of the crack, such that:

$$\tau_e = \mu\sigma_e = \mu(\sigma_n - \sigma_c)$$

where μ is the coefficient of sliding friction.

These frictional sliding stresses were then superimposed on the stress field around the flaw (from the solution derived by Griffith) and gave rise to a relationship between principal stresses at failure of the form:

$$\mu(\sigma_1 + \sigma_3 - 2\sigma_c) + (\sigma_1 - \sigma_3)(1 + \mu)^{\frac{1}{2}}$$
$$= 4T \sqrt{1 - \frac{\sigma_c}{T}}. \qquad (1.66)$$

If the flaws are long, narrow spaces which close under very low compressive stresses, so that $\sigma_c = 0$, then the relationship between principal stresses at failure given in Eq. (1.66) reduces to:

$$\mu(\sigma_1 + \sigma_3) + (\sigma_1 - \sigma_3)(1 + \mu)^{\frac{1}{2}} = 4T. \qquad (1.67)$$

This equation represents a linear relationship between principal stresses at failure, which has a Mohr's envelope for compressive stresses of the form:

$$\tau = 2T + \mu\sigma_n. \qquad (1.68)$$

This relationship differs from the Navier–Coulomb failure criterion only by the fact that the cohesive strength is replaced by $2T$.

Closure of the cracks will not take place in a tensile stress field and therefore the shape of the envelope in tension will be determined by Eq. (1.65), so that the complete envelope will be of the form represented in Fig. 1.54, where the junction (X) between the two criteria occurs where the slopes of both envelopes are identical.

If the reasonable assumption is made that cracks close at various levels of compressive stresses, it is possible to derive relationships between principal stresses at failure where N (a power law exponent) may have any value between 1.0 and 2.0. This covers the range of experimentally determined relationships between principal stresses at failure.

Failure when the least principal effective stress is tensile

In the preceding paragraphs, emphasis has been placed on the development of shear failure, when the least principal stress has been compressive (i.e. positive). However, in crustal conditions the fluid pressure (p) may sometimes be high and greater in magnitude than the total least principal stress (S_3), so that the least effective stress (σ_3) is tensile (negative). Depending upon the magnitude of the value of the least principal stress and also the value of the differential stress, relative to or expressed in terms of, the tensile strength of the rock (T), failure may occur in one of two ways.

(i) If $(S_3 - p) > T$ and also $(S_1 - S_3) < 4T$ failure will occur in tension and one or more fracture planes will develop perpendicular to the axis of least principal stress as the result of *hydraulic fracture*.

(ii) If $0.8T < (S_3 - p) < T$ and also $4T < (S_1 - S_3) < 5.5T$ hybrid extension and shear forms of failure may result.

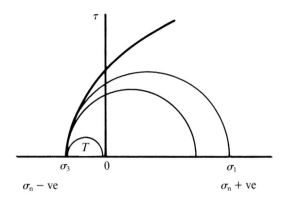

Fig. 1.55. Tensile failure conditions, when $\sigma_1 - \sigma_3 < 4T$.

We shall now deal with these two forms of failure and indicate how the stress conditions indicated above are derived and why they are so important.

Tensile failure
The stress criteria for tensile failure, cited in (i) are straight forward. The first requirement, that:

$$(S_3 - p) > T \qquad (1.69(a))$$

merely means that the least effective principal stress must attain a magnitude sufficiently great to overcome the inherent resistance of the rock to fail in tension. The second requirement that:

$$(S_1 - S_3) < 4T \qquad (1.69(b))$$

is based on the geometry of the parabolic failure envelope predicted by the Griffith failure criterion. It will be inferred from Fig. 1.55 that stress circles which touch the failure envelope at the coordinates $(0, -T)$ cannot have a diameter greater than $4T$. Stress circles with a diameter greater than $4T$ must touch the envelope at some other coordinate and so will give rise to some form of failure in shear rather than in tension.

The tensile strength of rock will be commented upon and possible or probable magnitudes of this quantity (T) will be proposed in several of the chapters of this book. Consequently it is apposite to discuss at some length the problems of determining this tensile strength using laboratory and/or theoretical techniques.

It can be inferred from the Griffith criterion (see Eq. (1.64)) that the tensile strength of a rock (which obeys this criterion) is $\frac{1}{8}$ of the rock's strength in uniaxial compression (when $\sigma_3 = 0$). If the rock in question obeys the combined Griffith/Navier–Coulomb criteria, it may be inferred from Fig. 1.54 that the ratio of uniaxial compressive to tensile strengths will be a little greater than 8:1. However, many rocks are neither homogeneous nor isotropic (fundamental requirements of the failure criteria) so that, for such rocks, the ratios quoted above are no

guide to the likely values of T. Moreover, from perusal of a reference dealing with rock properties (such as Clark, 1966) the reader may infer that the ratio of compressive to tensile strength may exceed 30:1. Such strength ratios may be real when dealing with slates, phyllites or other rocks with obvious anisotropy. However, even some massive sediments or igneous rocks are quoted to have ratios of more than 20:1. The latter rock types may be approximately homogeneous and isotropic, so such large strength ratios are suspect, and are probably related to a record tensile strength which is too low. As we shall see, such a low recorded value can often be attributed to the use of an unsuitable technique of ascertaining the tensile strength.

By far the largest amount of experimental effort in rock mechanics has been directed to determining the behaviour of rocks in wholly compressive stress systems. This, in part, represents the interests of the various investigators, but in part is a measure of the experimental difficulty of obtaining a meaningful value for the tensile strength of a rock specimen. To illustrate this latter point, we present, in a very abbreviated form, the results of a series of tests and analyses carried out by the senior author, over thirty years ago, when rock mechanics testing was in its 'formative years'. These results, which were not published, indicate how the 'strength' of one specific rock type (Pennant Sandstone) is related to the mode and technique of testing. Several different techniques were employed, which are represented diagrammatically in Fig. 1.56, and the results obtained, together with comments on each test and analysis, are listed in Table 1.1.

Some of these tests gave obviously spurious results. However, tests described in 1(d), 2(b), 3 and 4(b) and (c) gave results which were thought to represent, with reasonable accuracy, the tensile strength of this one particular rock type. Yet even these 'acceptable' values indicate a range of tensile strengths between 290 and 658 bar. The differences in strength values obtained for tests 4(b) and (c) are, at first sight, particularly vexing. This is a point to which we shall return.

At this point the investigation ceased, so that the best, and possibly most accurate, method of establishing the tensile strength was not included in the investigation. This omission was the result of a basic and general ignorance among rock mechanicists, at that time. Indeed, it was not until several years later that the important paper by Hubbert & Willis (1957) brought the concept of *hydraulic fracturing* to the attention of geologists.

The hydraulic fracture mechanism of brittle failure
For many years the development of veins and extension gashes presented a paradox, in that geologists

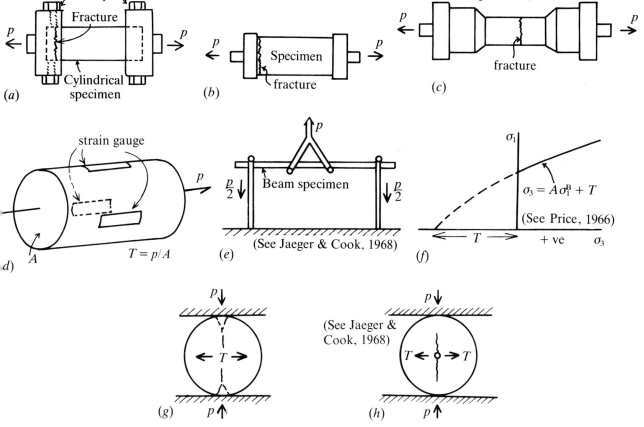

Fig. 1.56. Various methods of 'Tensile Strength Tests', with equations for determining the value of 'T'. See Table 1.1.

Table 1.1.

	Strength		Strength
1. *'Straight pull' tests*		2. *Bending beam tests*	
a) *Pull-test with clamped ends (Fig. 1.56(a))* (Failure occurred within clamp – result meaningless)	230 bar	a) *Classical beam test (Fig. 1.56(e))* (Strength estimated from equation shown in this figure assumes Young's Modulus (E) to be equal in compression and extension.)	350 bar
b) *Pull-test with cemented end toggles (Fig. 1.56(b))* (Failure occurred at junction between 'clean' rock and cement impregnated rock – result better, but still extremely dubious.)	150 bar	b) *Same test as above* (E in compression is greater than in extension. Strength recalculated accordingly.)	370 bar
c) *Pull-test on 'dog-bone' shaped specimen (Fig. 1.56(c))* (Failure occurred in the neck area, so result would appear to be good. However, to check this a test (d) was conducted to calibrate for 'straightness' of pull')	180 bar	3. *Extrapolation from compressive strength data* An equation was established for the relationship between principal stresses (see Fig. 1.56)*f*). *T* represents the tensile strength (when $\sigma_1 = 0$).	310 bar
d) *Calibration of straight pull (Fig. 1.56(d))* (Stress–strain curves were obtained for the three attached gauges. The ratio of least to greatest strains at failure was 3:1 (straight pull should give 1:1). The estimated value of strength cited was obtained from the 'mean' stress–strain curve which was extrapolated to the highest recorded strain. Even this strength estimate is almost certainly low.)	250 bar	4. *Brazilian Disc Test*	
		a) *Test of solid disc (Fig. 1.56(g))* (It was demonstrated that wedges were formed at the plattens and split open the disc, as indicated, so for this particular rock type the test results are meaningless.)	150 bar
		b) *Disc with 4 mm diameter hole (Fig. 1.56(h))* (The central hole gives a six fold increase in the tensile stresses, as indicated, and ensures tensile failure.)	465 bar
		c) *Disc with 2 mm diameter hole* (Comments as for 4(*b*). Smaller hole, greater tensile strength!)	658 bar

recognised that they were a form of 'tensile' failure,
but that stresses in the crust were compressive. This
paradox was not resolved until it was realised that the
stresses which are always (or almost always) compressive are the total stresses and that, provided the
fluid pressure (p) becomes sufficiently large, the least
principal effective stress may become tensile. If the
magnitude of the tensile, effective stress becomes
greater than the tensile strength (T) of the rock, tensile
failure by the hydraulic fracture mechanism takes
place. That is, hydraulic fracture occurs when
$(S_3 - p) > T$. This simple relationship, derived by
Hubbert & Willis, was applied to fracture development in crustal conditions by Secor (1965), who
demonstrated that there was no limiting depth in the
crust where this mechanism could not apply.

Indeed, the mechanism has a wide range of
application which includes, for example, the
emplacement of large-scale features such as dykes and
sills and possibly the initiation of some diapirs. At the
macro- and mesoscopic level, the mechanism is
responsible for the generation of veins, some barren
fractures (joints) and fracture cleavage: and on the
microscopic level it gives rise to microfracturing and
rock disaggregation leading, possibly, to cataclastic
flow.

In certain situations, it is apparent from the
geological evidence that the fluid is derived from
outside the system (as, for example, the emplacement
of dykes and sills), while in others the internal fluid
pressure already exists within the system, giving rise to
the formation of fracture cleavage and many other
minor extension fractures. When the fluid is initially
external to, and is introduced into, a system at high
pressure, then it is permissible to use Eq. (1.69(a)) to
describe the hydraulic fracturing conditions. (This
means that one can use this process of failure to
establish experimentally the tensile strength of the
rock with accuracy, Fig. 1.57.) However, when the
fluid pressure is generated internally, the situation is
not so clear cut.

This problem was addressed by Secor (1968),
who used the conceptual model (Fig. 1.58) in which a

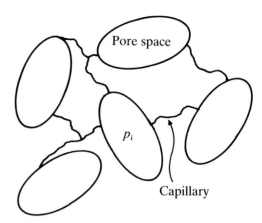

Fig. 1.58. Secor model for explaining the development of hydraulic
fractures.

homogeneous and isotropic body contains randomly
oriented, and relatively widely spaced, elliptical flaws
that are interconnected by a network of capillary
channels. Initially, all void spaces are filled with fluid
at pressure, p. Secor makes the point that an elliptical
flaw, in which there exists a fluid pressure, such that
the flaw is subjected to an effective tensile stress
(Fig. 1.59(a)), is analogous to, and will generate tip
stresses which are identical to, those generated in the
Griffith flaw (Fig. 1.59(b)) subjected to total tensile
stresses. It is these tip stresses that will be able to
rupture bonds within the rock mass and permit the
elliptical flaw to propagate.

Secor then argues that the elliptical flaw with the
most appropriate orientation (i.e. aligned perpendicular to S_3) and with greatest eccentricity (i.e. with the
greatest length to width ratio) will begin to propagate
when the fluid pressure exceeds the total, least principal stress (S_3) plus the tensile strength (T). When the
flaw propagates, it increases in volume so that, if the
volume of fluid within the flaw remains constant, the
flaw fluid pressure drops to a value p_f. Because the
external pressure $p < p_f$, Secor argues that fluid will
flow into the 'preferred' flaw, so that the fluid pressure

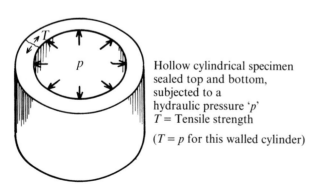

Fig. 1.57. The hydraulic fracture mechanism as a means of determining accurately the tensile strength of rock.

Hollow cylindrical specimen
sealed top and bottom,
subjected to a
hydraulic pressure 'p'
T = Tensile strength

($T = p$ for this walled cylinder)

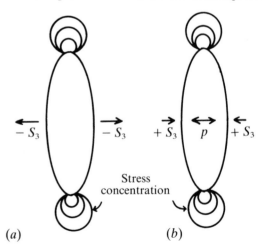

(a)   (b)

Fig. 1.59. Stresses at the tip of an ellipse derived by (a) total tension
and (b) negative effective stresses.

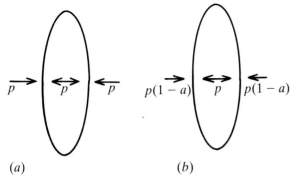

(a)   (b)

Fig. 1.60. If $\sigma = S - p$, flaw exhibits no tendency to propagate (a). However, flaw, or ellipse, propagation is possible if $\sigma = S - p(1 - a)$ (b).

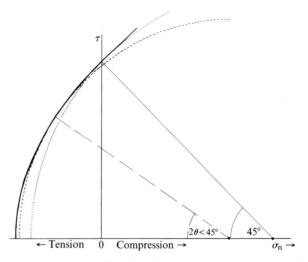

Fig. 1.61. Stress conditions for failure in the negative quadrants of the Griffith failure envelope and the predicted angle (2θ) between conjugate *hybrid extension/shear fractures*.

will increase and the flaw will undergo further extension.

However, Secor uses the effective stress relationship:

$$S - p = \sigma. \tag{1.50}$$

In order that this relationship may operate, the influence of the fluid pressure must be *completely uniform throughout the model*. Hence, as may be inferred from Fig. 1.60(a), any tendency for flaw propagation (accompanied by a pressure drop to p_f) would be resisted by the effect of the fluid pressure, p, external to the flaw.

There is another, and more general, relationship between fluid pressure and stresses, where:

$$\sigma = S - p(1 - a) \tag{1.70}$$

where a is a material constant which, for many rocks, has a value of 3–5 per cent.

Skempton (1961) suggested and Nur & Byerlee (1971) demonstrated that:

$$a = K_b/K_m \tag{1.71}$$

where K_b and K_m are the bulk moduli of the whole rock and the minerals respectively. The relationship generally used in rock mechanics (Eq. 1.50) is clearly a special case of Eq. (1.70), when $a = 0$.

The $(1 - a)$ term in Eq. (1.70) means that the action of the fluid pressure, p, is not 100 per cent effective within a rock mass. Fluid pressure inside the flaw would, however, be 100 per cent effective on the walls of a critical flaw. It can be inferred from Fig. 1.60(b), therefore, that the Secor mechanism works, provided it is assumed that the effective stresses satisfy the conditions of the general relationship expressed in Eq. (1.70) rather than the special and restricted relationship of Eq. (1.50). Experimental data which confirms the argument set out above are quoted in Fyfe, Price & Thompson (1978).

It has been noted that tensile failure is only possible provided the differential stress does not exceed the value $4T$. As already noted, this is one of the requirements of the Griffith criterion of failure and relates to the fact that any stress circle (Fig. 1.55) with a diameter of less than $4T$ can only touch the failure envelope at the origin where, of course, it would give rise to tensile failure.

Hybrid extension and shear failure

Let us now consider failure when $S_1 - S_3$ exceeds $4T$ by a small amount, and when the least effective stress is tensile and relatively large but is smaller in magnitude than the tensile strength of the rock. Such stress conditions are represented in Fig. 1.61. In this example, the stress circle touches the failure envelope in the negative, tensile sector. Mohr suggested that the Navier–Coulomb condition for predicting the angle (θ) between the failure surface and the axis of maximum principal stress could, by analogy, be extended to non-linear failure envelopes. Accordingly he suggested that if, at the point where the stress circle touches the failure envelope, a tangent is drawn to the envelope, then the angle which a line drawn perpendicular to this tangent makes with the normal stress axis defines the angle (2θ) between conjugate shear surfaces (Fig. 1.61). Mohr's suggestion, which was based on empiricism, was later demonstrated analytically by Griffith to be generally valid.

For the stress and failure conditions represented in Fig. 1.61, it can be inferred that the normal stress acting on the fracture plane will be tensile, moreover a shear stress acts along the failure plane, so that it will also experience shear displacement. Consequently, the failure plane will be a hybrid extension/shear fracture.

It follows from the geometry of the failure envelope (Fig. 1.61) that should conjugate hybrid failure planes develop, the acute angle between the planes will be less than 45°. Price (1975) carried out a series of graphical constructions and obtained

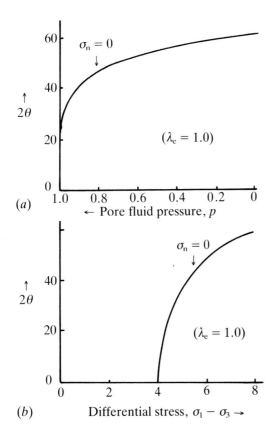

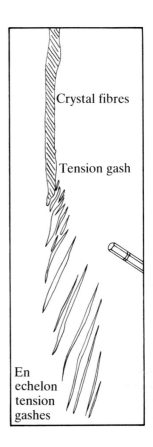

Fig. 1.62. (*a*) Relationship between 2θ and pore fluid pressure for hybrid fractures. (*b*) Differential stress, expressed in terms of *T*, necessary to cause conjugate hybrid fractures with acute angle of intersection 2θ. λ_e is the ratio of fluid pressure to the total min. principal stress.

Fig. 1.63. Example of hybrid extension and shear fracture, running laterally into a shear zone.

relationships between the acute angle 2θ between the conjugate planes and (i) the pore fluid pressure (expressed as $\lambda = 1.0 + KT$, where *K* is in the range 0 to 1.0) and (ii) the differential stress (expressed in multiples of *T*), Fig. 1.62(*a*) and (*b*).

These relationships are based wholly on the theoretical predictions of the Griffith theory of brittle failure. This particular aspect of the theory, concerned with failure when σ_3 is negative, has not been verified experimentally, since experiments of this kind are extremely difficult to conduct. However, other aspects of the Griffith theory concerned with tensile failure and failure when σ_3 is compressive have been substantiated experimentally. Moreover, field evidence exists which demonstrates that such hybrid fractures have developed. In some situations, it may be argued that a fracture formed in extension and was subsequently sheared or, alternatively, developed as a shear and later opened in extension and, of course such sequential events do take place. However, in the example illustrated in Fig. 1.63, it is clear that the planar fracture developed in hybrid extension and shear failure. This follows from the fact that the planar fracture passes laterally into a shear zone with en echelon extension fractures and also that the infilling

quartz in the planar fracture exhibits 'growth fibres' which are parallel to the direction of the opening of the fracture (i.e. from the orientation of the fibres with respect to the vein walls it can be inferred that the vein opened with a component of both extension and shear). These points are discussed in greater detail in Chap. 2 and elsewhere in the book. In the meantime, if the reader is not familiar with the interpretation of en echelon fractures and fibres, the conclusion reached above may be 'taken on trust'.

Temperature, time and scale effects on rock strength
The concepts and criteria so far presented relate to fluid and confining pressures. Other parameters of importance regarding the strength of rock in the crust include (i) temperature and (ii) time.

The effect of *temperature* has been established in a range of experiments by several workers. The results they obtained are not surprising, in that they accord with everyday experience and expectation of material behaviour. Thus, as temperature is increased, but all other parameters remain constant, the strength of rock decreases. In the example shown in Fig. 1.64, the rock behaviour is ductile for all the tests represented. The ductility in these experiments relates to the fact that the rocks were at a relatively high confining pressure of 1.5 kbar. An increase in temperature can

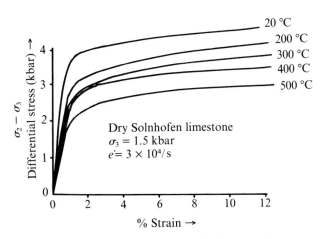

Fig. 1.64. Ductile stress–strain curves obtained at elevated temperatures and constant strain-rates. (After Rutter, 1970).

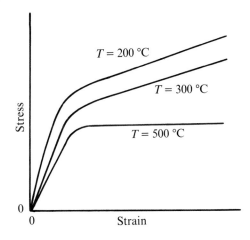

Fig. 1.65. Strain-hardening of rock at low temperatures.

also influence the mode of behaviour, so that a rock which is brittle at low temperature will become ductile as the melting temperature is approached. At the melting temperature the rock, of course, begins to take on obvious characteristics of a liquid. However, when the second parameter *time* (which is so important in geological processes) is taken into consideration, it becomes apparent that rocks may have the characteristics of liquids at temperatures considerably lower than their melting point.

Experiments which can be used to demonstrate this form of behaviour are conducted at elevated temperatures and confining pressures and the rock specimens are deformed at *constant strain-rates* ($\dot{e}$). Here we take the simple definitions of strain (e) as:

$$e = \frac{\mathrm{d}l}{l_0} \qquad (1.72)$$

where $\mathrm{d}l$ is the change in length and l_0 is the original length. So that strain-rate $\dot{e}$ is given by:

$$\dot{e} = \frac{e}{t} \qquad (1.73)$$

where t is the time in which a specific strain is induced.

The pioneering work in the use of the 'strain-rate technique' in rock deformation was conducted by Heard (1963). The technique involves obtaining a whole series of stress–strain relationships for a range of temperatures and strain-rates. In his investigation, Heard conducted experiments in the temperature range 25–500 °C (he also reported experiments conducted at temperatures up to 800 °C) and strain-rates from 10^{-1}/s to 10^{-8}/s. At the fast strain rate 10^{-1}/s, a strain of 10 per cent (10^{-1}) was induced in a second, while at 10^{-8}/s, 10 per cent strain is induced in 10^7 seconds (approximately four months). At low temperatures and/or high strain-rates, the specimens often did not exhibit a well-defined yield strength, but showed strain-hardening (Fig. 1.65). Consequently,

for convenience, Heard took the 'strength' or yield point of the various tests to be the differential stress which the specimen supported at a strain of 10 per cent. These strength data were then plotted on a log strain-rate base as shown in Fig. 1.66, to which Heard then fitted an Equation of State, which is based on the theory of steady-state, diffusion forms of deformation. The equation he used was:

$$\dot{e} = \text{Const.} \exp(-Q/RT_K)\sinh(S_1 - S_3) \quad (1.74)$$

where $\dot{e}$ is strain-rate, Q is the apparent heat of activation, R is the Boltzmann constant, T_K is temperature in degrees Kelvin and, as previously defined, $(S_1 - S_3)$ is the differential stress.

Heard and others later considered it more suitable to use a related equation of the form:

$$\dot{e} = \text{Const.} \exp(-Q/RT_K)(S_1 - S_3)^n$$
$$(1.74(a))$$

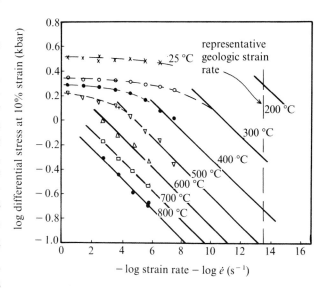

Fig. 1.66. Log differential stress/log strain-rate for Yule marble; a representation of the type of data indicated in Figs. 1.64 and 1.65. (After Heard, 1963.)

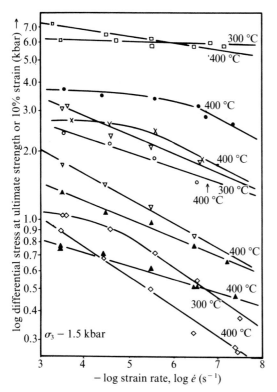

Fig. 1.67. Data, comparable with that shown in Fig. 1.66, for various types of carbonate rock types. (After Rutter, 1970.)

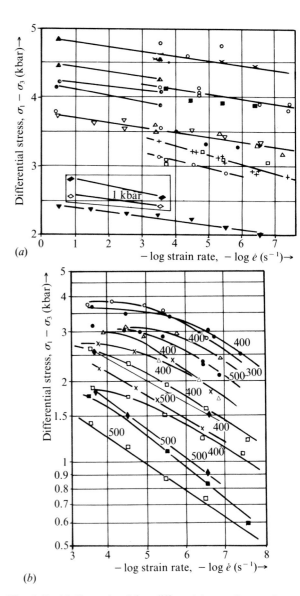

Fig. 1.68. (*a*) 'Low slope' log differential stress/log strain-rate relationships. (*b*) 'High slope' log differential stress/log strain-rate relationships. (After Rutter, 1970.)

where the exponent n can be empirically determined for a specific rock, or may be theoretically related to one or other of the diffusion mechanisms. (We shall comment later on these micro-mechanisms.)

It will be seen that the curves calculated from the equation of state only fit those data points relating to experiments conducted at high temperatures and/or slow strain-rates. That is because these data relate to experiments which exhibit relatively little strain-hardening, so that deformation was approaching steady-state. From Fig. 1.67, it can be seen that a compilation of data from various sources reveals that carbonate rocks exhibit high and low slope forms (Rutter, 1970).

An important feature of the equations of state is that they describe the behaviour of liquids, which can reasonably be applied to rocks at high temperatures. But from the practical, rather than the philosophical view point, it is often more reasonable to treat competent rocks, in low temperature environments, as 'solids'.

It can be inferred from Fig. 1.68 that, for the low slope relationships, the strength of the rock decreases by about 2 per cent for every 'decade' decrease in strain-rate. Therefore, if this slope can reasonably be extrapolated to geological strain-rates of 10^{-14} to 10^{-16}/sec, the test rocks would still exhibit a very significant strength. However, an extrapolation made on a log-time scale can be deceptive, so that the question regarding the strength of rock at low tem-

perature, but at slow strain-rate, cannot be answered by having recourse to data obtained from constant strain-rate experiments. Price (1966) and in Fyfe, Price & Thompson (1978) argues that a definitive answer can, however, be obtained from a study of creep test data. Briefly, he states that, in tests designed to establish the rheological behaviour of rocks at room temperature and in uniaxial compression, the time–strain data indicate that rocks have a long-term strength. (See Fig. 1.69 and refer to the section on creep tests.) However, the most compelling evidence comes from tests when the specimens actually expanded while under a heavy deadweight load (Fig. 1.70(*a*) and (*b*)). Price interpreted this behaviour as the result of the release of residual stresses which reflect stress conditions experienced by the rock some 10^8 years ago. It was inferred that the average magni-

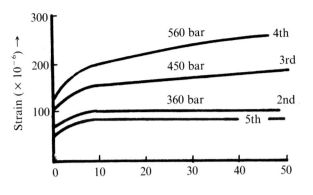

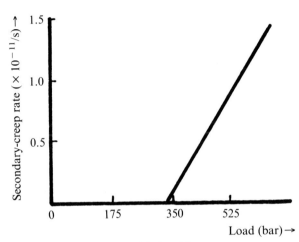

Fig. 1.69. 'Long term' strength inferred from creep tests.

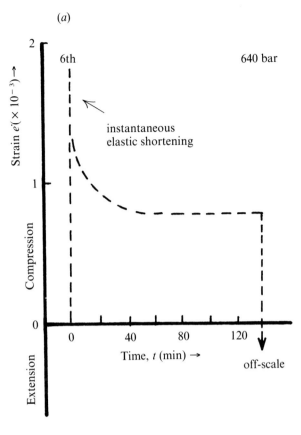

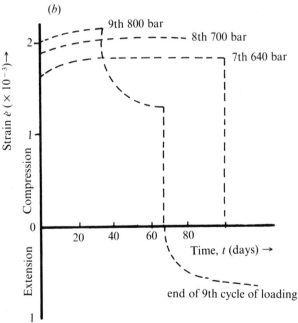

Fig. 1.70. (*a*) and (*b*) Examples in which a rock specimen expanded while under a uniaxial compression.

tude of the residual stresses released in the experiments represented in Fig. 1.70(*a*) and (*b*) is 600–700 bar.

If rock behaved as a liquid, these stresses would have dissipated by a process of relaxation, where the relaxation time (t_R) is given by:

$$t_R = \frac{\eta}{G} \, \text{s} \tag{1.75}$$

where η is the coefficient of viscosity in poise and G is the elastic shear modulus in dynes cm². The value of G for rock is approximately 10^{11} dynes cm², so that if rock were assumed to have a viscosity to 10^{22} poise (a very high value) then, by substituting for η and G in Eq. (1.75), the relaxation time would be only 3200 years. From the creep data alone, therefore, it can be concluded that competent rock at low temperatures behaves as a solid.

The high slope relationships shown in Fig. 1.68(*b*) are compatible with steady-state diffusion processes. Such experimentally based relationships can be combined with theoretical predictions regarding specific deformation mechanisms to produce deformation maps such as those in Fig. 1.72.

Such maps indicate, for a specific grain size, the interrelationships between differential stress, temperature and strain-rates for various mechanisms. We shall use data obtained from such diagrams from place to place throughout the book.

It is not our intention to deal with the various

mechanisms of crystal deformation mentioned in Fig. 1.72(*a*) and (*b*) and elsewhere in any detail. However, for the convenience of the reader, we give a brief outline of the micro-mechanisms in the final section of this chapter.

Scaling factor

Almost all rock specimens tested in the laboratory have dimensions which fall in the 1–10 cm range. In addition, the specimens are usually carefully selected to be free from observable flaws and fractures. It is, therefore, pertinent to enquire what relationship can be expected to exist between laboratory tests on small, continuous (i.e. unfractured) specimens and extremely large, discontinuous (fractured and/or bedded) rock masses in the crust. This is a topic on which there is little direct or experimental evidence. It is, however, worth commenting on the range of tensile strengths quoted earlier which relate to one reasonably homogeneous and isotropic rock type. It is suggested that the range of strengths reported is related to the volume of rock subjected to a tensile stress capable of initiating a fracture, for this was different in the various tests. The smallest volume of rock tested was in the disc test with the 2 mm hole (where the volume of rock in tension would be only a few cubic millimetres). The next smallest volume would be in the disc test with a 4 mm hole. The volume of rock subjected to the highest tensile stress in a bending beam test is difficult to ascertain with accuracy, but was probably in excess of 10 mm³. The largest volume of rock tested, or inferred, was approximately 4×10^4 mm³ for the cylindrical straight pull (calibrated or extrapolated) tests. The data are not sufficient to permit precise formulation of the relationship between strength and volume of rock, but enables the conclusion to be reached that the smaller the volume of rock tested, the larger its measured strength.

The rock types that contain a range of fractures and which has been investigated with a view to establishing the scale–strength relationship are coal and clay-rocks. Over the range of specimen sizes tested by them, (from a few millimetres to tens of centimetres) Evans & Pomeroy (1958) established that the uniaxial strength of a specific type of coal was given by the power law:

$$S_0 = kL^n \tag{1.76}$$

where k is a constant, L the length of side of the cubical specimens tested and n was a constant for different coals with values between 0.17 and 0.32. This type of power law has been verified in tests conducted underground on coal pillars, where it was established that the decrease in strength ceased for specimen sizes with L greater than about 1.0 m. Other strength/scale data (compiled by Kojima, 1983) are indicated in Fig. 1.71.

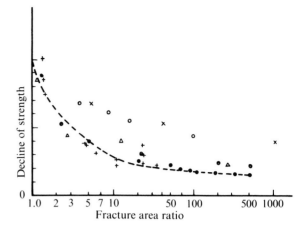

Fig. 1.71. Strength/scale data. (From Kojima, 1983; after Yoshinaka, 1976.)

As regards crustal scale, arguments have been put forward by Jeffreys (1952) that the crust has sufficient strength to support, possibly for millions of years, stress differences induced by topographical relief. Such stresses rarely exceed 500 bars. Other arguments and evidence have been adduced by Fyfe, Price & Thompson (1978) which indicate that the average crustal, differential stress is of the order of a few hundred bars, though, as we shall see in Chap. 5, the stronger units in a sequence may be significantly stronger than this latter figure would indicate.

Clearly, evidence regarding the scaling effect is limited, but it would appear that data from rock mechanics tests indicate rock strengths significantly higher than those which are likely to obtain in similar rocks in the crust. In general, it seems probable that in crustal conditions, the strength of a specific rock is not likely to exceed 50 per cent of that ascertained or inferred from laboratory experiments (where specimens are deformed under comparable environmental conditions) and in some circumstances this ratio may be as little as 10 per cent. Moreover, as we shall see in Chap. 4 and elsewhere, under special circumstances (when residual stresses become significant) it is necessary to infer that rock actually attains zero tensile strength.

Mechanisms of crystal deformation

A variety of micro-mechanisms operate during the bulk creep of rocks. These include *cataclasis, pressure solution, Coble creep, Nabarro–Herring creep, dislocation creep and dislocation glide.* The contribution of any single micro-mechanism is determined by a variety of parameters including *pressure, temperature, strain-rate and grain size* and the processes compete with one another, each dominating under a specific range of physical conditions. This concept can be represented diagrammatically on a deformation map (Fig. 1.72), which was first proposed by Ashby (1972). The axes of the 'maps' are commonly temperature and

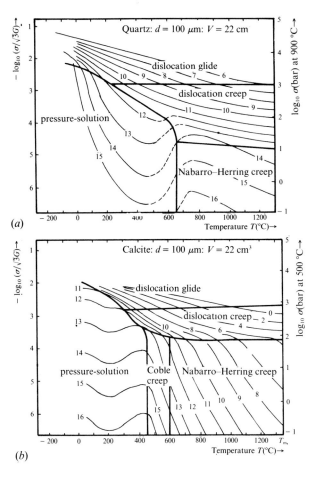

Fig. 1.72. Deformation maps for (*a*) Quartz and (*b*) Calcite. Curves for strain-rates between 0 and 10^{-16}/S are shown. (After Rutter, 1976.)

stress and each map is contoured for strain-rate. If all other parameters, e.g. grain size, are kept constant then, given a particular condition of temperature and stress, the dominant mechanism of crystal deformation and the associated strain-rate can be determined from the map.

A discussion of the micro-mechanisms associated with the creep of crystals is outside the scope of this book and only a brief definition of the various mechanisms is given. The reader who is particularly interested in this topic is referred to Nicolas & Poirier (1976) and Poirier (1985) for clear discussions of the deformation of crystals and to Frost & Ashby (1982) for the treatment of deformation maps.

Some of the mechanisms of crystal deformation listed above (e.g. Coble creep and Nabarro–Herring creep) are based on well established physical models which involve the migration of crystal defects, such as vacancies and dislocations, through and around crystals. The models successfully predict the constitutive equations that characterise the micro-mechanism. Other mechanisms (e.g. pressure solution and cataclasis) are less well understood.

Burton (1977) points out that when a stress is

applied to a polycrystalline material at elevated temperatures, creep can occur mainly by processes involving *dislocation movement* (dislocation creep and glide), or by the stress-directed *diffusion of vacancies* (Coble creep, Nabarro–Herring creep and pressure solution). The former group is characterised by a strong dependence of creep-rate upon stress and the latter, known collectively as *diffusional creep* tends to be the predominant deformation mode at lower stress levels (Fig. 1.72).

Diffusional creep can be controlled by the diffusion of vacancies *through* grains, known as Nabarro–Herring creep, during which the strain-rate is inversely proportional to the *square* of the grain size (*d*), or by diffusion *around* grains, i.e. along grain boundaries, known as Coble creep, where the strain-rate is inversely proportional to the *cube* of the grain size. Because of its stronger grain size dependence and because boundary diffusion has a lower activation energy, then lattice diffusion, Coble creep, predominates at lower temperatures and in fine-grained materials (Fig. 1.72(*b*)).

The geologically important mechanism of crystal deformation, termed *pressure solution*, also involves the transport of material along grain boundaries and, in this respect, is exactly analogous to Coble creep. However, for pressure solution to occur it is assumed that a 'fluid film' exists along the grain boundary which facilitates the migration of material. Thus, pressure solution occurs at a lower temperature than Coble creep (Fig. 1.72(*b*)). Although the mechanisms of solution, migration and reprecipitation of minerals in this film of fluid along grain boundaries are not fully understood, the process is thought to be analogous to the solution of material in a solvent, the migration of the solute along concentration gradients and the precipitation of the material at sites of relatively low stress. (See Chap. 17.)

The mechanisms of crystal deformation involving movement of dislocations through a crystal have been sub-divided into dislocation creep and dislocation glide. In both these types of deformation, strain of the crystal is induced by the glide of dislocations. However, the strain-rate is controlled by different parameters. In dislocation creep, the obstacles to dislocation movement are large compared with the size of the dislocations and the strain-rate is determined by the ease with which the dislocations can circumvent the obstacles, by the processes of dislocation slip and cross-slip. In dislocation glide, the obstacles to dislocation motion have dimensions on the same scale as the dislocations (i.e. a few interatomic distances) and these can be overcome by thermal agitation, helped by an applied effective stress. Dislocation glide is controlled by the glide velocity of dislocations, which is related to the properties of the crystal lattice and not to obstacles and defects within the lattice.

Cataclasis is characterised by brittle microfailure which gives rise to loss of cohesion of the material along grain boundaries and across individual grains and/or crystals. A considerable amount of intergranular slip (grain boundary sliding) occurs. Other mechanisms of crystal deformation play a much less important rôle. This deformation mechanism is often associated with relatively high strain-rates and low temperatures. We consider that cataclasis is the major mechanism involved in crystal and grain deformation in rocks in the upper levels of the crust. Even at relatively high temperatures, cataclasis may be responsible for a high proportion of the total strain. However, it is reasonable to infer that evidence for cataclasis may often be destroyed by later diffusion mechanisms, which possibly only account for a small proportion of the total strain, so that, unfortunately, the earlier assertion cannot be proved. To date, cataclasis has yet to attract the attention of the majority of experimenters. The usefulness of deformation maps to geologists interested in the micro-mechanisms involved in crustal deformation will be greatly enhanced once the extent of the field of cataclasis on such maps is clearly established.

References

Ashby, M. (1972). A first report on deformation-mechanism maps. *Acta Metall.*, **20**, 887–97.

Brun, A. (1981). Instabilités gravitaires et deformation de la croute continentale. Thesis d'Etat, University of Rennes, France.

Burton, B. (1977). Diffusional creep of polycrystalline materials. *Diffusion and defect monograph series.* Aedermansdorff: Trans. Tech. Pub., 119 pp.

Clark, S.P. (1966) (Ed.). *Handbook of physical constants.* Geol. Soc. Am. Mem., 97.

Colback, S. & Wiid, B. (1965). The influence of moisture content on the compressive strength of rocks. *Proc. Symp. on Rock Mech. Ottawa*, 66–83.

Dunnet, D. (1969). A technique of finite-strain analysis using elliptical particles. *Tectonophysics*, **7**, 117–36.

Evans, I. & Pomeroy, C.D. (1958). The strength of cubes of coal in uniaxial compression. *Mech. Prop. Non-metallic Brittle Materials*, ed. W. Walton, pp. 5–28. Butterworths.

Flinn, D. (1962). On folding during three-dimensional progressive deformation. *Q.J. Geol. Soc. London*, **118**, 385–433.

Frost, H.J. & Ashby, M.F. (1982). *Deformation-mechanism maps.* Oxford: Pergamon.

Fyfe, W., Price, N.J. & Thompson, A.B. (1978). *Fluids in the Earth's crust.* Amsterdam: Elsevier.

Griffith, A.A. (1925). The theory of rupture. *1st Int. Cong. Appl. Mech. Proc. Delft*, pp. 55–63.

Heard, H. (1963). Effects of large changes in strain-rate in the experimental deformation of Yule Marble. *J. Geol.*, **71**, 162–95.

Hubbert, M.K. & Rubey, W. (1959). Role of fluid pressure in mechanics of over-thrust faulting. Pts I and II. *Geol. Soc. Am. Bull.*, **70**, 115–205.

Hubbert, M.K. & Willis, D.G. (1957). Mechanics of hydraulic fracturing. *Trans. A.I.M.E.*, **210**, 153–68.

Jaeger, J.C. & Cook, N.G.W. (1968.) *Fundamentals of Rock Mechanics.* London: Methuen.

Jeffreys, H. (1952). *The Earth.* Cambridge University Press.

Kojima, K. (1983). In *Geological structures*, ed. T. Uemura & S. Mizutani. New York: Wiley.

Lisle, R.J. (1977). Clastic grain shape and orientation in relation to cleavage from the Aberystwyth Grits, Wales. *Tectonophysics*, **39**, 387–95.

Lisle, R.J. (1985). *Geological strain analysis – a manual for the R_f/ϕ technique.* Oxford: Pergamon.

McClintock, F.A. & Walsh, J.B. (1962). Friction on Griffith Cracks under pressure. *Fourth U.S. Nat. Cong. Appl. Mech. Proc.* 1015–21.

Means, W.D. (1976). *Stress and strain.* New York: Springer–Verlag.

Murrell, S.A.F. (1958). The strength of coal under triaxial compression. In *Mech. Prop. Non-metallic Brittle Materials.* ed. W. Walton, pp. 123–53. Butterworths.

Nicolas, A. (1984). *Principes de tectonique.* Paris: Mason.

Nicolas, A. & Poirier, J.P. (1976). *Crystalline plasticity and solid state flow in metmorphic rocks.* London: Wiley Interscience.

Nur, A. & Byerlee, J.D. (1971). An exact effective stress law for elastic deformation of rocks with fluids. *J. Geophys. Res.*, **76**, 6414–28.

Poirier, J.P. (1985). *Creep of crystals.* Cambridge University Press.

Price, N.J. (1960). The compressive strength of coal measure rocks. *Colliery Eng.* (London), 106–18.

Price, N.J. (1966). *Fault and joint development in brittle and semi-brittle rocks.* Oxford: Pergamon.

Price, N.J. (1975). Rates of deformation. *J. Geol. Soc. London*, **131**, 553–75.

Ramsay, J.G. (1967). *Folding and fracturing of rocks.* New York: McGraw-Hill.

Ramsay, J.G. & Huber, M.I. (1983). *The techniques of modern structural geology. Vol. 1. Strain analysis.* Academic Press.

Ramsay, J.G. & Wood, D. (1973). The geometric effects of volume changes during deformational processes. *Tectonophysics*, **16**, 263–77.

Rehbinder, P.A. & Lichtman, V. (1957). Effects of surface active media on strain and rupture in solids. *Proc. 2nd. Int. Cong. on Surface Activity*, **3**, 563.

Rutter, E. (1970). An experimental study of the factors affecting the rheological properties of rock in simulated geological environments. Unpublished Ph.D. Thesis, University of London.

Rutter, E. (1974). The influence of temperature, strain-rate and interstitial water in the experimental deformation of calcite rocks. *Tectonophysics*, **22**, 311–30.

Rutter, E. (1976). The kinetics of deformation by pressure solution. *Phil. Trans. Roy. Soc. London*, **A283**, 203–13.

Secor, D. (1965). Role of fluid pressure in jointing. *Am. J. Sci.*, **263**, 633–46.

Secor, D. (1968). Mechanics of natural extension functioning at depth in the Earth's crust. *Geol. Surv. Paper Can., Pap.*, **68**, 52.

Skempton, A.W. (1961). *Effective stress in soils, concrete and rocks. Pore-pressure and Suction in Soils.* London: Butterworth.

Terzarghi, K. van (1923). Die Berechnung der Durchlassig-keitziffer des Tones aus dem Verlauf der hydrodyna-mischen Spannungerscheinungen. *Sber. Akad. Wiss. Wien*, **132**, 105.

Yoshinaka, R. (1976). Size effects of the strength of bedrock. *Seko Gijutsu*, **9**, 8, 58–60.

Zing, T. (1935). Beitrag zur Schotteranalyze. *Schw. Miner. und Pet. Mitteilung*, **15**, 39–140.

2 Minor fractures – their nomenclature and age-relationships

Introduction

From the view point of hydrology, engineering, mining and hydrocarbon geology, fractures are among the most important of geological structures. Yet, paradoxically, they tend to be neglected by most geologists and so are probably the least well studied and documented of geological features. The geological processes which interact to produce fractures of various types and sizes are extremely complex, for many of these interacting controls, parameters and effects change with time as the rocks in a region experience geological evolution. In general, however, fracture developments are related to three main geological processes, which are:

(i) Deformations resulting from Orogenic processes,
(ii) Deformations resulting from Epeirogenic processes, and
(iii) 'Shrinkage' caused by cooling or dessication.

A specific rock sequence may, of course, be subjected to any combination of these fundamental processes, all of which will be considered in the appropriate sections and chapters of this book.

As one of the topics which bedevils fracture interpretation is that of nomenclature, we shall encounter this aspect as a recurring theme throughout this chapter. However, the main subject to be dealt with here concerns the criteria that can be used to assess the age relationships of fractures. This seemingly academic task is at the heart of many practical engineering and industrial problems.

It will be shown here (and also in Chaps. 3, 4, 9 and 14) that fractures may form throughout much of the history of the evolution of a given rock unit. Consequently, it is desirable to establish when a specific fracture type developed. Many fractures form in the near-surface environment as the result of weathering processes. It is, therefore, necessary to recognise such fractures so that, if the geologist is concerned with engineering, hydrological or hydrocarbon migration problems well below the zone of weathering, such surface 'noise' can be 'filtered out'. A forecast may then be made, from the surface patterns, regarding the fracture types and orientations, coupled with the frequency of fractures, thereby making the geological input to the problem more realistic.

Various criteria which can be used to establish the age relationships of fractures are presented and discussed in subsequent sections of this chapter. However, it is necessary that we first consider the basic types of fractures which may develop.

Basic fracture types

The majority of minor fractures (perhaps >95 per cent) that form in sedimentary rocks are oriented within ten degrees of normal to the bedding plane. Individual fractures may be either planar, with a straight fracture trace on the bedding, (i.e. *systematic*) or, with a curved fracture trace (i.e. *non-systematic*) (Fig. 2.1(a)). when systematic fractures develop paral-

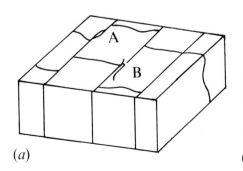
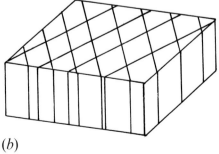

Fig. 2.1. (*a*) Orthogonal systematic and non-systematic fractures. Parallel systematic fractures form a set. Some of the ways in which individual fracture traces reveal how separate fractures join, or over-lap and die out are indicated at A and B. (*b*) Two sets of systematic fractures. These are often orthogonal or, as shown here, intersect at an acute angle, when it is common to interpret one or perhaps both sets as 'shear fractures', even though evidence for displacement is not clearly discernible (see text for the arguments justifying such practice). It can be seen that the orientation of the exposure surface may greatly influence apparent spacing of fractures.

(*a*)   (*b*)

0.1 mm

Fig. 2.2. Photomicrograph of veins of quartz cut and displaced by a vein of calcite. The quartz veins are dilational (i.e. extension) features and the calcite vein is a hybrid extension/shear fracture.

lel to one another they form a *set*. Often non-systematic fractures are arranged in a roughly orthogonal pattern with such a systematic set and this probably represents the simplest fracture pattern. Indeed, it has been suggested (Nickelsen & Hough, 1969) that this is the 'natural unit' of minor fracture development. Usually, a reasonably large exposure contains more than one set of systematic fractures (this, somewhat confusingly, is known as a *system* of fractures, Fig. 2.1(*b*)).

In the main, we shall concern ourselves in this chapter (and elsewhere in the book) with systematic fractures. Non-systematic, minor fractures, it will be argued, are most likely to develop in the zone of weathering. From the geometry of the fracture surface one may infer that such fractures are mainly *dilational* features, for the curved form of the fracture obviates the possibility of significant amounts of shear. Even a small amount of shear displacement on such curved fractures would result in the development of void spaces intermittently along the fracture: a feature which is rarely observed.

There is no such constraint when dealing with systematic fractures. Accordingly, Griggs & Handin (1960) classified fractures as *faults* if they exhibit shear displacement and as *joints* if they are dilational features which exhibit no shear. From the philosophical view-point one cannot take exception to this classification. However, from the practical point of view it may lead to the misinterpretation of real geological data. The shortcoming of this definition is that the scale at which the observation is made is not specified.

In the field, it may prove extremely difficult to demonstrate that small shear displacements have taken place. Take, for example, the intersecting micro-veins shown in Fig. 2.2. From this photomicro-

graph, it is clear that the calcite-filled vein exhibits a shear displacement of the quartz veins of 0.025 mm. This is indeed a very small displacement and can only be inferred because of the infill of the fractures. Were all three fractures barren, then it would prove virtually impossible to demonstrate that the 'horizontal' fracture has experienced shear displacement. In the field, weathering of the fractures would probably preclude seeing the displacement, even with the use of a powerful hand lens. Moreover, the barren fractures would be planes of weakness, so that were a thin section of the rock to be prepared, then, unless great care were taken, the specimen would separate along the fracture planes.

The inherent danger of the Griggs & Handin definition is that geologists who are unable to demonstrate a shear displacement on a fracture will be tempted to classify such a fracture as a 'joint' resulting only from extension. This is a point to which we shall return (Chap. 9). Here it is sufficient to note that we would argue that it is legitimate and necessary to use other evidence (e.g. the orientation of fracture planes with reference to other structural features, such as fold axes, or known or inferred axes of maximum principal stress) to conclude that certain barren fractures with no obvious signs of shear displacement may, nevertheless, be reasonably classified as shear fractures. Moreover, the Griggs & Handin classification ignores a third and, from the interpretive point of view, important group of fractures. These are the 'hybrid' fractures which exhibit both extension and shear. Indeed, it can be inferred that the calcite vein in Fig. 2.2 is such a hybrid feature.

To understand how these various fracture types develop and how they may be interpreted, in terms of crustal stress and fluid pressure conditions, it is necessary to consider the criteria of failure generally applied to rocks. For a detailed treatment of these criteria, the reader is referred to Chap. 1. Here, it is merely necessary to refer to Fig. 2.3, which shows the stress conditions necessary to induce the three basic types of fractures. It may be inferred from Fig. 2.3 that for the formation of the hybrid shear/dilation fractures the effective normal stresses acting across the fracture planes are negative (i.e. tensile). Hence there is a tendency for these fracture planes to open, which can, of course, be attributed to the relatively high fluid pressure necessary to induce effective tensile stresses in the Earth's crust. If the temperature, pressure and chemical conditions are suitable and there is also a sufficient quantity of water draining through the fractures, then the circumstances are conducive to the emplacement of vein material. That is, shear-vein emplacement can occur during, or immediately following, fracture development. It will be noted from the geometry of the fracture envelope of Fig. 2.3 that the acute angle (2θ) between conjugate hybrid shears is less than 45°, or $\theta < 22.5°$.

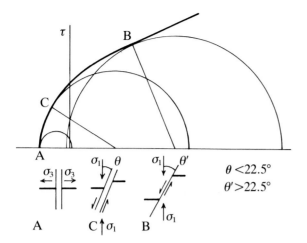

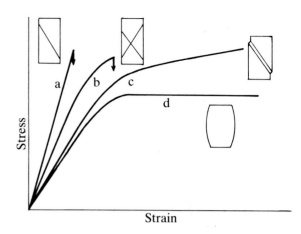

Fig. 2.3. Mohr's stress circles, representing rock stress conditions that will give rise to A. tensile failure B. shear failure or C. hybrid extension/shear fractures. The sense of movement exhibited by the fracture planes and their orientation with respect to the axes of greatest and least principal stress are also shown.

Fig. 2.4. Diagrammatic representation of (*a*) brittle, (*b*) semi-brittle, (*c*) semi-ductile and (*d*) ductile stress–strain curves, with corresponding modes of shear failure.

In the field, one may occasionally encounter fracture sets, which are classified as conjugate shear fractures, that make an acute angle with each other of greater than 45°, which also contain vein minerals. In such instances, it is reasonable to conclude that (because of the large acute angle between the fractures, i.e. $2\theta > 45°$) these fractures were originally closed and barren but were subsequently opened up in response to high fluid pressures and that vein material was deposited at this later time. An example of such behaviour, when the infill is by magmatic rather than gangue material, is cited in Chap. 3. There may be a hiatus between fracture development and the emplacement of vein material if the acute angle between shear planes is less than 45°: however, if the angle exceeds 45°, this hiatus is mandatory.

Before proceeding further, we must first define the most basic of terms used, namely that of 'fracture' itself. The definition of this word in the context of its application to geological phenomena is by no means simple. Perhaps the meaning of the word 'fracture' most usually held is that proposed by Griggs & Handin (1960), who defined a material as being fractured when, as the result of applied stresses, there is a total loss of cohesion of the material along some plane. This definition is not as satisfactory as it may at first appear. Although some 'fractures' in rock may, from visual inspection, seem to have lost cohesion, nevertheless, the rock material across the fracture plane can still retain a certain degree of cohesive or tensile strength. One may take as an example a hand specimen that is completely cut by a fracture, but which still possesses sufficient strength to resist one's efforts to pull the blocks bounding the fracture apart. Without doubt, the joint represents a plane of weakness relative to the unfractured rock material, but equally certainly it has not completely lost cohesion.

Indeed, in unweathered rock, it is probable that relatively few minor fractures completely lose cohesion.

Another criticism of this definition is that lack of cohesion can be related to the scale of the phenomenon being observed. For example, on the macroscopic scale, shear failure may appear to have taken place without loss of cohesion. Yet, at the microscopic level, if the shear zone was the result of cataclastic flow, it may be ramified by a host of small-scale fractures, each exhibiting local loss of cohesion.

It was noted in Chap. 1 that there is a continuum of stress–strain relationships from brittle, through semi-brittle to ductile failure. These types of stress–strain relationships with the types of failure planes which correspond with the different modes of deformation are illustrated diagrammatically in Fig. 2.4. Because there is such a gradation between the various types of shear planes, it is difficult to set out an unequivocal definition of what constitutes a 'fracture'. For the reasons outlined, the word 'fracture' will be used here to cover both the cohesive and non-cohesive forms of planar failure and adjectives such as ductile and brittle will be used to differentiate between the various forms. We shall also use the word to describe any obvious planar feature which had at one time lost most, or all, its cohesion through failure, but which is now healed and completely coherent.

The failure criteria we have been discussing here (and in Chap. 1) should, strictly, be applied only to brittle behaviour, Fig. 2.5.(*a*). However, it may not be necessary to be completely 'purist' in approach, for it appears that the brittle concepts may be applied to rocks which have not deformed in a brittle manner. Examples of what may be termed semi-ductile shear are shown in Fig. 2.5(*b*). From the geometry of the features shown, it can be seen that there are obvious points of similarity between brittle and semi-brittle or even more ductile behaviour; and few geologists

(a)

(b)

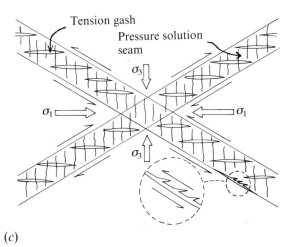

(c)

Fig. 2.5. (a) Brittle conjugate shears. (b) Conjugate ductile shear zones with associated en echelon quartz-filled tension gashes and pressure-solution seams from which the quartz in the gashes has been derived. (c) Interpretation of conjugate shear system in terms of the orientation of the stresses which gave rise to the main shear zones. The inset shows a commonly observed variation in which the zone of en echelon fractures is bounded on one side by a quartz-filled fracture. In these (inset) situations, the array of en echelon gashes are termed 'feather fractures' and can be regarded as flight feathers on an arrow, which reveal the direction of shear movement.

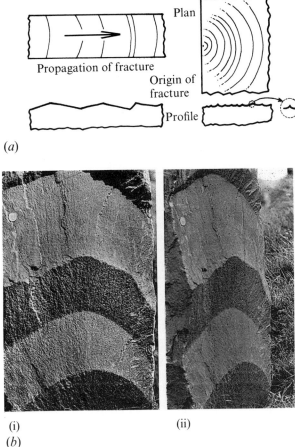

(a)

(i)   (ii)

(b)

Fig. 2.6. (a) Examples of plan and profile through the type of fracture surface features known as rib-marks. (b) Example of large scale rib-marks on one of the Gorsedd Stone, marking the site of the National Eisteddfod in Aberystwyth, Wales. (i) oblique view; (ii) plan view.

would hesitate to define the general orientation of principal stresses which gave rise to the shear zones, as though they originated as brittle shears, (see Fig. 2.5(c)). A discussion of the secondary features associated with these ductile zones of shear is given in Chaps. 6 and 18.

Because the mode of shear failure is transitional, it is not always easy to define with precision the mode of formation of a particular fracture (e.g. whether it should be considered as semi-brittle, or perhaps semi-ductile). One aspect which may help the geologist in this respect is to determine whether or not the fracture exhibits surface features.

Fracture surface features

A detailed study of individual fracture faces which exhibit well-defined morphological features was first made by Woodworth (1896). More recently, they have been studied by Hodgson (1961) in the U.S.A. and Roberts (1961) in Britain. By a remarkable coincidence, these works were conducted independently and submitted to the same journal, where they are now to

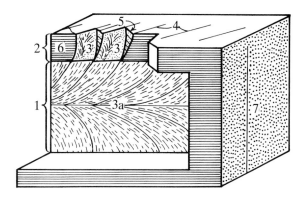

(*a*)

Fig. 2.7. Block diagram showing details of surface features on a fracture plane (after Hodgson, 1961). 1. Main fracture face 2. Fringe 3. Primary plumose or hackle structures 3a. Secondary plumose structures 4. en Echelon fractures 5. Minor cross fractures joining members of 4., 6. Shoulder and 7. Trace of main fracture face.

be found as companion papers. Bankwitz, working in East Germany, was carrying out a more extensive study, but did not publish his work until 1965 and 1966.

The surface features fall into two main groups which are generally termed (a) *Rib-marks* and (b) *Plumose* or *Hackle-marks*. Examples of rib-marks are shown in Fig. 2.6, where it will be seen that they form arcuate features. In profile, they may form cuspate, wave-forms or they may have a more rounded crestal form. When these rib-marks developed in coal, it was noticed by German miners that the central zone of the fracture was circular or elliptical. These zones, which they termed 'augen' (eyes), were about 1–2 cm in diameter. Examples of rib-marks are frequently found on the edge of shattered glass. However, their development is not restricted to such obviously strong brittle material, for they may also develop in gelatine (jello). Nor are they restricted to sediments (as illustrated in Fig. 2.6), but may develop, on a much larger scale, as features of exfoliation sheets in granite, (where they may only be revealed by careful mapping (Jahns, personal communication). It is known that rib-marks develop in relatively fast-propagating fractures which are cutting through a solid medium in which the stress field is vibrating. The ribs may represent changes in direction of the fracture as the stress field changes. Alternatively, the wave-form profile may result from a fracture developing in a plane within a vibrating solid, so that the rib-marks (especially the ones of smaller amplitude) may be 'frozen' displacements of such vibrations. It is possible that some of the rib-marks frequently seen at the edge of broken glass may have formed in this way. It is also suspected that rib-marks may develop slowly, possibly as the result of stress corrosion at the tip of the growing fracture. The more planar profiles shown in Fig. 2.6(*b*) may result from yet another mechanism, in which conjugate shears (or one set of shear and one

(*b*)

Fig. 2.8. (*a*) Plumose mark on a fracture surface in a siltstone. Note also the poorly developed rib-marks. (*b*) Hackle-marks on a fracture surface in unconsolidated silt in a tailings pond of a mine in Sardinia.

set of extension fractures) interact so that the lines of intersection define arcuate ribs. From these various mechanisms it may be inferred that the direction of development of the fracture plane was approximately normal to the trend of the rib-marks.

The morphology of *Plumose* and *Hackle-marks* and their associated features is complex and can best be shown in diagrammatic form (Fig. 2.7). These features are usually associated with material such as well-indurated rock, which would readily be described as brittle (Fig. 2.8(*a*)). However, they may also develop on fracture surfaces in only slightly consolidated sediments, such as the partially dried silt of a mine tailings pond (Fig. 2.3(*b*)). Indeed, readers may wish to demonstrate for themselves that plumose marks can readily be induced in 'bouncing' putty, a material that is so weak that it will flow under its own weight (see Chap. 1).

The *shoulder* features are less frequently seen than the plumose or hackle-marks. Indeed, good examples of the type which enabled Bankwitz and others to compile the characteristics shown in Fig. 2.7 are relatively rarely observed. The best shoulder features which have been seen by the senior author were developed in granite. The *fringe faces* for these

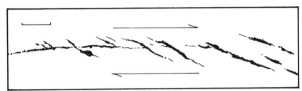

Fig. 2.9. En echelon barren features exposed on an exfoliation surface in granite (see Chap. 4) at Peggy's Cove, Nova Scotia, Canada. (See item 4 in Fig. 2.7).

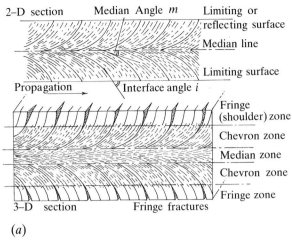

(a)

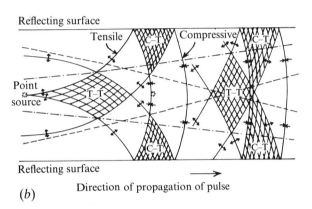

(b)

Fig. 2.10. (*a*) Hackle (or Chevron) mark geometries and descriptive terms. (After Syme–Gash, 1971.) (*b*) Illustrating a pulse mechanism of direct and reflecting elastic waves which interact to form hackle or plume structures. The diagram represents the form of an advancing compressive pulse with reflected tensile pulses in its wake. (C–T = zone of intersecting compression and tensile stresses. T–T = zone of intersecting tensile pulses. Dashed lines = limit of C–T zone, dot-dashed lines = limit of T–T zone). (After Syme–Gash, 1971.)

features were more dominant than those represented in Fig. 2.7. Moreover, the shoulder features could best be seen where the vertical fractures terminated against a flat-lying exfoliation plane and were exposed as an array of minor fractures similar in geometry to barren, en echelon extension gashes. Indeed, because of the similarity in geometry it is tempting to accord them the same general explanation. That is, the fringe-faces shown in Figs. 2.7 and 2.9 result from shear, or hybrid shear and extension failure on the main fracture face.

The senior author suggested to Gash that hackle-marks (Fig. 2.10(*a*)) may possibly be the result of a reflected elastic wave (the elastic wave would be triggered by the initiation of a brittle fracture) which interfered with the fracture surface as it advanced, as indicated in Fig. 2.10(*b*). This suggestion was taken up and augmented by Syme-Gash (1971), who differentiates between plumose structures and hackle-marks. Plumes he takes to be the result of interference of elastic waves originating from a single centre. Hackle-marks, he suggests, result from running fractures with an advancing shock source. As with rib-marks, the morphology of the plumose or hackle-marks permits the direction of fracture propagation to be inferred (Fig. 2.10(*b*)).

Other workers (such as Bahat, 1979 and Pollard *et al.*, 1982) prefer to explain the development of surface features by reference to *Fracture Mechanics*. Bahat explains surface features in terms of Mode I fractures, the term used in Fracture Mechanics to describe extension failure, while Pollard *et al.* explain the shoulder features in terms of Mode I and Mode II fractures, so that they consider the shoulder features to be a combination of extension and shear. *However,*

there is general agreement that surface features develop in the brittle or semi-brittle mode of failure. It is emphasised that the present authors believe that the various mechanisms proposed to explain the mode of development of these surface features which are seen in the field result from a combination of extension and shear, it will prove difficult, or even impossible, to decide from field evidence alone, whether surface features are the result of either extension or shear. For further information on this subject the reader is referred to the monograph on 'The application of fractography to core and outcrop fracture investigation' Kulander *et al.* (1979).

Timing of fracture development

Fractures develop in rocks at many different times and for many different reasons. When and what type

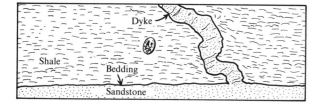

Fig. 2.11. (*a*) Neptunian (Sedimentary) dyke of sandstone intruded into overlying shales. The buckling of the dyke probably occurred as the result of compaction of the shale during diagenesis. The parent sandstone bed is visible along the lower edge of the photograph. Northcote Mouth, N. Cornwall, England. (*b*) Fractures, on the bedding plane of a siltstone unit, into which mud-rock has been injected. Milook, N. Cornwall, England.

of fracture develops depends upon the condition of a particular rock and the environment in which it exists. Igneous rocks which are emplaced as molten masses do not begin to fracture until they have cooled sufficiently to take on some of the characteristics of solids, but thereafter may continue to fracture in different ways as the result of cooling and/or orogenic or epeirogenic processes and weathering until they are exposed at the surface. (This aspect is discussed in Chap. 4.) One of the most characteristic forms of fractures which result from cooling of tabular intrusions is that which gives rise to polygonal columnar blocks. This mode of fracture is discussed in Chap. 3 where it is indicated that recently deposited sediments

may also develop polygonal fracture patterns which result from dessication and shrinkage while the sediments are at the surface. These fractures are often destroyed by subsequent events, but sometimes they remain well preserved throughout the geological evolution of the sediments. Other features which may develop, when the sediments are non-indurate, are Neptunian Dykes and other minor faults and fractures. These features usually result from the early phase of dewatering of the sediments before they have attained any marked degree of induration. Sand which has been injected into clay rock is shown in Fig. 2.11(*a*). At a somewhat higher degree of induration of the sand or siltstone it is the adjacent, and still mobile, muds or clays which can be injected (Fig. 2.11(*b*)).

Sediments may become metamorphosed and/or may be subjected to orogenic deformation, with the associated development of such features as dykes, veins and a variety of faults and barren fractures. Subsequently, further fracturing may develop in the phase of uplift and exhumation. Finally, many more fractures develop in the near surface environment as the rocks become weathered. As the unravelling of these various phases of fracture development has obvious importance for the geologist involved in projects that require the assessment or prediction of fracture system development at depths in the crust, various environments in which fractures can develop will be considered at length in the appropriate chapters of this book.

Fracture nomenclature and age relationship criteria

Classification and nomenclature are the bricks and mortar of scientific thought. It is unfortunate in the extreme, therefore, that classification and nomenclature of geological fractures have received so little attention. The terms available to describe such fractures include *fault, joint, fissure, gash, shear* and *vein*. Each of these words can be given a general definition and one can cite specific examples and state that these structures are 'typical' faults, joints etc. Unfortunately, there are many small-scale geological fractures which do not fit neatly and 'ideally' into one or other of these categories. It is with these 'non-ideal' fractures (and they form a large proportion of the types observed in outcrop) that difficulty and confusion arise. In the field, where two or more geologists stand before an outcrop containing, or cut by, fractures, discussion will minimise this difficulty. Unfortunately, the terms noted earlier are often used to describe fractures in an ambiguous, confusing and sometimes contradictory manner. The word 'joint' is particularly singled out for this negligent treatment. We shall, therefore, concentrate here on the terminology relating to mesoscopic structures (megascopic fractures and those of regional extent will be discussed later) and attempt to clarify the nomenclature problem. Paradoxically, this may be accomplished, in part, by considering the various ways in which a geologist may establish the age relationships of fractures.

A number of criteria are given here which permit relative age relationships of fractures to be assessed and determined. They are listed below and will be discussed in subsequent sections.

Criteria

(i)   Presence or absence of infilling material or staining.
(ii) Interaction of fractures of different sets.
(iii) Morphology of fractures.
(iv) Aspect ratio of fractures.
(v)   Fracture separation and bed thickness.

The first three of these criteria need relatively little comment, but the remaining two will be treated at some length.

Presence or absence of infilling material or staining
If some fractures of one set contain infilling vein material, such as quartz or calcite, and there are barren fractures which are parallel, or sub-parallel, to the infilled fractures, then the barren fractures, with little doubt, are younger than the veins; for otherwise one would expect all the sub-parallel fractures to contain some vein material.

Even when vein minerals are absent from a fracture, the fluid flow through the fracture may have left evidence of its passage. Staining of the walls of the fracture may sometimes take place and the presence of staining permits one to infer that fluid pressure in the fracture was higher than in the country rock. The presence, or absence, of staining of the fracture walls may also be used to differentiate between the ages of barren fractures within one set. However, one must use such evidence with extreme care, especially if the staining is related to near-surface flow of water, for the staining may merely indicate 'preferred' flow paths.

In addition, it is possible to deduce the geological environment (temperature and pressure), in which the veins developed, by determining the *homogenisation temperature* (at which the liquid/gaseous phases in a primary fluid inclusion, enclosed within the vein minerals, are seen to merge into a single fluid phase, when observed on a heated microscope stage). The emplacement temperature can also be determined by the O^{16}/O^{18} isotope ratio of the vein material (Fyfe, Price & Thompson, 1978). Kerrich (1974) used both these techniques on material from the same vein and by so doing was able to establish the geothermal gradient that obtained when the vein was emplaced.

Ladeira (1978) used only the bubble method of estimating temperature for adjacent veins and was able to show that, from the marked differences in the homogenisation temperatures obtained from the two sets of veins, they must have been of very different ages.

Interaction of fractures of different sets
This is the criterion most usually used by geologists to determine the relative ages of fractures. In general, the criterion has been used to 'date' shear fractures which obviously cut and displace earlier fractures. When dealing with 'joints', shear displacements, if they exist, are so small that they are difficult to assess in the field. Moreover, 'apparent' shear displacements can sometimes be erroneously used as a criterion. For example, consider the intersecting fractures shown in Fig. 2.12(*a*) and (*b*), both of which show the traces of sets of fractures on the upper surface of a turbidite unit. Fig. 2.12(*b*) is a close-up of representative fractures of the various sets shown in Fig. 2.12(*a*). It can be seen that the fractures may be grouped into three sets A, B and C (as indicated in Fig. 2.12(*c*)).

These illustrations have been used during teaching by one of the authors (N.J.P.) for several years. Students are asked to study the examples and then indicate the probable age relationships of the various sets. A table (Table 2.1) with eight options is set up and an 'opinion poll' compiled. (The reader may like to carry out this exercise now before reading further.) A reasonable proportion of the students choose option 7. A few consider that the problem is not amenable to solution (option 8). However, the majority (usually more than 70%) nominate option 2. The reasons given by those who decide on this option are usually as follows. Consider Fig. 2.12(*d*) and (*e*), which are representations of the fracture traces shown in Fig. 2.12(*a*) and (*b*); it is argued that traces 1 and 1a and 2 and 2a were originally a single line which was displaced by fracture 3a and, therefore, fractures in Set A are older than those of C which 'displace' Set A. They further argue that as the fractures of Set B cut through the other fracture set and are not displaced, those of Set B must be the youngest.

Unfortunately, these arguments are based on incomplete observation and incorrect inferences. Thus, fracture 4a in Fig. 2.12(*d*), cuts through fracture 3a without apparent displacement. Moreover, at point 5 there appears to be a movement on a fracture in Set C which is contrary to that at 1–1a. It cannot be expected that the displacement on 3a should decrease from 7–8 cm to zero (at 4a) in a distance of only 30 cms. Furthermore, shear offset on parallel fractures usually occurs in the same sense. Consequently, one can conclude that fractures of Set A are not, in fact, offset by fractures of Set C and that fractures 1 and 1a were not ever a single plane but are two separate fractures which are later in age than 3a and

(a)

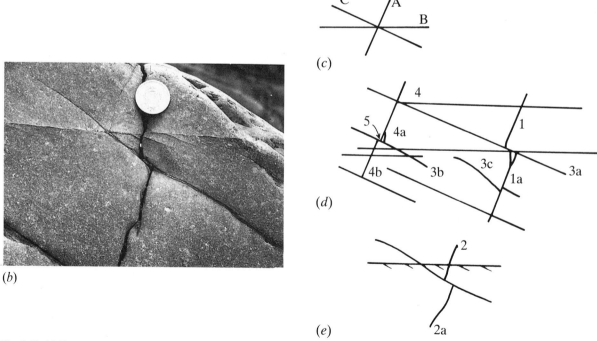

(b)

(c)

(d)

(e)

Fig. 2.12. (a) Fracture-traces on a bedding plane, of a siltstone exposed on a wave-cut beach platform, showing three main trends A, B and C which are represented diagrammatically and in detail in 2.12 (c)–(e). (b) Details of members of the fracture sets A, B and C, from an area adjacent to that shown in 2.12 (a). (c) Generalised trends of the fractures shown in (a) and (b). (d) Line diagram of fractures shown in (a), the numbers are referred to in the text. (e) A line diagram of the fractures shown in (b).

Table 2.1. *Opinion poll table*

Option	1	2	3	4	5	6
Oldest	A	A	B	B	C	C
	B	C	A	C	A	B
Youngest	C	B	C	A	B	A

Option 7 – Don't know
Option 8 – Impossible

that 1 and 1a stopped developing when they reached fracture 3a. Fracture 4a, however, may have crossed fracture 3a, or stopped and bifurcated, as it approached 3b.

Let us now turn to an inspection of Fig. 2.12(*b*). It will be noted that the fracture shown in this figure which belongs to Set B, is, in fact, a vein (the infilling is quartz). As we have seen, the feather fracture (Fig. 2.5(*c*)) associated with this vein indicates that it is a hybrid extension shear fracture where the upper block (with the coin) has moved to the right relative to the lower block. Such small feather fractures are in evidence in the field, but are only barely perceptible in the photograph, on all the (parallel) veins (i.e. Set B) shown in Fig. 2.12(*a*). Thus, all the veins of Set B exhibit shear. However, it is evident that they do not displace fractures of Sets A and C. (The open fracture, on which the coin in Fig. 2.12(*b*) is placed, is the result of weathering and wave action, so that the effects upon the interaction of fractures should be disregarded.) Consequently, the fractures of Set B pre-date those of Sets A and C, and the answer to the question appears to be option 4, i.e. Set B is older than Set C which is, in turn, older than Set A.

If the reader has reached this conclusion, he will have built into his reasoning the tacit assumption that one should compare 'like with like'. The choice of option 4 results from comparing the interrelationships of the main fractures of Sets C and A. However, it will be noted that the minor fracture (3c) which is subparallel to the trend of Set C abuts against, and stops at, fracture 1a. Using our previous reasoning, we would argue that fracture 1a is, therefore, older than 3c.

Anyone who objected to the setting up of the first six options would now seem to have a degree of justification on his or her side. However, it is usually possible to indicate the age relationships of certain fractures or, in one instance in our example, of a set of fractures (Set B). However, the question was based on a tacit and incorrect assumption, namely that all fractures of a parallel set are of the same age. Clearly, from our analysis it can be concluded that fractures which form a set are not necessarily of the same age. *The longer, more dominant fractures in any set tend to*

(a)

(b)

Fig. 2.13. (*a*) Semi-ductile, poorly defined minor thrusts formed in thick turbidite unit. Refraction of these shear zones occur at the sandstone–shale interface. Thin grit bands in the shales form drape-folds across the shears. When viewed on bedding planes such sequences of drape folds may have the appearance of ripple marks. (*b*) Brittle fracture plane in a unit of the same lithology which strikes parallel with and is situated within a few metres of the fractures seen in 2.13(*a*).

be the oldest ones belonging to that set and the progressively shorter ones tend also to be progressively younger representatives of the set. Consequently, although the fractures of Set C which formed earliest pre-date the earliest formed fractures in Set A, fractures in both sets continued to develop, with the result that contradictory evidence regarding the relative ages of the sets became apparent. Thus, in such an exercise, we must state the tacit assumption noted above and compare 'like with like'; i.e. determine the age relationships of the longest, that is, the oldest, fractures in the sets.

Morphology of fractures

From a study of the morphology of a fracture, it is sometimes possible to infer its mode and environment of development. For example, fractures may exhibit characteristics which permit them to be termed ductile, semi-ductile, semi-brittle or brittle. Such a classification must depend, to some extent, upon

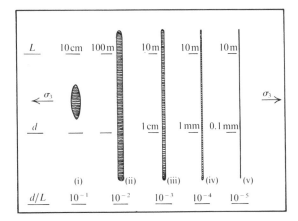

Fig. 2.14. Diagrammatic representation of a hypothetical set of parallel extension fractures with a range of aspect ratios of dilation (d) to length (L).

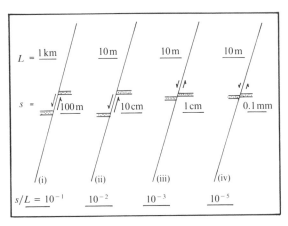

Fig. 2.15. Diagrammatic representation of a hypothetical set of parallel extension fractures with a range of aspect ratios of shear displacement (s) to length (L).

subjective assessment. Nevertheless, the exercise is well worthwhile.

Intense ductile deformation of rock has usually taken place (i) at considerable depth in the Earth's crust where the temperature and confining pressures are high or (ii) before the sediments have been completely indurated. By the time that rocks are studied by geologists they are usually at, or near, the surface, so that intensely deformed rocks have become indurated and/or, in the meantime, have passed through an environment of relatively low temperature and confining pressure. If fractures develop in both the *prograde* and *retrograde deformation* environments (here we use the terms relating to deformation in the way they relate to metamorphism: Prograde deformation is associated with downwarp and/or orogenesis and retrograde deformation occurs in the phase of uplift and exhumation), it follows that the more ductile type of fracture will pre-date the brittle type of structure. This simple sequence of events may, of course, be complicated by multiple phases of deformation. However, if brittle fractures existed prior to a ductile phase of deformation they would, in general, show evidence of this later deformation or even become transferred into ductile fractures. Hence, it may be concluded that because the fracture illustrated in Fig. 2.13(*a*) (from its appearance) can be classified as semi-ductile, it is older than the brittle fracture (Fig. 2.13(*b*)) that is exposed only a few metres distant in the same rock type, but which developed when the rock was completely indurated.

Aspect ratio

To illustrate what we mean by aspect ratios of fractures, let us consider, as examples, the series of extension fractures shown in Fig. 2.14. This diagram represents sections of the various fractures together with some typical dimensions. This, and other series of fractures, which we shall discuss later, throw light on the problems of nomenclature, as well as constituting one of the criteria by which the relative ages of fractures may be assessed.

The first of these structures (i) would be called an 'extension gash' by most geologists. The fractures, diagrammatically represented by (ii) and (iii), would be classified as veins. Many geologists may describe the fourth structure (iv) as a thin vein, although many might, in the field, describe it as a joint, or possibly an 'infilled joint'. The final example of extension fractures (v) would be classified by most, if not all, geologists as a 'joint'. Those who called example (iv) an 'infilled joint' would probably differentiate between the two types by referring to example (v) as a 'barren joint'. It will be seen, therefore, that for types of fractures represented in Fig. 2.14, the limited terminology available to the geologist can be made to work, albeit with some ambiguity. However, when one realises that there must a complete gradation between all four examples cited and that few geologists would be prepared to define specific limits for the terms they use to describe these structures, one can understand the difficulties which arise in practice.

Shear fractures can be represented diagrammatically by a similar series (Fig. 2.15). The first of these diagrammatic examples (i) represents a 1 km long fracture with a maximum displacement of 100 m. Such a fracture would invariably be referred to as a fault. The second example (ii) represents a 10 m long fracture with a maximum displacement of 10 cm (such a fracture would probably be classified as a minor fault). The third example (iii) depicts a fracture which some geologists might term a 'minor fault', or perhaps a 'minor shear fracture', while others might classify it as a major shear joint. Geologists belonging to the latter group would, therefore, probably not differentiate between examples (iii) and (iv), of Fig. 2.15, which they would possibly also term shear joints.

Again it must be emphasised that there will be a

complete continuum linking the examples of shear fractures cited. As with extension fractures, the number of terms which one may use to describe these structures is limited, and the geologist is faced with the same problem regarding the precise definition of the various terms. The problems of nomenclature are further compounded when it is recalled that the two groups of examples which illustrate extension (or dilation) and shear fractures are themselves end-members of a continuum, for a whole host of geological fractures exist which exhibit both shear and dilation.

It is clear that words fail as a means of providing an unambiguous nomenclature system. Therefore, it is suggested that geologists would be advised to use numbers or, more specifically, ratios to help them provide an adequate description of natural fractures. Thus, when dealing with dilation fractures, one can indicate the ratio of the maximum amount of dilation (d) to the length of the fracture (L). The d/L, or aspect, ratios for the cited fractures are as given in Fig. 2.14. Similarly, when considering shear fractures, one may give the ratio of the maximum shear displacement (s) to the length of the fracture (L) (Fig. 2.15).

It will be noted that the dimensions of the structures represented in Figs. 2.14 and 2.15 are realistic and that the end members of the series presented in these figures approximate to the maximum and minimum possible ratios which are likely to occur in the field. The aspect ratio not only provides factual and unequivocal information about specific fractures, but also, it is suggested, may provide information regarding the relative times of development of the structure, i.e. whether, for example, the fracture is syn-tectonic or post-tectonic. To demonstrate this point, it is necessary to consider increments of movement (e.g. shear displacement) on a fracture plane. If, for the sake of presenting the following argument, one assumes that all the fractures in a given rock type are brittle or semi-brittle and that the range of aspect ratios represented is wide, it is of interest to enquire how the fractures developed.

During movements in 1930, the maximum offset along the Tana Fault, in Japan, was 2.5 metres. During this particular increment of movement, the length of the fault actually reactivated was estimated to be 25 kilometres. Thus, for this single increment of movement, the s/L aspect ratio was 1×10^{-4}. Similarly, for the San Andreas Fault, in California, the s/L ratio for the 1906 increment of movement was 1.4×10^{-4}. These regional faults reach the surface and also extend to great depth (possibly to a zone of detachment). However, most minor faults, when and where they are developing, neither reach the surface nor an underlying zone of decollement. Therefore, because such fractures are totally constrained peri-

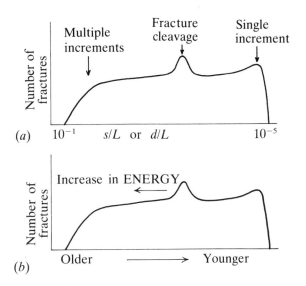

Fig. 2.16. (*a*) Diagram representing relationship between number of fractures and their aspect ratios for a hypothetical geological situation and (*b*) indicating a tentative age-relationship of various fractures based on their aspect ratios.

pherally, there will be a corresponding increase in restraint to shear movement along the fault plane. When one considers the natural data already quoted and the great increase in restraint which must develop around most mesoscopic shear fractures, it is not unreasonable to expect the s/L aspect ratio for a single increment of movement along such a fracture plane to be smaller than the figures already quoted for the Tana and San Andreas faults and they would be probably in the range $1-5 \times 10^{-5}$.

Let us now consider a hypothetical example relating to fractures within some given area, where the number of fractures with given aspect ratios varies in a manner similar to that indicated in Fig. 2.16. It is emphasised that the following argument is in no way affected by the configuration or form of the curve shown in this figure. It has been argued that a single increment of movement will result in an s/L aspect ratio of $1-5 \times 10^{-5}$. Hence, it follows that the fractures which occur with aspect ratios which approximate to this latter figure are the result of a single movement (Fig. 2.16). Fractures with these aspect ratios would be termed, in the strict sense, *joints*. At the other end of the s/L or d/L aspect ratio scale, it follows that fractures with ratios of about 10^{-2} are the result of hundreds of increments of movement, after each of which there is a significant drop in the intensity of the differential stress which initiated movement. In order that further movement may take place the differential stresses must be replenished. Such a repeated build-up or replenishment of stress could take place only during active tectonic deformation: i.e. such fractures are *syn-tectonic*.

At the other end of the aspect ratio scale, where the single increment fractures are located, it follows that, because there has been only one increment of movement, the stresses that caused the initiation of such fractures have not been replenished. As we shall see later, many fractures which develop near the Earth's surface form in response to *residual* and *remanent* stresses which are largely dissipated by the fractures and are not replenished. This argument, which will be presented at some length in Chap. 4, coupled with the fact that many of these single movement fractures also exhibit surface features, which themselves are evidence of brittle failure (i.e. such fractures formed at high levels in the crust) points unequivocally to the conclusion that such fractures with aspect ratios approaching 10^{-5} are *post-tectonic*. Further, it is reasonable to suggest that between the extreme values of aspect ratios indicated in Fig. 2.16 one may *very tentatively* use the aspect ratios to indicate the relative age relationships of structures with intermediate aspect ratios. It has been assumed in this discussion that the fractures were brittle, or semi-brittle. However, it will be appreciated that the reasoning may, with care, be extended to fractures which have developed in ductile or semi-ductile forms.

When fractures are less than 10 metres or so in length, especially careful observation is needed before attempts at classification may be made: for unless these fractures cut some obvious marker, a shear displacement of a few millimetres, or even a centimetre or two, may pass unnoticed. Thus, a fracture with an aspect ratio of 10^{-3} could easily be placed in the same category as fractures with aspect ratios of 10^{-5}. This lumping together of fractures which represent different ages and environments of development is one of the reasons why attempts at fracture analysis have frequently proved to be unrewarding, if not confusing or positively misleading. In this context, it is interesting to note that the term *joint* used in its widest (and most confusing) sense would be extended by some geologists to include fractures with an aspect ratio of about 10^{-3}. The use of this term to include fractures with such an aspect ratio is one of the reasons why there is debate about whether joints are syn- or post-tectonic.

When dealing with shear fractures it is often difficult to determine the precise displacement. However, because we are only interested in determining the general order of magnitude of the aspect ratios, a two or threefold error in estimating the displacement should not significantly affect the classification of an individual fracture.

When fractures are observed at the surface, weathering may greatly increase the apparent dilation. One must therefore guard against making erroneous classifications of such open, barren fractures. In addition, the d/L ratio is likely to change from a maximum at the time of the initiation of such barren extension fractures to a lower value, as the fluid pressure, which caused the fracture, decays.

As regards the hypothetical *fracture cleavage* which gives rise to the peak in Fig. 2.16, the energy involved in deformation may manifest itself by causing multiple displacements on a few fractures or, in some circumstances, may cause only a limited dilation on an extremely large number of fractures. Hence, from energy considerations, fracture cleavage would relate to an earlier tectonic deformation than its position on the s/L–d/L aspect ratio continuum might seem to indicate.

From the previous comments, it is apparent that the difficulties of applying these concepts are considerable, for they require detailed and often time-consuming observation. Moreover, it may not always be possible to find sufficient, significant field data when one is trying to establish the aspect ratios in poorly exposed terrain. In the light of these difficulties, the use of aspect ratios to indicate the relative age of a fracture must be used with extreme caution. Nevertheless, because careful observation is a desirable habit for any geologist to develop in any structural problem in which fractures play an important role, the benefits which result from the application of these concepts can make the effort involved well worth while.

Fracture separation and bed thickness

The spacing between fractures, especially those in the mesoscopic size range, is often of great relevance. For example, it determines the maximum size of block that can be obtained in quarrying. Fracture spacing, combined with orientation, is important in the design and stability of engineering and mining projects, such as major cuts, open pits, as well as large mines and other underground openings and chambers. Clearly, as fracture spacing plays an important role in all aspects of fluid transport, it is of import in hydrogeology, the hydrocarbon industry and nuclear waste disposal problems. Thus, if fractures are large, interconnected and closely spaced they may present problems in those situations (e.g. reservoir construction) where it is desirable to prevent leakage. Conversely, the relative absence of fractures in rocks may cause problems (e.g. the migration of hydrocarbons into a well may be so hindered by the absence of natural fractures that they need to be artificially induced). Thus, the problems relating to fracture spacing are of such importance, in their own right, that they warrant the applied geologist's closest attention. Spacing also permits some decisions to be made regarding the age of fractures.

It has long been established that the lithology and thickness of competent (i.e. strong) beds influence the spacing of fractures that cut, and are contained by, them. It was indicated by Bogdanov (1947), Novikova (1947) and Kirillova (1949) that the spacing (W_0) between fractures varied as:

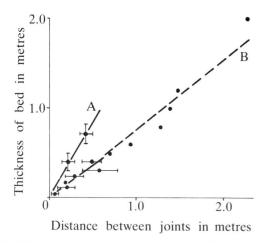

Fig. 2.17. Approximately linear relationship between fracture separation and bed thickness for two different lithologies. (From data by Bogdanov, 1947; Novikova, 1947 and Kirillova, 1949.)

$$W_0 = k' a \qquad (2.1)$$

where a is the bed thickness and k' is some constant that is related to the lithology of the bed (Fig. 2.17). It has also been established that the spacing of fractures is influenced by the degree of orogenic deformation. Harris *et al.* (1960) concluded that, for the Goose Egg Dome (Wyoming, U.S.A.), the largest number of fractures, per given area, developed in those parts of the structure that exhibited the greatest degree of curvature, whether measured in the dip or the strike directions (Fig. 2.18).

The relationship of fracture separation to bed or unit thickness has been investigated by model experiments and also by theoretical analysis. The experimental work by Sowers (1973) indicates that there is a linear relationship between fracture separation and the thickness of the units used in his model. As regards theoretical treatments various analyses have been carried out (e.g. Price (1966), Hobbs (1967), and Sowers (1973)) all of which predicted that there should be a linear relationship between fracture spacing and bed thickness.

Here we shall consider only the simple theoretical model used by Price (1966) which comprises a uniform competent bed, thickness a, with tensile strength T, set between two incompetent units (Fig. 2.19(a)). The competent bed is subjected to a uniform effective tensile stress (σ_T) which causes the competent unit to fracture at AA, (Fig. 2.19(b)). If the bedding planes had no frictional restraint, this fracture would open (and thus form a considerable cavity) until the tensile stresses in the rock were relieved over a wide area. In general, such cavities do not form because a shear stress ($\tau = \mu\sigma_n$, see Eq. (1.60)), plus, perhaps, a small degree of cohesion between beds, will tend to resist such bedding plane slip and widening of the initial fracture. At the free surface created by this fracture, the horizontal effective stress falls to zero.

However, because of the frictional restraint to movement along the bedding plane, the effective tensile stress in the competent unit increases in magnitude away from the fracture until, at distance W from AA, it attains its original magnitude. This distance W represents the minimum distance at which a second tensile fracture may develop, but it is probably also close to the actual fracture separation (i.e. $W \simeq W_0$). The equilibrium condition for this portion of the competent bed, length W, represented in Fig. 2.19(c), is given by:

$$\sigma_T a = \tau W \quad \text{or} \quad W = \frac{\sigma_T a}{\tau} \qquad (2.2)$$

where τ is the combined average shear stress acting along the top and the bottom of the bed, throughout the distance W.

Consider now the equilibrium of a second, nearby, competent unit which is also subjected to the same tensile stress σ_T. Let this second competent bed and its adjacent layers of incompetent material have the same properties as in the first model, but let the thickness of the second competent layer be ka. Then, as for the first layer, equilibrium is given by:

$$\sigma_T k a = \tau W'$$

or

$$W' = \frac{\sigma_T k a}{\tau}. \qquad (2.3)$$

Then W'/W can be obtained by dividing Eq. (2.2) by Eq. (2.3), so that:

$$\frac{W'}{W} = k. \qquad (2.4)$$

That is, there will be a linear relationship between fracture separation and bed thickness, for a given material, or lithology. The analyses proposed by both Hobbs and Sowers are more sophisticated, but point to the same conclusion. Thus, the field data so far cited, the experimental models and the mathematical analyses all indicate that there is a linear relationship between fracture separation and bed thickness. Consequently, despite a warning by Norris (1958) and Mastella (1972) who suggested that the relationship in some instances was non-linear, the linearity of the relationship is still widely accepted.

However, even a casual inspection of the fracture spacing which is exhibited by massive competent beds, when these are well exposed (Fig. 2.20) indicates that the fractures are relatively closely spaced: and this should immediately lead one to suspect the general validity of the simple linear relationships represented in Fig. 2.17 and by Eq. (2.4). The reason for this apparent paradox results mainly from the fact that the early field investigations, on which the linear relationships are based, were conducted in sedi-

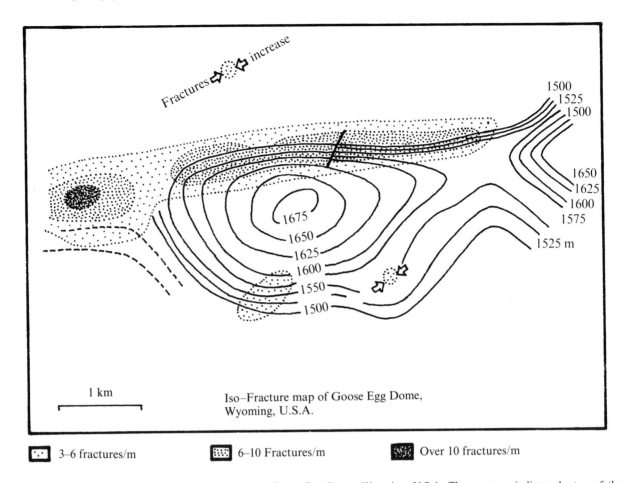

3-6 fractures/m   6-10 Fractures/m   Over 10 fractures/m

Fig. 2.18. Distribution of fracture intensity about the Goose Egg Dome, Wyoming, U.S.A. The contours indicate the top of the Pennsylvanian. (After Harris *et al.*, 1960.)

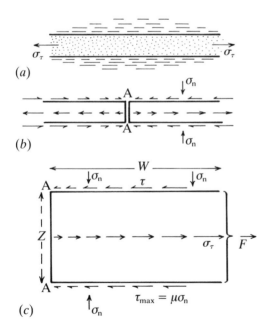

Fig. 2.19. (*a*) Single competent unit subjected to a uniform effective tensile stress. (*b*) Diagrammatic representation of reduction of lateral effective tensile strength and increase of bedding parallel shear stress that occurs in response to the generation of a single tensile fracture. (*c*) Equilibrium conditions for portion of competent unit, length *W*, adjacent to the tensile fracture.

mentary sequences in which bed thickness did not exceed 1.5 m.

Measurements in Carboniferous turbidites and Jurassic limestones, in which the thicker beds exceeded 1.5 m are reported by Ladeira & Price (1980). These measurements enabled the relationship between fracture spacing and bed thickness to be established for two markedly different lithological types. In addition, a study of the development of fracture spacing in thick units of Asmari limestone had previously been conducted by McQuillan (1973) and these data augment those reported by Ladeira & Price. The significance of McQuillan's data is not immediately apparent because he expressed fracture frequency in terms of the number of fractures per unit length of traverse and plotted this number against bed thickness expressed on a log-scale. The curves that fit the data reported by Ladeira & Price and McQuillan are represented in Fig. 2.21(*a*) and (*b*). It was suggested by Ladeira & Price that these various data can be represented by two straight line relationships (Fig. 2.21(*c*)) which are the results of two different mechanisms. The first of these two relationships, indicated by line OA in the latter figure, is the linear relationship forecast by the various theoretical

Fig. 2.20. Relatively closely spaced fractures in massive sediments exposed in the cliffs above the Temple of Hatshepsut, Luxor, Egypt.

treatments already cited, in which fracture separation is influenced by traction at the competent-incompetent rock interface. This relationship need not be further commented upon.

The second relationship for which fracture separation in beds of a given lithology is independent of bed thickness, is represented by line BC in Fig. 2.21(*c*). This relationship, it is suggested, results from the hydraulic fracture mechanism (see Chap. 1), when:

$$p > S_3 + T \tag{2.5}$$

where p is the fluid pressure, S_3 is the least principal stress and T is the tensile strength of the rock. This mechanism will strictly operate only in the development of extension fractures. However, the influence of fluid pressure will also be extremely important in the formation of the hybrid extension and shear fractures which develop in sediments. As far as we are aware, the proportion of extension and hybrid fractures which develop in thick sedimentary units, relative to shear fractures, has not been quantified. However, from studies of fracture traces and published data (e.g. Norman *et al.* (1977) and that presented in Chap. 9) we suggest that extension and hybrid fractures often appear to be dominant. In any event, in some environments, they probably constitute a very important

proportion of the total number of fractures that exist in thick layers.

With few exceptions, total stresses in the crust will be compressive. However, as we have seen, extension fractures develop when the effective stress is tensile. This situation exists when the fluid pressure is higher than the least principal total stress (S_3). Fluid pressures are most frequently abnormally high in the *prograde* phase of deformation, i.e. when rocks are undergoing active burial or orogenic deformation. Arguments supporting this statement have been put forward by Price, in Fyfe *et al.* (1978). Consequently, fractures which fall on the vertical line BC in Fig. 2.21 are relatively old fractures which developed perhaps at considerable depth.

The mechanism which Ladeira & Price put forward to explain the development of such fractures is outlined below. They point out that hydraulic fracture occurs when the least effective principal stress exceeds the tensile strength (T) of the rock, as indicated in Chap. 1. However, from 'fracture mechanics' concepts it is known that:

$$T = \frac{\pi K_{IC}}{2C} \tag{2.6}$$

where, as before, T is the tensile strength, C is the half length of the fracture and K_{IC} is a material property

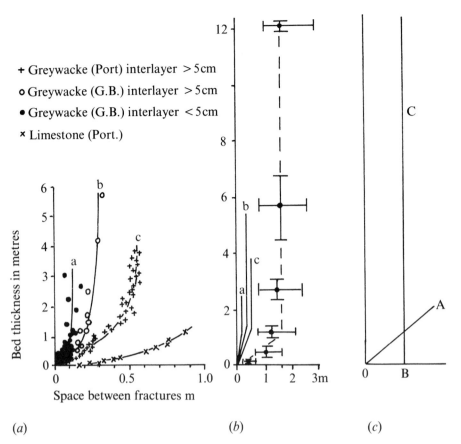

+ Greywacke (Port) interlayer >5cm
o Greywacke (G.B.) interlayer >5cm
• Greywacke (G.B.) interlayer <5cm
× Limestone (Port.)

Fig. 2.21. (*a*) Relationship between fracture separation and bed thickness, for thick units using the data from Ladeira & Price, 1980. (*b*) The trends obtained from the data in 2.21(*a*) are represented as curves a, b, and c, together with a compilation of data obtained by McQuillan. (*c*) Possible interpretation of curves represented in 2.21(*b*) in terms of two mechanisms of development (OA it is suggested is related to frictional constraint on the bedding plane (Fig. 2.19) and BC to hydraulic fracturing).

known as the critical stress intensity factor. Thus, the tensile strength is not a constant, but depends upon the length of the fracture. From Eqs. 2.5 and 2.6, it may be inferred that for a given value of least principal (total) stress (S_3), the magnitude of the fluid pressure (p_p) required to propagate a developing fracture is smaller than the fluid pressure (p_1) required to initiate the fracture (i.e. when the fracture length is extremely small). Consequently, when the fracture is developing, there will be a fluid pressure gradient from the relatively low pressure (p_p) in the fracture to the higher fluid pressure (p_i) which exists in the rock at some distance (L) from the fracture, as indicated in Fig. 2.22. Therefore, at a distance L from the fracture, the rock is 'unaware' of the existence of that fracture. At this distance from the fracture, the original conditions for hydraulic fracture exist and so a second fracture is able to develop. Fracture separation in such a set would therefore be closely linked to the distance L which, in turn, is clearly related to the gradient of the fluid pressure dp/dx, as indicated in Fig. 2.22, but is independent of bed thickness. In its turn, the fluid pressure gradient will largely be determined by: (i) the rate of propagation of the fracture (which will be related to the strain-rate) and (ii) the permeability (K_p) of the unfractured rock, which, of course, will be related to the lithology. With reference to the second of these factors, although data are

lacking, it is reasonable to infer that the fractures are more closely spaced in the turbidites than in the limestones because the clay mineral content renders the turbidites more impervious. By the same token, one would expect fractures in thick, porous, coarse-grained sandstones (with a low value of K) to be yet more widely spaced.

As has already been indicated, these fractures will develop at depth, and will be classified subsequently as 'early' fractures. Because they are most clearly seen in thick beds or units, it follows that the fractures must themselves be reasonably extensive. Thus, these findings completely corroborate the conclusion reached earlier, when using the criterion of fracture interaction, namely that 'the oldest fractures in a set are usually the longest'.

The arguments regarding mechanisms outlined above are qualitative and more data are required before a rigorous and quantitative analysis can be attempted. However, no matter what our state of understanding may be, we suggest that the empirical relationships indicated in Fig. 2.21 are valid and relevant in many problems of practical and economic importance.

The data of McQuillan and Ladeira & Price were obtained from areas which had experienced orogenic deformation. As we shall see, differential stresses, in such orogenic environments, are often

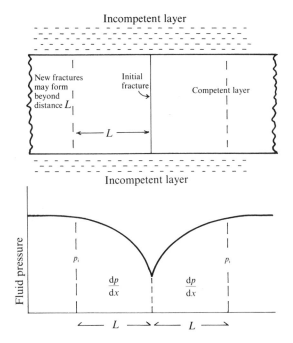

Fig. 2.22. Mechanism proposed by Ladeira & Price, 1980, whereby spacing of hydraulic fractures may be influenced, or even controlled, by the permeability of the rock unit.

sufficiently large to induce shear fractures. Also, it will be argued, the fluid pressures can reasonably be expected to be high, so that many fractures will result from the hydraulic fracturing mechanism. However, as we shall see (Chap. 9), in non-orogenic environments it is likely that many near-vertical fractures form by a mechanism other than hydraulic fracturing. Consequently, the mechanism proposed by Ladeira & Price, which related specifically to the hydraulic fracturing mechanism, will not be applicable; and it will be necessary to seek some other explanation for the spacing of major fractures in thick sedimentary units.

Commentary

The use of the various criteria, outlined above, which can be used to evaluate the relative ages of fractures requires careful and intelligent application. In some situations, the application of one or other of the criteria may not give rise to significant information. Nevertheless, when used in conjunction with each other they should provide sufficient evidence on which to establish a sequence of the relative ages of fracture sets and even of individual fractures within sets.

References

Bahat, D. (1979). Theoretical considerations on mechanical parameters of joint surfaces based on studies on ceramics. *Geol. Mag.*, **116**, 81–92.

Bankwitz, P. (1965). Über Klüfte, Beobachtungen im Thüringischen Schiefergebirge. *Geologie*, **14**, 242–53. (1966). Über Klüfte II. *Geologie*, **15**, 896–941.

Bogdanov, A. (1947). The intensity of cleavage as related to the thickness of the bed (Russian Text). *Sov. Geol.*, **16**.

Fyfe, W., Price, N.J. & Thompson, A.B. (1978). *Fluids in the Earth's Crust*. Amsterdam: Elsevier.

Griggs, D.T. & Handin, J. (1960). Observations on fracture and a hypothesis of earthquakes. In *Rock Deformation*. Geol. Soc. Am. Mem., 79, eds. D. Griggs & J. Handin, pp. 347–73.

Harris, J.F., Taylor, G.L. & Walper, J.L. (1960). Relation of deformational fractures in sedimentary rocks to regional and local structures. *Am. Assoc. Petrol. Geol. Bull.*, **44**, 12, 1853.

Hobbs, D.W. (1967). The formation of tension joints in sedimentary rocks. *Geol. Mag.*, 550–6.

Hodgson, R.A. (1961). Classification of structures on joint surfaces. *Am. J. Sci.*, **259**, 493–502.

Kerrich, R. (1974). Aspects of pressure solution as a deformation mechanism. Unpublished Ph.D. Thesis, University of London.

Kirillova, I.V. (1949). Some problems of folding (Russian Text). *Trans. Geofian.*, **6**.

Kulander, B.R., Barton, C.C. & Dean, S.L. (1979). *The application of Fractography to core and outcrop fracture investigation*. Morgantown, West Virginia: U.S. Dept. Energy. Morgantown Energy Tech. Center.

Ladeira, F.L. (1978). Relationship of fractures to other geological structures in various crustal environments. Unpublished Ph.D. Thesis, University of London.

Ladeira, F.L. & Price, N.J. (1980). Relationship between fracture spacing and bed thickness. *J. Struct. Geol.*, **3**, 179–84.

Mastella, L. (1972). Interdependence of joint density and the thickness of layers in the Podhale Flysch. *Bull. de L'Adac. Pol. des Sci.*, **20**, 3, 187–96.

McQuillan, H. (1973). Small-scale fracture density in Asmari Formation of Southwest Iran and its relation to bed thickness and structural setting. *Am. Assoc. Petrol. Geol. Bull.*, **57**, 12, 2367–85.

Nickelsen, R.P. & Hough, V.N.D. (1969). Jointing in the Appalachian Plateau of Pennsylvania. *Geol. Soc. Am. Bull.*, **78**, 924–31.

Norman, J., Price, N.J. & Peters, E.R. (1977). Photogeological fracture trace study of controls of kimberlite intrusions in Lesotho Basalts. *Trans. Instn. Min. & Metall.*, B78–80.

Norris, D.K. (1958). Structural conditions in Canadian coal mines. *Bull. Geol. Surv. Can.*, **44**, 1–53.

Novikova, A.C. (1947). The intensity of cleavage as related to bed thickness (Russian Text). *Sov. Geol.*, **16**.

Pollard, D.D., Segall, P. & Delaney, P.T. (1982). Formation and interpretation of dilitant echelon cracks. *Geol. Soc. Am. Bull.*, **93**, 1291–302.

Price, N.J. (1966). *Fault and joint development in brittle and semi-brittle rocks*. Oxford: Pergamon.

Roberts, J.C. (1961). Feather-structures and the mechanics of rock-jointing. *Am. J. Sci.*, **259**, 481.

Sowers, G.M. (1973). Theory of spacing of extension fractures. *Engl. Geol. Case Histories*, **9**, 27–53.

Syme-Gash, P. (1971). A study of surface features relating to brittle and semi-brittle fractures. *Tectonophysics*, **12**, 349–91.

Woodworth, J. (1896). On the fracture system of joints, with remarks on certain great fractures. *Boston Soc. Nat. Hist. Proc.*, **27**, 163–84.

3 Concordant and discordant intrusions

Introduction

When igneous material is intruded into flat-lying or gently inclined sediments to form extensive sheets, as indicated in Fig. 3.1(*a*), the intrusions may be *concordant* with, or *discordant* to, the bedding. Near-parallel-sided concordant sheets in this situation are termed *sills*, while discordant sheets, which commonly cut orthogonally across the bedding, are termed *dykes* (or *dikes*, using the American spelling). Other forms of concordant intrusions, some of which will be commented upon in this chapter are shown in Fig. 3.1(*b*).

A situation which is commonly encountered is represented in Fig. 3.1(*a*) but, even when bedding is far from horizontal, geologists often continue to term steeply inclined or vertical igneous sheets as dykes. (Indeed, even when such an intrusion is not discordant, some geologists may still refer to it as a dyke.) An example of such a sheet which is concordant with vertical foliation in schists is represented in Fig. 3.2. In Fig. 3.2(*a*) the situation is given as it may be encountered in the field and recorded on a mapping slip or sketched in a field note-book. The manner in which such information may be presented on the final map is indicated in Fig. 3.2(*b*). In a report based on the map and field observations, the inexperienced

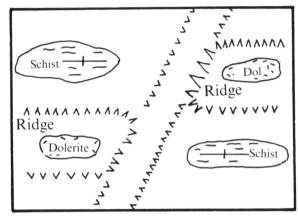

(*a*)

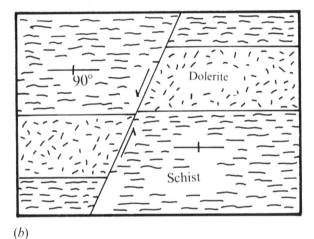

(*b*)

Fig. 3.2. (*a*) Schematic representation of field data showing dolerite intruded into schists, with a vertical foliation, which have been cut by a fault. (*b*) Possible (and common) interpretation of the field evidence shown in (*a*).

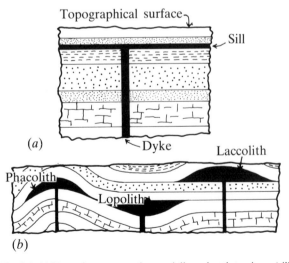

(*a*)

(*b*)

Fig. 3.1. (*a*) Near-planar concordant and discordant intrusions, (sill and dyke) in flat-lying sediments. (*b*) Types of non-planar concordant intrusions: a laccolith may have a flat base but an upward concave roof; a lopolith may have saucer-shaped or approximately flat roof and downward-curved (sagging) floor and a phacolith has a concavo-convex form. Phacoliths when they are ore bodies are termed saddle-reefs.

geologist would almost certainly express the opinion that the dyke was cut and displaced by the inferred fault. Such a statement may, of course, be true. However, on the available evidence, the relative ages of the intrusion and the fault cannot be known. The experienced geologist or, as we shall see, one who understands the mechanisms by which intrusions develop, will realise that there is an alternative interpretation.

However, this is not the only reason why dykes, sills and other related forms of intrusion are worthy of study by the structural geologist. These bodies result from an intrusion of magmatic material when in a molten state, which then cools and solidifies *in situ*.

Hence, by studying these igneous bodies, insight can be obtained into the way in which other fluids, such as water and brines, may produce fractures and escape from the rock mass leaving no evidence of their passage, other than sets of barren fractures. Furthermore, a study of these concordant and discordant intrusions enables geologists to make inferences about the stress environment obtaining at the time of intrusion and the manner in which the country rock responds to that intrusion. In this chapter, therefore, we shall first outline the various factors and parameters which pertain to the intrusion of such concordant and discordant bodies. Furthermore, we shall attempt to quantify and/or set limits on these parameters. The manner in which dykes may pass into sills and how, in turn, sills may give rise to dykes higher in the crust will also be considered.

Other topics to be discussed include: (i) the way in which laccoliths develop and the inferences that may be drawn about the intruding material and also the mechanics of deformation of the roof of the intrusion; (ii) types of country rock deformation induced by, or responsible for, the form of intrusions; (iii) the development of the polygonal fractures which are so commonly a feature of sills and dykes. However, we shall first consider the general mode of development of planar intrusions.

Development of planar intrusions

Igneous materials which are intruded as dykes and sills at high levels in the crust are derived from either primary or secondary magma chambers. The processes of partial melting and segregation of lower crustal, or mantle, material, that form a primary magma chamber will not be discussed here. However, as it is likely that such chambers comprise a porous mesh of unmolten material, it is only the pore spaces, constituting perhaps 30 per cent of the chamber, which contain liquid magma. Secondary magma chambers may develop higher in the crust (by processes which will be described later) in the form of thick and relatively extensive sills (though for much of their active life, because of rapid cooling at their margins, the molten material may be restricted to the central portions of such intrusions) and these major bodies may, before they cool and solidify, subsequently give rise to the development of dykes and smaller sills yet higher in the crust.

The magmatic pressure (P_m) in the primary magma chamber may be somewhat less than that of the overburden pressure (σ_z). However, at depths in the crust where primary magma chambers exist, the rocks are hot and unable to withstand high differential stresses, so that the horizontal stress is unlikely to be much smaller than the vertical stress. Consequently, until P_m becomes approximately equal to

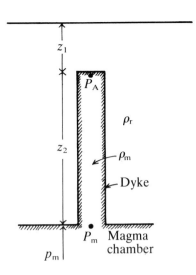

Fig. 3.3. Symbols and dimensions used to estimate magmatic pressure (P'_A) within a dyke.

σ_z, significant upward migration of magma in sheet intrusions is unlikely to take place. As regards the magmatic pressure in secondary magma chambers which forms sills, it is apparent from the geometry of these bodies that P_m cannot be significantly less than σ_z. Indeed, as we shall see, because of the flexural rigidity of the upper layers of the crust, P_m may exceed σ_z. However, it is sufficient at this time to assume that at the initiation of a dyke which penetrates high into the upper levels of the crust, $P_m = \sigma_z$.

Let us assume that a vertical dyke is intruded from a primary magma chamber, as indicated in Fig. 3.3. (The stress conditions that must be satisfied for such an intrusion to be initiated will be discussed later.) If $P_m = \sigma_z$, the values of P_m at depth z_m is given by:

$$P_m = \rho_r g(z_1 + z_2) \tag{3.1}$$

where ρ_r is the average density of the cover rocks, g is the constant of gravitational acceleration and $z_m = z_1 + z_2$.

If the fluid in the intrusion is at rest, the fluid pressure (P_A) at point A (depth z_1) at the top of the intrusion is given by:

$$P_A = P_m - \rho_m g z_2 \tag{3.2}$$

where ρ_m is the average density of the magma and z_2 is the height of the dyke. From Eqs. (3.1) and (3.2), it follows that the excess of magmatic pressure over geostatic pressure (S'_z) at depth z_1 is given by:

$$P_A - S'_z = z_2 g(\rho_r - \rho_m). \tag{3.3}$$

This relationship is the well-known buoyancy equation which, as we shall see in Chap. 4, has been applied to explain the emplacement of salt diapirs. In order that there be an excess of magmatic pressure over vertical load, and therefore a tendency for the magma to rise in the crust, it follows from Eq. (3.3) that two conditions must be met. They are: (i) $z_2 \neq 0$ and (ii) $\rho_r > \rho_m$. It is emphasised that the symbols ρ_r and ρ_m refer to average densities of the cover rocks and column of magma respectively. In natural environments there is a tendency for both the density of the country rock and of the magma to decrease, albeit at different rates, as the surface is approached. It is, therefore, not a simple task to determine the precise depth at which an intrusion will cease to rise. However, as will be apparent from data presented in Chap. 4 (Fig. 4.15), a dyke will be unable to penetrate far into cover rocks which have a density significantly smaller than that of the rising magma.

There is further limitation regarding the applicability of Eq. (3.3): namely that this equation is based on the assumption that the magma has ceased to flow. While the magma is still mobile and moving upwards in the crust, it exerts a viscous drag on the wall rock of the intrusion and some part of the initial magmatic pressure is needed to overcome this resistance to flow. For these conditions, therefore, Eq. (3.3) gives too large a value of the excess magmatic pressure. The viscous drag of the magma on the intrusion walls is in part related to the velocity of the upward flow of the molten material. It is now necessary to consider the time during which the intruding magma remains fluid so that, knowing the quantity of magma to be emplaced and the width of the intrusion, the minimum velocity of emplacement can be estimated. An estimate of the maximum time available for the magma to be intruded can be obtained by assessing the rate at which cooling of such a hot fluid takes place. The maximum time available for emplacement will then be some period shorter than that in which the hot 'fluid' magma cools to become a 'solid'.

Cooling of planar intrusions

The mobility of magma is extremely dependent upon its temperature. This may be inferred from the manner in which extruded basic lava flows freely when it is very hot, near the vent, but becomes progressively more turgid (even if the slope down which the lava flows remains constant) as the lava cools. More specifically, it has been demonstrated (Shaw *et al.* 1968, and Williams & McBirney, 1979) that basalt acts as a Newtonian liquid in the temperature range 1200 °C to 1100 °C, when the coefficient of viscosity is about 10^4 poise (10^3 Pa s). However, below a temperature of 1100 °C the flow behaviour becomes non-linear and markedly more viscous. Indeed, the onset of crystallisation renders the magma so viscous that it effectively arrests further flow.

A mathematical analysis of the manner in which a sheet of hot material (which has been introduced into a relatively cold, solid medium) cools has been presented by Jaeger (1959). For mathematical con-

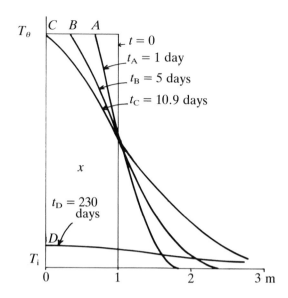

Fig. 3.4. Theoretical time/cooling curves for an intrusion. (After Jaeger, 1959.)

venience, he found it necessary to assume that the hot material, at temperature T_θ, is injected instantaneously into a uniform medium that is at an initial uniform temperature T_i and which extends for an infinite distance on either side of the simulated intrusion. The change in temperature which ensues with the passing of time in the 'intrusion' and in the adjacent 'country rock' for a dyke with a half width (x) of 1 metre is indicated in Fig. 3.4. The cooling is expressed in terms of the difference between the intrusion and original ambient temperatures. The timing of the cooling process represented in this figure is dependent upon the values of thermal conductivity chosen for intrusion and country rock. Moreover, in real situations, a dyke, or sill, is often intruded into porous, water-laden country rock. The heat lost from a relatively thick intrusion (where $x \gg 1$ m) may cause convection resulting in a more rapid transfer of heat away from the intrusion. Nevertheless, although cooling may sometimes be more rapid than is indicated in Fig. 3.4, the times represented there can be taken as sufficiently accurate for our purposes.

The times cited in this figure refer specifically to an intrusion with a half-width (x) of 1 metre. A more general relationship can be derived (Jaeger, 1959) in which the time (t_c) required for a dyke or sill to attain the temperature distribution shown by curve C in Fig. 3.4 is given by:

$$t_c = \frac{x^2}{4kF^2} \tag{3.4}$$

where x is the half width of the intrusion, k is the thermal diffusivity of the magma and the country rock (which is assumed to be the same) and F is a dimensionless constant. The interested reader is referred to Jaeger (1959) for the derivation and methods of

Table 3.1.

Half thickness	t_s
0.01 m	15.6 mins.
0.1 m	2 h 36 mins.
1.0 m	10.9 days
2.0 m	43.6 days
10.0 m	1090 days (3 years)
100.0 m	300 years

Table 3.1a. *Time to cool to 10 per cent of original temperature*

Half thickness	$t_{10\%}$
0.1 m	2.3 days
1.0 m	230 days
10.0 m	66 years

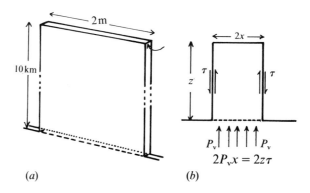

Fig. 3.5. (*a*) Schematic representation of a 1 cm thick 'sliver' of a dyke. (*b*) Stress/viscous drag relationship for such a sliver.

utilising this equation. Here, it is sufficient to note that from Eq. (3.4) it follows that t_c, and the other times required to attain the other corresponding cooling curves shown in Fig. 3.4, vary as the square of the half thickness of the intrusion.

The temperature drop in a basalt before its viscosity becomes too high for rapid flow is relatively small (about 100 °C). Hence, if the temperature in a basaltic intrusion followed curve C (Fig. 3.4), then much, or even all, of the dyke would have lost its ability to flow rapidly. Indeed, much of the dyke would have solidified. Consequently, the times given in Table 3.1, which may best be applied to basaltic intrusions, can be expressed as the time to solidification (t_s). This gives an outside limit for the rate of emplacement of an intrusive sheet of a known thickness and extent, provided the basic assumptions of the analysis are met (i.e. that emplacement is instantaneous and that magmatic material no longer flows along the intrusion). If flow continues, then longer cooling times would be required.

If a volume of magma Q is emplaced to form a parallel-sided intrusion of width $2x$ in time t, then, assuming the rate of flow is a maximum at the base of the intrusion and falls to zero at the highest level the dyke attains, the *average* rate of flow (v) is given by:

$$v = \frac{Q}{kxt} \tag{3.5}$$

if Q is given in cm^2 and x is in cm, v is in cm/s. Let us consider a vertical, 10 km high dyke with an average width of 2 m. From Fig. 3.4 and Table 3.1, the time to solidification (t_s) for such a sheet is 10.9 days. However, from the argument presented above, it can

be concluded that the intrusion would probably be emplaced in a time span of 1–10 per cent of t_s represented by curve C in Fig. 3.4 (and cited for various thicknesses of intrusion in Table 3.1). Hence, the emplacement time for the specific example represented in Fig. 3.5 is probably in the range 10^4–10^5 s (2 hr 45 mins–1 day). Let us arbitrarily assume that $t = 5 \times 10^4$ s, then for the values cited for Q and x, it follows from Eq. (3.5) that the average velocity of emplacement will be 10 cm/s. Again it is emphasised that this analysis does not apply to feeder dykes, in which flow is likely to be protracted.

It can be shown that a Newtonian liquid (i.e. a liquid exhibiting linear stress/strain-rate behaviour) which flows between parallel plates, at an average velocity v, gives rise to a viscous drag (τ) on the walls of the channel given by:

$$\tau = \frac{k\eta v}{x^2} \tag{3.6}$$

where η is the coefficient of viscosity of the liquid and k is a numerical constant (the value of which will be commented upon later). Let us now apply this equation to the model represented in Fig. 3.5(*b*). It will be seen that the total viscous resistance to emplacement, of the 1 cm wide sliver of the intrusion, requires the action of an element of the pressure in the magma (P_v) such that:

$$P_v 2x = 2\tau z$$

or

$$P_v = \frac{\tau z}{x} \tag{3.7}$$

Hence from Eqs. (3.6) and (3.7) it follows that:

$$P_v = \frac{k\eta v z}{x^3}. \tag{3.8}$$

This relationship (Eq.(3.8)) has been verified experimentally. The value of k, however, depends upon the walls of the fissure, i.e. their degree of 'roughness', departure from parallelism and constance of width. It is probable that k ranges in value from 12–20. Here we shall take $k = 16$.

The buoyancy excess pressure given by Eq. (3.3) is reduced by the pressure P_v. For the values cited previously of $z = 10$ km, $2x = 2$ m, an average velocity $v = 10$ cm/s and $\eta = 10^4$ poise, then the value of P_v necessary to overcome the viscous resistance to emplacement is approximately 0.008 bars.

The model element presented above is extremely simple and does not represent the dimensions likely to be encountered throughout the plan of a dyke. Clearly, such low values of P_v do not significantly influence the buoyancy effect. However, the viscosity of more acid rocks (at their appropriate melt temperature) may be at least six orders of magnitude greater than that of basalt at 1100 °C. (We exclude from this statement such materials as pegmatites when water and other volatiles play important roles.) For such a high viscosity and similar geometries of intrusion, the corresponding values of P_v would be 8 kbar. This is an impossibly high values, so that the geometry of an intrusion envisaged in Fig. 3.5(a) (a dyke of 2 m width and 10 km height) would not develop with such magmas. If, however, the width of the intrusion were to be increased by at least an order of magnitude, then the possible values of P_v become feasible.

Emplacement of dykes

Igneous material may be emplaced as a dyke either (i) by flowing relatively passively into a pre-existing open fissure, or (ii) by means of forceful injection. Open, empty fissures cannot exist at considerable depths in the crust, so the first of these possible modes of emplacement can only have local significance, in near-surface environments in the crust. The main mechanism of emplacement of dykes must include at least a component of forceful injection; and this occurs as the result of hydraulic fracturing. If, in this simple treatment, we ignore the possible influence of the tip stresses which may be associated with an elliptical or wedge-shaped advancing intrusion, then the conditions necessary for the formation of such a fracture are given by:

$$P_m > S_3 + T \qquad (3.9)$$

where P_m is the magmatic pressure, S_3 is the total least principal stress and T is the tensile strength of the rock in which the fracture develops. For ease of presentation, we shall relate the stress system to the vertical and horizontal. Consequently, it follows that for the emplacement of dykes:

$$P_m > S_x + T_\parallel \qquad (3.10)$$

where S_x is the least principal horizontal stress and $T_\parallel$ is the tensile strength of a given rock type tested in extension parallel to the bedding (i.e. fracture would occur perpendicular to the bedding).

If the magmatic pressure P_m is exactly equal to the right hand side of Eq. (3.10), then this is just sufficient to initiate hydraulic fracturing. If the magmatic pressure remains unchanged, it then exceeds the least principal horizontal stress (σ_h) by the magnitude of $T_\parallel$ and so would push back the walls of the intrusion to form a dyke of finite width. Clearly, it is desirable to establish the magnitude of the excess pressure of the magma relative to that of the least horizontal principal stress.

Estimates of the magnitude of this excess pressure can be obtained by applying an elastic analysis coupled with a knowledge of the width/length aspect ratio of one or more dykes determined in the field. The analysis on which such estimates are based relies on assuming that the rock mass is homogeneous and isotropic, and that the stress orientations and magnitudes are constant, within the area of interest. The fracture or dyke is taken to be elliptical in plan and is assumed to open as the result of an effective stress, (P_e) which is the excess of the magmatic pressure relative to the least horizontal principal stress.

It can be shown (Gudmundsson, 1983) that the effective tensile stress ($-\sigma_3$), where ($-\sigma_3 = -P_e$) is given by:

$$-P_e = \frac{m^2 E}{2(m^2 - 1)} \cdot \frac{W_{max}}{L} \qquad (3.11)$$

where E and m are the Young's modulus and Poissons number of the country rock, W_{max} is the maximum width of the dyke or fissure and L is its length in plan. Gudmundsson obtained length and width data for 68 fissures or dykes of the Vogar swarm of the Reykjanes Peninsula of Iceland and established that the average aspect ratio of width to length was 1:650. He took the reasonable value of $E = 4.8 \times 10^4$ bars (4.8×10^3 MPa) and $m = 4.0$ for the elastic constants of the country rock. Putting these figures in Eq. (3.11), it follows that $-P_e = 40$ bars (4 MPa). This, he noted, is less than the tensile strength of basalt, as determined in the laboratory, and suggested that the low value could be attributed to the fact that the country rock is mainly formed of lavas which contain numerous columnar fractures that would significantly reduce the *in situ* strength of the lens. It may be noted that the scaling effect on strength mentioned in Chap. 1 would also contribute to this relatively low value.

Because the excess magmatic pressure is so small in this particular example, it is reasonable to expect that quite minor changes in the strength of the country rock (which would reflect the orientation and extent of specific columnar fractures, or groups of fractures) would influence the degree of development and even

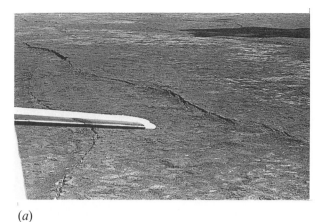

(a)

(b)

Fig. 3.6. (a) Aerial view of sinuous pattern of fissures in S.W. Iceland. (b) Morphology of fissure in the Thingvellir area, Iceland.

induce local changes in direction of·the fissures or dykes. Fissues in Iceland are thought to develop most frequently above dykes which do not quite reach the surface (though they may also form as the result of normal faulting). The non-linear and slightly irregular outcrop pattern of fissures in S.W. Iceland, is shown in Fig. 3.6(a) and (b).

Dykes sometimes occur as single, isolated features; but more frequently, as in Iceland, they occur in groups of parallel, or sub-parallel, units forming what is termed a swarm. The parameters which determine the conditions that give rise to single dykes, as opposed to dyke swarms, are two fold: (i) the quantity of dyke material which is available and (ii) the degree of lateral extension of the crust. As we shall assume here that there is always sufficient magmatic material, we shall be concerned only with the question of crustal extension. As we shall see, the ultimate in forceful injection must be invoked when the crust exhibits zero lateral extension. In such a situation, it will be shown that the ratio of dyke thickness to dyke separation is usually significantly smaller than 1 per cent, so that individual dykes will be relatively widely spaced and so will not be generally regarded as forming a swarm. Dyke swarms, therefore, require lateral extension in the crust; and the greater that extension is, the more dense the dyke swarms are likely to be. It was the opening of the Atlantic in Tertiary times that gave rise to some of the dyke swarms of Scotland, and the continued widening of the Atlantic has been a contributory factor in the formation of dyke swarms and continuing volcanic activity in Iceland. An example of a dyke which was intruded in an environment of 'ocean floor spreading' is shown in Fig. 3.7. This dyke cuts pillow lavas of the ophiolite complex of the Troodos Mountains, Cyprus.

Dyke swarms are not necessarily simple. For example, Speight & Mitchell (1979) determined the orientation, the frequency of occurrence and the ratio to country rocks of Permo-Carboniferous dykes in areas of Scotland, Fig. 3.8(a). It may be noted that individual dykes, especially in the north of the mapped area, make a marked angle with the trend of the contoured zone. One possible explanation of such a distribution of orientation and quantity of dykes is indicated in Fig. 3.8(b), which shows that the regions of high intensity in the number of dykes may be related to a 'fin' in a magma chamber which also marks a shear zone of left hand strike-slip motion. Unfortunately, no geometric control of the spreading is available in such a model.

The Tertiary dyke swarm which occurs in E. Greenland is very definitely controlled by the geometry of the flexure of the basement rocks (Wager & Deer, 1938 and Fig. 3.9). This geometrical control mainly gives rise to a dyke swarm which fans upwards and strikes parallel to the coast as indicated in Fig. 3.10. However, there is a kink in the trend near Kangerdlugssuak, which has caused horizontal extension resulting in a fan of vertical dykes (as indicated).

There is qualitative control of the problem in the Greenland example, but information regarding dyke thickness and numbers of dykes is unavailable. However, an example where dyke emplacement can be

Fig. 3.7. Some irregular dykes intruded into pillow lavas in the Troodos Mountains, Cyprus. The irregular walls of the intrusion may indicate that the magmatic intrusions were formed relatively passively, in country rock already near tensile failure, rather than by forceful injection.

related to lateral extension in a quantitative manner is in the Idwal and Tryfan areas of Snowdonia (Fig. 3.11), where the fold axes are curved and change from the horizontal, near Cwm Tryfan, to plunge 15° S–SW to the east of Glyderfach and so form a pericline. It can be seen from the map (Fig. 3.11) that six major dykes cut through the Tryfan ridge. These have an average width of 15 m at the top of the ridge, thinning downwards to zero in the valleys on either side of the ridge, in a vertical relief of about 700 m. The simple model on which the following calculations are based is shown in Fig. 3.12(*a*). From the dip of the fold axes and the dimensions of the fold on the map, it follows that the *fold axis* has a radius of curvature (*R*) of about 7 km (Fig. 3.11). If it is assumed that the pinching-out of the dykes in the downward direction occurs near the neutral surfaces of R_c (see Fig. 10.26), then the extension d*L*, which occurs at the top of the ridge 700 m above the neutral surface is given by:

$$dL = \frac{dR\theta}{180°}. \qquad (3.12)$$

Consider the portion of the fold from A to B (Fig. 3.11), where the change in dip of the fold axis is 15°. Therefore, taking d*R*, and *θ*, in Eq. (3.12), to be 700 m and 15° respectively, then d*L* = 58 m. In this traverse from A to B, four main dykes are

encountered, so that, if the average thickness is taken as 15 m, the extension required is 60 m (Fig. 3.12(*b*)). The correlation between the value of d*L* and the total thickness of dykes is almost exact.

Forceful injection

The most extreme cases of forceful injection by magma occur when the crustal rocks, prior to intrusion, exhibit approximately zero lateral strain, at least in the direction parallel to the least principal horizontal stress (σ_3). As was noted earlier (Eq. (3.9)), the magmatic pressure must exceed the least horizontal stress plus the tensile strength of the rock to initiate hydraulic fracture. Moreover, we have noted that the maximum value P_m can have for dyke intrusion is close to that of the total vertical stress (σ_z), for if P_m is only a little greater than σ_z, sills will form. Forceful injection will be best developed when the horizontal stress has the minimum value in the conditions postulated above. The question which arises, therefore, is what is the largest differential stress between the vertical and least horizontal stresses (i.e. $\sigma_z - \sigma_x$) that is likely to occur at any given level in the crust?

A point that has been made in Chap. 1, but not yet taken into consideration here, is that Eq. (3.9) is only valid provided $\sigma_1 - \sigma_3 < 4T$ (where *T* is the tensile strength of the rock unit). If $\sigma_1 - \sigma_3 < 4T$, then

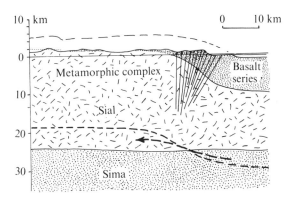

Fig. 3.9. Section through dyke-swarm, E. Greenland. (After Wager & Deer, 1938.)

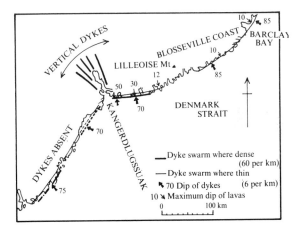

Fig. 3.10. Map showing trends of dyke-swarm, E. Greenland. (After Wager & Deer, 1938.)

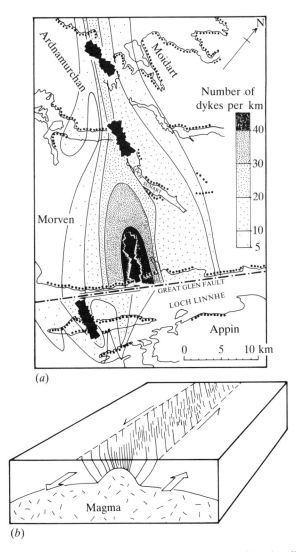

(a)

(b)

Fig. 3.8. (a) Permo-Carboniferous dyke-swarm of Northern Argyll: Contour map illustrating the intensity-distribution of the number of dykes per kilometre, (dots indicate location of traverse). The off-set and difference in intensities of the dykes on either side of the Great Glen Fault can be attributed to post-dyke-swarm, oblique displacement with a 7–8 km component of dextral slip and a substantial down-throw to the S.E. (After Speight & Mitchell, 1979.) (b) Schematic representation of a mechanism that could give rise to the intensity-distribution and dyke orientation illustrated in (a).

shear failure will take place. However, it will be recalled that hybrid extension/shear failure can take place for a range of values such that $4T < (\sigma_1 - \sigma_3) < 5T$. These hybrid fractures would, of course, generally result in the development of inclined, rather than vertical, dykes. Alternatively, the element of shear may make itself manifest in the development of en echelon dykes, which may, at depth, be connected to a single planar 'feeder' dyke, possibly experiencing strike-slip motion (see Anderson, 1951). So the question posed above now revolves around the probable maximum value of T that can exist in the crust.

From the theories of brittle failure discussed in Chap. 1, it follows that the uniaxial compressive strength (S_0) is predicted to be about eight times greater than the tensile strength (T). When determined by laboratory measurements, this ratio of strengths often exceeds 20:1, rather than 8:1. However, as we have seen in Chap. 1, the measurement of tensile strengths of rock in the laboratory is notoriously difficult and the results obtained are often unreliable. So, provided the rock is situated at depth in the crust, does not contain minor barren fractures and any veins that may exist are as strong as the host rock, then in some metamorphic units, massive sandstones or similar competent massive units, it is probably reasonable to assume that $T \approx 0.125 \, S_0$. From rock mechanics data and applying a probable scaling factor, the value of S_0 will rarely exceed 1.6 kbar. Hence, because $\sigma_1 - \sigma_3$ may be as large as $5T$, the largest differential stress ($\sigma_z - \sigma_x$) will be about 1.0 kbar. If we assume that the magmatic pressure does not greatly exceed the magnitude of the vertical stress, at any specific level in the crust which is under consideration, then the maximum magmatic overpressure will also be about 1.0 kbar, though, we emphasise, such a high value is *rarely* likely to be attained.

Consider now the simple, even naive, model

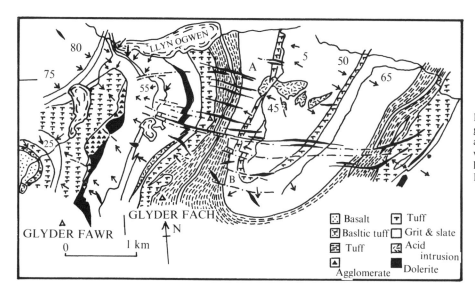

Fig. 3.11. Sketch map of the geology of the Idwal and Tryfan area, N. Wales, showing dykes which increase in thickness at higher levels. (After Williams & Ramsay, 1959.)

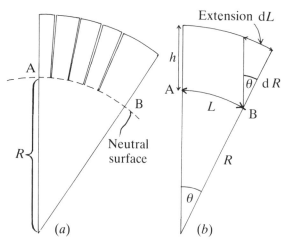

Fig. 3.12. Simple graphical representation used to quantify the increase in thickness with height of dykes represented in Fig. 3.11.

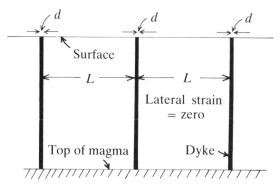

Fig. 3.13. Model used to estimate the ratio of dyke to country rock that can result from forceful injection.

$$d_w = \frac{L(P_m - S_n)}{E} \tag{3.13}$$

where E is the average Young's modulus of the sediments between the dykes, and σ_x is the original value of the horizontal stress. Using the simple model represented in Fig. 3.13, and assuming that $E = 10^5$ bars and that the magmatic over-pressure $(P_m - \sigma_x)$ has a value of 1.0 kbar, it follows that the ratio d_w/L, for such pressure conditions and elastic moduli, will be 1.0 per cent. This probably represents the maximum ratio of dyke material to country rock which would result from the forceful injection of the magma into a crustal environment of zero horizontal strain. The value of magmatic over-pressure of 1.0 kbar used in this calculation is probably the maximum value ever likely to be attained in the crust. More often, the magmatic over-pressure is likely to be significantly lower than this value. Hence, in general, when dyke material is intruded into a crust which has *not* been subjected to horizontal extension, the ratio of dyke material to country rock is likely to be significantly less than 1 per cent.

This argument has been based on magmatic intrusion, but applies equally to intrusion of non-

shown in Fig. 3.13, which represents conditions in the crust of zero extension but into which a series of dykes is emplaced. These dykes are separated one from another by a distance L. At some specific depth, the magmatic pressure in the magma chamber is assumed to be equal to the gravitational load. The dykes are intruded vertically and the magmatic pressure, which at every level in the crust is assumed equal to the vertical pressure, acts against the horizontal pressure in the crust at that level. If we suppose that at some depth (say 8 km), where the total vertical and horizontal stresses are about 2.0 kbar and 1.0 kbar respectively, then the magmatic over-pressure in the dyke will push back the walls of the intrusion until they have a width d_w. This results in elastic compressive strain (d_w/L) of the country rock and so increases the horizontal pressure. This lateral compression of the country rock may continue until σ_x almost equals σ_z. From this it follows that the width of a dyke d_w is given by:

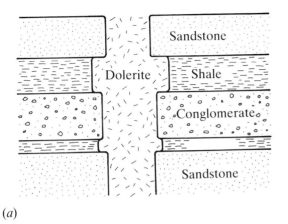

(a)

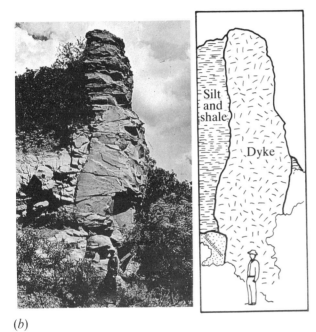

(b)

Fig. 3.14. (a) Diagram of steps in walls of a dyke related to the lithology of the country rock. (b) Photograph showing the same relationship as in (a).

magmatic material. For example, the Sylvinite Dykes of Unita Basin in the U.S.A. have thicknesses of 1–2 metres and are separated by an average of at least 5 km of country rock, so that these non-magmatic dykes are almost certainly the result of intrusion into a crustal condition of approximately zero strain.

Forceful injection which results in the pushing back of the walls of the intrusion can give rise to a number of features in the adjacent country rock, which we shall now consider.

Associated features in the wall rock

As not all layered rocks have the same elastic moduli, one may expect dyke walls to exhibit steps (as indicated in Fig. 3.14) when the elastic moduli, and the horizontal stress in adjacent beds, are markedly different. These steps may be wholly mechanical, or they may have a chemical–mechanical explanation. Let us

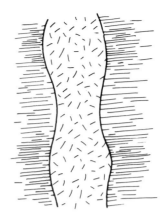

Fig. 3.15. Pseudo pinch-and-swell developed in a dyke intruded at a deep crustal level by forceful injection.

first examine the mechanical effects. The values of Young's moduli for weak rocks, such as shales and mudstones, are commonly as much as an order of magnitude smaller than those of stronger rocks such as sandstones. Thus, provided the magmatic over-pressure is the same, strain induced in the weaker rocks would tend to be larger than that in the stronger rocks. Hence, where the walls of an intrusion are of shale they could be pushed back further than those portions of the walls composed of strong rocks. There are, however, two factors which militate against this tendency. Firstly, let us assume that we are here concerned with flat-lying sediments not subjected to horizontal tectonic compression. Then, as we shall see in Chap. 9, the horizontal pressures in weak rocks tend to be greater than they are in adjacent strong rocks. Consequently, the magmatic *over-pressure* will not be constant, even throughout a small vertical section of a dyke. The over-pressure acting against a layer of strong rocks will be considerably larger than it is against a layer of shale. Secondly, because of the effects of gravitational loading, there would usually be a frictional resistance to differential movement between adjacent strong and weak rocks. Thus, the mechanical problem is not amenable to a simple solution, so we shall defer a discussion of this latter point until later and comment first on what happens if the elastic limit of the wall rocks is exceeded.

A representation of the type of wall rock geometry of a dyke intruded into high-grade, meta-morphic rocks, which is known to form in some basement complexes, is shown in Fig. 3.15. We argue that the intrusion occurred while the country rocks were deep-seated and 'hot'. The total ductile strain in the wall rock may be as little as 10 per cent (i.e. 10^{-1}). From the cooling curves shown in Fig. 3.4 and the times given in Table 3.1, it may be inferred that the temperature of the country rock adjacent to a moderately thick dyke would be significantly higher than its initial temperature for several months (say 10^7 s); so that in such a situation the average strain-rate would

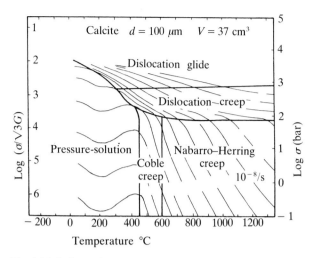

Fig. 3.16. Deformation map for calcite. (After Rutter, 1976.)

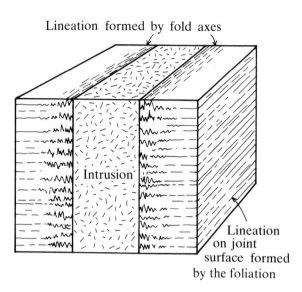

Fig. 3.17. Schematic representation of deformation of wall-rock as the result of forceful injection. Such features occur in Namibia. (Halbich, personal communication.)

be about 10^{-8}/s. Let us assume that certain layers in the country rocks contain significant amounts of calcite. We make this assumption because rock mechanics data exist for this mineral. Then, from the strength/deformation chart for calcite (Fig. 3.16) it can be inferred that, if the country rock included a reasonably high proportion of calcite, and was, for most of the cooling period at or above a temperature of 800 °C, deformation could take place by dislocation creep at 10^{-8}/s, provided the differential stress is at least 100 bar. As we have seen, the excess magmatic pressure may exceed several hundred bars. Consequently, it is suggested that the contact metamorphic effect of the intrusion gave rise to differential weakening of different layers adjacent to the intrusion, so that the magmatic over-pressure was able to push back some parts of the wall rock further than others, thereby giving rise to a pseudo-boudinage effect. The situation presented above has been related to rocks containing calcite. When more is known about the strength/deformation behaviour of other minerals, it will be possible to quantify the parameters involved in natural rock deformation in comparable geological situations.

A further example of wall rock deformation brought about by forceful injection is shown in Fig. 3.17, where a dolerite dyke is intruded into finely laminated, acid country rock. We suggest that the metamorphic aureole has been greatly weakened by the local production of water by dehydration reactions and the finely laminated material of the country rock adjacent to the dyke has been folded. Instances of such wall rock deformation in Namibia are reported by Halbich (personal communication).

An example of larger scale deformation of country rock by an intrusion of diorite, elliptical in plan, in the Atacama Desert of N. Chile has been described by Skarmeta & Price (1984). This deformation has resulted in a narrow belt of spectacular

mesoscopic folds (Fig. 3.18), in an otherwise flat-lying sequence, comprising poorly compacted conglomerates and clay-rocks with gypsum intercalations. Fig. 3.19 shows an oblique view of one of the folds. The clay-rock units in these folds are cut by a large number of fractures, which are infilled with fibrous anhydrite and secondary gypsum (Fig. 3.20(a)), while some of the fractures near the crest of the folds have been gently folded. From a study of the orientation and geometry of the fractures and the fibre forms and by 'unfolding' the structure using a stereographic technique (Fig. 3.20(b)), it was possible to establish the history of the country rock deformation.

The first event was the initiation of a system of minor thrusts at a time when the beds were essentially flat-lying. From the orientation of the fractures, it was shown that the axis of maximum principal stress (σ_1) was inclined to the horizontal at 14°, Fig. 3.21(a). That is, at this time, a shear stress acted parallel to the bedding. Moreover, from a stress analysis, it was established that during this phase of deformation the rocks were essentially dry. This latter conclusion is, of course, extremely reasonable, because the sediments were at a shallow depth of only a few hundred metres and, as already mentioned, are now, and were at the time of intrusion, in a desert environment. Towards the end of this phase of deformation, elastic buckle folds with an infinitesimal amplitude were initiated. Finite fold development did not take place until heating of the country rock by the intrusion caused the dehydration reaction from gypsum to anhydrite to occur, which released large quantities of water at a pressure (p) comparable with the overburden pressure. The mechanical requirements that dictate the need for such high fluid pressures before finite fold

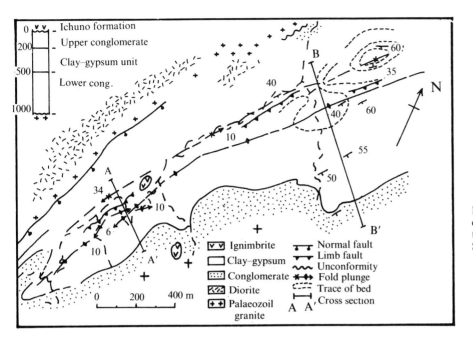

Fig. 3.18. Geological map and cross sections of the area cut by the Quinchamale Gulch in the Sierra de Moreno, N. Chile.

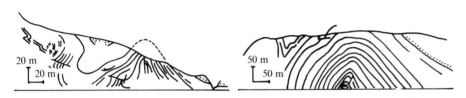

development can take place are dealt with at length in Chap. 14, so will not be considered here.

The high fluid pressures also opened up the shear fractures to form veins. This was accomplished preferentially, in that those barren fractures which were most nearly horizontal, in the earliest phase of finite fold development, were the first to develop as veins. As will be seen in Fig. 3.22, these were the 'short' shear fractures in the west limb (V1W), but were the 'long' shear fractures in the east limb (V1E). Later in the phase of finite fold development, the changes in rock stress permitted the fluid pressure to open up the conjugate fractures in the E. and W. limbs which became veins (Fig. 3.22(b) and (c)). After buckling ended and the fold 'locked-up', post-buckle flattening occurred. In this flattening phase, minor thrusts and reverse faults developed (Fig. 3.22(d)). The final phase of deformation occurred as the country rock cooled and experienced lateral shrinkage. This resulted in a reversal of the stress-field (Fig. 3.22(e)) so that normal, dip-slip shear developed on several of the minor fractures of suitable orientation.

One other aspect regarding this paper by Skarmeta & Price merits comment. Because of the excellent exposure (Figs. 3.18 and 3.19), the stratigraphy and depth of burial of the folds could be accurately assessed (the depth of cover ranged from 200–500 m). From the acute angle ($2\theta = 70°$) between the initial

shear planes (which, of course, eventually became veins) and the inferred orientation of the axis of greatest principal stress, which dipped at 14° (Fig. 3.21(a)), it was possible to determine qualitatively the stresses which gave rise to the initiation of the shears and elastic buckling (Fig. 3.21(b)).

The history of deformation outlined above is, of course, justified by Skarmeta & Price, and the various aspects and phases of folding mentioned here are dealt with at length in Chaps. 10 to 15. The example has been quoted here because it illustrates so well the processes of forceful injection which can give rise to folds in the country rock. (The same process of high fluid pressure development must also be invoked to explain the development of the smaller crinkles illustrated in Fig. 3.17: and also to permit detachment of bedding planes and so allow differential movement of beds to occur, as shown in Fig. 3.14.) Moreover, the examples cited above illustrate how necessary it is for the field geologist to be familiar with most, if not all, the processes involved in rock deformation, even when dealing with such an apparently 'simple' problem as intrusions. It must be emphasised that the various examples quoted earlier merely serve to illustrate some of the forms of country rock deformation induced by forceful injection. The larger the intrusion, the greater and more complex the country rock deformation is likely to be. Indeed, we argue in Chap. 7 that many of the major overthrusts in Canada, U.S.A. and

Fig. 3.19. Oblique views of structures represented in Fig. 3.18.

Inclined dykes

It has been noted that if $(\sigma_1 - \sigma_3) > 4T$ then hybrid extension/shear fractures may develop. Hence, if we continue the previous assumption that the greatest principal stress acts vertically, the dykes would be inclined and could attain a minimum dip of 67.5°. This follows from the fact that the maximum interfracture angle between conjugate, hybrid extension/shears is 45° (Chap. 1). Alternatively, if normal faults have already developed in the rock mass, the magma may intrude into these existing planes of weakness rather than generate new vertical fractures by hydraulic fracturing. For such behaviour, it follows from Eqs. (3.9) and (1.16) which, respectively, are the hydraulic fracture criterion and the normal stress acting on a shear plane, that:

$$(S_3 + T) > P_m$$
$$> \frac{(S_1 - S_3)}{2} - \frac{(S_1 - S_3)}{2} \cos 2\theta. \tag{3.14}$$

The limiting conditions are met when:

$$(S_3 + T) = \frac{(S_1 + S_3)}{2} - \frac{(S_2 - S_3)}{2}. \tag{3.15}$$

From the latter relationship it can be shown that the differential stress $(\sigma_1 - \sigma_3 = \sigma_z - \sigma_x)$ compatible with Eq. (3.15) varies with the angle of dip (θ) in the manner indicated in Fig. 3.23. It can be inferred from this graph that magma materials can quite readily intrude along faults, or fractures, which dip at angles of as little as 60° and may even intrude into more gently dipping planes of weakness.

Should the necessary stress conditions be met and an inclined dyke is initiated, either as a hybrid extension/shear fracture or by infilling by magma of an existing normal fault, noticeable shear displacement of the country rock should ensue. Skarmeta (1979) studied dykes, in the Mejillones Peninsular of N. Chile, which exhibit a fabric, with the geometry shown in Fig. 3.24. This fabric is clear evidence that the fault was initiated or, more probably, reactivated, by the intrusion of the magma and that slip movement occurred while the intrusion was in the initial stages of cooling. The reader will have noted that the form of the fabric is not that which is usually associated with a shear zone. In this example, the fabric in the central part of the dyke is relatively flat-lying. Relatively intense shear has occurred only near the margins of the intrusion. It is interesting to note that, by careful determination of the strain distribution across the 'sheared dyke', Skarmeta was able to establish that the rheology of the magma approximated closely to that of a Newtonian liquid, and so substantiates the conclusions, cited earlier,

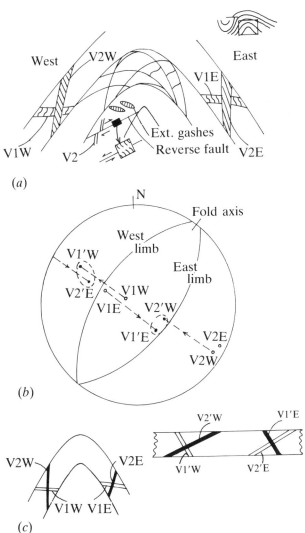

(a)

(b)

(c)

Fig. 3.20. (a) Diagrammatic representation of the orientation of sets of veins 1 and 2, relative to an anticline, together with the fibre directions. Note the curved veins in the core of the anticline and the reverse faults that displace V1 veins. Fibre growth directions are shown by thin parallel lines. (b) Stereogram showing rotation of veins of Set 1 and 2 from the western (V1W, V2W) and eastern (V1E, V2E) limbs to the horizontal position. Open circles denote veins in the folded state and black circles veins in the unfolded state. Arrows indicate the rotation direction for unfolding. (c) Schematic representation of the position of the vein Sets V1 and V2 (in E. and W. limbs) in the folded and unfolded state.

reached by Shaw *et al.* (1968) and Williams & McBirney (1979).

So far we have considered only the structures which are associated with forceful injection of vertical sheets when viewed in a vertical profile normal to the intrusion. Forceful injection may also give rise to effects seen in plan view (particularly if the injection is centred upon a vertical pipe-like intrusion) when radiating dykes may form.

Radiating dyke patterns

The dykes in the vicinity of the Spanish Peaks, Colorado, U.S.A. are of Eocene age and exhibit an axisymmetric form of radiating pattern (Fig. 3.25) with the line of symmetry trending N80°E. However, the dykes are not of equal development throughout the pattern. Those to the north, south and east of the Peaks can be traced for 15 km or more: but to the west they are usually less than 5 km in extent and terminate in the vicinity of the Sangre de Cristos Mountains. As will be seen from Fig. 3.26, these dykes sometimes weather to form impressive walls which can be traced across country for many kilometres.

An interesting mathematical analysis of these dykes was conducted by Odè (1957), in which he assumed that the Sangre de Cristos Mountains formed a rigid boundary, or *Mountain Front*, as indicated in Fig. 3.27(a). The concepts encountered in the derivation of stress functions and their uses in determining the orientation and magnitude of principal stresses, for given boundary conditions, are discussed in Chap. 5. Here, we merely note that Odè used a stress function, expressed in polar coordinates, for his elastic analysis which he coupled with a useful mathematical model 'dodge' which enabled him to take in to account the restraint of the rigid Mountain Front. He calculated the tensile stress trajectories which resulted from this analysis and presented half of a symmetrical pattern, as shown in Fig. 3.27(a). He then made the reasonable assumption that prior to intrusion there existed a regional horizontal compressive stress field, with trajectories normal and

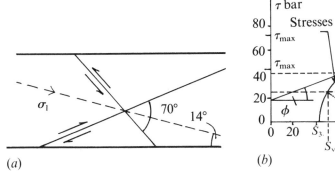

(a)

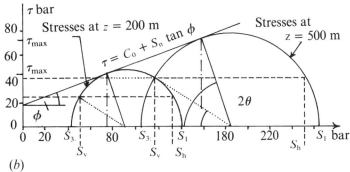

(b)

Fig. 3.21. (a) Orientation of maximum principal stress inferred from the pre-buckle orientation of V1 and V2 in Fig. 3.20. (b) Quantitative stress analysis of the development of veins represented in Figs. 3.18–3.20 and 3.22, using the Navier–Coulomb criteria for stress data inferred from field measurements and observations.

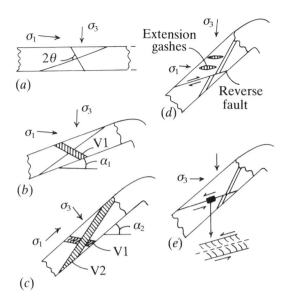

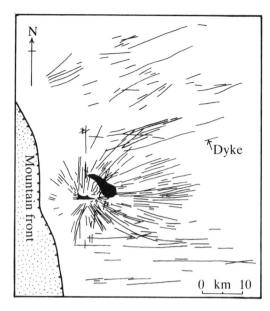

Fig. 3.22. Diagram showing sequence of fracture formation and vein infilling at different stages of deformation. (*a*) Pre-buckling fractures: (*b*) elastic buckling and infilling of veins 1: (*c*) inelastic fold amplification and infilling of veins 2: (*d*) and (*e*) post-buckling stages ((*d*) shows reverse faulting stage and a late normal faulting phase related to cooling). The infilling of V1 and V2 continues into the inelastic and post-buckling phases respectively. The estimated stress orientations are also shown.

Fig. 3.25. System of dykes surrounding the Spanish Peaks. (After Knopf, 1936.)

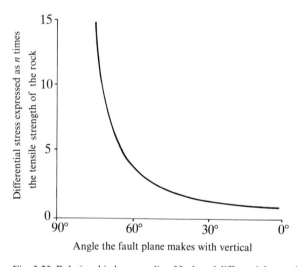

Fig. 3.23. Relationship between dip of fault and differential stress in terms of multiples of the tensile stress (*nT*) compatible with magmatic intrusion along the fault.

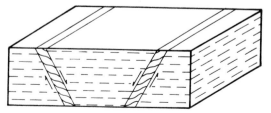

Fig. 3.24. Schematic representation of foliation in dykes induced by shear displacement concomitant with intrusion. (After Skarmeta, 1979.)

parallel to the Mountain Front. The final orientation of the tensile stress trajectories (Fig. 3.27(*b*)) was obtained by superimposing the calculated tensile field on the assumed compressive field. It will be seen that the degree of correspondence between the tensile stress trajectories and the observed distribution of dykes is reasonably good.

It may be noted, however, that the Spanish Peaks intrusions are sited near a pre-existing synclinal axis. We shall be dealing with the general fracture patterns which are commonly associated with folds in Chap. 14. It is necessary here merely to note that the trends likely to be associated with the major Spanish Peaks Syncline are as indicated in Fig. 3.28. In addition to these trends, Badgely (1965) notes that the area is also cut by fractures which pre-date the folding. Individual fractures parallel to the indicated trends may be only a few tens or hundreds of metres long. By running from one fracture to another of the same, or different, trend, suitably curved dykes, when viewed on the regional scale, could readily develop. The possible steps and changes in direction which such dykes could exhibit are grossly magnified and exaggerated in Fig. 3.28 to illustrate the proposed mechanism. Such a mechanism does not, of course, invalidate the general conclusions and methods of analysis adopted by Odè.

We have already noted that dykes do not necessarily form simple linear features when exposed at the surface. For example, outcrops of fissures in Iceland (which are considered to be manifestations of dykes that have not quite reached the surface) are often markedly curved (Fig. 3.6(*a*) and (*b*)). Also, dykes may replace each other in an en echelon manner (Fig. 3.29(*a*)). Such an array has been attributed

Fig. 3.26. Wall of igneous rock exposed through differential weathering of one of the Spanish Peak dykes.

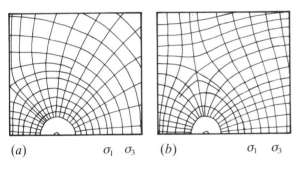

(a) σ_1 σ_3 (b) σ_1 σ_3

Fig. 3.27. (*a*) Pattern of principal-stress trajectories (caused by two sources of equal sign). (*b*) Pattern of principal-stress trajectories from (*a*) superimposed on a regional-stress system. (After Odè, 1957).

(Anderson, 1951) to the upward fingering of a planar dyke into a near-surface level, where the orientation of the least principal stress is slightly oblique to that at lower levels (Fig. 3.29(*b*)).

Similar geometrical relationships have been noted by Delaney & Pollard (1981) who indicated that the geometry may be even more complex. Moreover, they infer that the direction of propagation of the fracture (Fig. 3.30(*a*)) and also the direction of flow of the magma (Fig. 3.30(*b*)) can be inferred. In the example shown in Fig. 3.30(*b*) the direction of flow is vertical. However, it should not be assumed that all dykes are fed by vertical magmatic flow, for Gudmundsson (1984) has argued convincingly that many

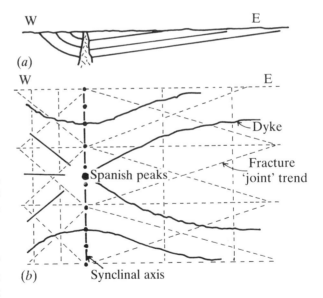

Fig. 3.28. (*a*) Profile of syncline through the Spanish Peaks area. (*b*) Plan view of fracture trends likely to be associated with the syncline and indicating how these trends may have been utilised to result in the observed dyke pattern.

of the dykes in Iceland are emplaced and developed as the result of horizontally flowing magma.

Delaney & Pollard, following Anderson, also infer that the en echelon, or off-set dykes are the result of a rotation of the stress field as the surface is approached. An alternative explanation for such an

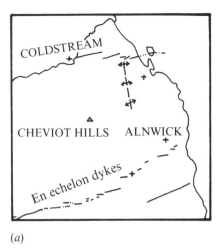

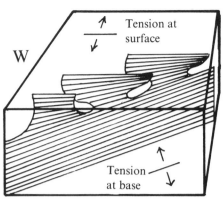

W

Tension at surface

E

Tension at base

Fig. 3.29. (*a*) Outcrop pattern of en echelon dykes in the north-east of England. (After Anderson, 1951.) (*b*) Postulated change of stress direction with depth which could explain the development of en echelon dykes developing from a single planar sheet at depth. (After Anderson, 1951.)

(*a*)   (*b*)

n echelon, or off-set, array is that the underlying planar dyke is the result of extension and shear. Thus, the en echelon and off-set features could possibly be likened to the fringe structures around a joint (Chap. 2), (see also Fig. 3.8(*a*) and (*b*), for which a comparable explanation is proposed).

We have noted that the orientation of dykes may be influenced by pre-existing joints or other fractures which form planes of preferential weakness. In the example cited by Delaney & Pollard they noted that, at the current level of exposure, joints do in fact run parallel to the dykes. However, they emphasise that such joints do not represent a regional trend but occur only in the immediate vicinity of the intrusion. Consequently, they conclude, these joints are related in some way to the mechanism of dyke emplacement. Similar observations regarding the relationship between dykes and parallel joint sets in the San Moreno area of Chile have been made by Skarmeta (1983) who also concluded that the joints were generated as the result of the intrusion mechanism. (It may be noted in passing that the above observations are not incompatible with the possible existence of a regional set of fractures at a deeper level in the crust which may have permitted the emplacement of a planar dyke by a combination of extension and shear.)

Emplacement of sills

For the emplacement of sills, the conditions which must be satisfied are given by:

$$P_m = S_z + T_\perp \tag{3.16}$$

where S_z is the total vertical pressure and $T_\perp$ is the strength of a bed or sequence of beds tested perpendicular to the bedding plane. The tensile strength of a pile of sediments will be determined by the strength of the individual bedding planes and $T_\perp$ would therefore be approximately zero. Here, we neglect the stress concentrations which may exist at the tip of the sill as it is emplaced. We shall consider this effect later.

Sills must be fed. The feeder channel may possibly be a pipe, but in many instances (possibly the majority of cases) the sill is fed through a dyke. At the junction of the feeder dyke and the sill, the conditions required by both Eqs. (3.10) and (3.16) must be satisfied simultaneously (Fig. 3.31). Hence, it follows (by subtracting Eq. (3.10) from (3.16)) that:

$$(S_z - S_x) < (T_\parallel - T_\perp). \tag{3.17}$$

We have noted that $T_\perp$ is approximately zero. The tensile strength of most unjointed or unfractured rocks is not likely to exceed 100 bar and could be considerably smaller. If, as is likely, the rock mass prior to sill emplacement contains joints the tensile strength parallel to the bedding could be very small indeed. Therefore, we can conclude that sills are only emplaced when the differences in magnitude of the horizontal and vertical stresses are very small. As we shall see, this is an important conclusion when dealing with the question of how a dyke becomes a sill and/or a sill terminates and changes into a dyke.

Let us now consider how a dyke turns into a sill. From the section of a dyke represented in Fig. 3.32(*a*), it will be seen that P_m acts in a hydrostatic sense. If the dyke has a 'knife edge', then (because Force = Stress × Area) the upward force at that edge, generated by the magmatic pressure, approaches zero, so that the ability of the magma to lift off the superincumbent strata and so form a sill would also appear to be zero. An elegant solution to this paradox was obtained by Pollard (1969), who demonstrated analytically and also by model work how a vertical dyke could develop into a symmetrical sill. The action of the magmatic pressure causes an elastic extension in the horizontal direction in the adjacent strata immediately above the dyke and there is a concomitant downward pull (as a result of the Poisson ratio effect, see Chap. 1). As the dyke approaches a bedding plane of near-zero strength, and provided the pressure conditions of Eq. (3.16) are fulfilled, this downward pull will cause detachment along the bedding plane (Fig. 3.32(*A*)),

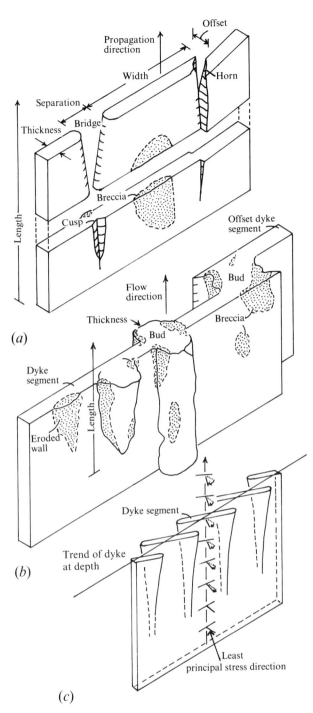

(a)

(b)

(c)

Fig. 3.30. (a) Terms used to describe discontinuous and off-set dykes. (b) Further morphological features of off-set dykes. (c) Postulated change of orientation of stresses with depth necessary to cause the development of en echelon dykes. (a), (b) and (c) after Delaney & Pollard, 1981.)

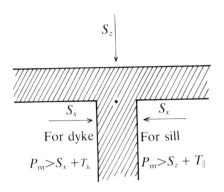

Fig. 3.31. Stress conditions compatible with dyke and sill intrusion.

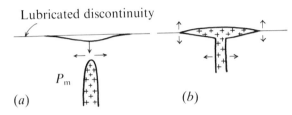

Fig. 3.32. (a) Schematic representation of a plane of weakness in advance of a developing dyke enabling a sill (b) to form. (After Pollard, 1969.)

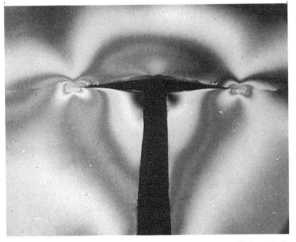

Fig. 3.33. Photo-elastic model demonstrating the validity of the mechanism indicated in Fig. 3.30. (After Pollard, 1969.)

thereby providing a void into which the magma can flow and so generate a symmetrical sill, Fig. 3.32(b). Results of a photo-elastic model which confirms the correctness of this analysis and prognosis are indicated in Fig. 3.33.

So far, we have assumed that the dyke was emplaced in a vertical plane and cut through the layered sequence by hydraulic fracturing of each successive bed, or possibly was emplaced along an existing vertical fracture such as a wrench fault or 'master joint'. However, the dyke could utilise an existing plane of weakness, such as a normal fault, and be emplaced oblique to the strata. If the necessary conditions are then obtained and the dyke turns into a sill, it may be seen from Fig. 3.34(a) and (b) that the resulting sill is asymmetrical. This asymmetry is caused by the action of the magmatic pressure which, as can be seen in Fig. 3.34(b), acts in such a way that the bedding plane is kept closed to the right hand side of the diagram, but tends to open to the left hand side of the diagram.

We are now in a position to reassess the question

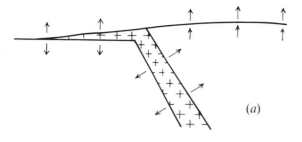

(a)

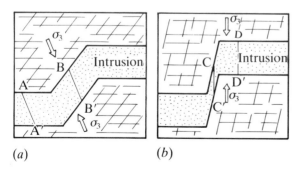

(b)

Fig. 3.34. (a) Mechanism whereby an intrusion advancing along an inclined plane of weakness will give rise to an asymmetrical concordant intrusion (sill). (b) Photo-elastic model demonstrating mechanism shown in (a). (After Pollard, 1969.)

regarding the age relationships of intrusion and fault shown in Fig. 3.2(a) and (b), in which both the foliation and the fault represent planes of weakness. The foliation is likely to have a profound influence on the orientation of the rock stress. Nevertheless, the orientation of the principal stresses may be slightly oblique to the foliation. Therefore, the opening of the concordant intrusion need not be perpendicular to the foliation, but may be oblique (Fig. 3.35). This direction of opening may also permit the intrusion to follow the fault plane indicated in Fig. 3.2. The answer to the question concerning the relative age of the fault and the dyke posed in an earlier section is therefore indeterminate. From the available fault exposure indi-

Fig. 3.35. Geometry of an intrusion along oblique planes of weakness in relation to the orientation of the axis of least principal stress.

cated in Fig. 3.2(a), it is not possible to give the age relationship of the intrusion on the fault. Further evidence is needed. It would be necessary to excavate and expose the fault plane before the question could be answered. However, it must be noted that if the extension direction is exactly parallel with the fault (and the fault plane is slightly irregular) it will not be possible to resolve this problem without extensive investigation of the site.

An example which illustrates how an intrusion follows one type of weakness plane (bedding) and then is deviated by a second set of weakness planes (faults) is indicated in Fig. 3.36(a), which shows the intrusion of dolerite into tilted strata and along normal faults in the coal measure rocks in the Midland Valley of Scotland. As we have seen in an earlier section, such intrusions were emplaced in a relatively short time and therefore the magma 'floated off' the superincumbent strata. If one can find a direction in the different parts of the intrusion for lines which are parallel to, and of the same length as, each other, then the direction of opening and hence the orientation of S_3 can unequivocally be established (Fig. 3.35), provided, of course, that the $S_1 S_3$ plane can be defined. An example of a sharp change in direction in a 'dyke' is shown in Fig. 3.36(b) and a sequence of such 'kinks', comparable with those represented in Fig. 3.36(a) can be seen (Fig. 3.36(c)) in veins, indicating that the concepts outlined above may also be applied to aqueous solutions. When dealing with this latter situation care must be taken by the observer to ensure that the off-set of the type shown in Fig. 3.36(c) is not the result of pressure solution.

Symmetry of sills and the development of off-shoot dykes

When considering the emplacement of a dyke, it is reasonable to assume that the long axis of the dyke is usually central to the intrusion and so forms an axis of symmetry. That is, in general, the magmatic overpressure pushed back the wall-rock by an equal amount on either side of the intrusion. However, when dealing with the emplacement of sills, such as simplifying assumption may not, in general, be valid.

The majority of sills are emplaced under relatively shallow depths of cover. Also, such shallow intrusions frequently exhibit a lateral extent which is many times greater than the depth of cover which existed at their time of emplacement. As we shall see, if the ratio of the depth of cover to the horizontal extent of the sill is equal to, or smaller than, 1:10, it will be easier for the magmatic overpressure to cause an upward displacement of the roof, rather than induce an equal strain in floor and roof rocks. Consequently, it is to be expected that a high proportion of sills will exhibit asymmetry with respect to the horizontal plane. However, sills commonly exhibit a ratio of

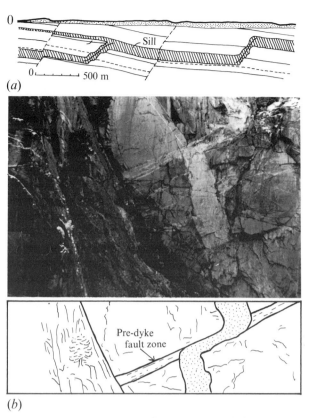

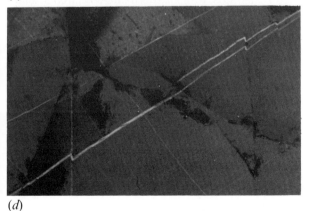

Fig. 3.36. (*a*) Stirling dolerite 'sill' in the Midland Valley of Scotland. (after Macgregor, M. & Macgregor, A., 1936.) (*b*) Step or kink in tabular intrusion into a granite, Colorado, U.S.A. (*c*) Steps or kinks in quartz veins exposed on a bedding-plane, which is comparable in morphology to that shown in (*a*). (*d*) Detail of (*c*).

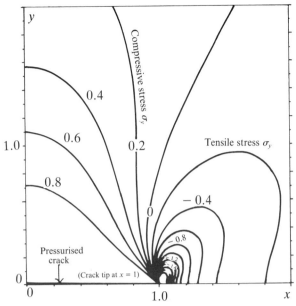

Fig. 3.37. Distribution of stresses in a quadrant adjacent to a pressurised crack. (After Delaney & Pollard, 1981.)

height to horizontal extent in excess of 100:1. Consequently, when one takes into account the imperfections of geological exposure in most terrains, it will usually be extremely difficult to detect such asymmetry. Moreover, the degree of asymmetry may vary along the intrusion. For example, if the intrusion is emplaced in weak, or brecciated rock, the termination of the sill, in profile, may be well rounded, so that it approximates to the tip of an ellipse. In such a situation the sill may locally exhibit almost perfect symmetry. Nevertheless, near the central portions of the sill, it is to be expected that the roof will be displaced upward to a far greater extent than the floor has been depressed; so that in the central zone the sill will be asymmetrical.

This question of symmetry is important, for it determines the stress conditions that may develop around the termination of a sill and this, in turn, controls the orientation of any dyke which may develop as an off-shoot from that sill.

The contours of stress component acting normal to a pressurised crack in a quadrant adjacent to that crack is represented in Fig. 3.37 (Delaney & Pollard, 1981). (This analysis was used to determine the stress distribution around a dyke. However, provided it is assumed that the crack is completely symmetrical and that variations in stress which result from body-weight can be neglected, the stress distribution indicated can also be applied to the ends of some sills.) It can be inferred from this figure that if tensile failure occurs near the end of the crack, then an off-shoot, tensile failure plane can develop at an angle of about 60°. (Such off-shoots, resembling a 'rhinoceros horn', are known to exist.)

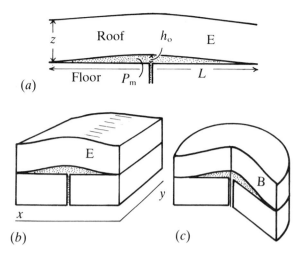

(a)

(b)   (c)

Fig. 3.38. Geometry used in elastic analysis of concordant intrusions. (a) Two dimensional section. (b) Example in which the section shown in (a) is constant in the y direction. (c) Example with circular symmetry.

However, if the sill is intruded along a bedding plane between competent rocks (Fig. 3.38) there is no comparable change in pattern and intensification of the stresses at the tip of the sill as indicated in Fig. 3.37. Instead, the changes in horizontal stresses near the tip of such an asymmetrical sill can be inferred from an analysis by Johnson (1970). In this analysis he applied elasticity theory to establish the deflection curve of the roof of a concordant intrusion in response to a magmatic over-pressure (P_m) that is assumed to be constant throughout the intrusion. He further assumed that the concordant body was supplied from a central source, as shown in Fig. 3.38(a), and that the floor of the body was rigid, so that only the roof deflected in response to this magmatic over-pressure. This figure represents a section through a concordant intrusion of width L and maximum height h_0 which is set at a depth z within the crust. The elastic modulus B, is the rigidity modulus, where $B = m^2 E/(m^2 - 1)$ and m is Poisson's number. He showed that the maximum displacement h_0 (i.e. the maximum thickness of the intrusion) is given by:

$$h_0 = P_m L^4/32Bz^3 \qquad (3.18)$$

or

$$P_m = 32h_0 Bz^3/L^4. \qquad (3.18(a))$$

It is emphasised that this relationship is based on displacement and flexing of a *thin cover unit* in which the ratio of L/z is greater than 10:1.

From this analysis it can be inferred that the decrease in horizontal stress at points A and B in Fig. 3.39(a) are equal to $2P_m$, where P_m is the magmatic over-pressure. It has been noted that in order for a sill to develop from a dyke, the horizontal and vertical stresses must be approximately equal. Consequently, if the horizontal stresses at points A and B in

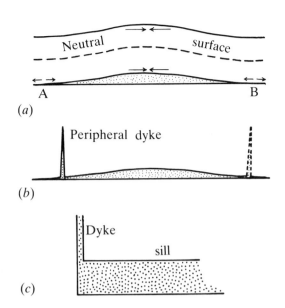

(a)

(b)

(c)

Fig. 3.39. (a) Representing variation in lateral stress caused by flexing of an elastic cover rock. (b) End or peripheral dykes. (c) Detail of a rectilinear transfer from a sill to a dyke.

Fig. 3.39(a) are reduced by an amount $2P_m$, it is possible that the stress conditions conducive to the emplacement of dykes may result, with the orientation shown in Fig. 3.39(b).

Sills may be circular or elliptical in plan. Therefore, because the stress reductions are related to the geometry of the roof at the edges of the intrusion, then, if the horizontal stress has approximately radial symmetry, the resulting off-shoot dykes, which will follow the edge of the sill, will also be curved or arcuate in plan. Occasionally, even in areas which dominantly contain dykes with linear traces, dykes with curved or even arcuate traces may be observed. Such traces were observed by Norman *et al.* (1977) in a photogeological study of the Lesotho basalts (Lesotho, S. Africa). These traces were attributed to dykes which were initiated at the periphery of major, deep-seated sills, with an approximately circular plan. It is envisaged that such peripheral dykes would define blunt, or even rectilinear, ends to the sills (Fig. 3.39(c)). Such terminations of sills and their junction with vertical dykes are known to occur in Tasmania (see *Dolerites: a Symposium*, edited by Carey, 1958).

Laccoliths

Laccoliths are concordant intrusions which are markedly asymmetrical about the horizontal plane and also exhibit a very much smaller aspect ratio of length (L) to maximum height (h_0) than sills; for whereas sills frequently fall in the aspect ratio range of 100–1000:1, laccoliths have L/h_0 ratios of about 3–15:1. The geometry in plan of a laccolith is often approximately circular, though, as can be seen in Fig. 3.40(a), if the intrusion is built up from a number of separate intrusive events, the outline may be irregular. The profiles of laccoliths, which are circular in plan, commonly approximate to the form represented in Fig. 3.38(*c*), as may be inferred from the view of a laccolith in E. Iceland (Fig. 3.40(*b*)).

Examples of groups of laccoliths that have been much studied are to be found in the Henry Mts. of S.E. Utah, U.S.A. These were reported and analysed

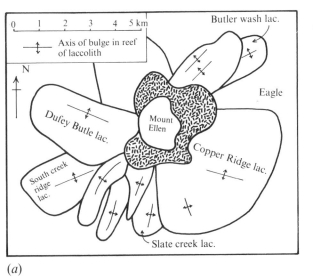

(a)

by Gilbert as early as 1877 and more recently have been the subject of study by Johnson (1970) and Pollard & Johnson (1973). These laccoliths are intruded in two groups, the upper and smaller of which comprise intrusions with an average diameter of about 2 km, while the lower group have an average diameter slightly in excess of 4 km. The depth of emplacement of these groups of laccoliths is thought to be approximately equal to their respective mean diameters.

Johnson (1970) used the elastic analysis, based on the model shown in Fig. 3.39(*a*), and dealt with in the previous section, to derive a dimensionless expression that represents the deflection of the roof which is induced by a uniform excess magmatic pressure. This expression is:

$$h(2Bz^3/P_m L^4) = (x/L)^4 - 2(x/L^3) + (x/L)^2$$

$$(3.19)$$

where *h* is the vertical uplift at a distance *x* from one end of the intrusion, and the other terms are as previously defined.

When *h* is plotted against *x* using this expression, the resulting curve can be impressively and persuasively similar to the laccolith profile represented in Fig. 3.40(*b*). However, Eq. 3.19 relates to

(b)

Fig. 3.40. (*a*) Plan of laccoliths fed from a central (Mt. Ellen) stock in the Henry Mountains, Utah, U.S.A. (After Hunt, 1953.) (*b*) View of laccolith on the East Coast, Iceland, showing the same general form as in Fig. 3.38(*b*) and (*c*).

conditions in which the ratio of the depth of cover to the length of the intrusion is equal to, or less than, 1:10. However, the ratio for laccoliths in the Henry Mts. is about 1:1.

A further analysis and discussion of the mechanics of growth of the Henry Mts. laccolithic intrusions was presented by Pollard & Johnson (1973). This in part, was meant to obviate the problem regarding the fact noted above that the ratio of horizontal extent of the intrusion to its depth of cover is about 1:1. This paper, which is mainly concerned with the analysis of a *conceptual model*, also includes results of *physical model* studies in which it is obvious that the models reached and exceeded their elastic limits before the aspect ratio comparable with that of the Henry Mts. laccoliths was attained. Indeed, the maximum extension in the roof of both the physical models and real intrusions sometimes exceeds 20 per cent; and this is greatly in excess of possible elastic strains.

A major feature of the conceptual model used in this paper was to assume that the cover rocks above the intrusion were layered and can be considered to have an *effective thickness* significantly smaller than the real thickness (z). However, in order that the layered unit may exhibit an effective thickness which is considerably smaller than the real thickness, slip must take place on a number of bedding planes in the rock forming the roof of the intrusion. The conditions necessary to induce slip on such planes are not considered by Pollard & Johnson.

Even if such planes exhibit no cohesion, it is necessary to overcome the frictional resistance to sliding before movement may take place. This is given by Amonton's Law (Chap. 1), which for the situation under consideration can be expressed as:

$$\tau = S_z(1 - \lambda)\tan \phi \qquad (3.20)$$

where τ is the shear stress acting on the bedding, or slip, plane, λ is the ratio of the water pressure in the rocks to that of the total vertical stress (S_z) and ϕ is the angle of sliding friction. We are not aware of any evidence that these rocks, at the time during which the intrusion formed, were generally overpressured. Hence, the value of λ is likely to have been about 0.4 to 0.5. The cover rocks include massive sandstones, with a value of sliding friction of $\phi = 30°$. Substituting these values in Eq. (3.20), it follows that the shear stress required to cause sliding would range from about zero, near the surface, to about 250 bar at a depth of 4 km. Therefore, slip could take place on planes near the surface and also on planes very close to the intrusion (where the heat of the magma may temporarily raise the ambient fluid pressures).

It could be argued that the angle of sliding friction on some planes in the sedimentary sequence may be less than 30°. Even so, it would appear likely that slip on the majority of bedding planes would be inhibited by the frictional resistance, so that the effective thickness would not be substantially less than the real thickness of the cover rock above the intrusion.

However, it can be inferred from field evidence that bedding plane slip has occurred on some of the planes in the roof rock (Pollard, personal communication). So it is necessary to consider how such slip may have been able to occur.

Pollard & Johnson suggest that the final development of the laccoliths is by plastic deformation of the sandstone roof, though they do not indicate how this may be brought about at the shallow depths and the relatively low stresses which must have obtained when the intrusions were emplaced. Price (1975) suggested that this problem can be resolved if the effects of the heat of intrusion are taken into account, for this would heat up the fluids in the country rock and generate high fluid pressures. If, as is likely, the sandstone forming the roof obeys the special law of effective stress (Chap. 1), this high fluid pressure will cause the sandstone to disaggregate and become a sand which will readily deform, mainly by 'ductile' shearing. Later, as the intrusion and roof rocks cool, the sands become reconstituted as a sandstone, as grain to grain cohesion takes place in the presence of hot fluids, which would aid diffusion and cementation processes. (We admit that we have not studied thin sections of the rock and cannot therefore adduce evidence to support this suggestion. However, the ductile shears and faults shown in Fig. 3.41(*a*), which are of the roof rock of one of the Henry Mts. intrusions are compatible with this explanation. This process can, of course, also be invoked to explain the intrusion breccias which are sometimes associated with concordant and discordant intrusions, Fig. 3.41(*b*).)

Finally, it should be noted that laccoliths are usually composed of rocks which are more acid than the dolerites and gabbros that so often form sills. Acid rocks, at their melting temperature, are many orders of magnitude more viscous than basaltic magma so that the component of magmatic pressure required to overcome viscous drag between the intrusion and the adjacent rock is correspondingly higher. This means that there will be a significant fall-off in magmatic pressure away from the supply vent. It is this effect (also discussed by Pollard & Johnson) which, we suggest, causes the observed roof displacement geometry of acidic laccoliths and that the mechanism includes the local development of bedding plane slip which, Pollard & Johnson have argued, is so important.

It should be noted that the distance to which temperature effects, which are related to magmatic intrusion, extend into the country rock, during a

(a)

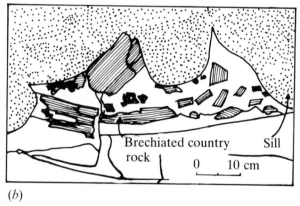

(b)

Fig. 3.41. (a) Ductile shear zones and fault in sandstones forming the roof of one of the Henry Mt. laccoliths. (b) Intrusion breccia, Henry Mt. laccoliths.

specific time interval, is largely related to the thickness of the intruded body. Thus, the heating of the country rock around a developing laccolith will be limited about the peripheral zone where the intrusion is thin, but will have a greater importance above the central zone where the intrusion is thicker. This means that bedding plane slip will be relatively enhanced in the roof above the centre of the intrusion. So it is above the central zone that the flexural rigidity will be most reduced, as the result of bedding plane slip brought about by the heating of the interstitial water in the roof rock, and upward curvature of the roof with further thickening of the intrusion will be enhanced.

We suggest that it is this effect, which tends to concentrate thickening of the central zone of the intrusion, coupled with the local weakening of the roof by disaggregation and failure by ductile normal faults and shear zones (Fig. 3.41) that helps determine the final form of a laccolith.

Polygonal fractures

One of the common features of sills and dykes is that they often contain a host of polygonal fractures which define arrays of columns which may be both spectacular and picturesque; as illustrated in Figs. 3.42(a) and (b) which show *columnar jointing* near the bottom of the Grand Canyon, Arizona, U.S.A. and in a sill in S. Iceland. When the fractures develop in vertical dykes the colums tend to be horizontal. Gudmundson (1984) describes an example of an Icelandic dyke which is composed of rows, or sheets, of columns which he infers, represent separate phases of intrusion. An individual intrusion may contain as many as a dozen such sheets. If a late phase of such a multiple intrusion dyke occurs while the material in the dyke is fractured but still hot and ductile, it may be bent by the later intrusive phase (see the junction between sheets 1 and 2 in Fig. 3.43).

These polygonal columns are bounded by 3 to 12 planar, 'facet' fractures (Fig. 3.44). Most frequently the number of longitudinal, bounding facets is between 5 and 8, with the preponderance of columns exhibiting 6 or 7. It can be seen in Fig. 3.45 that the columns may also be cut by ladder fractures which are approximately orthogonal to the facet fractures. These ladder fractures are usually orthogonal to the facet surfaces on the outside of the column, but are often convex or concave away from the edges of the column and are sometimes seen to have originated at or near an impurity or bleb in the basalt.

These various fractures are associated with the cooling and consequent shrinkage of the solidifying igneous rock mass. Thus, a basaltic magma is emplaced at a temperature (T_i) of about 1200 °C and will eventually cool to the initial temperature (T_c) of the country rock. The shrinkage cooling strain ($-e_c$) will be tensile and given by:

$$e_c = \alpha(T_i^\circ - T_c^\circ) \qquad (3.21)$$

where α is the coefficient of thermal expansion (approximately 5×10^{-6}/°C); so that $e_c = 0.6$ per cent.

Shrinkage, as the result of desiccation, can also take place in sediments. Spectacular polygonal fractures in the salt flat, Salar de Atacama, in Northern Chile are shown in Fig. 3.46(a) and fractures in the dried-out sediments of a tailings pond of a mine in Sardinia are illustrated in Fig. 3.46(b). These latter fractures are worthy of close scrutiny, for they are

Fig. 3.42. (*a*) Columns in basalt near the base of the Grand Canyon, Arizona, U.S.A. (*b*) a sill in South Iceland.

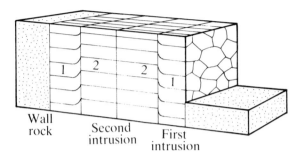

Fig. 3.43. Schematic morphology of horizontal, polygonal columns in a dyke which has experienced two phases of emplacement and exhibits a central suture. A later phase of intrusion may cause distortion of earlier polygons.

(a)

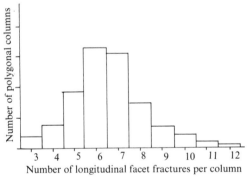

(b)

Fig. 3.44. (a) Close up of polygon from the Giant's Causeway showing convex and concave cross fractures. (b) Histogram showing type of distribution of number of facet fractures per column commonly observed.

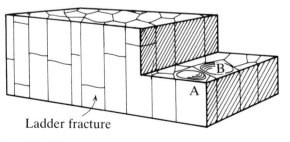

Ladder fracture

Fig. 3.45. Schematic representation of polygonal columns in a sill with ladder fractures. Dome and basin form of the ladder fractures are shown at A and B respectively.

instructive as regards the probable mode of development of polygonal fractures in basalts. It may be inferred that there are different generations of fractures. Initially, tensile fractures, probably of random orientation, spread to interact and define a reasonably extensive block. As desiccation (or cooling in basalts) proceeds, the block contained within the first generation of fractures eventually generates tensile stresses of sufficient magnitude for fractures internal to the block to develop. Because many of the fractures appear to form with random orientation, the number of fractures which eventually bounds a polygonal column of rock is probably dictated by chance.

It is interesting to note that an early generation of fractures may define irregular units which are clearly destined to be split (Fig. 3.47(*a*)). If it is assumed that shrinkage is directed toward two centres (SC$_1$ and SC$_2$) within the unit which is eventually to be split into blocks, as indicated in Fig. 3.47(*b*), this gives rise to a component of shear along the potential line of division. This shear component results in an array of en echelon, sigmoidal fractures (Fig. 3.47(*c*)). Such 'shear zones' are also known to occur in basalts and may give rise to 'walking sticks', perhaps a metre long with a cross section as shown in Fig. 3.47(*d*)).

As regards the ladder fractures in basalts, it is probable that their formation may be related to one of two mechanisms. The first of these is that which gives rise to exfoliation fractures, which will be discussed in Chap. 4. However, it is more probable that they are best explained in terms of the second mechanism which is dependent upon alteration effects to the 'skin' of the polygonal column. As the column develops, it is free to shorten along its length as it cools. This longitudinal shrinkage would obviate the development of longitudinal tensile stresses. Hence, there would be no potential or necessity for ladder fractures to develop. In most near-surface crustal environments, however, soon after the polygonal fractures developed, they would be penetrated by ground or connate water and hydration reactions would occur at the surface of these fractures and also to some small distance into the column. These reactions would tend to cause expansion, or at least slow down the shrinkage process, of the (relatively thick) skin of the column. Thus, the shrinkage caused by the cooling of the column is slowed or even reversed in the skin area but continues unimpaired in the interior. It is this differential shrinkage, we suggest, that gives rise to the formation of the ladder fractures. The internal sites of initiation of these cross fractures may often be determined by inclusions (of somewhat different elastic properties to the matrix) which causes them to be 'stress raisers'. (The reader is referred to a text on rock mechanics and stress measurements (e.g. Jaeger & Cook, 1969) for a treatment of the concept of stress raisers.)

Number of polygonal columns

3 4 5 6 7 8 9 10 11 12
Number of longitudinal facet fractures per column

Wall rock · Second intrusion · First intrusion

(a)

(b)

Fig. 3.46. (a) Polygonal cracks in salt flats, Salar de Atacama, North Chile. (Courtesy of J. Skarmeta.) (b) Mud cracks in tailings pond, S.W. Sardinia.

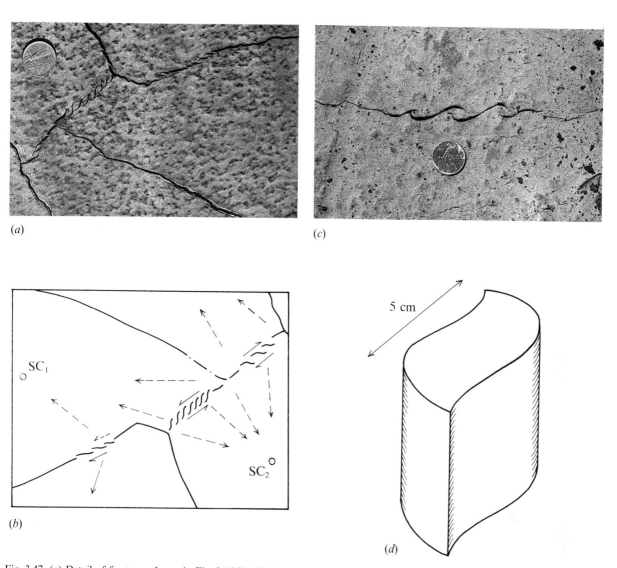

Fig. 3.47. (*a*) Detail of fractures shown in Fig. 3.46(*b*). (*b*) Interpretation of development of fractures represented in (*a*). (*c*) Detailed morphology of curved fractures which form as the result of shear. (*d*) Sketch of portion of 'walking stick' basalt.

Finally, it must be emphasised that these polygonal fractures only take place in intrusions which occur in the upper levels of the crust and, more particularly, in relatively thin intrusions (or extrusions). At depth in the crust, and particularly in thicker intrusions, the rate of cooling is relatively slow. The rate of shrinkage will be commensurately slow. The vertical shrinkage can be accommodated by a slight thinning of the intrusion and the lateral shrinkage may be offset by flow of the intrusive material probably by one of the diffusion mechanisms. In an area where some thin intrusions contain polygonal fractures while other thicker intrusions of the same age and composition do not, then given data as regards depth of burial, it should be possible to quantify the rheological properties of the hot, crystallised, igneous material. As far as we are aware, no such exercise has been attempted.

Commentary

The study of minor intrusions is too often neglected by structural geologists. The analysis of their emplacement not only provides invaluable insights into such problems as the mode of fluid flow-paths in the crust, but it also relates to the factors and processes involved in deformation of the wall or roof rocks. Moreover, it enables one to see clearly the dangers of accepting the conclusions of theoretical analyses before these conclusions have been subjected to the critical test of quantification. The proportion of dyke material to country rock also provides information regarding the environment in which the magma was intruded. Finally, in certain environments, it may be possible to establish the rheological properties of the magmatic material and/or make a quantitative estimate of the stress regime of the rock mass at the time of intrusion.

References

Anderson, E.M. (1951). *The dynamics of faulting*. Edinburgh: Oliver & Boyd.

Badgley, P.C. (1965). *Structural and tectonic principles*. New York: Harper & Row, and Tokyo: John Weatherhill.

Carey, S.W. (1958). *Dolerites. A symposium*, University of Tasmania, Hobart, Australia. 1–274.

Delaney, P.T. & Pollard, D.D. (1981). *Deformation of host rocks and flow of magma during growth of minette dikes and breccia-bearing intrusions near Ship Rock, New Mexico*. U.S. Geol. Surv., Prof. Paper 1202.

Gilbert, G.K. (1877). *Report on the geology of the Henry Mountains*. U.S. Geog. & Geol. Surv. Rocky Mountain Region.

Gudmundsson, A. (1983). Stress estimates from the length/width ratios of fractures. *J. Struct. Geol.*, **5**, 623–6.

(1984). A study of dykes, fissures and faults in selected areas of Iceland. Unpublished Ph.D. Thesis, University of London.

Hunt, C.B. (1953). *Geology and geography of the Henry Mountains region, Utah*. U.S. Geol. Surv. Prof. Paper 228.

Jaeger, J.C. (1959). Temperature outside a cooling sheet. *Am. J. Sci.*, **257**, 44–54.

Jaeger, J.C. & Cook, N. (1969). *Fundamentals of rock mechanics*. London: Methuen.

Johnson, A.M. (1970). *Physical processes in geology*. San Francisco: Freeman Cooper & Co.

Knopf, A. (1936). Igneous geology of the Spanish Peaks Region, Colorado. *Geol. Soc. Am. Bull.*, **68**, 1727–84.

Macgregor, M. & Macgregor, A.G. (1936). *The Midland Valley of Scotland*. British Regional Geology. Inst. Geological Sciences, H.M.S.O.

Norman, J., Price, N.J. & Peters, E.R. (1977). Photogeological trace study of controls of kimberlite intrusions in Lesotho basalts. *Trans. Instn. Min. – Metall.*, **86**, B78–90.

Odè, H. (1957). Mechanical analysis of the Dike Pattern of the Spanish Peaks area, Colorado. *Geol. Soc. Am. Bull.*, **68**, 567–78.

Pollard, D.D. (1969). Aspects of the mechanics of sheet intrusions. Unpublished M.Sc. dissertation, University of London.

Pollard, D.D. & Johnson, A.M. (1973). Mechanics of growth of some laccolithic intrusions in the Henry Mountains, Utah, II. Bending and failure of overburden layers and sill formation. *Tectonophysics*, **18**, 311–54.

Price, N.J. (1975). Rates of deformation. *J. Geol. Soc. London*, **131**, 553–75.

Rutter, E.H. (1976). The kinetics of rock deformation by pressure solution. *Phil. Trans. Roy. Soc. London*, **A283**, 203–13.

Shaw, H. (1969). Rheology of basalts in the melting range. *J. Petrology*, **10**, 510–35.

Shaw, H., Wright, T.L., Peck, D.L. & Okamura, R. (1968). The viscosity of basaltic magma: analysis of field measurements in Makaopuhi Lava Lake, Hawaii. *Am. J. Sci.*, **266**, 225–64.

Skarmeta, J.J. (1979). Analisis Estructural de Diques Deformados en la Peninsula de Mejillones, Norte de Chile. *Revista Geol. de Chile*, **9**, 3–16.

(1983). The structural geology of the Sierra de Moreno, Northern Chile. Unpublished Ph.D. Thesis, University of London.

Skarmeta, J.J. & Price, N.J. (1984). Deformation of country rock by an intrusion in the Sierra de Moreno, northern Chilean Andes. *J. Geol. Soc. London*, **141**, 901–8.

Speight, J.M. & Mitchell, J.G. (1979). The Permo-Carboniferous dyke-swarm of northern Argyll and its bearing on dextral displacements on the Great Glen Fault. *J. Geol. Soc. London*, **136**, 3–11.

Wager, L.R. & Deer, V.A. (1938). A dyke swarm and crustal flexure in East Greenland. *Geol. Mag.*, **75**, 39–46.

Williams, S.H. & McBirney, A.R. (1979). *Volcanology*. London: Freeman.

Williams, D. & Ramsay, J.G. (1959). *Geology of some classic British areas: Snowdonia*. No. 28. Geol. Ass. Guide.

4 Diapirs, related structures and circular features

Introduction

Diapirs are features which result when a body of relatively low density pierces and rises through overlying rocks of higher density. The diapirs so formed may be composed of ice (giving rise to pingos) peat, coal, mud and overpressured clays and shales. However, the largest and most important diapirs are composed either of salt (or other evaporites) or igneous rock. We shall concentrate wholly on these major diapiric forms and in the first section of this chapter we discuss some of the classical concepts and experiments which have been used to explain major diapirism. It will be shown that some of these concepts leave much to be desired when applied to the emplacement of salt diapirs. We shall describe the morphology of diapirs and discuss how they may have been emplaced. It will be argued, however, that the classical concepts of diapirism may reasonably be applied, as a first approximation, to the emplacement of igneous intrusions. Consequently, in the second section of this chapter we shall deal with the use of fractures and other features to deduce the geometry of such intrusions and their emplacement. The development of barren fractures in these intrusions in the near-surface environment is a vexing problem which is considered at some length and which throws light on the mechanism by which 'joints' develop in other rock types.

Major igneous intrusions are not uncommonly accompanied by intrusive sheets which are arcuate or circular in plan. The possible mode of development of some of these circular features is briefly discussed. The diapirs themselves and the associated curved sheets together form an important group of circular features; however, what may prove to be the most important group of circular features are those which result from meteoritic impact. Impact Tectonics and the structures which result, are largely or completely ignored in texts on the development of geological structures. We shall, therefore, consider this topic at some length in the third and final section of this chapter.

SECTION I

Diapiric intrusions

Diapiric intrusions may take many forms, and we shall describe the characteristics exhibited by many such bodies later in this chapter. However, one of the features common to diapirs, composed either of igneous or evaporite rocks, is that they are intruded as distinct and separate bodies which may be related as a suite in space and time. Thus, an empirical relationship has been established that individual volcanoes and plutons are, in some areas, so spaced from each other that their average separation is approximately equal to the depth of the magma source layer. Such spacing of igneous diapiric intrusions, in Baja California, has been studied by Rickard (1984) (Fig. 4.1), who indicated that many of the observed spacings could be compared favourably with data of magma depth deduced from published crustal profiles in that area, but that the relationship is partially masked by a secondary effect related to ponding of magma beneath an upper, 6 km thick, brittle layer.

Examples of spacing of salt domes, plugs or stocks in N.W. Germany are reported by Trusheim (1960) and Sannemann (1968), (Fig. 4.2). Turcotte &

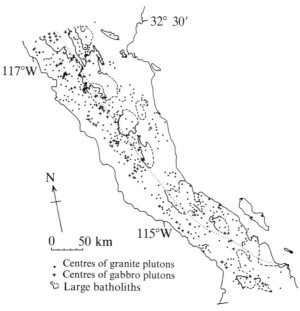

Fig. 4.1. Sketch map of distribution of intrusions in Baja, California. (From Rickard, 1984; after Gastil *et al.*, 1975.)

North
sea

Denmark

Holland

Hamburg

Germany

Hannover

Elbe

Munster   0   50 km

Fig. 4.2. Sketch map of salt wall and domes in N.W. Germany. (After Trusheim, 1960.)

Schubert (1982) note that these domes probably emanated from a salt layer which was at a depth of about 5 km and that the domes of plugs have, on average, a separation of 10–15 km. By using a simple conceptual model, they predicted that the separation (*D*) between intrusions should be:

$$D = 2.568z \qquad (4.1)$$

where *z* is the depth to the salt layer. They suggest that this relationship (accurate to three decimal places?) and the data for N.W. German salt domes are in good agreement.

The conceptual model noted above is ultimately based on experimental model investigations of diapiric intrusions. An example of the growth of a single model diapir is indicated in Fig. 4.3 (Price, 1975). This and similar models are based on a two layer system of liquids in which the more dense layer overlies the less dense material, so that they are initially at rest but in unstable equilibrium. As the result of some perturbation which is sometimes accidentally (but more usually deliberately) induced, the equilibrium condition is upset and the less dense material rises through the more dense liquid. The Turcotte & Schubert relationship (Eq. (4.1)) is based on a theoretical representation of such a model situation, (Turcotte & Schubert, 1982), in which a *Rayleigh–Taylor* instability develops (Fig. 4.4). The conditions represented in this figure are extremely simple. More complex theoretical conditions have been considered by Ramberg (1972 and 1981), which have been checked and shown to represent adequately physical models of some natural igneous intrusions.

Magma is a viscous material. Deep crustal rocks at elevated temperatures can also be regarded as viscous (though usually non-Newtonian) liquids. Indeed, even at intermediate levels in the crust, the

heat of a major intrusion can modify the rheological behaviour of the country rocks so that they flow relatively freely. Hence, the Rayleigh–Taylor instability, or better, the more complex, and geologically more realistic, models used by Ramberg may reasonably be applied to predict the morphology of individual diapirs of the type indicated in Fig. 4.5, as well as sometimes describing their separation from each other. Only at the higher levels in the crust, where the rocks are sufficiently cool and behave as brittle solids, does the model break down. However, the same cannot, in general, be said for the intrusion of salt diapirs, for these originate at relatively shallow depths in the crust, within, or possibly just below, the zone in which many sedimentary rocks behave as brittle solids. Hence, except in special cases (which we shall discuss later), it is unrealistic to analyse diapiric emplacement of salt domes in terms of viscous liquid systems; so that the degree of correlation between observed separation of salt domes and that predicted by Eq. (4.1) is possibly more apparent than real. In the following section, we therefore consider in some detail possible modes of initiation and intrusion of salt diapirs into rocks which behave as brittle and/or as ductile plastic solids.

Emplacement of salt diapirs and related structures
Diapirs have considerable economic importance not only for the salt itself (which in Europe, Asia and Africa has formed an important element in trade for many thousands of years) but also in more recent times, because the intrusion of the salt frequently creates conditions which are favourable for the development of hydrocarbon reservoirs in the adjacent sediments.

The term *diapir* was proposed by Mrazec (1915) to describe salt bodies in the Carpathians that occupied the cores of elongated domes and anticlines and which had locally broken through the crests of the folds, or had been intruded along faults. Such structures had earlier been recognised in Transylvania by Posepny (1871) who represented them in the diagrammatic form shown in Fig. 4.6.

In areas such as N.W. Germany, the morphology of the intrusions is simpler and tends to occur as *salt pillows, salt walls* or individual *salt domes* or *stocks*. Examples of such structures are given in the composite diagram by Trusheim (1960) (Fig. 4.7). The main salt walls are thought to be associated with major faults.

The morphology of isolated domes, even when intruded into flat-lying sediments, exhibits considerable variety. The form of such intrusions and their associated internal and external structures are indicated in Fig. 4.8. The reader interested in obtaining further information and greater detail is referred to other publications, such as the text by Halbouty

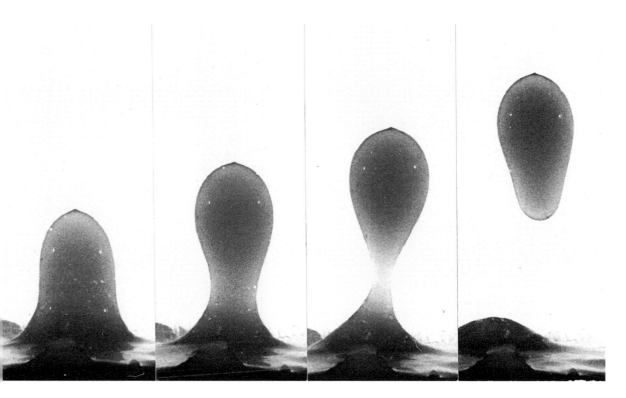

Fig. 4.3. Sequences in diapiric development in which a low density liquid rises through a higher density liquid. (After Price, 1975.)

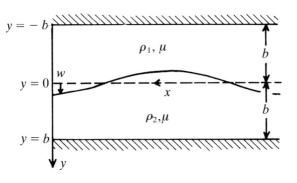

Fig. 4.4. Diagrammatic representation of the initial stage of a Rayleigh–Taylor instability.

(1967) and *Diapirism and Diapirs* (edited by Braunstein & O'Brien, 1968). Here we shall consider only the generalised and simplified morphology and associated features of the types of individual intrusions represented in Fig. 4.9. At their deeper levels, *bollard* type plugs (type A in Fig. 4.9) tend to be elliptical in plan; at intermediate levels they are more nearly circular in section with parallel (cylindrical walls), and only at the higher levels do the walls diverge to form an overhang or mushroom shape.

As has already been indicated, the diapirism of salt has been attributed to its general mobility as a 'liquid' of relatively low viscosity (coupled, of course, with its low density). We suggest that 'mobility' of salt (halokinesis) is a somewhat vague term, for it includes both *intrinsic* and *extrinsic* elements. We consider that intrinsic mobility is related solely to the rheological

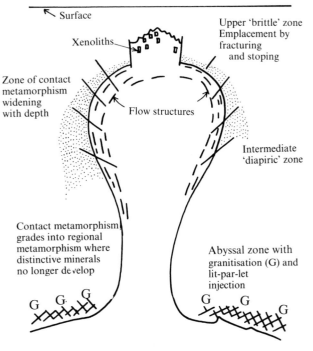

Fig. 4.5. Schematic representation of an igneous diapir with ductile and brittle deformation of the country rock at the lower and higher levels respectively.

properties of the salt which is controlled by the precise composition of the material, temperature and moisture content. Extrinsic mobility is related to the geometry of the salt body and, in particular, to the thickness of the *mother salt layer*, from which the salt plugs subsequently develop.

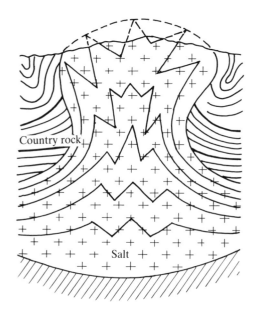

Fig. 4.6. Early representation of a salt diapir in Transylvania. (After Posepny, 1871.)

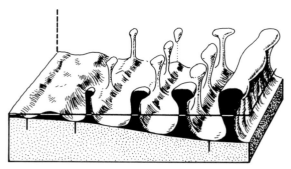

Fig. 4.7. General morphology of salt pillows, walls and domes. (After Trusheim, 1960.)

Intrinsic mobility

An interesting assessment of the intrinsic mobility of salt (and other aspects of diapirism) was presented by Gussow (1968), who pointed out that, from the dimension of chambers and other working spaces in salt mines (see for example Fig. 4.10), it can be inferred that cold salt is as strong as some limestones. He further argued that high mobility of salt is only achieved when the mother salt approaches an ambient temperature of about 300 °C, which in the U.S. Gulf Coast area occurs at a depth of about 10 km.

For a quantitative approach, we must turn to Heard (1972) who conducted experiments on polycrystalline dry halite and established its equation of state (see Chap. 1):

$$\dot{e} = -4.8 - \frac{25.2}{RT} + 5.4 \log S \qquad (4.1(a))$$

where $\dot{e}$ is strain-rate, R is Boltzmann's constant, T is temperature in degrees Kelvin and S is the differential stress. From this equation, the stress–strain-rate

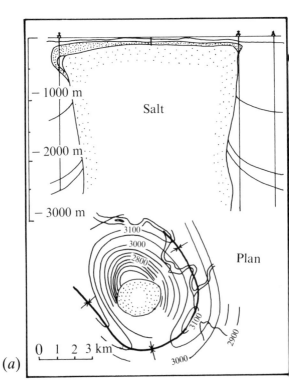

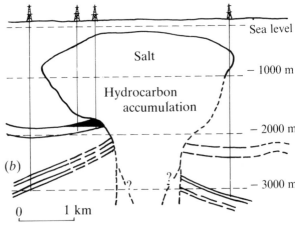

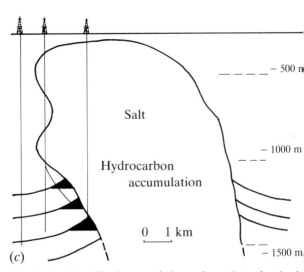

Fig. 4.8. Profiles indicating morphology of a variety of real salt domes or plugs (a) Zanapa Dome, Mexico. (b) Bethel Dome, Texas, U.S.A. (c) Cote-blanche Dome, Louisiana, U.S.A. ((b) and (c) after Halbouty, 1967).

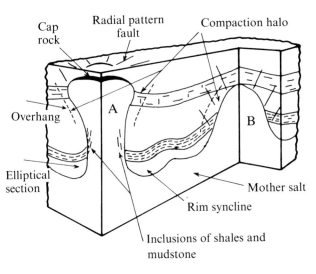

Fig. 4.9. Diagrammatic representation of characteristics and features associated with (*A*) bollard-type and (*B*) bell-type salt plugs.

Fig. 4.10. Sketch of vertically plunging folds revealed in the galleries of a salt mine. (After Balk, 1953.)

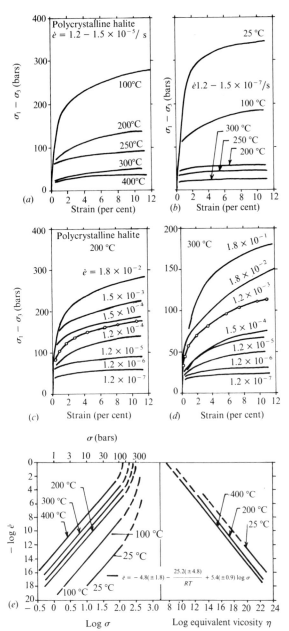

Fig. 4.11. Stress–strain relationships of polycrystalline halite at different temperatures (*a*) and (*b*) and strain-rates (*c*) and (*d*). (*e*) Viscosity, differential-stress and strain-rate relationships inferred from an equation of state (Eq. (4.1(*a*))) derived from the types of data shown in graphs (*a*)–(*d*). (After Heard, 1972.)

relationship, and hence the viscosity of salt, can be established for probable geological strain-rates and temperatures (Fig. 4.11). These results generally support Gussow's conclusions for, at room temperature, dry salt has a yield strength of several hundred bar and only at a temperature of 300 °C and probable geological strain-rates will it flow under a differential stress of the order of 1.0 bar.

Salt mines are usually extremely dry places, largely because salt is hygroscopic. However, natural salt is rarely completely dry, for it usually contains brine in intracrystalline inclusions. In addition, when salt is adjacent to water-saturated sediments, it is certain that the salt in the contact zone will contain water. Moreover, as we shall see, in certain situations, the original salt deposit itself may contain significant amounts of brine. Unfortunately, the rheology of wet salt has not been studied to the same extent as dry salt, but Urai (1983) has shown that, at a confining pressure of 280 bar and a temperature of 60 °C, wet salt can support a differential stress only 30 per cent of that which can cause yielding in dry salt under similar conditions. Hopper (1974) carried out some pilot experiments, at a higher temperature of 250 °C, in which he established that wet salt has a yield strength which is markedly less than that of dry salt tested under similar conditions. His results show a similar strength reduction to that observed by Urai for tests conducted at low temperatures. However, he established that, in the temperature range 200–300 °C, a transition point exists above which wet salt deforms by the mechanism of pressure solution at very low differential stresses. Indeed, these stresses are so low that they are comparable with the strength of the thin copper jacket in which the salt specimen is sheathed, so that it is difficult to establish with accuracy, the differential stress at which the salt, alone, deforms. For example, in one experiment, conducted at a strain-rate of 4×10^{-5}/s, at 300 °C and a value of fluid pressure (p) such that the ratio of fluid pressure to confining pressure was 0.8:1 when the confining pressure was 2.0 kbar, he established that the combined yield strength of copper jacket and salt specimen was only 5 bar.

Here we shall be more concerned with the behaviour of salt at high temperatures, although, as we have seen, the data relating to the behaviour of wet salt under these conditions are scarce. We shall therefore be guided by the results obtained by Hopper and, as a first approximation, we shall take the differential stresses that can be supported by wet polycrystalline salt, for specific strain-rates and temperatures, as not more than 10 per cent of those shown in Fig. 4.11.

Extrinsic mobility

Let us consider a horizontal layer of mother salt (thickness z_s) at a temperature of 250–300 °C. Under these conditions, as we have seen from the previous section, salt cannot support a significant differential stress, so that the stresses in the salt are essentially hydrostatic. In order that salt may migrate laterally and so feed any salt pillow, wall, or plug, a horizontal hydraulic gradient must be induced in the salt layer. This is most easily accomplished by a lateral variation in vertical loading on the salt layer, as indicated in Fig. 4.12. Such lateral variations of loading may be induced by changes in surface, or sub-surface topography, or lateral changes in density of the overlying sediments which result from facies changes or, of course, any combination of these factors. For simplicity, it as assumed in Fig. 4.12 that the change in vertical load varies in a linear fashion and so gives rise to a constant shear stress (τ) which will induce lateral migration of the salt. For equilibrium conditions, in Fig. 4.12, it follows that:

$$S_{ex}z_s = 2\tau L \tag{4.2}$$

where S_{ex} is the excess horizontal pressure acting at AA′, relative to BB′ and the linear dimensions z_s and L are as indicated.

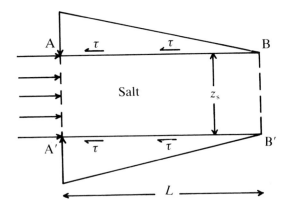

Fig. 4.12. Boundary stresses and fluid (Newtonian) flow between parallel layers.

It can be shown that the average velocity (v_s) of salt, assuming it to be a Newtonian fluid, flowing between parallel plates, in response to a shear stress (τ) is:

$$v_s = \frac{\tau z_s^2}{k\eta} \tag{4.3}$$

where η is the coefficient of viscosity and k is a numerical constant (see Chap. 3 for comment). Then from Eqs. (4.2) and (4.3):

$$v_s = \frac{S_{ex}z_s^3}{2\eta k L}. \tag{4.4}$$

The volume (Q) of salt that will migrate in planar flow, in a specific time, is:

$$Q = v_s z_s$$

so that

$$Q = \frac{S_{ex}z_s^4}{2\eta k L}. \tag{4.5}$$

That is, the extrinsic mobility of a salt layer is related to the fourth power of its thickness. By using Eq. (4.5), with the data and conclusions related to the intrinsic mobility and realistic (low) values for S_{ex}, it can be shown (and the reader is invited to tackle the exercise) that the salt layer needs to be of the order of 500 m thick before significant lateral migration can be achieved, even when the salt is at a temperature of 250 °C. Hence, layers of salt which are only perhaps 100 m thick will not develop into salt domes. This finding is, of course, completely in keeping with the quantity of salt required to form an individual dome or plug and also their separation one from another. The average diameter of a typical cylindrical, or bollard-shaped plug, in the U.S. Gulf Coast areas, of the type represented in Fig. 4.9(a), is about 6 km. If we assume that the plug has a height, in this area, of 10 km and is composed only of salt, it follows that the volume of salt required to form such a structure is only a little less than 300 km³.

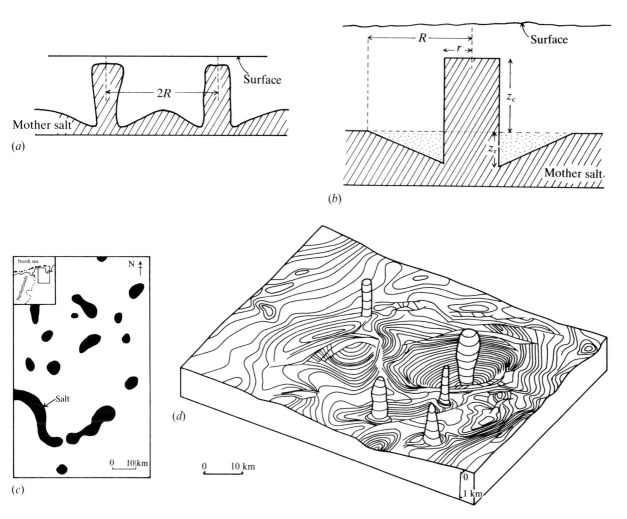

Fig. 4.13. (*a*) Schematic representation of section through two adjacent, bollard-type plugs. (*b*) Simplified geometry of salt layer and plug. (*c*) Distribution of Salt Domes in a small area of N. Germany. (After Turcotte & Schubert, 1982.) (*d*) Block diagram of salt plugs piercing the top of the Woodbine whose structure contours are shown. (After Jackson & Seni, 1983.)

Let us now reconsider the question of the factors which control the average separation of such salt domes in N.W. Germany. Here we shall assume that a single layer of mother salt gives rise to adjacent domes in the manner indicated in Fig. 4.13(*a*). Further, we shall simplify the geometry so that the roof of the mother salt subsides by a maximum of z_r, as the plug develops in a linear fashion, as indicated in Fig. 4.13(*b*). The volume of salt displaced by the subsiding roof (stippled in Fig. 4.13(*b*)) will be approximately equal to the (stippled) column (z_c) of the salt plug. That is, the volume of mother salt (V_m) displaced is:

$$V_m = \frac{\pi z_r (R^3 - 3Rr^2 + 2r^3)}{3(R - r)}. \tag{4.6}$$

Similarly, the volume of the shaded portion of the plug (V_p) is:

$$V_p = z_c \pi r^2. \tag{4.7}$$

Thus, if we take $V_m = V_p$, it follows from combining and simplifying Eqs. (4.6) and (4.7) that:

$$z_r = \frac{3z_c(Rr^2 - r^3)}{(R^3 - 3R^2 + 2r^3)}. \tag{4.8}$$

Let us apply this relationship to the salt plugs of N.W. Germany shown in Fig. 4.13(*c*). Their shapes are somewhat irregular, so we shall take the average radius, r, to be 2.0 km and (after Turcotte & Schubert, 1982) the depth to the mother salt is taken to be 5 km. The relationship between the amount of subsidence (z_r) of the roof of the mother salt and the average separation ($2R$) between salt domes, based on Eq. (4.8), is represented in Fig. 4.13(*a*) and (*b*). For an average separation of $2R = 15$ km (the upper figures used by Turcotte & Schubert) the required roof subsidence of the mother salt would be 1.0 km. If we use the lower figure of $2R = 10$ km, the required roof subsidence would be 2.2 km, which probably exceeds the highest possible value.

The model used in these calculations is, of course, unrealistically simple; moreover, it is based on the assumption that the plug contains only salt. This assumption is probably incorrect. However, to off-set

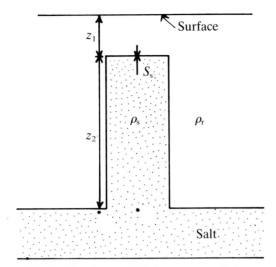

Fig. 4.14. Model used to establish the buoyancy effect.

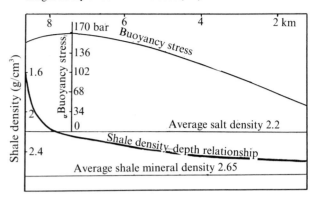

Fig. 4.15. Relationship between buoyancy, pressure and thickness of overburden for probable density distributions. (After Gussow, 1968.)

the error incurred in this assumption, it should be noted that we have ignored the likelihood that a significant proportion of the mother salt may be lost to the system and carried away by circulating brines which will flow along the contact between plug and country rock. Despite these shortcomings, it would appear that the results of the simple analysis set out above permits us to suggest that spacing of salt plugs is not, in general, controlled by the Rayleigh–Taylor viscous instability, but is related to the dimensions of the salt plug and the permissible limits of subsidence of the roof rock. This is a point to which we shall return.

In this simple treatment we have assumed that the 'catchment area' of the mother salt is symmetrically distributed around the plug. In a more realistic treatment, Jackson & Seni (1983) indicate that, when the mother salt is inclined, a greater volume of salt flows from the down-dip direction into the salt plug, so that the resulting 'rim-syncline' is not symmetrical. They present in their paper an interesting isometric block diagram of the salt diapirs in the East Texas area, part of which is shown in Fig. 4.13(*d*).

Emplacement mechanisms

In this section, we shall consider how isolated salt plugs are emplaced. The arguments and concepts we shall present we hold to be valid. However, we realise that to some extent they are contentious, and therefore the reader should be prepared to exercise critical appraisal.

In the past, the emplacement of diapirs has been explained in terms of the *buoyancy effect*. This effect, which has already been dealt with in a different context in Chap. 3, is related to the potential excess pressure that can develop in a column of low density fluid which intrudes into a higher density cover. Thus, as can be inferred from Fig. 4.14, in the static situ-

ation, there is a tendency for the upward pressure in the salt column (S_s), at the interface at depth z $(z_1$ in Fig. 4.14), to exceed the downward gravitational load (S_z) by the amount:

$$S_s - S_z = z_2 g(\rho_r - \rho_s) \qquad (4.9)$$

where the buoyancy effect is given by $(S_s - S_z)$, ρ_r and ρ_s are the densities of cover rock and intruding salt respectively and the dimensions z_1 and z_2 are as indicated in the figure. This equation is based on constant (or average) values of density of salt and country rock and results in a linear increase in excess pressure with height of column (z_2). A more realistic situation, which reflects probable variations in density of the sedimentary sequence into which the salt intrudes, results in an excess-pressure–column-height relationship of the type represented in Fig. 4.15.

It is emphasised that there is no tendency for excess pressure to develop when the column has zero height. Therefore, although the buoyancy effect is of paramount importance in the further development of a diapir, once it has become well established, the buoyancy effect plays no role whatsoever in initiating the diapir. Consequently, we must first establish some trigger mechanism(s) that will give rise to intrusion initiation.

In the following paragraphs, therefore, possible trigger mechanisms will be briefly outlined and we shall indicate how the subsequent suggested intrusion mechanisms result in the salt plugs with the bollard geometry. This is followed by a brief consideration of the conditions of intrusion that can give rise to bell-shaped plugs (Type B in Fig. 4.9); a geometrical form that is commonly encountered in some areas.

Trigger mechanisms

As we have seen, it is necessary for some trigger mechanism to be invoked, whereby a sufficient amount of intrusion can take place, so that the buoyancy effect can become sufficiently important,

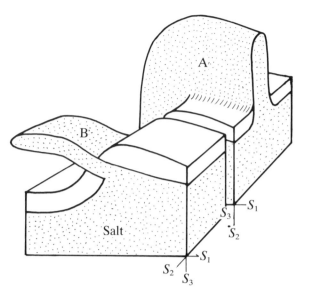

Fig. 4.16. Salt wall (A) and sill (B) triggered by the folding of a salt layer and the surrounding country rock.

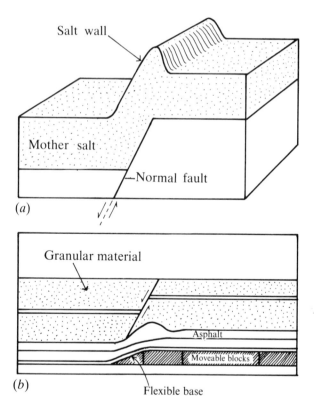

Fig. 4.17. (*a*) Representation of salt wall triggered by a fault in the basement, below the mother salt (stippled). (*b*) Bulging of asphalt against a normal fault observed in an experiment. (After Parker & MacDowell, 1951.)

thereby ensuring that the diapir will evolve. Some such trigger mechanism is often necessary for the development of model diapirs. This triggering may take the form of 'model earthquakes' induced by the gentle tapping or, more vigorously, the kicking of the model apparatus. Alternatively, some foreign body such as a nail, or tack, is dropped into the type of two

liquid-layer model previously described and such a body, on reaching the interface between the fluids, sets up a local instability that initiates a diapir.

In an orogenic environment, folds or faults readily form trigger mechanisms, (Figs. 4.16 and 4.17, respectively). If competent units above a layer of mother salt are folded and the axis of least principal stress (S_3) acts parallel to the fold axis, hydraulic fracture can occur normal to the fold axis and break-through may take place to form a salt wall (A in Fig. 4.16). Alternatively, if S_3 acts vertically (B in Fig. 4.16), small thrusts can develop, along which salt may be intruded. In the latter event, with S_3 acting vertically, the salt intruded along the thrust would soon develop into a salt sill (see Chap. 3). Full diapiric development is likely to be delayed to a post-folding phase when the lateral stresses become smaller and the fluid pressure of the salt could push back the country rock. In the instances represented in Figs. 4.16 and 4.17, these forms are unlikely to be preserved, for they would give rise to conditions in which the buoyancy effect would become important. This would happen more readily when the vertical stress becomes S_1.

In an environment of extension tectonics, the development of salt walls can be induced by a major fault that cuts through and displaces the salt–rock interfaces (Fig. 4.17(*a*) and (*b*)), and so causes a vertical relief in the mother salt to develop. However, as has been pointed out by Fyfe, Price & Thompson (1978), it is not necessary for a fault which develops in the cover rock to cause significant displacement of the cover–salt interface in order to provide a suitable trigger mechanism. The difference in magnitudes of stress in the overlying cover rock relative to those which obtain in the mother salt may itself be a sufficient trigger, as soon as the fault plane actually reaches the interface plane.

As we shall see in Chap. 9, normal faults are likely to develop in competent sedimentary units, even in an environment in which the beds are merely subjected to vertical subsidence. Hence, the example of a trigger mechanism outlined below will apply even in areas which have experienced no obvious deformation.

An example of the stresses that could give rise to such faulting in a competent unit at a depth of 10 km, and the corresponding appropriate stresses in the mother salt adjacent to the interface, is indicated in Fig. 4.18(*a*) and (*b*). The horizontal and vertical stresses in the salt are equal because the salt behaves as a fluid. However, the horizontal stresses in the limestone are significantly smaller than the vertical stress because, at the stipulated depth, limestone behaves as a solid (Chaps. 1 and 9). It will be seen from the magnitudes of the various stresses and pressures represented in this figure that there is a large potential difference in the magnitudes of stress between the salt,

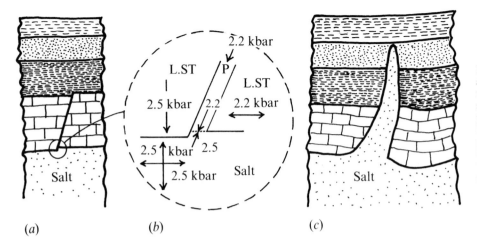

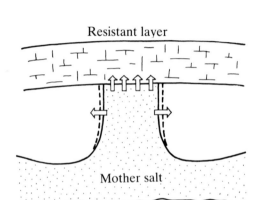

Fig. 4.18. (*a*) Fault initiation in a competent layer above salt as the result of the differential stress shown in inset (*b*). The potential stress imbalance between the salt and the fluid pressure in the fault, where the fault meets the salt, gives rise to the early stage of salt intrusion shown in (*c*).

the fluid pressure in the fault plane and the lateral stress in the faulted cover rock. From rock mechanics experiments on salt, it is known that specimens which are at a temperature of 300 °C will deform at a strain rate of about 10^{-2}/s, when subjected to a differential stress of about 300 bar. Consequently, it can be inferred that in the hypothetical natural situation represented in Fig. 4.18, the salt would be explosively injected into the fault zone at initial strain rates of 10^{-2}/s or faster. We mentioned earlier that model diapirs can be initiated by 'model earthquakes'. It is interesting to note, therefore, that in nature, the initiation of a salt plug may actually give rise to an earthquake. Indeed, small magnitude earthquakes which originated off the coast of New Zealand have recently been attributed to the initiation of 'diapiric folds'.

Development of bollard-type diapirs

The very high strain rates, cited in the previous paragraph, could not be long sustained. Gradually the walls of the fault zone would be pushed back and the cover rock would subside as the mother salt, in the vicinity of the fault, is transferred to the rising and expanding intrusion. Because the salt intrusion would be initiated along a fault line, the lower levels of the intrusion are likely to be elliptical rather than circular in plan.

Because the trigger fault is assumed to be restricted to a particular competent horizon, as the upward penetration of the salt progresses it may, or is even likely to, encounter a resistant horizon that will temporarily halt upward progress and enable lateral spreading to take place (Fig. 4.19). The upward pressure on the cover rock can be likened to an inverted example of the loading of a foundation. The probable shear directions that may develop in such situations are indicated in Fig. 4.20. Other configurations of shear patterns will be discussed later in this chapter. However, because the salt will have risen quite rapidly, the upward loading of the interface

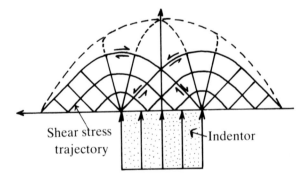

Fig. 4.19. Salt intrusion spreads laterally beneath an initially impenetrable layer. Breakthrough may occur when lateral spreading is sufficiently large.

Fig. 4.20. The Prandtl model showing maximum shear stress trajectories and general distribution of stress intensities induced in the homogeneous plastic material by loading of 'indentor'. (After Tapponnier & Molnar, 1976.)

shown in Fig. 4.19 will cause relatively rapid strain rates in the rock above the interface. This will cause local strains to develop in the rocks above the advancing salt intrusion. These strains will tend to deform

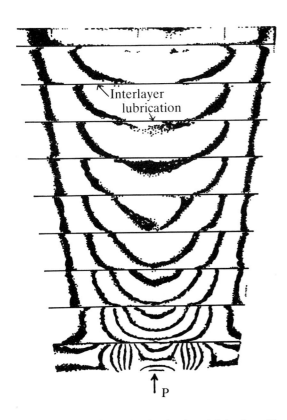

Fig. 4.21. Photo-elastic model showing how lubrication of layers 'channels' the compression rather than the dispersion which occurs in the model represented in Fig. 4.20.

and close void spaces, and so cause a concomitant development of relatively high fluid pressures. Consequently, the shear stresses on the bedding planes will be low, which will ensure that the principal stresses in the cover rocks will act approximately vertically and horizontally. The higher the fluid pressures generated, the more closely will the axes of stress approach vertical and horizontal. An example of how lubricated surfaces influence the transmission of stresses is given in Fig. 4.21. This figure is a photo-elastic representation of a uniaxial compression test and shows how excellent uniformity of stress may be attained using this technique.

The loading situation that results when the salt configuration is as shown in Fig. 4.19 is comparable with the intrusion of a sill (see Chap. 3). Consequently, hydraulic fractures will tend to develop perpendicular to the axis of least principal stress. Salt will intrude along these vertical fractures. The body-weight of the roof may then readily cause detachment along bedding planes, so that *stoping* takes place. This mechanism is represented in Fig. 4.22(*a*) and (*b*) and a natural example with major xenoliths of country rock is shown in Fig. 4.22(*c*).

The area of fracture plane that defines such an upward migration of salt is a minimum if it has a circular plan. The energy needed to initiate and propagate a fracture will be related to the surface area

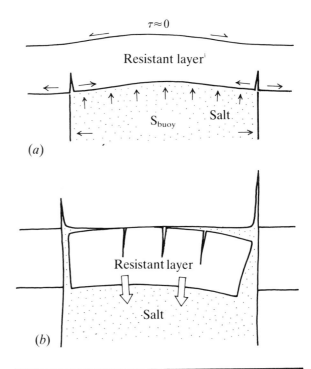

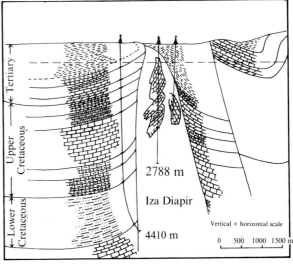

(*c*)

Fig. 4.22. (*a*) and (*b*) show stages in the breakthrough of a hitherto impenetrable layer (because of the effect illustrated in Fig. 4.21) leading to the development of stoping: a natural example of which (the Iza Diapir of N. Spain) is shown in (*c*).

and will be at a minimum when the fracture surface area is smallest. It follows, therefore, that the circular form of bollard type diapirs, in their upper levels, is determined by minimum energy requirements: and that the upward growth of the cylindrical fracture is ensured by the type of stress distribution shown in Fig. 4.21.

From Fig. 4.15, it will be seen that, ideally, a buoyancy effect of approximately 170 bar will develop at a distance of 8 km above the mother salt. However, this ideal figure does not take into account the reduction in the buoyancy stresses which takes

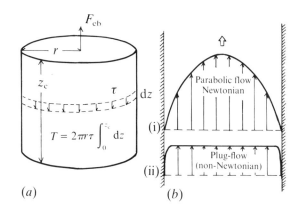

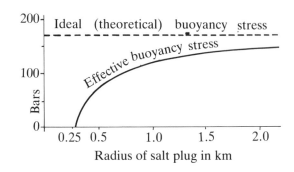

Fig. 4.23. (a) Dimensions and geometry used to estimate boundary traction resisting upward migration of salt. (b) (i) Parabolic (Newtonian) flow and (ii) plug (non-Newtonian) flow.

Fig. 4.24. Theoretical buoyancy stresses, based on Fig. 4.15, and effective buoyancy for a cylindrical plug, 8 km high, and an average shear stress of 3 bar, acting on the salt–country rock interface.

place as the result of the intrusion of the salt. If we assume that the salt moves by plug flow (Fig. 4.23(b) (ii)), so that the internal deformation of the salt can be ignored, then it is only necessary to consider the influence of the viscous drag between the salt and the country rock. Let us assume that the salt plug is exactly cylindrical throughout its lower and intermediate levels and that the average viscous (shear) resistance per unit area of contact is τ; then the total resistive force (T_r) against upward movement of the salt plug is given by:

$$T_r = 2\pi r z_c \tau \qquad (4.10)$$

where z_c is the height of the cylindrical salt plug (Fig. 4.23(a)). The value of the shear resistance (τ) is not, of course, constant, but (because the rheological properties of salt are temperature dependent) this resistance increases from a very small value at the lowest levels of the intrusion to values of several bars, or even tens of bars, near its upper level. It is difficult to be precise as regards the average value of τ. However, here we shall assume that it has what we consider to be a reasonable value of 3.0 bar.

With this average value of τ, we can establish the theoretical or ideal and effective buoyancy forces (F_{ib} and F_{eb}) for different values of radius of cylindrical salt plugs. The effective buoyancy force is given by $F_{eb} = F_{ib} - T_r$ so that the effective buoyancy stress (S_{eb}):

$$S_{eb} = \frac{F_{ib} - T_r}{\pi r^2} \qquad (4.11)$$

can readily be established for different values of r. This relationship is shown graphically in Fig. 4.24. It will be seen that high effective buoyancy stresses are only encountered in diapirs of large radius, though clearly the effective buoyancy stress is significantly less than the ideal buoyancy stress. It will also be noted that the effective buoyancy stress decreases to zero for a plug radius of about 300 m. Such a plug

would therefore exhibit no further tendency to rise. Because this figure is based on an assumed value of $\tau = 3.0$ bar, the relationship indicated is specific rather than general. Nevertheless, it is clear that the ability of a diapir to progress upward is enhanced if its radius of curvature is large. For any natural example, the plug with the largest radius will rise furthest; the limit in nature will be set by the available supply of salt from the mother layer.

Let us now consider how the effective stress may deform the cover rock into which the salt plug intrudes. If the salt is sufficiently mobile in the upper levels and encounters weak ductile rocks, in which the horizontal stresses are comparable with the vertical load, then conditions conducive to sill-like intrusion of the salt are encountered. For example, the Wienhausen salt dome of N.W. Germany has a 'mushroom' profile (Fig. 4.25), the overhanging salt having a maximum thickness of about 400 m. It may be noted in passing that an alternative interpretation to the formation of the overhanging salt is that it represents fossil salt glaciers.

At high levels in the plug, where it is relatively cool, the salt may be able to sustain a differential stress of a few tens of bars. Hence, the effective lateral

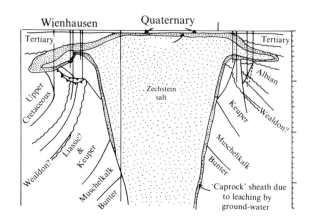

Fig. 4.25. Section through the Wienhausen Salt Plug, N.W. Germany. (From Gussow, 1968; after Schott, 1956.)

pressure in the salt will be less than that which acts vertically. However, as we shall see in Chap. 9, the horizontal stresses in flat-lying sediments are often less than the vertical stress at the same level. This means that the pressures in the salt will often have a comparable, or even greater, effect in the horizontal direction than in the vertical direction. In order that the horizontal stresses in the salt and in the country rock may attain balance, the country rock will be pushed back. This may be accomplished by flexuring of the country rock. Alternatively, or in combination with flexuring, the pore spaces in the country rock may be reduced, or closed, to form around the plug what is termed a *compaction halo*. This action, which gives rise to a widening of the plug, tends to become progressively more pronounced as the surface is approached, where the cover rocks are commonly weaker (Fig. 4.25).

If the near-surface sediments are not recent and have become well compacted or cemented, the near-cylindrical mode of intrusion may be maintained almost to the surface, where the effective vertical buoyancy stresses may eventually cause a break-through along faults and so lift up an inverted, truncated cone of the cover rock. However, as may be inferred from the data in Fig. 4.24, because the effective buoyancy stress will be relatively small, the thickness of cover rock that may be so uplifted will not generally exceed a few hundred metres.

The models presented above are extremely simplistic; nevertheless, although they are based on idealised geometries and simplified rheological behaviour, and also tacitly assume a single continuous phase of salt emplacement, they are sufficient to indicate how the main geometrical aspects of bollard type diapirs may develop. They do not, however, provide any insight into the bell-shaped intrusions (Fig. 4.9(*b*)); a problem to which we now turn.

Bell-shaped intrusions

The similarity in shape between the bell-shaped intrusions and the first stage of development of the model intrusion represented in Fig. 4.3 is such that one is persuaded that the mode of emplacement must also be similar. We have noted that the model was composed of two layers of liquid and that sedimentary rocks in the upper levels of the crust usually behave as solids. There is, however, one situation in which rocks have such low yield strengths that they become ductile and approximate to liquid behaviour when subjected to low values of differential stress. Such behaviour can be exhibited by over-pressured clay and mud rocks of high porosity.

If the range of porosities of such rocks is 30–40 per cent, the corresponding bulk density of these rocks will be 2.15–1.98 g/cm^3, which is less than that for halite (2.2 g/cm^3). In such circumstances there

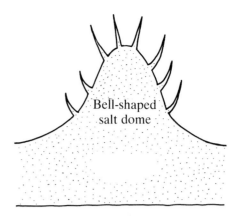

Fig. 4.26. Hypothetical fracture patterns associated with the development and 'freezing' of bell-shaped diapirs.

would be no inherent instability. Indeed, at depths and temperatures at which such over-pressured clays could exist, dry salt would probably be significantly stronger than the cover rocks: which would also contribute to the inherent stability of the situation.

However, if the underlying mother salt was also highly porous, with the void spaces filled with brine, the situation could be completely transformed. Thus, even if the porosity of the mother salt was as little as 20 per cent, its bulk density would be only 1.96 g/cm^3; and the unstable element of a layer of relatively high density overlying low density salt would be reintroduced. Moreover, because of the physical, and possibly chemical, action of the brines in the salt, the mobility of the mother layer would be greatly increased. Such natural situations would approximate closely to the liquid–liquid model represented in Fig. 4.3 and an instability of the type shown in that figure could then develop.

It is emphasised that the mobility of both the cover rock and the mother salt in the postulated situation is related to the quantity of water in the system and its degree of overpressuring. As diapirism is initiated, and the diapiric structure develops, it pushes aside the country rock, which will become ruptured and faulted (Fig. 4.26). Relatively rapid migration of the fluids from both the salt and the cover rock into, and along, these fractures and faults will ensue. As the salt and cover rock become depleted of water and brines, they lose their high mobility and eventually the intrusion 'freezes': for at the relatively shallow levels in the crust where the conditions described above can obtain, the salt will be cool, so that once it becomes reasonably dry it has a low intrinsic mobility and will not then penetrate further through the cover. (However, should the sediments in which the bell-shaped intrusion occurs become more deeply buried and reach a temperature of about 300 °C, the intrusion may become rejuvenated and may even develop into a bollard-type intrusion. Hence, bell-shaped intrusions could also act as a trigger mechanism for the development of bollard-type diapirs.)

These conditions which, it is suggested, are necessary for the development of bell-shaped intrusions, approximate closely to those which are assumed by Turcotte & Schubert, Hence, as they suggest, the Rayleigh–Taylor instability may well control the spacing of bell-shaped domes. Moreover, because bell-shaped domes could eventually develop into bollard-shaped structures, the Rayleigh–Taylor instability may also influence the separation of these latter types of intrusions. However, as argued earlier, the potential supply of salt (which is related to the thickness of the mother layer) must ultimately determine the degree of development, and hence the separation, of the structures at high levels in the sedimentary sequence.

It has been argued (Fyfe, Price & Thompson, 1978) that pillows of bell-shaped salt intrusions can possibly be attributed to the fluids generated by dehydration reactions below the mother salt. The vast quantities of fluids so generated can initially collect below the impervious mother salt, but will eventually breach the salt layer barrier (by progressive solution) and so introduce vast quantities of high concentration brines into the cover rock, where much of the salt will come out of solution. However, this mechanism is conjectural and will not be further considered.

The arguments presented above are quantitative and also, to some extent conjectural. Nevertheless, we consider them essentially sound and reasonable explanations for the development of bell-shaped domes. However, it will be clear from the various facts, concepts and hypotheses presented and discussed in the preceding sections that there is still a great deal to be learned regarding the emplacement of salt and other non-igneous diapirs.

SECTION II

Major igneous intrusions

Large-scale igneous bodies have been mapped and studied for many reasons. At the simplest level, they have been considered merely as a rock type, the boundaries of which were to be delineated on a map. The geochemistry and petrology of major bodies have received, and continue to receive, considerable attention for academic and economic reasons. Granites around the world were intensely studied as the result of the 'migma-versus-magma' controversy regarding their formation, which raged during the middle decades of this century. This controversy eventually subsided when it was widely recognised that granites could be formed by both mechanisms and that there were 'granites and granites'. In some of these studies, use was made of the earlier work of Cloos and his

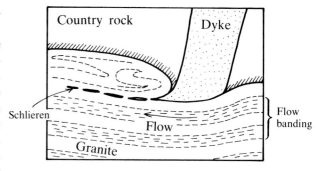

Fig. 4.27. Diagrammatic representation of foliation, flow banding and schlieren.

co-workers (reported by Balk, 1937) who indicated ways in which flow structures, faults and fractures were often systematically related to the shape of the intrusion and its methods of emplacement. One can expect a resurgence of interest in the structural mapping of major ingeous bodies because of the possibility of such bodies being used as the repositories for the disposal of toxic waste material. Consequently, we shall briefly consider the structural mapping methods advocated by Cloos and also address the type of problem facing geologists investigating the incidence of fractures in granites at depths of 1 km or more.

Structures in large igneous intrusions

Fractures and related structures which develop in major igneous intrusions are, in part, controlled by the environment of the intrusive body. If, as for example, in S.W. England, granite bodies occur in an orogenic belt with well-defined structural trends, many of the fractures in the major intrusions can be directly correlated with those fracture patterns which occur elsewhere in the orogenic belt. In addition, the individual intrusions will exhibit fracture patterns which can sometimes be perceived to relate to the mode of emplacement of the specific igneous body. The pioneering work which enables such relationships to be established was conducted by Hans Cloos and his co-workers.

The philosophy expressed by Cloos relates to the development of a large igneous intrusion, particularly granite, at intermediate and high levels in the crust, and involves the intrusion of igneous material in a fluid state. This fluid slowly crystallises so that, in some localities within the body, one is concerned with the flow of a crystalline 'mush'. Because the magma is at a higher temperature than the country rock, the intrusion cools most rapidly adjacent to the country rock. At this time, *flow structures* develop, usually adjacent to the margins of the intrusion. Such structures include *flow banding* or *foliation*, and sometimes include *schlieren* (smeared-out xenoliths of country rock) as may be seen in Fig. 4.27. These structures

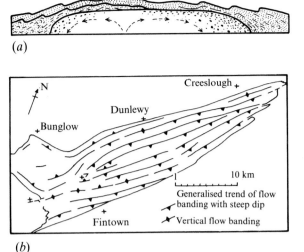

(a)

(b)

Fig. 4.28. Flow direction and banding or foliation in two intrusions: (a) The Dalbeattie Granite, Scotland. (After Phillips, 1956.) (b) The Derryveagh Granite, Eire. (After Pitcher & Read, 1959.)

may permit the geologist to infer the direction of flow of the fluid material within the chamber of the intrusion (Fig. 4.28(a)) and, to some extent, the general shape of the intrusion, because the foliation tends to be parallel to the intrusion/country-rock interface. However, care must be exercised as, for example, in the Derryveagh Granite (Fig. 4.28(b)), where the foliation is seen to transgress the walls of the intrusion.

As the margins of the intrusion become less mobile, they begin to behave as a solid which becomes progressively stronger, and even brittle, as time passes. Meanwhile, in the central parts of the intrusion the magma is still fluid. The central molten magma acts against the solidified, intrusive material and, if the intrusion is continuing to grow, disrupts the outer parts of the intrusion, and also the country rock, to form a series of *primary fractures*, some of which form *auto-intrusions* filled with new 'granitic' material (Fig. 4.29(a) and (b)). In any one intrusion, there usually exists a continuum from that represented in Fig. 4.29(a), where the boundaries of the auto-

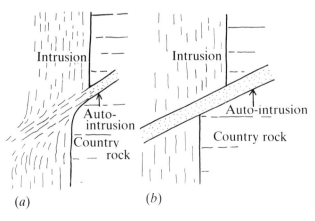

Fig. 4.29. (a) Ductile and (b) 'brittle' auto-intrusions.

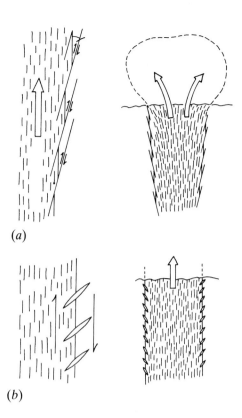

(a)

(b)

Fig. 4.30. Examples of how the geometry of an intrusion may be inferred from (a) thrusts and (b) extension fractures which develop near the interface between intrusion and country rock.

intrusion are diffuse and poorly defined, to that in Fig. 4.29(b), where the boundaries of the auto-intrusion (dyke) are sharp and clearly defined and the dyke material is obviously different, either in grain size and/or composition. From the characteristics represented in Fig. 4.29(a), one would infer that this structure formed relatively early in the history of the intrusion when there was little difference in physical properties between the intruded and intruding material. The structure in Fig. 4.29(b) developed later, when the foliated material had become relatively brittle and so cool that it had induced very rapid cooling of the intrusion, resulting in an aplite (relatively fine-grained) dyke.

These intrusions along primary fractures are not necessarily wholly dilational features but may also exhibit a considerable element of shear displacement. Where such dilation/shear fractures develop and show a consistent shear direction when they cut through the outer zone of the major intrusion into the country rock, they may be used to adduce the general three-dimensional form of the intrusion. For example, from the sense of thrust shown by the extension/shear fractures represented in Fig. 4.30(a), we can infer that, at the level of observation, the intrusion has a 'neck' and that it balloons upwards, as indicated: although, of course, the precise form of the upper and eroded parts of the intrusion must remain conjectural. Frac-

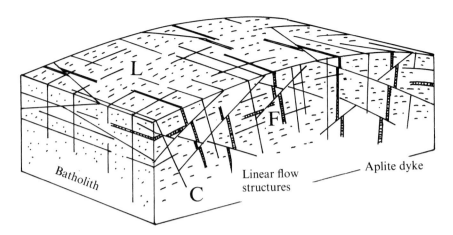

Fig. 4.31. Typical orientations and interrelationships between primary joints and foliation, C, cross joints, L, longitudinal joints and F, flat-lying joints. (From Balk, 1937; after Cloos, 1922.)

tures with orientations which are similar to those represented in Fig. 4.30(*a*) are shown in Fig. 4.30(*b*). However, the structure shown differs in two important respects from those in Fig. 4.30(*a*); namely, (i) they tend to make a somewhat larger angle (typically 45°) with the foliation and (ii) they are extension fractures and *not* hybrid extension and shear features. From the configuration of these extension fractures and foliation, one can infer that the walls of the intrusion (at the level of observation) are essentially parallel-sided; and that the extension fractures were induced by the shearing couple which the upward flow of the inner fluid magma generates as it moves past the country rock and the solidified outer zone of the intrusion.

Cloos (1922) correlated the orientation of the fractures with that of the planes of foliation in the intrusion. He defined four main classes which can sometimes be seen in the steeply-dipping foliation, but are, however, best developed, and seen, in the roof of intrusions where the foliation is flat or only gently inclined. The classes are: (i) Cross Joints (ii) Longitudinal Joints (iii) Flat-lying Joints and (iv) Diagonal Joints; the orientation of which relative to foliation is illustrated in Fig. 4.31.

Cross 'joints' form perpendicular to the foliation and, where they can be defined, to the flow lines (as defined by schlieren etc.). According to Cloos, these fractures are among the earliest to form. They are frequently occupied by aplite, or else their walls are almost invariably coated by hydrothermal minerals. Moreover, the fracture surfaces commonly exhibit 'slickensiding'. Hence, these fractures are commonly hybrid extension and shear fractures. Cloos regarded cross 'joints' as extension fractures either as the result of 'drag' of the fluid core against the walls and roof, or as the result of the continued expansion of the intrusion. To these possible causes, one could add the effects of shrinkage in the solidified parts of the intrusion as cooling continued.

Longitudinal 'joints' are planes which are usually approximately orthogonal to both foliation and to cross joints, so that they are steep planes which strike parallel to the flow lines. This type of fracture is rarely filled with aplite or other dyke material. When they are not barren, their mineral content is usually different from that found in other forms of primary fractures. Moreover, shear movement on these fractures is rarely observed. Hence, although they are often geometrically related to cross joints, they are later features. As we shall see, it is probable that these fractures were generated by the cooling of the igneous body coupled with subsequent uplift and lateral stretching. The orientation of these fractures is possibly related to the basic anisotropy of the strength of the intrusive material which, in turn, is a reflection of the mode of emplacement of the intrusion. (Quarry workers have long been aware of the strength anisotropy in granite and refer to orthogonal planes of increasing difficulty of rock-splitting as *rift*, *grain* and *hardway*.)

Primary, flat-lying 'joints' tend to develop near the apex, or dome, of an intrusion where the foliation is (or was) horizontal. These fractures may be filled with aplite or hydrothermal minerals; and so would be primary fractures. When they are barren, it would be difficult to distinguish them from other barren flat-lying fractures which, in areas of little relief, would be better termed 'exfoliation' features. These will be discussed later in this chapter.

The development of primary, flat-lying fractures is perhaps difficult to interpret on dynamic grounds. Cloos (op cit) and Balk (1937) suggest that they form when the centre of the intrusion shrinks as the result of cooling. It is difficult to envisage why, under gravitational loading, this suggested mechanism should give rise to flat-lying 'extension' fractures. We believe that these flat-lying features reflect variations in internal magmatic pressure. During forceful intrusion, magmatic pressure must slightly exceed that which is required to balance gravitational loading. However, as the intrusion develops and extends its boundaries, possibly by stoping, through the formation of sheet intrusions external to the main body (by a process

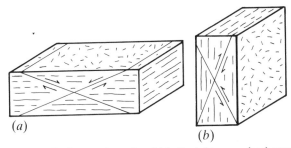

Fig. 4.32. Conjugate shears in which the 'obtuse wedges' move inwards.

Fig. 4.34. Examples of sub-horizontal exfoliation fractures in a granite in the Yosemite National Park, California.

analogous to that shown in Fig. 4.22), or by the generation of cone sheets or ring dykes, or even by erupting at the surface, these various events and processes would result in an abrupt, if transient, decrease in magmatic pressure. The flat-lying, solidified granite would be more dense than the fluid magma and could part along a weak plane of foliation and start to sink gently into the magma. Magma would flow into the parting to form an 'aplite sill'. Alternatively, flat-lying fractures could result from the active injection of magma by the hydrofracture mechanism: a form of 'embryonic stoping'.

Diagonal 'joints' form at 45° or less to the foliation. Displacement along the features indicates that they are shear phenomena which result from compression normal to the foliation, and extension in the foliation parallel to the flow lines. These fractures, which are also commonly filled with aplite or hydrothermal minerals, are not necessarily restricted to flat-lying foliation (where they would have the orientation of low angle, 'normal' faults with *movement along the shear planes into the obtuse angle*, (Fig. 4.32(*a*)). They also develop in steeply dipping foliation (e.g. in the Aar Massif) as indicated in Fig. 4.32(*b*), where they would be classified as reverse faults (but again the shear movement is into the obtuse angle). Clearly, these conjugate fractures cannot be interpreted by using the Brittle Criteria of failure. We contend that

they are related to the *characteristic* directions which may develop in anisotropic material compressed perpendicular to the planes of anisotropy. However, we shall leave a discussion of such features until Chaps. 8, 13 and 18.

It is apposite at this time to comment on the use of the word 'joint' in the context of these various primary fractures. It will be noted that many of the classes of fractures are infilled with aplite or hydrothermal material and that they often exhibit evidence of considerable shear movement. The use of the word joint to describe these is probably now too firmly embedded in the literature to be eradicated; consequently, as here, we strongly advocate that the word be used in inverted commas to indicate that the structures are not what would normally be termed 'joints'. (See the discussion given in Chap. 2.) Certainly, fractures which could correctly be termed joints do occur in granites and we shall now discuss the orientation, morphology and mode of development of these types of fractures.

Barren and exfoliation fractures in granites

In south-west England and in many other localities in the world, granite weathers to form 'tors', which are upstanding and often dramatic piles of joint-bounded blocks (Fig. 4.33). One of the features of these tors is that of 'flat-lying' fractures which, in this example, are sub-parallel to the topography. Superb examples of these topography-parallel features, which are termed *exfoliation*, or *sheeting fractures*, or joints, are to be seen in the Yosemite National Park, California (Fig. 4.34). In the exposure shown in Fig. 4.34, it is apparent that the exfoliation fractures are dominant

Fig. 4.33. Sketch of 'Rough' Tor, Bodmin Moor, showing the remains of horizontal exfoliation fractures.

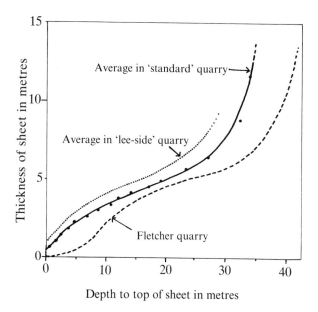

Fig. 4.35. Plot of separation between exfoliation fractures against depth. (After Jahns, 1943.)

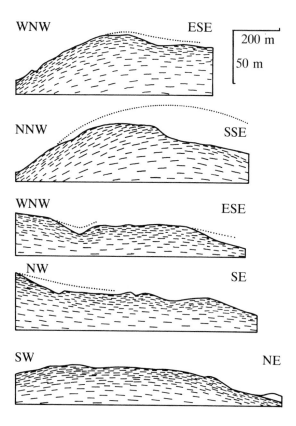

Fig. 4.36. Representative cross sections of Oak Hill and adjacent hills, in the Chelmsford granite N.E. Mass., U.S.A. showing the relationship between the exfoliation fractures and the topography. Dotted lines show pre-glacial topography. (After Jahns, 1943.)

and that other fractures perpendicular to the exfoliation have not developed to the extent that they have in Fig. 4.33. We infer, therefore, that exfoliation fractures tend to develop first and that the majority of barren fractures which form perpendicular to the exfoliation develop later.

Exfoliation fractures in granites in New England, U.S.A., have been studied by Jahns (1943), who demonstrated that these fractures are closely spaced near the surface, but that spacing or separation between the exfoliation fractures increases with depth (Fig. 4.35). At depths beyond about 20–30 m, these fractures are rare or may even cease to form. These features have also been studied by Chapman & Rioux (1958), who suggested that they are largely related to pre-glacial topography. One may also infer this from Fig. 4.34. In a few localities, however, sheeting was formed parallel to ice-cut surfaces and in these instances may be post-glacial. Exfoliation fractures, where they crop out at faces in quarries are usually non-planar and consequently cannot be attributed to shear, but must be considered as tensile fractures.

It is instructive to consider how exfoliation fractures may be generated in granites (and other rock types), for they apparently represent a paradoxical situation. They have the morphology of extension fractures and indeed they occur approximately perpendicular to the inferred direction of least principal stress. However, a small but positive (i.e. compressive) stress induced by gravitational loading must have existed (and in most instances continues to exist) across the fracture surfaces.

It is interesting to note that exfoliation fractures are not well developed in all granite quarries. For example, in several of the granite quarries in central

Minnesota, barren fractures, in quarry faces hundreds of feet in extent, are conspicuous by their absence. The granite in some of these quarries contains *in situ* horizontal stresses which, as we shall see, may attain magnitudes in excess of 150 bar, (C. Nelson, personal communication).

At low levels in the cited quarries in New England the granites which are cut by exfoliation fractures may also contain high *in situ* stresses. This may be inferred from the feature shown in Fig. 4.37, which shows a circular cavity of approximately 2.5 m diameter on the vertical granite face. Several tonnes of

Fig. 4.37. Scar left by spalling of granite from a quarry face.

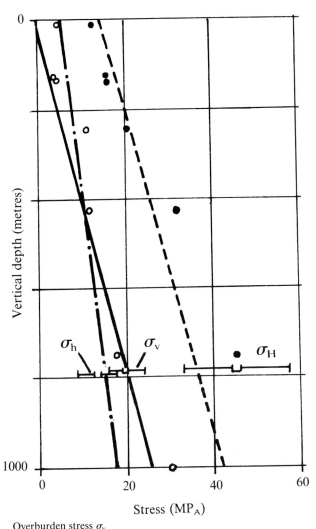

Overburden stress σ_v
Hydrofracturing
Min. horizontal stress σ_h
Max. horizontal stress σ_H
Overcoring
Average stresses +
1 standard deviation

Fig. 4.38. Relationship between magnitude of remanent stresses and depth in the Carnmenellis Granite, Cornwall, England. (After Pine *et al.*, 1983.)

decrease in the magnitude of the lateral stress at shallower levels, so that, just below the surface, the horizontal stress is unlikely to exceed 400 bar.

Horizontal stresses of 400 bar near the surface and 625 bar at depths of 1 km are geologically realistic, or even conservative. For example, near Elliot Lake, Ontario, at depths between 300 and 700 m, horizontal stresses between 210 and 370 bar have been recorded. At Timmins, the horizontal stresses at 700 to 850 m were in the range of 530 to 725 bar; while at depths of 1200 to 1700 m in the Sudbury Basin, the near horizontal stresses attained values between 800 and almost 1300 bar.

Pine *et al.* (1983) have published *in situ* stress data (Fig. 4.38) which relate to the horizontal stresses which exist through a range of depths in the Carnmenellis Granite of Cornwall, England. It will be seen that the magnitude of the stresses present in this granite, at a depth of 1 km, are in reasonable agreement with those forecast by Price (1979). The near surface stresses are somewhat lower than those which were forecast. This is probably attributable to a degree of stress release which took place as the result of small movements on pre-existing fractures. This possible mechanism was not taken into account in the analysis by Price (1979).

Let us now consider how exfoliation fractures which develop parallel to topography and other related, orthogonal, fractures may develop. As already noted, these exfoliation fractures are particularly intriguing. It can be demonstrated that they have the characteristics of extension fractures, yet they develop in a situation where the stress normal to the fracture surface is compressive, and we cannot appeal to high fluid pressures to render the normal stress effectively tensile. To understand how these fractures develop, we must first establish what forms *in situ* stresses may take and also define some of the terms relating to types of stress that we shall encounter in the ensuing discussion.

There are many terms used to indicate various types of stress. Some, such as *total, effective, shear* and *normal* are well defined and have been dealt with in Chap. 1. Other terms such as *regional* and *in situ* are not so clearly defined. We shall define what we mean and understand regarding the use of these latter and related terms in the following paragraphs.

Regional stresses are usually deduced by geologists from an interpretation of the structural patterns, such as are formed by the distribution of fold axes or disposition of faults, dykes, veins or other fractures. When, as is usually the case, a geologist is dealing with an ancient orogeny, it is the orientation of the palaeo-regional stress system that is inferred. This ancient regional stress may have little in common as regards the orientation of the current regional stress pattern. This latter system would best be established from

granite have spalled from the face and were ejected across the quarry to impact on a parallel face some 15–20 m away.

Before we can proceed further with the question of how exfoliation and other barren fractures may form in the near surface environment, we need to consider *in situ* stresses more closely. Price (1979) made a theoretical evaluation of the range of magnitudes of *in situ* stresses likely to be encountered in a granite at a depth of 1 km and also near the surface. The methods he used in this analysis will be discussed in detail in Chap. 9, so here we shall merely quote the results obtained. He concluded that horizontal stresses, at a depth of 1 km could be expected to reach values of about 625 bar (62.5 MPa). It is inherent in his analysis that there would be an almost linear many measurements of *in situ* stresses, preferably obtained from many stations throughout a region.

There are a number of different stress types which can individually and collectively be termed in *in situ* stresses. The type of such stress, measured or inferred, depends upon the geological environment in which the measurement is made. If, for example, the measurements were made in an area of the world which is currently experiencing orogenic (Plate Boundary) deformation, then one would be measuring the *in situ orogenic* or *tectonic* stresses. In other areas, where active orogeny or active subsidence is not going on, one would probably be measuring *in situ* stresses which are composed in part of *remanent stresses* and in part of *residual stresses*. The differences between remanent and residual stresses are not always clearly understood, but it is possible to indicate important differences between them. As with many natural phenomena, however, the distinction between them is not absolutely clear cut, especially when considering small domains which approach grain size.

Remanent Stresses are those stresses which are inferred from an *in situ* strain-measuring technique such as overcoring. If a block is isolated and removed from its former position in the crust (i.e. it is removed from a borehole, mine or quarry) then (ignoring body-weight effects) the boundary stresses which previously existed and acted on the block are completely dissipated. Specifically, the term remanent stresses should be restricted to those *in situ* stresses in rocks which are now at shallow depths or exposed at the surface and which have experienced cooling, uplift and exhumation; that is, the rocks have experienced *retrograde deformation*. The horizontal stresses measured by Pine *et al.* (1983) (Fig. 4.38) are examples of remanent stresses. Thus, the relatively high lateral stresses which exist at the surface are the remainder of the stresses which the rock contained when it was at a depth of 1 km, while the larger lateral stresses which now exist at a depth of 1 km are a remanent of those higher magnitude stresses which the rock contained when it was buried at greater depths in the crust.

Residual Stresses are a somewhat surprising phenomenon. It was noted that when a block of rock is isolated and removed from its *in situ* position in the rock mass, it experiences elastic strains which very rapidly result in the dissipation of the *in situ* or remanent stresses. Although the boundary stresses have been removed, this does not mean that the block is necessarily free from stresses. For example, overcoring tests conducted on such a recently isolated block of granite by Friedman (1972) revealed that the block contained *residual stresses*.

We believe that the first experiments to demonstrate the existence of these residual stresses in rock (i.e. those stresses which remain after the dissipation of the remanent stresses) were conducted and com-

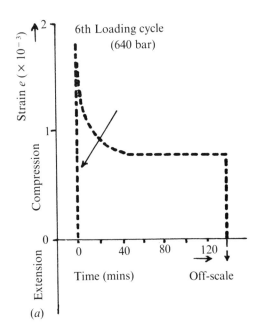

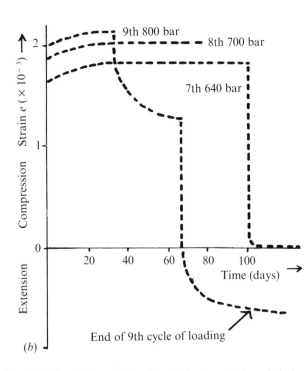

Fig. 4.39. Strain–time relationships indicating (*a*) the relatively rapid and (*b*) a slower release of residual stresses.

mented upon by Price (1964, 1966 and 1970). The data obtained relate to creep experiments conducted on specimens of Coal Measure greywackes prepared from fresh mine rock which *expanded*, or showed other anomalous behaviour, while actually being subjected to large *uniaxial compressive stresses*, as shown in Fig. 4.39. It was inferred from these tests that some rock specimens contained residual stresses, prior to test, which were of the order of 60 MPa, or 600 bar.

From the fact that the duration of some of the creep tests sometimes exceeded a year and that the

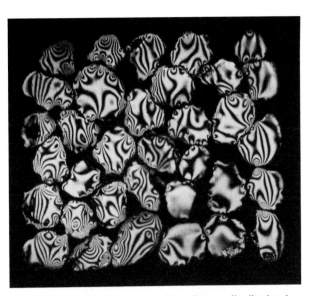

Fig. 4.40. Photo-elastic representation of stress distribution in a granular aggregate subjected to uniaxial compression.

stored residual stresses gave no sign of their existence until a certain level of applied stress was reached, it is clear that such residual stresses are metastable. However, the question remains: what are these residual stresses, how do they develop and how may they influence fracture development?

The mechanism proposed by Price (1966, 1970) to explain the distribution and mode of development of such residual stresses is based on the photo-elastic model shown in Fig. 4.40. This model comprises pieces of perspex fashioned to represent a section through a porous monomineralic rock. The stress conditions in individual grains which are induced during uniaxial loading (as indicated by the fringe patterns) are clearly complex. This model simulates a sedimentary rock, such as a quartzite, or the crystal-line mesh that would develop in an igneous rock prior to its complete solidification. Imagine now that such a model, *while still under external compression*, has epoxy resin poured into the void space. The matrix thus formed, when the resin sets, would be unstressed. If now the external compression is removed (this is equivalent to reducing the remanent stresses to zero), the compressive stresses in individual grains would not be completely dissipated because, as the individual grains, or network of grains, tried to attain their original dimensions they would be constrained by the matrix that would, in turn, be subject to complex tensile stresses. Thus, when the boundary stresses are completely released, the model would contain a complex array of small domains in which compressive stresses in grains are balanced by tensile stresses in the matrix. Clearly, the magnitude of these residual stresses increases as unloading progresses until their maximum development in the completely unloaded state is attained. Some, or even all, of these residual stresses can be released by any mechanism

which causes a breakdown in the rock on a grain to grain, or grain to matrix, basis. The larger the number of the internal boundaries that are disrupted, the more general will be the release of these residual stresses. In experiments, in which the specimen is subjected to high differential stresses, the release may be relatively rapid. In nature, the release may occur over a long period, as the result of stress corrosion and/or weathering processes.

Clearly, the model proposed above is an imperfect simulation of a natural system. However, we can infer that a polymineralic rock, such as granite, will be made up of a complex of grains with different elastic moduli and coefficients of thermal expansion. Hence, even if the original stress state in the rock, at depth in the crust, were completely uniform throughout the constituent grains of the rock mass, this state would be changed during cooling and uplift, so that residual stresses of the type envisaged in this model would result. There is, of course, a limit to the magnitude of these residual stresses and this is set, at the surface, in unconfined conditions, by the strength of the rock. Rock is approximately an order of magnitude stronger in uniaxial compression than in extension, hence it is the tensile strength which sets the level of the residual stress. Unfortunately, this limit cannot be clearly defined because the tensile strength of rock is influenced by a variety of factors which include linear scaling factors and time (see Chap. 1). In addition, when *in situ*, the rock mass is usually subject to total compressive stresses. Moreover, there is an inter-action between the measurable strength of rocks and the magnitude of residual stresses they contain. From the viewpoint of rock behaviour, the tensile component of the residual stresses *reduce the effective tensile (i.e. measurable) strength of the rock*. Thus, when an isolated block develops a crack in response to the residual stresses, the effective tensile strength of that part of the block has been reduced to zero, or may even become a positive quantity (for free standing blocks have been known to 'explode').

This concept of the reduction of the effective tensile strength of the rock enables us to understand how exfoliation (and other) fractures can develop in granites at shallow depths below the surface, for we can resolve the paradoxical situation in which exfoliation fractures (which have the morphology of extension fractures) can develop in rocks which contain *in situ*, gravitational and remanent stresses which are all compressive. If we assume that the topography is approximately flat-lying, then, at a depth of about 30 m, below which exfoliation fractures are rare (Fig. 4.35) the vertical, compressive, gravitational load is about 10 bar or 1 MPa. However, the development of horizontal extension fractures will be initiated at such a depth if pockets of tensile residual stress exist which exceed in magnitude the tensile strength of completely stress-free, isolated rock by about 10 bar. Pockets of tensile residual stresses with somewhat smaller excess values of stress (1–9 bar smaller) will initiate exfoliation fractures nearer the surface.

When such fractures are initiated they will tend to extend by stress corrosion so that they will interconnect with adjacent pockets of tensile residual stress or other minor fractures which have already formed. (For a treatment of the way in which such, residual stress induced, fractures propagate, the reader is referred to Johnson (1970).) At this stage, the development of the fractures will be almost wholly determined by the orientation of the *in situ* gravitational and remanent stress field. The extension fractures will most readily develop and propagate in a plane which is normal to the least principal stress. As we have seen, a granite may contain high lateral stresses near the surface. The gravitational loading, however, is determined wholly by the density of the rock and the depth of burial. Hence, in near-surface conditions, the least principal stress (S_3) will be oriented normal to the topographic surface, so that exfoliation fractures will form approximately parallel to the topography.

If one of the horizontal remanent stresses is small, or even slightly tensile, then the mechanism outlined above could first give rise to the formation of vertical barren fractures. These may exist at considerable depths, depending upon the position in the crust at which the horizontal stress approaches zero magnitude. However, let us revert to the original supposition that the least principal stress acts perpendicular to the topography. The development of the exfoliation fractures will relax only the residual tensile stresses acting in the vertical direction. Those which operate in the horizontal plane will still exist to influence the development of fractures yet to develop during the weathering process and which will be mainly orthogonal to the exfoliation fractures.

It is difficult to be specific about the way in which such fractures, which are perpendicular to exfoliation planes, will develop, for it depends upon the relative magnitudes of the horizontal remanent stresses. (Here, we continue to assume that the topography is approximately horizontal.) If one of the horizontal, remanent stresses is almost zero, vertical fractures may develop in the manner already outlined, or if the differential horizontal stress is sufficiently large, shear, or hybrid extension/shear fractures may form. The orientation of these late fractures will be determined by the orientation of the remanent stresses and these will have been inherited from ancient events. These stresses may cause sheets of granite, formed by the development of exfoliation fractures, to buckle and so induce tensile stresses at the upper surface, so that fractures may be formed with the orientation indicated in Fig. 4.41(*a*). An example, from the Yosemite National Park, California, in which an exfoli-

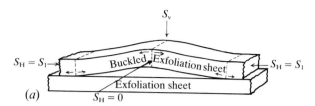

(*b*)

Fig. 4.41. (*a*) Possible mode of development of secondary fractures perpendicular to exfoliation. (*b*) Buckled exfoliation sheet, with fracture pattern. Yosemite National Park, California, U.S.A.

ation sheet has been buckled to form a minor dome, with associated, somewhat irregular fractures, is indicated in Fig. 4.41(*b*). (When the surface of this feature is stamped upon, the 'hollow sounding response' leaves one in no doubt that the sheet is unsupported on its lower surface.)

The arguments presented above regarding remanent and residual stresses and their influence upon the development of barren fractures, including exfoliation features in massive units, in near-surface environments apply with equal validity to sedimentary and metasedimentary rocks (Chaps. 9 and 14).

SECTION III

Circular features

When individual diapirs are eroded and the intermediate levels of the intrusion are exposed, they are often seen to exhibit an approximately circular plan, sometimes associated with other igneous features such as *ring dykes*, *cone sheets* and *radial dykes* which combine to form a *ring complex*. Such a system was first recognised in Skye, Scotland (Harker, 1904). Other examples occur at Ardnamurchan, Scotland (Fig. 4.42); in New Hampshire, U.S.A.; Norway; S.W. Africa; Victoria, Australia; and elsewhere. The various terms used above are defined in Fig. 4.43(*a*) in terms of simple 'idealised' geometries. The anatomy of such ring complexes and the form of real circular

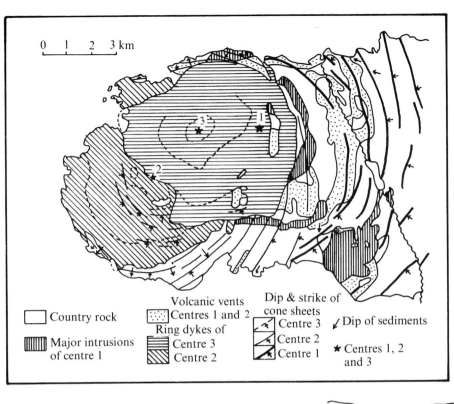

0 1 2 3 km

Country rock

Major intrusions
of centre 1

Volcanic vents
Centres 1 and 2
Ring dykes of
Centre 3
Centre 2

Dip & strike of
cone sheets
Centre 3
Centre 2
Centre 1

Dip of sediments

★ Centres 1, 2
and 3

Fig. 4.42. Generalised distribution of ring dykes and cone sheets in Ardnamurchan, Scotland. (After Rickey *et al.*, 1930.)

structures and their probable association with *caldera* are represented in Fig. 4.43(*b*) (after Cloos, 1936).

In Anderson's (1936) analysis of the formation of the arcuate features, he explained the formation of cone sheets and ring dykes as shear failures, generated by 'point dilation' and point downward push respectively. He assumed that these fractures were subsequently infilled by magma. Such mechanisms, particularly 'dilation of a point' do not appear particularly applicable to real process of forceful injection. However, Jeffreys (1936), commented that the postulated point push is closely imitated when a pointed tool is placed in contact with a block of flint and struck with a hammer, resulting in the generation of a conical fracture. Such pressure cones produced in flints and glass are spiral cracks (Robson & Barr, 1964) and that idea was taken up by Durrance (1967).

Phillips (1974) reviewed the various analyses and concluded that Anderson was correct as regards ring dykes, and that they developed as intrusions along shear fractures in the country rock which opened when the magma subsided. Radial dykes and cone sheets, Phillips argues, form as the result of increased (upward) action of magmatic pressure on the country rock. Radial dykes, (Chap. 3) are taken to be the result of simple hydraulic fracturing. However, cone sheets, he suggests, occupy shear fractures which formed as the result of dynamic stresses induced in response to rapid expansion of magma undergoing retrograde boiling. These failure orientations and mechanism are indicated in Fig. 4.44.

The mechanics proposed by Phillips is persuasive, and the importance he placed on transient

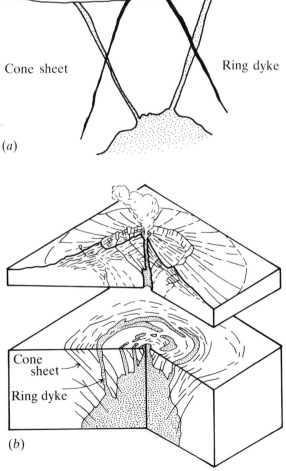

Cone sheet   Ring dyke

(*a*)

Cone
sheet

Ring dyke

(*b*)

Fig. 4.43. (*a*) Idealised geometry of cone sheets and ring dykes. (*b*) Ring structures in relationship to a major intrusion. (After Cloos, 1936.)

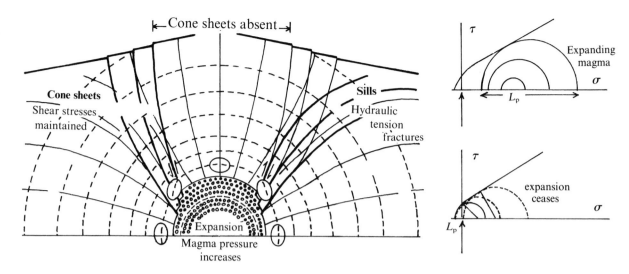

Fig. 4.44. (*a*) Stress trajectories, fracture orientation and generation of cone sheets through transient stresses resulting from the rapid expansion of a magma undergoing retrograde boiling. (*b*) High strain rate. High deviatoric stress. Low magma pressure on cone sheets. (*c*) Lower strain rate. Decreasing deviatoric stress. Increasing magma pressure on cone sheets. (After Phillips, 1974.)

variations in magmatic pressure is, we believe, correct. There is, however, a limit to the magnitude of the differential stresses that can be induced by this mechanism, which is set by the strength of the crustal rocks and the depth at which the event takes place. This, in turn, sets a limit to the size of structural assembly that can be generated. The mechanism may be applied to complexes which are a few tens of kilometres in diameter. However, for circular complexes which exceed a diameter of about fifty kilometres, it is necessary to consider some other form of generation. Many larger circular structures, we suggest, have formed in response to meteoritic impact.

Impact tectonics

The surfaces of the Moon, Mercury, Mars and other bodies of the Solar System bear testimony to the importance of impacts and the developing of cratering. The Earth could not have escaped this process. Indeed, the impact of solid bodies is the most fundamental process to have taken place on the terrestrial planets, which were almost certainly formed by such an accretionary process: the last stage of which is still proceeding, albeit at a very slow rate.

In the past, the impact mechanism has received scant attention from the majority of geologists, (this is not to say that the literature does not abound with articles or books on the subject) for geologists usually spend their professional lives trying to understand and explain geological phenomena wholly in terms related to terrestrial processes. Thus, terrestrial-based explanations are at the heart of our science. Unfortunately, there is a tacit corollary, held by some, that explanations of structures on Earth which are based on extra-terrestrial events are equated with non-science.

One of the reasons for this point of view is an uncritical application of the concept of Uniformitarianism. A really major impact on Earth has not occurred during the period of recorded history. Subconsciously, or even consciously, therefore, such events are categorised as rare or remote geological occurrences and tend to be dismissed from our thoughts. The two most important, practical reasons regarding the lack of acceptance of the importance of major impact structures were, firstly, the lack of a sufficiently large 'world view', so that circular features were not readily recognised (this difficulty has now, of course, disappeared with the arrival of satellite imagery), and secondly that even small features, which have now been shown to be of impact origin have, in the main, been previously attributed to other mechanisms.

The total accumulation of impact craters on the lunar surface as a function of age is shown in Fig. 4.45(*a*). The Earth will have experienced a similar impact history. However, because of their relative size, the probability of impact on the Earth is over 20 times greater than that for the Moon. (Unlike the Moon, the Earth has an atmosphere, so weathering can often rapidly remove craters, which are the surface expression of impact.) Clearly, from Fig. 4.45(*a*), the greatest impact activity took place in the early Pre-Cambrian. However, a significant number of impacts have occurred since this period (≈ 650 my). For example, the seismic profile in Fig. 4.45(*b*) shows a probable buried impact structure affecting the Jurassic rocks of central Montana. Indeed, the Earth continues to be bombarded; see, for example the recent crater in central Australia (Fig. 4.45(*d*)). The average impact energy received each year, at this present time, is about 3×10^{18}

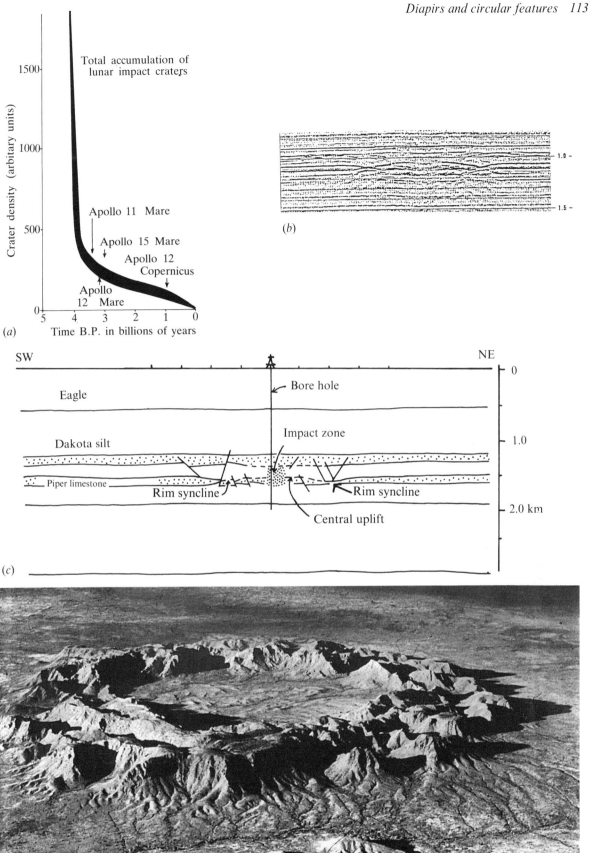

Fig. 4.45. (*a*) Frequency of impact events with time, for the Moon. (*b*) Seismic section showing a probable buried impact crater in the Jurassic rocks of central Montana. (After Plawman & Hager, 1983; in Bally, 1983.) (*c*) Interpretation of (*b*). (*d*) Impact structure in central Australia (courtesy Swiss Air).

joules. This energy mainly derives from the impact of small or micro-meteorites.

There is a group of asteroids that cross the Earth's orbit, which are called Apollo asteroids. This group, of which about 20 are currently known, were named after the first asteroid with such a track which was discovered in 1932. The closest observed approach of an Apollo asteroid to the Earth occurred in 1937 when 'Hermes' passed at a distance of 780 000 km. It will be appreciated that because the Apollo asteroids have only recently been discovered (and also because they have received scant attention from astronomers) there are certainly more than the known number. Indeed, Shoemaker (1977) has estimated that the number of such steroids with a diameter equal to, or in excess of, 0.7 km is 800 ± 400. The size of the larger known Apollo asteroids ranges up to 12 km diameter. However, among the hundreds of unknown asteroids, it is reasonable to expect that many will exceed this diameter. Indeed, they could be 1 to 2 orders or magnitude larger and approach the size of the largest known asteroid, Ceres, which has an approximate diameter of 450 km. Shoemaker has shown that the mean probability of an 'Earth-crossing' impact (per asteroid) is one every 2.2×10^8 yr and that the mean impact velocity is 24.6 km/s. Given that there are about 800 Earth-crossing asteroids with a diameter of 0.7 km or larger, the current rate of impact of such asteroids is about 3 $(\pm 1.5)/10^6$ years. This is in general accord with the geological record in the last few million years. As we shall see, even these relatively small asteroids with a diameter of 0.7 km can give rise to structures with a diameter of about 10 km. Hence, one would conclude that, during the Proterozoic, there have been over a thousand major impacts which could have given rise to craters with a diameter greater than about 10 km. Indeed, if one accepts that Earth-crossing asteroids with a diameter of only about an order of magnitude greater than the largest known Apollo asteroid collided with the Earth at the average impact velocity of about 25 km/s, then a feature with a crater diameter of 200 to 300 km would have resulted. Moreover, several such events are likely to have occurred during the Phanerozoic.

From Fig. 4.45 and the conclusion that the Earth received about 1000 impacts which gave rise to craters of 10 km or more in diameter in the last 600 my, one may infer that the number of events of similar magnitude that have occurred in the last 3.5×10^9 years is of the order of 3500. Moreover, in the period $4-3.5 \times 10^9$ years ago, the number of major impacts was of the order of 10 000. One may also reasonably infer that impacts from large asteroids were more probably in the early Pre-Cambrian, and that such a relatively large proportion of events would have resulted in craters with a diameter considerably

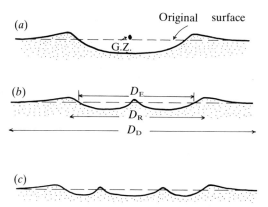

Fig. 4.46. Diagrammatic section through: (*a*) a simple 'basin' (*b*) central uplift and (*c*) ringed uplift craters. (G.Z. is ground zero, the point of impact or site of explosion.)

greater than 100 km. Unlike the Moon, however, the surface of the Earth has undergone continuous modification through the agents of weathering and Plate Tectonics. Consequently, all traces of the majority of the smaller craters (i.e. 10 to 30 km) resulting from impacts which occurred in the Pre-Cambrian are likely to have been obliterated. But what of the structures generated by high energy impacts which would have given rise to craters of 100 km or more in diameter? In order that this question may be answered, it is necessary to be aware of the morphology and the structural elements which may be exhibited by craters.

Morphology and structures associated with craters

The morphology of craters has been widely studied in natural examples on the Earth and the Moon and also by means of theoretical and physical models. The theoretical models and many of the physical models (especially those using small energy sources) have used 'target' material which was homogeneous and isotropic. The results of such calculations or experiments are usually in the form of a basin crater (Fig. 4.46(*a*)). Most natural craters on Earth with a diameter of less than about 3.5 km (e.g. Barringer Crater, Arizona) exhibit such a form, as do some of the craters which resulted from the Nuclear Test explosions. However, the size limits depend upon the strength of the target rock. For example, in an important experiment (codename Snowball), in which 500 tons of T.N.T. were set off on weak, lacustrine sediments, complex surface morphology was produced with a marked central uplift (Jones, 1977; Fig. 4.46(*b*)). Natural examples of craters (Dence *et al.*, 1977) which range in diameter from 3.5 to 35 km (e.g. Brent Crater) exhibit this central uplift. Natural terrestrial craters with a diameter of greater than 35 km tend to exhibit rings of alternating high and low topography, 'ringed uplift' (e.g. Manicouagan, Canada) as shown in Fig. 4.46(*c*). Ringed uplift was

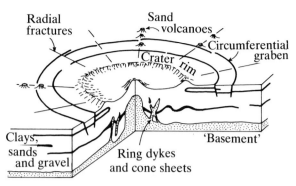

Fig. 4.48. Schematic synthesis of structures generated by a 500 ton T.N.T. explosion which resulted in a central uplift crater.

Fig. 4.47. Ringed uplift crater 75 m in diameter and 4 m deep, generated by a 500 ton T.N.T. explosion.

also developed in the crater formed by a second 500 ton T.N.T. explosion, (codename Prairie Flats) (Fig. 4.47). Hence, the size at which craters will exhibit ringed rather than central uplift characteristics appears to be related to target material and also the energy density of the explosion.

These structures which result from impact or explosion, can be divided into three circular lines of demarcation. The inner line (D_E) is the diameter of the excavated cavity (Fig. 4.46(b)). The crater rim (D_R) marks the second line of demarcation, while the outer limit (D_D) marks the limit to which rock is disrupted by the impact or explosion. The diameter of a crater rim of a newly formed crater is readily established, either on the ground or from aerial survey, while the limits of rock disruption can only be approximately established even after very considerable mapping and field examination. It is for this reason, of course, that the identification of an impact event hinges so heavily on identifying the crater rim. Moreover, it is because the crater rim and also the ringed or central uplift areas are so readily attacked and destroyed by the process of erosion, that impact events on the Earth dating back more than a few tens of million years are so difficult to detect.

There are, however, more easily identifiable features which will help in the recognising of impact structures. Based on descriptions provided by Jones, Price (1975) presented a schematic reconstruction of the main structural features which had developed as the result of the 'Snowball' 500 ton T.N.T. explosion, (Fig. 4.48). The circumferential normal faults and/or graben are clear manifestations of an extension phase, but do not, however, mark the limit of the compressional features. Minor folds, which are too small to show in Fig. 4.48, form beyond the zone of circumferential normal faults. Other features which developed beyond this zone were radial fractures which were subsequently utilised for the outflow of water, muds and sands from below to form 'volcanoes'. Other megascopic features of impact phenomena are the overturned units which form recumbent, isoclinal folds beneath the rim.

As a result of the energy of the explosive, or impact, event, some melting, or even vaporisation, of the rock mass takes place. In the 500 ton experiments, this was restricted to the production of ejected, hollow spherules of silica (i.e. silica 'popcorn'). In nuclear experiments of greater energy, tectites (solid glass globules) may be ejected and considerable quantities of glass are to be found in the crater rocks. In moderate-sized terrestrial craters, such as the Ries Structure of S. Germany, suevite (an impact breccia with a glass matrix) is a common feature.

Small-scale and microscopic structures also result from impact. The smaller-scale features which appear to be diagnostic of impact events are shatter cones: while on the microscopic scale there are forms of deformation and mineral assemblages which are a feature of shock metamorphism. Let us now look at the relationship between the energy of the event and the size of the structure produced.

Energy and size of structure

A 500 ton T.N.T. explosion (energy (E) approximately 5×10^{12} J) gives rise to a crater with a rim diameter of 70 m. Using this and other comparable experimental data obtained from nuclear explosions, one can employ theoretical relationships coupled with evidence adduced from the studies of natural, terrestrial craters to obtain scale-law relationships. Those

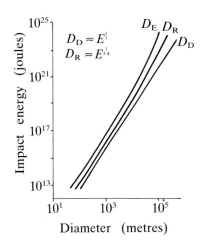

Fig. 4.49. Energy/diameter relationships for excavation (D_E), crater rim (D_R) and outer limit of disruption (D_D).

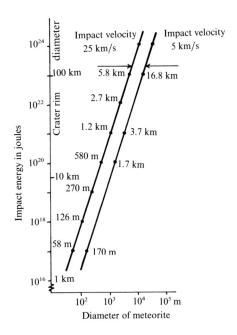

Fig. 4.50. Scaling relationships, for two velocities of impact, between energy and size of impacting body (the meteorite) and the diameter of the crater rim.

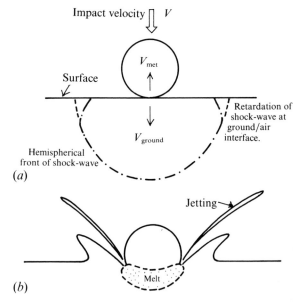

Fig. 4.51. (a) Shock wave velocities induced, in meteorite and ground, on impact. (b) Jetting phase.

produced by Dence *et al.* (1977) are shown in Fig. 4.49, where it will be seen that the crater rim diameter (D_R) = $E^{\frac{1}{4}}$ and that the total extent of disruption (D_D) = $E^{\frac{1}{3}}$. (We shall not consider the crater excavation relationship here.) Taking the earlier conclusion, that the average velocity of meteoritic impact is 25 km/s and the observation that 93 per cent of the meteorites are 'stony' with a density of about 3.5, one can estimate the size of the crater that would result from an impact on Earth of an extra-terrestrial body of a given size, simply by using the relationship $E_i = Mv^2/2$ (where E_i is the impact energy, M is the mass of the meteorite or asteroid and v is the velocity of impact) and the relationships indicated in Fig. 4.49. (Here we assume that the impacting body is sufficiently large, i.e. more than about 1000 tons, so that the energy lost by frictional heating in the atmosphere can be neglected.) The interrelationships between 'missile' diameter, energy and crater diameter for stony meteorites impacting at 25 km/s or 5 km/s are shown in Fig. 4.50.

Let us now consider cratering mechanisms. Upon the instant of impact, two shock waves are produced (i.e. pressure waves with an instantaneous increase), one of which is produced in the ground and the other in the meteorite (Fig. 4.51(a)). These shock waves initially propagate through the rock at supersonic velocities, but the ground pressure pulse, as it travels further from the source of energy, eventually changes in form, velocity and magnitude. The initial shock wave, with an instantaneous rise-time, degenerates into a plastic wave with a fast, but finite, rise-time. The plastic wave is slower than the shock wave but is still supersonic. Further from the source of the event, the plastic wave degenerates into an elastic wave form which, of course, travels thereafter at the speed of sound in rock, but continues to decrease in magnitude away from the energy source.

The maximum shock pressure and also the initial velocity of propagation depend upon the impacting and impacted materials and the particle velocity (Ahrens & O'Keefe, 1977). For stony meteorites and iron meteorites, both impacting on gabbroic anorthosite, the relationship between impact velocity and shock pressure is given in Table 4.1.

The initial velocity of propagation also depends upon the materials and velocity of impact. It may be seen from Table 4.1 that, for the quoted impact velocities, the peak impact stress exceeds the strength of rock by a factor of 10^3 to 10^4. At this stage, the

Table 4.1.

Impact Velocity (km/s)	Peak shock pressures	
	Stony Meteorite Pressure (Mbar)	Iron Meteorite Pressure (Mbar)
5.0	0.6	0.8
15.0	3.0	4.8
30.0	10.1	15.9
45.0	21.3	33.6

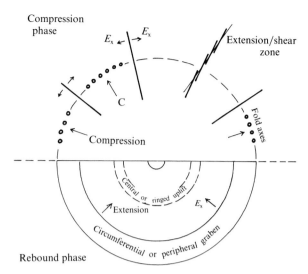

Fig. 4.52. Plan view of structures associated with the compression and rebound phases, shown in the upper and lower parts of the diagram respectively.

shock-deformed material forms a lens-shaped volume below the point of impact (Fig. 4.51(*b*)). The magnitude of the impact stresses so greatly exceeds the strength of the rocks that they begin to flow hydrodynamically (i.e. as a liquid). The process known as 'jetting' is initiated at this stage, in which material near the interface of the meteorite and the ground is squirted out in either a molten or vaporised plume.

As further penetration occurs, the shock wave pattern becomes more complex, with (on Earth) a coupled air blast and refracted compression waves on the ground. As compression waves reach a free surface, either at the ground–air interface or at the meteorite–space interface, they reflect and change phase to become 'rarefraction' waves of high stress intensity. The end of the compression stage near Ground Zero (G.Z.), i.e. near the point of impact, is reached when the shock wave reflects from the back side of the meteorite as a tensile pulse, which is then completely 'consumed' or disintegrated by the tensile, reflected, shock wave. The main mass of material (in the molten or solid state) is then ejected from the crater area in what is known as the 'excavation stage'.

The shock wave velocities in the meteorite and ground are identical for stony meteorites, and are from 19–24 per cent higher in the ground than in the meteorite when the latter is composed of iron. Hence, it can be inferred that when the compression stage near G.Z. is coming to an end, it is just beginning at a distance of 2–2.5 times the diameter of the meteorite away from G.Z. Thus, the phases of compression and rarefraction rip outwards in a band. It is this process which will not only cause the cratering (by refraction at the ground–atmosphere interface) but, at depth, results in the rebound phenomenon which gives rise to the inward and upward movement of material below the crater floor. As we have seen, this inward movement may result in central or ringed uplift of the crater floor. The main structures which develop in these compression and rebound phases are indicated in Fig. 4.52. The pressure wave gives rise to radiating fracture patterns which are only clearly seen beyond the crater rim and therefore develop at relatively low stress levels. These fractures are essentially brittle phenomena, even though they are contemporaneous with the ductile structures resulting from the radiating compressive pulse. (See folds in Fig. 4.48.) Clearly, the brittle fractures develop as a result of the circumferential tensile strains which are associated with the radiating compressive pulse.

Let us now consider the peak values and attenuation of the pressure curves. The peak values are indicated in Table 4.1, from which it is apparent that the magnitude is controlled by material properties and impact velocities. Ahrens & O'Keefe (1977) have shown that the attenuation of the stress wave is of the form indicated in Fig. 4.53(*a*), which shows that there is a 'near field' and a 'far field' behaviour, with different rates of attenuation. Thus, combining the data of Table 4.1 and Fig. 4.53(*a*), it is possible to quote a maximum impact stress and an appropriate mode of attenuation.

The area under the stress–strain curves (Fig. 4.53(*b*)) represents the work done on the rock mass. Usually, the unloading curve is different from that caused by the initial loading. This is known as the *hysteresis* effect. The difference in area of the loading and unloading curves represents the amount of work done and results in the heating up of the rock mass. Hysteresis associated with the high pulse-stresses gives rise to shock metamorphism, melting and even vaporisation. Regarding the pressure related to melting, Dence *et al.* (1977), who were concerned with relatively small craters (30–80 km) developed in rocks in the Canadian Shield, considered that, as a result of the hysteresis effect, the melting of these rock types (schist, gneisses and granites) would take place when the peak pressure attained a value of 600 kbar. From the pressure/attenuation relationships derived by

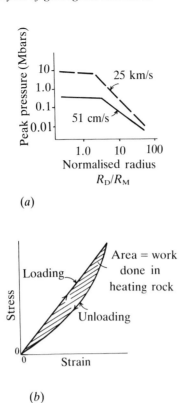

(a)

(b)

Fig. 4.53. (*a*) Decay of magnitude of stress wave generated by two velocities of impact expressed in normalised form (R_P/R_M) where R_P is the distance from ground zero (i.e. the radius of the pressure wave) and R_M is the radius of the impacting body. (After Ahrens & O'Keefe, 1977.) (*b*) Loading and unloading pressure relationships exhibiting hysteresis.

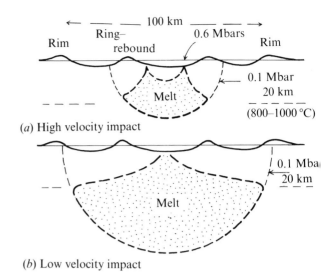

(a) High velocity impact

(b) Low velocity impact

Fig. 4.54. Qualitative indication of the degree of melting that may be generated by high and low velocity impact events with the same total energy. (*a*) Impact velocity 25 km/s and meteor radius 2.9 km. (*b*) Impact velocity 5 km/s and meteor radius 8.4 km.

Ahrens & O'Keefe, one may infer that, from a stony meteorite making impact at 25 km/s, the pressure pulse would decline from several Megabars to 600 kbar at a radius from ground zero (G.Z.) of about 6 times the radius of the impacting meteorite.

The figure of 600 kbar required to induce melting, which is used by Dence *et al.*, however, relates to an initially 'cold' near-surface rock. At lower levels in the crust, the rocks are already at an elevated temperature. In Pre-Cambrian times, the crust was thin and the geothermal gradient significantly higher than it is today, so that the rocks would be near their melting point at relatively shallow depths. Let us assume that at depths of about 50 km, melting results from a transient pressure pulse of about 100 kbar. This level of pressure pulse will occur at a distance of about 16 times the radius of the impacting meteorite at $v = 25$ km/s. Hence, the zone of melting will widen out from the surface in the form of an inverted mushroom, as indicated in Fig. 4.54(*a*). As may be inferred from Table 4.1, the peak shock pressure attained by a stony meteorite making impact at a velocity of only a little less than 5 km/s will not be sufficiently high to cause melting at the surface. It could, however, induce melting to take place in the sub-surface rocks which are already at an elevated

temperature. Moreover, the radius of a slow-moving meteorite (5 km/s) required to produce a 100 km diameter crater is almost three times greater than that of a 25 km/s meteorite. Hence, because the 100 kbar level of shock extends to a distance of 6 times the radius of the impacting body for $v = 5$ km/s, the zone of sub-surface melt is extremely extensive (Fig. 4.54(*b*)) and could possibly spread as far as, or even beyond, the limit of the crater rim.

If this molten unit is regarded as a sill, the cooling time for it to reach half the initial differential temperature ($T^{\frac{1}{2}}$) is given approximately by $T^{\frac{1}{2}} = 40d^2$ where d is half the sill thickness given in centimetres. If one attributes to this molten 'sill' an average thickness of as little as 10 km, the cooling time ($T^{\frac{1}{2}}$) will be of the order of a million years, a period of sufficient duration to enable 'differentiation' and 'diapirism' of the lighter elements to take place resulting in the emplacement of granite and other intrusives.

Synthesis

It is of interest to present a synthesis or 'scenario' that will enable geologists to recognise large and ancient structures which have been formed as the result of major meteoritic impacts.

We consider that the starting point of the scenario should be the data relating to the two 500 ton T.N.T. experiments, Snowball and Prairie Flats. Because they involve deformation of weak lacustrine deposits, they are the most realistic model experiments of crustal deformation that have so far been conducted in this field of investigation. We shall take linear scaling factors of 10^3, 3×10^3 and 10^4 and comment upon the magnitude of the various features and structures which one would expect to find in larger structures based on the information derived

Table 4.2.

Diameters of rim (D_R)	70 m	70 km	210 km	700 km
Depression of crater floor	4–5 m	4–5 km	12–15 km*	50–50 km*
Depression of asphalt. strip	2 m	2 km	6 km	20 km
Wavelength of folds (asym.)	1–5 m	1–5 km	3–15 km*	10–50 km*
Dip-slip extent of N. faults	>1 m	1 km	>3 km	10 km
Displacement of N. faults	>10 cm	>100 cm	>300 m	>1 km
Slumping into crater	Noticeable	Important	Important	Important
Melting	Insignificant	Noticeable	Significant	Extensive

*Limits may not be attained.

from the two 'high explosive' experiments. As the diameter of the crater rim, in both experiments, was close to 70 m, we shall consider larger structures with crater diameters of 70, 210 and 700 km respectively (Table 4.2).

Consider first the depression of the crater floor. In the T.N.T. experiments, the average depression is 4–5 m. If this is scaled-up geometrically, the resulting depression of floors in craters of 70–700 km diameter indicated in Table 4.2 is almost certainly too large. Indeed, Jones pointed out that in a series of experiments where craters were produced from charges ranging from 5 tons of T.N.T., the craters did not scale geometrically, but became progressively wider and shallower than the 'model ratio' prediction. Nevertheless, it is probable that the crater floor would be a topographic low and so would begin to infill by slumping and sedimentation. Deposition of sediments

adjacent to the rim and possibly to the ringed uplift would be particularly rapid. The resulting narrow zones of thick sediments which accumulate in such positions could eventually become deformed by the diapiric processes which were initiated in the underlying molten rocks (Fig. 4.55). Hence, these sediments would be lightly metamorphosed and could evolve into the narrow, arcuate and sometimes sinuous greenstone belts which are such a feature of Pre-Cambrian geology.

Elsewhere within the crater, the sediments would be thinner. However, because of the extremely high geothermal gradients that would exist above the molten rocks, burial metamorphism would occur in the sediments. This, coupled with any diapirism of the former crustal material, which had been depressed by the impact, could result in mantled gneiss domes or a complex of acid intrusions into metamorphosed sediments.

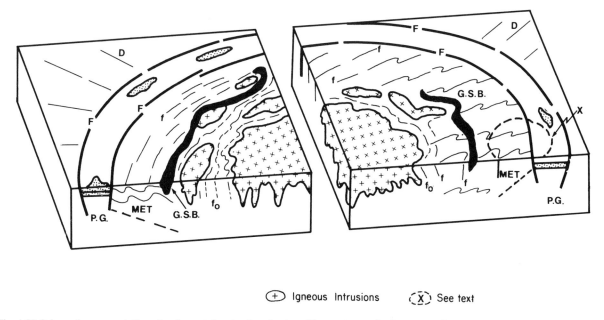

⊕ Igneous Intrusions   (X) See text

Fig. 4.55. Schematic representation of an impact site, that has developed into an 'astron', when crater-rim topography etc. has been eroded. The left hand unit shows the circumferential trends of fold axes which would develop if the impacted rock were flat-lying and initially unfolded. The right-hand unit represents the interference patterns that could develop if the impacted rock were originally folded. P.G. = peripheral graben, G.S.B. = Greenstone belt, F = fault, f = fold, f_0 = foliation, MET = metamorphosed rock and D = radial dykes.

In the 500 ton T.N.T. experiments, asphalt strips, used during the pre- and post-explosion surveys, provided excellent markers which exhibited two main common features. Firstly, these strips were depressed increasingly from the limit of disruption (D_D), so that below the crater rim they were approximately 2 m below their original level. The scaled-values for the depression which ranges from 40–50 km are again likely to be somewhat exaggerated. Nevertheless, significant depressions will develop beneath the crater rim. Such a depression will, as we have seen, be accompanied by some measure of shock-metamorphism, but this will subsequently be over-printed by thermal metamorphism, as the depressed sediments adjust to the ambient temperatures appropriate to the various depths of burial, which will obliterate much of the evidence of impact deformation so readily demonstrated in smaller terrestrial craters. The second feature of the asphalt strips, and also the adjacent sediments, is that they have buckled to form asymmetrical and outward-verging folds with wavelengths of 1–5 m. The equivalent fold wavelengths of the larger craters (see Table 4.2) would, therefore, have an estimated range of 10–50 km.

Hence, the central area of mantle-gneiss domes and/or intrusive masses would be surrounded by a 'shield' of metamorphosed rocks. If the rocks originally possessed flat-lying bedding or foliation, they would exhibit an array of asymmetrical folds verging outwards with fold axes tending to a circumferential trend. (See the left hand side of Fig. 4.55, which indicates the type of features that will be exposed after the crater and original surface morphology has been eroded away.) However, the impact area may originally have exhibited a marked 'grain', e.g. an existing fold belt. In such an event, the fold pattern would be more complex and would exhibit one of the interference patterns which is so typical of Pre-Cambrian metamorphosed rocks. (See the right hand side of Fig. 4.56, where the interference relationship can be seen.) However, if only a small part of the impact feature is now exposed (i.e. area X), the structural significance is not so readily inferred. The fracture pattern associated with the T.N.T. experiments is likely to be reproduced in large, natural events. As we have seen, they comprise radiating and circumferential fractures, many, or even most, of which are used as paths for migrating fluids. (See the sand volcanoes in Fig. 4.48.) Consequently, as the fractures produced by the impacts would form paths for the subsequent out-flowing of molten material, one would expect to find radiating dykes associated with some major impact structures.

The central uplift type of crater is also characterised by having a peripheral graben. The dip-slip extent of individual normal fault planes in the peripheral graben of the Snowball experiment certainly

0   20 km

Fig. 4.56. Relationship between circular features and mineralisation in Arizona; the black dots are known mines and prospects. (After Saul, 1978.)

exceeds 10 cm, so that these would represent major fault structures in the larger craters. Moreover, the displacement of about 10 cm would be an equivalent of a downthrow of about 1 km in the larger impact structures. In early Pre-Cambrian times the crust was thin: and in really large impacts the whole of the crust and part of the upper mantle could have been 'peeled' back to form the flat-lying isoclinal fold shown in Fig. 4.48. Part of the attenuated limb could be thrown down in the subsequent graben development and, if erosion was modest, preserved.

Hence, a complete impact feature would exhibit a reasonable, but not perfect, radial symmetry. However, if only a part of the structure is preserved, one must search to find the arcuate features. The synthesis indicates that impacts give rise to geological features which are found in many Pre-Cambrian 'cratons'. It has been suggested (Norman *et al.*, 1977) that many of these cratons are the result of meteoritic impact, and for this reason one of the present authors coined the word 'astron' to indicate the extra-terrestrial importance in the structural evolution of Pre-Cambrian crustal structures.

Circular features are often associated with mineralisation. The mechanisms involved are somewhat conjectural so that a discussion will be omitted. It will suffice to show the results of an investigation that illustrates this relationship between the sites of metalliferous mines and relatively small circular features in Arizona, U.S.A. (Fig. 4.56).

The influence of major impacts on early Earth evolution was enormous. However, even the relatively small number of major impacts (i.e. those producing craters of 50–100 km diameter) which have occurred since the Pre-Cambrian would have had far-reaching effects on the Earth's climate. A volcanic eruption such as Mt. St. Helens may affect weather trends for a year or two, but a major impact with 10^6 times the amount of energy of such a modest volcanic eruption would induce climatic conditions which could result in Ice Ages and/or have a pronounced influence on evolution (particularly of terrestrial life forms such as the dinosaur). However, discussion of such events is outside the scope of this book.

References

Ahrens, J. & O'Keefe, W. (1977). Equations of state and impact induced shock wave attenuation on the Moon. In *Impact and explosive cratering*, ed. Roddy *et al.*, Oxford: Pergamon.

Anderson, E.M. (1936). The dynamics of the formation of cone sheets, ring-dykes and cauldron subsidence. *Proc. Roy. Soc. Edin.*, **56**, 128–57.

Balk, R. (1937). Structural behaviour of igneous rocks. *Geol. Soc. Am. Mem.*, **5**.

Balk, R. (1953). Salt structure of Jefferson Island salt dome. *Am. Assoc. Petrol. Geol. Bull.*, **37**, 11, 2455–74.

Braunstein, J. & O'Brien, G.D. (1968). *Diapirism and diapirs: a symposium.* Am. Assoc. Petrol. Geol. Mem. 8.

Chapman, C.A. & Rioux, R.L. (1958). Statistical study of topography, sheeting and jointing in granite, Acadia National Park, Maine. *Am. J. Sci.*, **256**, 111.

Cloos, H. (1922). Tektonik und Magma. *Bd. 1 Abh. d. Preuss. Geol. Land*, **89**.

Cloos, H. (1936). *Einführung in die Geologie.* Berlin: Borntrager.

Dence, M., Greive, R. & Robertson, D. (1977). Terrestrial impact: principal characteristics and energy considerations. In *Impact and Explosive Cratering*, ed. Roddy *et al.*, Oxford: Pergamon.

Durrance, E.M. (1967). Photoelastic stress studies and their application to the mechanical analysis of the Tertiary ring complex of Ardnamurchan, Argyllshire. *Proc. Geol. Assoc.*, **78**, 289–318.

Friedman, M. (1972). Residual elastic strain in rocks. *Tectonophysics*, **15**, 297–300.

Fyfe, W.S., Price, N.J. & Thompson, A.B. (1978). *Fluids in the Earth's Crust.* Amsterdam: Elsevier.

Gastil, R.G., Philips, R.P. & Allison, E.C. (1975). *Reconnaissance geology of the state of Baja California.* Geol. Soc. Am. Mem., 140, 170 pp.

Gussow, W.C. (1968). Salt diapirism: importance of temperature and energy source of emplacement. In *Diapirism and Diapirs*, ed. J. Braunstein & G.D. O'Brien, pp. 16–52. Mem. 8 Am. Assoc. Petrol. Geol.

Halbouty, M.T. (1967). *Salt domes, Gulf Region – U.S. and Mexico.* Houston: Gulf Pub. Co.

Harker, A. (1904). *The tertiary igneous rocks of Skye.* Mem. Geol. Surv.

Heard, H.C. (1972). Steady-state flow in polycrystalline halite at pressure of 2 kb. *Am. Geophys. Un. Mon.*, **16**, 191–210.

Hopper, F.W.M. (1974). Salt Diapirism. Unpublished M.Sc. dissertation, University of London.

Jackson, M.P.A. & Seni, S.J. (1983). Evolution of Salt Structures, East Texas Diapir Province, 2. Patterns and rates of Halokinesis. *Am. Assoc. Petrol. Geol. Bull.*, **67**, 8, 1245–74.

Jahns, R.H. (1943). Sheet structures in granites: its origin and use as a measure of glacial erosion in New England. *J. Geol.*, **51**, 2, 71–98.

Jeffreys, H. (1936). Note on Fracture. *Proc. Roy. Soc. Edin.*, **56**, 158–63.

Johnson, A.M. (1970). *Physical Processes in Geology.* San Francisco: Freeman, Cooper & Co.

Jones, G.H.S. (1977). Complex craters in alluvium. In *Impact and explosive cratering*, ed. Roddy *et al.*, Oxford: Pergamon.

Mrazec, M.L. (1915). Les plis diapirs et le diapirisme en general: *Rumanian Inst. Geol. Comptes Rendus*, **4**, 226–70.

Norman, J., Price, N.J. & Shukwu-Ike, M. (1977). The Earth's oldest scars. *New Scientist*.

Parker, T.J. & McDowell, A.N. (1951). Scale models as a guide to interpretation of salt-dome faulting. *Am. Assoc. Petrol. Geol. Bull.*, **35**, 9, 2076–86.

Phillips, W.J. (1956). The Criffell–Dalbeatie granodiorite complex. *Q. J. Geol. Soc. London*, **112**, 2, 221–40.

Phillips, W.J. (1974). The dynamic emplacement of cone sheets. *Tectonophysics*, **24**, 69–84.

Pine, R.J., Tunbridge, L.W. & Kwaka, K. (1983). *In situ* stress measurements in the Carnamanellis Granite. *Int. J. Rock Mech. Min. Sci. & Geomech. Abstr.*, **20**, 51–62.

Pitcher, W.S. & Read, H.H. (1959). The main Donegal Granite. *Q. J. Geol. Soc. London*, **114**, 2, 259–305.

Plawman, T.L. & Hager, P.I. (1983). Seismic expression of structural styles. In A.W. Bally, *A picture and work atlas.* Vol. 1. Am. Assoc. Petrol. Geol.

Posepny, F. (1871). Studien aus dem Salinargebiete Siebenbürgens: Jahrbuch K.K. *Geol. Reichsanstalt*, **17**, 4, 475–516.

Price, N.J. (1964). The study of time-strain behaviour of coal measure rocks. *Int. J. Rock Mech. Min. Sci.*, **1**.

(1966). *Fault and joint development in brittle and semibrittle rock.* Oxford: Pergamon.

(1970). Laws of rock behaviour in the Earth's crust. In *11th Sympos. Roc. Mech. Berkeley*, ed. W.H. Somerton, pp. 3–25.

(1975). Rates of deformation. *J. Geol. Soc. London*, **131**, 553–75.

(1979). *Fracture patterns and stresses in granites.* Canadian Geoscience.

Ramberg, H. (1972). Theoretical models of density stratification and diapirism in the Earth. *J. Geophys. Res.*, **77**, 877–89.

(1981). The role of gravity in orogenis belts. In *Thrust and Nappe Tectonics*, eds. K. McClay & N.J. Price, pp. 125–40. Geol. Soc. London Special Publ. No. 9. Blackwell Scientific Publications.

Rickard, M.J. (1984). Pluton spacing and the thickness of crustal layers in Baja California. *Tectonophysics* **101**, 167–72.

Rickey, H., Thomas, H.H. *et al.* (1930). *The Geology of Ardnamurchan, North-west Mull and Coll.* Mem. Geol. Surv.

Robson, G.R. & Barr, K.G. (1964). The effect of stress on

faulting and minor intrusions in the vicinity of a magma body. *Bull. Volcanol.*, **27**, 315–30.

Sannemann, D. (1968). Salt-stock families in N.W. Germany. In *Diapirism and Diapirs*, ed. J. Braunstein & G.D. O'Brien, pp. 261–70. Mem. 8 Am. Assoc. Petrol. Geol.

Saul, J.M. (1978). Circular structures of large scale and great age in the Earth's crust. *Nature*, **271**, 5643, 345.

Schott, W. (1956). Das niedersächsische Becken östlich der Weser, in Symp. sobre yacimientos de petroleo y gas. *20th Int. Geol. Cong. Mexico*, **5**, 59–64.

Shoemaker, E.M. (1977). Astronomically observable crater-forming projectiles. In *Impact and explosive cratering*, ed. Roddy *et al.*, Oxford: Pergamon.

Tapponnier, P. & Molnar, P. (1976). Slip-line field theory and large scale continental tectonics. *Nature*, **264**, 319–24.

Trusheim, F. (1960). Mechanism of salt migration in N. Germany. *Am. Assoc. Petrol. Geol. Bull.*, **44**, 9, 1519–40.

Turcotte, D.L. & Schubert, G. (1982). *Geodynamics*. New York: Wiley.

Urai, J.L. (1983). Deformation of wet salt rock. Unpublished Ph.D. Thesis, University of Utrecht, Holland.

5 Faults – nomenclature, classification and basic concepts

Introduction

We shall first consider the scale at which faults occur and the nomenclature used to allude to structures of different magnitudes, attitudes and direction of displacements. As the classification of faults is commonly related to the direction(s) of slip along the fault plane, we shall discuss the problems inherent in establishing these slip directions. The Andersonian classification of faults will be outlined and the relative magnitudes of differential stress associated with the various Andersonian faults examined, after which we shall deal with some aspects and features which are related to these different stress magnitudes. The three main types of faults which form the Andersonian Classification, (i.e. *strike-slip*, *thrust* and *normal faults*) are dealt with in Chapters 6, 7 and 8 respectively, the remainder of this chapter will mainly be concerned with non-Andersonian aspects of faulting. These include conditions when: (i) at least two of the axes of maximum principal stress are inclined to the vertical (ii) the rock mass undergoing faulting is not homogeneous and isotropic, i.e. the rock contains a fabric or pre-existing planes of weakness and (iii) all three axes of principal stress are inclined to the fault plane.

Fault nomenclature and movement

The definition of the word *fault* given in the Tectonic Dictionary (Dennis, 1967) is that 'it is a fracture surface or zone along which appreciable displacement has taken place'. This definition is one which would be acceptable to many geologists but, as we shall see, it leaves much to be desired. The main problem is determining what is meant in this context by 'appreciable'.

As was pointed out in Chap. 2, an 'appreciable' displacement will be related to the linear dimensions (L) of the fault plane, and its aspect ratio (i.e. the s/L ratio, where s is the maximum shear displacement). For most faults which extend for at least a few hundred metres, even when they exhibit s/L ratios smaller than 10^{-2}, the displacement is still likely to be appreciable. However, fractures which extend for only a few tens of metres, or less, and which exhibit s/L ratios of 10^{-4} to 10^{-5} will have a maximum displacement of only a few millimetres. Indeed, the displacement, rather than being 'appreciable' may be completely overlooked; so that the field geologist may

well classify such fractures as 'joints'. Even if the shear displacement is detected, it may reasonably be concluded that it is so small as to be 'un-appreciable': and hence the fracture should not be called a fault but should rather be described as a 'minor shear fracture' or some such descriptive term. Thus, it is inherent in Dennis' definition that faults must, of necessity, be moderately large features, but that the point of change-over from fault to some minor shear fracture is not defined. This constitutes a problem in nomenclature for the field geologist who may frequently encounter shear fractures which fall into this 'twilight zone'.

Faults, or shear fractures, form a continuum, as regards the magnitude of the fracture plane, which extends from the microscopic to mega-structures which are hundreds, or even thousands, of kilometres long. The type of nomenclature that may be used to describe individual fractures which fall within this size range is indicated in Table 5.1. Such a set of limits, as is given in this table, need not necessarily be rigidly maintained. Indeed, the choice is completely arbitrary and will largely depend upon the experience, interest and area of ground studied. For example, for a mining geologist, a shear fracture with a displacement of a few metres and an extent of 100–300 m could have considerable economic significance and be considered as a 'large scale' structure. However, the same fracture may be considered of minor importance by a geologist concerned with regional fault patterns. Alternatively, in an area under investigation there may be a system of faults which range in length from 200 to 2000 m, together with a number of faults which extend from 8 to 12 km. In such an area it would be convenient to place the limit between small and intermediate-size, megascopic fractures at 2 km. Thus, the limits given in Table 5.1 are only presented for guidance. The important thing is that when such descriptive terms are used, the reader should be left in no doubt as to their meaning and quantitative limitations.

One of the earliest methods used to classify faults in sedimentary rocks was to relate the orientation of the *fault* or *fracture trace* (i.e. the line which the fault plane or zone makes when it intersects the topographic surface) to the disposition of the beds through which the fracture ran. If the fault trace was parallel, or sub-parallel, to the strike of the beds, it was termed a *strike-fault*. Similarly, if the trace was aligned approximately perpendicular to the strike it was termed a *dip-fault*;

Table 5.1. *Magnitude of faults*

	Description	Length of fracture
Mesoscopic	Minor (small scale)	$< 10^0$ m
	Intermediate	$10^0–10^1$ m
	Large	$10^1–10^2$ m
Megascopic	Minor	$10^2–10^3$ m
	Intermediate	$10^3–10^4$ m
	Large	$10^4–10^5$ m
Regional or continental		$> 10^5$ m

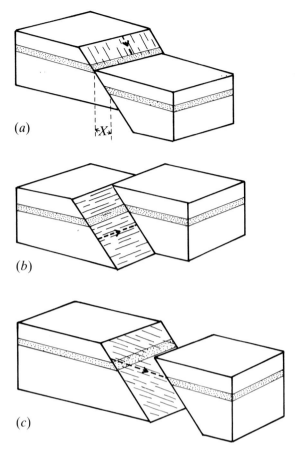

(a)

(b)

(c)

Fig. 5.1. Block diagram of a fault, showing (a) dip-slip (b) strike-slip and (c) oblique-slip. (X is the lateral separation caused by normal faulting).

while if the fault trace was parallel to neither the dip nor the strike the structure was termed *transverse* or sometimes an *oblique-dip fault*. Such terms, which do not take into account the orientation of the shear plane and the direction of shear displacement and are concerned only with a line exposed in an essentially two dimensional view, are clearly of very little significance and nowadays are rarely used.

In structural geology we are usually concerned with a three-dimensional interpretation of structures. As with other planar geological surfaces, the three-dimensional orientation of a fault can be defined in terms of its dip and strike. The modern classification of faults is based upon the orientation of the fault plane coupled with the known, or inferred, direction of shearing and the sense of movement along the fault plane. Thus, the direction of shear movement may be *dip-slip, strike-slip* or *oblique-slip* (Fig. 5.1). However, some of the terms which are used in classification evolved in mining fraternities and are of some considerable antiquity, dating probably from the 18th century, or even earlier. Fault planes encountered in British coal mines were frequently inclined and the rock mass adjacent to the shear plane was termed either a *footwall* or a *hanging wall*, according to whether the miner could stand and rest his foot on the fault plane or whether the rock mass hung above his head as he approached the fault plane. The shear direction of the majority of these faults exhibited a large component of dip-slip, with the hanging wall displaced downwards relative to the footwall (Fig. 5.1(a)). Such structures were called *Normal Faults* because they were the faults which were 'normally', or most frequently encountered in British coal mines. If the movement of the hanging wall was upward relative to the footwall the structure was termed a *Reverse Fault*, (Fig. 5.2(b)).

Because of the lateral separation of a geological horizon which results from normal faulting, such structures have been termed *extension faults* (Fig. 5.1(a)). Similarly, because reverse faults exhibit a repetition or overlap of a geological horizon, these structures have been termed *compression faults*. Van

Hise (1896) extended the concept that such faults may be the result of either extension or compression and accordingly suggested that normal faults and reverse faults should be classified as *gravity faults* and *thrusts* respectively, thereby introducing an element of mechanistic thinking into the classification.

It was noted that in relatively flat-lying sediments the structures which exhibited reverse movement could be divided into two groups. One group often dipped at about 60°, while the other group often dipped at a relatively low angle of about 25–30°. In order to differentiate between these groups it has been suggested that only those faults showing relative upthrow of the hanging wall which dip at angles of less than 45° should be termed thrusts, while those fractures which dip at greater than 45° should be classified as reverse faults. This distinction is, nowadays, not much used and it is usual to refer to all faults with the movement indicated as reverse faults. Indeed, the term thrust (or *overthrust*) is now usually restricted to describe those major structures in which the displacements can be measured in kilometres or tens of kilometres and whose shear surface is, or was originally, within a few degrees of horizontal.

There is another main type of fault, the plane of

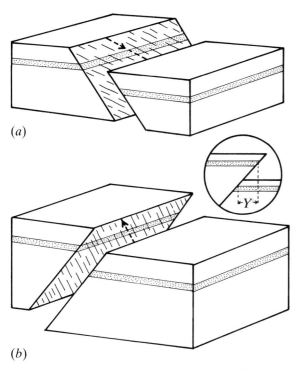

(a)

(b)

Fig. 5.2. Block diagrams which indicate the oblique slip nomenclature that may be used if the slip direction is known. (a) a left-normal slip fault and (b) a left-reverse-slip fault. (Y is the overlap of strata caused by reverse faulting.)

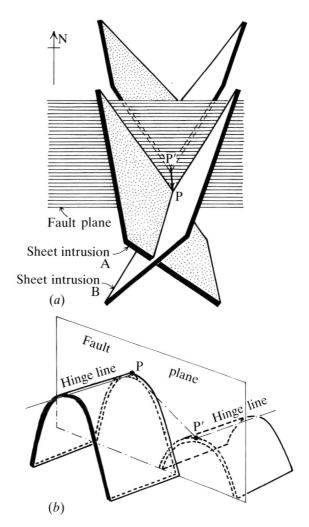

Fig. 5.3. Block diagrams showing how a point on either side of the fault plane may be identified by projecting a line that can be determined from structural units on either side of the fault, until it intersects the fault plane at points P and P' respectively. If it can be assumed that the structural elements predate, and were displaced by, the fault then the magnitude and direction of displacement can be accurately assessed. (a) Shows how this may be done by determining the lines of intersection of two sheet intrusions. (b) Indicates how the hinge line of a fold at a specific horizon may be utilised to determine the direction and magnitude of displacement.

which is close to vertical and which exhibits near horizontal, i.e. strike-slip movement. These faults received little attention until the turn of this century when they began to attract a large number of descriptive terms. Of the various synonyms the most common are *tear*, *transcurrent*, *wrench* and *strike-slip faults*. The sense of movement on such faults is defined by the view of a geologist facing the fault. If the fault block on the far side of the fault from the observer has been displaced to the left, the movement has been termed left-handed, sinistral or anticlockwise movement. Conversely, if the movement is to the right, the fault may be termed right-handed, dextral or clockwise. All these terms are to be found in the literature; however, the modern trend is to describe such faults as left-slip or right-slip. If the direction of movement on an inclined fault plane is oblique to its dip, the various terms noted above can be combined to provide a suitable description of the fault. For example, the structures in Fig. 5.2(a) and (b) would be described as 'left-normal-slip' and 'left-reverse-slip' faults respectively.

Determination of direction and sense of movement
Because knowledge regarding the direction and sense of slip is so important to the classification of faults, it is of interest to digress briefly to consider how these features may be determined. It must at once be admitted that the task of defining the direction and

precise displacement from data available in the field is usually a difficult, if not impossible, task. However, three main methods may be employed, either singly or in combination.

The first of these methods is represented in Fig. 5.3(a) and (b), where it will be seen that the direction, sense and magnitude of movement is determined by identifying two points on either side of the fault plane which, prior to fault movement, were coincident. The way in which such points are defined is to find a linear element which pre-existed fault movement and which was cut and displaced by that movement. The two examples represented in Fig. 5.3(a) and (b) are based on a line defined by two intersecting intrusive sheets and the hinge line of a fold in a specific bed respectively. Such a disposition

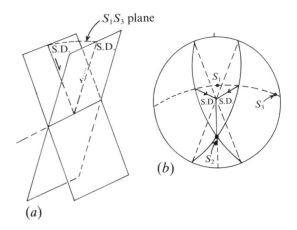

Fig. 5.4. Photograph showing fibres and imprint fibres on the adjacent rock.

of elements provides a wonderful basis for the compilation of problem maps for students. Unfortunately, structures and elements with suitable dispositions adjacent to a fault are somewhat infrequently encountered in nature.

As regards the second method: occasionally, when a fault plane is exposed, it is possible to infer the sense of movement from linear striations on the fault plane. There are two distinct types of such lineations. The first of these is called *slickensides* which are grooves and fine linear features on a 'slick', polished fault plane. Such features are the result of melting of some of the rock minerals abutting the fault plane coupled with the gouging action of more resistant minerals and rock particles. In its extreme form this leads to the development of pseudo-tachylite, which we shall discuss later in this chapter. Slickensides and pseudo-tachylites develop when the effective stress acting normal to the fault plane is high.

The second type of lineation on fault planes, which is sometimes erroneously called slickensides, is the result of *growth-fibres*. Mineral infill, commonly quartz or calcite, on the fault plane may exhibit characteristics which are comparable with slickensides. However, the growth fibres (which will be discussed more fully in Chaps. 10 and 15) form in conditions of high fluid pressure, so that the effective stress acting normal to the fault plane is low, or negative. Eventually, when the fluid pressure decays, growth-fibres may subsequently be subjected to higher normal stresses that result in an imprint of the fibres on the fault plane. It is then sometimes quite difficult to determine whether the features are 'true' or 'apparent' slickensides (Fig. 5.4). The evidence presented by slickensides and growth-fibres frequently indicates that a fault has moved more than once and in more than one direction.

The third method of determining the direction and sense of movement on a fault can be employed when conjugate pairs of mesoscopic shear planes or faults can be identified and their orientations estab-

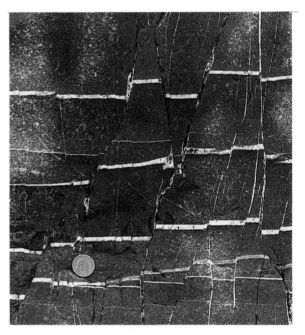

Fig. 5.5. (*a*) Example of conjugate, minor fractures from which initial stress orientations and direction of shear movements may be inferred by using the stereogrammetric technique indicated in (*b*). Unfortunately, such examples (*c*) are not commonly encountered in the field. (*b*) Stereographic representation of the conjugate shear planes shown in (*a*) and the interpretation showing orientation of stresses that caused the fractures and the direction (S.D.), but not the magnitude of the shear displacements.

lished (Fig. 5.5(*a*)). It is then possible to use the Navier–Coulomb theory of brittle failure, in conjunction with the field data, to infer the slip direction by using the stereographic technique (Fig. 5.5(*b*)). Thus, it may be inferred that the axis of intermediate principal stress acted parallel to the line of intersection of the two shear planes and that the movement along the two shear planes occurred in the plane which contains the axes of least and greatest principal stress. Such information can usually be best obtained from small-scale shear fractures such as those shown in Fig. 5.5(*c*). Once again the data which permit the use of this third technique are more frequently found in laboratory problem maps than in the field.

From the above, we can infer that the field

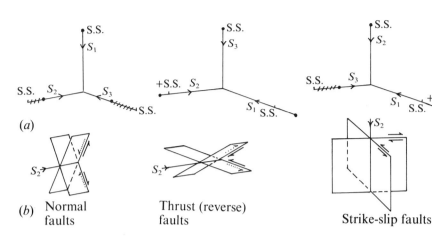

Fig. 5.6. (*a*) Showing the three possible ways in which the 'standard state' (S.S.) may be varied and (*b*) the corresponding types of faults (i.e. normal, thrust (reverse) and strike-slip) that can be generated.

(*b*) Normal faults   Thrust (reverse) faults   Strike-slip faults

geologist will only rarely be able to determine the direction and amount of total displacement on a fault. Even when this can be accomplished, what is determined is the average displacement direction. The actual incremental movement paths may never be known. It can be argued that many, if not most, faults exhibit some degree of oblique-slip motion. However, because precise evidence of such oblique-slip is usually absent, geologists, quite naturally, tend to place faults into one of the three types which figure in Anderson's dynamic classification, namely Normal, Strike-slip and Reverse Faults (or thrusts).

Anderson's dynamics of faulting

Anderson (1951) based his classification on the general observation that in many areas of the world, especially in those of low topographic relief, it can be inferred from studies of extension fractures, faults and dykes, that the axes of principal stress are close to the horizontal or vertical. He postulated a *standard state* of stress in the Earth's crust, which is equivalent to a 'hydrostatic state', in that the magnitude of the horizontal stresses, at any specific depth in the crust, is equal to that of the vertical *geostatic stress* induced at that depth by gravitational loading. Anderson pointed out that the magnitudes of the horizontal stresses, relative to that of the vertical geostatic stress, could change in one of three ways and (if the changes in the magnitudes of the stresses were sufficient) could cause faults to develop. The three basic ways are:

(i) both horizontal stresses decrease in magnitude, (but not by the same amount),
(ii) both horizontal stresses increase in magnitude, (but not by the same amount), and
(iii) one horizontal stress increases in magnitude while the other horizontal principal stress decreases in magnitude.

These variations in horizontal stress will result in the triaxial stress states represented in Fig. 5.6(*a*) which, if the differential stress is sufficiently large, will give rise to the three main types of faults (normal, reverse and strike-slip) represented in Fig. 5.6(*b*).

Anderson did not specify the magnitude of the stress changes required to initiate these three types of faults. However, Sibson (1973) estimated the relative magnitudes of the differential stress required to cause slip on these three sets of faults subsequent to their initiation. In his analysis, Sibson assumed that the axis of intermediate principal effective stress (σ_2) acted parallel to the plane of the fault and that the axis of maximum principal effective stress (σ_1) made an angle (θ) with the fault plane of:

$$\theta = 45° - \frac{\phi}{2} \qquad (5.1)$$

where $\phi = \tan^{-1}\mu$ and μ is the coefficient of sliding friction. He further assumed that sliding on the fault plane was governed by the frictional sliding criterion of Amonton's Law, namely:

$$\tau = \mu\sigma_n. \qquad (5.2)$$

This equation can be expressed in terms of principal stresses, so that, provided the axis of maximum principal stress makes an angle of $\theta = 45° - \phi/2$ with the fault plane, sliding will take place when:

$$\sigma_1 = k\sigma_3 \qquad (5.3)$$

where, as we saw in Chap. 1, (Eq. (1.60*a*)),

$$k = \frac{1 + \sin\phi}{1 - \sin\phi}.$$

By subtracting σ_3 from both sides of Eq. (5.3), we obtain:

$$(\sigma_1 - \sigma_3) = (k - 1)\sigma_3. \qquad (5.4)$$

It will be noted that Eqs. (5.2) to (5.4) are expressed in terms of effective stresses. That is, the effects of fluid pressure are automatically taken into account.

In a thrust regime, as defined by Anderson, the effective gravitational loading (σ_z) is such that $\sigma_z = \sigma_3$. Therefore, the conditions represented in Eq. (5.4) represent the slip conditions that must be met for frictional sliding on an Andersonian Thrust, i.e.:

$$(\sigma_1 - \sigma_3) = (k - 1)\sigma_z. \qquad (5.4(a))$$

In a normal fault regime σ_1 acts vertically and is therefore equal to σ_z. Hence from Eqs. (5.3) and (5.4) we may write:

$$(\sigma_1 - \sigma_3) = (k - 1)\frac{\sigma_1}{k} \qquad (5.5)$$

or

$$(\sigma_1 - \sigma_3) = (k - 1)\frac{\sigma_z}{k}. \qquad (5.5(a))$$

With regard to strike-slip fault regimes, the vertical stress (σ_z) may fall anywhere in the range between σ_1 and σ_3, so that the limiting conditions for sliding on strike-slip faults are those for normal and thrust faults respectively. Sibson took an arbitrary situation which he claimed, without justification, was 'representative', in which $\sigma_z = \sigma_2 = (\sigma_z + \sigma_3)/2$. For this stress condition it can be shown that, for frictional sliding on strike-slip faults:

$$(\sigma_1 - \sigma_3) = \frac{2(k - 1)}{(k + 1)}\sigma_z \qquad (5.6)$$

From Eqs. (5.4(a)), (5.5) and (5.6), it follows that, for a given depth (z), the ratios of the differential stresses required for frictional sliding on thrusts, strike-slip and normal faults are, respectively:

$$(k - 1):2(k - 1)/(k + 1):(k - 1)/k.$$

If $k = 4.0$ (a figure which obtains if the coefficient of friction (μ) is 0.75, and is the one assumed by Sibson) then the ratios of the magnitude of the differential stresses necessary for sliding on the three types of faults are 4:1.6:1.

Sibson related these ratios to the *strain-energy* of the different systems. In uniaxial compression (σ_1), the strain-energy (W) of a linear, elastic body, is:

$$W = \frac{\sigma_1 e}{2} \qquad (5.6(a))$$

where e is the elastic strain. The strain energy is therefore equal to the area beneath the stress–strain curve. Using Eq. (1.35) (i.e. $e = S/E$) we can rewrite Eq. (5.6(a)) as:

$$W = \frac{\sigma_1^2}{2E}. \qquad (5.6(b))$$

If the elastic body is subjected to biaxial stress it can be shown that:

$$W = \frac{(\sigma_1 - \sigma_3)^2}{2E}. \qquad (5.6(c))$$

Thus, the distortional elastic strain energy is related to the square of the differential stress: so the corresponding ratios for the energy for sliding on the three groups of faults are:

16 : 2.56 : 1.

As Sibson points out, these ratios of stress and strain energy are in good agreement with those which can be inferred from geophysical data and geological observations. Thus, more than 90 per cent of seismic energy is released from areas of compressive plate interaction, where thrust and strike-slip motions on faults dominate, whereas only 6 per cent of seismic energy is derived from oceanic rifting, where normal faulting will dominate. Sibson also notes that the extensive development of mylonites with thrusts and their rare association with normal faults is consistent with there being higher differential stresses during the development of thrusts than during the development of normal faults.

Once again it is emphasised that these conclusions are based on the assumption that the axis of maximum principal stress acts at the optimum angle of $\theta = 45° - \phi/2$ to the fault plane and that the intermediate principal stress acts parallel to the plane of the fault. As we shall see in Chap. 8, once a fault has been initiated the relative magnitudes of the principal stresses may be so changed that the erstwhile least principal stress becomes the intermediate principal stress. Moreover, the angle which the maximum principal stress makes with the fault plane may be other than the optimum value defined by Eq. (5.1). Nevertheless, in general terms, Sibson's analysis provides important quantification regarding the relative magnitudes of differential stresses which are associated with movement on the various Andersonian Faults.

Seismic faulting and associated events

Shearing may take place along fault planes or fault zones by slow, ductile processes. Such slow modes of deformation do not give rise to major seismic events. On the contrary, seismicity is associated with abrupt sliding on a fault plane which is completed in a time-scale which ranges from a few milliseconds to a few tens of seconds. The shorter durations of motion are related to small displacements of a millimetre or less and the longer durations to increments of movement of more than a metre.

Whether or not a particular rock type fails in a brittle or ductile manner depends upon the deformational environment (Chap. 1). Thus, Rutter (1974) showed (Fig. 5.7) that, for a given strain-rate, ductile deformation is favoured by high effective confining pressures and high temperatures. Indeed, as may be inferred from Fig. 5.7, brittle or seismic faulting can

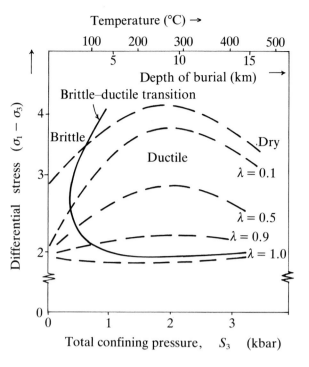

Fig. 5.7. Brittle–ductile transition for limestone. (After Rutter, 1974.)

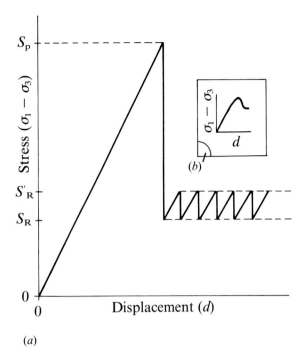

Fig. 5.8. (*a*) Brittle failure stress–displacement curve showing the 'residual strength' S_R of the shear plane that exists after fracture initiation. (*b*) Stress–strain curve obtained in a 'stiff' machine.

only occur at depths of more than about 5 km if the fluid pressure (p) approaches the magnitude of the total confining pressure (S_3) i.e. $\lambda_e = p/S_3 = 1.0$.

It is emphasised that the reader must distinguish between λ and λ_e, where $\lambda = p/S_z$ and $\lambda_e = p/S_3$. Only when the seismically active fault in question happens to be a thrust, so that S_3 will be approximately vertical and so will be close to S_z (the vertically acting *geostatic pressure*) will λ be equal to λ_e. For other types (i.e. normal and strike-slip faults) where $S_3 < S_z$, brittle failure will occur at values of $\lambda_e < \lambda$.

The idealised stress/displacement relationships for initial brittle failure and subsequent shear sliding on a fracture plane is represented in Fig. 5.8(*a*). Initial failure occurs at the peak stress (S_p) which the rock can sustain. This is followed by a sharp decrease in the ability of the rock to sustain a differential stress. Originally, it was thought that subsequent to the peak stress, failure was catastrophic with an instantaneous stress drop, but with modern stiff testing machines the stress decay can be followed and has been shown to be of the form represented in Fig. 5.8(*b*). If the rock mass is not subject to a confining pressure ($S_3 = 0$), the differential stress which the failure plane is able to support is exceedingly small. However, if a confining pressure exists such that there is a positive effective stress acting normal to the fracture plane, the rock mass is able to sustain a differential stress, the magnitude of which is dictated by the frictional resistance to sliding on the fracture plane. This differential stress (S_R in Fig. 5.8(*a*)) is termed the *Residual strength*.

Subsequent displacements on the fracture are not accomplished at a constant differential stress; but rather take place in a succession of jerks (Fig. 5.8(*a*)) in which the differential stress increases to a level S'_R, at which an increment of slip takes place and the differential stress once again falls to the magnitude of the residual strength (S_R). This *saw-tooth* stress/displacement relationship has been attributed to a *stick–slip mechanism*. It has been suggested that the peak sliding stress $S_{R'}$ is related to *static friction* while the lower level of stress S_R has been attributed to a lower value of *sliding friction*. It is not absolutely clear why there should be such a variability in the coefficient of friction. It has been suggested that it is possible that the difference can be related to the inertia of the system. However, it is more probable that the initial high resistance to sliding in rock mechanics experiments can be attributed to the resistance to shear displacement of minor imperfections, or *asperities*, on the plane of sliding, coupled with a small amount of 'spot welding' of the rock across this plane. In natural conditions, increments of fault movement may be separated by many tens of years, or even centuries. In the intervening periods, the fault plane can reasonably be expected to develop a small degree of cohesion as the result of percolating fluids, coupled perhaps with pressure solution. The change in stress in the system brought about by re-shear (i.e. reactivation of the fault) is, of course, what causes seismic activity. The magnitude of this stress drop, in nature, is usually in the range of 10–100 bars (Chinnery, 1963). As may be inferred from Fig. 5.8, the stress-drop associated with fracture initiation will be very much larger. However, faults are initiated only once and may re-shear many hundreds of times. Hence, it is the re-shear stress drop which is likely to be the one which is encountered or associated with seismic records.

Only a part of the energy of the system which is dissipated by fault movement takes the form of seismic elastic waves. A significant proportion of the work done in moving the rock mass along the fault plane results in the generation of heat. Indeed, it is the heat so generated that may give rise to the possible 'spot-welding', referred to above, regarding rock mechanics experiments, and which will be instrumental in natural faults developing a cohesive strength. Let us now consider this phenomenon.

Frictional heating

The problem of frictional heating and the quantity of heat which is generated by seismic faulting has been commented upon by several authors including Jeffreys (1942), Anderson (1951) and Price (1970). However, the most detailed and definitive studies are those carried out by Sibson (1973, 1975 and 1977).

Mechanical work is done during fault movement; and this work gives rise to the almost instantaneous generation of a specific quantity of heat (q) which, in turn, causes a rise in temperature ($\theta°$) of the rock mass adjacent to the fault zone. In some instances, the quantity of heat generated is sufficient to cause some of the rock mass to melt. This molten material cools quickly to form a glassy rock known as *pseudotachylite*.

The work done (W) per unit area of fault plane in overcoming frictional sliding resistance is:

$$W = d\tau \tag{5.7}$$

where d is the displacement and τ is the shear stress acting on the plane of sliding. The quantity of heat (q) generated by the work done is:

$$q = \frac{W}{J} = \frac{d\tau}{J} \tag{5.8}$$

where J is the mechanical equivalent of heat. (The authors are aware that this form of unit is now outmoded, but because these units still exist in the literature and J is used there, it is convenient to quote the old form here and later in the text). It is emphasised that d in these equations relates only to a single increment of movement, rather than the total displacement (s) on the glide plane.

If the heat, which is generated during the brief period of fault movement cannot be rapidly dispersed (and this must be the general case) the rock mass adjacent to the fracture plane will experience a rise in temperature ($\theta°$) such that:

$$q = MC\theta° \tag{5.9}$$

where M is the mass of the rock heated and C is its specific heat.

Price (1970) noted that pseudotachylites are rare rocks and do not occur on all faults. Indeed, he points out that most shear fractures do not even exhibit *polished* surfaces, which result from the melting of minute quantities of rock adjacent to the fracture. From this observation and the use of Eqs. (5.7)–(5.9) he concluded that the differential stress acting on most fractures undergoing re-shear will, in general, be less than 400–500 bars (Eq. (1.15), Chap. 1). This conclusion, however, does not apply to the stress conditions at the initiation of a shear fracture, (S_p, Fig. 5.8(a)) nor is it applicable to conditions in dry rock, where the differential stress required to cause shear and re-shear may be very large.

Sibson (1973, 1975 and 1977) who studied the Outer Isles Thrust in N.W. Scotland found that the shear dislocations which disrupted the amphibolites and granulite facies rocks of the Lewisian complex were, in some localities, frequently associated with layers and/or injections of pseudotachylite. He noted that many faults are not true planar features, but that shear takes place within a zone of finite thickness. The resistance to shear (τ) and movement is assumed to be evenly distributed throughout a fault zone of width w. Then, assuming no heat loss, it follows from Eqs. (5.8) and (5.9) that the temperature rise ($\theta°$) is given by:

$$\theta° = \frac{\tau d}{\rho J w C} \tag{5.10}$$

where d is the displacement, ρ is the rock density and C, the specific heat. Sibson took ρ and C as 3.0 g cm^{-3} and 10^7 erg g^{-1}°C^{-1} respectively.

In order to ascertain the conditions under which rock may melt as the result of seismic motion on a shear plane or zone, it is necessary to quantify the parameters in Eq. (5.10) and also to make assumptions regarding the environment which obtained when shearing took place, so that an estimate may be made regarding the temperature rise required to bring about melting of the rock. Sibson took the depth in the crust at which shearing took place to be 7 km and so estimated that the increase in rock temperature necessary to cause melting was about 800 °C. Coupling this conclusion with the assumption that there was a 'jerk' displacement of 1.0 m, he indicated that the width (W) of the shear zone which would melt at shear resistances of 100, 500 and 1000 bars would be 0.4, 1.9 and 3.8 cm respectively. It must be emphasised that these estimates relate only to dry rock. The mainly quartzo-feldspathic rocks studies by Sibson had experienced retrograde metamorphism and were almost certainly dry at the time of faulting, so that the application of the concepts outlined above are valid. Sibson (1975) later estimated the likely range of differential stresses which could have given rise to the

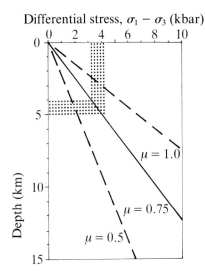

Fig. 5.9. Values of differential stress for given depths and coefficients of sliding friction (μ) that may be inferred from the generation of pseudotachylites. (After Sibson, 1975.)

shear-induced pseudotachylite in these rocks. The important stress parameter in Eq. (5.10) is that of the shear stress (τ). This is related to the normal stress (S_n) by:

$$\tau = \mu S_n$$

where μ is the coefficient of sliding friction. Hence, the differential stress which may obtain in such conditions will be determined by the value of μ. As may be seen in Fig. 5.9, at a depth of 4–5 km in the crust, the differential stress necessary to give rise to the generation of pseudotachylites in dry rock, with the value of $\mu = 0.75$ is 3.3–4.0 kbar. The reader may infer that for a value of $\mu = 0.5$ the differential stress may be as low as 2.0 kbar and for a value of $\mu = 1.0$ the differential stress acting at a depth of 5 km may be as high as 7.0 kbar. The compilation of data relating to rock friction indicates that the probable upper limit of differential stress is about 4.0 kbar. Moreover, it must be emphasised that such a differential stress is only likely to exist in 'cold and dry' rocks in a near-surface environment.

As Sibson points out, when the rocks contain water the analysis presented above needs to be modified. In particular, if water is present on the fault plane and the adjacent rocks are so impermeable that this water cannot quickly and easily migrate, then an increase in temperature ($\theta°$C) will cause a transient increase in the fluid pressure (dP). If this pressure change takes place at an approximately constant volume then:

$$\frac{dP}{d\theta°} \approx 1.8 \times 10^7 \text{ dynes cm}^{-2} \text{ s}^{-1}. \qquad (5.11)$$

This equation, which is derived from the steam tables, applies over a temperature range of several hundred

degrees, for water initially at a temperature of 200 °C and a pressure of 700 bar. If it is assumed that we are dealing with slip on a horizontal shear plane then the value of λ can be written as:

$$\lambda = \frac{(P + dP)}{S_z} = \lambda + \frac{dP}{S_z}. \qquad (5.12)$$

As soon as $\lambda = 1.0$, the effective normal stress on the shear plane falls to zero, so that further heat generation by frictional sliding is prevented. If, as Sibson assumes, at a depth of 7 km, the value of λ_0 is 0.3, then it follows from Eqs. (5.11) and (5.12) that sliding on the fracture plane will need to raise the temperature by only 80 °C before $\lambda = 1.0$. Sibson also considers the more general case in which dilation of the rock mass adjacent to the fault plane takes place. However, because the degree of dilation is not likely to be known for any specific instance, we shall not consider this aspect further.

The generation of high fluid pressures on the fault plane as the result of quite modest temperature rises, caused by initial frictional resistance to shear movement, means that (as p approaches S_3, so that $\sigma_3 = 0$) subsequent resistance to movement is so reduced that it is to be expected that the stress drop associated with such fault movement will be close to the differential stress that caused the initiation of that movement, whether it be for fracture initiation or re-shear. This is a point which has significance in the discussion presented in Chap. 6, relating to the development of second order fractures. If the rock mass adjacent to the shear plane is porous or is well fractured so that the fluid pressure generated by frictional sliding results in fluid migration away from the heat-producing shear plane, the problem becomes intractable. That such fluid migrations take place may be inferred from the distribution of quartz veins adjacent to some thrusts (personal communication C. Fabry, Chap. 7).

Sibson (1977) argues that heat will also be generated if a fault fails in a ductile mode. Indeed, he shows that such heating may, for a significant geological period, influence the distribution of isotherms about major faults, such as the Outer Isles Thrust, which exhibit both ductile as well as brittle deformation. The heat generated by such movement may be sufficient to cause local inversion of the isotherms and so give rise to a form of 'contact metamorphism'. Such an effect occurring below a thrust in the Haute-Savoie Alps, France, has been reported by Aprahamian & Pairis (1981).

Inclined principal stresses

As we have seen, the Andersonian classification of faults is concerned with the initiation of idealised, perfectly planar structures that develop in those geo-

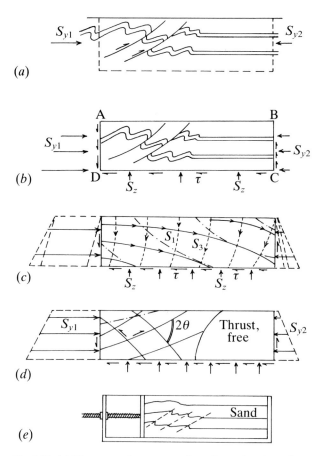

Fig. 5.10. (*a*) Diagrammatic representation of a section through a fold belt and foreland. (*b*) Qualitative boundary conditions which, it can be inferred, will give rise to the structures shown in (*a*). (*c*) Semi-quantitative and analytical solution, using stress functions, of the boundary conditions shown in (*b*). (*d*) Orientation and distribution of conjugate shears and zones of fractured and unfractured rock obtained by interpreting (*c*). (*e*) Representation of sandbox experiment simulating (*a*). (After Hubbert, 1951.)

logical environments in which one of the axes of principal stress acts vertically and the others act in the horizontal plane. However, one may infer from the geometry of geological structures in many areas of the world that two, or perhaps all three, axes of principal stress are inclined to the vertical and horizontal. Such structures can be seen in traverses from an orogenic fold belt into the relatively undeformed *foreland*, Fig. 5.10(*a*), which represents, in simplified form, the style and disposition of structures which may be encountered, for example, in a SE–NW traverse from the folds and thrusts of the Moines in the Scottish Highlands to the mainly undeformed Torridonian sediments of the foreland. Similar situations are encountered in the Canadian Rockies (e.g. in a profile from Banff to Calgary), in the Eastern Alps (e.g. in a traverse north from the Zillertal in Austria to the Molasse of Southern Bavaria) and in many other areas of the world. It will be noted that the folds represented in Fig. 5.10(*a*) are symmetrical and inclined and that they may be cut by fault planes which are curved (i.e. *listric*). From the fact that the

folds are asymmetrical and their axial planes are inclined, one may reasonably infer that the axis of maximum principal stress that obtained when these structures were initiated, and which existed when the folds were fully evolved, was inclined to the horizontal.

Because the degree of deformation represented in Fig. 5.10(*a*) grades from high to low, as a traverse is made from left to right, it was at one time customary to suggest that such deformation resulted from a 'one-sided' or an 'active' push from the left. Such a statement is, from the mechanistic viewpoint, completely unacceptable. A body subjected to a one-sided push (*F*) must, in accord with Newton's Second Law, result in a mass (*M*) undergoing an acceleration (*a*), so that:

$$F = Ma. \tag{5.13}$$

However, the rates of deformation of rock mass of the dimensions represented in Fig. 5.10(*a*) are, when averaged out, so exceedingly slow that orogenic deformation can be taken to develop in *static equilibrium*. There are, of course, exceptions to this statement which relate to meteoritic impacts, catastrophic landslides and the considerable ground accelerations associated with seismic faulting. Nevertheless, in general, it is to be expected that the rock mass represented in Fig. 5.10(*a*) was, at any one instant, in over-all equilibrium.

One can simplify the system represented in Fig. 5.10(*a*) as being delimited by arbitrarily selected boundaries subjected to *boundary stresses*, which are so chosen that they do not disturb the internal stress distribution. For example, we may choose the rectangle ABCD (Fig. 5.10(*b*)) as representing the boundary of the two dimensional problem to be studied. Because the problem is a static one, the boundary stresses must be in complete equilibrium. Thus, the body-weight of the material within the boundaries must be supported by a vertical stress (S_z) which acts on the lower, horizontal boundary BC (Fig. 5.10(*b*)). Because the rock mass in the left part of the enclosed unit is more intensely deformed than that to the right, it must be inferred that S_{y1} is greater than S_{y2}. Such a stress distribution is not in equilibrium, for the whole of the rock mass within the boundaries would move and accelerate to the right. In order to bring the system into equilibrium, we can postulate a shear stress (τ) which acts, as shown in Fig. 5.10(*b*), along the basal boundary so as to balance the inequilibrium between the boundary horizontal stresses S_{y1} and S_{y2}. In turn, this shear stress introduces a turning moment on the block so that it will attempt to rotate in a clock-wise direction. In order to counter such a turning motion, it is necessary to postulate shear stresses acting on the vertical boundary faces (Fig. 5.10(*b*)). The shear stress that can develop at the

Earth's surface is only that which can be generated by wind friction, and so can be equated to zero. Consequently, the shear stresses acting on the vertical boundaries must vary in intensity from zero at the upper limit of the vertical boundaries to that of the horizontal shear stress at the base of the model.

The boundary stresses so derived can, of course, only be considered in qualitative terms. Nevertheless, it is clear that the axes of principal stress can only be vertical and horizontal near the upper surface of the rock mass. As may be inferred from Fig. 5.10(c), the principal stresses elsewhere within the bounded mass will be inclined. However, to enable the precise orientation of the axes of principal stress to be ascertained, it is necessary to have recourse to a quantitative analysis. The method most usually adopted in such analyses is based upon the use of the *Airy Stress Function*.

As we have already noted, the geological problem in hand is essentially a static one, so that the boundary stresses must be in equilibrium. Let these boundary stresses be represented by a function ψ; then it can be shown that the conditions of static equilibrium are automatically satisfied if:

$$S_y = \frac{\partial^2 \Psi}{\partial z^2}, \; S_z = \frac{\partial^2 \Psi}{\partial y^2}, \; \tau = -\frac{\partial^2 \Psi}{\partial^2 \partial y} \quad (5.14)$$

when body forces are unimportant. However, in general, when gravity represents the only body-force, the S_z term in Eq. (5.14) is replaced by:

$$S_z = \frac{\partial^2 \Psi}{\partial y^2} + \rho g z \quad (5.14(a))$$

where ρ is the mean density of the rock, z is the depth and g the gravitational acceleration.

The stresses and strains induced in the deforming body are related by elastic constants and are, therefore, not independent of each other. Indeed, they must be connected by what is termed the compatibility relationship, which ensures that the material maintains its continuity during deformation and does not develop 'holes', or discontinuities. This compatibility relationship (Jaeger & Cook, 1969) can be written as:

$$\left(\frac{\partial^2}{\partial y^2} + \frac{\partial^2}{\partial z^2} \right)(S_y + S_z) = 0. \quad (5.15)$$

Both the equilibrium and the compatibility equations must simultaneously be satisfied before a solution of the static problem under consideration can be achieved. If Eqs. (5.14) and (5.15) are combined, a single equation, known as the *biharmonic equation*, results:

$$\frac{\partial^4 \Psi}{\partial y^4} + 2 \frac{\partial^4 \Psi}{\partial y^2 \partial_z^2} + \frac{\partial^4 \Psi}{\partial z^4} = 0 \quad (5.16)$$

Solutions to this equation automatically satisfy

Eqs. (5.14) and (5.15) and which are, therefore, valid solutions to the static problem under consideration. These solutions (known as Airy Stress Functions) will, of course, provide boundary conditions which are apposite to the problem. It will be noted that the stress function Ψ is a solution of a fourth order differential equation, so that it will contain four independent constants of integration. However, any stress system must satisfy the boundary conditions at the Earth's surface (i.e. $S_z = 0$ and $\tau = 0$ when $z = 0$), so that two of the constants must be used for this purpose. The remaining two constants, however, can be chosen to modify the boundary stresses. In any realistic geological situation, it is probable that more than two constants will be needed to define the total boundary conditions. This apparent difficulty is readily overcome for, from the *principle of superposition* (an important aspect of elasticity theory) it follows that if there are two stress functions Ψ_1 and Ψ_2, each of which represents a valid solution to the biharmonic equation (Eq. (5.16)), then the sum of the two functions (Ψ') is also a valid solution, so that:

$$\Psi' = \Psi_1 + \Psi_2. \quad (5.17)$$

Let us follow the analysis by Hafner (1951) to show how stress functions can be used to represent the states of stress that give rise to the Andersonian classification of faults. To do this we shall represent the *Standard State* of stress assumed by Anderson by the stress function Ψ_1 and superimpose upon this a second function Ψ_2 which represents what Anderson termed the *supplementary stresses*. To derive the Standard State, Hafner selected a stress function of the form of a third order polynomial:

$$\Psi_1 = a_1 y^3 + a_2 y^2 z + a_3 y z^2 + a_4 z^3 \quad (5.18)$$

which satisfies the biharmonic equation (Eq. (5.16)). The stress components are derived from Eqs. (5.14), (5.14(a)) and (5.18) and are given by:

$$S_y = \frac{\partial^2 \Psi_1}{\partial z^2} = 2a_3 y + 6a_4 z \quad (5.19(a))$$

$$S_z = \frac{\partial^2 \Psi_1}{\partial y^2} + \rho g z$$

$$= 6a_1 y + 2a_2 z + \rho g z \quad (5.19(b))$$

$$\tau = \frac{\partial^2 \Psi_1}{\partial_y \partial z} = (2a_2 y + 2a_3 z). \quad (5.19(c))$$

At the Earth's surface $S_z = 0$, and $\tau = 0$. It follows, therefore, from Eqs. (5.19(b), (c)) that two of the constants, a_1 and a_2 are zero. If a_3 is also set at zero and a_4 put at $\rho g/6$, then it follows from (Eqs. (5.19(a), (b)) that the stress conditions for the standard state are then given by:

$$S_y = S_z = \rho g z.$$

Hafner demonstrated how the supplementary

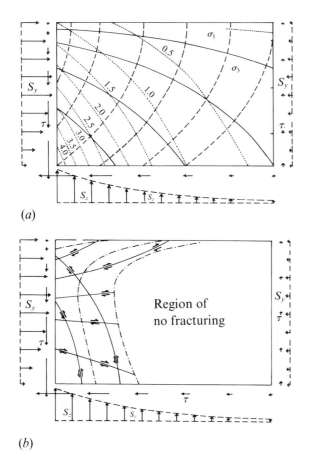

(a)

(b)

Fig. 5.11. Examples of (a) orthogonal stress trajectories and (b) the orientation of possible shear fractures, obtained by using different and more complex boundary conditions than those represented in Fig. 5.10. The dotted lines in (a) are contours of differential stress. (After Hafner, 1951.)

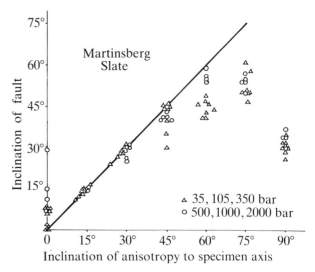

Fig. 5.12. Relationships between the angle of shear fracture (fault) and orientation of planes of weakness (anisotropy) in Martinsburg Slate. (After Donath, 1961.)

stresses could be taken into account. He also considered more complex boundary conditions. Once these boundary conditions have been established, the magnitude of the principal stresses at any point within the space delimited by the boundaries can be quantified. Moreover, the orientation of the axes of principal stress can also be calculated for any point within the block by using the expression:

$$\tan 2\alpha = \frac{2\tau}{S_y - S_z} \tag{5.20}$$

where α is the angle between the S_1 stress trajectory and the horizontal. The *stress trajectory* pattern (i.e. orthogonal lines to which, at any one point, the axes of principal stress, S_1 and S_3 are tangential) for the problem under consideration is shown in Fig. 5.10(d), and, for different boundary conditions, is represented in Fig. 5.11. Contours of differential stress are superposed onto the stress trajectory pattern (e.g. Fig. 5.11(a)) and it is clear from the values on these contours that the differential stress in certain portions of the block is small and insufficient to cause finite deformation. In those areas of the block where the

differential stresses are large enough to cause the generation of faults, the probable orientations of the fractures which may be induced can be inferred by selecting the appropriate value of θ (e.g. 30°) which is the angle between S_1 and a plane of shear failure, Fig. 5.10(d) and 5.11(b).

As has been pointed out by Odè (1960), once such fractures have been initiated, we are no longer dealing with a homogeneous and isotropic material, so that the assumptions, on which the application of stress functions are based, are no longer valid. Indeed, the geologist should always query whether the rock mass to be modelled, using this technique, can in fact be represented as a homogeneous and isotropic material. Even if the unit can be so modelled, only infinitesimal strains may be induced in the model before failure, and hence breakdown, both of the 'rock mass' and the 'model' takes place. It is for these reasons that this technique has been so infrequently applied to geological problems. One of the problems tackled using this technique has already been noted in Chap. 3, and relates to the emplacement of the Spanish Peaks dyke system.

Shear failure on existing planes of weakness

Shear failure and frictional sliding on a pre-existing single plane of weakness or along a plane of fabric such as exists in anisotropic rocks, e.g. slates, is not considered in the Andersonian scheme. This problem falls into two cases. The *special case* in which one of the axes of principal stress acts parallel to the weakness plane or fabric and the *general case* in which all three axes of principal stress are oblique to these planes. We shall deal with these two cases, in turn, in the following paragraphs.

The influence which the orientation of planes of weakness, relative to the axis of maximum principal stress, has upon the orientation of the plane along which shear actually occurs has been demonstrated by Donath (1961) in a classic series of rock mechanics experiments conducted upon specimens of Martinsburg Slate. In these experiments, cylindrical specimens were prepared with the cleavage making an angle (β) which ranged from 0° to 90°, in steps of 15°, to the long axis of the cylinder. The angle at which shear failure occured in these various specimens, when compressed along the axis, and the differential stress required to cause failure, are shown in Fig. 5.12. The confining stress in these experiments was axisymmetric, so that we can take these experiments to represent the special case in which the least and intermediate principal stresses were equal; and one of the axes was aligned parallel to the plane of weakness. Accordingly, we shall now discuss the significance of these results.

Special case

In this section we shall assume that one of the axes of principal stress acts parallel to the plane of weakness and so plays no part in the process of slip. It is usual to assume that this axis is that of the intermediate principal stress (σ_2). However, *provided there is a sufficiently large difference between the various principal stresses, any of the axes may be assumed to act parallel to the plane of weakness*. With this proviso in mind, we shall now assume that slip takes place under the action of the maximum and least principal stresses (σ_1 and σ_3).

The minimum differential stress necessary to induce shear displacement on a plane which exhibits no cohesion occurs when the axis of maximum principal stress (σ_1) acts at an angle (θ) to the weakness plane, such that:

$$\theta = 45° - \frac{\phi}{2} \qquad (1.61)$$

where, as before, ϕ is the angle of sliding friction.

If the failure condition is expressed in terms of principal stresses, then (it will be recalled from Eq. (5.3)) failure will take place for this very specific condition (i.e. $\sigma_0 = 0$) when:

$$\sigma_1 = k\sigma_3 \qquad (1.57)$$

where, $k = (1 + \sin \phi)/(1 - \sin \phi)$. For the value of $\phi = 30°$, $k = 3.0$. This specific angle of θ is, of course, the angle which the axis of maximum principal stress makes with the shear plane which is about to be initiated in a homogeneous and isotropic rock mass (Fig. 5.13(a)). Once failure has occurred, or when dealing with materials which possess one or more planes of weakness, there is no reason why θ should be restricted to this one very special value.

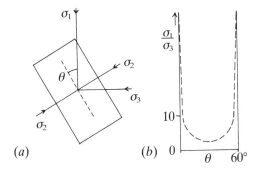

(a)   (b)

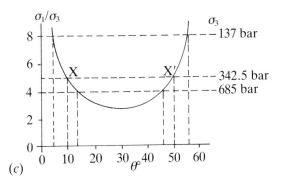

(c)

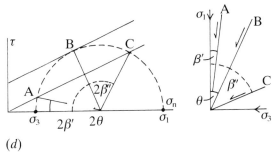

(d)

Fig. 5.13. (a) Showing existing fracture plane containing the axis of intermediate principal stress (σ_2). The axis of σ makes an angle θ with the plane. (b) Relationship between the ratio of σ_1/σ_3 and the angle θ that holds when slip on the fracture plane may occur. (c) For conditions in which $\sigma_3 \neq 0$, very high ratios of σ_1/σ_3 are not possible. The limiting values of θ for three values of σ_3 are indicated (the unusual numerical values of $\sigma_3 = 137$, 342.5 and 685 bar occur because the original calculations were conducted using σ_3 as 2000, 5000 and 10 000 lb. in² respectively). (d) Failure conditions for specimen containing two planes of weakness (A and C) which make an angle of β' and β'' with the σ_1 axis respectively. For these conditions, fresh fractures (B) may also form, which makes an angle θ with the σ_1 axis.

As we have noted (Eq. (5.2)) failure will take place on a plane of weakness (with zero cohesion) when:

$$\tau = \mu\sigma_n. \qquad (1.58)$$

This relationship can be expressed in terms of principal stresses by writing τ and σ_n in terms of the principal stresses (Eqs. (1.15) and (1.16)), so that:

$$(\sigma_1 - \sigma_3)\sin 2\theta = \tan \phi(\sigma_1 + \sigma_3) - (\sigma_1 - \sigma_3)\cos 2\theta$$

where $\tan \phi = \mu$.

It can be shown by rearranging and simplifying this expression that:

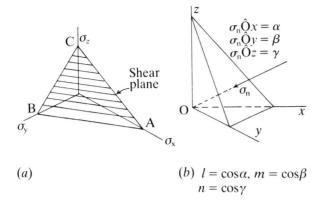

(a)

(b) $l = \cos\alpha$, $m = \cos\beta$
$n = \cos\gamma$

Fig. 5.14. (*a*) Shear plane which is oblique to all the axes of principal stress. (*b*) Cosine directions for a plane oblique to three orthogonal axes.

$$\frac{\sigma_1}{\sigma_3} = \frac{1+d}{1-d} \tag{5.21(a)}$$

where $d = (\sin 2\theta + \cos 2\theta \tan\phi)/\tan\phi$. $\tag{5.21(b)}$

If, as before, we let $\phi = 30°$, Eq. (5.21(a)) can be expressed graphically as a relationship between the ratio of the principal stresses (σ_1/σ_3) which must exist if slip is to take place on a plane of weakness, and the angle θ that the axis of maximum principal stress makes with the weakness plane (Fig. 5.13(*b*)). It will be seen that the ratios go to infinity at values of θ of $0°$ and $60°$.

The condition for failure in unbroken rock is given by:

$$\sigma_1 = \sigma_0 + k\sigma_3 \tag{1.57}$$

where σ_0 is the uniaxial strength of the rock. This failure relationship can be plotted on the graph in Fig. 5.13(*b*) for any value of σ_3. Plots for three different values of σ_3 are shown in Fig. 5.13(*c*) and define the limiting values of θ, between which slip will occur on the pre-existing plane of weakness, in preference to the generation of a new fracture plane. If the stress and angular relationships defined by points, X, X' etc. are satisfied, then the conditions are simultaneously met for the generation of a new shear plane and the concomitant shear along an existing plane of weakness, as indicated in Fig. 5.13(*d*). This is an aspect of fault movement and fault generation to which we shall return in Chap. 6, when dealing with the formation of *second order fractures*.

General case
When all three axes of principal stress act at an angle to a plane of weakness (Fig. 5.14(*a*)) it is possible to analyse the system and determine the direction of shear-sliding on that plane, *provided the orientation of the principal stresses relative to the plane and their magnitudes is known.* The inverse problem, with which

the analyst of field data will be faced, namely that of deducing the orientation and relative magnitudes of the principal stresses from the slip direction on a plane (or planes) of known orientation, is extremely difficult and may even be impossible. We shall first consider the theoretical problem, from which one may infer the scope of the problem facing the geologist who wishes to interpret such field data.

In a theoretical analysis, where the magnitudes and orientations of the principal stresses are known, then whether movement will take place on a specific plane of weakness of given orientation and properties (i.e. degree of cohesion of the plane and its coefficient of friction), and also the direction of slip, can be ascertained by analytical or graphical methods. Both methods require the use of the *direction cosines* (*l*, *m* and *n*) which are defined in Fig. 5.14(*b*). If one considers a plane ABC which cuts all three axes of principal stress, as shown in Fig. 5.14(*a*), where, for the sake of simplicity σ_1, σ_2 and σ_3 are allocated the *x*, *y* and *z* directions respectively, it can be shown that the normal stress (σ_n) on the plane is given by:

$$\sigma_n = l^2\sigma_1 + m^2\sigma_2 + n^2\sigma_3. \tag{5.22}$$

It can also be demonstrated that:

$$\tau_{max} = \frac{(\sigma_1 - \sigma_2)^2}{4} + n^2(\sigma_1 - \sigma_3)(\sigma_2 - 3\sigma_3)$$
$$- \left[\sigma_n - \left(\frac{\sigma_1 + \sigma_3}{2}\right)\right]. \tag{5.23}$$

Furthermore, Bott (1959) has shown that the *pitch* (θ) of the direction of action of τ_{max} is given by:

$$\tan\theta = n/lm[m^2 - (1 - n^2)$$
$$\times (\sigma_z - \sigma_x)/(\sigma_y - \sigma_x)]. \tag{5.24}$$

As it has been assumed that σ_1, σ_2 and σ_3 coincide with the *x*, *y* and *z* axes respectively, Eq. (5.24) can be rewritten as:

$$\tan\theta = \frac{n}{lm}\left[m^2 - (1 - n^2)\frac{(\sigma_3 - \sigma_1)}{(\sigma_2 - \sigma_1)}\right]. \tag{5.25}$$

In an analysis where σ_1, σ_2, σ_3, *l*, *m* and *n* are all known, one can determine the values of σ_n, τ_{max} and θ, and using the relationships:

$$\tau_{max} = \mu\sigma_n \tag{5.26}$$

or

$$\tau_{max} = C_0 + \mu\sigma_n \tag{5.27}$$

(Chap. 1, Eq. (1.59(*a*)), where C_0 is the cohesive strength of the rock) it can be ascertained whether slip can take place; and, if so, in which direction this slip will occur.

If a graphical solution to this problem is required, the stress state is represented as Mohrs Circles (Fig. 5.15); and it can be deduced from a given

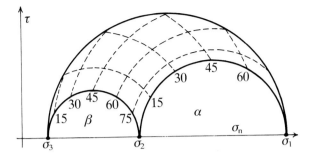

Fig. 5.15. Mohr's Circle representation of stress in three dimensions.

failure condition whether or not frictional sliding on a plane of weakness will occur. This was the method used by Jaeger & Cook (1969) to determine *domains of orientation* in which slip would occur on weakness planes, for a given orientation and magnitude of the principal stresses.

We have treated this aspect of stress analysis only in brief outline because relatively few geologists are likely to become involved in conducting such analyses: and these few will be, or will very rapidly become, familiar with the various texts (such as Love, 1892 or Jaeger & Cook, 1969) which deal with stress analysis in three dimensions.

The field geologist will be faced with an entirely different problem. Usually it is only the orientation of the failure plane and the pitch (θ) of the movement direction which can be established in the field: and even these data will only be available in rather rare situations. In general, the orientations of the principal stresses (and hence the values of l, m and n) will be unknown. Thus, only one of the variables (θ) contained in Eq. (5.25) can be ascertained from field observations made on a single failure plane. In these circumstances a solution is clearly not possible. Before this type of problem can be tackled, it is necessary to establish that there has been movement on a number of different failure planes of different orientations in what is assumed to be a uniform stress field. Alternatively, a solution is possible if there is evidence (such as fibre growth on infill material which was deposited in the fracture plane during shear movement) that permits the orientation of one or more of the principal stresses to be deduced (so that l, m and therefore n, as well as θ are known). Some estimate (i.e. inspired guess) may also be made of the ratios of two of the three principal stresses (so that the term $(\sigma_3 - \sigma_1)/(\sigma_2 - \sigma_3)$, which appears in Eq. (5.25), can be simplified). Then, by using the failure criteria (Eq. (5.26) or (5.27)) in conjunction with Eq. (5.25), a solution of the problem is theoretically possible. This difficult problem has been addressed by Carey (1979) and Angelier *et al.* (1982). Their basic hypothesis is that, although the fault planes studied may be of arbitrary orientation, the striae (fibre-growth) or

slickensiding accurately indicate the direction and sense of the maximum shear stress. The significance of the results obtained is therefore wholly related to the validity of this basic assumption. Only rarely will sufficient field data be available for this approach to be used with confidence.

Thus, as the reader will now appreciate, an estimate of the orientation and relative magnitudes of the principal stresses based on the field evidence of natural shear planes, which exhibit many increments of movement, is an extremely vexing problem, requiring very special geological information, probably coupled with 'special pleading' and dubious assumptions.

Commentary

In this chapter we have dealt with some of the fundamental aspects of fault initiation and re-shear along planes of weakness. The treatment is far from encyclopaedic, but is sufficient to indicate some of the problems likely to be encountered in an analysis of natural fractures. It can be argued that probably all major shear fractures, in some localities of the shear plane, will exhibit a component of oblique-slip movement. Such movement is not considered in the Andersonian classification of faults. However, field observations which can be used to deduce movement directions on major faults are rarely available. Consequently, in the absence of such evidence, by observing or inferring the orientation of the fault plane and the *apparent slip directions*, geologists are generally obliged to fit specific faults into the Andersonian classification. It is for this reason that the three subsequent chapters of this book deal with strike-slip, normal and thrust faults respectively.

References

Anderson, E.M. (1951). *The dynamics of faulting.* Edinburgh: Oliver & Boyd.

Angelier, J., Tarantola, A. & Valette, B. (1982). Inversion of field data in fault tectonics to obtain the regional stress – I. Single phase fault populations: a new method of computing the stress tensor. *Geophys. J.R. astr. Soc.*, **69**, 607–21.

Aprahamian, J. & Pairis, J.-L. (1981). Very low grade metamorphism with a reverse gradient induced by an overthrust in Haute-Savoie (France). In *Thrust & nappe tectonics*, eds. K. McClay & N.I. Price, pp. 159–66. Geol. Soc. London Special Publ. No. 9. Blackwell Scientific Publications.

Bott, M.P.H. (1959). The mechanics of oblique-slip faulting. *Geol. Mag.*, **96**, 109–17.

Carey, E. (1979). Recherche des directions principales de contraintes associées au jeu d'une population de failles. *Rev. Geol. dyn. Geogr. phys.*, **21**, 57–66.

Chinnery, M.A. (1963). Earthquake magnitude and source parameters. *Bull. Seismol. Soc. Am.*, **59**, 1969–76.

Dennis, J.G. (1967). *International tectonic dictionary*. Am. Assoc. Petrol. Geol. Mem. 7.

Donath, F.A. (1961). Experimental study of shear failure in anisotropic rocks. *Geol. Soc. Am. Bull.*, **72**, 985–9.

Hafner, W. (1951). Stress distributions and faulting. *Geol. Soc. Am. Bull.*, **62**, 373–98.

Hubbert, M.K. (1951). Mechanical basis for certain familiar geological structures. *Geol. Soc. Am. Bull.*, **62**, 355–72.

Jaeger, J.C. & Cook, N.G.W. (1969). *Fundamentals of rock mechanics*. London: Methuen.

Jeffreys, H. (1942). On the mechanics of faulting. *Geol. Mag.*, **79**, 291–5.

Love, A.E.H. (1892). *A Treatise on the mathematical theory of Elasticity*. New York: Dover Publ.

Odè, H. (1960). Faulting as a velocity discontinuity in plastic deformation. *Geol. Soc. Am. Bull.*, **79**, 293–321.

Price, N.J. (1970). Laws of rock behaviour in the Earth's crust. In *11th Sympos. Rock Mech. Berkeley*, ed. W.H. Swinnerton, pp. 3–25.

Rutter, E.H. (1974). The influence of temperature, strain-rate and interstitial water in the experimental deformation of calcite rocks. *Tectonophysics*, **22**, 311.

Sibson, R.H. (1973). Interaction between temperature and pore-fluid pressure during earthquake faulting – a mechanism for partial or total stress relief. *Nature Phys. Sci.*, **243**, 66–8.

Sibson, R.H. (1975). Generation of pseudotachylite by ancient seismic faulting. *Geophys. J. R. Astr. Soc.*, **43**, 775–89.

(1977). Fault rocks and fault mechanisms. *J. Geol. Soc. London*, **133**, 191–213.

Van Hise, C.R. (1896). Deformation of Rocks. *J. Geol.*, **4**, 449–593.

6 Strike-slip faults

Introduction

The faults of which this group are comprised have been given a number of names, such as *tear, wrench, transcurrent* and more recently *transform* faults. The term now most generally used, however, is that of the chapter heading, namely *strike-slip* faults. The characteristics of these structures are that the fault plane, or fault zone, is usually approximately vertical and the movement on the fault is dominantly in the horizontal (i.e. strike) direction.

Strike-slip faults are considered nowadays to be important in two main geological settings. Firstly, they may occur as relatively small structures, in the upper levels of the crust, with an extent not usually exceeding a few tens of kilometres. Two examples of strike-slip faults in relatively high level crustal environments are shown in Fig. 6.1(*a*) and (*b*), which represent the structures exposed in the Jurassic rocks of the Jura Mts. (France and Switzerland) and the Palaeozoic rocks of Dyfed (Pembrokeshire) in Wales. These structures are relatively large examples of the types of strike-slip faults that may develop in rocks in the upper levels of the crust. Such fractures may range down in size to features which can be readily seen in quarry or cliff sections, small outcrops and even hand specimens. These latter examples will be discussed under the appropriate heading of Minor Structures (Chap. 18).

The second and more important group is comprised of strike-slip faults which are of regional or continental extent. Many of these major features are thought to extend through the crust and to develop at present, or ancient, Plate Boundaries. Large displacements on some strike-slip faults had long been suspected, but it was not until 1946 that Kennedy adduced evidence which indicated that the Great Glen Fault of Scotland had a *sinistral* strike-slip displacement of about 150 km (Fig. 6.2). (It was noted in Chap. 3, that the off-set of dykes indicates that there was subsequent *dextral* strike-slip, combined with normal displacement along this fault.) It is worth recalling that during a lively debate at the Geological Society of London which followed the presentation of this thesis, H.H. Read remarked that Kennedy would long be remembered as the man who gave Scotland 'a lean and hungry look'. However, even this figure of 150 km was overshadowed when, Hill & Dibblee (1953) suggested that the San Andreas Fault system

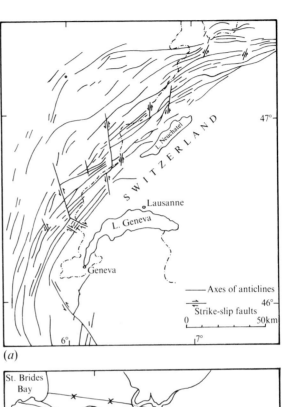

(*a*)

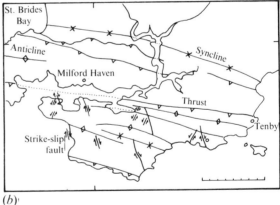

(*b*)

Fig. 6.1. (*a*) Sketch map of the Jura, showing fold trends and major, cross-cutting, strike-slip faults (after Heim, 1919). (*b*) Fold and fault trends in S. Dyfyd (Pembrokeshire) Wales. (After Anderson, 1951.)

had given rise to a horizontal displacement of about 500 km. The importance of this and other strike-slip faults in relation to the circum-Pacific environs was indicated by Allen (1962) (Fig. 6.3), in which the San Andreas, Atacama, the New Zealand-Alpine and Philippine Faults are represented on the same scale.

For many years, it was generally held that all major strike-slip faults exhibited the characteristics of the other groups of faults, namely that the maximum

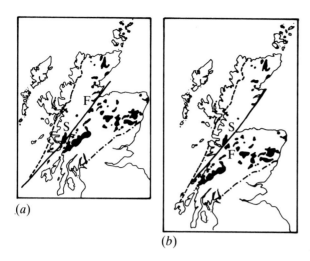

Fig. 6.2. (*a*) Geological sketch map of the Scottish Highlands, showing the present position of the Moine Injection Complex, the Strontian and Foyers granites (S and F) and the Moine Thrust-plane. (*b*) Reconstruction of the situation of the above mentioned geological features prior to the lateral displacement on the Great Glen Fault. (After Kennedy, 1946.)

displacement usually occurs in the central regions of the structure and that towards the ends of the faults the displacements gradually decrease and the main fault dies out or ends by passing into a complex array of secondary faults (e.g. the northern end of the Alpine Fault, Fig. 6.4). However, the studies related to *ocean floor spreading* have, more recently, led to the

Fig. 6.4. Array of second order faults at the northern termination of the Alpine Fault, New Zealand. (After Chinnery, 1966.)

recognition of a different type of major strike-slip fracture which has been termed a *Transform* fault. As illustrated in Fig. 6.5, transform faults are different from the classical type of faults in that the displacement on the transform fault remains approximately constant throughout its length. Transform faults are mainly associated with oceanic ridge systems. However, these submarine structures may, by crustal evolution, 'come ashore' thus entering the domain of the 'continental' structural geologist.

Movements along these major faults, whether they be transform or the more 'traditional' strike-slip structures, may give rise to a variety of secondary features, which range from sedimentary basins, tectonic sub-provinces and major folds, to many other

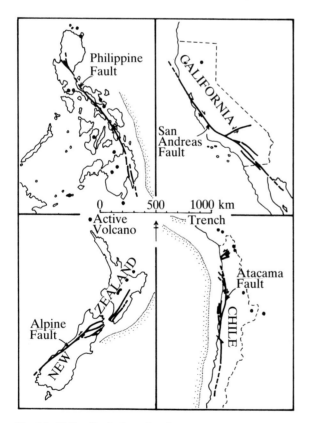

Fig. 6.3. Strike-slip faults, of regional extent, in four localities around the Pacific. (After Allen, 1962.)

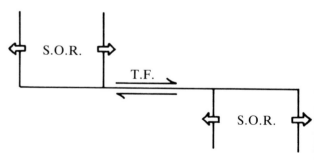

Fig. 6.5. Diagrammatic representation of a transform fault (TF) connecting two, off-set, spreading, oceanic ridges (SOR).

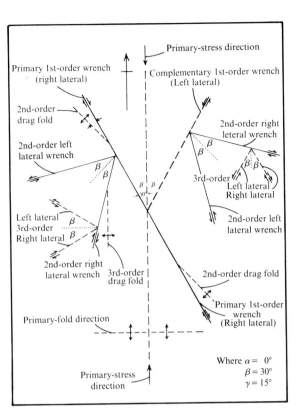

Fig. 6.6. Diagram of first-, second- and third-order faults and associated fold directions which may 'theoretically' result from a north–south primary compression. (After Moody & Hill, 1956.)

secondary fractures. Indeed, the complexity of these associated features induced Moody & Hill (1956) to categorise them (Fig. 6.6) and term the process that gave rise to them *Wrench-fault Tectonics*. These secondary features and structures, some of which develop in 'cover rocks' which overlie the 'basement' cut by the major strike-slip fault, will be discussed in the following sections. Special attention will be given to the various concepts which have been proposed to explain the range of secondary fractures associated with these relatively major *First Order* strike-slip faults.

Structures associated with strike-slip faults

The many structures associated with first order strike-slip faults depend upon the form, scale and environment in which a specific fault develops. As regards environment, large scale, first order structures may cut basement rocks which crop out at the surface. The types of secondary structures that develop depend upon their position with respect to the first order fault. For example, the orientation of *splay* faults, so commonly found towards the end of these major fractures, is shown in Fig. 6.7(*a*) and that of secondary strike-slip faults is shown in Fig. 6.7(*b*). It should be noted that the geologist can only distinguish clearly between these two forms when the sense of movement in the first and secondary fractures is known.

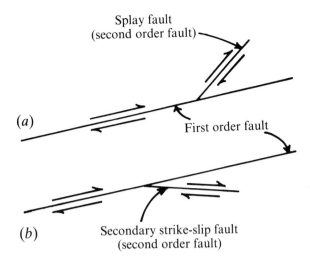

Fig. 6.7. Relationship between first order strike slip fault and (*a*) Splay fault and (*b*) 'Secondary' strike-slip fault.

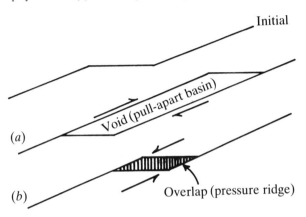

Fig. 6.8. (*a*) Potential void or (*b*) potential overlap which result from strike-slip movement on a vertical, curved or kinked fault plane.

It may be inferred from the sketch maps of Fig. 6.3, that some major strike-slip faults exhibit a curved, or even kinked, trace at the surface. Strike-slip movement on a major fault which exhibits such a non-linear form must obviously create space problems. Either a major gap may tend to develop or else two portions of opposing fault blocks try to occupy the same volume, i.e. an overlap situation develops (Fig. 6.8). Examples where both a void and an overlap have developed occur in the vicinity of the Dead Sea, (Freund, 1965). Indeed, the Dead Sea itself is considered to occupy the void space created by the strike-slip movement, while the Palmyra and other folds form in an area of potential overlap, Fig. 6.9.

The potential void space cannot be maintained to great depth in the crust for it will be filled as a result of slumping or normal faulting of the walls. Partial infill may result from igneous activity or some form of general ductile flowage of the deeper crustal wall rocks. However, the most probable manifestation of such a 'pull-apart' of the crust is the development of a *graben*. Braiding, or divergence and branching of

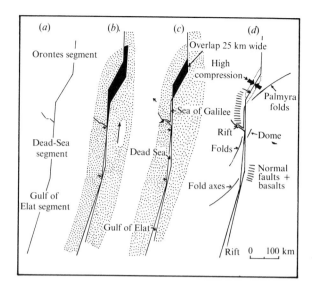

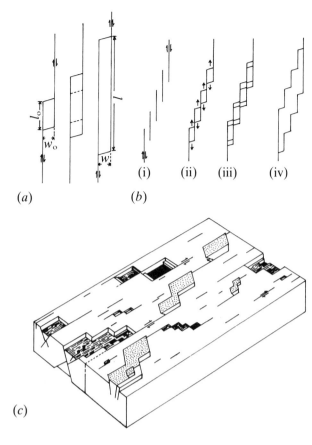

Fig. 6.9. An example in which potential void and overlap situations have given rise to the Dead Sea Graben area and folds, respectively, as the result of strike-slip movement on the Dead Sea Fault, the initial position of which is shown in (*a*). Because of the tendency to overlap, rotation takes place as indicated in (*b*) and (*c*). The secondary structures which form as a result, are shown in (*d*). (After Freund, 1965.)

strike-slip faults may result in the development of horsts, as well as grabens, Fig. 6.10. On a larger scale, it has been suggested that the gap and associated grabens that may be induced by such strike-slip movement give rise to downwarp areas which may be infilled, resulting in the creation of a significantly large sedimentary basin. However, grabens are usually associated with extension tectonics (Chap. 8).

A study of rhomb-shaped, pull-apart basins, together with the complementary rhomb-shaped horst areas which result from the over-lap situation, have been conducted by Aydin & Nur (1982). They point out that it would seem likely that the width of these basins (and also of the horsts) which are associ-

Fig. 6.11. (*a*) Simple model of a pull-apart basin. Length of basin increases (from l_0 to l) with increasing fault off-set, but the width (w_0) is constant. (*b*) The coalescence of rhomb grabens associated with en echelon strike-slip faults. The end product is a composite pull-apart basin. (*c*) Discontinuous strike-slip faults with different lengths and displacements interact to form basins and ridges of various size. (After Aydin & Nur, 1982.)

ated with strike-slip faults would be controlled by the initial fault geometry: where, as indicated in Fig. 6.11(*a*), the length of the basin or horst would be directly related to the displacement of the fault. Aydin & Nur analysed the shapes of 70 such rhomboidal structures, which ranged in length from a few tens of metres to over 80 km. Contrary to expectation, they found that the ratio of length to width of these various structures usually fell between 2 and 5:1, and that the maximum frequency value was between 3 and 4:1. From this study they concluded that smaller basins coalesce to form larger features, as slip of the major fault continues (see Fig. 6.11(*b*)). Hence, it may be inferred that the fault zone tends to increase in width in certain localities as displacement on the major structure continues. A block diagram of a broad zone of strike-slip movement, containing discontinuous strike-slip faults, of various lengths and displacements is shown in Fig. 6.11(*c*). As movement along the zone occurs, these faults interact to form basins (dashed or black) and ridges or horsts.

Where a strike-slip fault in the basement is covered by a sequence of sediments, movement along

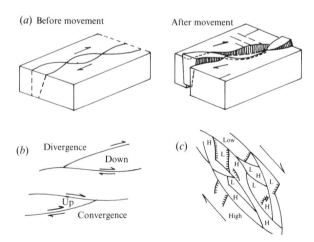

Fig. 6.10. Examples of braided strike-slip faults that produce associated extensional basins and grabens and compressional uplift blocks (horsts). (After Reading, 198o.)

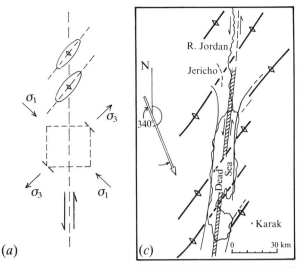

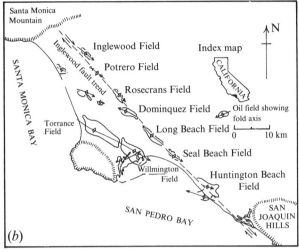

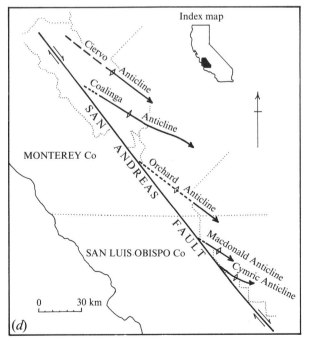

Fig. 6.12. (*a*) Shear in the basement induces compression in the cover rocks which produces en echelon folds such as are developed above (*b*) the Inglewood Fault (California) (*c*) the sinistral Dead Sea fault and (*d*) the San Andreas Fault. (After Moody & Hill, 1956.)

the fault gives rise to structures in these cover rocks. As may be inferred from Fig. 6.12(*a*), an induced shear in the cover rock will result in compression and extension. The compressive stress so induced in the cover rocks may cause a series of structures to develop which can include folds and a variety of different types of fractures. If the compression gives rise to the development of folds, they will initially have the orientation shown in Fig. 6.12(*a*). However, continued shearing of the basement can give rise to rotation of the fold structures so that they eventually exhibit orientations characterised by the well-known array of domes along the Inglewood Fault, S. California (Fig. 6.12(*b*)). Other examples of the association of folds with major strike-slip faults are shown in Fig. 6.12(*c*) and (*d*). The amount of shear strain required to rotate folds, which originally made an angle of 45°, to near parallelism with the first order fault trace must be very large. However, if the fault exhibits oblique-slip, then *drape* folds would tend to develop with their fold axes approximately parallel with the fault trace, in response to the dip-slip component of movement: while the strike-slip component of movement would ensure that the fold axes made an angle with the fault trace. The oblique-slip fault motion could thereby give rise to the orientation of fold structures represented in Fig. 6.12(*b*) by inducing a relatively small amount of horizontal shear strain. This is a point to which we shall return.

The complex array of secondary faults that may develop in basement rocks at the end of a major feature such as the Alpine Fault is indicated in Fig. 6.4, and in Fig. 6.13(*a*), while those which more commonly form adjacent to the first order fault in regions remote from the ends are indicated in Fig. 6.13(*b*). The geometrical relationship of these various structures to each other and the major first order strike-slip fault are represented diagrammatically in Fig. 6.13(*c*) and (*d*).

We have permitted the words cover and basement to slip into the text without definition. This is because the terms are not readily defined. However, it will be apparent from Fig. 6.14(*a*) in what sense the terms are used here.

We shall now attempt to show how such secondary fractures and structures may be interpreted and their mode of development understood by making reference firstly to Experimental Studies and secondly to Theoretical Analyses.

Experimental studies

The type of structures which may be generated in cover rocks as the result of strike-slip movement in the basement can be simulated by using very simple apparatus. Such apparatus may be made up of two rectangular rigid blocks which can be caused to move

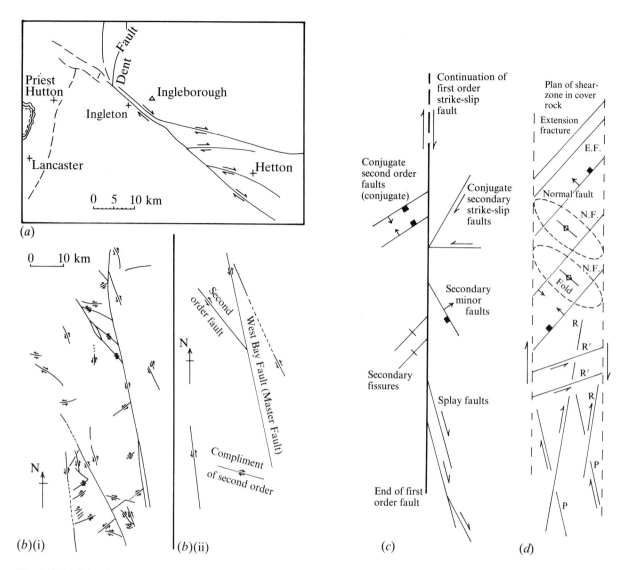

Fig. 6.13. (*a*) Splay faults at the end of a major first order fault, Ingleborough, N.W. England. (After Anderson, 1951.) (*b*) (i) Secondary strike-slip faults associated with a first order strike-slip fault, in the Yellowknife District, N.W.T., Canada. (ii) Idealised pattern of (i). (After McKinstry, 1953.) (*c*) Compilation showing types of secondary features which commonly develop adjacent to a first order fault in *basement rocks*. (*d*) Compilation showing secondary features which commonly form in *cover rocks* above a major strike-slip fault in the basement.

relative to each other, as indicated in Fig. 6.14(*a*), and so represent strike-slip movement of a basement fault. The cover rock in such experiments is represented by a slab of weak material, such as clay, putty or wax. As fault movement occurs in the basement, the traction which exists at the interface between basement and cover gives rise to stresses in the cover rock which result in their deformation. The types of structures which develop in the various models depend upon the properties of the materials, whether or not they contain horizontal planes of weakness, or banding (which would represent stratification of the cover rock) and the rate of strain which is induced in the model.

One of the earliest studies of the development of secondary fractures associated with movement on a main first order fault was conducted by Riedel (1929). Using thick, homogeneous tablets of clay to simulate the cover rocks, Riedel found that movement of the basement blocks induced a shear zone in the cover which, in profile, exhibited a V-shape, with the base of the V immediately above the basement 'fault', Fig. 6.14(*a*). The type of fractures, depending upon the strength of the clay, were either extension or shear fractures, as indicated in Fig. 6.14(*b*). The mechanism of development of these secondary fractures can now be inferred directly from Fig. 6.14(*c*). The shear fractures (R and R′) form a complementary system and are usually termed Riedel Shears. Of these two sets, R is the one which develops more frequently. When R′ also develops, it often experiences rapid rotation as the shear progresses on the basement fault and after a relatively small degree of rotation 'locks-up' and so becomes inactive.

Similar experiments have been conducted by Tchalenko (1968), in which, however, the clay tablet

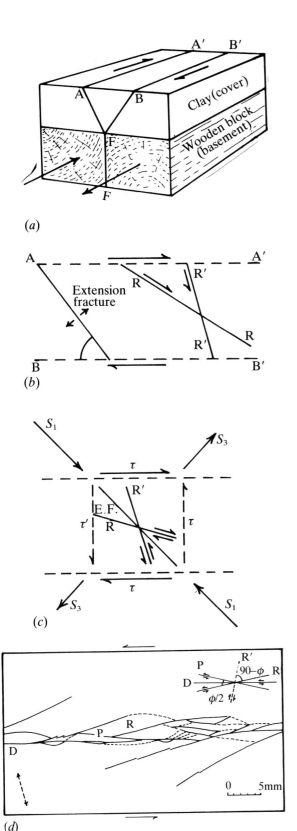

(a)

(b)

(c)

(d)

Fig. 6.14. (a) Model used by Riedel to simulate cover-rock deformation as the result of strike-slip movement on a fault FF' in the 'basement'. (b) Type of fractures which developed at the surface in the shear zone with the boundaries AA', BB' (see (a)). (c) Interpretation of the stress system that gave rise to the fractures shown in (b). (d) Terminology used by Tchalenko (1968) to describe shear discontinuities in a clay tablet subjected to uniform shear in direction D, these include conjugate Riedel shears (R and R'), what Tchalenko termed 'thrust-shears' (P) and structures influenced by stress concentrations at the edge of the shear box.

was subjected to a uniform shear. Such a mode of displacement simulates deformation of cover rock above a ductile shear zone in the basement. The fracture patterns which develop in these experiments are shown in Fig. 6.14(d), where D is the direction of principal shear (i.e. analogous to the first order shear direction) and R and R' are Riedel Shears. Other important secondary fractures which developed in these experiments are termed *P-shears*, the orientations of which are also indicated in Fig. 6.14(d). (Tchalenko originally termed these fractures 'thrust shears': however, this term is so obviously a potential source of confusion that we shall not use it.) The mode of development of these P-shears, which form after the R-shears was not understood by Tchalenko. Tentatively, he suggested that they were fractures which were related to kink-bands and, following the concepts which were held *at that time*, formed at an angle of 45° + $\phi/2$ to the axis of maximum principal stress where, as in the Navier–Coulomb criterion of brittle failure, ϕ is the angle of sliding friction. For brittle failure, it will be recalled, the angle between shear plane and the axis of maximum principal stress is given by $\theta = 45° - \phi/2$ (see Chap. 1). As will be inferred from the discussion of the development of kink-bands, which is presented in Chap. 13, this tentative suggestion in the light of current understanding can not be invoked to explain the formation of P-shears. Later in this chapter it will be shown that these shears can be explained in terms of transient stresses. Eventually, after considerable movement on the basement fault, a fracture with the same orientation as this basement fault will propagate upward into the cover rock and disrupt the earlier formed secondary features.

These experiments by Riedel and Tchalenko are informative in that they give rise to fracture patterns which simulate those which are known to develop in nature. However, the models have not been subjected to detailed analysis. For example, the stress pattern near the surface of the clay tablet will be markedly different from that which exists in the clay near the basement cover interface. Thus at, or immediately below, the surface of the clay tablet, the principal stresses will act horizontally, or vertically. At, or immediately above the interface, however, the axis of principal stress will be inclined as a result of the shear stresses which exist on this junction plane. Consequently, the orientations of the secondary structures immediately above the first order fault in the basement will be markedly different from those which occur at the surface. This is an aspect of the problem which appears to have been neglected by Riedel. This criticism does not, of course, apply to the work conducted by Tchalenko, for his underlying structure was a shear zone rather than a single strike-slip fracture.

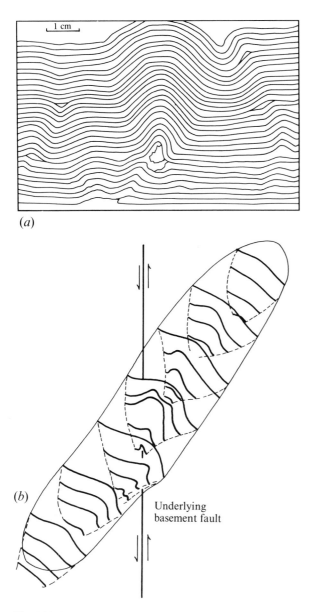

(a)

(b)

Underlying
basement fault

Fig. 6.15. (*a*) Examples of fold structures in layered wax, acting as cover-rocks, above a strike-slip fault in the basement. (*b*) Diagrammatic representation of such a single fold. The form of the structure is represented by a series of profiles. Note the reversal in symmetry of the fold on opposite sides of the basement fault. (Courtesy of D. Oliver, 1986.)

The type of apparatus illustrated in Fig. 6.14(*a*) has also been used to investigate the development of fold structures in cover rock which result from strike-slip fault movement in the basement. This problem has been addressed by several experimenters. However, we consider that the most realistic and elegant series of experiments have been conducted by Oliver (1987), in which he used thin layers of wax to simulate the sedimentary cover rocks. Unlike the V-shaped shear-zone, reported by Riedel, Oliver noted that the shear zone was U-shaped, so that it narrowed dramatically on approaching the basement fault. He produced a variety of realistic structural forms which include major, upright box-folds and

overturned, or even isoclinal, folds which may pass laterally into thrusts (see Fig. 6.15). In plan-view, the anticlines are mostly sinusoidal or arcuate: and, at increasing depth of cover, their fold axes tend to rotate towards the line of shearing. In section, he found that individual anticlines vary greatly in their geometry. At depth, a major structure often divided into two or more smaller, relatively tight folds. However, the major structures became simpler and gentler towards their ends. Oliver further noted that the folds on either side of the shear zone in the cover tended to be asymmetric and, in the upper levels of the cover rock, to verge towards the line of the basement fault. However, at deeper levels, the anticlines frequently verge away from the fault. Oliver suggests that these patterns of vergence and also the shape of the shear-zone itself are largely controlled by the distribution of shear-strain induced in the cover by the underlying basement fault.

In these experiments, Oliver studied the influence of a number of parameters which included: depth of cover, layer thickness and competence, lubrication of layers, number and spacing of basement faults, angle of oblique-slip and strain-rate. He makes the point that many of the observed model structures are comparable with natural examples described in the literature, but that some of the analogue structural patterns have yet to be identified in the field.

As we have noted, these types of experiments do not permit quantitative analysis. A kind of experiment which does, however, allow such an analysis is one based on the photo-elastic technique. Such experiments were conducted by Freund (1974), who noted that secondary fracture patterns were particularly well developed adjacent to curved, or kinked, portions of a first order fault, as indicated in Fig. 6.9. This geological situation was simulated in a photo-elastic model (Fig. 6.16). Freund determined the orientation of the axes of principal stress that would be induced in the vicinity of the 'kinked' area (Fig. 6.16(*b*)) and, by assuming that the angle of friction was 30°, he was able to construct the trajectories of potential strike-slip faults (Fig. 6.16(*c*)). It will be seen that there is a distinct resemblance between the experimental results and the natural distribution of secondary shear faults (Fig. 6.16(*d*)).

Freund's analysis holds only for those situations in which the first order fault is non planar. However, secondary shears and other forms of secondary structures can, and often do, occur adjacent to planar sections of first order faults as, for example in the Yellowknife District of the Yukon (e.g. Fig. 6.13(*b*)). In order to pursue this subject further it is necessary to turn to purely theoretical studies of the development of secondary features associated with first order, strike-slip faults.

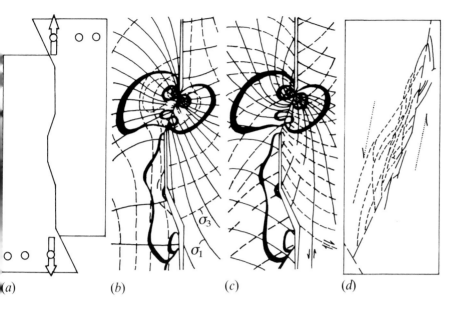

(a)   (b)   (c)   (d)

Fig. 6.16. (*a*) Representation of a photo-elastic model of a curved fault, arrows indicate the direction of movement. (*b*) Disposition of black isochromatic fringes from which the principal stress trajectories can be derived. Solid lines represent σ_1 (compressive) and dashed lines σ_3 (extensile) principal stresses. (*c*) Directions of potential strike-slip secondary faulting (angle of friction ϕ taken as 30°). (*d*) Strike-slip faults parallel to the southern part of the Dead Sea Rift. (After Freund, 1974.)

Theoretical analyses

In this section, various theoretical analyses relating to the development of secondary fractures adjacent to a first order strike-slip fault will be treated on a historical basis, starting with the analysis by Anderson and followed by those of McKinstry, Chinnery, Lajtai and Price.

Anderson's analysis

An early attempt to explain the development of secondary splay faults, which form near the end of first order strike-slip faults, was proposed by Anderson (1951). In this analysis, he assumed that a major strike-slip fault could be represented by a 'gigantic' elliptical 'Griffith' crack which reached from the surface to some deep layer of detachment. By using this model, he was able to consider the problem in terms of plane strain and he was also able to use mathematical relationships already employed by Griffith. Anderson assumed that principal stresses of $+P$ (S_1) and $-P$ (S_3) acted at 45° to the first order fault (as indicated in Fig. 6.17(*a*)), and that a single increment of movement occurred on the fault as the result of the applied stresses. It was then shown that the initiating stresses changed in orientation and magnitude because of the induced movement and that the resulting orientation of the stress trajectories were as represented in Fig. 6.17(*b*). It will be seen that near the ends of the first order fault, the principal stress trajectories are aligned approximately perpendicular and parallel to the fault plane. Anderson pointed out that the magnitude of the stresses near the ends of the faults increased relative to the initial values of $+P$ and $-P$. The degree of increase is related to the radius of curvature at the tip of the first order fault. He suggested that because the radius of curvature at the tip of the 'elliptical crack' must be very small, the local

increase in magnitude of the stresses would be so large that secondary faulting would take place. It may be inferred from Fig. 6.17(*c*) that splay faults would develop in quadrants A and C in which the maximum principal stress trajectories are aligned approximately parallel to the first order fault. Anderson omitted to note that, by the same token, secondary strike-slip fractures (FF) could be expected to develop in the other two quadrants (B and D) where the maximum principal stress trajectories are perpendicular to the first order fault. The splay and second order faults in Fig. 6.17(*c*) are drawn at 45° to the axes of principal stress. However, if, more realistically, it is assumed that the angle of sliding friction (ϕ) is 30°, then the orientation of secondary faults would be as shown in Fig. 6.17(*d*), where the orientation of the splay fault is in good agreement with those observed in nature, but that of the 'second order' fault is not.

Anderson also made the point that the magnitude of the stresses near the central part of the first order fault decreased almost to zero. Indeed, it is necessary to traverse a distance of 0.4L perpendicular to the central part of the first order fault (length L) before the original value of stress ($+P$ and $-P$) is reached. This reduction in stresses adjacent to the central regions of a first order fault is the reason which ensures that there is a discrete spacing between major fractures of the same set. It was also emphasised by Anderson that the stress drop in the central parts of the first order fault was so large that it completely inhibited the growth of secondary faults in the central region.

Anderson's analysis is, therefore, unable to account for the development of secondary strike-slip faults which, as we have seen, are observed in the central sectors of real first order strike-slip faults. Indeed, he maintains that they could not develop. It may be suggested that the major first order fault

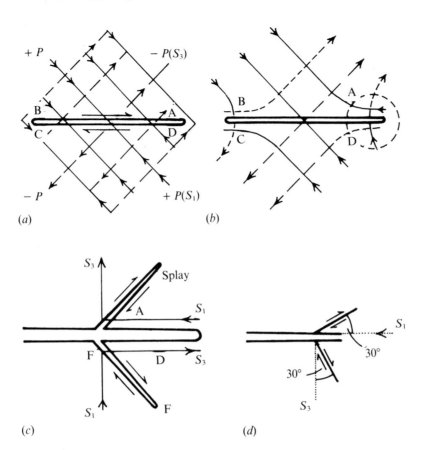

+ P   − P(S₃)

B
A
C
D

− P   + P(S₁)

(a)

B
A
C
D

(b)

S₃
Splay
A   S₁
F   D   S₃
S₁   F

(c)

S₁
30°
30°
S₃

(d)

Fig. 6.17. (a) Initial stress system assumed for the Andersonian model. (b) Stress system after an increment of movement. (c) Detail of stress orientation at the end of the first order fault, (dashed circle (b)) indicating the splay and associated secondary fault FF that could develop. (d) Showing the orientation of the splay and secondary strike-slip faults that will develop, according to the Andersonian analysis, if the angle of friction (φ) is 30°.

developed incrementally and that those secondary structures which developed at the ends of the first order fault at each increment of the development 'migrated' to the central portions of the structure as the first order fault 'grew'. However, it should be noted that splay faults are usually restricted to the ends of faults. If the incremental growth model were valid, one would expect both splay and other secondary strike-slip faults to be equally developed throughout the length of the first order fault.

There are obvious shortcomings to Anderson's basic model. For example, major wrench faults cannot be considered as gigantic and open elliptical 'cracks', along which frictional sliding is neglected. (Anderson's analysis pre-dates the modified analysis of the Griffith crack by McClintock & Walsh (1962) in which frictional sliding is included with the original Griffith analysis on which Anderson's treatment is based, (see Chap. 1).) Other shortcomings of this analysis will be noted later.

McKinstry's analysis

An analysis was presented by McKinstry (1953) in which he considered frictional sliding on two major parallel and adjacent strike-slip faults which moved in response to a uniaxial compressive stress (P) acting at 30° to the fault planes (Fig. 6.18(a)). He argued that the stress in the fault block between the first order faults was related to the normal stress (σ_n) and the shear stress (τ), where $\tau = \mu\sigma_n$ and μ is the coefficient

of sliding friction of the rock. An 'approximate' analysis was presented which purported to show that a refraction of stress occurred between the faults and that a stress P' acted at 60° to the faults (Fig. 6.18(b)). He pointed out that if secondary strike-slip faults (F_1, F_2) developed in response to the stress P' they would be inclined at about 30° and 90° respectively to the first order faults. Thus, the analysis seems to explain the development of the secondary strike-slip faults which are seen to develop adjacent to real first order, strike-slip faults. Unfortunately, the analysis can be dismissed as being incorrect, for McKinstry neglected to include the influence of the component of stress generated by P which acts parallel to the first order fault. Indeed, as is emphasised by Chinnery (1966), if this component parallel to the major faults is included in the analysis, the stress between the fault planes is refracted in the opposite sense, so that P' makes an angle of 11° with the first order faults. Moreover, Chinnery points out that the value of P' is only 0.80P. Consequently, because P is invoked to cause movement on an existing fracture, it may be inferred that the stress $P' = 0.80P$ would be insufficient to bring about the development of secondary shear fractures in hitherto unbroken rock.

Chinnery's analysis

Chinnery carried out an analysis in which an increment of movement on the fault was likened to the planar movement of a defect in a crystal. In this

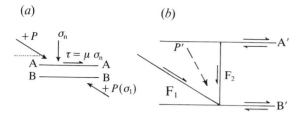

Fig. 6.18. (*a*) Initial external stress (+ *P*) at 30° to parallel faults, which, in the McKinstry model, are assumed to cause an increment of movement. (*b*) Calculated (but incorrect) orientation of internal stress (*P'*) which, McKinstry (1953) suggested, gives rise to the secondary faults F_1 and F_2.

analysis, model first order faults were subjected to (i) the 'pure-shear' stress system (i.e. $\sigma_1 = -\sigma_3$) used by Anderson (Fig. 6.17(*a*)) and (ii) the uniaxial compressive stress system (*P*) used by McKinstry. The results of his two analyses are represented in Fig. 6.19(*a*) and (*b*), which show the trajectories of maximum shear stresses that will develop near the ends of a first order, strike-slip fault after a single increment of movement.

It may be inferred from these figures that the most likely development is that the first order fault will continue to propagate. However, if such a propagation is inhibited, other fractures may develop. The possible fracture patterns which, Chinnery suggests, are compatible with the calculated patterns and intensities of the shear stresses are shown in Fig. 6.19(*c*) and (*d*). Chinnery points to the similarity between the forecast pattern shown in Fig. 6.19(*c*) and (*d*) and the fracture pattern which develops at the northern end of the Alpine Fault of S. Island, New Zealand (Fig. 6.4). However, although the Alpine Fault may originally have been a simple strike-slip fault, it now exhibits a marked component of reverse fault movement (Sibson *et al.*, 1981) with compression at a high angle to the fault plane. This latter compression may have influenced the development of the fault pattern, especially the cross faults (type D in Chinnery's model, Fig. 6.4).

Chinnery reaffirmed Anderson's statement that the stress drop near the central parts of the first order fault is so large that the development of secondary strike-slip faults in this area is not viable.

It must be pointed out that these three analyses contain an inherent inconsistency which is to do with the initiating stress field and the tacit, or assumed, failure condition. Both Anderson and Chinnery assume failure resulting from a 'pure-shear' stress condition, as indicated in Figs. 6.17(*a*) and 6.19(*a*). Similarly, both McKinstry and Chinnery assume failure in uniaxial compression, (Figs. 6.18(*a*) and 6.19(*b*)). The failure condition for sliding on a fault with zero cohesive strength is shown in Fig. 6.20, together with these two stress conditions. It will be immediately apparent that the stress conditions cannot be made to satisfy the failure criteria. Con-

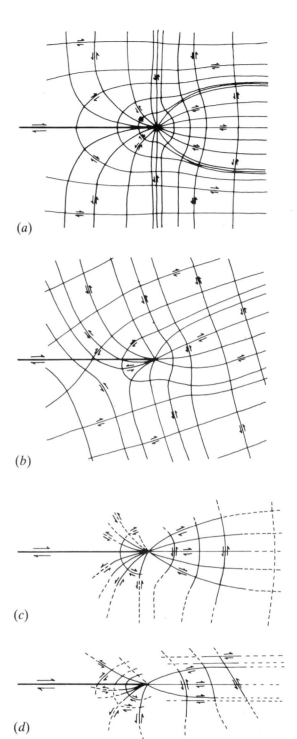

Fig. 6.19. Maximum shear stress trajectories obtained, using the Chinnery model, after an increment of movement on the first order fault subjected to (*a*) 'pure shear' (see text) and (*b*) uniaxial compression. Likely directions of secondary faulting compiled from the data given in (*a*) and (*b*) are indicated in (*c*) for pure shear and in (*d*) for uniaxial compression. In (*c*) and (*d*) the solid lines indicate likely locations of faults; less likely locations are indicated by dashed lines. (After Chinnery, 1966.)

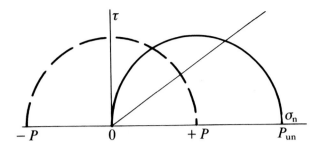

Fig. 6.20. Mohr's stress circles for pure shear ($-P$, $+P$) and uniaxial compression (0, P_{un}) and failure envelope for sliding on an existing first order fault with zero cohesion. Neither of the stress circles are compatible with the failure envelope (i.e. both give 'over-kill').

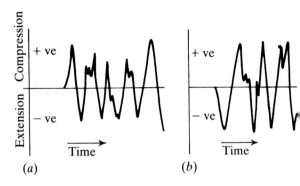

Fig. 6.21. Form of seismic trace obtained by instruments distant from the focus, showing (*a*) positive and (*b*) negative first motions.

sequently, the analyses presented by the three workers are not strictly correct. The uniaxial stress condition can be made a viable one by assuming that the first order fault exhibits a cohesive strength. However, it may be inferred that, unless the cohesive strength is large, the magnitude of the uniaxial stress (σ_1) will be small, and hence there is some doubt whether such a small stress can bring about conditions which will lead to the development of secondary or splay fractures in unfractured rock.

Even so, we would not wish to suggest that this criticism completely invalidates the analyses of Anderson and Chinnery, as regards their conclusions relating to the development of splay and other fractures which may develop at, or near, the ends of major first order fractures. These are interesting and broadly valid. However, it is emphasised that their conclusions, that the stress levels in the central parts of the first order fault fall to values so low that no secondary faults can develop, are at variance with field observations.

Lajtai's analysis

Lajtai (1968) presented a graphical stress and failure analysis which showed that, provided the stress component parallel to the fault is assumed to be small or zero, then failure and the development of secondary fractures would be possible. He, therefore, assumes the same stress conditions which McKinstry obtained in error. This is not, of course, any criticism of Lajtai's analysis. He made the basic assumption that, provided the component of stress parallel to the first order fault was suitably reduced, failure would follow. He did not attempt to explain how such a stress reduction might come about.

Price's analysis

At the same meeting as that attended by Lajtai, Price (1968) presented a paper which did indicate how such a stress reduction could be invoked. He argued that all the analyses presented and noted in the preceding paragraphs were 'static' analyses. That is, an initial

stress pattern was assumed to give rise to an increment of movement and, after movement was complete, there was a final static stress pattern which could, or could not, give rise to the development of secondary fractures.

As these static analyses gave only a partially correct answer regarding the development of splay and other fractures near the ends of first order faults and a completely erroneous conclusion regarding secondary fractures near the central portion of the fault, Price suggested that it would be better to consider a dynamic model. The arguments in the following paragraphs are based on this paper by Price, but are freely augmented and extended.

Brittle failure on a fault results in the development of elastic vibrations in the adjacent rock, which propagate on an approximately spherical front. At a point which is distant from the source of the 'earthquake', the seismic trace measured by a suitable instrument will be of the form represented in Fig. 6.21. The forms of this and other similar wave traces are highly complex: a complexity which largely results from interfering wave patterns that arise from internal reflections of bedding planes, different path lengths and the physical response of the recording equipment. Nearer the source of the earthquake, it is probable that the elastic disturbance has a simpler form. This may be inferred from the distribution of the *first movements* at the seismic stations.

The simple model which we shall use as regards the generation of elastic waves is shown in Fig. 6.22. In this model we shall assume that the wave-front close to the fault is approximately planar: we consider the effect of a spherical front later in the chapter. Initiation of re-shear occurs at some specific weak zone and is generated along the existing fault at a velocity (V_f). The release of elastic distortion indicated in Fig. 6.21 gives rise to P- and S-waves which propagate at velocities V_P and V_S, where $V_P > V_S > V_F$. As will be seen from Fig. 6.22, the P-waves are compressive ($+$ ve) in two opposing quadrants, and extensile ($-$ ve) in the other quad

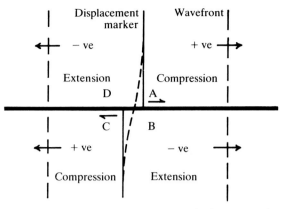

Fig. 6.22. Diagrammatic representation of planar wave-fronts generated by fault movement. Reverse polarity of first movements will be generated in the opposing fault blocks.

rants. Because the P-wave velocity is almost twice as fast as the S-wave velocity, the 'first arrivals' are of the positive or negative P-waves (Fig. 6.21(*a*) and (*b*) respectively). The distribution of the first arrivals forms a 'four-leaf clover', as indicated in Fig. 6.23, seismic stations in areas A and C recording + ve and B and D − ve first arrivals. It will be noted that there is a degree of ambiguity regarding the interpretation of the first arrivals in that they may be attributed to movement on either fault A or B. It will also be noted that along the trace of the faults, but at a considerable distance from the disturbance, there are no first arrivals (shaded areas, Fig. 6.23). This can be attributed to the fact that, close to the initiating movement, the P-waves, on opposite sides of the fault, will be of similar form but 180° out of phase. As the positive and negative waves spread along the line of the fault trace (Fig. 6.23) they will cancel out. For perfect cancelling

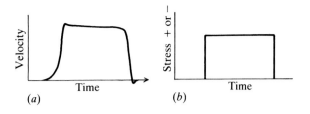

Fig. 6.24. (*a*) Probable displacement-velocity/time curve for fault movement. (*b*) Form of stress/time pulse (both + ve and − ve) assumed in the text.

it may be inferred that the wave form is likely to be simple.

As regards slip on the fault plane, the stress which gives rise to slip will cause an initial phase of accelerating displacement parallel to the fault, which may be followed by a period of approximately constant displacement velocity with a final phase of decelerating displacement (Fig. 6.24(*a*)). Here we assume that local strain, and therefore stress, follows a similar pattern, Fig. 6.24(*b*). As a first approximation, therefore, we shall consider the P-wave pulse to have a similar square-wave form.

Thus, we maintain that the stress situation adjacent to a first order, strike-slip fault is as shown in Fig. 6.25. The original static effective stresses which cause re-shear on the fault are σ_1 and σ_3, which give rise to the three indicated components of stress σ_x, σ_y and τ. In addition, there is a transient stress, $\pm P$ which travels through the rock. (Because the velocity of the S-wave is so much smaller than that of the P-wave, there is a brief period when we need only consider the effect of the P-wave in developing transient stresses.)

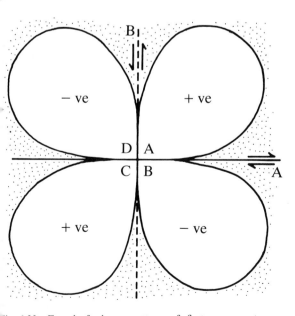

Fig. 6.23. Four-leaf clover pattern of first movement senses recorded in areas after movement on one of two faults (A or B). Note that ambiguity always exists when interpreting first movements.

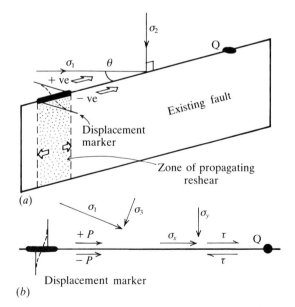

Fig. 6.25. (*a*) Oblique and (*b*) plan view, representing components of stress σ_x, σ_y and τ which result from the static stress field σ_1 and σ_3 and the pulse stresses + P and − P, which give rise to a transient stress system.

If, for the sake of simplicity, we initially assume that the P-wave adjacent to the fault has a planar front, then the P-wave is considered to influence only the normal stress (σ_x) which acts in the direction of the P-wave translation. Thus, at point Q, the initial static stress (σ_x) is changed by the transient P-wave to:

$$\sigma'_x = \sigma_x + P \qquad (6.1)$$

or

$$\sigma'_x = \sigma_x - P \qquad (6.1(a))$$

Now the values of σ_x, σ_y and τ are uniquely related to the magnitudes of σ_1 and σ_3 and the angles they make with the fault plane. Consequently, if σ_x changes to σ'_x, there will be a concomitant change in the values of the principal stresses and also the angles they make with the fault plane. These changes in the values and orientation of the principal stresses can be determined either analytically or graphically.

Graphical solutions

The graphical method by which we may determine the magnitude of the pulse stress (P) which is required to initiate second order fractures is indicated in Fig. 6.26(a). The original 'static' stress circle is that drawn through points σ_1 and σ_3 with centre O. The position of σ_2 is not specified in this figure. (The importance of the magnitude of σ_2 relative to those of σ_1 and σ_3 will be discussed later.) The normal and shear stresses acting on the fault (σ_y, τ) are given by point A and the 'normal' stress *parallel* to the fault and the shear stress acting normal to the fault (σ_x, $-\tau$) are given by point B (see also Fig. 6.25(b)). Using the planar wave-front concept outlined above, the transient stress situations are as indicated in Fig. 6.25(b) where, as we can see:

$$\sigma'_x = \sigma_x + P \qquad (6.1)$$

or

$$\sigma'_x = \sigma_x - P \qquad (6.1(a))$$

Consider the situation for the negative pulse, then σ'_x plots at point C in Fig. 6.26(b). The stresses represented by point A are unaffected by the pulse, so both points A and C are on the transient stress circle. Hence, we can construct this circle with centre at O' and radius O'A = O'C. From the intercept of this circle on the σ_n axis we can read off the values of σ'_1 and σ'_3. We can also obtain the angle θ' by taking half the angle marked $2\theta'$. This gives the angle which the axis of the maximum principal transient stress makes with the first order fault plane (in this instance that angle $\theta' = 60°$). It will be noted that the transient stress circle touches the failure envelope for unbroken rock and can, therefore, initiate second order fractures.

The orientation of these fractures depends upon

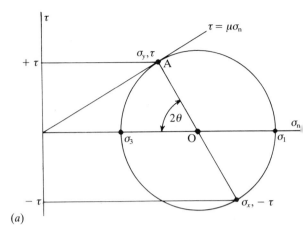

(a)

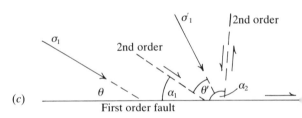

(b)

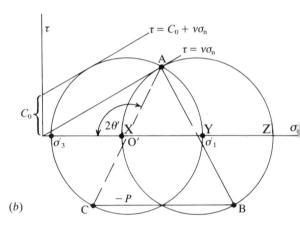

(c)

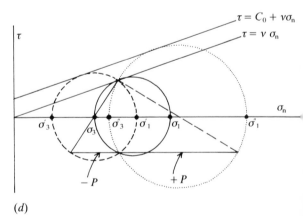

(d)

Fig. 6.26. (a) Static stresses σ_1 and σ_3 which cause movement on the first order strike-slip fault. (b) Static, fault parallel, stress is reduced from B to C by pulse $-P$ and gives rise to a transient stress system σ'_1 and σ'_3 that is compatible with movement on the first order fault and the generation of a secondary fracture in unbroken rock with cohesive strength C_0. (c) Plan view showing the orientation of static and transient stresses and potential planes of secondary strike-slip faulting. (d) Failure conditions for negative and positive pulse stresses.

he precise value of the intermediate principal stress σ_2). We may infer from Fig. 6.26(*b*) that if the magnitude of σ_2 lies in the range XY: the transient maximum principal stress (σ'_1) remains the greatest principal stress and the resulting second order fracture will be a strike-slip feature, with the orientation indicated in Fig. 6.26(*c*) and A in Fig. 6.27. However, if the static σ_2 (the vertical principal stress) has a value in the YZ range in Fig. 6.26(*b*), then the transient principal stress which we have so far termed σ'_1 is actually less than the static vertical stress. Hence, the vertical stress becomes the transient σ'_1 and the maximum horizontal principal stress then becomes σ'_2 which are those stress conditions that would give rise to normal faults with the orientations represented as B in Fig. 6.27,

It will be noted from Fig. 6.26(*d*) that the magnitude of the positive pulse required to initiate the corresponding set of second order fractures is significantly larger than that of the negative pulse. However, for any individual increment of fault movement, $+ P$ will have the same magnitude as $- P$. There will be a tendency, therefore, for second order fractures to be more readily associated with the negative pulse but, as we shall see, the positive pulse may also give rise to secondary fractures when conditions are favourable.

Analytical solutions

Price (1968) used the alternative approach and solved the problem outlined in the previous section analytically. This method has the potential advantage that the solutions can be derived with greater accuracy, but the method is less elegant than the graphical one: and certainly more difficult to envisage. Nowadays, however, the ease with which analytical problems can be solved using a computer makes it the method to be preferred if multiple solutions are required.

In outline, the analytical method used is as follows. Assumptions were made regarding the angle (θ) which the axis of maximum principal stress (σ_1) makes with the fault plane. It was also assumed that the axis of intermediate principal stress (σ_2) lies in the

fault plane, so the orientation of the least principal stress (σ_3) is also determined. The angle of sliding friction (ϕ) on the fault plane is given a specific value (30°) as is the 'uniaxial' strength (σ_0), of the unfractured rock. The conditions for failure (i.e. re-shear on the fault plane) and for the generation of secondary shear fractures are, respectively, given by:

$$\sigma_1 = k\sigma_3 \qquad (6.2(a))$$

and

$$\sigma_1 = \sigma_0 + k\sigma_3 \qquad (6.2(b))$$

[see Chap. 1, Eqs. (1.57) and (1.59)] where $k = (1 + \sin \phi)/(1 - \sin \phi) = 3.0$, when $\phi = 30°$. In addition, the confining pressure, or least principal stress (σ_3), is given a specific value.

From these assumed values and given relationships, the normal and shear stresses acting on the first order fault plane can readily be calculated using the stress equations:

$$\sigma_x = \frac{(\sigma_1 + \sigma_2) + (\sigma_1 - \sigma_3)\cos 2\theta}{2} \qquad (6.3(a))$$

$$\sigma_y = \frac{(\sigma_1 + \sigma_3) - (\sigma_1 - \sigma_3)\cos 2\theta}{2} \qquad (6.3(b))$$

and

$$\tau = \frac{(\sigma_1 - \sigma_3)\sin 2\theta}{2} \qquad (6.3(c))$$

(see Eqs. (1.15) and (1.16)).

It is then assumed that these components of static stress acting on the fault plane (Fig. 6.25), under the action of a planar pulse wave, become the transitory stresses:

$$\sigma'_y = \sigma_y \qquad (6.4(a))$$

$$\tau' = \tau \qquad (6.4(b))$$

and

$$\sigma'_x = \sigma_x + P \qquad (6.4(c))$$

or

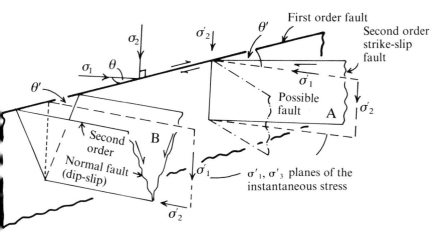

Fig. 6.27. Showing the orientation of (A) secondary strike-slip fractures, when the vertical stress is the intermediate principal stress throughout the duration of the pulse and (B) second order normal faults which may develop when the vertical stress is momentarily the greatest principal stress.

$$\sigma'_x = \sigma_x - P. \qquad (6.4(d))$$

The values of σ_y, σ_x and τ are uniquely related to σ_1, σ_3 and θ. Consequently, if one of these components of stress is changed it means that a completely new state of stress is brought into being, such that σ_1 becomes σ'_1, σ_3 becomes σ'_3 and θ becomes θ'. To obtain the values of σ'_1 and σ'_3 the values of σ'_x and τ' (Eqs. 6.4) are substituted into Eq. (1.9), to give:

$$\sigma'_1 = \frac{(\sigma'_x + \sigma'_y) + [(\sigma'_x - \sigma'_y)^2 + 4\tau^2]^{1/2}}{2} \qquad (6.5(a))$$

$$\sigma'_3 = \frac{(\sigma'_x + \sigma'_y) - [(\sigma'_x - \sigma'_y)^2 + 4\tau^2]^{1/2}}{2} \qquad (6.5(b))$$

The value of θ' can be obtained using Eq. (1.7):

$$\tan \theta' = \frac{(\sigma'_1 - \sigma'_x)}{\tau}. \qquad (6.5(c))$$

Price stipulated values of σ_3 (68.5 MPa), ϕ (30°), θ (30°) and σ_0 (68.5 MPa) and calculated the values of σ'_1 and σ'_3 for various values of $\pm P$, using Eqs. (6.4) and (6.5). (The results are shown graphically in Fig. 6.28.) He pointed out that given values of σ_0 and ϕ, the stress σ'_{F_1}, at which unfractured rocks fail, is determined by the magnitude of σ'_3, such that:

$$\sigma'_{F_1} = \sigma_0 + k\sigma'_3 \qquad (6.5(d))$$

(see Eq. (6.2(b)), where, as already noted, $k = (1 + \sin \phi)/(1 - \sin \phi)$.

The failure conditions (σ'_{F_1}) are also plotted in Fig. 6.28, using the appropriate values of σ'_3. It will be seen that if reshear on the existing first order fault gives rise only to P-waves of small magnitude, the transient stress (σ'_1) will not be sufficient to cause failure of the rock ($\sigma'_{F_1} > \sigma'_1$) and second order fractures will not be initiated, but will only form at higher magnitudes of P, when $\sigma'_{F_1} > \sigma'_1$. These points are clearly seen in Fig. 6.28, where the two curves for σ'_{F_1} and σ'_1 intersect at X and Y. Equivalent conditions were obtained using the graphical method, represented in Fig. 6.26, by means of trial and error. The analytical method is inherently more precise and systematic.

The angle (θ') which the axis of transient principal stress (σ'_1) makes with the first order fault, when the failure conditions for the negative pulse ($-P$), shown in Fig. 6.27 are reached, is 50°. As we have already pointed out, when discussing the graphical solution to this problem, the type and orientation of the secondary fractures depend upon the value of the static intermediate principal stress (σ_2). Thus, if the value of σ_2 lies between σ'_1 and σ'_3, strike-slip faults develop (A in Fig. 6.27). However, if σ_2 is greater than the maximum horizontal transient principal stress, σ_2 becomes σ'_1, so that normal faults will form (B in Fig. 6.27). However, as pointed out by Lajtai (1968)

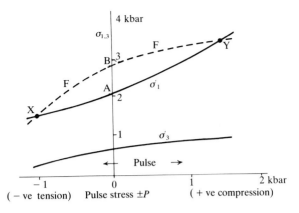

Fig. 6.28. Magnitudes of transient stresses (σ'_1 and σ'_3) for various values of pulse stresses (+ P and − P) and the corresponding failure curve FF. Failure for tensile and compressive pulses occur for conditions met at points X and Y respectively.

two other circumstances need to be considered. The Andersonian classification requires that for initiation of strike-slip faults σ_2 must act vertically and in the plane of the ideal, vertical, strike-slip fault. However once a first order, strike-slip fault has been initiated, it is feasible for the vertical principal stress to become the least principal stress (σ_3). If this stress remains smaller than both the horizontal, transient, principal stresses, thrusts may develop. Finally, if the effective confining pressure is small, the negative pulse may cause the rock unit adjacent to the fault to fail in tension (or possibly along hybrid shear/extension fractures). The orientation of these various fracture types relative to the first order fault, together with the appropriate stress conditions, are shown in Fig. 6.29.

The conditions discussed so far constitute a special case, in that the angle θ, which the axis of

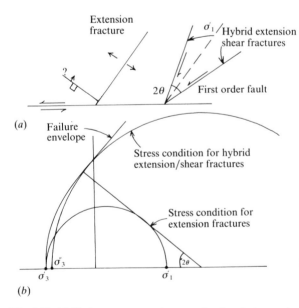

Fig. 6.29. (a) Various structures that may develop during re-shear on a first order strike-slip fault. (b) Stress systems that may give rise to second order extension or hybrid shear/extension fractures shown in (a).

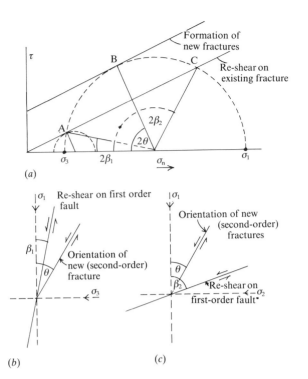

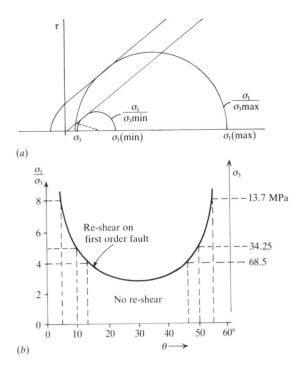

Fig. 6.30. (*a*) Conditions for re-shear on an existing fault and for the formation of new fractures (see also Chap. 5). (*b*) The orientation of new fractures with respect to the axis of maximum principal compression, when σ_1 is initially inclined at β_1 to it, and (*c*) when σ_1 is at β_2 to the first order fault.

Fig. 6.31. (*a*) Showing the limiting ratios of σ_1/σ_3 (for a particular value of σ_3) for re-shear on a fault without the formation of new fractures. (*b*) Ratios of σ_1/σ_3 and angle θ, which σ_1 makes with the first order fault plane that is compatible with sliding on that fault. For a particular value of confining pressure (σ_3) the range of values of β for which re-shear will occur can be read off. The range decreases with increase in value of σ_3.

maximum principal stress makes with the first order fault plane prior to re-shear, is constrained to be the same as that which initiated the first order fault, and is given by $\theta = 45° - \phi/2$. This angle θ is determined by the Navier-Coulomb criterion of brittle failure for fracture initiation. However, once shear has taken place, the angle θ is not then rigorously constrained and may fall within a certain range of values. This constitutes the *general case*, which we shall now consider. However, even in this general case, it should be noted that we shall continue to assume that the axis of intermediate stress lies in the plane of the first order fault.

The influence of planes of weakness, or anisotropy, upon brittle failure and/or frictional sliding has been considered at some length in Chap. 5. Here it will merely be necessary to remind the reader that the important feature is the orientation of the axis of maximum principal stress with reference to the plane of weakness. For convenience, the arguments given in Chap. 5 will be briefly recapitulated here. The conditions for failure of unbroken rock and for slip on a plane of weakness with zero cohesion are represented in Fig. 6.30(*a*). If the angle between the plane of weakness and the axis of maximum principal stress is less than β_1 or greater than β_2, then movement on the plane of weakness is not possible. If the differential stress is sufficiently large, as shown in Fig. 6.30(*a*), then a new fracture will be generated in hitherto

unbroken rock: and this will form at an angle $\theta = 45° - \phi/2$ to the axis of maximum principal stress.

If the angle which the plane of weakness makes with the axis of maximum principal stress is exactly equal to β_1 or β_2, so that the stress conditions are represented by points A or C in Fig. 6.30(*a*) then, for these two special conditions, shear may take place along the existing plane(s) of weakness and, *at the same time*, one or more new shear fractures may develop (Fig. 6.30(*b*) and (*c*)). If we take the plane of weakness in these two examples to represent a first order fault, then the new fractures which could develop (under stress condition A) would be splay faults with respect to the re-sheared first order fault (Fig. 6.30(*c*)). It will be noted that both these types of secondary fractures would form without the influence of a transient pressure pulse. However, as already emphasised, these secondary structures would only form for exact angular relationships between the axis of maximum principal stress and the plane of weakness, which must rarely be encountered in nature. In general, when slip along the plane of weakness is important, it is to be expected that the axis of principal stress will make an angle with the weakness plane which lies between β_1 and β_2. If this angle falls between these limits the influence of the plane of weakness dominates, so that the differential stress that can exist is less than that represented by the stress circle σ_1 and σ_3 in Fig. 6.30(a).

The limiting stress conditions which are compatible with shear on a plane of weakness (Fig. 6.31(a)) were discussed in Chap. 5, where it was shown that (Eqs. (5.21(a) and (b)):

$$\frac{\sigma_1}{\sigma_3} = \frac{1 + d}{1 - d} \tag{6.6}$$

where $d = (\sin 2\theta + \cos 2\theta \tan \phi)/\tan \phi$ and θ is the angle between the axis of maximum principal stress and the plane of weakness and ϕ is the angle of friction. This relationship is shown graphically in Fig. 6.31(b), where it can be seen that the permitted range of $\theta = \beta$ is in part determined by the magnitude of the least principal stress. For example, if the least principal stress is 137 bar (13.7 MPa) then, for $\phi = 30°$, the minimum ratio of principal stress (3:1) for re-shear on the first order fault to occur will be when $\theta = 30°$ and the maximum ratio of principal stresses for re-shear is 8:1, (Fig. 6.30(a)). This maximum ratio occurs when θ is either 5° or 55°.

The range of ratios of principal stresses which (as can be inferred from Fig. 6.31(b)) would give rise to an increment of movement on a first order fault was used by Price to determine the magnitude of the pulse stresses that could give rise to transient conditions capable of generating second order structures. This required obtaining solutions for σ'_1, σ'_3 and θ', by the method outlined in the preceding paragraphs, for the appropriate values of σ_1 and σ_3 in steps of 5° throughout the permissible range of θ. The relationships between the pulse stress and the angle θ for the conditions which can cause secondary failure are given in Fig. 6.32.

The positions of the horizontal lines AA' and BB' in this figure represent the limiting angles of θ'' which correspond to points A and C in Fig. 6.30(a). (It will be noted that we now apply the argument to θ', i.e. the angle which the transient stress makes with the plane of weakness.) For angles θ' between AA' and BB' (i.e. between β_1 and β_2) the influence of the plane of weakness is dominant and the transient principal stresses will be dissipated by shearing along the first order fault. For angles $\theta' > $AA' or $\theta' < $BB', slip on the first order fault will not occur and second order fractures will develop. It will be seen that positive pulses give rise to values of $\theta' > $BB' and $\theta' > $AA' (Fig. 6.32, curve cc'). Thus, adjacent to a first order fault, it is to be expected that only the negative pulse will give rise to second order fractures, (Fig. 6.32, curve dd'). However, as we shall see, for other situations it is reasonable to expect that the positive pulse may also give rise to the development of some secondary features.

It can be seen in Fig. 6.32 that the magnitude of the pulse stress that is required to give rise to second

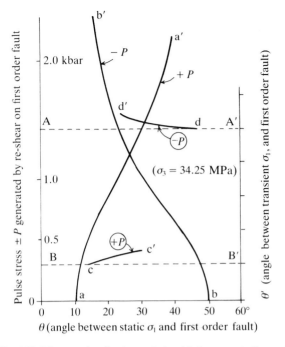

Fig. 6.32. Diagram showing interrelationship between θ_3 (the angle the static principal stress makes with the first order fault) and (i) the magnitude of pulse stresses ($+P$ and $-P$) required to cause second order failure (curves aa', bb') and (ii) the angle θ' which the transient principal stress σ'_1 makes with the first order fault (curves cc', dd').

order fractures is influenced by the value of θ'. As can be seen from this figure, when $\theta' = \beta_1$ or β_2 (the angles defined in Fig. 6.30(a)) the required pulse stress is zero. The magnitudes of the pulse stresses required to produce second order fractures for other angles of θ'' are given as Fig. 6.31. The question which obviously must be posed is whether, as the result of the initiation, or during an increment of brittle re-shear on a first order fault, the pulse stress is likely to be realised?

Magnitude of pulse stresses

It is reasonable to assume that the magnitude of the pulse stresses generated by the initiation of a fracture, or by re-shear along an existing fracture plane cannot exceed the magnitude of the differential stress that gives rise to the failure. Indeed, it is probable that the magnitude of the pulse stress is limited by the *stress-drop* that accompanies fracture initiation or re-shear. From rock mechanics tests it can be inferred that the stress-drop that occurs with brittle failure and subsequent *stick-slip* movement along the failure plane is of the general form shown in Fig. 6.33(a) (see also Chap. 5). The largest stress-drop is experienced when the fracture is initiated. The stick-slip phase of re-shear is accompanied by a relatively small stress-drop. The magnitude of the stress-drop is also related to the mode of failure. As rock behaviour becomes progressively less brittle, the associated stress-drop decreases until, for completely ductile failure, it falls to zero.

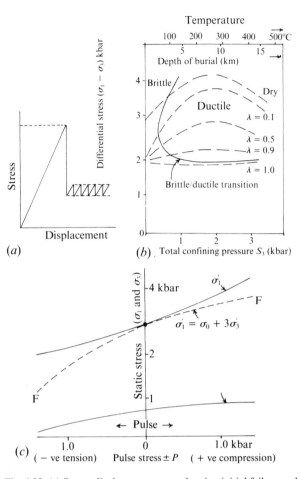

Fig. 6.33. (*a*) Stress–displacement curve showing initial failure and subsequent stick–slip behaviour (see also Chap. 5). (*b*)Brittle–ductile transition for limestone. (After Rutter, 1974: see Chap. 5.) (*c*) Transient stresses which develop ahead of a propagating first order fault, cf. Fig. 6.28.

It has been demonstrated in rock mechanics experiments that brittle failure can occur at considerable depths in the crust, provided the effective confining pressure (σ_3) is low, even though the total confining pressure (S_3) may be high. This situation can obtain only if the fluid pressure acting within the rock mass is high, so that $\sigma_3 = S_3 - p$. For this relationship, it will be recalled that we define the ratio of the fluid pressure to the total least principal stress (p/S_3) as λ_E. As may be seen from Fig. 6.33(*b*), brittle failure at depths in the crust exceeding a few kilometres can take place only when $\lambda_E \approx 1.0$. Thus, the rock mass will experience an effective confining pressure which will be close to zero.

The uniaxial compressive strength of dry rock, as determined in the laboratory at relatively low temperatures ($< 200\,°C$) and fast strain-rates of about $10^{-4}/s$, may exceed several kilobar. However, when one takes into consideration the influence of lower strain-rates, the mechanical and chemical effects of fluids at pressure and also the influence of the scaling factor (that is the size of the test specimen relative to the much larger rock mass in the crust, see Chap. 1) it

is probable that the uniaxial strength of the strongest competent rocks is only about 1.0 kbar (100 MPa). Hence, the maximum stress-drop associated with the intiation of a brittle first order fault, in such a strong rock type will have similar limits. So we may conclude that the maximum magnitude of pulse stress which can develop is unlikely to exceed 1 kbar.

As we have noted, once the first order fault has developed, the stress-drop that can occur thereafter is significantly smaller. The magnitude of such a stress-drop will be related to the differential stress that acts during re-shear. Let us assume that a fracture plane undergoing successive increments of re-shear possesses zero cohesion and is subjected to a small but positive confining pressure (σ_3). It is difficult to specify the magnitude of the differential stress and hence of the possible stress-drop in such a general situation. However, it seems unlikely that the magnitude of the differential stress, and hence of the stress-drop, will exceed 25 per cent and probably is less than 10 per cent of that stress-drop that can be experienced when the fracture is initiated.

Thus, from a consideration of the magnitude of the stress-drop, it would seem probable that the initiation of second order features occurs when the first order fracture is first formed. Price (1968) considered the influence of pulse stresses generated during fracture initiation (Fig. 6.33(*c*)), where it will be seen that the addition of a pulse stress (either positive or negative) ensures that failure conditions will also be reached adjacent to, and in advance of, a propagating first order fault. However, Price considered the action of pulses which followed a sinusoidal magnitude–time form. During such a form of pulse generation, the axes of transient principal stress will be continually changing in orientation and magnitude. Because failure could occur for all values of pulse stress (Fig. 6.33(*c*)), it follows that only microfractures could develop adjacent to the first order fault, and such microfractures would have a reasonably wide range of orientations. Consequently, Price concluded that organised, planar, second order fractures would not develop as the result of first order fracture initiation.

This conclusion, which is predicated upon the assumption regarding the sinusoidal form of the magnitude–time relationship of the pulse, we now consider to be incorrect. As has been noted earlier, we are of the opinion that the pulse magnitude–time relationship approximates to a square wave-form, (Fig. 6.24). Such a form would permit transitory stress conditions to be maintained which would be constant in orientation for a length of time which is sufficient to permit planar secondary fractures to develop. In support of this contention we can refer to the small-scale pinnate, or 'feather', fractures associated with relatively minor first order shear fractures which have experienced only one (or at most, a few) increments of movement.

Furthermore, the same mechanism can be invoked to explain the 'fringe' structures which are associated with some 'joints', (Chap. 2). In the light of the mechanism outlined above, these shoulder features can be interpreted as either extension or hybrid shear/extension fractures which developed in the first and only increment of movement on a somewhat larger, minor, first order fracture, a joint.

Given that secondary features are initiated at the same time as a first order fracture is generated, then, because a major, first order fault is likely to experience several hundred increments of re-shear, each increment of which can cause the growth and extension of already existing second order fractures, it is to be expected that the re-shear phase will have an influential and even dominant role in determining the magnitude of most secondary fractures.

It has been noted that the magnitude of the stress-drop and hence of the pulse stress that arises when a first order fault re-shears is probably in the range 1–200 bar. It is of interest, therefore, to estimate the likely strength of rock in which second order fractures may form if subjected to such relatively small pulses. With reference to the stress conditions shown in the graphical solution of the problem represented in Fig. 6.26(*b*): if the negative stress pulse $(-P)$ is taken to have the value of 100 bar then, on the same scale, the cohesive strength C is about 30 bar. It can be shown (Chap. 1, Eq. 1.59(*b*), (*c*)) that the uniaxial strength (σ_0) is given by:

$$\sigma_0 = 2Ck^{\frac{1}{2}} \tag{6.7}$$

where $k = (1 + \sin\phi)/(1 - \sin\phi)$, so that when $\phi = 30°$, $k = 3.0$: and therefore $\sigma_0 \approx 100$ bar.

If the average negative pulse stress were only 10 bar, then the strength of the rock in which secondary features may be induced would be commensurately smaller. The situation represented in Fig. 6.26, it will be recalled, represents the special case in which the axis of maximum principal stress makes an angle of 30° with the first order fault plane. For the general case, in which this angle may be as low as 5° or as high as 55°, the pulse stress required to cause second order failure may be smaller than that for the special case (curve bb', Fig. 6.32). Consequently, for a pulse stress of −100 bar the rock in which second order fractures may develop may be proportionally stronger. However, it is clear from these arguments that a greater degree of development of second order fractures is to be expected in the weaker brittle rock adjacent to a first order fault, where the magnitude of the pulse stress is related to the stress-drop which results from the *strongest* unit involved in the re-shearing of that fault.

Width of zone of secondary features

It has so far been assumed that the direction of propagation of the pressure pulse is parallel to the

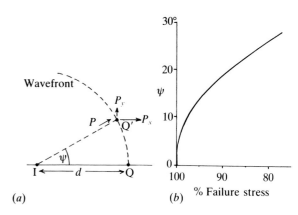

Fig. 6.34. (*a*) Schematic representation of pulse stresses for a circular wave-front when the pulse is resolved parallel and perpendicular to the first order fault. (*b*) Relationship between ψ and failure stress for a value of pulse stress just sufficient to cause failure when $\psi = 0$.

trend of the first order fault and that the pressure pulse has a planar wave-front. This assumption holds as a close approximation immediately adjacent to the main fault plane. However, the area of strain in which the pulse is initially generated may be small compared with the extent of the main fault. The pulse-front, in the horizontal plane, resulting from such a relatively small source of disturbance will be approximately circular. If we assume that the focus of the disturbance (I) in Fig. 6.34(*a*) is at a distance d from elements Q, Q' etc., such that the pulse-front at these various elements can be taken as planar, with the direction of the pulse stress acting along the radius centred at I: then at point Q', the component of pulse stress acting parallel and perpendicular to the first order fault will be:

$$P_x = P\cos^2\psi \tag{6.8(a)}$$

and

$$P_y = P\sin^2\psi. \tag{6.8(b)}$$

If it is assumed that the pulse stress generated by movement on the first order fault is just able to cause second order fracture development at point Q, immediately adjacent to the first order fault: then the stress intensities at points Q', Q'' etc. can be obtained by introducing the values of P_x and P_y from Eqs. (6.8(*a*), (*b*)) into Eqs. (6.5(*a*), (*b*)). The stresses so derived can be expressed as a percentage of that required for failure at point Q. These data are plotted in Fig. 6.34(*b*) against ψ (where ψ is the angle IQ', IQ'' etc. makes with the first order fault plane (Fig. 6.34(*a*)). It will be seen that there is a rapid fall-off in the intensity of the stresses available to generate secondary features away from the first order fault. This rapid reduction can be invoked to explain why second order fractures occur in relatively narrow zones on either side of a first order fault.

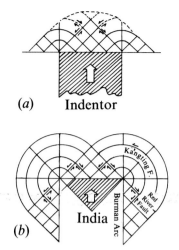

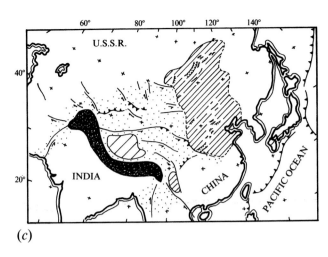

Fig. 6.35. (*a*) Shear stress trajectories in uniform plastic material induced by a square-ended indentor. (From Prandtl; after Tapponnier & Molnar, 1976.) (*b*) and (*c*) Strike-slip structures induced when, it is suggested, India acted as an indentor and Tibet and neighbouring areas acted as the 'plastic' body into which the indentor was pushed.

Wrench-fault and indentation tectonics

As we have seen, major strike-slip fractures are among the most important of tectonic structures and have a wide range of associated secondary and even tertiary structures. Indeed, as was noted earlier (Fig. 6.6) the associated features which include basins, folds and flexures are of sufficient complexity to have caused Moody & Hill (1956) to coin the descriptive term *Wrench-fault Tectonics*.

Another situation in which strike-slip faults may be associated with large scale structural or orogenic development of the Earth's crust has been termed *indentation tectonics*. It was noted in Chap. 4 that Prandtl calculated the distribution of shear stress trajectories which would develop in a plastic material (soil) when loaded by a square ended indentor (which simulated the load generated on its foundation by some heavy engineering construction).This distribution of trajectories is shown in Fig. 6.35(*a*) and it has been suggested (Tapponnier & Molnar, 1976) that it resembles the fracture traces of the strike-slip faults to the north of the Himalayas (Fig. 6.35(*b*) and (*c*)). It is proposed by them that these strike-slip faults result from the northward movement of the Indian subcontinent, relative to the main body of Asia and that the northern (somewhat square-ended) flank of India acted as an 'indentor'. The phrase 'indentation tectonics' was coined to describe this proposed mechanism.

This concept is interesting and may even have some validity. However, it must be emphasised that it is based on a semi-qualitative 'look-alike' evaluation of structural geometry. The Prandtl analysis is based on a two-dimensional model in the vertical plane. This model has been applied to interpret structures as viewed in the horizontal plane, where gravitational loading is important. This means that at shallow depths, the vertical stress will tend to be the least rather than the intermediate principal stress, so that

thrusts will tend to develop rather than strike-slip faults. Furthermore it is important to recognise other difficulties in applying the Prandtl model on a regional scale. In the model the indentor has exact rectangular geometry and the plastic material is assumed to be a homogeneous and isotropic plastic. Neither of these conditions are likely to be met in nature on the scale envisaged in indentation tectonics.

Commentary

Major strike-slip faults are profoundly important structures. However, as will be apparent from this chapter, it is the secondary fractures associated with the first order faults which have received most detailed, theoretical analysis. The range of secondary fracture types is such that no single mechanism is capable of satisfactorily explaining how the various types are initiated. One group of secondary fractures appears to be related to the geometry of the walls of the first order fracture. Another group can probably be related to the changes in stress that occurs towards the ends of first order faults. Yet a third group, which develop in sections remote from the ends of the first order faults, in which the main fault is planar, may be attributed to transient stresses which develop when the main fracture undergoes seismic faulting.

References

Allen, C.R. (1962). Circum-Pacific faulting in the Phillipines–Taiwan Region. *J. Geophys. Res.*, **67**, 4795–812.

Anderson, E.M. (1951). *The dynamics of faulting*, 2nd edn. Edinburgh: Oliver & Boyd.

Aydin, A. & Nur, A. (1982). Evolution of pull-apart basins and their scale independence. *Tectonophysics*, **1**, 91–105.

Chinnery, M.A. (1966). Secondary Faulting. *Can. J. Earth Sci.*, **3**, 2, 163–90.

Freund, R. (1965). A model of the structural development of Israel and adjacent areas, since Upper Cretaceous times. *Geol. Mag.*, **102**, 189–205.

(1974). Kinetmatics of transform and transcurrent faults. *Tectonophysics*, **21**, 93–134.

Heim, A. (1919–22). *Geologie der Schweiz.* Bds. I–III, Leipzig.

Hill, M.L. & Dibblee, T.W. (1953). San Andreas, Garlock and Big Pine fault, California. *Geol. Soc. Am. Bull.*, **64**, 443–58.

Kennedy, W.Q. (1946). The Great Glen Fault. *Q.J.G.S.*, **102**, 41–76.

Lajtai, E.Z. (1968). Brittle fracture in direct shear and the development of second order faults and tension gashes. *Proc. Conf. on Research in Tectonics*, eds. A.J. Baer & D.K. Norris, Geol. Surv. Can. GSC Paper 68-52, pp. 96–112.

McClintock, F.A. & Walsh, J. (1962). Friction on Griffith Cracks in rocks under pressure. *Proc. 4th U.S. Nat. Cong. Appl. Mech.* (Berkeley, California).

McKinstry, H.E. (1953). Shears of the second order. *Am. J. Sci.*, **251**, 401–44.

Moody, J.D. & Hill, M.J. (1956). Wrench fault tectonics. *Geol. Soc. Am. Bull.*, **67**, 1207–46.

Oliver, D. (1987). The development of structural patterns above reactivated basement faults. Unpublished Ph.D. Thesis, University of London.

Price, N.J. (1968). A dynamic mechanism for the development of second order faults. *Proc. Conf. on Research in Tectonics*, eds. A.J. Baer & D.K. Norris. Geol. Surv. Can. GSC Paper 68–52.

Reading, H.G. (1980). Characteristics and recognition of strike-slip systems. In *Sedimentation in Oblique-slip Mobile Zones*, ed. P.F. Ballance and H.G. Reading, pp. 7–26, *Spec Publ. Int. Ass. Sediment.*

Riedel, W. (1929). Zur Mechanik geologischer Bruchers-cheinungen. Zbl. Miner. *Geol. Palaeent.*, B 354.

Rutter, E.H. (1974). The influence of temperature, strain-rate and interstitial water in the experimental deformation of calcite rocks. *Tectonophysics*, **22**, 311.

Sibson, R.H., White, S.H. & Atkinson, B.K. (1981). Structure and distribution of fault rocks in the Alpine Fault Zone, New Zealand. In *Thrust & nappe tectonics*, eds. K. McClay & N.J. Price, pp. 1197–210. Geol. Soc. London Special Publ. No. 9. Blackwell Scientific Publications.

Tapponnier, P. & Molnar, P. (1976). Slip-line field theory and large-scale continental tectonics. *Nature*, **264**, 319–24.

Tchalenko, J.S. (1968). The evolution of kink bands and the development of compression textures in sheared clays. *Tectonophysics*, **6**, 159–74.

7 Overthrusts and thrust nappes

Introduction

It is difficult to give an unequivocal and generally acceptable definition of an overthrust and the associated thrust nappe, thrust sheet or thrust block. These structures, which are of regional extent, do not conform to the simple Andersonian classification of faults (Chap. 5). The overthrust, that is the fault plane, is usually a feature which, when initiated, was flat-lying or of low dip (i.e. less than about 5°), though, of course, the orientation of the thrust plane may be subsequently altered by folding. The thrust plane is sometimes a simple, geometrical surface, as shown in Fig. 7.1, but often comprises a layer whose thickness is commonly less than a few metres.

The thrust nappe, or sheet, consists of an allochthonous, extensive, but thin, block of rock which has been displaced on the thrust plane. It must be emphasised that the use of the words 'thrust' or

Fig. 7.1. Portion of the thrust surface of the Keystone-Muddy Mountain Thrust. (Courtesy of M. Johnson.)

'overthrust' does not necessarily have a generic connotation for, as we shall see, an overthrust may be a true thrust, but equally it may be a low-angle normal fault plane – a form of slide.

Major overthrusts were recognised in Europe by pioneer geologists. Examples include the Moine Thrust of Scotland, and the Glarus and other thrust nappes in the Alps. Large overthrusts have been reported in the Himalayas, the islands of Indonesia and on the eastern side of the Peruvian and Argentinian Andes and in many other orogenic areas elsewhere in the world.

Overthrusts, especially when they are well exposed, are impressive structures. The Glarus Thrust as seen from the Cassons Grat near Flims in Switzerland, is shown in Fig. 7.2(*a*). The section through the main structure from north to south is shown in Fig. 7.2(*b*). (Fig. 7.2(*c*) shows the original, historic, double nappe interpretation held by Escher van der Linth in 1841.) However, even the Alpine structures become less awesome when compared with those which have developed in North America; for probably some of the best developed and most intensely studied overthrusts are to be found on the eastern side of the Canadian and U.S. Rocky Mountains and the western side of the Appalachians. A view of the front of the McConnell Thrust can be seen in Fig. 7.3(*a*) and (*b*). The scale of thrust belts can be inferred from the sections in Fig. 7.4 where the view shown in Fig. 7.3 can be identified.

Recent interest in, and intensive study of, these and other overthrusts result from the discovery of important hydrocarbon reserves in the Eastern Canadian Rockies and this has led to a more detailed understanding of the geometry of these structures; (see for example Bally *et al.*, 1966; Norris, 1972; Price *et al.*, 1972; Dahlstrom, 1970; Boyer & Elliot, 1982; and Butler, 1982).

Geometry and nomenclature

The essential geometry and nomenclature of overthrusts and associated features is summarised in Fig. 7.5. The *sole* thrust plane, although often subhorizontal for considerable distances, can *climb* up through the stratigraphic succession. In areas where climb has taken place, the sub-horizontal portions of the thrust plane are termed *flats* and the parts that cut through the succession are known as *ramps*: together,

(a)

(b)   (c)

Fig. 7.2. (a) View of a portion of the Glarus Nappe, Switzerland. (b) Profile through the Glarus Nappe. (c) Original 'double nappe' interpretation of the Glarus structure.

they define a *staircase* (Fig. 7.5(a)). The block overriding the thrust plane is termed the *hanging wall* and the underlying block the *footwall*. The relative positions of the ramp in the hanging wall (H.W.R.) and footwall (F.W.R.) after movement has occurred on the thrust are indicated in Fig. 7.5(b). It will be noted that an anticline forms above a ramp. Also, it may be noted that a ramp, which often exhibits a dip of about 30°, satisfies the Andersonian definition of a thrust.

From the three-dimensional geometry of the thrust plane (Fig. 7.5(c)) it can be inferred that a thrust may climb in the direction of main thrust movement (on *frontal ramps*), at right angles to thrust movement (on *lateral ramps*), or in any other direction (on *oblique ramps*). The dip of lateral ramps may range from about 30° to 90°. When they are approximately vertical, they are termed *tear* faults (*T* in Fig. 7.5(c)). It was noted that antiforms develop over ramps (Fig. 7.5(b)): the effect of lateral ramps on these antiforms is to off-set the ramp anticlines to form an en echelon array of folds (Fig. 7.5(d)–(f)). This spatial disposition of folds is similar to that observed in both experimentally and naturally produced buckle folds (e.g. Fig. 10.39).

The ancillary thrust planes associated with the frontal portions of a major overthrust are often numerous and form a stack of *imbricate* or *schuppen* structures (Fig. 7.4), thought to form sequentially by *footwall collapse* (Fig. 7.6). A block completely bounded by thrust planes (Stage 1) is termed a *horse*. The flat fault at the base, which is common to adjacent ramps, is termed a *floor* or *sole* thrust while the flat at the top, which is common to adjacent ramps, is called a *roof thrust*, (Fig. 7.6, Stage 3). A series of horses (a herd?), together with their bounding thrusts collectively form a *duplex* (Dahlstrom, 1970). The profile shown in Stage 3 (Fig. 7.6) is also represented in Fig. 7.7, where it is related to the three-dimensional geometry of a typical duplex system. The reader interested in further nomenclature related to the intersection and junction of two thrust surfaces (termed a *branch line*) is referred to Hossack (1983).

Individual thrust structures are sometimes so large that one must, of necessity, think of them in terms of their Plate tectonic setting. As Bally (1981) has pointed out, there are currently three schools of thought relating to the tectonics of mountain building. They are (i) the 'fixists' (e.g. Beloussov, 1977) who advocate that mountain building is largely diapiric and that thrust nappes result either from gravitational sliding or gravitation spreading; (ii) those who adhere to the Expanding Earth concept (Carey, 1977), who also require that thrust nappes are formed by gravitational gliding or gravitational sliding; and (iii) the adherents of Plate Tectonics who would envisage that thrust nappes may be formed not only by gravitational spreading or gravitational sliding, but may also be engendered by lateral compression which

(a)

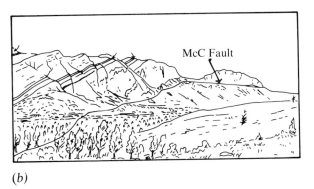

(b)

Fig. 7.3. (a) View of the McConnell fault as seen from the Calgary-Banff highway. (b) Sketch and interpretation of (a). (After Norris, 1972.)

results from plate collisions. Thus, there are three basic mechanisms which are thought to give rise to thrust nappes: *horizontal compression, gravitational sliding* and *gravitational spreading*.

As will already be apparent from these brief introductory remarks, overthrusts have received considerable attention from geologists and geophysicists as regards their geometry and morphology. This interest in establishing the geometry is being matched by the efforts of various geologists to explain how these large sheets of rock can be emplaced. Consequently, the bulk of this chapter is given to discussing the three main mechanisms noted above, namely, horizontal compression, gravitational sliding and gravitational

spreading, but we shall first consider the means of estimating the magnitude of displacement associated with these overthrusts.

Displacement and strain in overthrust nappes

The problem relating to determining the direction and amount of total movement exhibited by a fault has been considered in Chap. 5. However, a different and interesting approach to establishing the displacement of overthrusts has been suggested by Elliot (1976) who noted that thrusts often crop out along a well-defined, arcuate form (Fig. 7.8). He likened this to a bow and the movement direction and displacement (which reaches its maximum at the centre of the bow) to an arrow and demonstrated that there is a systematic relationship between the length of the 'bow string' and the displacement 'arrow' (Fig. 7.9) where, for the Rocky Mountain thrusts, it will be seen, that maximum displacement (u) is 7 per cent of the length (L) of the bow string between the extremities of the thrust.

More usually, geologists have used the method of the *balanced cross-section* to estimate the displacement or orogenic contraction which results from thrusting and associated imbrication and/or folding. This technique also provides a check on whether the geological cross-section constructed on the basis of field observations and/or seismic and borehole data is

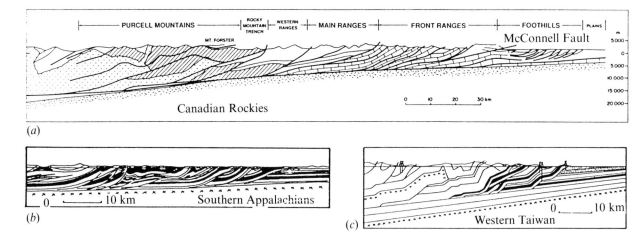

Fig. 7.4. Cross section of three foreland fold-and-thrust belts: (*a*) Canadian Rockies. (After Bally *et al.*, 1966.) (*b*) Southern Appalachians. (After Roeder *et al.*, 1978.) (*c*) Western Taiwan. (After Suppé, 1980.)

geometrically possible. However, it is emphasised that the solution is not necessarily unique and for any given set of data there are generally several 'possible' solutions.

Cross-sections are generally oriented so that they are normal to fold axes, or parallel to the direction of thrusting. Accepting the assumptions and rules of section balancing, outlined below, a complex structural cross-section can be restored to its undeformed state. If this restoration can be achieved without leaving any gaps, or overlaps, in the strata, then the proposed cross-section is held to be possible (thought still not necessarily correct), and the magnitude of the contraction, or extension, associated with the deformation can be determined.

In constructing a balanced section it is generally assumed that no volume changes took place during deformation. In addition, it is assumed that the conditions of plane strain obtain, i.e. no deformation occurs in the direction normal to the cross-section, so that there are no area changes in the section. It is further assumed that any folds represented in the section are *parallel* folds formed by *flexural slip* folding (these latter terms are defined in Chap. 10), so that the length of the beds in the cross-section remained constant during the deformation.

Based on these assumptions, two techniques of balancing sections have been developed. These techniques, which are outlined below, are *line length* and *area* balancing, respectively. (For a more detailed discussion, the interested reader is referred to the excellent review by Hossack, 1979.) The first step in the use of either technique requires that the stratigraphical thickness, prior to deformation, be determined. This can usually be achieved by examination of the stratigraphy of the adjacent, undeformed foreland. The undeformed section, known as the *template*, may reveal the beds to be parallel (the so-called,

layer-cake template) or possibly wedge-shaped. The next step is to find a reference position in the area where no interbed-slip (either by folding or thrusting) has occurred. This position, known as the *pin-line* (Fig. 7.10(*a*)), may be the hinge surface of a fold, or it may be located in the foreland.

In line length balancing, once a template and pin-line have been established, the length of a particular stratigraphic marker such as the boundary between layers A and B (Fig. 7.10(*a*)) across the section, is measured, i.e. the sum of the distances 1 to 2, 3 to 4, 5 to 6 and 7 to 8 in Fig. 7.10(*a*). (It is often found convenient to determine these bed lengths by using a piece of string, which is carefully placed along the various portions of the section.) This is repeated for a number of marker horizons and the section is restored (Fig. 7.10(*b*)). Ideally, lengths should be measured between two reference pin-lines, one on either side of the deformed zone. In order that the section be balanced, all the marker horizons should have the same length in the restored section. Generally, however, when this technique is applied to an area of thrust tectonics, the cover rocks are found to be 'too long' for the basement. From this it may be inferred that the cover rock and basement are decoupled on a decollement horizon which is termed a *sole fault*.

In some cross sections, for example when cleavage is widely developed in the rocks, the assumption that the beds maintained a constant thickness during deformation is unreasonable. In such situations the section can be balanced using the technique of area balancing.

The technique of area balancing was first used by Chamberlain (1910, 1919), to estimate the depth to a decollement plane beneath folds. The simple geometric concept used, which is based on the assumption that deformation of the section is by plane strain, is

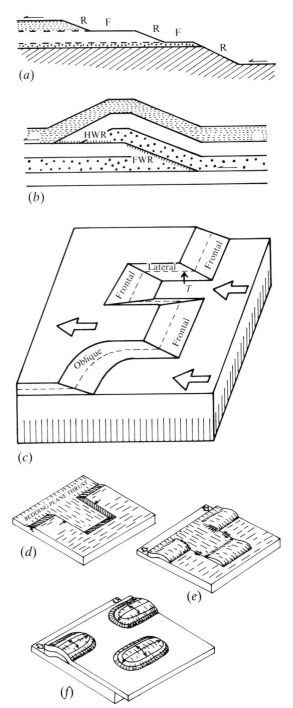

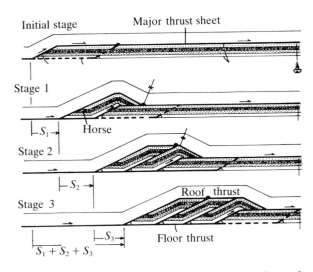

Fig. 7.6. The formation of a duplex by the progressive collapse of a footwall ramp. The roof thrust sheet undergoes a sequence of folding and unfolding. (After Boyer & Elliot, 1982.)

illustrated and the method indicated in Fig. 7.10(*c*). It can be seen, however, that if the depth to the decollement (t_1) is known from data obtained from exposure, borehole or seismic tests, the process and logic represented in Fig. 7.10(*c*) can be used to determine the shortening associated with deformation. The bed AB, originally at height BC above the decollement surface, is folded into a new position A'B'. The original and final lengths of the section are (AB = l_0) and (AO = l_1) respectively and OB is the shortening.

If plane strain is assumed then:

$$A'B'C'D = ABCD.$$

Fig. 7.5. Geometry and nomenclature of thrusts. (*a*) Trajectory of a staircase in a thrust, composed of ramps (R) and flats (F). (*b*) Relationship between hanging wall (HWR) and footwall ramps (FWR). (*c*) Block diagram of footwall topography giving ramp nomenclature: arrows indicate direction of thrust transport. (*d*), (*e*) and (*f*) Block diagrams showing the effect of lateral ramps and tear faults on the upper part of a thrust sheet. (*d*) shows the footwall boundary (*e*) a bedding surface in the lower half of the thrust sheet and (*f*) a bedding surface in the upper part of the thrust sheet. Note that the hanging wall anticlines are displaced to form an en echelon array. (*a*), (*b*) and (*c*), after Butler, 1982: (*d*), (*e*) and (*f*) after Dahlstrom, 1970.)

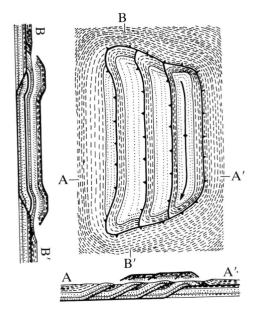

Fig. 7.7. A window showing the three dimensional geometry of a duplex. The lateral termination of the window occurs at lateral ramps in the sole thrust, which the roof and floor thrusts rejoin both along and across the strike. (After Boyer & Elliot, 1982.)

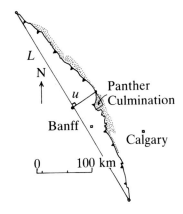

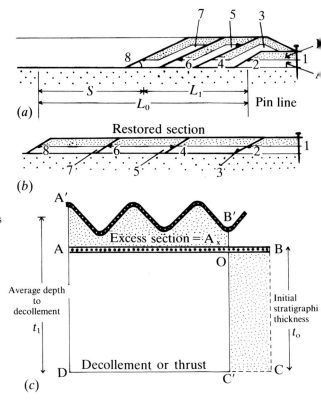

Fig. 7.8. Sketch map of the McConnell thrust, strike length (*L*) is 410 km and maximum displacement (*u*) is 45 km.

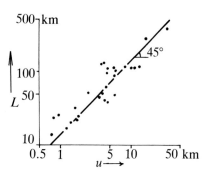

Fig. 7.9. Log–log plot of *L* against *u* for 29 thrusts in the Foothills and Front Ranges of the Canadian Rockies. Slope of line indicates a linear relationship of $u = 7\% \ L$. (After Elliot, 1976.)

Fig. 7.10. (*a*) Section through a thrust, indicating pin-line in the undeformed foreland. (*b*) Line balanced section of undeformed state. (*c*) Chamberlain's equal area model for calculating the depth of decollement beneath folds. The model can, however, be used to determine the tectonic shortening across the section if the depth of decollement is known. (See text.)

It follows from the geometry of Fig. 7.10(*c*) that the area of the excess section A_x is:

$$A_x = OBCC' = AA'B'O.$$

The initial depth to the decollement (t_0) is:

$$t_0 = \frac{A_x}{OB} \qquad (7.1)$$

or the shortening along the section is:

$$OB = \frac{A_x}{t_0}. \qquad (7.2)$$

Within the area A'B'C'D, the rocks may deform in any manner (e.g. folding or faulting).

As with line balancing it is necessary to establish a template and a pin-line in the undeformed part of the section. The area of a particular unit or group, in the deformed part, is then measured (preferably by using a planimeter). As the thickness of the unit is known from the template, its undeformed length can be determined and the section restored to its undeformed state. If all the restored units have the same length, the section balances.

From the assumption of plane strain deformation it follows that, in adjacent cross-sections, the amount of shortening along a specific horizon between two comparable pin-lines will be approxi-

mately the same, unless there is an intervening fault. This does not, however, preclude a gradual change in shortening along the strike of a fold, or thrust belt. Moreover, there may be a change in the manner in which shortening is achieved in different sections. For example, folds may pass laterally into an imbricate thrust zone. Compatibility between adjacent cross sections provides a further check on the validity of an individual cross-section.

Such determinations of strain and displacement, which are limited by the basic assumption of plane strain, lead to a minimum value of the estimated shortening. Reduction in the area of a balanced cross-section can result from compaction during lithification, pressure solution or elongation along the orogenic strike. Hossack (1979) points out that these may change the area by 15–45 per cent. Pressure solution effects and changes of strain along orogenic strike are particularly important in the more deeply buried structures. This has led to studies of strain in major nappes: for example, the work of Milton & Williams (1981) above a major thrust fault in Finmark (N. Norway) and Coward & Kim (1981) in a study of the Moine Thrust zone.

These and similar studies result from protracted investigations to determine the local strains, using one

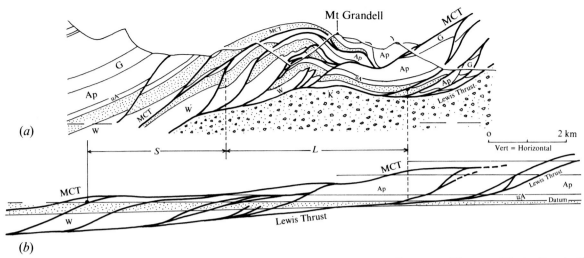

Fig. 7.11. (*a*) Structural profile through structures which have developed above the Lewis Thrust, near Waterton, Alberta, Canada. (*b*) Balanced cross-section of the structures represented in (*a*). (W) Waterton, (uA) Mid and Upper Altyn, (Ap) Appekunny, (G) Grinwell, comprising a Pre-Cambrian Belt supergroup thrust over (K) Cretaceous Siliclastics. *L* is current length and *S* is shortening. MCT = Mt. Crandell Thrust. (From Boyer & Elliot, 1982.)

or other of the methods outlined in Chap. 1 where it was indicated that strain determinations are usually first approximations and the attendant errors must automatically be transferred to conclusions regarding estimates of shortening based on the use of balanced sections. Nevertheless, the balanced-section construction remains an important technique for structural interpretation. An example of how complex structures are unravelled by this method is shown in Fig. 7.11. A synoptic diagram showing the stages in the evolution of the Moine Thrust zone in the region of Loch Eriboll, is shown in Fig. 7.12.

Mechanics of emplacement

Push-from-behind

Considerable effort has been directed to the problems of understanding the mechanics of the emplacement of these huge, thin overthrust sheets of rock. Indeed, for many decades these structural features were considered to represent a paradox in that, from the mechanistic viewpoint, they appeared to be 'impossible': a point of view that was somewhat marred by the fact that overthrusts only too clearly exist.

An early study of the overthrust problem was carried out by Smoluchowski (1909) who represented an overthrust block as a simple rectangular prism of dimensions *x*, *y* and *z* which was caused to move over a flat, dry surface, as shown in Fig. 7.13(*a*). The resistance to movement under the influence of force *F* is given by:

$$F = (xyz)w\mu \tag{7.3}$$

where *w* is the weight per unit volume and μ is the coefficient of sliding friction. In his analysis, Smoluchowski took $\mu = 0.15$ (the coefficient of friction of

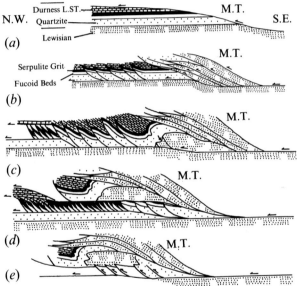

Fig. 7.12. Evolution of the Moine Thrust zone at Loch Eriboll. (After Coward & Kim, 1981.)

iron sliding on iron), and inferred that a 100 km long block would require a stress (given by F/yz) that would be far greater than any rock would sustain.

Let us now consider a somewhat more realistic loading condition, (Fig. 7.13(*b*)). We select an angle of sliding friction (ϕ) of 30°, which is appropriate to sandstone, (where, as we saw in Chap. 1, the coefficient of sliding friction $\mu = \tan \phi = 0.577$). It has been noted that the strength of rock, as represented by the relationship between principal stresses at failure, is given by:

$$S_1 = S_0 + kS_3 \tag{7.4(a)}$$

(see Chap. 1, Eqs. (1.57) and (1.59(*b*)) where S_0 is the uniaxial strength and $k = (1 + \sin \phi)/(1 - \sin \phi)$. It will be recalled that for thrusting, S_1 and S_3 act

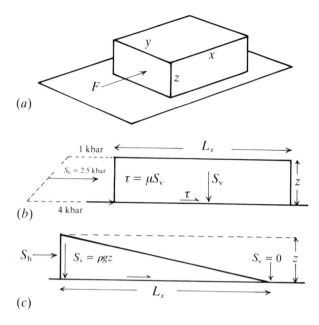

(a)

(b)

(c)

Fig. 7.13. (a) 'Push-from-behind' Model analysed by Smolu-chowski (1909) for the emplacement of an overthrust. (b) Boundary stress conditions for modified 'push-from-behind' model used by Hubbert & Rubey, (1959). (c) Corresponding boundary conditions for the wedge model.

horizontally and vertically respectively. Using the strength of the rock (S_1), determined from Eq. (7.4(a)), to represent the maximum stress that can be applied to the end of the block (Fig. 7.13(b)), the maximum length (L_x) of block that can be moved, can be calculated.

The confining stress (S_3) at the base of the block is given by:

$$S_3 = \rho g z \qquad (7.4(b))$$

where ρ is the density of the rock, z the thickness of the block and g is the gravitational constant. The value of ρ depends upon the mineralogical composition of the rock and its porosity. However, a wide range of sediments contain minerals of density 2.65 g cm^{-3} and, if the porosity is about 9 per cent, then $\rho = 2.5$ g cm^{-3}. Taking the value $\rho = 2.5$ g cm^{-3} and $z = 4$ km, it follows that $S_3 = 1.0$ kbar. It can be seen from Eq. (7.4(a)) that the strength of a block depends upon S_3 and that the maximum horizontal stress that can be sustained at the top of the block (where $S_3 = 0$) is the uniaxial strength (S_0), which here we take to be 1.0 kbar, and at the bottom of the block (where $S_3 = 1.0$ kbar) is 4.0 kbar. (Hence the average horizontal stress is 2.5 kbar, Fig. 7.13(b)).

We can recast Eq. (7.3) as:

$$F = S_1 zy = S_3 L_x y \mu \qquad (7.4(c))$$

which shows that the process of block sliding is independent of the width y. Rearranging Eq. (7.4(c)) gives:

$$L_x = \frac{S_1 z}{S_3 \mu} \qquad (7.4(d))$$

where L_x is the maximum length of thrust block that can move in response to the average horizontal stress S_1. Using the derived values of S_1 and S_3 (2.5 kbar and 1.0 kbar respectively) and a value of $\mu = 0.577$, then it follows from Eq. (7.4(d)) that $L_x = 17.5$ km.

If, instead of representing the thrust block as a rectangular prism, we take the block to have the form of a wedge-shaped prism, as indicated in Fig. 7.13(c), the length of the block which can move in response to an average stress S_x is double that for the rectangular prism. This follows from the fact that the average vertical stress S_z for a wedge which is 4 km high at the rear is 0.5 kbar (rather than 1.0 kbar for the rectangular block). Even a wedge with a length of 35 km is only a small fraction of the length of many major overthrusts. If we increase the scale of the feature and take the back wall of the wedge to have a vertical height of 12 km, then $S_x = 5.5$ kbar and $S_z = 1.5$ kbar. It follows from Eq. (7.4(d)) that $L_x = 72.26$ km. Hence, the length of the block, in which the dimensions have been made as favourable as possible, is only about 30 per cent of that inferred from the field evidence for some major structures.

This paradox was apparently largely resolved by the classic papers of Hubbert & Rubey (1959), who pointed out that the analysis by Smoluchowski of the type presented in the preceding paragraphs relates to the sliding of a block of dry rock on a dry thrust plane whereas, they emphasised, sedimentary rocks in the Earth's crust are mainly saturated with fluids.

The hydrostatic fluid pressure (p) which will be generated in a well (depth z) is given by:

$$p = \rho_w g z \qquad (7.4(e))$$

where ρ_w is the density of the water (which remains close to unity in the crust). Similarly, the vertical stress in rock $(S_v$ or $S_z)$ generated by gravitational loading is also given by:

$$S_v = S_z = \rho g z \qquad (7.4(b))$$

where ρ is the density of the wet rock.

In such saturated rocks, with void spaces and pores freely interconnected, we would expect that the ratio of pore fluid pressure to rock pressure (i.e. $p/S_z = \lambda$) to be approximately 0.4. The ratio λ is, of course, a measure of the ratio of the density of water and saturated rock. Fluid pressures determined in deep bore holes however, often exceed the expected ratio (Fig. 7.14). As we have seen in Chap. 1, the mechanical influence of pore fluid pressure on rock properties in general, and frictional sliding in particular, is profound. For dry rock, the normal and shear stresses are S_n and T respectively, but for wet rocks exhibiting a fluid pressure (p) (assuming the special law, see Chap. 1), the effective normal stress (σ_n) is:

$$\sigma_n = S_n - p \qquad (7.5)$$

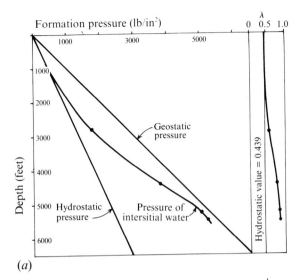

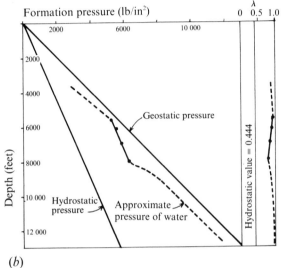

(a)

(b)

Fig. 7.14. Variations of pressure and corresponding values of λ with depth: (a) in the Khaur Field, Pakistan (after Keep & Ward, 1934) and (b) in Chia-Surkh, Iraq. (After Cooke, 1955.)

and

$$\tau = T. \tag{7.6}$$

In the situation of thrusting on a horizontal plane, $S_n = S_z$ and $\lambda = p/S_z$, so that:

$$\sigma_n = S_z(1 - \lambda). \tag{7.7}$$

Clearly, as p approaches S_z, i.e. as λ approaches 1.0, the normal effective stress becomes progressively smaller and approaches zero. That is, the fluid pressure (p) reduces the effective normal stress acting on any potential slide plane, but has no effect upon the shear stress.

The condition for frictional sliding is given by Amonton's Law, i.e.

$$\tau = \mu \sigma_n = \sigma_n \tan \phi \tag{7.8}$$

where the coefficient of friction $\mu = \tan \phi$ and ϕ is the angle of sliding friction. If, as the result of high fluid pressure, the effective normal stress (σ_n) on a plane approaches zero, then it follows from Eq. (7.8) that the resistance (τ) to frictional sliding also approaches zero.

If the rectangular model depicted in Fig. 7.13(b) is taken, then the relationship between the maximum length of the block which will slide and values of λ are as indicated in Table 7.1.

If we consider a wedge-shaped, thrust block as represented in Fig. 7.13(c), the maximum length of block is double the number shown above, so that it would apparently be feasible to push a 350 km long wedge, provided $\lambda = 0.9$, along the horizontal sole.

Gravitational sliding

Hubbert & Rubey also emphasised the importance of high fluid pressures in the situation in which a rectangular block slides down an inclined slope under its own body-weight. For sliding to take place in the situation represented in Fig. 7.15, $T = S_n \tan \theta$. Using Eqs. (7.5), (7.6) and (7.8), this can be rewritten as $\sigma_n \tan \phi = S_n \tan \theta$. Also, from Eqs. (7.5) and (7.7), $\sigma_n = S_n - p = S_n(1 - \lambda)$, therefore:

$$(1 - \lambda)\tan \phi = \tan \theta. \tag{7.9}$$

Table 7.1.

For $z = 4.0$ km							
Maximum λ	0	0.4	0.5	0.6	0.7	0.8	0.9
Length of block (km)	17.5	29.2	35.1	43.9	58.3	87.5	175.0

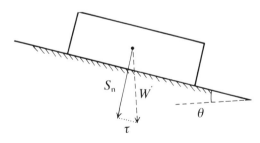

Fig. 7.15. Gliding of a rectangular block down an inclined plane.

Table 7.2.

λ	1 − λ	tan θ	θ (°)
0.0	1.0	0.577	30.0
0.4	0.6	0.346	19.1
0.6	0.4	0.231	13.0
0.8	0.2	0.115	6.6
0.9	0.1	0.57	3.3

Table 7.3.

λ	θ (°)	$\sin \theta$	L_x (km)
0.0	30.0	0.500	16.0
0.4	19.1	0.327	24.5
0.6	13.0	0.225	35.5
0.8	6.6	0.115	69.6
0.9	3.3	0.058	137.93

Table 7.4.

	θ	
λ_2	$(\lambda_1 = 0.95)$	$(\lambda_1 = 1.0)$
0.60	4.7(°)	3.1(°)
0.80	4.1	2.5
1.00	3.6	1.9

From this equation it follows that the angle of slope (θ), down which the block will slide depends upon the pore-fluid pressure. The relationship between θ and λ is presented in Table 7.2.

Slopes of 3.3° of unlimited extent cannot exist on the Earth's surface; and the difference between the altitude of the top and bottom of the slope must be realistic. Hubbert & Rubey took this limiting height to be 8 km and, therefore, the longest possible slope (L_x) will be given by $L_x \sin \theta = 8$ km. Hence, the lengths of glide block which can slide under their own body-weight, at various values of $\lambda = p/S_z$, can be determined and are as listed in Table 7.3.
Hubbert & Rubey also calculated the length of block which could be pushed down an inclined slope. We shall not, however, consider that aspect here because, admirable though the original treatment was, the block-line model proposed by Hubbert & Rubey is thought to be too simple.

More realistic conceptual models were proposed by Raleigh & Griggs (1963), in the first of which (Fig. 7.16(a)), they terminated the downhill side of the glide block with a simple 'toe' on a ramp which dipped, against the angle of movement, at 30°. They took the fluid pressure/overburden-pressure ratio under the main block to be λ_1, and under the toe the ratio was λ_2; while the thickness and length of the main block were taken as 6 and 100 km respectively. The relationships between θ (the inclination of the thrust plane necessary for sliding) and λ_1 and λ_2 are presented in Table 7.4.

In a second, more elaborate model which simulates the Pine Mountain Overthrust, they considered four elements (Fig. 7.16(b)). It was demonstrated that if, for simplification, $\lambda_1 = \lambda_2 = \lambda_3 = \lambda_4$, then the slope θ, required for gravity gliding for various values of λ, is as listed in Table 7.5.
It is evident from these studies that conditions at the front end of the glide block put serious constraints on its ease of movement. Moreover, it should be noted that Raleigh & Griggs assumed that no additional energy was needed to bring about flexuring of the main block at the ramp or toe.

However, it would appear that an even more important restraint which must exist at the initiation of movement of a natural glide block is the cohesion across the potential slide plane that must be broken before sliding can take place. As Hsu (1969) pointed out, the sliding condition assumed by Hubbert & Rubey (and also by Raleigh & Griggs) was:

$$\tau = \mu\sigma_n. \tag{7.8}$$

However, for the *initiation* of a glide surface, the correct condition is given by:

$$\tau = C_0 + \mu\sigma_n \tag{7.10}$$

where C_0 is the cohesive strength on the shear surface prior to any shear movement.
The value of C_0 may be several tens of bars, so that the stress needed to initiate the movement of a block on a horizontal plane, or the angle of inclination of a potential glide plane, must be very significantly higher than may be inferred from the analysis of either Hubbert & Rubey or Raleigh & Griggs.

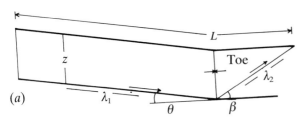

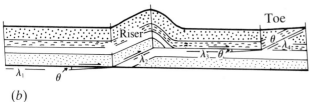

(b)

Fig. 7.16. (a) Gravity gliding of a rectangular block with a prismatic toe and (b) a more complex ramp and toe geometry. (After Raleigh & Griggs, 1963.)

Table 7.5.

$\lambda_1 = \lambda_2 = \lambda_3 = \lambda_4$	θ (°)
0.91	6.8
0.95	5.0
1.00	3.0

Despite these various criticisms, modifications and amendments, one cannot over-emphasise the importance of the contribution made by Hubbert & Rubey, if only for the way in which they demonstrate the existence and the mechanical significance of high fluid pressures in the development of these structures. They, themselves, were reasonably conservative in their assessment. Later authors postulate very high values, so that $\lambda \approx 1.0$. Such conditions may, in fact, obtain; however, such high values of λ which are sometimes considered necessary to bring about gliding, as, for instance, in the Raleigh & Griggs analysis (Tables 7.4 and 7.5), may not always be realised.

For example, Aprahamian & Pairis (1981) have demonstrated a degree of metamorphism and temperature inversion associated with an overthrust in the Haute Savoie (France). To account for the observed relationship, they proposed that heat was produced by friction along the thrust plane, from which one can infer that $\lambda < 1.0$. In such instances, the thrust plane acts as a 'heat engine' and may induce local convection of the fluids contained in the rock mass. Such local inversion of the geothermal gradient may result in a general metamorphism. An example of this occurs adjacent to some high level parts of the Moine Thrust, where such local 'metamorphism' gives rise to the development of a quartz depletion zone, adjacent to and immediately below the thrust plane, which is, in turn, underlain by a zone of quartz veins which decrease in importance (i.e. number and thickness of veins) away from the thrust plane (C. Fabry personal communication).

The energy and stresses needed to bring about flexuring of the thrust block at a ramp and/or toe have been analysed by Wiltschko (1981). His analysis was also modelled on the geometry of the Pine Mountain thrust block of the Southern Appalachians, and he assumed that the thrust block behaved as a linear (i.e. Newtonian) viscous sheet. He concluded that the flexure associated with the ramp, although not the most important energy sink, constituted an important source of resistance to movement of the thrust block. Certainly, flexure at the ramp is ductile deformation, but it may not be, and for the environment of the Pine Mountain structure was probably not, the result of viscous deformation.

An alternative method of analysing the movement of an over thrust block is to treat it as a plastic body. This approach has been used by Mandl & Crans (1981). In their analysis of gravity glide in a hypothetical delta, they were able to demonstrate how listric thrusts develop in the toe region of the delta. However, even when using the 'plastic' model, Mandl & Crans found it necessary to invoke over-pressured sediments in the deeper layers beyond the proximal portions of the delta. In this deltaic environment it is,

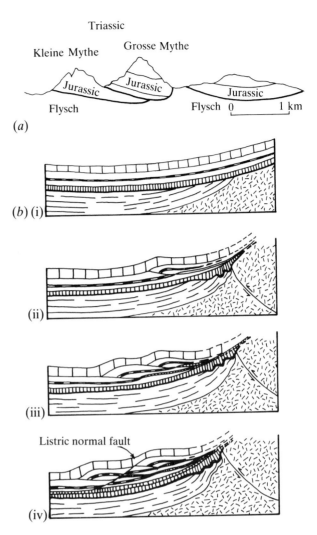

Fig. 7.17. (*a*) View of the Mythen and Rotenfluh outliers (Switzerland) emplaced by gravity gliding. Thick lines represent main glide planes. (*b*) Gravity glide model for Tinee Nappe, in the Maritime Alps, France. (After Graham, 1981).

of course, eminently reasonable to expect high fluid pressures and 'thrusts' to develop by gravitational gliding. Similarly, the higher tectonic units of a mountain belt, such as the Pre-Alps, and Klippen, e.g. the Mythen and Rotenfluh of the Swiss Alps (Fig. 7.17(*a*)) or the Tinee Nappes of the Maritime Alps, France (Fig. 7.17(*b*), see Graham, 1981), can reasonably be attributed to gravity glide.

Unfortunately, the frictional sliding hypothesis gives rise to 'catastrophic' failure. That is, once the slip conditions are realised, movement on the glide plane will take place at a relatively high velocity. Certainly, some landslides are literally catastrophic. One cannot remain unimpressed by the scars left by the infamous Frank's Slide, Alberta, Canada, in which huge boulders (many greater than 5 m in diameter) were transported a distance of about 2 km in a time estimated to be of the order of 100 s. However, equally certainly, some gravity glide phenomena are slow moving and non-catastrophic as, for example,

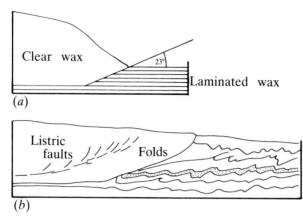

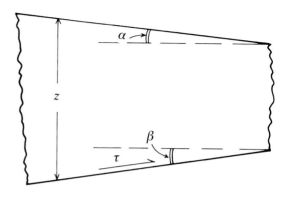

Fig. 7.18. Gravitational Spreading Model. (*a*) Before and (*b*) after deformation. (After Bucher, 1956.)

Fig. 7.19. The Nye Gravitational Spreading Model used to explain movement of glaciers.

the large blocks and rafts of limestone that are slowly moving down the slopes of the spine of Italy. The rate of progress of these blocks is controlled by the 'creep' of the 'argille scagliosi' (shaly clays) on which the rocks rest.

Moreover, as will be seen from the sections of the eastern structures of the Canadian Rockies and of the western overthrust of the Appalachians (Fig. 7.4(*a*) and (*b*)), the basement and the sole thrusts may dip *against* the direction of slip. Hence, gravity gliding in such a situation is out of the question. For movement of such thrust masses, we must take into account not only the sliding friction (and possible cohesion) but also the down-slip component of the body-weight of the thrust mass. As we shall see, these requirements pose serious problems for the frictional slip model. It is for this reason that other models and mechanism of thrust development have been envisaged.

Gravitational spreading
This mechanism is based on the concept that a ductile material which has gravitational potential energy will spread laterally under its own weight. As an example, we may take such material as 'bouncing' or 'potty' putty which slowly changes shape and spreads out laterally on a flat surface (see Chap. 1). A much more impressive experiment was conducted by Bucher (1956) (Fig. 7.18) in which a mass of wax, constructed as shown, abutted onto a lower level of laminated wax. The massive wax subsequently flowed under its own weight and developed a series of folds within the laminated wax and the trace of a listric fault and 'schuppen' developed within the massive wax.

Elliot (1976) took a mathematical model, first developed by Nye (1952) to explain glacial flow, and argued that, given a configuration of a surface and glide plane which form the boundaries of a wedge (as shown in Fig. 7.19), then, if the stresses in the 'spreading' mass are near hydrostatic, the shear stress at the base is given by:

where, surprisingly, α is the slope of the surface (in radians) of the spreading mass.

Ramberg (1981) used a comparable mathematical model which lends support to his well-known diapiric models produced in a centrifuge, to indicate that gravitational spreading can be of importance in the development of fold nappes. However, let us here restrict ourselves to considering the development of thrust nappes. The treatment by Nye relates to the deformation of ice in a glacier, where the material is near its melting point and is, therefore, very weak. Similarly, the experiment by Bucher was conducted on wax (with a melting temperature of about 40 °C) at a temperature of 20 °C.

Metallurgists use the concept of homologous temperature (T_H) such that:

$$T_H = \frac{T_D}{T_m}$$

where T_D is the temperature of the material at deformation and T_m is its melting point. Both T_D and T_m are measured in degrees Kelvin (K). It has been shown, that metals with vastly different melting temperatures behave in a comparable manner when the homologous temperatures are the same. The melting temperature of the wax used by Bucher (T_m) was about 40 °C or 313 K, and the temperature at which it was deformed was 293 K. Thus, the ratio $T_D = 0.94\,T_m$. Using this argument, acid crystalline rock will only behave like wax or ice when it, too, exhibits a comparable homologous temperature, i.e. when the rock temperature is approaching, or in excess of, 600 °C.

Not all rocks undergoing deformation, however, are of crystalline material, 'Soft' sediments in their unindurated state, especially if the interstitial fluid pressures were high, could deform under their own body-weight. Hence, 'gravitational' spreading could be an important component mechanism in delta tectonics especially in the development of structures near the toe of the delta.

$$\tau = \rho g z \alpha \qquad (7.11)$$

It has also been suggested that, because a reduction in strength is associated with a scaling effect (Chap. 1), it is incorrect to use laboratory-determined data on rock strength as a direct indication of the strength of an overthrust of such rock. However, the uniaxial strength of rock is not reduced to zero by the scaling factor, or even to such a low value as to allow the rocks to flow under their own weight. This is substantiated by the existence of residual stresses which may be as large as 600 bar (Chap. 4) and which could not exist if the rock strength fell to almost zero. Hence, it follows that indurated brittle rock, in the cooler regions of the upper crust, will not exhibit a strength/scaling effect that would satisfy the conditions required by gravitational spreading, namely that the strength would fall almost to zero so that the rock mass would flow under its own body-weight.

Consequently, we can conclude that the mechanism of gravitational spreading will be viable only when the rock mass is very weak: a condition which is realised when indurated rocks are at elevated temperatures and near their melting point, or when the sediments are non-indurated and exhibit a high interstitial fluid pressure. This mechanism will, therefore, appear to have played an insignificant role in the emplacement of such major structures as are exposed and inferred in the Eastern Canadian and U.S. Rockies and elswhere in the world: so let us return to the sliding elastic-block model.

Push-from-behind (Smith model)

From the arguments presented in the previous sections, it may be inferred that the general, and at this point vague, mechanism which can best be invoked to explain the emplacement of the majority of major overthrusts is that in which a wedge of rock is pushed from behind. In some circumstances such as, for example, the westward overthrusts in the Appalachians (Fig. 7.4(*b*)), thrust movement may possibly be attributed directly to the collision of plate boundaries. However, in other circumstances, such as in the eastern zones of the Canadian and U.S. Rockies, the thrusts are thought to have developed at the same time as the emplacement of the major granitic batholiths to their rear (Fig. 7.20). Smith (1981) suggests that it is difficult to envisage the transmission of the stresses which develop as the result of plate boundary collision through the liquid magmatic welt of the batholith. To obviate this problem, Smith further proposes that the emplacement of the granite intrusions, which went to form the batholith, were themselves responsible for the development of the overthrusts. He adduced the data to support his thesis: and inferred from the volumes of the thrust wedges and the batholithic material emplaced, combined with the rates of accretion, that the data are compatible (Figs. 7.20 and 7.21).

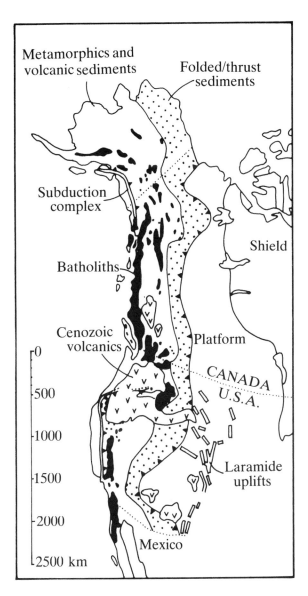

Fig. 7.20. Distribution of late Mesozoic and early Cenozoic subduction complexes, batholiths and fore-land folding/thrusting in the N. American Cordillera. (After Smith, 1981.)

Smith also presented a qualitative model of the mechanics of the 'fluid push' mechanism (represented in Fig. 7.22). In Fig. 7.22(*a*) and (*b*), it is assumed that the fluid welt extends throughout the full height of the back boundary of the thrust wedge; and in Fig. 7.22(*c*), the generally more realistic model is taken, in which there is a solid cap above the fluid welt. It is assumed that a weak layer exists at the base of the wedge, along which frictional slip will take place.

Consider the model represented in Fig. 7.22(*a*). The mass (*M*) of the wedge, of unit thickness parallel to the strike of the thrust plane is:

$$M = \frac{tL\rho}{2} \qquad (7.12(a))$$

where *t* and *L* are as indicated in Fig. 7.20(*a*) and ρ is

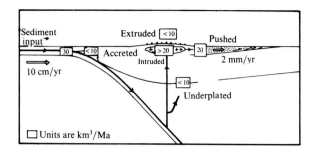

Fig. 7.21. Tectonic flow diagram linking subduction, batholithic intrusion and foreland thrusting. The numbers indicate the increase in volume of the continental plate by various processes. Units are in km³/Ma. (After Smith, 1981.)

the density (which Smith takes to be equal to that of the fluid welt).

The down-slip component of force (F) exerted by this mass is given by:

$$F = \frac{Lt\rho g \sin \beta}{2}. \qquad (7.12(b))$$

The sliding resistance (F_r) to movement on the basal plane of weakness is given by:

$$F_r = \frac{\tau L}{\cos \beta}. \qquad (7.12(c))$$

From Fig. 7.22(a), it is clear that the magmatic pressure (P_m) at the top of the welt is zero and at the base is given by:

$$P_m = \rho_m g t. \qquad (7.12(d))$$

Hence, the average magmatic pressure is $P_m = \rho_m g t/2$. A force (F_m) is exerted by this pressure of:

$$F_m = P_m t = \frac{\rho_m g t^2}{2}. \qquad (7.12(e))$$

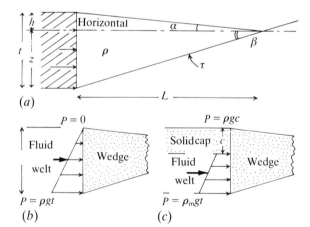

(a)

(b)   *(c)*

Fig. 7.22. Model proposed by Smith (1981) for the emplacement of thrusts as a result of fluid pressure associated with the emplacement of a batholith (see text). (a) Idealised fluid welt model (shaded) which has the same density (ρ) as the wedge. The yield stress of the decollement horizon is τ and the horizontal arrows indicate the fluid pressure at the rear of the wedge. (b) Pressure at the rear of a wedge driven by a completely fluid welt. (c) Pressures when fluid welt is capped by rock.

If this average magmatic pressure results in movement of the wedge up the plane, it has to overcome the down-slope component of the body force and the 'frictional', or other, resistance to sliding on the weak layer. Hence,

$$P_m t = \frac{\rho g t^2}{2} = \frac{tL\rho g \sin \beta}{2} + \frac{\tau L}{\cos \beta}. \qquad (7.12(f))$$

For small values of β, $\sin \beta = z/L$, where z is the vertical thicknes of the wedge downward from the horizontal. Also, for small angles, $\cos \beta = 1.0$, so that Eq. (7.12(f)) reduces to:

$$\frac{\rho g t^2}{2} = \frac{tL\rho g z}{2L} + \tau L. \qquad (7.12(g))$$

Rearranging and dividing by L:

$$\tau = \frac{\rho g t^2}{2L} - \frac{\rho g t z}{2L} = \frac{\rho g t(t-z)}{2L}. \qquad (7.12(h))$$

But $t - z = h$ and $h/L = \sin \alpha = \alpha$, in radians, for small values of α. Hence,

$$\tau = \frac{\rho g t \alpha}{2} \qquad (7.13)$$

or, for α in degrees,

$$\tau = \rho g t \alpha \frac{\pi}{360}. \qquad (7.13(a))$$

This relationship is comparable with that derived by Nye for gravitational spreading.

Let us now take reasonable values and substitute them in this equation to obtain the magnitude of the shear stress (τ). Consider a wedge where $\alpha = 1.5°$, $t = 10$ km and $\rho_m = 2.3$ g cm^{-3}, then ($2300 \times 1.5/7 \times 22/360 = 30$ bar). For the second model (Fig. 7.22(c)) with a solid cap of thickness $c = 2$ km, the value of τ would be reduced to about 24 bar.

The geometry assumed by Smith is simple. The frictional sliding mechanism is invoked, but he does not take into account any cohesion (C_0) across the thrust plane. The addition of a ramp to the Smith model will further reduce the proportion of the shear stress available to produce sliding. Indeed, Smith did not consider how any of the structures which develop at the front of major thrusts would influence his proposed mechanism of thrust emplacement.

Let us defer further discussion of thrust emplacement and first consider possible mechanisms involved in the development of some of the structures associated with the frontal regions of a major thrust.

Mechanics of development of 'frontal' structures

As may be seen from Figs. 7.4(a), (b) and (c), the main structures which develop towards, and at the front of, major overthrust structures are ramps, schuppen (imbricate) structures or fold belts such as the Jura

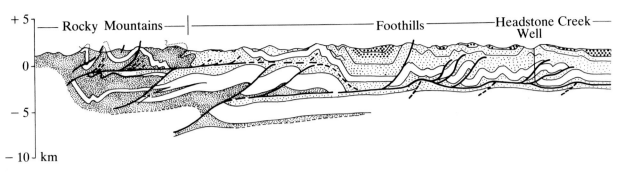

Fig. 7.23. Blind thrust (i.e. a thrust which does not crop out) beneath folds and reverse faults in the Laurier Section of the Canadian Rockies. (After Thompson, 1981.)

Mountains which form on a plane of decollement. The existence of a 'blind thrust' (i.e. one which does not crop out) may be inferred from such folds. For example, Thompson (1981) inferred such a thrust beneath the folds of the Laurier Section of the foothills of the Canadian Rocky Mountains (Fig. 7.23).

The association of folds and minor thrusts at the front of a major thrust can occur in at least two very different ways. The formation of a ramp will produce a hanging wall anticline in the manner shown in Fig. 7.6 (stages 1 and 2) or the amplification of a fold may lead to the development of a thrust as shown in Fig. 7.24 (this latter process is discussed in more detail in Chap. 10). The geometry of the final structures in Fig. 7.6, stage 2, and Fig. 7.24(*c*) are very similar and both may continue to develop by further slip on their thrust. Alternatively, movement may cease as new ramp thrusts are developed by the process of footwall collapse (Fig. 7.6, stages 2 and 3).

Field observations show that the relatively minor frontal thrusts when they form, whether they are single ramp features or one of a series forming an imbricate zone, are usually listric and sometimes sigmoidal in profile, Fig. 7.25.

The listric, or curved, kind of fault was attributed by Hubbert (1951) and Hafner (1951), in their analyses based on the use of stress functions (Fig. 7.26) to variations in orientation of the stress trajectories. However, it can readily be shown (Price, 1977) that the approach used by Hubbert cannot explain the 'asymptotic' junction between the listric fault and the flat-lying, or gently inclined, sole thrust. Price suggested that the listric faults originated and formed near the sole as a hybrid extension and shear fracture (of the type discussed in Chaps. 1 and 2). The proposed mechanism relates to the observation that interest in high fluid pressure has focused only on the fluid pressure beneath the glide block. Price noted that if a high value of $\lambda = p/S_z$ exists below the thrust block, then a high fluid pressure must also exist within the thrust block in a zone adjacent to the sole: though, as shown in Fig. 7.27(*a*)), this high fluid pressure will decay upwards from the sole. He then noted that in an

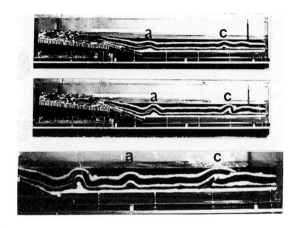

Fig. 7.24. Laboratory model, showing stages in the development of a fold which then evolves into a thrust.

air-dry test specimen, the stress–strain relationship at low stress levels is as shown in Fig. 7.27(*b*), where the markedly non-linear relationship is attributed to the closing of the pores and the void spaces in the specimen. If these void spaces are filled with water, as in the thrust block, and the block then undergoes deformation, then the fluid pressures will build up and conditions could be attained near the sole when λ exceeds unity (Fig. 7.17(*c*)).

If the fluid pressure attains a value such that $p = S_V + T_\perp$ (where S_V is the vertical total pressure and $T_\perp$ is the tensile strength of the rock in the thrust block (perpendicular to the bedding) and if $S_1 - S_3 < 4T$ then horizontal hydraulic fractures will develop. If, however, the conditions for hydraulic fracturing are not attained $s_1 - s > 4T - 5T$, a hybrid shear and extension fracture can develop at low angle to the sole thrust (Fig. 7.28). The shear resistance on a hybrid shear-extension fracture is zero. This causes an increase in the intensity of differential stress higher in the thrust block (as indicated by the convergence of the σ_1 stress trajectories, in Fig. 7.29). In this part of the thrust block the fluid pressure is lower than it is near the sole thrust. These two changes give rise to a shear with a steeper angle of dip, as indicated in Fig. 7.28. If high fluid pressures communicate with

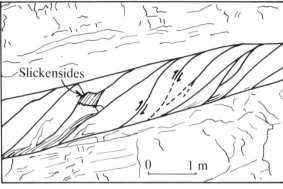

(a)

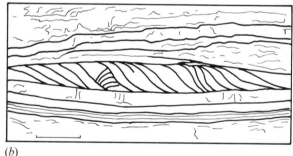

(b)

Fig. 7.25. (*a*) A series of sigmoidal thrusts in a sandstone bed, Widemouth, N. Cornwall, England. (*b*) A small scale thrust duplex from Northcott Mouth, N. Cornwall, England.

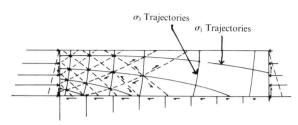

Fig. 7.26. Listric faults (dashed) which are inferred from a theoretical analysis based on boundary stress conditions and stress functions. (After Hubbert, 1951; see also Chap. 5.)

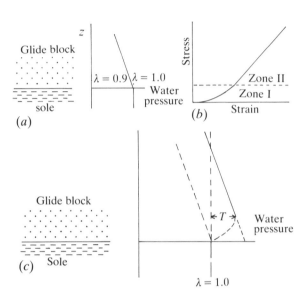

Fig. 7.27. (*a*) Distribution of λ in rocks below and above the sole thrust plane. (*b*) Stress–strain relationship exhibited by dry porous rock subjected to small compressive stress. (*c*) Distribution of λ for the situation represented in (*a*) immediately after the glide block above the sole has been subjected to horizontal compression. (After Price, 1977.)

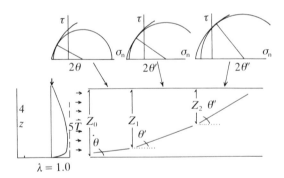

Fig. 7.28. Development of listric fault in a thrust block, which results from variations in λ and in differential stress. (After Price, 1977.)

the resulting listric fault, the shear stress on the listric fault becomes very low so that the principal stresses are reoriented, as indicated in Fig. 7.29. Thus, the listric fault locally takes on the role of the sole thrust and secondary listric imbrications can then develop. Although the arguments presented by Price related to a gravity glide model, the same concepts can be applied to a 'push from the rear' model.

As regards the high fluid pressure, Gretener

(1981) points out that the development of one listric ramp loads the footwall and so generates high fluid pressures below (Fig. 7.30(*a*)). Indeed, this mechanism can give rise to a 'piggy-back' arrangement of listric faults (with a possible concomitant isostatic response) as indicated in Fig. 7.30(*b*). It may be inferred from this mechanism that listric faults

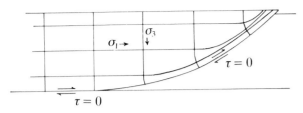

Fig. 7.29. Qualitative representation of the redistribution of stress trajectories that may arise in a thrust block as the result of the development of a listric fault. (After Price, 1977.)

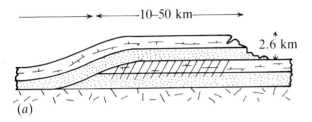

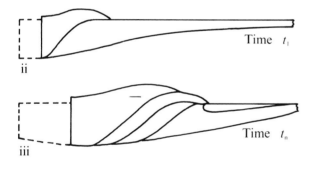

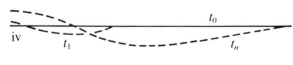

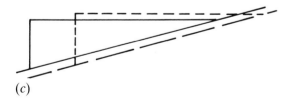

Fig. 7.30. (*a*) Major overthrusting results in the development of rapid loading which leads to high fluid pressures in the shaded area. After a delay, this effect may be enhanced by 'aquathermal pressuring'. From the dimensions indicated, isostatic response of the crust is inevitable. (*b*) Stages in the development of a 'piggy-back' thrust complex. The attendant and 'inevitable advancing isostatic bulge' is shown at times t_0, t_1 and t_n in (*b*) (iv). (After Gretener, 1981.) (*c*) Wedge thrusting with isostatic adjustment, dashed.

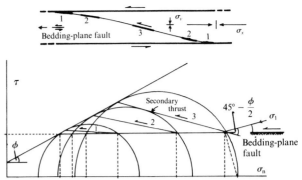

Fig. 7.31. Pore-pressure control of curved ramp, or step(flat–ramp–flat)-bedding-plane thrust. (After Mandl & Shippam, 1981.)

would form first at the rear and migrate forward, by the process of footwall collapse. Gretener introduces an important element in this figure relating to the question of thrust emplacement. An increase in vertical loading, which takes place as the thrust develops, results in isostatic adjustment and depression of the thrust plane (Fig. 7.30(*b*) and (*c*)). In these circumstances, the work done in thrusting the block forward may be somewhat less than that previously estimated, because the centre of mass of the thrust block is less elevated as a result of depression of the sole thrust.

It may be noted from Figs. 7.30(*a*) and (*b*), that the listric fault, on reaching the top of the ramp, flattens out and thus, in profile, becomes sigmoidal. Mandl & Shippam (1981) used a plastic model and showed how, with suitable variations in fluid pressure, sigmoidal-shaped shears could occur internally within a bed or unit. This form of shear requires that the fluid pressure is higher at the upper and lower limits of the bed or unit (Fig. 7.31). This diagram is specifically included because the stress analysis is achieved by use of the 'pole' method associated with the Mohr's stress circle. An introduction to this method is given in the paper by Mandl & Shippam, to which the interested reader is directed for further information. It has already been noted in the preceding paragraphs that treating the unit as a brittle body with a similar distribution of very high fluid pressures would give rise to a similar sigmoidal shear.

In their interesting analysis, Mandl & Shippam showed that individual frontal thrust units are usually bounded by front and rear thrusts, as indicated in Fig. 7.32(*a*). They also demonstrated that imbricates would develop from the rear towards the front of a thrust block (Fig. 7.32(*b*)). Moreover, they showed that once the value of α has been determined (using the equation in Fig. 7.32), or estimated from observed surface separation of neighbouring faults, the length of the unfaulted region (L_5) and even the length of the original thrust sheet (L_i) can be calculated by means of the relationship expressed in Fig. 7.32.

Let us now return to the discussion of emplacement of major thrust nappes.

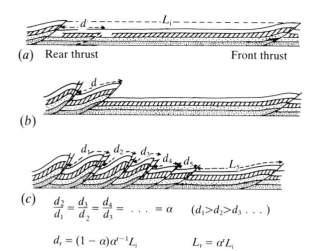

Fig. 7.32. (*a*) Thrust unit, length L_i, with bounding front and rear thrusts. (*b*) Imbrication thrusts generated at the rear of thrust unit. (*c*) Progressive thrust sheet imbrication. (After Mandl & Shippam, 1981.)

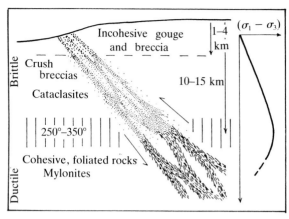

Fig. 7.33. Representation of the transition from brittle failure in the near surface portions of a thrust to ductile deformation at depth. (After Sibson *et al.*, 1981.)

Mechanics of thrust emplacement – continued

As has been noted, Smith (1981) did not take into consideration the multiplicity of frontal structures which frequently develop in the frontal zones of major overthrusts. All these various structures, one may infer, require an abundance of energy to be available in this frontal zone. However, Smith's model (which is perhaps the most pertinent one as regards emplacement of Rocky Mountain-type structures) clearly 'runs out of steam' as the front of the thrust wedge is approached. Yet, as the frontal structures attest, it is here that high stresses and large reserves of energy are necessary.

The impression was formed earlier in this chapter that the elastic model, pushed from behind, is barely feasible as a mechanism for overthrust emplacement even when all the various geological and physical parameters are optimised. When the requirements of the frontal structures are then taken into consideration, it is clear that the sliding block model in its simple form cannot explain the emplacement of overthrusts. It is for such reasons that some geologists have suggested that thrusts move in a 'rippling', caterpillar-like manner, a mechanism that is the large scale equivalent of slip on a glide plane, within a crystal, by the process of dislocation migration. This process produces a unit cell displacement of the crystal on one side of the slip plane with respect to the other. Because it is not necessary to rupture all the bonds across the slip plane simultaneously, this process has the advantage of requiring only a small amount of energy at any one time.

To consider this problem further, it is pertinent to note that many overthrusts immediately overlie a relatively thin layer, or zone, of fine-grain fault rock which is sometimes termed 'gouge' or, if it is suffi-

ciently fine-grained, 'mylonite'. Such a zone of fault rock can be seen in Fig. 7.1, immediately beneath a 'brittle' shear plane. Such a layer, which is so widespread beneath the Glarus Nappe in the Swiss Alps, is sufficiently important to warrant the appellation 'Lochseitenkalk'.

Sibson (1977) studied the Outer Isles Thrust of the Outer Hebrides of Scotland. Because of differential uplift along the outcrop of this thrust, the structure is exposed at different depths. Consequently, he was able to demonstrate from the associated rock products, that displacement on this major structure took place, at depth, along a narrow ductile zone and that deformation became progressively more brittle towards the forward, shallower areas of the thrust. Sibson *et al.* (1981) have shown that the rock products associated with the Alpine Fault, New Zealand, permit a similar conclusion to be reached (Fig. 7.33). This change in mechanism from ductile to brittle has important connotations as regards earthquake hazards associated with movement on specific faults. Elliot & Johnson (1980) drew similar conclusions regarding the Moine Thrust in Scotland.

It is not immediately clear how a narrow zone of ductile failure develops in the environment described above. However, in an elegant experiment, Mandl & De Jong (1977) demonstrated that a homogeneous aggregate of granular particles subjected to large strains, where the material is constrained to fail in simple shear, does, in fact, eventually break down to form a narrow shear zone. The materials in this narrow shear zone finally exhibit a markedly finer grain size than that which initially formed the uniform aggregate. In this example, the breakdown in particle size can only be attributed to cataclasis, which can be produced in the experimental deformation of rock samples at moderately high confining pressure. In the Earth's crust, cataclastic flow, which is a method causing volume increase, will most probably develop

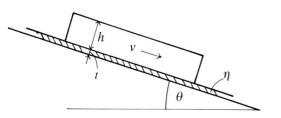

Fig. 7.34. Gravity gliding of a block underlain by an inclined viscous layer.

n rocks which exhibit high fluid pressures. As has been argued by Fyfe, Price & Thompson (1978), such conditions will readily develop at depths in the crust as the result of dehydration reactions associated with prograde metamorphism. Once cataclasis has been initiated and grain dimensions in the shear zone have been reduced, other deformation processes, such as pressure solution, Coble, Nabarro–Herring and Dislocation creep and dislocation glide may take over (see Chap. 1), and the textural evidence of cataclasis may be destroyed.

From studies of the Lochseitenkalk, Schmid (1975) and Schmid *et al.* (1977) conclude that, during thrusting, this material behaved as a 'super-plastic' material. It should be noted that this term 'super-plastic' relates to the micro-mechanisms involved in the deformation of the material and *not* to the stress–strain behaviour. Indeed, the Lochseitenkalk behaves as a slightly non-linear, viscous material (i.e. a Power-Law material, Fig. 1.39) with the Power-Law exponent (*n*) only a little greater than unity.

Kehle (1970) took such a distribution of material into consideration when analysing the model of a rigid block underlain by a material, with liner viscous properties, free to glide down a gentle slope (Fig. 7.34). It can be shown that the glide velocity (*v*) is given by:

$$v = \frac{(\rho g \sin \theta)}{\eta} (2ht - t^2) \qquad (7.14)$$

where ρ is the density of the gliding rock mass, of thickness *h*, η the coefficient of viscosity of the underlying material, of thickness *t* and θ is the angle of slope (Fig. 7.34).

Kehle postulated a hypothetical model of a 1 km thick unit of massive shale with a viscosity which he took to be 10^{17} poise (10^{16} Pa/s) overlain by a 1 km thick rigid block, on a slope of 5°. If these values are substituted in Eq. (7.14), it will be seen that the glide velocity approaches 10 m/yr.

The equation (7.14) set out above can, of course, be used to determine the viscosity of clay-rock, evaporites or other 'viscous' material, of known thickness, beneath a slide of known thickness, which moved at a known velocity. This type of model has considerable relevance to the gliding of massive blocks on thick

layers of weak materials (such as the scaly clays of Italy, mentioned earlier) or on thin layers which underlie some high-level nappes (such as the Tinee Nappe (Graham, 1981)). However, the model does not impinge upon the movement of the many major overthrusts that, it may be inferred, were pushed *up* a gentle slope.

This problem was addressed by Price & Johnson (1982) who included and combined both viscous and elastic behaviour as well as brittle frictional sliding in their analysis. They based their study on the Keystone-Muddy Mountain thrust of Nevada, U.S.A. This structure is one of a number of thrusts in the Sevier orogenic belt that has attracted the attention of geologists (e.g. Burchfield & Davis, 1972 and 1975). This thrust has a *toe* consisting of Palaeozoic–Mesozoic unmetamorphosed sediments which have been displaced over a footwall of Jurassic Aztec Sandstone. It provides remarkable evidence of thrust transport for tens of kilometres across an erosion surface. Johnson (1981) holds that the wedge-shape of the toe can be attributed to subaerial erosion and considers that the toe attained a 'steady state' condition, in that the thrust front remained approximately stationary as erosion kept pace with the continuous advance of the thrust sheet. A displacement of at least 25 km is envisaged, though one in excess of 40 km is thought not to be unrealistic.

The Aztec Sandstone which forms the footwall for the toe is a highly permeable, aeolian sandstone. It has been suggested that such rocks will be dry, so that one cannot have recourse to the concept of high fluid pressure, which sometimes appears necessary, if one is to explain the emplacement of these major thrust sheets. However, it should be noted that the footwall sandstone is Jurassic, but that movement on the thrust occurred mainly in the Upper Cretaceous.

Apart from indications that the palaeoclimate in the Upper Cretaceous in Nevada was warm and humid, there is clear evidence in the Aztec Sandstone for the circulating of ground water. Brock & Engelder (1977) demonstrated that Liesegang weathering patterns ('rings') are associated with thrust-related macrofractures. In addition, they report fractures which contain sand gouge and injection infill (clastic dykes), derived from the footwall in the hangwall rocks. These clastic dykes could be attributed to high fluid pressures; though Brock & Engelder prefer to interpret the injections as a passive response of the footwall to the opening of fractures in the hangwall. However, from the orientation of vein systems near the thrust plane and the angle between conjugate sets of veins, Price & Johnson concluded that a high fluid pressure, equal in magnitude to the overburden pressure (i.e. $\lambda = 1.0$) obtained during some phase of thrust activity. They point out that these high fluid pressures are probably transient effects attendant

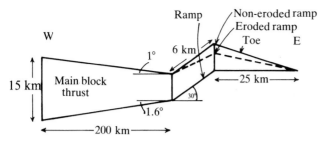

Fig. 7.35. Schematic representation of the three elements (Toe, Ramp and main thrust block) which comprise the Keystone-Muddy Mountain thrust sheet. (After Price & Johnson, 1982.)

upon the heat generated by overcoming the frictional restraint to thrust movement, at a time when the value of λ was considerably less than 1.0. However, it is emphasised that high fluid pressures are not an essential element of the thesis proposed by Price & Johnson. The above arguments are outlined here merely to indicate that geologists must guard against inferring zero, or low, fluid pressures on the basis of plausible but erroneous assumptions.

Price & Johnson presented the Keystone-Muddy Mountain thrust as a simplified and idealised model comprising three parts; namely, (i) the toe, (ii) the ramp and (iii) the main thrust block (Fig. 7.35). They considered the possibility of erosion and non-erosion of the ramp and toe and made assumptions regarding the fluid pressures beneath the ramp and below the toe. Furthermore, they followed a proposal set out by Smith (1981) discussed earlier, which assumed that the driving mechanism for the emplacement of the thrust sheet is provided by the intrusion of the Sierra Nevada Batholith. By making assumptions regarding the density of the country rock and magma, they quantified the magmatic pressure which was assumed to give rise to the thrusting. Then, by using concepts and principles outlined earlier in this chapter, they presented a mechanical analysis of the thrust, assuming frictional sliding throughout the whole of the three thrust units.

The analysis is incomplete because they did not estimate the stresses necessary to induce flexure at the bottom of the ramp: they merely refer to Wiltschko (1981) who indicates that the stresses necessary to induce flexuring are far from negligible. Further, they assume that the fluid pressure on the ramp was equal to the overburden pressure and hence there was no resistance to sliding on the ramp. Moreover they also assume that cohesion along the thrust plane was zero. Even by making all these 'favourable' assumptions, Price & Johnson determined that the model was barely able to satisfy the conditions of brittle sliding. If one took into consideration (i) a modest cohesion on the potential slide plane, (ii) lower fluid pressures beneath the ramp and toe or (iii) a high degree of resistance to flexuring: then, any one of these factors

may (and taken in combination they certainly would) render the brittle-sliding model an infeasible method of emplacing this major overthrust. Moreover, Price & Johnson emphasised that the brittle-sliding model may be realistically applied only to competent units at modest temperatures (probably not exceeding 150 °C) and hence to structures which occur at modest depths. For the Keystone thrust, it is reasonable to use the brittle-sliding model for the toe, probably for the ramp and possibly for the front end of the main block. However, it is completely unrealistic to use this model to explain movement towards the end of the block where, at a depth of 15 km and adjacent to the granite, the rock temperature may exceed 500 °C. At such temperatures, the rocks will approximate to viscous material. Because of the upper 'brittle' zone, it would not be apposite to use a completely viscous model (see Wiltschko, 1981). Consequently, a composite model of an elastic layer above and a viscous one below is used. The approach used was not comparable with that adopted by Kehle (1970), which has already been mentioned; but was based on a phenomenon (relating to slow-moving elastic waves) reported by Blay *et al.* (1977) and independently analysed by Bott & Dean (1973). This phenomenon will be considered later, in Chap. 10, in relation to its influence on fold development. However, as Price & Johnson showed, it also constitutes a vital element in the development of overthrusts and indicates how a 'caterpillar-like' motion' of a thrust unit may be achieved. Their composite model is therefore considered below at some length.

The conceptual model used by Bott & Dean and the physical models used by Blay *et al.* are indicated in Fig. 7.36. A feature of the physical models, noted by Blay, was that a slow moving wave travelled along the model in response to compression applied at one end. This surface wave was subsequently shown to be the manifestation of a slow moving 'stress front'. This was demonstrated by a photo-elastic experiment (Fig. 7.37) in which a layer of thick, 'stiff' gelatine which represented the elastic layer in Fig. 7.36, rested upon a thin, weak viscous layer. The model was deformed by applying a load through a platten (shown to the left in Fig. 7.37) which advanced at a very slow velocity (≈ 10 cm/hr). The stressed portion of the elastic layer shows 'white' in Fig. 7.37, while the unstressed area remains isotropic and 'black'.

We may infer from the position of the moving platten and the stress-front relative to the reference dots in Fig. 7.37, that the velocity of the stress-front decreases with distance from the platten. This relationship follows directly from the analysis of Bott & Dean (which, in turn, is based on the heat-flow equations derived by Carlslaw & Jaeger, 1959) which shows that a stress wave travels through a distance λ in time (t) where:

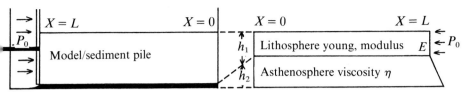

Sheared viscous base

Fig. 7.36. The conceptual model used by Bott & Dean (right) and the equivalent model representation used by Blay *et al.*, (left).

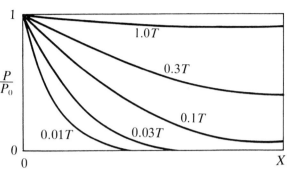

Fig. 7.38. Stress state throughout the model at different times, expressed as a fraction of *T*, the time required for the stress state at the far end of the model to reach 90 per cent of the applied end-load. (After Blay *et al.*, 1977.)

Fig. 7.37. Photo-elastic display of an advancing stress wave showing high differential stresses behind the wavefront (light area) and almost zero differential stress (dark area) ahead of the wave front. (After Blay *et al.*, 1977.)

$$t = \frac{\eta X^2}{E h_1 h_2} \qquad (7.15)$$

where η is the viscosity of the underlying layer, thickness h_1 and E is the Young's modulus of the elastic layer of thickness h_2. The transmission of a stress (P_0) applied at one end of the model (Fig. 7.36(*b*)), is shown graphically in Fig. 7.38. Each curve corresponds to the stress distribution at a particular time expressed as a fraction of the time (T) taken for the stress value at the far end of the model to reach $0.9\,P_0$.

The velocity (v) of the wave front is given by $\mathrm{d}x/\mathrm{d}t$, which can be obtained by rearranging and differentiating Eq. (7.15).

$$v = \frac{\mathrm{d}x}{\mathrm{d}t} = \frac{E h_1 h_2}{2 \eta X}. \qquad (7.16)$$

It is indicated by Price & Johnson that the model (Fig. 7.36) simulates a portion of an elastic overthrust block moving over a thin, weak viscous underlayer which could represent the thin layer of fault rock, or mylonite, seen to be associated with many thrusts. In order to show the likely range of stress-wave velocities and how this velocity varies with distance (X) from the source of the 'push', they took a rectangular block with arbitrarily chosen dimensions and properties. Thus, with $h_1 = 1.0$ m, $h_2 = 10$ km and

$E = 10^{11}$ dynes/cm^2 (10^5 bar): the relationship between the velocity (v) of the stress-front and the distance X from the source of the initial application of load and displacement can be plotted for different values of viscosity (η), as shown in Fig. 7.39(*a*). It will be noted that, for values $\eta = 10^{20}$ poise, the stress-front velocity falls below the inferred average displacement velocity of such structures as the Keystone Thrust. (The position of the curves shown in Fig. 7.39(*a*) are, of course, dependent upon the chosen values of h_1, h_2 and E.)

Price & Johnson then set up a model which is more representative of the Keystone-Muddy Mountain structure and took what is considered to be a representative value of viscosity for the thin layer, which ranges from 10^{18} poise for the high temperature environments to 10^{20} poise for the near surface environment. They took the dimensions of the main thrust block, as represented in Fig. 7.35, as a guide, and defined the dimensions of h_2 changing from 15 km (at $X = 1$ km) to $h_2 = 5.8$ at $X = 200$ km. Below this elastic layer, (again with $E = 10^{11}$ dynes cm^2 or 10^5 bar) it is assumed that the viscous layer similarly tapers from 1.5 m at $X = 1$ km to 0.6 m at $X = 200$ km. Moreover, they assume that Eq. (7.16) holds for every point value of h_1 and h_2, and can represent the variation in velocity (v) with distance (X) in the main thrust block for different values of viscosity (η), as indicated in Fig. 7.39(*b*).

In this model, h_2 is reasonably well defined: h_1 is defined at the top of the ramp, so the error at $X = 200$ km (6 km from the top of the ramp) is not likely to exceed a factor of 2 or 3. The value of h_1 may

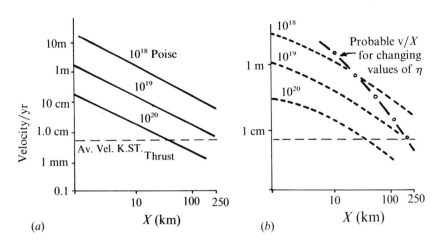

Fig. 7.39. (*a*) Relationship between velocity (*v*) of wavefront and *X* for different values of viscosity, assuming the thrust block to be (*a*) rectangular and the viscous layer to be of constant thickness. (*b*) Relationship between *v* and *X* for a tapering block and tapering viscous layer. (After Price & Johnson, 1982.)

increase very rapidly and attain a value considerably greater than 1.5 m. A factor of ten increase in this parameter would increase the velocity (*v*) by a corresponding amount. However, as we shall see, errors in estimates of the exact velocities at positions along the thrust block, where values of *X* are small, are unimportant. What is important is establishing the position along the thrust-block where the *velocity* of the stress-wave falls to the average displacement rate of the frontal portion of the thrust, namely the point which, they suggest, represents the change-over in mechanism controlling displacement of the thrust. Between the origin and the change-over point (at distance X_{crit}) displacement will be determined by the velocity of advance of the stress-wave; i.e. it will be associated with the coupled elastic/viscous layering of the structure. Beyond the change-over point to the leading edge of the wedge, or toe, at the front of the thrust, the rock mass as a whole is travelling faster than the stress-wave. Consequently, the stress-wave does not play an important role in the frontal part of the structure, which moves by brittle sliding of the frontal block as a whole, at a velocity determined by the velocity of the wave front at point X_{crit}.

Clearly, in order to establish the viability of this concept, it is necessary to determine the position of the change-over point as precisely as possible and to show that the available stresses are sufficient to cause frictional sliding of the frontal part of the structure. Using the values noted above and in Fig. 7.39(*b*), they obtained a cross-over point *X* = 160 km.

They considered the stability of their composite model and first established the value of the horizontal stress, which is generated by the stress-wave at various distances from the origin. Based upon the assumption regarding the density of the country rock and magma, they estimated the average magmatic pressure (P_{m}) for the model to be 1745 bar. However, this acts over a vertical height of only 13 km; the remaining upper 2 km are comprised of cover-rock that will make a

much smaller contribution to the horizontal stress (and which was taken to be zero) so that $P_0 = P_{\text{m}} \times 13/15$, or 1512 bar.

As can be seen in Fig. 7.38, the stress condition in the thrust block varies with time. Consider a time (t') at which the stress wave gives rise to a stress of 0.9 P_0 at a point 160 km along the block. The stress (P_X) at X_{crit} = 160 km (in a rectangular block) would be:

$$P_X = 0.9\, P_0$$

$$= P_{\text{m}}\left(\frac{13}{15}\right)\left(\frac{90}{100}\right) = 0.78\, P_{\text{m}}. \qquad (7.17)$$

However, the block is tapering, so that there will be a stress amplification which will be proportional to h_0/h_z (where h_0 = 15 km and h_z the block thickness at some distance *X*), so that for the wedge section:

$$P_X = P_{\text{m}}\left(\frac{13}{15}\right)\left(\frac{90}{100}\right)\left(\frac{15}{h_z}\right)$$

$$= \frac{11.7\, P_{\text{m}}}{h_z}. \qquad (7.18)$$

This relationship between P_X and *X*, in the tapering block shown in Fig. 7.35, is represented graphically in Fig. 7.40 where it will be seen that after an initial distance of about 30 km, in which P_X remains approximately constant, there is a significant increase in the magnitude of P_X with an increase in *X*. (At *X* = 160 km, P_X = 2666 bar.)

Price and Johnson then considered the equilibrium conditions for the forward portion (i.e. *X* = 160 km to *X* = 200 km) of the main thrust block. In this calculation it was assumed that the stress required to drive the thrust up the ramp, cause flexure and move the toe is 1500 bar. This is 50 per cent greater than the minimum figure quoted in the brittle, sliding-block analysis presented by them and, therefore, probably represents a realistic requirement for thrust movement forward of the main thrust block.

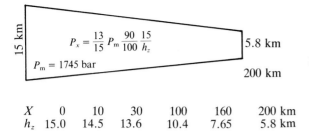

$$P_x = \frac{13}{15} P_m \frac{90}{100} \frac{15}{h_z}$$

$P_m = 1745$ bar

15 km | 5.8 km | 200 km

X	0	10	30	100	160	200 km
h_z	15.0	14.5	13.6	10.4	7.65	5.8 km

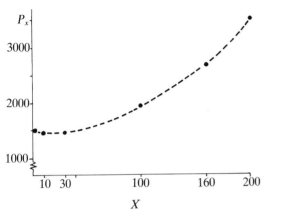

Fig. 7.40. Increase of average horizontal stress P_x in a tapering wedge with distance X. P_m = magmatic pressure, h_z = height of block at some distance X. (After Price & Johnson, 1982.)

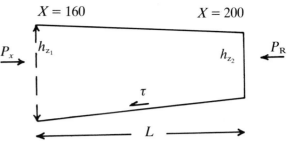

Fig. 7.41. Stability of the forward portion of the main thrust-block at a distance of 160–200 km from the origin. (After Price & Johnson, 1982.)

The stability requirements are represented in Fig. 7.41 where, it will be seen that sliding of the forward portion of the main thrust block and forward units will take place if cohesion is zero and λ (i.e. p/S_z) exceeds 0.6. The average shear stress (τ) available for sliding of the forward portion of the main thrust block is 486 bar. As the value of λ is extremely modest and the value of the shear stress is large, it can be concluded that sliding conditions will readily be met and that the hybrid stress-wave/sliding model satisfies the requirements for 'viscous' displacement and sliding on structures such as the Keystone Thrust.

It may be argued that one cannot necessarily be specific as regards the citing of the change-over point (at $X = 160$ km); depending as it does on specific values of η, E, h_1 and h_2. We have also completely neglected the cohesion of the glide plane which must first be broken down before frictional sliding may take place. To these criticisms Price & Johnson would reply that, hitherto, they had considered a stable, almost steady state condition of displacement, and had not been concerned with the initiation of frictional sliding, when the cohesive strength is an important factor.

Let us now consider the initiation of frictional sliding. This possibly took place prior to the development of the ramp. However, in order to maximise the stress requirement, let us assume that the initial configuration is as in Fig. 7.38 and that the plane which is to be sheared exhibits a cohesive strength of 100 bar. Then, if we take the value of the stress P_R required to

cause frictional sliding and flexuring of the ramp and toe to be 1500 bar, a further horizontal stress of 535 bar must be added to this to overcome the cohesive resistance; i.e., the break-down stress necessary to initiate frictional sliding will be 2035 bar. The stress-wave would reach the bottom of the ramp at a magnitude of over 3500 bar, and hence the stress-wave would initiate sliding at some distance before the bottom of the ramp. This would represent the change-over point at the instant of initiation of sliding friction. However, as soon as the cohesive strength of the basal slip plane has been broken down, there is a change in equilibrium conditions and the change-over point (X_{crit}) will migrate some distance towards the rear of the thrust block.

Thus, although the estimated change-over point stipulated in the previous argument is probably a reasonable one, the mechanism is not sensitive to this value. In reality, the change-over point would automatically be selected by the stress conditions, dimensions of the thrust unit and the physical characteristics of the rocks involved.

To summarise: the hybrid model presented in the preceding paragraphs combines the initiating influence of the batholithic intrusion with the displacement/stress-wave mechanism associated with a stressed, coupled elastic and viscous layer; which in the frontal portions of the wedge-shaped thrust body, eventually gives rise to frictional sliding. The model provides a sufficiently large intensity of stress to account for the possible proliferation of frontal structures observed to be associated with many major thrusts. Moreover, this hybrid model is compatible with rock mechanics data and field evidence.

Commentary

It will be apparent to the reader that we have only touched on the literature regarding field geometry and the mechanisms of development of major thrust nappes. A more extensive coverage of field data and further reference can be found in McClay & Price (1981).

Even though the treatment given in the sections

on the mechanics of emplacement of thrust nappes has been developed at some length, it will be further apparent that the coverage is far from complete. The literature on this subject is extensive and often openly, or tacitly, critical and contradictory. The object of the presentation here has been to lead the reader through the more important aspects of thrust nappe emplacement, so that he or she will be better able to assess the mechanism which should be employed to explain the emplacement of a specific structure. It is evident that much of the controversy in the literature has arisen because of (i) oversimplification of the conceptual model on which a particular mechanism is based, (ii) the choice of rheological model (elastic, plastic or viscous) and (iii) the tendency for workers to apply a single, specific mechanical concept to explain the emplacement of a natural example when the geometry and rheology are more complex than, or do not coincide with, that which can be encompassed by a single model. Indeed, it will now be apparent that a single, simple model is often not able to represent the emplacement of a single, large structure. The multiple, or complex, model approach is the one more likely to represent, satisfactorily, natural conditions of thrust nappe emplacement. We can, however, conclude by repeating the final remarks in the introduction of *Thrust and nappe tectonics* (Price & McClay, 1981) '... our knowledge of the geometry and the mechanics of thrust and nappes still needs to be improved. Nevertheless, it is apparent that the problem is tractable and we have the elements for its solution within our grasp'.

References

Aprahamian, J. & Pairis, J-L. (1981). Very low grade metamorphism with a reverse gradient induced by an overthrust in Haute-Savoie (France). In *Thrust & nappe tectonics*, eds. K. McClay & N.J. Price, pp. 159–66. Geol. Soc. London Special Publ. No. 9. Blackwell Scientific Publications.

Bally, A.W. (1981). Thoughts on the tectonics of folded belts. In *Thrust & nappe tectonics*, eds. K. McClay & N.J. Price, pp. 13–32. Geol. Soc. London Special Publ. No. 9. Blackwell Scientific Publications.

Bally, A.W., Gordy, P.L. & Stewart, G.A. (1966). Structure, seismic data and orogenic evolution of southern Canadian Rocky Mts. *Bull. Can. Petrol. Geol.*, **14**, 337–81.

Beloussov, V.V. (1977). Gravitational instability and the development of the structure of continents. In *Energetics of Geological Processes*, eds. Saxena & Battacharji, pp. 3–18. New York: Springer.

Blay, P., Cosgrove, J.W. & Summers, J.M. (1977). An experimental investigation of the development of structures in multilayers under the influence of gravity. *J. Geol. Soc. London*, **133**, 329–42.

Bott, M.P.H. & Dean, D.S. (1973). Stress diffusion from plate boundaries, *Nature Phys. Sci.*, **243**, 339–41.

Boyer, S.E. & Elliot, D. (1982). Thrust systems. *Am. Assoc. Petrol. Geol. Bull.*, **66**, 9, 1196–230.

Brock, W.G. & Engelder, T. (1977). Deformation associated with the movement of the Muddy Mountain overthrust in the Buffington windows, S.E. Nevada. *Geol. Soc. Am. Bull.*, **88**, 1667–77.

Bucher, W.H. (1956). Role of gravity in orogensis. *Geol. Soc. Am. Bull.*, **67**, 1295–318.

Burchfield, B.C. & Davies, G.A. (1972). Structural framework and evolution of the southern part of the Cordilleran orogeny, Western United States. *Am. J. Sci.*, **272**, 97–118.

Burchfield, B.C. & Davies, G.A. (1975). Nature and controls of cordilleran orogenesis, Western United States: extension of an earlier synthesis. *Am. J. Sci.*, **275A**, 363–96.

Butler, W.H. (1982). The terminology of structures in thrust belts. *J. Struct. Geol.*, **4**, 3, 239–45.

Carey, S.W. (1977). *The Expanding Earth: Developments in Geotectonics*. Series 10, Amsterdam: Elsevier.

Carlslaw, H.S. & Jaeger, J.C. (1959). *Conduction of Heat in Solids*. Oxford: Clarendon Press.

Chamberlain, R.T. (1910). The Appalachian folds of central Pennsylvania. *J. Geol.*, **18**, 288–51.
(1919). The building of the Colorado Rockies. *J. Geol.*, **27**, 225–51.

Cooke, P.W. (1955). Some aspects of high weight muds used in drilling abnormally high pressure formations. *4th World Petroleum Cong. Proc., Rome, Sec II*, 43–57, Discussion 57–8.

Coward, M.P. & Kim, J.H. (1981). Strain within thrust sheets. In *Thrust & nappe tectonics*, eds. K. McClay & N.J. Price, pp. 275–91. Geol. Soc. London Special Publ. No. 9. Blackwell Scientific Publications.

Dahlstrom, C.D.A. (1970). Structural geology in the eastern margin of the Canadian Rocky Mts. *Bull. Can. Petrol. Geol.*, **18**, 332–406.

Elliot, D. (1976). The energy balance and deformation mechanism of thrust sheets. *Phil. Trans. Roy. Soc. London*, **A283**, 289–312.

Elliot, D. & Johnson, M.R.W. (1980). Structural evolution in the northern part of the Moine Thrust Belt, N.W. Scotland. *Trans. Roy. Soc. Edin. Earth Sci.*, **71**, 69–96.

Escher van der Linth, A. (1841). *Gegirgsprofil von St.-Triphon*. N. Jahrb: Leonhard und Bronn. 342–46.

Fyfe, W., Price, N.J. & Thompson, A.B. (1978). *Fluids in the Earth's Crust*. Amsterdam: Elsevier.

Graham, R.H. (1981). Gravity sliding in the Maritime Alps. In *Thrust & nappe tectonics*, eds. K. McClay & N.J. Price, pp. 335–54. Geol. Soc. London Special Publ. No. 9. Blackwell Scientific Publications.

Gretener, P.E. (1981). Pore pressure, discontinuities, isostacy and overthrusts. In *Thrust & nappe tectonics*, eds. K. McClay & N.J. Price, pp. 33–40. Geol. Soc. London Special Publ. No. 9. Blackwell Scientific Publications.

Hafner, W. (1951). Stress distribution and faulting. *Bull. Geol. Soc. Am.*, **62**, 373–98.

Hossack, J.R. (1979). Use of balanced cross-sections in the calculation of orogenic contraction: a review. *J. Geol. Soc. London*, **136**, 705–11.
(1983). A cross-section through the southern Norwegian Caledonides constructed with the aid of branch line maps. *J. Struct. Geol.*, **5**, 2, 103–12.

Hsu, K.J. (1969). Role of cohesive strength in the mechanics of overthrust faulting and land sliding. *Geol. Soc. Am. Bull.*, **80**, 927–52.

Hubbert, M.K. (1951). Mechanical basis for certain familiar geological structures. *Geol. Soc. Am. Bull.*, **62**, 355–72.

Hubbert, M.K. & Rubey, W.W. (1959). Role of fluid pressure in mechanics of overthrust faulting. Pts. I & II. *Geol. Soc. Am. Bull.*, **70**, 115–205.

Johnson, M.R.W. (1981). The erosion factor in the emplacement of the Keystone thrust sheet (S.E. Nevada) across a land surface. *Geol. Mag.*, **118**, 501–7.

Keep, C.E. & Ward, H.L. (1934). Drilling against high rock pressure with particular reference to operations conducted in the Khuar field. *Punjab: Inst. Petroleum (London)*, **20**, 990–1013.

Kehle, R.O. (1970). Analysis of gravity gliding and orogenic translations. *Geol. Soc. Am. Bull.*, **81**, 1641–64.

Mandl, G. & Crans, W. (1981). Gravitational gliding in deltas. In *Thrust & nappe tectonics*, eds. K. McClay & N.J. Price, pp. 41–54. Geol. Soc. London Special Publ. No. 9. Blackwell Scientific Publications.

Mandl, G. & De Jong, L.N.J. (1977). Shear zones in granular material. *J. Rock Mechanics*, **9**, 95–144.

Mandl, G. & Shippam, G.K. (1981). Mechanical model of thrust sheet gliding and imbrication. In *Thrust & nappe tectonics*, eds. K. McClay & N.J. Price, pp. 79–98. Geol. Soc. London Special Publ. No. 9. Blackwell Scientific Publications.

McClay, K. & Price, N.J. (eds.) (1981). *Thrust & nappe tectonics*, Geol. Soc. London Special Publ. No. 9. Blackwell Scientific Publications.

Milton, N.J. & Williams, G.D. (1981). The strain profile above a major thrust fault, Finmark, N. Norway. In *Thrust & nappe tectonics*, eds. K. McClay & N.J. Price, pp. 235–40. Geol. Soc. London Special Publ. No. 9. Blackwell Scientific Publications.

Norris, D.K. (1972). *Guide-book*. Field Excursion XXIV Int. Geol. Cong. Montreal.

Nye, J.F. (1952). The mechanics of glacial flow. *J. Glaciol. London*, **2**, 82–93.

Price, N.J. (1977). Aspects of gravity tectonics and the development of listric fault surfaces. *J. Geol. Soc. London*, **133**, 311–27.

Price, N.J. & Johnson, M. (1982). A mechanical analysis of the Keystone-Muddy Mountains thrust sheet in S.E. Nevada. *Tectonophysics*, **84**, 131–50.

Price, N.J. & McClay, K. (1981). Introduction. In *Thrust & nappe tectonics*, eds. K. McClay & N.J. Price, pp. 1–6. Geol. Soc. London Special Publ. No. 9. Blackwell Scientific Publications.

Price, R.A., Balkwill, H.R., Charlesworth, H.A.K., Cooke, D.G. & Simony, P.S. (1972). *The Canadian Rockies and Tectonic Evolution of the S. Eastern Cordillera.* Guidebook, Field excursion AC 15, pp. 77–85. XXIVth Int. Geol. Cong. Montreal.

Raleigh, B. & Griggs, D. (1963). Effect of the toe in the mechanics of overthrust faulting. *Geol. Soc. Am. Bull.*, **74**, 819–38.

Ramberg, H. (1981). The role of gravity in orogenic belts. In *Thrust & nappe tectonics*, eds. K. McClay & N.J. Price, pp. 125–40. Geol. Soc. London Special Publ. No. 9. Blackwell Scientific Publications.

Roeder, D., Gilbert, O.C. & Witherspoon, W.D. (1978). *Evolution and macroscopic structure of Valley and Ridge thrust belt, Tennessee and Virginia.* Studies in Geol. 2, Univ. Tenn., Dept. Geol. Sci., 25.

Schmid, S.M. (1975). The Glarus overthrust: field evidence and mechanical model. *Eclog. Geol. Helv.*, **68**, 247–80.

Schmid, S.M., Boland, J.N. & Paterson, M.S. (1977). Superplastic flow in fine-grained limestone. *Tectonophysics*, **43**, 257–91.

Sibson, R.H. (1977). Fault rocks and fault mechanisms. *J. Geol. Soc. London*, **133**, 191–213.

Sibson, R.H., White, S.H. & Atkinson, B.K. (1981). Structure and distribution of fault rocks in the Alpine Fault Zone, New Zealand. In *Thrust & nappe tectonics*, eds. K. McClay & N.J. Price, pp. 197–210. Geol. Soc. London Special Publ. No. 9. Blackwell Scientific Publications.

Smith, A.G. (1981). Subduction and coeval thrust belts, with particular reference to North America. In *Thrust & nappe tectonics*, eds. K. McClay & N.J. Price, pp. 111–24. Geol. Soc. London Special Publ. No. 9. Blackwell Scientific Publications.

Smoluchowski, M.S. (1909). Some remarks on the mechanics of overthrusts. *Geol. Mag. new ser.*, **5**, 204–5.

Suppe, J. (1980). A retrodeformable cross section of northern Taiwan, *Geol. Soc. China*, **23**, 45–55.

Thompson, R.I. (1981). The nature and significance of large 'blind' thrusts within the N. Rocky Mts. of Canada. In *Thrust & nappe tectonics*, eds. K. McClay & N.J. Price, pp. 449–62. Geol. Soc. London Special Publ. No. 9. Blackwell Scientific Publications.

Wiltschko, D.V. (1981). Thrust sheet deformation at a ramp: summary and extenstion of an earlier model. In *Thrust & nappe tectonics*, eds. K. McClay & N.J. Price, pp. 55–64. Geol. Soc. London Special Publ. No. 9. Blackwell Scientific Publications.

8 Normal faults and associated structures

Introduction

In this chapter we shall deal with those various aspects of normal faults and their association with other structures which are not considered in other chapters. As we have seen, from the Andersonian classification of faults, referred to in Chap. 5, it follows that normal faults develop in an environment of lateral extension, when the maximum principal stress acts vertically. Such conditions are met in a variety of geological environments. For instance, lateral extension can be engendered during diapirism and also by meteoritic impact (Chap. 4).

Lateral extension and concomitant vertical loading can develop during the down-warp and uplift of sediments associated with epeirogenic events. (This aspect will be considered in Chap. 9.) Similarly, normal faults are also associated with the development of major folds and these will be discussed in Chap. 14. In addition, we have seen that normal faults are associated with the evolution of major strike-slip faults (Chap. 6). Consequently, we shall consider here some of the remaining environments in which normal faults develop. These include the extension regimes, to the rear of sliding blocks, associated with gravity and delta tectonics, and the sliding of blocks. We shall also deal with the regional, extensional environment (Plate Tectonics) which gives rise to the formation of grabens. Such extension may result from the processes of Plate Tectonics and Orogenesis. The final topics of this chapter will be concerned with the intimate relationship between normal and reverse faults and the distribution of sets of fractures. However, we shall first deal with the changes in the stress fields around a normal fault and the secondary, minor features that can result from normal fault development.

Stress patterns and secondary structures associated with normal faults

As we have noted, normal faults develop in an environment in which the rocks are usually experiencing some degree of extension. Many geologists expect that the axis of least principal stress (σ_3) will be oriented perpendicular to the line of intersection of conjugate fault planes and will bisect the obtuse angle between these fault planes. Or, if the beds are flat-lying and cut by only one, or a single set, of normal

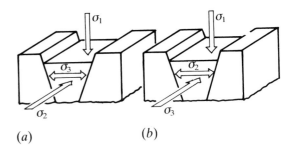

Fig. 8.1. (a) Interpretation of stress orientation that would be inferred to be associated with a graben, using the Andersonian classification. (b) Stress orientations determined by *in situ* measurements in Iceland.

faults, σ_3 is assumed to be approximately perpendicular to the strike of the fault (s): and will also probably be contained in the plane of the bedding. Indeed, such an interpretation is inherent in the Andersonian approach to the dynamics of faulting (Fig. 8.1(a)).

However, measurements of the magnitude of the horizontal stresses that develop (at depths of several hundred metres) in grabens associated with the active zones of rifting and vulcanism in Iceland reveal that the axis of minimum horizontal stress frequently acts along the length of the graben rather than perpendicular to the strike of the faults. This leads one to infer that the axes of least and intermediate principal stress act parallel and perpendicular to the strike of the normal faults respectively, rather than the opposite relationship, which is inherent in the Andersonian interpretation, (Fig. 8.1(b)).

The volcanic zones of Iceland are active and undergoing slow but continual extension. Hence, the measured stresses do not relate to an ancient palaeo-stress that has subsequently been modified; but represents the present situation. However, the difference in the inferred and measured stress distribution can be resolved. The Andersonian interpretation of the stress distribution is based on conditions at fault *initiation*. Shear failure conditions for fault initiation are represented by the combined Griffith/Navier–Coulomb failure envelope as indicated in Fig. 8.2, which also shows the failure envelope for a fault plane with zero cohesion. The theory of brittle failure is not, in general, concerned with the magnitude of the intermediate principal stress (σ_2) except to stipulate that it lies in the shear plane and has a magnitude somewhere between σ_1 and σ_3.

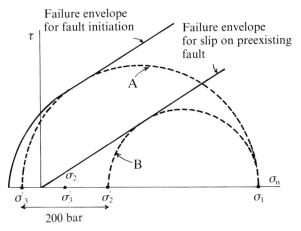

Fig. 8.2. The Andersonian classification relates only to the generation of new fractures and so must satisfy the stress magnitudes indicated by σ_1, σ_3 of circle (A). After shear failure, and provided the orientation of the stress axes remains unchanged, the stress circle (B) is sufficient to continue fault movement.

The failure condition for unfractured rock (i.e. that which gives rise to fault initiation) is represented by the stress conditions σ_1 and σ_3 in Fig. 8.2. However, once failure has been initiated, subsequent increments of movement on the fault plane (which is here assumed to possess zero cohesion) need only satisfy the failure conditions represented by line OA. The original stress conditions are incompatible with stability for the failure envelope (OA) of the fractured rock. Gravitational loading will remain virtually the same before and after fault initiation, so that in general σ_1 will remain unchanged. In order that the stresses may be compatible with this envelope (OA), the erstwhile stress σ_3 must increase until it attains the magnitude represented by point B. If, as indicated in Fig. 8.2, the original value of σ_2 was only a little larger than σ_3 (and this is completely realistic for many geological environments in which normal faults may develop), then the erstwhile σ_3 becomes σ'_2, while the original σ_2 becomes σ'_3. The σ'_2 now acts perpendicular to the strike of the fault.

As the result of faulting, the stresses will not only change in magnitude, but can also change in orientation. Let us assume a situation in which a normal fault develops as a totally enclosed system, i.e. it does not reach the topographic free surface. We shall now derive in qualitative terms the redistribution of strains and stresses which result from shear displacement on such a totally enclosed normal fault. To this end, we shall first consider the profile section through the normal fault, and then deal with the secondary effects parallel and adjacent to the fault plane.

The stress pattern which will develop around a fault may be inferred in general terms from the displacement pattern on the fault plane. It is suggested, that in the ideal case, the displacement along

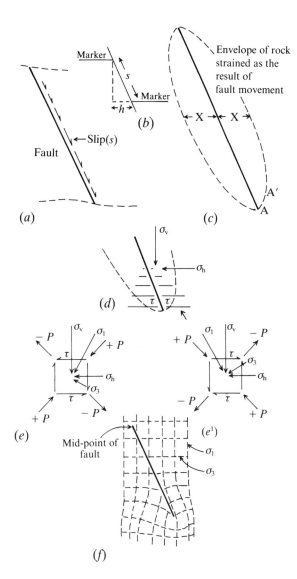

Fig. 8.3. (a)–(e) Various diagrams by which it is shown (see text) that the approximate stress distribution associated with normal fault development is as indicated in Fig. 8.3(f). The stresses $\pm P$ in (e) are those equivalent to the shear stress τ generated near the fault tip (d).

the fault plane will increase from zero at the upper end of the fault plane to a maximum in the central portion of the fault and then decrease to zero at the lower limit of the fault plane (Fig. 8.3(a)). It is clear from Fig. 8.3(b) that the displacement (s) on the normal fault results in a horizontal movement (h) of the rock mass. Clearly, h will also range from zero at the upper end of the fault, through a maximum in the central portion of the fault plane to zero at the lower end of the fault. In turn, the horizontal displacement h will induce a zone of strain around the fault (Fig. 8.3(c)) which reaches a maximum extent where s and h are greatest. Thus, slip on the fault will cause stress changes in a section of rock that is related to the pattern of displacement. For convenience, we shall assume that this section, in which stress changes occur, is approximately elliptical. (The increases in stress intensity which, according to the Griffith Theory of brittle failure, occur near the tip of the fracture (Chap. 1) are neglected in this simple analysis.)

It may be inferred that where the elliptical envelope of strain makes a marked angle with the fault plane (e.g. near the end of the fault as indicated by the portion of the ellipse AA' in Fig. 8.3(c)) significant shear strains and associated shear stresses develop parallel to the bedding. When this angle is large, the shear strains and stresses tend to be large (e.g. sector AA'): but where the elliptical envelope indicated in Fig. 8.3(a) is approximately parallel to the fault plane (i.e. throughout much of the central portion of the fault) the shear stress parallel to the bedding will be negligibly small. The magnitude of the shear stresses, which develop parallel to the bedding near the ends of the fault, are represented diagrammatically in Fig. 8.3(d). Equilibrium conditions require that there are shear stresses of comparable magnitudes acting at right angles to the bedding, as shown in Fig. 8.3(e). These stresses are compatible with normal stresses ($+P$ and $-P$) which must be added to the existing normal stresses given by σ_z and σ_x. The process of stress addition results in a rotation of the principal stresses. As may be inferred from Fig. 8.3(e), because the shear strains, and therefore shear stresses, on either side of the fault plane have different signs, they will give rise to a clockwise rotation of the maximum principal stresses on one side of the fault, and an anticlockwise rotation on the other, in the vicinity of the fault end. Because there are no horizontal shear stresses in the central portion of the fault, the principal stresses do not exhibit this rotation. A qualitative indication of how the orientation of the principal stresses will change adjacent to the lower half of a normal fault in a reasonably homogeneous material is shown in Fig. 8.3(f). The rock mass adjacent to the upper half of the normal fault would exhibit a similar but reversed pattern. The distribution of the axes of principal stresses indicated in Fig. 8.3(f) is in good agreement with that for a strike-slip fault obtained by Anderson, who used a mathematically rigorous analysis (Chapter 6).

As was noted above, the principal stresses remain undeflected near the central portion of the fault, but the horizontal stress at the fault plane increases from its original value (σ_3) to σ'_2. It may also be inferred that the magnitude of this horizontal stress decreases from σ'_2 at the fault plane to its original value (σ_3) at the limits of the ellipse of strain imposed on the country rock by the shear displacement on the fault (Fig. 8.3(f)).

It is of interest to estimate the general limits of influence of the fault, as represented by this ellipse. To do this, it is necessary to quantity:

(i)   the dip and the dimensions and displacement of the fault,

(ii) the elastic properties of the rock surrounding the fault,

(iii) the failure envelope for the intact and sheared rock, and

(iv) the original stresses that gave rise to the fault.

Here we shall take a hypothetical example in which the fault dips at 60°, has a length in the dip direction of 500 metres and exhibits a maximum dip-slip displacement of 0.5 metres. The Young's modulus is taken to be $E = 10^5$ bars (10^4 MPa) and the failure envelopes and effective stress conditions are as represented in Fig. 8.2, from which it can be inferred that the stress increase given by $\sigma'_2 - \sigma_3$ is 200 bars. Also from Fig. 8.3(b), a slip of 0.5 m on the fault plane gives rise to a lateral displacement (h) of 0.25 m.

Let us assume that h is shared equally by the rocks in both blocks abutting onto the fault plane and that for a single isolated fault, the increase of 200 bar in the lateral stress decreases outwards from the fault plane in a linear fashion until it reaches a value of zero at a horizontal distance X from the fault plane. This latter is a questionable assumption, but will suffice for our simple analysis. From this, it follows that the average strain (e_h) is given by:

$$e_h = \frac{h}{2X} = \frac{\sigma}{E}$$

or

$$X = \frac{hE}{2\sigma} \tag{8.1}$$

where σ is the average increase in stress (i.e. 100 bar). From the values given above, $X = 125$ m. That is, the influence of the fault plane extends to a distance of 125 m on either side of the fault. Clearly, this figure has no fundamental significance, as it depends upon the many parameters involved in the calculation. Nevertheless, it indicates the general type of influence which a minor normal fault with the dimensions, orientation and displacements cited above could impose on the stress field that gave rise to the structure. The limit of influence in Fig. 8.3 is represented as an ellipse. The outline of the limit will mainly be controlled by the amount of displacement on the fault. If the fault has experienced several phases of movement in which different portions of the fault moved at different times, then the limit outline will be correspondingly complex. Here we have assumed that the displacement at the ends of the fault are small and that

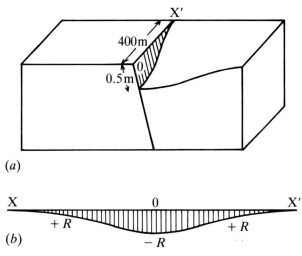

(a)

(b)

Fig. 8.4. (a) Block diagram showing form of displacement of normal fault discussed in text. (b) Diagram showing the form taken by displacement, perpendicular to the strike and in the plane of the fault.

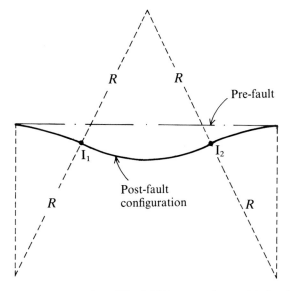

Fig. 8.5. Representation of Fig. 8.4(b) in terms of arcs of circles of radius $\pm R$ joined at inflection points I_1 and I_2.

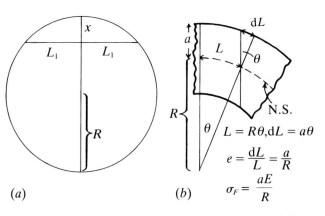

(a)   (b)

Fig. 8.6. (a) Euclidian theorem used to determine the value of R in Fig. 8.5. (b) Derivation of equation used to calculate fibre stresses (σ_f). N.S. = neutral surface.

there is a smooth and progressive increase in displacement along the fault, which reaches a maximum in its central portions. Because the horizontal stresses are increased within the limits of the ellipse, the stress increase will militate against the development of a second parallel fault within a distance of about 100 m of the first fault.

It is emphasised that the limits of influence cited in the previous paragraph are based on a linear decrease in stress away from the fault. If, as is likely, the stress follows an approximately exponential form, the real limits of influence would be about a quarter, or at most a third, of the dimensions cited above for a linear decrease.

Let us now consider secondary effects on the stresses that result from fault development. In the previous argument, we were concerned only with the stresses, in a plane at right angles to the fault, which are related to the shear stability of that fault. Strain and concomitant stress changes are also induced parallel to the fault plane. As far as the authors are aware, this problem has not been subjected to a mathematically rigorous and geologically realistic analysis. The concepts which are considered below are based on a simple model which enables a qualitative, and even a semi-quantitative, evaluation of the problem to be presented. Here, it is necessary to change the basic model from one which is reasonably homogeneous to one which more closely simulates sedimentary sequences.

As with folding (see Chap. 12) the development of a fault in a sedimentary sequence, composed of layers of different thickness and material properties, will be *controlled* by one or more strong units. A diagrammatic representation of the general form of the displacement of a normal fault is presented in Fig. 8.4(a). The displacement on the fault plane

between X and X' is indicated in Fig. 8.4(b) as a series of circular arcs with an upward (+ ve) radius of curvature (R) at the outer portions of the fault plane and a downward (− ve) radius of curvature in the central regions. (Note that the radii of curvature lie in the plane of the fault.) The order of magnitude of the radius of this curvature can be estimated in the following manner. Thus, if I_1 and I_2 in Fig. 8.5 are inflexion points, then R can be estimated from the Euclidean theorem which gives:

$$x(2R - x) = L_1^2 \qquad (8.2)$$

where the geometry and symbols are as defined in Fig. 8.6(a).

Let us now consider a minor normal fault with a maximum extent in the strike direction of 400 m and a maximum dip-slip displacement of 2.0 m, so that $2L_1$ for the distance between inflexion points in the central

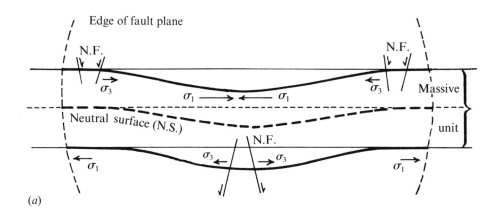

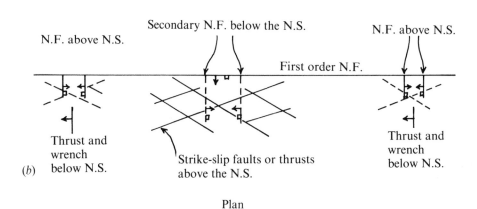

Fig. 8.7. (*a*) Indication of the compressive and extensional stresses induced by flexuring of a massive unit which, when added to the original ambient stresses may give rise to minor, secondary normal faults (N.F.) in the positions shown. (*b*) Plan of orientations of minor normal, strike-slip or thrust faults which may possibly be associated with a major normal fault.

portion of the fault plane is 200 metres and the value of $x = 1.0$ m. Substituting these values in Eq. (8.2), it follows that $-R$ is 5 km. Similarly, the upward positive curvature ($+R$) at the ends of the fault is $+5$ km. If the displacement line shown in Fig. 8.6(*b*) represents the neutral surface (N.S.) of the control unit, then it may be shown from the theory of flexing elastic plates that the extensile and compressive strains will develop below and above the neutral surface. As indicated in Fig. 8.7(*a*), there will be corresponding local increases and decreases in the stresses acting parallel to the fault plane. Engineers refer to these flexure-induced stresses as *fibre stresses*, (σ_f) the magnitude of which is given by:

$$\pm \sigma_f = \pm \frac{aE}{R} \qquad (8.3)$$

(Fig. 8.6(*b*)) where *a* is the distance of some bed (or fibre) above or below the neutral surface which has a radius of curvature *R* and *E* is Young's modulus.

The extensile and compressive stresses in the control unit, which act parallel to the fault plane are indicated in Fig. 8.7(*a*). Clearly these stresses, if they become sufficiently large, could permit the development of minor normal faults with the orientation and general position indicated in Fig. 8.7(*a*) and (*b*). In order to ascertain whether or not it is possible for such

secondary structures to develop, it is necessary to quantify any given specific situation. If we take the situation noted above such that $R = 5$ km and the value of $E = 5 \times 10^5$ bar $(5 \times 10^4$ MPa), then it follows that at a modest distance of 10 metres above or below the neutral surface the fibre stresses are 1.0 kbar. It may be inferred that in most situations in the crust, rocks could not sustain a decrease of 1.0 kbar parallel to the fault plane and would fail by forming normal faults. As we have seen, the trend of these secondary fractures would be at right angles to the strike of the main normal fault and, as is indicated in Fig. 8.7, would be located either near the central portion of the main fault, or towards its ends. If the degree of flexuring on the main fault is very small, then, of course, the secondary normal faults may never develop. Consequently, it is to be expected that from one to three minor grabens may develop which trend perpendicular to the major normal fault. However, even if these secondary faults do develop, they may not be observed in the field if the level of erosion cuts the main fault at some inappropriate distance from the neutral surface.

As regards the positive, compressive fibre stresses, one may infer that these could possibly give rise to minor secondary strike-slip faults, or thrusts, as indicated in Fig. 8.7(*b*). However, such secondary

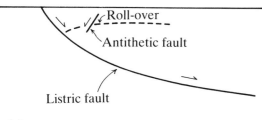

(a)

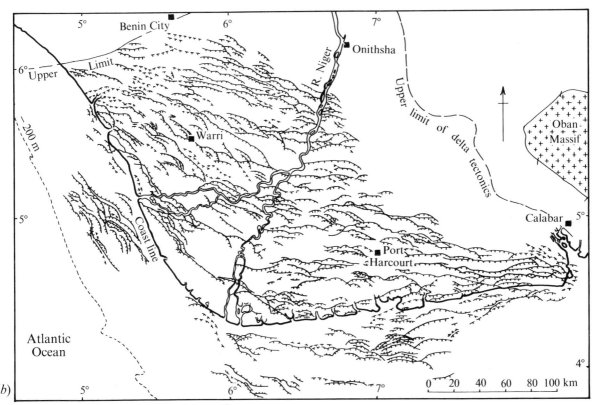

(b)

Fig. 8.8. (a) Section through listric normal fault showing associated roll-over structure and antithetic normal faults in the down-throw block. (b) Growth Faults in the Niger Delta. (After Evamy *et al.* 1978.)

features would usually require higher levels of stress (see Chap. 5) and are likely to be less frequently observed.

Normal faults in delta tectonic environments

As was noted in the Introduction, normal faults are likely to develop towards the rear of any block which is sliding, or about to undergo sliding, down a slope under its own body-weight. Such faults, often termed *growth faults*, are particularly important in the proximal sediments of major deltas (Fig. 8.8(*a*)); see, for example, the faults in the Niger Delta area (Fig. 8.8(*b*)). A common feature of these structures is that the upper portions of the normal faults may exhibit a dip of 45° and at depth the faults may be inclined at an even lower angle, i.e. the structures are *listric* (or spoon shaped). Associated with these major, listric normal faults, which exhibit a downthrow in the same general direction as the glide block, sets of minor

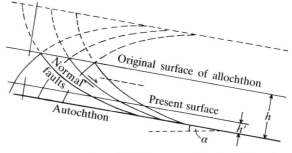

Fig. 8.9. Interpretation of development of listric surfaces in a gravity glide block, based on plasticity theory. (After Hose & Danes, 1973.)

normal faults that exhibit the complementary, or antithetic, movement sense are frequently observed.

Explanations have been proposed for these listric structures which are based on the use of plasticity theory. In a treatment presented by Hose & Danes (1973), the glide unit was assumed to be a homogeneous and isotropic block of plastic material. As a detailed treatment of this analysis is beyond the

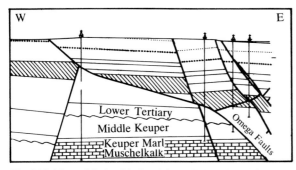

Fig. 8.10. Normal fault with 'knee' from the Landau oilfield.

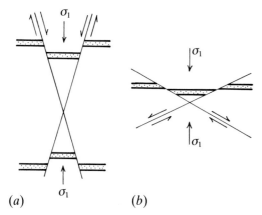

Fig. 8.11. (*a*) Relationship between shear fractures in which the wedge forming the acute angle moves inwards. This form of shear fracture relationship (most commonly observed in competent rock) is usually interpreted as the result of brittle failure. (*b*) Relationship between shear fractures in which the block forming the obtuse angle moves inwards. This form of shear failure has usually developed in incompetent rock or in crystalline basement rock undergoing ductile deformation.

scope of this book, it is sufficient here to note that the family of shear surfaces which are predicted by this analysis are as indicated in Fig. 8.9.

A second plastic analysis, presented by Mandl & Crans (1981) is based on a somewhat more geologically realistic model. Although they too assume that the glide block is homogeneous and isotropic in its fundamental, physical properties, they postulate (and also adduce evidence and arguments in support) that deltaic sediments are frequently over-pressured and that the degree of over-pressuring, which will vary throughout the glide block, influences the angle at which shear failure will take place. Because the failure angle is related to the degree of over-pressuring, the approach by Mandl & Crans is more versatile and can be adapted to explain situations when the geometry of the normal fault does not form a simple listric curve. For example, some normal faults exhibit a marked knee bend (Fig. 8.10). The Mandl & Crans analysis could be adapted to explain such a knee by postulating a marked variation in fluid pressure with depth in the glide block. It would, however, be difficult to make this adaptation compatible with the basic

assumption of homogeneity and anisotropy, without some rather special pleading.

This problem of low-angle faults is allied to a wider problem which has bedevilled geologists for many years. It has long been recognised that shear failure in competent rocks can be interpreted in terms of brittle failure. As we have noted, when brittle rocks fail, to form conjugate shears (Fig. 8.11(*a*)), the axis of maximum principal stress bisects the acute angle between the shear planes. However, conjugate shears exist in some weak or semi-ductile rocks where the movements on the shear plane indicate that the maximum principal stress bisected the obtuse angle between the shears (Fig. 8.11(*b*)). Towards the end of the last century, a concept was introduced by a geologist named Becker, which was based on the strain-ellipsoid and purported to explain the development of these features. Because these features developed in rock which exhibited considerable strain, they would often exhibit the form of 'ductile shears' (Chap. 2). This concept was widely used for almost half a century. However, as it can be demonstrated that there are many fundamental errors in Becker's thesis, his erroneous concept was gradually abandoned over the next couple of decades. For the last thirty years or so, geologists have therefore been without any generally acceptable basis for the interpretation of shears in semi-ductile material, where the angle between the shear planes (with the movement senses as shown in Fig. 8.11(*b*)) exceed 90°. As a consequence, observations regarding such structures have almost completely disappeared from the literature (though, of course, the structures have not disappeared from the outcrop).

The low angle portion of listric normal faults may be interpreted as such semi-ductile shears, for as can be seen from Fig. 8.8 the angle between the synthetic and antithetic fractures, when the movement sense is taken into account, indicates that the axis of maximum principal stress bisects the obtuse angle, which in this instance is about 100°.

Where it is apparent that the maximum principal stress (S_1) bisects the obtuse angle between two conjugate shear planes, it can sometimes be argued that the inter-shear angle was originally acute, but was increased during subsequent deformation. For example, if a normal fault forms in poorly compacted sediments with a porosity of 38 per cent and these sediments are then compacted until they have zero porosity, then the fault plane will be rotated, as indicated in Fig. 8.12.

Another explanation for the formation of such acute angles between conjugate shear planes is that they formed in rocks which are both inhomogeneous and anisotropic. These requirements, it is suggested, are more apposite to those glide blocks which are comprised of layered sediments of various bed

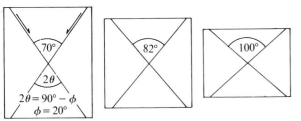

Fig. 8.12. Rotation of normal fault by compaction. (*a*) 0% compaction, (*b*) 20% compaction, (*c*) 38% compaction.

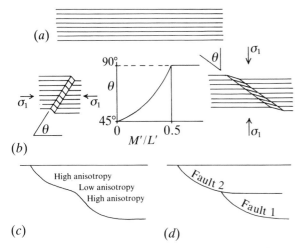

Fig. 8.13. (*a*) Anisotropic material compressed normal to fabric (or bedding). (*b*) Relationship between the material anisotropy (M'/L') and the angle (θ) the 'characteristic' makes with the compression direction, see Chap. 13. (*c*) Possible interpretation of knee in normal listric fault in terms of variations in anisotropy. (*d*) Alternative interpretation for formation of the knee.

thickness and lithology. As the treatment of failure of anisotropic material, which is based on the analysis by Biot (1965), is given in some detail in Chap. 13 we shall merely outline the concept here.

Consider an anisotropic unit (Fig. 8.13(*a*)) which is subjected to compression, that acts normal, or at a high angle, to the plane of anisotropy. It will be shown in Chap. 13 that planar shear instabilities develop along conjugate *characteristic* directions which make an angle θ with the direction of compression; and that this angle is determined by the degree of anisotropy of the compressed unit. The degree of anisotropy is given by the ratio M'/L', where M' and L' are respectively, the normal and shear moduli of the anisotropic material in the direction of the principal compression. When M'/L' tends to infinity (or zero) the angle θ goes to 45°. However, for relatively isotropic material (i.e. the M'/L' ratio goes to unity) the angle θ tends towards 90°, (Fig. 8.13(*b*)). (The reader is referred to Chap. 13 for a more detailed discussion of the concept of anisotropy and characteristics.) Clearly, if variations in the degree of anisotropy are postulated to occur within a sedimentary sequence, it may be inferred that the characteristics and hence, potentially, the shear surfaces could curve

in a complex manner. One such distribution of anisotropy which could give rise to the knee in a listric growth fault is represented in Fig. 8.13(*c*).

In passing, it may be noted that, if a significant degree of displacement takes place on the growth fault, the existence of such a knee poses considerable space problems. These knees are most usually inferred from seismic profiles, which, if the definition were sufficiently fine, may indicate that some of these knees represent the junction between two listric growth faults, as indicated in Fig. 8.13(*d*); in which event, the movement/space problem is largely obviated. Such a double listric fault is not, of course, dependent upon the anisotropic model for its explanation. It could also be explained by the Mandl & Crans model. Clearly, further detailed information and perhaps more refined models are required before the growth fault problem is completely understood.

Another possible explanation, which has been proposed to explain the existence of low angle normal faults, relates to bulk rotation of the fault blocks: and is sometimes referred to as the 'book-shelf' mechanism, Fig. 8.14. This geometry, first proposed by Lovering (1928) as a means of explaining fracture development in incompetent rocks (i.e. fracture cleavage development), is indicated in Fig. 8.15. It will be seen that, depending upon the angular relationship between the bounding shear (the 'bookshelf') and the internal shears (the planes between individual 'books'), the bulk shear may give rise to either dilation or compression of the unit between the bounding shears. The conditions in which no thickening of the bed occurs and the amount by which extension, parallel to the bounding shears, takes place are also given in Fig. 8.15. In the dilational mode, the rotating faults tend to open up and possibly become zones in which mineral infilling takes place (Beach, 1977). The second mode, which is sometimes referred to as the 'domino style' is the one of particular interest in our present discussion. This mode occurs over a wide range of scale. It can result from glacial action, or through flexural slip during folding, or arise from the development and rotation of normal faults which are related to gravity gliding in deltas and possibly, in the Basin and Range Province of the U.S.A., as the result of crustal extension. It may even occur as the result of strike-slip faultings in the Massif Central (Letourneur, 1953) or in the Najd fault complex of Saudi Arabia (Moore, 1979).

An extremely interesting mechanical analysis of both the dilational and the domino model has been presented by Mandl (1985). The interested reader is urged to make reference to the original paper for the full analysis. Here, we shall only mention the treatment of the domino model in which he demonstrates the mode of development, by the use of Mohr's stress circles and the linear Navier–Coulomb failure

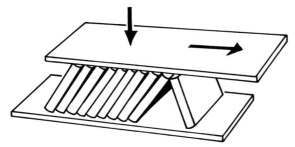

Fig. 8.14. Bookshelf kinematics. (After Mandl, 1985.)

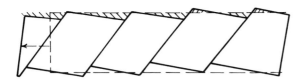

Fig. 8.15. Bookshelf mechanism and dilation of a sheared unit. (After Lovering, 1928.)

criterion. He reaches the somewhat surprising conclusion that the shear stress required to operate the domino bookshelf mechanism (Fig. 8.16(c)) is smaller than that required to produce the 'card-deck'-type shearing (Fig. 8.16(a)).

Roll-over and related structures

It will be noted in the seismic section shown in Fig. 8.17(a) that the down-throw block associated with a listric fault exhibits a larger component of vertical slip adjacent to the fault plane than it does at some distance from the fault plane, so that a *roll-over* anticline develops. These roll-over structures are of particular interest to the hydrocarbon industry as they may provide potential sites for the entrapment of gas and/or oil. The roll-over geometry is considered to be the inevitable result of movement on a listric fault (Hamblin, 1965). Movement on such a curved fault plane will tend to generate a 'gap' between the hanging wall and footwall (Fig. 8.17(b)) which it is held will be filled by collapse of the hanging wall in either a ductile or brittle manner. More usually, deformation occurs as a result of a combination of both these modes of behaviour to form a roll-over with associated structures (Fig. 8.17(a)). This suggestion, which is based purely on geometrical considerations, is extremely plausible. It is far from clear, however, what mechanical concepts can be advanced to justify the method.

For reasons which are not immediately apparent, the roll-over may exhibit little or no curvature (Fig. 8.18(a)) and such features may be termed *half-grabens*. Sometimes only a portion of the half-graben remains planar and one or two flexures, often with relatively low amplitude and considerable wavelength, develop in the zone adjacent to the main, bounding fault (Fig. 8.18(b)). In addition reverse

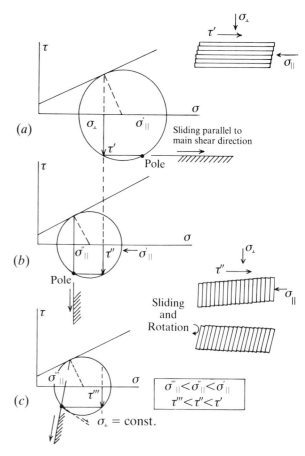

Fig. 8.16. Stress conditions and failure envelope indicating how failure stresses can be established for (a) the card-deck model and (b) and (c) the 'domino' model. (After Mandl, 1985.)

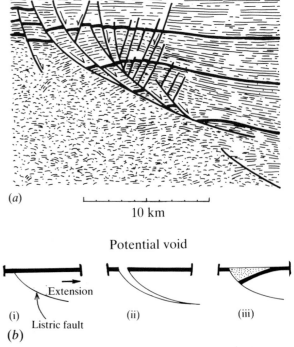

Fig. 8.17. (a) Seismic section of Growth Fault in Tertiary sediments of the S. Texas part of the Gulf Coast basin. (In Harding & Lowell, 1978.) (b) Roll-over structure caused by movement on a listric fault. (After Hamblin, 1965.)

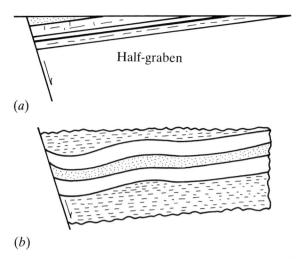

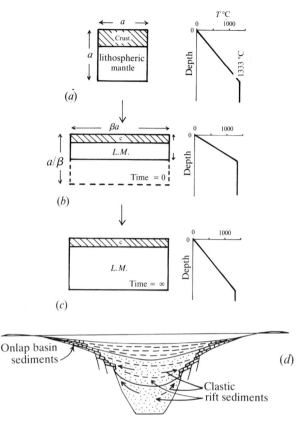

Fig. 8.18. (*a*) Half-graben. (*b*) Low amplitude buckling in half-graph.

Fig. 8.19. Lithospheric attenuation model. (After McKenzie, 1978.)

faults or thrusts may develop in the wedge of sediments abutting the boundary fault.

Attempts have been made to link the geometry of the folds such as those represented in Fig. 8.18 (*b*), to variations in orientation and attitude of the boundary fault. However, it can be shown that these are implausible from the mechanistic view point. Alternatively, it is possible to consider such folds not as roll-over structures but as buckle phenomena which, together with the thrusts and/or reverse faults are manifestations of a phase of compression. The development of the half-graben will be related to a phase of extension, so that the 'compressive' structures are said to have resulted from *tectonic inversion*. This 'inversion' is not necessarily the result of deep-seated crustal compression, but can develop in the cover rocks as the result of gravitational instability of the wedge. A quantified justification of this statement is based on unpublished analyses by the senior author and is beyond the scope of this book. However, it is sufficient to note that instabilities in deltas can occur on slip surfaces which are inclined at angles as low as 2°, and that dips of sediments in half-grabens may often exceed 5°. Consequently, if the sediments in the half-graben include rock types and fluid pressures comparable with those thought to develop in deltas, then gravitational instability is to be expected. If the width of the half-graben is in excess of about 17 km, it can be shown that sufficient gravitational potential exists within such a wedge to ensure the development of the compressive features, such as thrusts, reverse and strike-slip faults and buckle flexures.

Crustal extension

Many of the various aspects of normal fault development noted above are frequently encountered in regions which have experienced crustal extension. A frequently cited model of such extension is that of

Lithospheric Attenuation proposed by McKenzie (1978) in which stretching (by a factor β) causes a reciprocal amount of lithospheric thinning (a/β), (Fig. 8.19(*a*) and (*b*)). The lithosphere is taken to include the crust and part of the upper mantle, down to the 1333 °C isotherm, the peridotite solidus. During the stretching phase, fractures (usually planar, or listric, normal faults) develop in the upper (brittle) part of the crust; whilst the lower crust and upper mantle deform in a more uniform ductile manner. As stretching occurs the isotherms are packed more closely together and heat flow is increased (Fig. 8.19(*b*)). Following the stretching phase, thermal equilibrium is re-established and the 1333 °C isotherm returns to its original depth (Fig. 8.19(*c*)). It can be seen from this figure that because of the assumed crustal thinning that took place in the stretched region, the mean density of the lithosphere is greater here than in the adjacent, unstretched area. This results in subsidence of the attenuated region. The thermal subsidence basin that results overlaps the fault-bounded rift to produce what is sometimes termed the 'steer-head' or 'Texan Long-horn' basin profile (Fig. 8.19(*d*)). It should be noted, however, that such bovine characteristics are purely the result of vertical exaggeration.

It will be seen that this model is concerned with *uniform* ductile stretching of the lower crust and upper mantle. Because the lithosphere is a layered system,

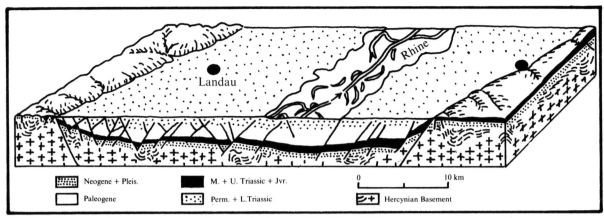

Fig. 8.20. Diagrammatic cross-section through the Rhein Graben. (After Illies, 1977.)

such homogeneous behaviour resulting in uniform stretching is unlikely to occur. As will be apparent (Chap. 16), stretching of ductile layers often results in the development of pinch-and-swell structures. Hence, layering in a lithosphere undergoing extension will probably result in a non-uniform deformation, so that narrow zones, of more marked thinning (pinched regions, see Chap. 16) will develop approximately at right angles to the extension direction. This regional scale pinch-and-swell structure, may determine the position and magnitude of a major brittle structure in the upper crust. This type of structure, the *graben*, is discussed at length in the following section.

Grabens

Graben, the German word for a ditch or trench, was used by copper miners in Thuringia to describe these geological structures as early as the 18th century and was introduced into the geological literature, towards the end of the last century, to describe large-scale rift features. Grabens are physically defined by two conjugate and converging dip-slip faults, so that, in profile, they contain between them a down-thrown, wedge-shaped fault block. Perhaps the best known example of such a feature is the African Rift System which can be traced throughout much of East Central Africa northward, via the Red Sea and the Gulf of Akaba, to the Jordan Valley. However, the *Rhine Graben* and associated structures have received especially intensive attention so that it is this unit which we shall take as our example. A diagrammatic profile through this structure is given in Fig. 8.20, which shows a specific feature of grabens; namely, that these major structures exhibit an impressive development of smaller faults, many of which are antithetic; so that the wedge-shaped block itself is disrupted by subsidiary grabens and horsts.

The dip of the major, bounding normal faults of the Rhine Graben is 60–65° and its width is about 35 km. Consequently, if the boundary faults retain

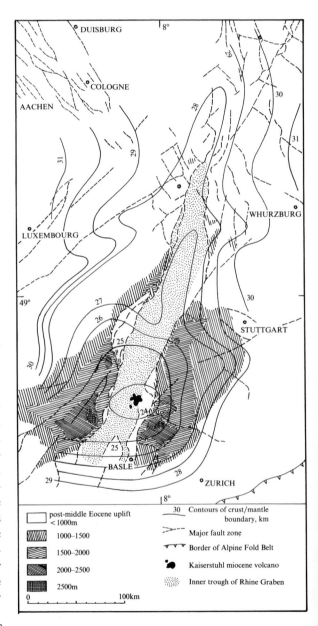

Fig. 8.21. Map showing amount of uplift and also the depth to the crust–mantle boundary in the area around the southern part of the Rhine Graben. (After Illies, 1977.)

their inclination at depth, the Rhine Graben and other comparable structures must have their roots in the upper mantle. It may be inferred from current or ancient volcanic activity associated with major grabens, that these structures do, indeed, extend to great depths. Volcanic activity varies from example to example. Some features, for instance the floor of the Afar Depression, or Graben, in Ethiopia, has been almost flooded by volcanic intrusions. However, the volcanic activity associated with the Rhine Graben is modest and is mainly found in the neighbourhood of the Kaiserstuhl. Indeed, it is most interesting to note that seismic studies have demonstrated that, in this area of the Baden-Rhineland Plain, the 'Moho' approaches to within 24 km of the surface (Fig. 8.21). Although there is no current activity, the sub-surface temperatures indicate that this area still exhibits a marked geothermal anomaly. Hundreds of square kilometres of the Rhine Graben between Strasbourg and Heidelberg exhibit geothermal gradients which range from $55\,°C/km$ to something in excess of $80\,°C/km$.

These geothermal anomalies are mainly attributed to hydrothermal convection along fractures. As was noted in Chap. 4, these high geothermal gradients greatly facilitate the development of diapiric intrusion of evaporites. The salt structures of the North Sea are almost certainly manifestations of such geothermal anomalies, for, as may be inferred from Fig. 8.22, the Rhine Graben, Lower Rhine Embayment and the graben complex of the North Sea Basin form a sinuous, and somewhat discontinuous, major crustal structure. It will be seen that the Rhenish Shield forms the major discontinuity and exhibits the preponderance of volcanic activity. Moreover, although a discrete rift valley is absent in the Rhenish Shield area, it is cut by a belt of seismic activity.

Illies (1981) suggests that the discontinuity can be attributed to a change in crustal behaviour (Fig. 8.23) where the Rhenish Shield is depicted as behaving in a relatively incompetent and ductile manner. A feature of the environs of the Rhine Graben, and other major graben structures, which is indicated in Fig. 8.20 and 8.23, is that the flanks of the graben are uplifted and tilted and that the graben itself occupies the 'keystone' position in an elongated arch.

Model studies of graben formation and development

Many attempts have been made to reproduce the graben structures by means of models. The pioneer work by Cloos (1928) and his co-workers is well known. In this chapter, therefore, we shall make reference to the more recent work of Elmohandes (1981) who made a three-dimensional model study using the apparatus indicated in Fig. 8.24(*a*) and (*b*). It will be seen from these figures that the lateral

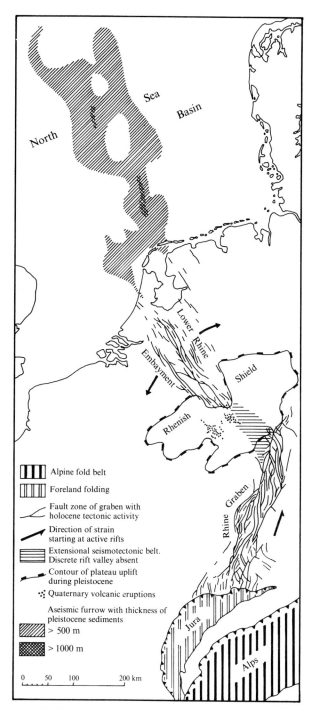

Fig. 8.22. Map of a major crustal structure reaching from the Jura to the North Sea Basin. (After Illies, 1981.)

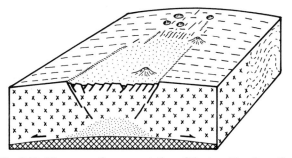

Fig. 8.23. Diagrammatic representation of the transition from the Rhine Graben to the Rhenish Shield. (After Illies, 1981.)

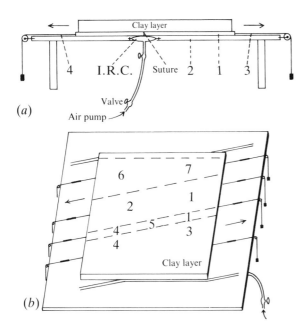

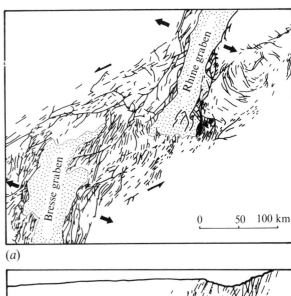

(a)

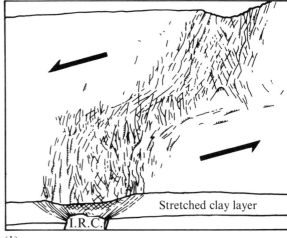

Fig. 8.24. (a) Diagrammatic representation of the apparatus (side view) showing (1) the upper plate with a cut, representing the suture; (2) lower plate with medial gap for the inflatable rubber case (I.R.C.); (3) and (4) spring balances. Arrows show extension direction. (b) Surface view of experimental set-up. Numbers 1–7 represent the small, wooden units forming the upper plate, permitting study of the transform faulting. (After Elmohandes, 1981.)

extension is, in part, generated by the weights acting via pulleys and, in part, by an air bag which can be made to expand and so cause uplift and arching of the clay tablet. A model graben with the profile shown in Fig. 8.25 results from this upwarping and concomitant lateral extension. Elhomandes also modelled the structures which connect the Rhine and Bresse Grabens. The real and model structures are represented in Fig. 8.26(a) and (b) respectively. As can be seen, the correlation is persuasive.

These connecting structures which are referred to as the Bresse Transformation, are markedly different in geometry from the oceanic transform faults which connect offset portions of mid-ocean ridges.

Fig. 8.26. (a) Fracture patterns linking the Bresse and Rhine Grabens. (After Illies, 1981.) (b) Fracture patterns evolved in clay tablet as the result of an induced 'transform'. The similarity between the model and natural fracture patterns is persuasive. (After Elmohandes, 1981.)

Illies suggests that the difference between the geometries of the oceanic and continental transforms can be attributed to the fact that the spreading rate of continental grabens is an order of magnitude slower than ocean floor spreading. However, we suggest that the difference is perhaps more probably attributable to the 'cover' of crustal rock. A single, deep transform fault in the basement would readily give rise to a shear zone in weaker cover rocks (c.f. the shear zone with Riedel and other shear fractures discussed in Chap. 6).

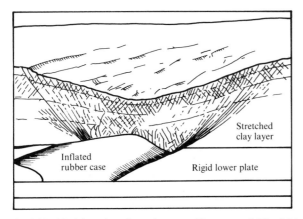

Fig. 8.25. Model graben formed over rubber case of Fig. 8.24. (After Elmohandes, 1981.)

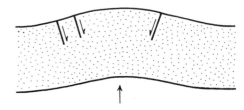

Fig. 8.27. Fractures in sand induced in a model in which the displacement of the basement followed a sinusoidal curve. (After Sanford, 1959.)

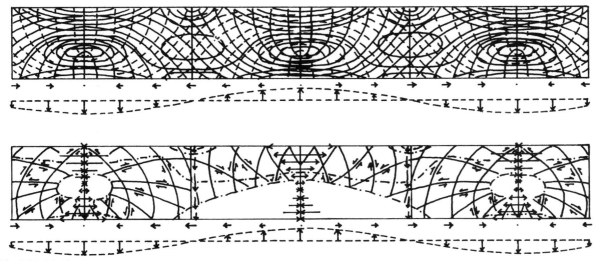

Fig. 8.28. (*a*) Theoretically derived stress trajectories that develop in a cover overlying a basement which has undergone a sinusoidal displacement as indicated below. Solid trajectories σ_1, dashed σ_3. (*b*) Shear fractures that may develop in cover rocks, as the result of the stress pattern shown in (*a*). (After Sanford, 1959.)

It will be noted that the profile through the clay model exhibits a very large number of shears. A lower fracture density is obtained if a graben is caused to develop in a sand-box experiment. The profiles shown in Fig. 8.27 were obtained by Sanford (1959) in experiments in which up-doming of the basement was induced. These experiments were conducted to verify the conclusions which resulted from theoretical analyses. These will now be considered.

Theoretical models of graben development

It will be recalled that stress functions (Chap. 5) can be used to determine the magnitude and orientation of stresses within a model, provided the boundary stress conditions are correctly defined. Sanford used a technique which is related to that used by Hafner (1951), but instead of expressing the boundary conditions entirely in terms of stress, Sanford's analysis is based on boundary displacements. The two methods can give similar results when applied to some problems but it is apparent that the boundary displacement method used by Sanford is particularly suitable for the study of the faulting of cover rocks which results from movement in the basement. Moreover, the analysis can be compared with the results obtained from the sand-box or other physical models in which the same prescribed basal displacements are imposed.

Sanford considered two main types of displacements. In the first of these examples, the elastic cover layer is assumed to undergo a sinusoidal displacement. The displacements and the resulting stress trajectories of this analysis are indicated in Fig. 8.28(*a*). In order that the whole analysis remains within the elastic range of materials, the vertical movements envisaged in Sanford's analyses are very small. Thus, if the sedimentary cover is taken as 5 km, the maximum movement in the basement will be of the

order of 10 m. The stress distribution and potential shear planes in the cover material, which result from sinusoidal basement movement, are indicated in Fig. 8.28(*b*).

To determine the regions where fractures are likely to form, or where material is likely to yield by plastic deformation, Sanford calculated values for the distortional strain energy (E_d) throughout the model. This he did by using the expression:

$$E_d = \frac{1}{12G}\left\{[\sigma_y - \sigma_z]^2 \right.$$
$$+ \left[\sigma_z - \frac{1}{m}(\sigma_y + \sigma_z)\right]^2$$
$$\left. + \left[\frac{1}{m}(\sigma_y + \sigma_z) - \sigma_y\right]^2 + 6\tau_{yz}{}^2\right\} \quad (8.4)$$

where G is the shear modulus and m is Poisson's number. It may also be inferred that the shear fractures which may develop in the vicinity of the crust would have the orientation of normal faults.

As noted above, since this analysis relates to a rather thin cover and minor basement movements, the displacement on the normal faults should be only a few metres. However, the extent and displacements of the normal faults which bound major grabens greatly exceeded the limits postulated in Sanford's analysis. As has been seen, the bounding faults which define a major graben probably maintain their orientation, so that, for example, the boundary faults of the Rhine Graben are rooted below the crust. Moreover, from the thickness of the accumulated sediments within the graben, it may be inferred that the down-throw is several kilometres in extent. Consequently, the conclusions of the analysis must be applied to real situations with great care. Moreover, subsequent events may change the stress conditions about real grabens.

For example, the development of the Alps (and possibly, also the formation of the Atlantic) subsequently influenced the evolution of the Rhine Graben. As a result the regional stresses adjacent to the graben are no longer consistent with the mechanism outlined above but, as may be seen from Fig. 8.29, they are oblique to the limits of the graben and, indeed, are evidence of the influence of the Alpine orogeny far into its foreland.

We have touched only lightly upon the topic of graben morphology and possible modes of formation. The literature abounds with contributions on these subjects. The interested reader is advised to consult the special issue of Tectonophysics, (1981, vol. 73) on the *Mechanism of Graben Formation*, for several articles on the subjects discussed above and further references to the pertinent literature. It is worthwhile to quote Illies (1981) who wrote the following warning in the epilogue of this special issue, '. . . an overall model of graben formation cannot be found because the different prototypes of continental rift valley. . . . the destructive process of continental rifting appears as multifarious as its constructive antagonistic partner, the orogeny'.

Normal and reverse faults related to basement movement

Reverse faults commonly dip at about 60–70°. That is, they usually have the orientation of normal faults but exhibit the reverse sense of movement. It is immediately apparent, therefore, that many reverse faults probably originated as normal faults and were subsequently rejuvenated in a stress field which gave rise to the reverse sense of movement. This aspect will be considered later. However, here we shall consider how reverse faulting at, or near, the surface may be induced by fault movement in the basement. We have already made reference to the 'displacement' method used by Sanford, when the formation of grabens was discussed. He also presented a second type of displacement, which approximated to a step-like uplift of the basement (Fig. 8.30(*a*)).

The stress trajectories and distribution of distortional strain energy for the step model are indicated in Fig. 8.30(*b*) and (*c*) respectively. It will be seen that the deformation of the 'cover' is almost completely restricted to a vertical zone above the 'fault line' in the basement. Shear failure develops at an angle to the axis of maximum principal stress. The precise angle, as we have seen, depends upon the values of the effective least principal stress, which ranges from less than zero, when 'near-tensile failure' occurs, through an increasingly larger angle for the hybrid extension and shear zone to $\theta = 45° - \phi/2$ (where ϕ is the angle of friction) for conditions when the effective least principal stress is compressive. (Here we assume that

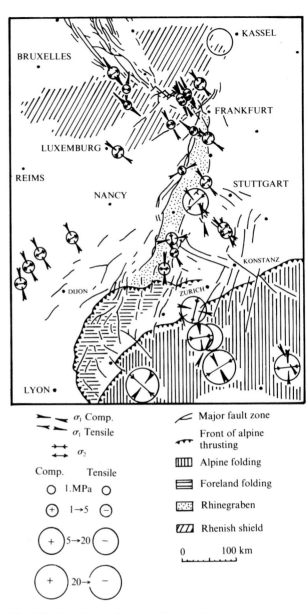

Fig. 8.29. Direction and magnitude of maximum horizontal compression in the W. Alps and their northern foreland. (After Baumann, 1981.)

the differential stress is greater than $4T$, see Chap. 1.) If such compressive stresses exist in the conditions represented in Fig. 8.30, then it may be inferred that a fracture which develops in the cover, immediately above the step displacement in the basement, may not be vertical. If, indeed, it is not vertical, the fault in the cover will run away from the projection of the fault line in the basement into a zone in which the stress trajectories are curved and, consequently, the fault will also be curved.

As has already been noted, this method of analysis is particularly amenable to verification, for displacements of the type used in the analysis can be applied in sand-box and other types of model experiments. Typical results obtained by Sanford, using a simple sand-box for the basement fault analysis, are

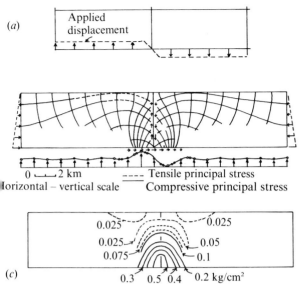

0 ⊢—⊣ 2 km
Horizontal – vertical scale

---- Tensile principal stress
—— Compressive principal stress

(c)

Fig. 8.30. (a) Type of step-like uplift of the basement discussed by Sanford. (b) Distribution of compressive and tensile principal stress engendered in the cover by the type of basement uplift indicated. (c) Distribution of distortional strain-energy in the deformed cover. (After Sanford, 1959.)

represented in Fig. 8.31, where it will be seen that the reverse fault in the cover material is initiated at the fault line but curves away from the zone towards the 'down-throw' side. As we can see, the geometry of the faults in the cover above the step graben (Fig. 8.31) is different from the classic Rhine Graben geometry. Indeed, it is interesting to note that if the graben in the basement results from very slight extension, the resulting reverse faults at the surface would almost certainly be interpreted in terms of a phase of horizontal compression.

The reverse faults in the cover, generated during these experiments, were curved fractures. It was found, empirically, that the degree of curvature was influenced by the thickness of the layer of cover material (the thinner this layer, the greater was the

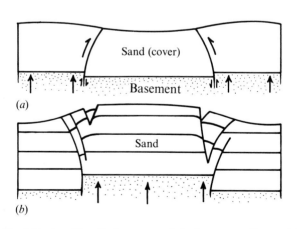

Fig. 8.31. Step-like displacements of the basement of a model experiment produced curved reverse faults in the sand representing the cover. (After Sanford, 1959.)

curvature of the fault plane) but was independent of the amount of displacement of the basement. Both these points may be inferred from a comparison of the curvature and displacement shown in Fig. 8.31. If one takes the angle of friction for sand to be $\phi = 28°$, then the coincidence between the experimentally determined fractures and those which may be inferred from the analysis is almost exact.

The sand-box experiments indicated in Fig. 8.31, were conducted using a vertical basement fault. Other experiments modelling movement on an inclined shear have been conducted by Parker & McDowell (1951) (Chap. 4). Horsfield (1977) carried out a series of sand-box experiments in which the angle of inclination of the basement fault, and hence the direction of movement, varied from 30–90° in steps of 15°. As would be expected, it was found that the inclination of the basement fault profoundly influenced the orientation of the faults which subsequently developed in the cover. The apparatus used (Fig. 8.32(a)) is larger and more versatile than most apparatus used in similar experiments. Sand was filled into the box by special equipment designed to give reproducible, homogeneous packing density. In the course of the experiment, the sand surface was kept roughly level by removing uplifted material using a modified vacuum cleaner or by filling troughs with extra sand.

Photographs were taken at regular intervals throughout the experiments. Off-set marker layers in the sand indicated the rough position of the faults, or shear zone (Fig. 8.32(b)). However, these were much more accurately defined when pairs of successive photographs were studied through a stereo viewer. A pseudo-stereo effect is seen because of the displacement of individual sand grains. Sharp variations in apparent height in the stereo-image represent fault planes, or narrow shear zones. Reverse faults in the cover developed only when movement in the basement was along faults dipping at 80 or 90°. For angles smaller than 80°, the features which developed in the cover rocks were normal faults which formed grabens.

The results of the experimental work have been applied to explain structures in the field. For example, Prucha *et al.*, (1965) discussed the Wyoming Province of the Rocky Mountain foreland, U.S.A. and Phipps & Reeve, (1969) have applied these experimental results to explain the structures which occur in the Malvern, Abberley and Ledbury Hills of England (Fig. 8.33(a) and (b)).

Bending of the lithosphere

A problem which is beginning to attract attention is the mathematical modelling of the lithosphere. Some of these analyses are based on elastic theory and are conducted to explain the observed bathymetric pro-

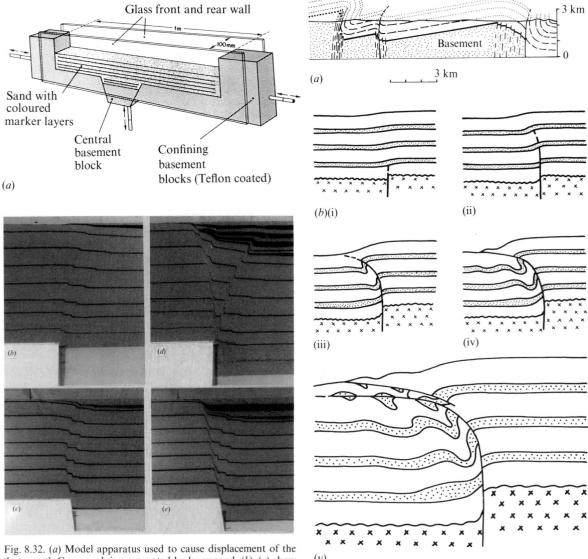

Fig. 8.32. (*a*) Model apparatus used to cause displacement of the 'basement'. Cover rock is represented by loose sand. (*b*)–(*e*) show two experiments illustrating the structures induced in the cover by dip-slip movement on basement faults. (*b*) and (*c*) over a vertical fault and (*d*) and (*e*) over a fault dipping at 78° to the right. (After Horsfield, 1977.)

Fig. 8.33. Structures in (*a*) the Sangre de Cristos area, New Mexico, U.S.A. (after Prucha *et al.*, 1965) and (*b*) in the Malvern, Abberly and Ledbury Hills, England, (after Phipps & Reeve, 1969) which are thought to have been caused by step-movements in their basements.

files associated with such features as Oceanic Islands, Island Chains and Arcs and Ocean Trenches, (Turcotte & Schubert, 1982). The models are essentially two-dimensional and it is assumed that a significant portion of the lithosphere can be treated as a homogeneous and isotropic, linear elastic layer, floating on a viscous mantle.

A mathematical model of the bending of the elastic portion of the lithosphere at an Ocean Trench is based on the geometrical situation represented in Fig. 8.34(*a*), where the deflecting load (V_0) and the Bending Moment (M_0) result from contact between two plates. The mathematical analysis of this situation is not given here. However, one of the important results of the analysis is the derivation of an expression which defines the *universal flexural profile*: a profile that is valid for any elastic flexure induced by

loading of the type represented in Fig. 8.34(*a*). Turcotte (1979) fitted this universal flexural profile to the *bathymetric profile* across the Mariana Trench, as shown in Fig. 8.34(*b*). The fit does indeed appear impressive.

By choosing suitable and reasonable elastic moduli ($E = 7 \times 10^5$ bar and $m = 4.0$) it can be calculated from this analysis that the thickness of the elastic lithosphere, in this instance, is 28 km; and this, Turcotte and Schubert claim, is in good agreement with the estimated thickness of the elastic lithosphere estimated from deflections which result from loading of the lithosphere by an Oceanic Island. Turcotte and Schubert noted one caveat: namely that the largest bending stress obtained from this analysis occurred 20 km seaward of the trench axis where it attained a

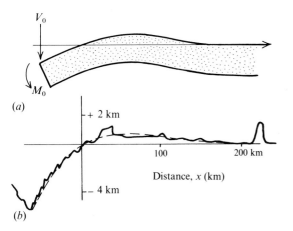

(a)

(b)

Fig. 8.34. (a) Boundary conditions and symbols used to obtain the 'universal flexural profile' of an elastic crust. (b) The profile derived by using the conditions indicated in (a) (dashed line) fitted to the bathymetric profile (solid line) across the Mariana Trench. (After Turcotte & Schubert, 1982.)

magnitude of 9.0 kbar. They comment that it is 'doubtful that the near-surface rocks have sufficient strength in tension' to withstand such a deviatoric stress. Indeed, it may be inferred that high extensional stresses will tend to develop near the surface, throughout most of the Trench and Island Arc profile. For example, it will be seen that the curve between $x = 0$ and $x = 100$ km in Fig. 8.34(b) is close to a circular arc (as emphasised by the dashed line). As noted earlier, from this geometry and the relationships shown in Fig. 8.6(a) and (b) and Eqs. (8.2) and (8.3), it can be shown that if one takes the elastic lithosphere to have a thickness (cited above) of 28 km, and the values for E and m already quoted, then, throughout the whole of the Island Arc, the outer (uppermost) fibres of the elastic lithosphere would tend to experience a tensile stress of almost 3.0 kbar. These bending stresses would be superimposed on the existing rock stresses. For example, the lithosphere may be in a state of general horizontal compression. It has been argued by Fyfe, Price & Thompson (1978) that in such a general compression, the average differential stress is not likely to exceed 1.0 kbar. Hence, throughout almost 200 km of the profile shown in Fig. 8.34(b), the upper levels would experience a significant lateral reduction in stress that has the potential of attaining a magnitude of total tensile stress of 1–2.0 kbar. Turcotte and Schubert's warning was not sufficiently strong. *Near-surface rocks could not possibly sustain such stresses without failure.* Indeed, the upper levels of the lithosphere in the vicinity of the Island Arc would develop normal faults, before the lateral stresses actually became tensile.

Just as the outer fibres of the elastic lithosphere would undergo extension, the inner fibres would experience compression and tend to develop a compressive differential stress that also could potentially

reach almost 3.0 kbar. If, in addition, the lithosphere was experiencing a general horizontal compression, these bending stresses would augment the general level of the differential stress. From the geometry of the bathymetric profile and an assumed rate of plate advance (say 10 cm/yr) the average strain-rate in the inner fibres (at a depth of 28 km) must be of the order 10^{-15}–10^{-16}/s. Even from the meagre pertinent rock mechanics data that are currently available, one may infer that, at temperatures of perhaps 700 °C and a strain-rate of less than 10^{-15}/s, the rocks forming the inner fibres would yield in a ductile manner, long before the differential stress attained its maximum potential value.

Where deep crustal rocks have been uplifted and exposed by erosion, it is commonly noted that significant deformation has resulted from the generation of ductile shear zones. Even though this type of failure is clearly non-brittle, nevertheless, from the acute angle between conjugate shears coupled with the sense of displacement, it is frequently possible to interpret some of these shear zones in terms of principal stress orientation in exactly the same way in which one would interpret brittle shears, (Sturt, 1969). (This is discussed further in Chaps. 2 and 18.) Consequently, it is reasonable to conclude that the ductile deformation of the lower levels of the so-called, elastic layer, in the situation under consideration, would probably result (in part at least) in the development of ductile shear zones. Price & Audley-Charles (1983) applied this concept to the collision of the Australian continental plate with the southward curving portion of the Banda Arc. From the geometry of the collision line they concluded that such shear zones, in this locality, would have the orientation of thrusts.

The development of thrusts and also of any associated pervasive general (as against localised) ductile deformation would result in thickening of the lithospheric plate. It is known from the recent retreat of the Polar Ice-Sheets that the isostatic response time is of the order of 10^4–10^5 years. Consequently, thickening of the lithospheric plate beneath an Island

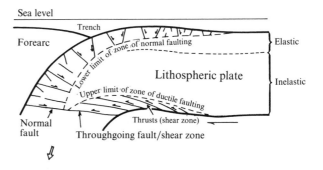

Fig. 8.35. Diagrammatic representation of the development of normal faults in the upper zones, and thrust shears in the lower zones of an oceanic, lithospheric plate, which may lead to through-cutting planes of weakness. (After Price & Audley-Charles, 1983.)

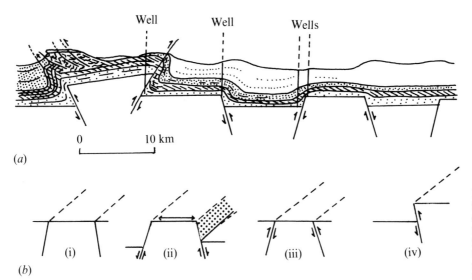

Fig. 8.36. (*a*) Reverse faulting
in the basement in the rocks of
central Tierra del Fuego. From
the orientation of the basement
faults, it is reasonable to infer
that they result from rejuvenation as indicated in (*b*). (After
Winslow, 1988.)

Arc would have ample time in which to give rise to isostatic adjustment. Such an adjustment and its influence on the bathymetric profile is not considered in the elastic analyses noted above. In addition, the failure at the outer and inner fibres, by brittle normal faulting and ductile shear respectively causes thinning of the elastic portion of the lithosphere.

The *flexural rigidity D* of the elastic plate is given by:

$$D = \frac{Eh^3 m}{12(m^2 - 1)} \tag{8.5}$$

where E and m are Young's modulus and Poisson's number respectively and h is the thickness of the elastic plate. Thus, because D is proportional to the cube of the thickness (h), a small reduction in h results in a large reduction in the flexural rigidity. If the boundary conditions are maintained constant, this will result in a sharper curvature of the lithospheric plate, as the Trench is approached. In turn, this will give rise to a tendency for yet higher fibre stresses to develop, so that the normal faults and thrusts approach each other. Should the elastic portion of the lithospheric plate become vanishingly thin, an individual thrust may run into an individual normal fault which, as it descends into the Trench, now has a much lower dip (Fig. 8.35) and so results in a listric structure. If the plate as a whole is experiencing horizontal compression, such a break-through, from ductile thrust to normal fault, could result in a reversal of movement on the normal fault. Moreover, it is argued by Price & Audley-Charles that in specific conditions, weak sediments of the continental shelf area may be forced into these 'breakthrough' fractures and so enable the downward moving slab to become detached.

The arguments presented above are qualitative, but are based on reasonable mechanical premises.

They indicate that, although extremely interesting, analyses of lithospheric deflections in terms of an elastic plate are, in some situations at least, only first approximations. However, because of the intrinsic difficulty of the problem, it will probably be several years before a more representative model, which incorporates progressive failure of the lithosphere coupled with isostatic adjustment, can be successfully analysed.

Reverse faults

It has been noted that normal fault displacement of basement rocks can give rise to reverse faulting (or even thrusting) in the cover rocks (see Figs. 8.31 to 8.33). Indeed, it is not infrequently observed that fractures which have the orientation usually associated with normal faults prove, on inspection, to exhibit a reverse displacement. In many such situations, it is difficult to avoid the conclusion that the fractures first developed as normal faults during a phase of lateral extension, and that they were subsequently rejuvenated with reverse movement induced by lateral compression. An example, from Tierra del Fuego, of such a group of structures is illustrated in Fig. 8.36 (Winslow, 1981).

Normal faults with the ideal, Andersonian orientation are initiated when the stresses are as shown in Fig. 8.37(*a*). Rejuvenation and reverse mode re-shear may be expected when the maximum principal effective stress (σ_1) acts horizontally (Fig. 8.37(*b*)). However, this is a purely qualitative interpretation. It has been noted in Chap. 6, that re-shear along an existing plane that contains the axis of intermediate principal effective stress (σ_2) will result when:

$$\frac{\sigma_1}{\sigma_3} = \frac{1 + d}{1 - d} \tag{8.6}$$

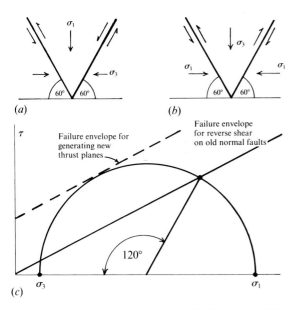

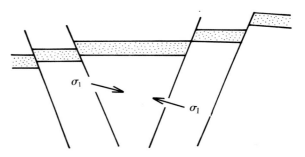

Fig. 8.38. Re-shear on one set of normal faults to form reverse faults is more easily accomplished if the maximum principal stress is inclined to the horizontal.

Fig. 8.37. (*a*) Stress configuration for the initiation of normal faults. (*b*) Stress configuration often proposed for reactivation of normal faults. (*c*) Mohr's circle and failure criteria which enable one to infer that stress magnitudes required to cause normal faults to re-shear to form reverse faults are likely to approach (or even exceed) those required to form thrusts dipping at about 30°.

where

$$d = \frac{\sin 2\theta + \cos 2\theta \tan \phi}{\tan \phi}. \tag{8.7}$$

Let us assume that $\phi = 30°$: then the ideal Andersonian normal faults would dip at 60°. Hence if, for potential re-shear, it is assumed that the axis of maximum principal stress is horizontal, then, as can be seen in Fig. 8.37(*a*), the angle $\theta = 60°$. It may be inferred from Eqs. (8.6) and (8.7), that if $\phi = 30°$ and $\theta = 60°$, the ratio of σ_1/σ_3 goes to infinity (see Chap. 5). Consequently, one may conclude that thrusts, dipping at 30° would develop long before the maximum principal stress could attain a value which would permit the rejuvenation and reverse shear of the erstwhile normal faults. Hence, such reverse re-shear on the ideal, Andersonian normal faults could only take place if the angle of sliding friction (ϕ_r) were considerably lower than the angle of friction (ϕ) of the solid unbroken rock (as would occur if the fault planes contained clay fault-gouge) and/or the axis of maximum principal stress is somewhat inclined to the horizontal.

If the latter alternative obtains, it may readily be inferred from Fig. 8.38 that the inclined principal stress will greatly enhance the probability of reverse re-shear on one set of normal faults. However, because σ_1 clearly acts at a high angle to the conjugate set of normal faults, the normal stress on these latter fractures will attain a high value which will completely inhibit re-shear on this second set.

The inclined axes of principal stresses may be attributed to a number of possible situations (see

Chap. 5 relating to Stress Functions). However, there is one very common possible cause for such inclined axes. As geologists usually study structures in surface outcrops, one may conclude that they are generally concerned with rock masses which have experienced uplift and denudation (or exhumation). Differential uplift is one of the situations which may give rise to inclined axes of principal stress. Thus, a differential uplift which gives rise to a tilting of the strata by 0.057° (0.001 radians) would result in a shear strain γ of 0.001. The shear modulus (G) of elastic competent rocks is commonly about 10^5 bar. The resulting shear stress (τ) that is associated with such a shear strain γ is given by:

$$\tau = \gamma G. \tag{8.8}$$

For the values of γ and G given above, the induced shear stress could attain a value of 100 bar. Let us assume that prior to this phase of quite minute, differential uplift, the principal stresses (σ_1 and σ_3) acted horizontally and vertically, respectively; and possessed the values of 200 and 500 bar. Then, as may be inferred from Fig. 8.39, the induced shear stresses will change the magnitudes of the least and principal

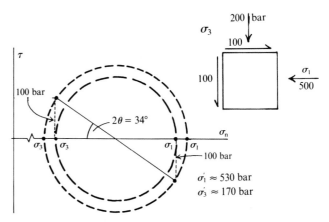

Fig. 8.39. Shear-strains induced by differential uplift result in the hitherto horizontal maximum principal stress becoming inclined and also increased in magnitude. The stress circles shown represent the original stresses of 200 and 500 bar (the inner circle) and the subsequent conditions when shear stresses of 100 bar are superimposed. The orientation of the planes on which the shear stresses develop are horizontal and vertical.

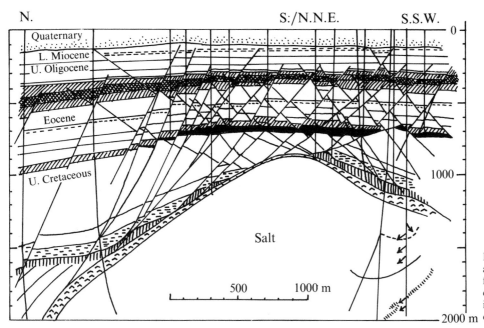

Fig. 8.40. Section showing interaction of conjugate, normal faults above the Reitbrook salt-dome of northern Germany inferred from seismic and drill data. (After Horsfield, 1980.)

stresses (σ'_1 and σ'_3) to 170 and 530 bar. Moreover σ'_1 will act at 17° to the horizontal. Such changes in magnitude and orientation of the principal stresses will greatly enhance the probability of reverse re-shear on one set of normal faults. If the sense of differential uplift is maintained, then reverse movement can take place only on one set of normal faults. However, the amount of differential uplift cited above (0.001 radians) is extremely small so that, in the process of prolonged uplift, the sense of differential uplift may change; with the result that reverse re-shear may occur on both sets of erstwhile normal faults.

Arrangement of sets of shear fractures

In this final section we shall discuss briefly the disposition of conjugate normal faults, though the comments are also likely to apply to the geometrical and spatial relationships of one set of shears to another set which, together, are thought to form a conjugate system, whether they be normal, strike-slip or thrust shears.

The intersections of two conjugate shears are only infrequently exposed in the type of cliff or quarry section which affords the field geologist the opportunity of close study of the detailed geometry of the mode of intersection. Indeed, even when sets of fractures with suitable senses of shear are exposed, it is not always possible to ascertain whether the sets are conjugate or, possibly, *synthetic* and *antithetic* features, where the latter may be secondary to the larger, primary synthetic faults. Relatively large normal faults above a salt dome (Fig. 8.40) may be inferred from drilling and seismic data; but these do not usually permit close study to be made of the mode of intersection of the conjugate shears.

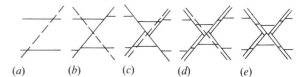

Fig. 8.41. Mechanism to produce irrotational plane strain by the alternate operation of conjugate faults. (*a*) Incipient fault (dashed line): (*b*) slight displacement on the first fault and initiation of conjugate fault: (*c*) first fault off-set by the conjugate fault: (*d*) new fault, parallel to first, off-sets conjugate: (*e*) new conjugate fault forms. (After Freund, 1974.)

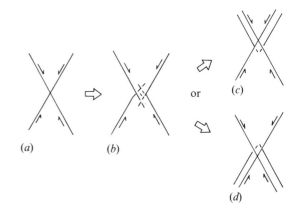

Fig. 8.42. Mechanisms proposed by Horsfield (1980) of contemporaneous movement along crossing faults. (*a*) Initial formation of conjugate pair. (*b*) Movement along each fault causes off-set of both faults. Further displacement leads to increasingly small-scale deformation in the cross-over region. (*c*) and (*d*) The original fault breaks through to form new extensions. External constraints on the fault positions may determine which of the alternatives (*c*) or (*d*) actually develops. Continued displacement will generate an increasing number of new fault extensions.

A mechanism which gives rise to irrotational plane strain by the development of conjugate shears has been proposed by Freund (1974). He suggested that an incipient fault (Fig. 8.41(*a*)) causes slight

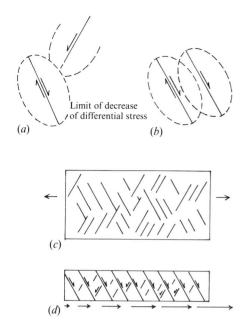

Fig. 8.43. (*a*) Stress changes induced by the first fault tend to inhibit the development of intersecting, conjugate fractures. (*b*) Similarly, stress changes ensure that parallel shear fractures of the same set are spaced. (*c*) Random development of shears of one or other of the two sets are to be expected in uniform, pure shear. (*d*) Progressive extension from one side is likely to develop one dominant set of shears.

displacement of marker horizons (Fig. 8.41(*b*)) and that the first fault is then off-set by a conjugate fault and that the two faults become locked (Fig. 8.41(*c*)). New faults which are parallel and off-set these first faults (Fig. 8.41(*d*)) eventually give rise to the form shown in Fig. 8.41(*e*), which illustrates a mechanism based on the *alternating* crossing and off-set of shear planes.

Horsfield (1980) presented an alternative mechanism which was based on the continuous crossing and off-setting of conjugate faults. This work was based on the observed behaviour of normal faults which develop in a sand-box and gave rise to the pattern of fractures indicated in Fig. 8.42 and which are seen to have formed in the sediments exposed in the cliffs of Heligoland.

As has already been said, intersecting conjugate shears are only infrequently encountered. It is far more usual for sets of shears to occur either as 'packets' (the interaction between conjugate packets is not seen, Fig. 8.43(*a*)) or else one of the conjugate sets dominates so that the other, if it exists at all, occurs in the form of minor fractures which may be confused with second order features.

The development of packets of shears of one set can be explained in terms of the static stress changes induced by shear displacement discussed by Anderson (see Chap. 6) and earlier in this chapter. The first shear (normal fault) to cut through a section of rock would have caused a significant reduction in the magnitude of the differential stress around the fault.

This change in the stress field extends for a significant distance from the fault plane. (For example, in Anderson's analysis, the stresses are affected to a distance of 0.4 times the length (*L*) of the fracture.) As may be inferred from Fig. 8.43(*a*), the existence of this field of reduced differential stress associated with movement on this first fault would inhibit the development and propagation of a conjugate shear. However, provided other fractures of the same set as the first fracture are sufficiently distant (Fig. 8.43(*b*)) they may readily develop to form a packet of one set of shears.

If a rock mass is subject to uniform and distributed lateral extension, it is probably a matter of chance whether fractures of Set *A* or Set *B* of a conjugate system develops in any one locality (Fig. 8.43(*c*)). However, when the initial fracture in one locality has developed then, by the mechanism outlined above, this will result in the formation of packets of shears of either Set *A* or *B*. However, if a rock mass is subjected to progressive extension, from one side (Fig. 8.43(*d*)), as may happen in gravity or delta tectonics, there will be a strong tendency for only one major set of shears to develop. These will be the synthetic set. The antithetic faults which may be conjugate, or secondary features are likely to be of relatively minor extent.

Commentary

In the last three chapters we have dealt with various aspects of strike-slip, thrust and normal faulting. It will have become apparent to the reader that such a treatment is, to some extent at least, artificial and arbitrary. Certainly, the field geologist may find specific examples and environments which relate to one of the three main types of faults. However, when dealing with many faults, especially when they are of regional extent, it is not uncommon to encounter two or more of the main types in association. For example, in Chap. 6 we noted the interrelationship between secondary normal, strike-slip and thrust or reverse faults adjacent to a major strike-slip fault. The various folds, flexures, fractures and ramps associated with the frontal portions of many major thrusts described in Chap. 7 will produce a comparable grouping of the three main fault types. Furthermore, as has been pointed out, it is usually extremely difficult to determine the precise direction of transport of larger faults. It is a matter of convenience, therefore, to designate a particular structure as normal, reverse or strike-slip, when in actual fact the structures may have exhibited oblique-slip characteristics. Consequently, it sometimes proves convenient to express some groups of major faults and their associated structures as hybrid forms which result from *transpression* (a combination of transcurrent and compressive movements) or *transtension* (a combination of transcurrent and extensional movements).

References

Anderson, E.M. (1951). *The dynamics of faulting*, 2nd edn. Edinburgh: Oliver & Boyd.

Baumann, H. (1981). Regional stress field and rifting in Western Europe. *Tectonophysics*, **73**, 105–12.

Beach, A. (1977). Vein arrays, hydraulic fractures and pressure solution structures in deformed Flysch sequence, S.W. England. *Tectonophysics*, **28**, 245–63.

Biot, M.A. (1965). *Mechanics of Incremental Deformations*. New York: Wiley.

Cloos, H. (1928). Experimente zur inneren Tektonik. *Centralbl. f. Mineralogie*, Abt. B, 609–71.

Elmohandes, S.E. (1981). The central European graben system: Rifting imitated by clay modelling. *Tectonophysics*, **73**, 69–78.

Evamy, B.D., Haremboure, J., Kameling, P., Knaap, W.A., Molloy, F.A. and Rowlands, P.H. (1978). Hydrocarbon habitat of Tertiary Niger Delta. *Am. Assoc. Petrol. Geol. Bull.*, **62**, 1, 1–39.

Freund, R, (1974). Kinematics of transform and transcurrent faults. *Tectonophysics*, **21**, 93–134.

Fyfe, W., Price, N.J. & Thompson, A.B. (1978). *Fluids in the Earth's Crust*. Amsterdam: Elsevier.

Hafner, W. (1951). Stress distributions and faulting. *Geol. Soc. Am. Bull.*, **62**, 373–98.

Hamblin, W.K. (1965). Origin of 'reverse drag' on the down-throw side of normal faults. *Geol. Soc. Am. Bull.*, **76**, 1145–64.

Harding, T.P. & Lowell, J.D. (1978). Structural styles, their Plate Tectonic habitats and hydrocarbon traps. In *Petroleum provinces. Am. Assoc. Petrol. Geol. Bull.*, **63**, 7, 1016–59.

Horsfield, W.T. (1977). An experimental approach to basement-controlled faulting. *Geol. Mijn.*, **56**, 363–70.

   (1980). Contemporaneous movement along crossing conjugate normal faults. *J. Struct. Geol.*, **2**, 3, 305–10.

Hose, R.K. & Danes, Z.F. (1973). The development of the late Mesozoic to early Cenozoic Structures of the Eastern Great Basin. In *Gravity tectonics*, ed. K.A. de Jong & R. Schotten, pp. 429–41, New York: Wiley.

Illies, J.H. (1977). Ancient and recent rifting in the Rheingraben. *Geol. Mijn.*, **56**, 329–50.

   (1981). Mechanism of graben transformation. *Tectonophysics*, **73**, 249–66.

Letourneur, J. (1953). Le Grand Sillon Houiller du Plateau Central Français, **238**, Paris.

Lovering, T.S. (1928). The fracturing of incompetent beds. *J. Geol.*, **36**, 709–17.

McKenzie, D. (1978). Some remarks on the development of sedimentary basins. *Earth & Plan. Sci.*, **40**, 25–32.

Mandl, G. (1985). Tectonic deformation by rotating parallel faults – the 'bookshelf' mechanism. *Shell Research B.V. Publication*, **712**.

Mandl, G. & Crans, W. (1981). Gravitational gliding in deltas. In *Thrust & nappe tectonics*, eds. K. McClay & N.J. Price. Geol. Soc. London Special Publ. No. 9. Blackwell Scientific Publications.

Moore, J. McM. (1979). Tectonics of the Najd transcurrent fault system, Saudi Arabia. *J. Geol. Soc. London*, **136**, 441–52.

Parker, T.J. & McDowell, A.N. (1951). Scale modelling as a guide to interpretation of salt dome faulting. *Am. Assoc. Petrol. Geol. Bull.*, **35**, 9, 2076–86.

Phipps, C.B. & Reeve, F.A.E. (1969). Structural geology of the Malvern, Abberley and Ledbury Hills. *J. Geol. Soc. London*, **125**, 1–34.

Price, N.J. & Audley-Charles, M.G. (1983). Plate rupture by hydraulic fracture resulting in overthrusting. *Nature*, **306**, 572–5.

Prucha, J.J., Graham, J.A. & Nickelsen, R.P. (1965). Basement-controlled deformation in Wyoming Province of Rocky Mountains Foreland. *Am. Assoc. Petrol. Geol. Bull.*, **49**, 966–92.

Sanford, A.R. (1959). Analytical and experimental study of simple geological structures. *Geol. Soc. Am. Bull.*, **76**, 19–52.

Sturt, B.A. (1969). Wrench fault deformation and annealing recrystallization during almandine amphibolite facies regional metamorphism. *J. Geol.*, **77**, 391–32.

Turcotte, D.L. (1979). Flexure. *Adv. Geophys.*, **21**, 51–86.

Turcotte, D.L. & Schubert, G. (1982). *Geodynamics: application of continuum physics to geological problems*. New York: Wiley.

Winslow, M.A. (1981). Mechanism of basement shortening in the Andean foreland belt of southern South America. In *Thrust & nappe tectonics*, eds. K. McClay & N.J. Price. Geol. Soc. London Special Publ. No. 9. Blackwell Scientific Publications.

9 Development of systematic fractures in slightly deformed sedimentary rocks

Introduction

One of the features of horizontally, or near-horizontally, bedded sedimentary series which have been subjected to little or no tectonic deformation is that, despite a lack of apparent finite deformation, the sediments are, nevertheless, cut by many well-defined sets of fractures. Striking examples of such sets of fractures are exhibited by sediments (from Cambrian to Cretaceous) in which the Grand Canyon, Colorado, U.S.A. has been cut (Fig. 9.1(*a*) and (*b*)). As may be seen in these figures, the fracture planes in such little-deformed sediments are predominantly perpendicular to the bedding. Individual fractures in any set rarely exhibit noticeable, or even discernible, field evidence of shear displacement; so that these fracture sets are usually referred to as 'joints' (cf. Chap. 2).

In the absence of tectonic features, such as major faults or folds, it is extremely difficult to interpret observed fracture patterns in such flat-lying expanses of sedimentary rocks. Although relatively few studies have been made of such patterns, there is one area which has been repeatedly surveyed. Consequently, in this chapter, we shall first outline and comment upon various attempts to interpret fracture patterns in the little-deformed sediments of this particular area, the Appalachian Plateau of the U.S.A. It will be shown that attempts at analysis of the fracture patterns contained in these sediments have not resulted in general interpretive agreement. In the light of these difficulties and differences in interpretation, we shall deal with some general relationships established in less extensive localities and then consider how stress patterns and fracture systems may be expected to develop and evolve in certain situations, so that the field geologist and fracture pattern analyst may be better able to understand the possible modes of development of real geological fracture systems.

Fracture systems in the Appalachian plateau

As we shall see in Chap. 14, the development of fracture patterns associated with well-developed folds often permits the geologist to establish the stress fields in which the fractures develop. Moreover, these studies even provide an important insight into mechanisms of development of the folds in which the fractures occur. However, in regions where folds are

(*a*)

(*b*)

Fig. 9.1. (*a*) View of part of the Grand Canyon, showing the horizontal beds cut by a large number of vertical fractures. (*b*) Fracture traces on bedding planes in the Grand Canyon.

poorly developed, or non-existent, the interpretation of fracture systems is a task which rarely leads to clear-cut, unequivocal conclusions regarding the stress history of an area. Indeed, the literature dealing

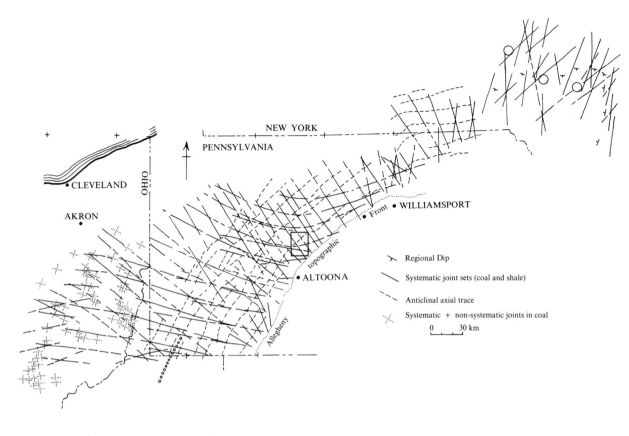

Fig. 9.2. Map of the Appalachian Plateau showing outline of state boundaries and general orientation of the fracture patterns in the various localities.

with such attempted interpretation often leaves the reader with an impression of confusion regarding the type and age relationships of the fracture studied; so much so that different observers in a given region may reach contradictory conclusions.

Such a region is exemplified by the Appalachian Plateau of Pennsylvania and New York, which has been studied by many workers, among whom are Parker (1942), Nickelson & Hough (1967) and, more recently, Engelder & Geiser (1980). In the maps and diagrams presented by these various workers, the area is seen to be traversed by a succession of 'folds'. The relief between adjacent structures decreases northwards from the Alleghenny Topographic Front to the Appalachian Basin which passes, with a NE–SW trend, through Pittsburgh. It is emphasised, however, that the surface rocks throughout much of the region are essentially flat-lying, for dips rarely exceed 1–2°. Indeed, the widely-spaced fold axes, shown in Fig. 9.2, are defined mainly by structural contouring. With such small angles of dip and large distances (20–30 km) between fold traces, it is extremely doubtful that these 'folds' can be attributed to a buckling mechanism. Consequently, it would be unwise to argue that, because there are 'fold axes', these structures must automatically trend normal to the axis of

greatest principal stress. Other evidence must be adduced before such a stress orientation can be inferred.

The fracture patterns which have developed in this area, as reported by the various workers, appear to exhibit quite a marked consistency in their pattern (Fig. 9.2). Parker distinguishes three groups of fractures ('joints'), all of which are essentially perpendicular to the bedding planes. The first of these groups comprises two sets (Ia and Ib) with an average acute angle between them of 19°. It is noted by Parker that these two sets of fractures are remarkably planar, in that they cut cleanly through concretions in relatively weaker strata and pass through cross-bedded rocks without deviation. The median line of this group of fractures (i.e. the trace of this plane bisecting the acute angle between Sets Ia and Ib) trends approximately perpendicular to the fold axis. Parker considers this group of fractures to be 'shear joints'.

The second group of fractures (Set II) is mainly perpendicular to the median line of Group I and, hence, is approximately parallel to the trend of the fold axis. Parker notes that the surfaces of the fractures in Set II tend to be curved and irregular with a rough, torn appearance and are held to be 'tension joints'. The third group (Set III) seems to be indepen-

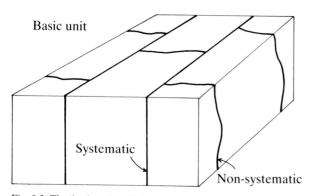

Fig. 9.3. The 'basic unit' of joint development used by Nickelson & Hough (1967) which comprises a set of systematic, planar fractures and the approximately orthogonal set of non-systematic, curved fractures.

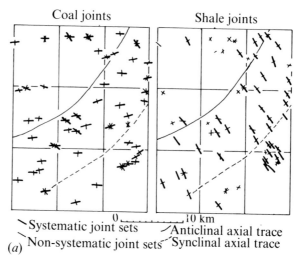

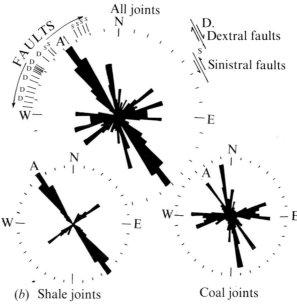

Fig. 9.4. (*a*) Map of the fractures in the Houtzdale quadrangle (from Nickelson & Hough, 1967). (*b*) Rose diagram of the fractures shown in (*a*). Rose diagrams are compiled representing the number of fractures which have a trace that falls within a 5° or 10° arc. The data used to compile such rose diagrams are weighted to take into account the length of individual fractures. For example, a 10 m long fracture may be represented as 1 mm on the diagram, so that a 50 m long fracture would be represented by 5 mm.

dent of the other groups of fractures, for it does not follow the change in trend of the fold axes. This third set, which trends N60°E ± 5° throughout the whole region, is attributed to a stress system unrelated to those that give rise to Sets Ia and b and II, and is of unknown age.

Nickelson & Hough (1967) studied the fractures in the coals and shales of the region and are not in agreement with all the conclusions reached by Parker. Their approach was different from that of Parker, in that they considered the 'basic unit' to be the *fundamental joint system*, which comprises two approximately orthogonal sets. The fractures of one of these sets is planar and systematic, while the fractures of the other set are curved and non-systematic (Fig. 9.3).

They conclude that systematic 'joints' are grossly transverse to fold axes. However, because the fold axes bend N30°E to N70°E, no single set of 'joints' remains perpendicular to the folds throughout the region. Rather, they suggest, the regional pattern consists of overlapping 'joint' trends which exhibit remarkably uniform orientation; each trend locally being dominant (i.e. showing greatest size, frequency and persistence) where it is most nearly perpendicular to the fold axis. They correctly emphasise that the 'joint' pattern of the region is cumulative, presenting a record of events which have produced effective stresses of sufficient magnitude to induce fracture.

Although they remark that systematic 'joints' may be polygenetic, they conclude, from the morphology of the 'joint' planes and particularly from the lack of lateral movement, that the joints are of extensional origin. They further note that only 6 per cent of the areas studied contained two acutely intersecting sets of systematic 'joints'. Nickelson & Hough are not of the opinion that these occurrences are conjugate shears, but believe that they represent overprinting of extension 'joints' by a slightly dominant trend of extension features which are better developed in an adjacent area.

Vertical fractures with horizontal 'slickensiding' (i.e. strike-slip faults) are reported to occur in some areas of the Appalachian Plateau. However, only in the Houtzdale quadrangle was there sufficient data for Nickelson & Hough to attempt clarification of the interrelationship between 'joints' and 'faults'. A map of the orientation of the fractures relative to fold axes in this rectangle together with rose diagrams of the data collected are presented in Fig. 9.4(*a*) and (*b*). They note that the strike-slip faults form a conjugate system which would result from compression when the greatest principal stress acted N50°W, i.e. approximately perpendicular to the fold axis trend. (See earlier comment.) They also note that Set A of the systematic 'joints' is approximately parallel to the sinistral faults; but comment (without further elucidation) that their parallelism in this area does not necessarily imply similar time or mechanics of genesis.

Engelder & Geiser (1980) also studied the fractures of the Appalachian Plateau. The basic premise of their paper is that 'vertical joints propagate normal to the least principal stress and thus follow the trajectories of the stress field present at the time of propagation'. Following the conclusions reached by Bahat (noted in Chap. 2), they further state that 'the existence of plumose structures, as opposed to slickensides on (the main fracture) Sets Ib joints and Ia joints leave little doubt that the joints formed normal to σ_3 as extension fractures rather than shear fractures'. They observe that Sets Ib and Ia differ in trend by as much as 30° and that this divergence must be interpreted as resulting from the rotation of the stress field by a comparable 30°.

Rose diagrams are used by all these cited groups of workers, for they are a convenient means of expressing data on fracture trends over a specific area. However, the grouping together of such data may obscure important relationships. For example, it is common to combine data on fractures from different lithologies in the same area. If two major trends exist in such a diagram separated by a small angle (say 20°–30°), then it is important to know whether or not these fracture trends occur together in individual beds. Once a set of extension fractures have developed in a bed, it is extremely difficult, or even impossible, to form a second set of extension fractures which make a small angle with the first set. (There is, of course, no great difficulty in permitting a second set of fractures to form at a small angle, provided they are the result of shear failure.) However, if the two postulated trends do not occur together in the same bed, then overprinting of one extension set upon another is possible.

Thus, Engelder & Geiser make the point that, in their area of study, the fracture trends (Ia and Ib), which are separated by an angle of about 30°, do not occur in the same units. They argue that fracture trend Ia developed first in the competent units, and after a hiatus and rotation of the stress field a second set of fractures (Ib) developed in the incompetent units. This second phase of fracturing would, of course, cause a second phase of opening in the Ia fractures coupled with a component of shear movement. That is, the first formed fractures would then exhibit some characteristics of shear fractures. It is emphasised that the second phase of opening and shear would usually prove exceedingly difficult to establish.

Clearly, the interpretation of real fracture patterns, especially those which develop throughout a region, which has possibly experienced more than one phase of deformation, is an extremely difficult and vexing task. For this reason, we shall merely note the

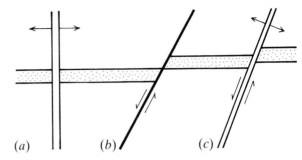

Fig. 9.5. The three types of fractures (*a*) extension, (*b*) shear and (*c*) the hybrid extension/shear fractures.

main types of fractures (Fig. 9.5) (Chaps. 1 and 2) and the 'typical' relationships which may exist between them. This is followed by an analysis of the development of a 'hypothetical' sedimentary basin, from which we indicate how and why such patterns *may* develop in some *strong* rocks.

Relationship between fractures in unfolded sediments

Fold belts are frequently flanked by a 'foreland' (Fig. 9.6(*a*)), in which the sediments do not exhibit obvious evidence of orogenic deformation. Examples of such forelands are to be found in North America (the Appalachian Mts, the Ouachitas and the foothills of the Rocky Mountains), Europe (the Alps and the Scottish Highlands) and elsewhere. An intuitive analysis of the stresses which could give rise to such a distribution of folded and undeformed sediments is given by Hubbert (1951), while a rigorous analysis based on the application of stress functions is given in a companion paper by Hafner (1951). (See also Chap. 5.) In this analysis, Hafner considered the situation in which the horizontal stresses perpendicular to the section represented in Fig. 9.6(*b*) are the intermediate principal stresses (σ_2). With such a stress system, one would expect minor thrusts to develop in the foreland sediments which are adjacent to the fold belt.

If, however, there is a slight extension in the foreland, or possibly in the fold belt, such as would occur if the trend of the fold belt were slightly convex towards the foreland: then the horizontal stress perpendicular to the section shown in Fig. 9.6(*b*) may become the least principal stress (σ_3). It will be noted that the magnitude of the differential stress decreases steadily from left to right in the illustrated section. Consequently, in those sediments of the foreland which are adjacent to the fold belt, the differential stresses will be higher than in the same sediments which are at comparable depths, but more distant from the fold belt. Hence, provided $\sigma_1 - \sigma_3 > 4T$ the fractures likely to develop in those sediments close to the fold belt will be strike-slip structures (as indicated in Fig. 9.6(*c*)); while in the sediments more distant

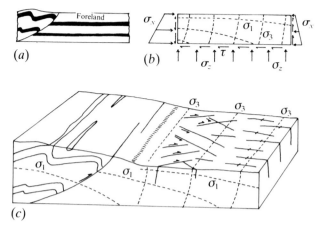

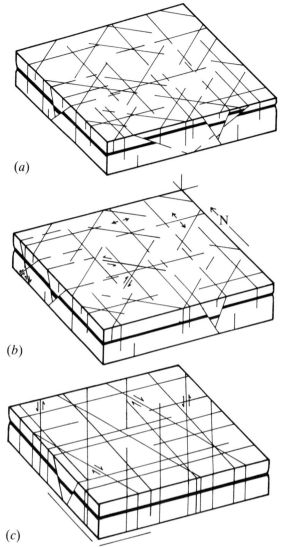

Fig. 9.6. (*a*) Schematic representation of fold and thrust belt abutting a lightly deformed foreland. (*b*) Boundary condition and internal stress trajectories compatible with the type of deformation represented in (*a*). (*c*) Block diagram showing fracture patterns which are likely to develop in the foreland area in response to the deformation and stress patterns represented in (*a*) and (*b*).

Fig. 9.7. (*a*) Fracture patterns which may be expected to form in unfolded rock. (After Price, 1966.) (*b*) Fracture patterns in the generally flat-lying sediments of the Cotswold Hills, England. (After Hancock, 1969.) (*c*) Fracture patterns deduced by Price (1974), from hypothetical basin analysis.

from the fold belt (where $\sigma_1 - \sigma_3 < 4T$) vertical fractures may form which are perpendicular to, but distant from, the fold belt. In a sedimentary sequence comprised of rock units of different physical properties, it is to be expected that there will be a zone in which both extension and shear fractures may develop. In addition, fractures of other types and orientations can be expected to develop if downwarp and subsequent uplift of the sediments in the foreland takes place. (Here, we ignore the possibility of flexure of the foreland as it passes beneath the thrust-, or fold-, belt. Such flexure may give rise to normal faults, or extension fractures, which trend approximately parallel with the junction between the foreland and the fold-belt.)

Theoretical considerations of this type, coupled with personal observations in the field and examples culled from the literature, led Price (1966) to present a generalised relationship of the various types and sets of fractures (faults and joints) that could develop throughout phases of an orogenic cycle in flat-lying sediments, not subjected to a compression sufficiently large to generate folds (Fig. 9.7(*a*)). Price noted that in many (but not, of course all) field examples, the inferred axes of principal stress appeared to remain remarkably constant in orientation throughout the generation of the different sets of fractures; but that the magnitude of the stresses may change. That is, the σ_1 axis of principal stress, during the thrusting phase, may be the σ_2 or σ_3 axis, and so on, during the development of normal faults and some joints. It will be recalled that we have seen in Chap. 8, how movement on a normal fault can locally engender such changes in the axes of principal stress.

The sets of fractures which have developed in the essentially flat-lying, Cretaceous limestones of the Cotswold Hills of S. England, have been mapped by

Hancock (1969) and presented in a block diagram (Fig. 9.7(*b*)). It will be noted that this pattern is remarkably similar to that shown in Fig. 9.7(*a*). Only thrusts are missing from Hancock's diagram, when compared with the relationships proposed by Price. The incidence of thrusts in flat-lying sediments tends to be low, but, as has been reported by Harper & Szyrmanski (1983), thrusts with complex morphology and displacements of at least two metres are known to develop in unfolded sediments.

The third block diagram (Fig. 9.7(*c*)) shows the fracture pattern which Price (1974) derived by considering the evolution of a simple and idealised, hypothetical basin, through a single phase of downwarp and subsequent uplift and exhumation. Again, the similarity between the pattern of fractures represented in Fig. 9.7(*c*) and those in (*a*) and (*b*) of the same figure, is striking. It is of interest, therefore, to follow this analysis in outline. As will become clear, the analysis is based on an extremely simple model for a sedimentary basin. It does not take into account the possible complexities of sedimentation or the probable fundamental mechanisms which could give rise to downwarp and subsequent uplift. Moreover, Price considers only average strain-states in the basin and uses stress–strain equations which are approximations (though the errors incurred by doing so are usually less than a few per cent). Furthermore, he ignores the influence of temperature upon his analysis. These various criticisms are not trivial. Nevertheless, it is still instructive to follow (and also correct some of the deficiencies of) the general argument and so establish how the fracture patterns represented in Fig. 9.7(*c*) may have developed.

A theoretical appraisal of fracture development during downwarp

In this treatment we assume that the flat-lying sediments have experienced only vertical i.e. epeirogenic movements. In general, such sediments will undergo vertical downwarp and, because the rocks are assumed to be now at the surface, they will have subsequently experienced uplift and exhumation. In a real situation, one may infer that a sequence of rocks (such as, for example, those exposed in the Grand Canyon which comprise limestones, shales, marls and dune-bedded sandstones), have often undergone more than one cycle of subsidence and uplift. Here, however, we shall examine the simplest situation of a series of sediments which have undergone only one cycle of downwarp and subsequent uplift. We shall consider an analysis of the infinitesimal lateral strains induced during subsidence in an idealised hypothetical basin, of elliptical plan, and the stresses and subsequent fracture pattern which can result.

There are two points that must be made. Firstly, because compaction results in permanent strain, the elastic limits of the rocks are frequently (or even continually) exceeded in downwarp. Hence, the behaviour can, at best, be regarded as pseudo-elastic. We shall nevertheless assume a linear stress–strain behaviour (even when the strain is irrecoverable) which can adequately be described by using elasticity theory equations. Secondly, the analysis throughout this section is based on the assumption that the competent rocks behave as simple, linear-elastic bodies, and that any stresses induced do not experience significant relaxation by any diffusion-controlled processes. Clearly then, we shall be concerned with, and postulate, events which occur at low temperatures. Thus, temperatures of less than about 150 °C must be considered for siliceous sediments, during the events under discussion (though, of course, the same

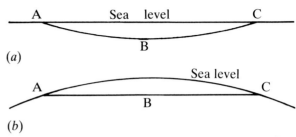

Fig. 9.8. (*a*) Simplistic representation of a sedimentary horizon ABC in a basin, of the 'Flat Earth' type most frequently and unthinkingly adopted by geologists. (*b*) Basin represented in (*a*) showing geometry it may attain in section, when the curvature of the Earth's surface is taken into account. This diagram, of course, shows excessive, vertical exaggeration, another common error in graphical representation of geological profiles.

rocks may experience higher temperatures at other times). Consequently, we shall be restricting the discussions in this section to depths from the surface of about 4–5 km (depending on the local thermal gradient) for siliceous rocks; and at somewhat shallower depths for carbonate rocks (in which pressure solution can take place at less than 150 °C).

Several simplifying, but reasonable, geological assumptions are made regarding the size the geometry of a hypothetical sedimentary basin and its history. Estimates are then presented for the development of the vertical and horizontal stresses which tend to arise in a number of rock types possessing different elastic constants during downwarp and subsequent uplift. It is demonstrated that when pore-fluid pressure is taken into consideration, the effective rock stresses must, on occasion, lead to the formation of fractures, which thereafter modify the history of stress development in the rock mass. From the analysis, the orientation of the fractures which are likely to develop can be determined. In the ensuing argument, the average strains which develop during subsidence are computed for the directions parallel to the axes of a hypothetical elliptical basin.

Geologists frequently represent sections of sedimentary basins by diagrams of the type shown in Fig. 9.8(*a*). Sea-level is represented as a horizontal, straight line and the base of the basin as a downward curving arc. One might infer from such sections that, because the arc distance ABC is greater than the sea-level distance AC, the sediments in the basin experience progressive extension as they are buried. Such a conclusion is approximately correct, provided the basin has a depth of many thousands of metres and the distance AC does not exceed a few tens of kilometres. However, if the basin is of greater extent, it is necessary to take the curvature of the Earth's surface into consideration.

A sectional representation of the basin, of surface length *L*, taking into account the curvature of the Earth, is indicated in Figs. 9.8(*b*) and 9.9. It may be inferred that, in such a basin, the sediments down to a

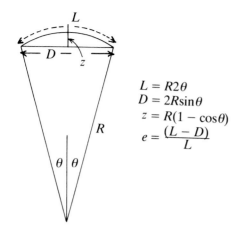

Fig. 9.9. Relationship between θ, R (radius of the Earth) and length L) and maximum depth (z) of basin, such that an axis of the basin defines a straight line D.

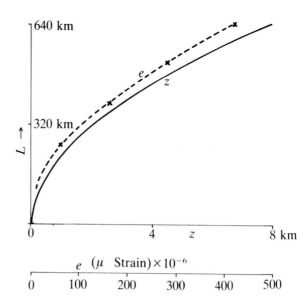

Fig. 9.10. Quantified relationships between length (L) of the basin and depth (z) so that the basin axis is straight, and the corresponding, average compressive strain along the major axis of the basin that results from burial to depth z.

depth z are in lateral compression; and this results purely from the true geometry of the Earth's crust. The depth z of any such zone of compression relative to the surface dimensions of the basin, and the lateral strains induced during subsidence to depth z, may readily be calculated. Thus, from Fig. 9.9:

$$L = 2R\theta \tag{9.1}$$

where R is the radius of the Earth and 2θ is the angle subtended at the centre of the Earth by the surface length, L. The depth z of the compression zone is given by:

$$z = R(1 - \cos \theta) \tag{9.2}$$

The average lateral strain (e) induced by depressing sediments to the linear dimension (D) of the basin at depth z is given by:

$$e = \frac{(L - D)}{L} \tag{9.3}$$

The relationship between the surface dimensions of the basin and the thickness z of the compression zone, and also the magnitude of the lateral strain induced by downwarping are shown in Fig. 9.9.

It is now necessary to attribute specific dimensions to the hypothetical sedimentary basin, in order to continue with the analysis. Let us select a maximum depth of burial for a specific bed of 4.0 km. Let it also be assumed that the length of the hypothetical basin is such that the base of the compression zone is reached when z equals 4.0 km. From the graph in Fig. 9.10, it follows that the length, AB, represented in Fig. 9.10, is 450 km. The lateral compressive strain (e_y) during downwarp to depth z of a basin with such a length is $.225 \times 10^{-3}$.

Some natural sedimentary basins are elongated, with a length/width ratio of 3–4:1. Here, the width of the hypothetical basin is arbitrarily fixed at 128 km $L/W = 3.5:1$). From reference to Fig. 9.11, it will be

seen that the zone of compression in the CD, or x, section extends only to the depth of $z = 0.3$ km. Thereafter, for the next 3.7 km of subsidence, the sediments are in extension in the x direction. A simple calculation (cf. Chap. 6) enables one to demonstrate that, at a depth of 4 km, the average lateral strain (e_x) throughout the sediments is about -1.2×10^{-3}. Using values calculated in this way for the average lateral strains and taking mean values for the elastic moduli of the rocks during the phase of downwarp, the stresses at the base of the basin can be obtained. By using the basic equations of linear elasticity developed in Chap. 1, the interrelationship between the stresses and strains in three-dimensional space can be established. (This will be demonstrated in the next section.) If, however, one makes the assumption that the 'cross strain' effects can be neglected, it is possible to use the much simpler, but approximate, relationships used by Price (1974), where:

$$S_x = \frac{S_z}{m - 1} - e_x E \tag{9.4(a)}$$

and

$$S_y = \frac{S_z}{m - 1} + e_y E. \tag{9.4(b)}$$

In these equations, compressive strain is taken as positive, and extension as negative. The errors incurred by using these approximate relationships will, in general, be less than 3 per cent and so will suffice for our present purposes. To quantify the horizontal stresses which can obtain in the hypothetical basin, it is now merely necessary to choose (i.e. guess) specific values of Young's modulus (E) and Poisson's number

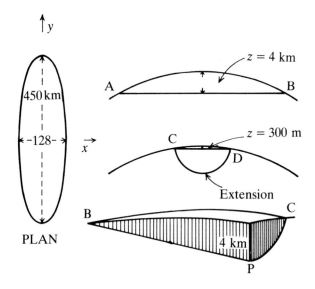

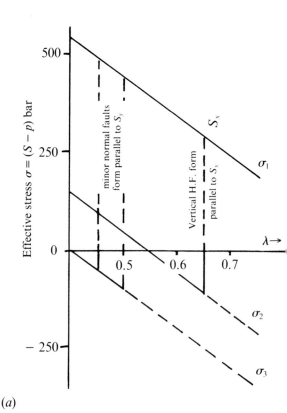

Fig. 9.11. Shape and dimensions of the hypothetical basin in plan and in section through the long and short axes, AB and CD respectively.

(m) which, it is hoped, represent the average elastic characteristics of the rocks throughout subsidence.

By making guesses regarding these values, Price indicates that the values of lateral stress (S_x and S_y) at point P in the basin (Fig. 9.11), for strong and very strong sandstone (R_1 and R_2) respectively for the given strains and the possible range of pore-fluid pressure likely to obtain are as shown in Fig. 9.12. It may be inferred from this diagram that the effective stresses ($S - p$) in the rock mass may be compressive or tensile. By plotting the effective stresses which tend to develop in the two rock types, as indicated in Fig. 9.13(a) (and (b)) respectively, it will be seen that even if λ (the ratio of the fluid pressure to the vertical total stress) is as little as 0.5, the differential stresses in the 'very strong' (high modulus) rock are sufficient to cause shear failure. The fractures which would

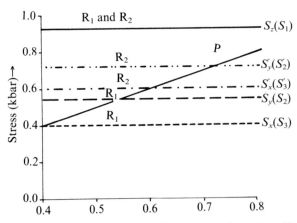

Fig. 9.12. Total horizontal stresses for rock types R_1 and R_2 with selected values of E and m are shown as horizontal lines (as labelled). The total vertical stress S_z is common to both rocks and is also shown at 920 bar. The oblique line represents the fluid pressure for values of $\lambda = 0.4$ to 0.8.

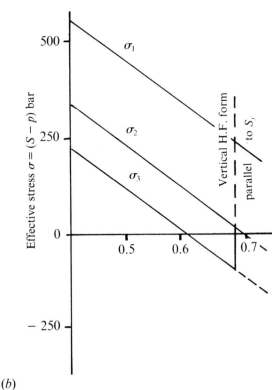

(*b*)

Fig. 9.13. These figures represent the data shown in Fig. 9.12 plotted in terms of effective stresses for (*a*) a very strong sandstone R_1 and (*b*) a strong sandstone, R_2.

develop in this stress field are normal faults striking parallel to the long axis of the hypothetical basin. The development of these faults would of course greatly influence the subsequent development of the lateral

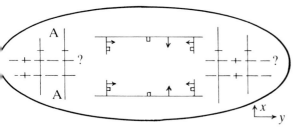

Fig. 9.14. Trends of fractures which may develop in an elliptical basin in downwarp. The major (and minor) normal faults will be initiated in the strongest rocks. The vertical, hydraulic fractures (trend AA) will develop in rocks of intermediate strength. The trend of hydraulic fractures represented by dashed lines will develop only if the fluid pressure reaches values approaching that of the gravitationally induced vertical stress (S_z).

stress σ_x (c.f. Chap. 8). If, however, the value of λ attained, or just exceeded, 0.65, further failure is possible. In this instance, the differential stress may not be sufficient to cause shear failure. Instead, the rock mass is likely to develop hydraulic fractures which would be vertical and perpendicular to the σ_y direction; i.e. they would develop parallel to the short axis of the basin.

It may be inferred from Fig. 9.13(*b*), that, because $(\sigma_1 - \sigma_3 < 4T)$, only the vertical hydraulic fractures are likely to develop in the less strong rock, and these will only develop provided λ attains a value of about 0.7, or higher. Fractures which trend parallel to the long axis of the basin may develop at the figure quoted, but λ would have to exceed 0.8 before the complementary set of vertical hydraulic fractures could develop parallel to the short axis of the basin. Thus, it is interesting to note that the higher the pore-water pressure in the crust, the larger the number of rock types in which fracture patterns will develop during the prograde deformation.

As we have seen in Chap. 8, minor fractures may also develop as a result of the stress changes which are attendant upon the development of the normal faults. The orientations of these minor fractures are indicated in Fig. 8.7(*b*). Hence, the main fracture types and trends, together with the associated minor fractures that can develop in the prograde, downwarp phase are as depicted in Fig. 9.14.

Uplift and exhumation

The physical properties of rocks during uplift are different from those which existed during subsidence, when the sediments were undergoing progressive induration. For the prograde subsidence phase, the analysis was 'pseudo-elastic' and there was an element of guesswork regarding the values of the elastic values which were attributed to the sediments. Indeed, it is because of the uncertainty regarding the values of the moduli that another factor (namely the effects of temperature change) has been tacitly ignored: for the temperature effect can be accommodated by choosing relatively small values for the elastic moduli E and m.

After such competent rocks have reached their maximum depth of burial and subsequently undergo uplift, it is reasonable to assume that, in general, the sediments in question attain their full degree of induration and that, during uplift, the rocks possess elastic properties which are closely related to those which one may later measure in laboratory tests. Therefore, before we continue further with the argument we shall develop the correct relationships between stress and infinitesimal elastic strain and also consider some experimental data which are pertinent to the subsequent argument.

Further aspects of elasticity, theory and practice
It was shown in Chap. 1 (Eq. (1.40)) that the basic linear–elastic relationships between stresses and strains in three-dimensions are:

$$e_z = \frac{1}{E}\left[\sigma_z - \frac{1}{m}(\sigma_y + \sigma_x)\right] \tag{9.5}$$

$$e_y = \frac{1}{E}\left[\sigma_y - \frac{1}{m}(\sigma_z + \sigma_x)\right] \tag{9.6}$$

$$e_x = \frac{1}{E}\left[\sigma_x - \frac{1}{m}(\sigma_z + \sigma_y)\right]. \tag{9.7}$$

From these equations it can be inferred that one horizontal stress (e.g. σ_x) is influenced by the magnitude of the other horizontal stress (σ_y) and vice versa. To eliminate this effect from calculations, one can rearrange the equations: thus, by transferring (cross multiplying) E to the left hand side of the equation, the right hand side of Eq. (9.6) can be substituted for σ_y in Eq. (9.7). From this operation we obtain:

$$\sigma_x = \frac{m^2}{m^2 - 1}\left[e_x E + \frac{e_y E}{m}\right.$$
$$\left. + \frac{\sigma_z(m + 1)}{m^2}\right]. \tag{9.8}$$

Similarly:

$$\sigma_y = \frac{m^2}{m^2 - 1}\left[e_y E + \frac{e_x E}{m}\right.$$
$$\left. + \frac{\sigma_z(m + 1)}{m^2}\right]. \tag{9.9}$$

Changes in lateral stress are not only related to changes in the geometry of uplift and the attendant lateral strains. During uplift and exhumation the rock mass becomes cooler and this gives rise to lateral 'shrinkage' strains which, in turn, form a component which contributes to the total reduction in lateral stresses during uplift. The relationship between thermal stresses (σ_t) and strains (e_t) is given by:

$$\sigma_t = e_t E = \alpha_t T° E \tag{9.10}$$

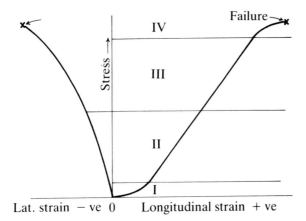

Fig. 9.15. Zones of behaviour of a cylindrical test specimen subjected to axial compression when under zero or small confining pressure.

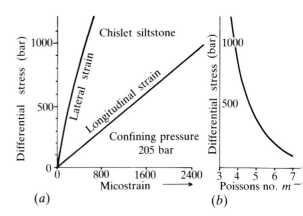

Fig. 9.16. Stress–strain curves indicating that longitudinal strain is approximately linear over a large range of differential stress, but lateral strains are slightly non-linear and give rise to variations of m with applied stress. (After Price, 1958.)

where α_t is the coefficient of linear thermal expansion (per °C) and $T°$ is the change in temperature (in °C) experienced by the rock mass. The thermal stress effect of Eq. (9.10) can be incorporated with Eqs. (9.8) and (9.9) to give the general relationships:

$$\sigma_x = \frac{m^2}{m^2 - 1}\left[(e_x + e_t)E \right.$$
$$\left. + (e_y + e_t)\frac{E}{m} + \frac{\sigma_z(m + 1)}{m^2}\right] \quad (9.11)$$

and:

$$\sigma_y = \frac{m^2}{m^2 - 1}\left[(e_y + e_t)E \right.$$
$$\left. + (e_x + e_t)\frac{E}{m} + \frac{\sigma_z(m + 1)}{m^2}\right]. \quad (9.12)$$

Clearly, to use these equations it is necessary to assign values of elastic moduli to the rocks being simulated in the uplift model: so it is pertinent to enquire whether rocks satisfy the basic assumptions of linear elastic theory, that the material is homogeneous and isotropic. Experimental determination of the elastic behaviour reveals that, in fact, rocks do not completely satisfy these conditions. It has been shown (Brace *et al.*, 1966) that the elastic stress–strain curve for rocks can be divided into four zones (Fig. 9.15). Zone I and Zone IV are non-linear. However, the extent of Zone I is small and can sometimes be reduced by 'cycling' (i.e. submitting the test specimen to a series of loading and unloading cycles). Here we need not consider Zone IV, which occurs as the failure stress is approached. The major part of the longitudinal deformation, throughout Zones II and III is, however, usually very close to linear. Indeed, many of the stronger, or more competent, rock types exhibit stress–strain curves of the type shown in Fig. 9.16 (after Price, 1958). This figure shows that at modest confining pressures and differential stress (205 bar

(3000 lb in^{-2})   and   1000 bar   (14 600 lb in^{-2}), respectively) the longitudinal strain is quite linear for this rock; so that the Young's modulus has a single value of 3.8×10^5 bar. However, it will be seen (Fig. 9.16(*b*)) that the lateral stress–strain relationship is not linear, and this means that Poisson's number (m) does not have a single value. Indeed, Price (1958) showed that the value of m is dependent not only upon the differential stress ($S_1 - S_3$) but also upon the confining pressure (S_3). The relationships between m and the differential stress and confining pressure for two coal measure rocks are shown in Fig. 9.17(*a*) and (*b*).

It was shown in Chap. 1 that if the lateral strains in the crust are zero, then the gravitational loading (S_z) gives rise to a horizontal stress

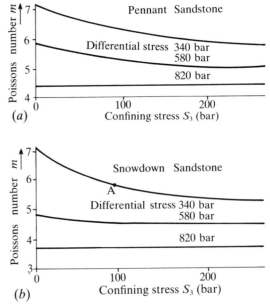

Fig. 9.17. Relationship between Poisson's number (m), differential stress and confining pressure. The unusual values for differential stress in this figure reflect the fact that the experiments were conducted using apparatus calibrated in lb in^{-2}. (After Price, 1958.)

(S_h) where these stresses are given by:

$$S_z = \rho g z \qquad (9.13)$$

and

$$S_h = S_x = S_y = \frac{S_z}{m - 1}. \qquad (9.14)$$

Price (1959) used the data represented in Fig. 9.17(*b*) to derive the relationship between the vertical principal stress and the horizontal stress $(S_x = S_y = S_3)$, shown in Fig. 9.18. This latter figure was constructed by choosing points on the curves in Fig. 9.17(*b*) which satisfy Eq. (9.14). For example, for point A in Fig. 9.17(*b*), the differential stress $(S_1 - S_3 = S_z - S_h)$ is 340 bar and the confining pressure $(S_3 = S_h)$ is 85 bar; hence $S_1 = S_z = 425$ bar. Also, it will be noted that $m = 6.0$. Putting these values in Eq. (9.14) we have:

$$S_x = S_y = \frac{S_z}{m - 1} = 425/(6 - 1) = 85 \text{ bar.}$$

It is then possible to construct the curve shown in Fig. 9.18 by choosing a number of such points in Fig. 9.17(*b*). It will be noted that m varies from about 4.0 at a depth of 7 km to about 7.0 near the surface.

Price (1974) used these variations in the value of m when estimating the changes in horizontal pressure in the crust during uplift and exhumation. It has been noted that in this latter paper he used the approximate

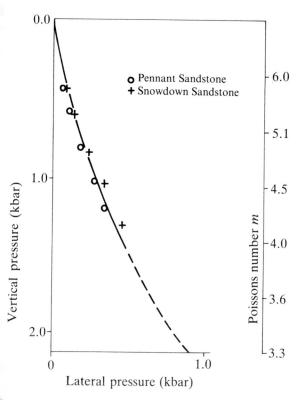

Fig. 9.18. Variations of horizontal pressure with depth calculated from data given in Fig. 9.17 and assuming lateral strain $e_x = e_y = 0$.

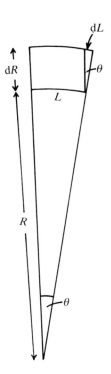

Fig. 9.19. Parallel uplift model. (See text for discussion).

relationships of Eq. (9.4) and also ignored the influence of changes in temperature. In the arguments that follow we shall use the Eqs. (9.11) and (9.12), which are not approximate and also take into account the effect of temperature changes. Two conceptual models will be discussed. These are (i) the parallel uplift model and (ii) a specific example of non-parallel uplift.

Parallel uplift

If, as is assumed here, a sedimentary sequence (of average relative density of 2.5) experiences erosion which exactly keeps in step with the amount of uplift, some specific reference point in the rock mass will experience a gradual reduction in the vertical stress. From Eq. (9.13) one may infer that, for the assumed average relative density, this reduction in vertical stress will be 250 bar/km of uplift.

The parallel uplift model (Price, 1959, 1966 and 1974) is represented in Fig. 9.19. From this diagram it follows that:

$$L = R\theta \qquad (9.15(a))$$

and

$$dL = dR\theta \qquad (9.15(b))$$

where θ is some arbitrary angle, L is an arbitrary arc length, R is a radius of curvature (in this context, the radius of the Earth) and dL is the increase in length of the flat-lying unit L after an uplift of dR. By dividing Eq. (9.15(*b*)) by (9.15(*a*)) one obtains:

$$\frac{dL}{R} = e_h = \frac{dR}{R} \qquad (9.16)$$

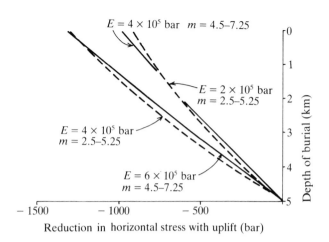

Fig. 9.20. Potential reduction in horizontal pressure with uplift from a depth of 5 km to the surface for four groups of E and m, of four typical competent rocks.

where e_h is the horizontal strain induced by an uplift dR. Here we shall take the value of R to be constant and equal to 6400 km. Hence, parallel uplift of 1 km (i.e. $dR = 1.0$ km) will give rise to an extensive strain (e_h) in the horizontal direction of -1.56×10^{-4} (where, in this instance, $e_h = e_x = e_y$).

In non-orogenic environments uplift and exhumation is a slow process which would provide ample time for the rock mass to adjust to, and equilibrate with, the geothermal gradient. Thus, if the geothermal gradient is 25 °C/km and the coefficient of linear expansion $\alpha_t = 5 \times 10^{-6}$/°C then, after an uplift of 1.0 km, the rock mass at the level of the reference point will cool by 25 °C, so that there will be a lateral cooling strain of $e_t = -1.25 \times 10^{-4}$.

These various components which give rise to changes in the horizontal and vertical strains and stresses can be substituted in Eqs. (9.11) and (9.12), together with the appropriate value of elastic moduli: and so enable the variations in potential reduction in horizontal stress, with uplift and exhumation, to be estimated. Four such exercises are represented in Fig. 9.20, which show the reductions in lateral stresses which would accompany an uplift and exhumation of 5 km, for four rock types with different representative elastic properties. The values of E and the range of values of m are indicated in the diagram. It is emphasised that these diagrams refer only to the potential reduction in lateral stress with uplift. In order to estimate the real rock stresses that exist throughout this period of uplift, it is necessary to know, or make assumptions regarding, the magnitudes of (i) the fluid pressures which obtain during uplift and (ii) the initial magnitude of the horizontal stresses when the reference point in the rock mass was situated at a depth of 5 km. Let us now address these two problems.

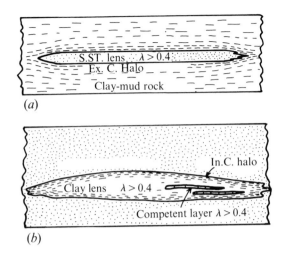

Fig. 9.21. (a) Sandstone lens forming a closed system, completely surrounded by clay–mud rock which forms an external compaction halo. (b) Clay lens with an internal compaction halo completely surrounded by pervious rock. The lens contains competent layers with $\lambda > 0.4$.

Fluid pressure during uplift

In a rock mass containing interconnecting pores and void spaces it has been noted that the hydrostatic head of pressure is given by:

$$p = \rho_w g z \qquad (9.17)$$

where ρ_w is the density of water (which remains close to unity throughout the depth range considered here, so that the hydrostatic pressure increases by about 100 bar for every kilometre of depth). As we have noted, because the density of compacted sedimentary rocks is often close to 2.5 g cm^{-3}, the ratio (λ) of p/S_z for such a situation is about 0.4. In an open system, of the type envisaged in order that Eq. (9.17) may hold, the fluid pressure during uplift will then fairly rapidly settle to a value of $\lambda = 0.4$. The value of fluid pressure in a *closed system* would not necessarily be given by Eq. (9.17) and would depend upon the strength of the rock material within the limits of the closed system. For example, the closed system may comprise a lens of porous but competent sandstone, completely surrounded by impervious weak clay or mud-rocks (Fig. 9.21(a)). Alternatively, the closed system may comprise a lens of weak clay or mud-rocks (Fig. 9.21(b)).

Let us consider the first of these alternatives. The bulk modulus of water is about an order of magnitude lower than that of strong rock: while the coefficient of expansion of water is about an order of magnitude larger than that for most rocks. Consequently, decreases in temperature and mean pressure will have a much more profound influence on the enclosed water than it will on the volume of the rigid pores contained in a strong sandstone lens. During uplift, pressure and temperature effects will very rapidly give rise to a reduction in fluid pressure.

and in the extreme state could leave pores and void spaces only partially filled by water. (An example of such an extreme state is the liquid and vapour contained in a primary inclusion in quartz or other vein minerals. When the vein was formed, the liquid in the primary inclusion was, of course, at the ambient temperature and pressure of the vein forming fluid.)

Initially, water would tend to flow out of the surrounding clays and muds into the sandstone lens and so nullify the effects of the temperature and mean pressure reduction. However, the water would be derived from the clays and muds immediately adjacent to the sand lens. This would eventually tend to give rise to a compaction halo of almost impervious low porosity clay-rock around the sandstone lens, so rendering the system 'closed' and thereby allowing abnormally low fluid pressures to develop in the lens as the temperature and mean pressure decreases during uplift. Such sand lenses with measured ratios of p/S_z of 0.2 or less are encountered in some oil and gas fields of Alberta.

The alternative situation is of an enclosed lens of clay as represented in Fig. 9.21(b). Because of the intrinsic weakness of the clay material, any reduction in pore-pressure would result in compaction; so there would be a strong tendency for an original, abnormally high fluid pressure to be maintained. Water would initially be drained from the outer portions of the lens which would result in an 'internal compaction halo' and so render the clay lens a closed system. Because the resistance to compaction of a clay-rock is a function of its porosity and the interstitial fluid pressure, any reduction in the fluid pressure will permit a further increment of compaction. Consequently, compaction of the clay-lens may occur during uplift and this effect could maintain a fluid pressure significantly higher than the hydrostatic head of pressure which one would estimate from the use of Eq. (9.17).

Initial horizontal pressure prior to uplift

It is difficult to be specific as regards the value of the horizontal stress acting at a depth of 5 km after a period of subsidence. However, some general arguments can be put forward which permit some tentative conclusions to be reached. For example, it is sometimes assumed in soil mechanics that low strength materials such as 'soils' and clay-rocks fail continually along micro, or small-scale shear planes as they are buried to greater depths. (Such a mechanism could help explain the development of some forms of fissility in clay-rocks.) Indeed, stronger rock types may also fail along larger scale normal faults. Consequently, for a wide range of circumstances, the lateral stresses, at any given depth, may be controlled, or strongly influenced by, such a mechanism.

Let us assume that failure is essentially by fric-

Table 9.1.

(a)	ϕ	$k = S_3/S_1$
	10°	0.70
	20°	0.53
	30°	0.33
	40°	0.22

(b)	ϕ	λ					
		0.4	0.5	0.6	0.7	0.8	0.9
	10°	0.80	0.85	0.88	0.91	0.94	0.97
	20°	0.68	0.77	0.81	0.86	0.91	0.95
	30°	0.54	0.67	0.73	0.80	0.87	0.93
	40°	0.47	0.61	0.69	0.77	0.84	0.92

tional sliding and that there are sufficient numbers of micro- or macro-shear planes with the necessary orientation, such that the relationship between effective principal stresses (Chap. 1) is given by:

$$\sigma_1 = \sigma_3 \frac{(1 + \sin \phi)}{(1 - \sin \phi)} \tag{9.18}$$

where, as before, ϕ is the angle of sliding friction. If we assume that σ_1 and σ_3 act vertically and horizontally, respectively, then the ratios (k) of horizontal to vertical effective stresses, for a range of values of ϕ, at a fault plane, are as given in Table 9.1(a).

However, we are here concerned with the ratios of total stresses (S_3/S_1) and these will be influenced not only by the relationship given in Eq. (9.18), but also by the maximum value of the fluid pressure and in particular by the value of λ (where, as before, $\lambda = p/S_z$). If it is assumed that $S_z = 1250$ bar (the vertical pressure that would develop in many sediments at a depth of about 5 km), then from the relationship given in Eq. (9.18) and because

$$\sigma = S - p$$

it can be shown that the ratios (k') of horizontal to vertical total stresses ($S_x/S_z = S_3/S_1$) are as given in Table 9.1(b), for the specified values of λ and ϕ.

Many weak clay-rocks have angles of sliding friction between 10° and 20° (here we take the angle for 'residual' conditions, see Chap. 1). It will be seen that even for modest values of fluid overpressure (e.g. $\lambda = 0.6$–0.7) the total horizontal stress will certainly exceed 80 per cent and may even be over 90 per cent of the total vertical pressure. At higher values of λ the horizontal pressure will closely approach the magnitude of the total vertical stress (Table 1)(b) and Fig. 9.22).

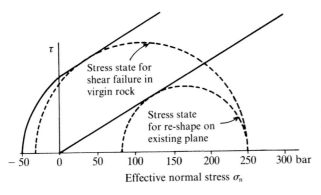

Fig. 9.22. Mohr's circles representing the initial stress conditions and those which obtain after shear movement has taken place for a value of $\lambda = 0.8$ and a depth of 5 km.

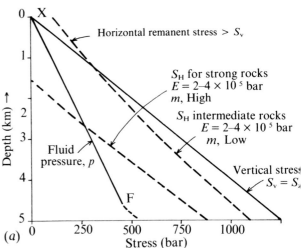

(a)

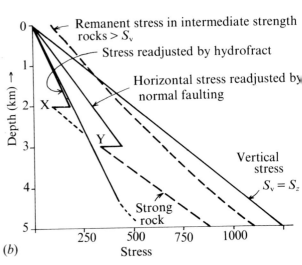

(b)

Fig. 9.23. (a) The curves from Fig. 9.20 which show that the potential reductions in horizontal stresses S_H during uplift have been superposed upon probable initial, horizontal stresses and fluid pressures. (b) These curves indicate the probable variation in horizontal stresses during uplift showing abrupt breaks in the potential curves when failure conditions are met. These are most likely to occur in strong rocks with high elastic moduli.

If we make the reasonable asumption that the majority of strong sediments will exhibit a value of ϕ between 30° and 40°, then it can be inferred from Table 9.1(b) that the horizontal stress will only exceed 90 per cent of the total vertical stress for high values of λ (between 0.8 and 0.9). Alternatively, if values of λ remain low (i.e. 0.5–0.6) the horizontal stress will be significantly smaller than the vertical stress.

Because the estimates of lateral stress are based upon shear failure, any ratio of the total horizontal to vertical stresses which develop at a high fluid pressure will not change if the fluid pressure subsequently decreases. However, as we noted in Chap. 8, the high value of the horizontal stress quoted above will only develop immediately adjacent to the normal fault. At some considerable distance away from the fault, the horizontal stress will be close to that required for fault initiation. Hence, for the stipulated depth and failure conditions, the mean value of the horizontal stress is likely to range from 900–1100 bar, provided the value of λ exceeds 0.7. That is, the horizontal stresses, for such a rock type, at the initiation of uplift, will be 72–88 per cent of the vertical stress. (These initial stresses would, of course, be commensurately lower if λ were less than 0.7).

Similar conclusions may be reached regarding the ratio of horizontal to vertical stresses by considering rocks which fail as the result of hydraulic fracturing. However, unless there is offset on the fractures or they are infilled with vein material, so that the extension fractures cannot return to their original closed position, when the excess fluid pressure that gave rise to fracturing decays, the rocks will regain the stresses they experienced prior to the generation of high fluid pressure. Consequently, conclusions regarding the horizontal stresses in a rock mass cut by hydraulic fractures must of necessity be rather vague.

From these arguments, we conclude that the horizontal stresses in flat-lying sediments at a depth of about 5 km depends upon the rock type and the class of fractures by which they are cut, but of greatest

importance is the maximum level of fluid pressure which the rocks experienced. The lowest magnitude of the horizontal stress tends to develop in the stronger units. But even in these situations, the value of the horizontal stress is likely to be about 75 per cent of the vertical stress. In the weaker rocks, the horizontal stress is likely to attain higher value and may exceed 90 per cent of the total vertical stress.

Let us now use these *inferred* initial stress conditions, together with the estimated fluid pressures which may develop in uplift, and with the calculated reductions in horizontal stress shown in Fig. 9.20, to obtain the probable rock stress and failure conditions which will prevail during parallel uplift.

Stress and fracture development during parallel uplift
The data shown in Fig. 9.20 have been grouped to represent 'strong' (solid lines) and 'intermediate

strength' (dashed lines) competent rocks. Using the conclusions of the arguments presented in the previous paragraphs, these curves are set, in Fig. 9.23(*a*), at the appropriate 'average' initial horizontal stress to be expected to obtain in these rock types at the specified depth of 5 km. The decrease in the vertical stress and the probable fluid pressure with uplift and exhumation is also represented by lines AX and FX respectively.

If we assume that the horizontal stresses in rocks of intermediate strength are initially about 85 per cent of the total vertical stress, then it will be seen that the horizontal stresses in the intermediate strength rocks become greater than the vertical stress when they are uplifted to a depth cover of about 1.2 km and that the near-surface horizontal stress is of the order of 100 bar. As regards the lateral stress reduction in the stronger rocks (where it is assumed the horizontal stresses are initially about 70 per cent of the total vertical stress) these become lower in magnitude than the fluid pressure at a depth of 2.7 km and have the tendency to become totally tensile at a depth of about 1.5 km. This tendency is, of course, not realised because new fractures will form, or pre-existing fractures of suitable orientation will be widened.

Only if the rocks of intermediate strength are associated with a 'closed system' lens, in which the fluid pressure is abnormally high will this conclusion not hold. One may infer from Fig. 9.23 that the horizontal stresses in these rocks remain close to the magnitude of the vertical stress and because both greatly exceed the normal hydrostatic head of fluid pressure, *fractures are unlikely to develop in these rock types, as a result of this mechanism, during the uplift phase.*

The remanent horizontal stresses which remain in the rocks of intermediate strength as they approach the surface, eventually become dissipated by the development of the multitude of systematic and, more usually, non-systematic fractures which are induced in the zone of weathering (as discussed in Chap. 4).

With reference to the stronger rocks: it may be inferred from the stress conditions shown in Fig. 9.24 (which is based on the information regarding stresses at a depth of 3 km given in Fig. 9.23(*a*)) that, provided normal faults already exist in the rock mass, these will be rejuvenated. The changes in horizontal stress which would result from such re-shear are represented in Fig. 9.23(*b*), level Y. At some distance from a normal fault, the horizontal stress will have its lowest value. Continued uplift may possibly give rise to the development of hydraulic fractures in these areas of relatively low stress when the rocks reach level X in Fig. 9.23(*b*). The subsequent variations in horizontal stress which result in the continued opening of these extension fractures, at higher levels in the crust, are also shown in Fig. 9.23(*b*).

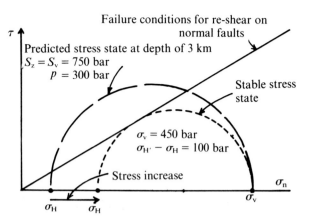

Fig. 9.24. Stress conditions represented by level Y in Fig. 9.23(*b*) will cause reshear on existing normal faults as shown.

In this latter figure only one value of horizontal stress is shown for a given rock type. However, in a situation in which normal faults have developed, it is most unlikely that the horizontal stresses will exhibit radial symmetry, but rather that there will be a maximum and a least horizontal stress. The parallel uplift conditions will maintain these initial stress differences. Hence at, or close to, level X, it is possible that an orthogonal system of hydraulic fractures will develop in strong rock and that the orientation of these fractures will be controlled by the orientation of the original horizontal stresses, at a depth of 5 km: which in turn are related to the downwarp conditions which determine the fundamental stress orientations in the basin.

Non-parallel uplift and fracture development
Parallel uplift is unique. However, the possible variety of forms of non-parallel uplift is large. Clearly, as it is not feasible to consider such a potentially wide variety, we shall be content, here, merely to consider briefly one form of non-parallel uplift in which the 'hypothetical basin' reverses its mode of vertical movement, as indicated in Fig. 9.25 (Price, 1974). That is, the long axis of the basin undergoes extension during uplift, while the short axis undergoes compression. Here the degrees of extension and compression are arbitrarily set at -2×10^{-3} and $+5 \times 10^{-3}$ respectively. These are infinitesimal strains which may readily be experienced during uplift but, as can be inferred from Fig. 9.26(*a*), the potential effects upon the horizontal stresses in strong rocks with a high value of Young's modulus are dramatic. The horizontal stress (S_3 at the original depth) parallel to the long axis of the basin decreases quite rapidly with uplift and becomes less than the fluid pressure at a depth of about 2 km. (This could, of course, result in the development of hydraulic fractures parallel to the short axis of the basin at higher levels in the crust, as shown in Fig. 9.26(*b*)).

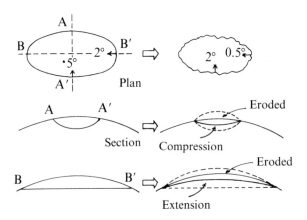

Fig. 9.25. Model of non-parallel uplift used by Price (1974) in which vertical movements assumed to take place during downwarp are reversed during uplift and exhumation. Bed dips are reduced during uplift.

The horizontal stress (S_2 at a depth of 5 km) acting parallel to the short axis of the hypothetical basin increases with uplift and becomes the maximum principal stress (S_1) at a depth of about 4 km. The vertical stress which hitherto had been the maximum

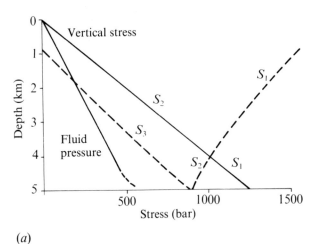

(a)

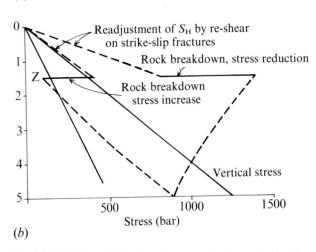

(b)

Fig. 9.26. (a) Potential horizontal stresses that may develop in strong rocks during uplift of type represented in Fig. 9.25. (b) Probable changes in horizontal stresses during uplift taking into account re-adjustments that occur as a result of shear failure.

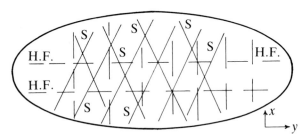

Fig. 9.27. Orientation of shear fractures (S) which can develop in non-parallel uplift and hydraulic fractures (H.F.) which may form in parallel uplift.

principal stress becomes the intermediate principal stress (S_2) at this point. Thus, the rocks at and above this level are subject to a stress state that could give rise to strike-slip fractures. At the level of 1.5 km, the differential stress is approximately 1200 bar (Fig. 9.26(b)), so that the development of strike-slip fractures at this, or at a slightly higher level is indeed probable, with the orientation represented in Fig. 9.27. The subsequent magnitude of the horizontal stresses will therefore undergo a significant change as the strike-slip fractures develop (Fig. 9.28) and will thereafter undergo continual readjustment with subsequent uplift, as re-shear occurs on these faults (Fig. 9.26(b)). It will be seen that in such circumstances, at least one of the horizontal stresses will be significantly larger than the vertical stress, as the surface is approached. In this near-surface environment, such high lateral remanent stresses, coupled with the *in situ* residual stresses could occasionally give rise to minor thrusting.

The combination of the fractures which develop during downwarp (Fig. 9.14) and those which may form in uplift (Fig. 9.27) give the pattern derived by Price (1974) shown in Fig. 9.7(c).

It is of interest to note that the ideas presented here were originally developed not for geologists, but for mining engineers. Indeed, several of the ideas were

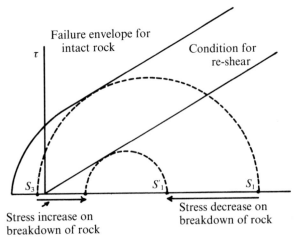

Fig. 9.28. Indicating changes in intensities of the horizontal stresses when shear fractures are induced at level Z in Fig. 9.26(b).

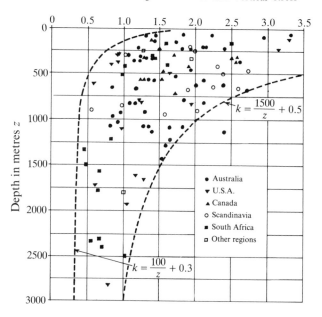

k = Ratio of average horizontal and vertical stress

$$k = \frac{1500}{z} + 0.5$$

Depth in metres z

- Australia
- U.S.A.
- Canada
- Scandinavia
- South Africa
- Other regions

$$k = \frac{100}{z} + 0.3$$

Fig. 9.29. Showing ratios of average horizontal stress/vertical stress against depth, as determined by *in situ* stress measurements in many localities throughout the world. (After Brown & Hoek, 1978).

formulated by the senior author in 1971, when he conducted research into the *coal-burst* problem, at the University of Minnesota. The object of that particular exercise was to demonstrate that high horizontal stresses (which it was held were of importance in the development of coal-bursts) could be expected to obtain in rocks at 'mining depths'. The necessity for such arguments no longer exists for, in recent years, *in situ* stress measurements have demonstrated that in *rocks undergoing uplift*, even at depths of as great as 3 km, the magnitude of the horizontal stress may exceed that of the vertical stress (Fig. 9.29). This is a general vindication of the type of uplift analysis presented above. The orientation and types of fractures which are predicted by this analysis, and which are in such good agreement with the orientation and type of fractures observed to develop in flat-lying sediments, *appear* to lend even greater support to the general type of analysis outlined above.

Throughout the rather simplistic analyses presented in the previous paragraphs, it has been necessary to make assumptions regarding the elastic moduli which have been used in the various calculations. These values have been representative of a wide range of rock types; but obviously the calculations cannot be of universal applicability. Also, it may be noted, in order to obtain the degree of agreement between the patterns predicted by the analyses and field observations, it was necessary to combine the parallel and non-parallel uplift models. This is, however, not a serious objection, for it is likely that an extensive unit of flat-lying sediments will indeed have experienced

both parallel and non-parallel uplift. This merely means that some of the fracture types and orientations will be present in some areas and missing in others. This is what we observe in the field. Indeed, the criticism may be turned on its head and we suggest that the mode of data presentation in Fig. 9.7(*b*) should not be used to represent real data, for it obscures or neglects important regional aspects which must be considered in fracture interpretation.

It will be apparent from the arguments presented in the preceding paragraphs that the mechanism outlined by Price (1974), and presented here in modified form, is capable of giving rise to sets of fractures with orientations which are in accord with some field observations. However, it is emphasised that the mechanisms work best for strong rocks (with high values of E and m) which do not exhibit too high a ratio of horizontal to vertical total stress. Moreover, in order to cause failure by the hydraulic fracture mechanism, it is necessary to postulate the existence of high values of λ, and the weaker the rock, the higher the fluid pressure must be to induce failure.

High fluid pressures are known to develop at considerable depths (i.e. >3 km) in basins undergoing sedimentation. However, there is little evidence for values of λ (p/S_z) greater than 0.5 in sediments, even clay-rocks, which are buried to depths of less than 3 km. Indeed, MacKenzie *et al.* (1986) argue that the porosity and permeability of clay-rocks at such shallow depths of cover, when covered at geologically realistic rates of sedimentation, will permit a degree of over-pressuring which does not exceed 10 per cent of the hydraulic head. That is, they conclude that at these shallow depths the value of λ will rarely exceed a value of 0.55.

There is abundant evidence that weak sediments (even clay-rocks) which have never experienced more than 2–3 km of burial, but which are now seen at the surface, are cut by one or more well defined sets of fractures. It can be inferred from the preceding arguments that it is not feasible to attribute the generation of such fractures to the hydraulic fracture mechanism, so they cannot be categorised as 'extension' fractures. Also, for clay-rocks and other sediments which have been buried to only shallow depths, the mechanism of non-parallel uplift which could give rise to lateral stresses of the relative magnitudes represented in Fig. 9.26 is extremely unlikely to be operative. This follows from the fact that the Young's modulus of clay-rocks which have been uplifted from a depth of 2 km (rather than a depth of 5 km, as shown in Fig. 9.26) is likely to be about 3×10^4 bar and the uplift strains and temperature changes will probably be less than 50 per cent of those used to estimate the lateral stresses shown in Fig. 9.26. (It could be argued that strike-slip fractures may develop in the final stages of uplift and exhumation, when the effective

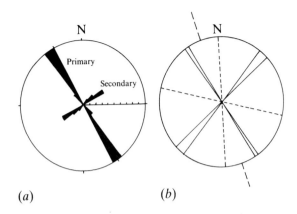

Fig. 9.30. (*a*) Orientation and frequency of primary and secondary major joints at Saxon Pit, Lincolnshire, England. (*b*) Orientation of planar fissure and regional faults (dashed lines). (After Burland *et al.*, 1977.)

confining pressure is very small. We shall comment on this aspect later. It is sufficient to note here that some clay-rocks exhibit *orthogonal* fracture sets and it would not be feasible to apply a late-stage, brittle, strike-slip shear mechanism, which requires an acute angle between such sets, to explain the development of such a joint system.)

We are therefore faced with the problem that the traditional methods of explaining joint development (while they work for competent rocks with high elastic moduli, uplifted from considerable depths and having experienced high fluid over-pressures) cannot convincingly be applied to explain the development of joint sets or systems which have developed in incompetent rocks which have never been buried to depths greater than 2–3 km and have never experienced high, fluid over-pressures.

This is a very serious short-coming as far as structural interpretation of joints is concerned. Before proceeding to rectify this situation, it is apposite to comment on some field observations regarding the orientation and morphology of joint sets which develop in a clay-rock in one specific locality.

Field observations

The joints and fissures exposed in the Oxford Clay at Saxon Pit, operated by the London Brick Company, have been described by Burland *et al.* (1977). The trend and frequency of major joints are indicated in Fig. 9.30(*a*). In general, those joints oriented normal to the 29 m high face remain tightly closed, though, it is reported, they are eventually more readily seen as a result of weathering and shrinkage of the clay mass. 'In general, the major joints dip within a few degrees of the vertical, though local deflection from the vertical sometimes occurs . . . The joints are both planar and continuous over distances of over 100–200 m . . . Separation of the joints varies from 2–24 m with 8–10 m being most common.'

Burland *et al.* also report a study of what they term 'fissures', which are small-scale fractures ranging in size from 1 cm to 1 m in extent. Fifty-two per cent of such fractures are systematic (i.e. planar) and the remainder are non-systematic (curvi-planar). Many are approximately parallel to the bedding, while the more planar, near-vertical fissures exhibit orientations which are very similar to those of the major joints (Fig. 9.30(*b*)).

In addition to orientation and systematic or non-systematic forms, fissures are also noted to be open or closed with their surfaces ranging from rough, to smooth and slickensided (though, unfortunately, the orientation of the lineation is not reported).

The interpretation of these structures by Burland *et al.* is contradictory and confusing. In part, this is the result of the intrinsic difficulty of analysing fractures which, from their description, have clearly formed as the result of more than one mechanism. Burland *et al.* accept the viewpoint that many of the fissures occurred in proximity to the ground surface, when the horizontal stresses may exceed the vertical stress (Fig. 9.29), and take place 'when the maximum shear stress in the plane becomes equal to the shear strength'; i.e. they are the result of shear failure. Yet it is also suggested that the *in situ* stresses that gave rise to fissures and joints were the same. Further, from their assessment of the nature of major joint surfaces and because the major fractures form orthogonal sets, Burland *et al.* categorise these fractures as 'tension joints' (and quote the results of the analysis by Price (1974) to support this contention). There is implicit contradiction in these statements.

Certainly, some of the 'fissures', namely those which would be better termed 'minor joints', have the appearance of shear surfaces (Fig. 9.31). Moreover, from the planar morphology, we would also classify fractures belonging to the primary, major joint set, as 'shear fractures' (Fig. 9.32).

It is interesting to note that the near-vertical fractures which trend approximately parallel to the primary, major joint set exposed in Saxon Clay Pit occur in a variety of rock types throughout much of south-east England and also in north-west France. The 'traditional' interpretation of the fractures which belong to this trend would be that they are the result of 'tension', or 'extension', in a direction perpendicular to the plane of the fractures.

An alternative interpretation of 'joint' development

One important parameter which has been largely neglected in the theoretical interpretation of fractures in sediments is that of shear-strain. Price (1979) indicated the importance of this parameter in relation to the potential for the development of fractures in granite (Chap. 4) and the formation of reverse faults

Fig. 9.31. Minor planar joint in Oxford Clay, Saxon Pit. From the smoothness of the fracture surface, relative to the fissility of the clay, it can be inferred that the joint is the result of shear failure. (Courtesy of the London Brick Co.)

(Chap. 8). Let us now consider the importance of this parameter regarding the development of fractures in relatively weak sediments.

If an angular shear (ψ) (which, for infinitesimal strain, is also equivalent to the shear-strain) is introduced into a series of elastic sediments by differential vertical movements, then shear stresses (τ) are induced, as indicated in Fig. 9.33, so that (cf. Eq. (1.43), Chap. 1):

$$\tau = G\psi \qquad (9.19)$$

where G is the elastic shear modulus of the rock. If the sediments are acted upon by total vertical and horizontal stresses (S_z and S_x respectively) then, prior to differential vertical movements, these normal stresses will be principal stresses, such that $S_z = S_1$ and $S_x = S_2$ or S_3. Even if S_z and S_x remain unchanged, the development of shear strains (which in the situation under consideration are best induced by differential uplift) completely changes the stress situation, in that the differential stress ($S_1 - S_3$) is increased and the axes of principal stress are rotated through an angle (α) which, of course, is half the angle in Fig. 9.34.

It can be inferred that if the rocks are dry (so that effective stresses do not have to be considered) and the shear-stress is sufficiently large, then the induced stress circle S_1, S_3 can touch a failure envelope and so induce shear failure. This example will serve to indicate how fractures may develop in dry rock, such as granite, but in sediments which contain fluids, it is necessary to take into account the influence of the fluid pressure (p) which reduces the 'effectiveness' of the normal stresses, so that the effective vertical and normal stresses (σ_z and σ_x) are given by:

$$\sigma_z = S_z - p \qquad (9.20(a))$$

and

$$\sigma_x = S_x - p. \qquad (9.20(b))$$

The graphical construction that can be used to determine the value of the shear stress (τ_{crit}) which must be induced to cause failure in sediments acted upon by a fluid pressure is shown in Fig. 9.35. In order to obtain this value for a given rock type, it is necessary to specify:

(i) the initial principal stresses (S_z) and (S_x),
(ii) the value of the fluid pressure (p), and
(iii) the cohesive strength (C_0) and the angle of friction (ϕ).

Here, we assume that the rock type under consideration has a linear failure envelope. Hence, from

Fig. 9.32. A primary major joint surface in the Saxon Pit, which forms a relatively smooth surface with facets. (Courtesy of the London Brick Co.)

the Navier–Coulomb criterion of failure, it follows that the failure stress circle will touch the failure envelope on a line which is normal to that envelope and also runs through the stress point $(\sigma_1 + \sigma_3)/2$, $\tau = 0$ (i.e. line xx').

It can be seen that the shear stress (τ) that must develop to cause a sufficiently large differential stress, such that the radius of the stress circle will be xx',

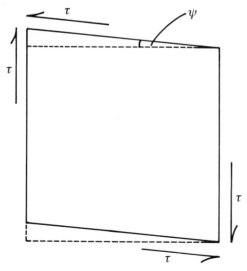

Fig. 9.33. Relationship between shear-stress and shear-strain.

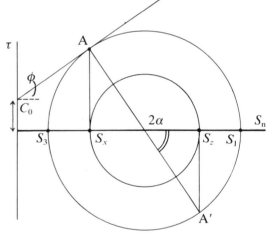

Fig. 9.34. Showing change of magnitude of differential stress and orientation of S_1 as the result of a shear-stress (A, S_x) superimposed upon vertical and horizontal stresses (S_z and S_x). Shear failure occurs when the stress circle touches the failure envelope at point A. Points A and A' represent the stresses in the horizontal and vertical planes respectively, after differential uplift has occurred.

is given by TT'. From Fig. 9.35, it is apparent that the position of T' is dictated by the value of S_x, while that of X' is determined by the value of C_0, ϕ and $(S_1 + S_3)/2$. In general, therefore, T' and X' will not be coincident. The angle XX' makes with the normal

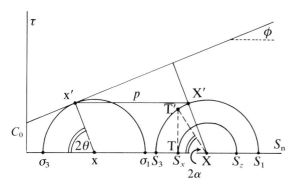

Fig. 9.35. Graphical construction used to determine the critical value of the shear stress TT′, necessary to cause failure for specific values of S_z, S_x, p, C_0 and ϕ. The steps in the construction are:

1. Draw stress circle S_z, S_x.
2. Drop normal from failure envelope to X.
3. By trial and error, locate a horizontal line X′x′, of length p, which touches the failure envelope at x′ and then draw in the normal to the failure envelope at x′.
4. Draw a stress circle S_1 S_3, with radius XX′.
5. Draw a perpendicular from S_x until it meets the stress circle at T′.
   The required shear stress is TT′.
   Angle X′X S_3 = x′x σ_3 = 2θ and angle T′XT = 2α.

stress axis (S_n) is 2θ, where:

$$2\theta = 90° - \phi \qquad (9.21)$$

and 2θ is the acute angle between conjugate planes of shear failure (Fig. 9.36(a)).

The angle (2α) which the line T′X makes with the normal stress axis, is twice the angle (α) through which the shear stress (τ) causes the axis of principal stress (S_1) to rotate from the vertical (Fig. 9.36(b)). We can now combine Figs. 9.35 and 9.36(a) and (b) to determine the orientation of the potential shear planes which will develop in the rock mass. This combination of conditions is represented in Fig. 9.36(c). It can be inferred that, because we are dealing with esssentially

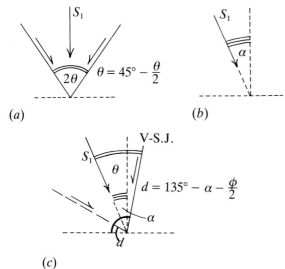

(a)

(b)

(c)

Fig. 9.36. (a) Acute angle 2θ between conjugate shear planes. (b) Orientation of S_1 which makes an angle α with the vertical. (c) 'Dip' ($d°$) of vertical-shear joint (V-S. J.) is as shown.

flat-lying sediments, the shear plane which is most nearly perpendicular to the bed, which we term the *Vertical-Shear Joint* (V-S.J.), will cut through a smaller linear dimension (L) of rock than the conjugate shear plane, before it reaches the limits of the rock unit under consideration. The energy expended in forming a fracture is related to its area (L^2). Therefore, according to the principle of 'least energy', only fractures parallel to the V-S.J. are likely to be widespread. From Fig. 9.36(c), it follows that the angle of 'dip' ($d°$), which in our treatment can exceed 90°, is given by:

$$d° = 135° - \alpha - \frac{\phi}{2}. \qquad (9.22)$$

The value of τ for any given value of S_z, S_x, p, C_0 and ϕ can be obtained by a series of graphical constructions of the type shown in Fig. 9.35. However, this is a tedious and (when ϕ has a low value) somewhat inaccurate procedure. Accordingly, it is more convenient to tackle the problem analytically.

It can be shown that for specific values of S_z, S_x, p, C_0 and ϕ, the critical value of the shear stress (τ_{crit}) necessary to induce failure is given by:

$$\tau_{crit} = \sqrt{\left\{ \left[S_zA - (S_zA + pD^2 - C_0)\left(\frac{D^2 + 1}{D^2}\right) \right]^2 + \left[S_zA - \left(\frac{2 - D^2}{1 - D^2}\right) \times (S_zA + pD^2 - C_0D) - S_z - B \right]^2 \right\}}$$

$$(9.23)$$

where $A = (1 + k)/2$ and $B = (1 - k)/2$ (in which $k = S_x/S_z$), $D = \tan\phi$ and other symbols are as previously defined.

Similarly, it can be shown that the angle (α) which the axis of maximum principle stress makes with the vertical is given by:

$$2\alpha = \tan^{-1}\left(\frac{\tau_{crit}}{S_zB}\right)^{\frac{1}{2}} \qquad (9.24)$$

where τ_{crit} is obtained from Eq. (9.23), so that the angle of 'dip' ($d°$) is given by:

$$d° = 135° - \alpha - \frac{\phi}{2}. \qquad (9.22)$$

There are restrictions to the validity of these equations which define three fields of behaviour. Thus, because we are dealing with approximately flat-lying sedimentary rocks which contain bedding planes, in order that Eq. (9.23) may be valid, the shear stress (τ_{crit}) necessary to induce fracture must have a

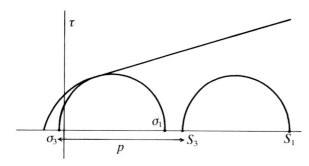

Fig. 9.37. Condition for 'traditional' (Andersonian) normal faults, $(\sigma_1 - \sigma_3) > 4T$.

value which is less than that which can exist on the bedding planes (τ_b), where:

$$\tau_{\text{crit}} < \tau_b = \tan \phi (1 - \lambda) S_z. \tag{9.25}$$

Moreover, as can be inferred from Fig. 9.37, above a specific value of k ($= S_x/S_z$), the initial differential stress is so large that, for the specified value of fluid pressure (p), normal faults will be induced without having recourse to the need to increase the differential stress by the generation of shear stresses.

Yet another limitation may be set by the possibility of hydraulic fracturing, i.e. if $\sigma_1 - \sigma_3 < 4T$, (Fig. 9.38). As we have seen, for the values of p (and hence λ) which we shall consider here, such conditions are not likely to be encountered in clay-rocks in the upper 2–3 km of the crust, so will not be considered further.

If we specify the value of the vertical stress S_z, the values of C_0 and λ then, by setting these quantities in Eqs. (9.23)–(9.25), two tables can be constructed. The first of these (Table 9.2(a)) gives the values of the shear stresses required to cause shear failure for values of angle of friction (ϕ) from 10–45° (in steps of 5°) and values of k from 0.5–1.00 (in steps of 0.05), for a specified total vertical stress (S_z) and cohesive strength (C_0). These ranges cover the probable values of ϕ and k likely to be applicable for all clay-rocks and also a large number of other rock types. The bottom line of Table 9.2(a) gives the limiting value of shear stress (τ_b)

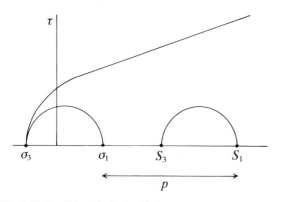

Fig. 9.38. Conditions for hydraulic fracture $(\sigma_1 - \sigma_3) > 4T$.

at which bedding plane slip will be induced (here we assume that cohesion on the bedding plane is zero).

It will be seen that the three 'fields' mentioned above can be clearly defined in Table 9.2(a). The first of these, which is shown by the 'print out' ERROR, delimits the initial stress conditions that will give rise to traditional normal faults, so that it is not necessary to have recourse to shear strains to induce failure. The second field occurs in the lower left hand corner of the table and relates to those values of ϕ and k for which the shear stresses necessary to induce failure in the rock mass are greater than that (τ_b) necessary to bring about bedding plane slip. In this field, bedding plane slip will occur at a level of shear stress below that necessary to induce vertical-shear fractures and so will inhibit the development of such fractures. The third and, for the conditions specified in Table 9.2(b), largest field is that in which vertical-shear fractures, with general orientations similar to that shown in Fig. 9.36(c) will develop. Table 9.2(b) represents the 'dip' of the shear fractures which are closest to the vertical (i.e. $d° = 90°$), for the commensurate data in Table 1. The combined data of Tables 9.2(a) and (b) can be represented in a single diagram (Fig. 9.39) which shows the three main fields and also (within the field of specific interest to us) the dip of the near-vertical fractures. It will be seen that, for a wide range of values of ϕ and k, shear fractures will develop which are within 10° of perpendicular to the bedding

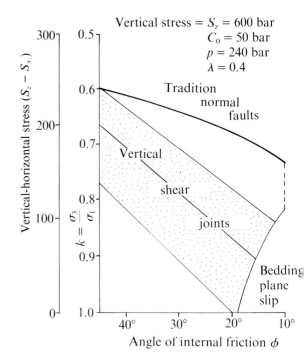

Fig. 9.39. Fields of modes of deformation for ranges of values of initial differential stress ($S_z - S_x$), k and ϕ. The development of vertical-shear fractures (which are assumed to dip within 10° of perpendicular to the bedding) occur for conditions which fall within the shaded area.

Table 9.2. *(a) Value of shear stress (τ) required to induce shear failure for various values of k and ϕ at the specified values:*
$S_z = 600$ bar, $p = 240$ bar, $C_0 = 50$ bar

$k = S_x/S_z$	Phi (ϕ)							
	10°	15°	20°	25°	30°	35°	40°	45°
0.50	E	E	E	E	E	59.6	86.8	106.3
0.55	E	E	E	38.6	77.8	103.3	123.4	139.9
0.60	E	E	47.5	84.5	110.8	132.3	150.6	166.3
0.65	E	45.1	83.6	111.4	134.7	155.0	172.8	188.4
0.70	33.7	76.6	106.4	131.6	153.9	173.9	191.8	207.6
0.75	64.2	96.3	123.5	147.8	169.9	190.2	208.4	224.7
0.80	81.7	110.7	137.0	161.3	183.8	204.4	223.2	240.1
0.85	93.7	121.8	148.0	172.7	195.6	217.0	236.5	254.1
0.90	102.2	130.3	157.0	182.3	206.1	228.3	248.6	267.0
0.95	108.1	136.8	164.3	190.5	215.3	238.4	259.6	278.9
1.00	111.8	141.5	170.1	197.5	223.3	247.4	269.7	289.9
BPS.	63.5	96.4	131.0	167.9	207.8	252.1	302.1	360.0

(b) Angle of 'dip' ($d°$) of vertical-shear joints, for conditions as given in Table 9.2(a).

$k = S_x/S_z$	Phi (ϕ)							
	10°	15°	20°	25°	30°	35°	40°	45°
0.50	E	E	E	E	E	106.7	99.9	94.8
0.55	E	E	E	114.5	105.0	98.8	93.8	89.5
0.60	E	E	114.2	104.9	98.6	93.6	89.3	85.4
0.65	E	115.9	105.7	99.1	93.9	89.6	85.6	82.1
0.70	119.7	107.3	100.1	94.7	90.1	86.2	82.6	79.2
0.75	BPS	101.4	95.6	90.9	86.9	83.2	79.9	76.7
0.80	BPS	BPS	BPS	87.7	84.0	80.7	77.5	74.5
0.85	BPS	BPS	BPS	BPS	81.5	78.3	75.4	72.5
0.90	BPS	BPS	BPS	BPS	79.1	76.2	73.4	70.7
0.95	BPS	BPS	BPS	BPS	BPS	74.3	71.6	69.0

E = Error, BPS = Bedding Plane Slip.

and (because the beds are flat-lying) are, therefore, near-vertical.

The fractures which are delimited by the 80° and 100° dip contours (the shaded areas in Fig. 9.39) are grouped together as *vertical-shear joints*. Fractures which dip from 80–90° in this group would be the result of reverse shear, while those which dip from 90–100° result from normal shear. It is emphasised that the angle and senses of shear noted here are based on an assumed sense of shear strain. These angles may be used only in conjunction with field data if the sense of shear of the vertical-shear joints can be deduced from field evidence (Fig. 9.40).

We have demonstrated how vertical-shear joints can occur as a result of shear stresses induced by differential vertical movements. It has also been shown that these fractures can form in response to a wide range of values of λ and ϕ. However, we have looked at only one specific set of values of S_z, λ and C_0. Let us now see how variation of these three parameters will influence the extent of the range of k and ϕ in which vertical-shear joints can develop.

Influence of depth, fluid pressure and strength upon the development of vertical-shear joints

In the following paragraphs, the influence is assessed of:

(i) the depth of burial (as represented by the total vertical stress (S_z)),

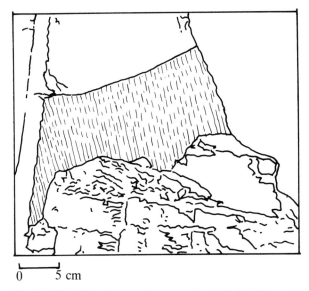

|_____|
0   5 cm

Fig. 9.40. Faint lineations on fracture surface in Oxford Clay at the Saxon Pit, which indicate vertical-shear motion in the 'reverse' mode. (Courtesy of London Brick Co.)

(ii) the strength of the rock or, more specifically, the cohesive strength (C_0) in determining the size of the field in which vertical-shear joints may form, and

(iii) the fluid pressure (p), where p is given in terms of its ratio to the total vertical stress, such that $\lambda = p/S_z$.

The cohesive strengths (C_0) which will be considered range from 12.5 to 200 bar and are approximately 25 per cent of the uniaxial strength (σ_0) of a rock. Taking into consideration the influence of time and scale (Chap. 1), these values of C_0 correspond to values of S_0 which range from about 50 to 800 bar and represent rocks which range from 'moderately weak' to 'very strong'.

The assessment of the influence of these various parameters can be carried out by constructing a series of diagrams of the type shown in Fig. 9.39. In this way, the variation in areal extent of the different

fields, that is, (i) traditional normal faults, (ii) bedding plane slip and (iii) vertical-shear joints, can be readily seen. It is this last field in which we are particularly interested. Consequently, we wish to determine those conditions most likely to give rise to the widest range of values of k and ϕ in which vertical-shear joints will develop. That is, we are looking for the largest extent of this specific area (shaded in Fig. 9.39).

The influence of:

(i)   the magnitude of the vertical stress (S_z),
(ii) the magnitude of the cohesive strength (C_0), and
(iii) the value of λ

are represented in Fig. 9.41, along the z, x and y axes respectively. From the graphical representation of the size of the fields, and particularly the stippled areas which represent the conditions in which vertical-shear joints may form, the following conclusions can be reached regarding the development of such joints.

(i)   Their development is enhanced as the magnitude of S_z increases: though the effect, for values of $S_z > 400$ bar, is not large.
(ii) Vertical-shear joints will develop most readily in those rocks with low strength (as represented by the value of C_0).
(iii) Their development is favoured more by low values of fluid pressure (i.e. $\lambda = 0.4$ to 0.5), than when conditions of over-pressuring obtain.

From these three trends, it follows that the conditions particularly conducive to the development of vertical-shear joints occur in weak rock, not subjected to over-pressure, when at depths of 2–4 km.

'Joint' displacement

The authors could be taken to task in categorising the structures formed by this vertical-shear mechanism as 'joints', without first demonstrating that they would exhibit a suitably small amount of differential slip. In order to estimate the amount of slip that would be associated with the fractures seen in the Saxon Clay Pit it is necessary to quantify the *in situ* characteristics of Oxford Clay that would have existed at a depth of 2 km. (The estimate of depth has been derived from the measured porosity of a specimen of clay taken from the clay pit and from the known depth/porosity relationships for clays in the region.) These various parameters have never been measured directly, so here we shall present values which are based upon laboratory measurements and which are 'adjusted' for time and scale effects (Chap. 1).

The various parameters which we quantify, for Oxford Clay are:

(i)   Young's modulus (E), taken as 32.5×10^3 bar
(ii) Poisson's number (m), taken as 2.5

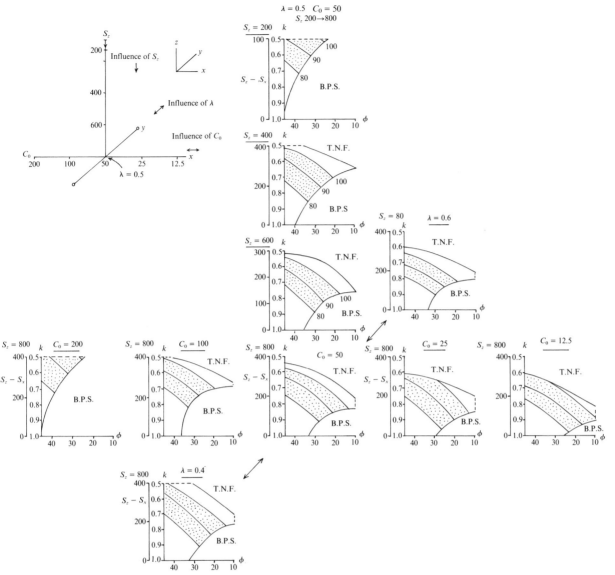

Fig. 9.41. Compilation of diagrams similar to Fig. 9.39 which permit the influence upon the likelihood of V-S Joint development to be assessed of:

(i)   the magnitude of the vertical stress (S_z),
(ii)   the magnitude of the cohesive strength (C_0) and
(iii)   the value of λ

These parameters are represented along the z, x and y axes, respectively. The stippled areas indicate the conditions for the development of V-S Joints.

(iii)   Cohesive strength (C_0), taken as 10 bar
(iv)   Peak and residual angles of friction, taken as 27.5° and 13° respectively.

The latter two parameters are needed to estimate the stress-drop that would be associated with the generation of vertical-shear fractures in Oxford Clay. Two possible failure conditions are indicated in Fig. 9.42. In this diagram it is assumed that the total vertical stress is 400 bar, which is the value of S_z at a depth of approximately 2 km, and that $\lambda = 0.5$, consequently, the effective vertical stress is 200 bar.

It will be seen that if shear failure is controlled by the residual friction angle (ϕ_r) failure envelope, then, on the generation of a shear fracture stress circle

ZC would change to ZD with a stress-drop of 17 bar. The largest stress-drop that could occur would be brought about if failure were controlled by the ϕ envelope (with a stress circle ZA) which then reduced to stress circle ZD with a resulting stress-drop of 65 bar. It can be argued that the residual envelope represents the 'long-term', i.e. geological, strength of clays, at strain-rates of 10^{-20}/s and is likely to be close to 17 bar. At strain-rates pertinent to the formation of fractures (say 10^{-14}–10^{-16}/s) the stress-drop is likely to be larger, and is taken here to be 30 bar. This stress-drop is compatible with failure occurring at an angle of sliding friction of $\phi = 20°$.

The stress-drop inferred from the failure envelopes represented in Fig. 9.42 holds only at the actual

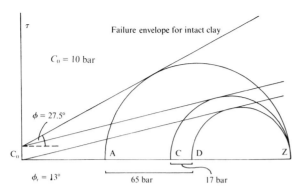

Fig. 9.42. Stress-drop conditions associated with failure of Oxford Clay.

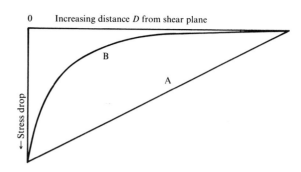

Fig. 9.43. Linear (A) and exponential (B) increase in differential stress (and decrease in stress-drop) away from a shear plane.

fractures plane. At some distance D (which is likely to be closely related to fracture separation) the stress-drop will be zero. If the fall-off in stress-drop from the fracture plane were linear (curve A in Fig. 9.43) the average stress-drop for the cited example would be 15 bar. However, from elastic analyses of stress changes adjacent to holes, or inclusions, in an elastic medium it is known that the intensity of the stress changes with distance in a non-linear, manner (curve B, Fig. 9.43). Assuming an exponential decrease in stress-drop away from the shear plane, the average stress-drop for this specific example is likely to be about 5 bar.

It must at once be admitted that the extent and influence of a stress-drop induced by shear failure is not well understood. In addition, little is known regarding the factors that control separation between parallel shear planes. Consequently, for simplicity, we shall assume that the stress-drop influences strain only on one side of a shear fracture. Unsatisfactory though this assumption is, it permits an estimate to be made of the displacement that would be associated with a stress-drop. Let us, therefore, consider the situation (Fig. 9.44) in which the axis of maximum principal stress (σ_1) makes an angle θ with respect to a set of parallel shear fractures, which have a mean separation D. From this geometry, an estimate can be made of the shear displacement (d) on each fracture, which is compatible with the stress-drop (σ_d) when the rock between the fractures has an elastic modulus E.

From the geometry represented in Fig. 9.44, it follows that the displacement d is given by:

$$d = \frac{\sigma_d D}{E \cos\theta \sin\theta}. \tag{9.26}$$

The angle θ is, of course, determined by the brittle-failure criterion, so that:

$$\theta = 45° - \phi_r.$$

The value of ϕ_r for *in situ* failure of Oxford Clay we have established to be 20°, so that $\theta = 35°$. If we substitute $\theta = 35°$, $\sigma_d = 5$ bar, $D = 10$ m and $E = 32.5 \times 10^3$ bar in Eq. (9.26) it transpires that the

shear displacement $d = 3$ mm. Such a small displacement could be detected in particularly favourable field exposures if they were subjected to the closest scrutiny. In the majority of field situations, and assuming the normal degree of scrutiny, the displacements would go undetected and *the vertical-shear fractures would be classified as joints*.

If such vertical-shear joints develop in stronger rocks (and, as may be inferred from Fig. 9.41, they certainly can) the associated stress-drop may be several times larger than the figure cited for Oxford Clay. However, to off-set the influence of this larger stress-drop, the value of Young's modulus may be over an order of magnitude higher than that of Oxford Clay. Moreover, the separation between joints in the Oxford Clay is large (i.e. 10 m). In stronger rocks, the separation may be of the order of 1.0 m. The nett result of these effects is that the probable displacement on vertical-shear joints in strong rock is likely to be small fraction of a millimetre and, therefore, may not be detectable in the field.

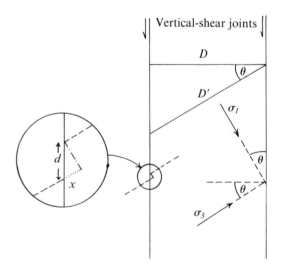

Fig. 9.44. Dimensions and geometry used to estimate the shear displacement (d) on a vertical-shear joint.

Magnitude of shear-strain required to initiate vertical-shear joints

It is now necessary to ascertain what critical value of shear strain must be induced in order to give rise to the necessary magnitude of shear stress, and to determine the changes in dip that are required to bring about these critical shear strains.

As already noted, the elastic relationship between shear stress and shear strain is given by:

$$\tau_{crit} = G\psi \qquad (9.19)$$

where τ_{crit} is the critical value of the shear stress, ψ is the angular shear and G is the shear modulus, which can also be written as:

$$G = \frac{mE}{2(m+1)} \qquad (9.27)$$

where m is Poisson's number (the reciprocal of Poisson's ratio).

The mean value of Poisson's number for Oxford Clay is taken to be $m = 2.5$. Therefore, substituting this value of m, together with a reasonable value of $E = 32.4 \times 10^3$ bar in Eq. (9.27), gives for Oxford Clay a value of $G = 11.60 \times 10^3$ bar.

Let us now ascertain the value of shear stress that must be induced, by the differential movement of sediments, before near-vertical-shear fractures will develop.

The vertical-shear joints exposed in the Oxford Clay at Saxon Pit are, on average, vertical (i.e. $d° = 90°$). Moreover, failure may be initiated when $\phi = 27.5°$, or possibly when $\phi_r = 13°$ or some intermediate angle of sliding friction.

From Table 9.3(a) and (b) it can be inferred that the value of shear stress (τ_{crit}) required to induce fractures which dip at 90° is approximately 65.0 bar when $\phi = 20°$. This limiting condition occurs when $k = 0.87$.

In order that vertical-shear joints can develop in response to such a shear stress (of 65 bar) an angular shear (ψ) must be induced, which is given by:

$$\psi = \tau_{crit}/G,$$

or

$$\psi = \frac{65}{12.5 \times 10^3} = 5.2 \times 10^{-3} \text{ rads} = 0.3°.$$

Thus, an extremely small change in dip would induce fracture (for the specific ratio (k) of S_x/S_z, of 0.87) provided the sediments in question are buried at a depth of about 2 km. Here, it is assumed that the sediments took on the elastic properties used in the calculation (which, thereafter, during uplift, are assumed to be constant). In nature, it is to be expected that, as a result of compaction and induration, there will be a progressive increase in the values of the

Table 9.3. *(a) Value of shear stress (τ) in bars required to induce failure for various values of k and ϕ at the specified values:*
$S_z = 400$ bar, $p = 200$ bar, $C_0 = 10$ bar

$k = S_x/S_z$	Phi (ϕ)				
	13°	15°	20°	25°	37.5°
0.65	N	N	N	N	N
0.70	N	N	N	32.5	42.5
0.75	N	N	34.4	52.4	60.0
0.80	22.2	31.8	50.1	65.4	72.4
0.85	37.5	44.5	60.5	75.1	82.1
0.90	BPS	52.6	68.0	82.8	89.8
0.95	BPS	BPS	BPS	88.8	96.1
BPS	46.2	53.6	72.8	93.3	104.1

(b) Angle of 'dip' (d°) of V-S. Joints for conditions specified in Table 9.3(a).

k	Phi (ϕ)				
	13°	15°	20°	25°	27.5°
0.65	N	N	N	N	N
0.70	N	N	N	108.3	103.6
0.75	N	N	107.7	99.3	97.2
0.80	114.0	108.3	99.3	93.2	90.7
0.85	102.8	99.5	91.2	88.4	86.6
0.90	BPS	92.9	88.2	84.3	82.5
0.95	BPS	BPS	BPS	80.7	78.8

N = Normal Faults, BPS = Bedding Plane Slip.

elastic moduli. If differential, vertical movements occurred continuously from the time the sediments were deposited, there would be a continual build-up in shear stresses throughout the history of subsidence. If it is assumed that the increase in elastic moduli is linear, then a change of 1° in dip as the sediments are buried from the surface to a depth of 1.5 km would generate a shear stress sufficient to initiate failure, provided the value of k is appropriate.

If the clay-rocks reached a depth of 2 km untilted, and hence unfractured, any subsequent, differential, vertical movement of less than 0.3° (whether it was caused by further burial or uplift) would certainly give rise to the development of vertical-shear joints.

Hence, it is concluded that clay-rocks, at quite modest depths (i.e. 1–2 km) will almost invariably be cut by near-vertical-shear joints, if the dip is in excess of 1°, or the sediments have experienced significant uplift which induced a change in dip of about 0.3°.

Clearly, vertical-shear joints may develop at

depths other than 2 km (on which depth the calculations above are based). However, in order to ascertain
at what precise depth such fractures actually develop
in a specific clay-rock, it is necessary to know the
physical properties of that particular clay-rock as it
becomes indurated and also to be able to produce
isopach maps of the area of interest, for the specific
sedimentary horizon. If such data are available, then it
is possible to forecast the orientation and also the
ratio of the horizontal to vertical total stresses which
existed when the vertical-shear joints developed.

Morphology of vertical-shear joints
An important feature of the fracture surfaces, which is
consistent with the dip-slip motion, are the 'facets'
(Fig. 9.45) which occasionally occur on the joint surfaces. These major joints may cut through considerable thickness of rock which is unlikely to be completely homogeneous throughout. Thin layers or
lenses will be encountered with physical properties (in
particular a value of the residual angle of friction ϕ_r)
which are different from the rest of the unit. Such a
change in the value of ϕ_r would result in refraction of
the vertical-shear joint surface, when viewed in
section, which would, of course, cause the development of a facet on the surface (Figs. 9.32 and 9.45).
These facets, we suggest, are diagnostic features of
vertical-shear joints.

In previous 'traditional' analyses, it has been
concluded that the majority of joints are the result of
extension and, consequently, these fractures would
constitute potential pathways for fluid migration.
However, we have established here that vertical joints
which occur in clay-rocks at modest depth are best
attributed to vertical-shear failure and that such
'joints' developed when the maximum and least principal stresses were inclined at quite large angles to the
bedding. It will be seen from Fig. 9.42 that, following
failure, the normal effective stress acting on the failure
plane (i.e. the horizontal stress perpendicular to the
joint) is positive and large. For the conditions represented in Fig. 9.42 the normal stress is 153 bar. That
is, the fracture planes are completely and tightly
closed and would not form channels of easy fluid
migration. (Indeed, in order to open these fractures,
assuming total stresses remained constant, it would be
necessary to increase fluid pressure by over 153 bar, so
that the fluid pressure would become 353 bar and λ
would be in excess of 0.88.)

Multiple sets of vertical-shear joints
So far we have considered the development of only a
single set of V-S Joints and demonstrated that they
could form in a unit of clay-rock which experienced an
angle of tilt of as little as 0.33°. This angle is so small
that it is to be expected that many, if not most,
sediments will experience several periods in their

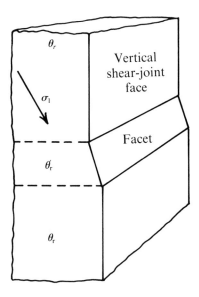

Fig. 9.45. Facet on V-S joint surface.

history when they undergo tilting, of this magnitude,
in different directions. For example, the earliest phase
of tilting may be associated with simple, progressive
sedimentation as the basin deepens. A different direction of tilt could be induced by the local deposition of
sediments, such as a delta or a facies change, vertically
above the unit in question, which, by isostatic
readjustment, causes differential sinking of the clay-
rock unit. Similarly, in uplift and exhumation, further
tilting may take place on an axis which is unrelated to
those caused by earlier events. If the various axes of
tilt are separated by a sufficiently large angle, then
several sets of V-S joints could develop. When the
angle between the tilt axes is 90°, it is obvious that the
earlier set of fractures would have no influence upon
the subsequent development of an orthogonal set.
However, if the angle between the tilt axes is small,
conditions may favour re-shear on the existing fractures rather than permit the generation of a new set.
As we saw in Chap. 5, it is not a simple task to
ascertain this limiting condition where a second set of
fractures is able to form. However, if we make the
simplifying assumption that the two smaller principal
stresses are equal (i.e. $\sigma_1 > \sigma_2 = \sigma_3$) then we can follow
the analysis presented by Jaeger & Cook (1969). For
these conditions, the stress in a rock mass can be
represented by a single stress circle (Fig. 9.46(a)),
where the mean stress (σ_m) is given by:

$$\sigma_m = \frac{(\sigma_1 + \sigma_3)}{2} \qquad (9.28)$$

and the radius of the stress circle (τ_m) is:

$$\tau_m = \frac{(\sigma_1 - \sigma_3)}{2}. \qquad (9.29)$$

Let us assume the existence of a plane of weakness

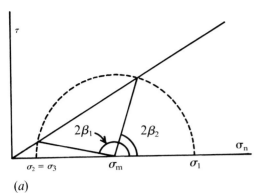

(a)

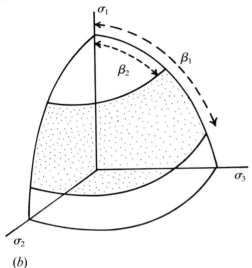

(b)

Fig. 9.46. (*a*) Failure conditions along a plane of weakness and the critical angles β_1 and β_2 for the special situation in which $\sigma_2 = \sigma_3$. (*b*) Region (shaded) in which the normal to the plane must fall for sliding to be possible.

(with zero cohesion). Failure by frictional sliding occurs when:

$$\tau_m = \sigma_n \tan \phi \qquad (9.29(a))$$

where σ_n is the normal effective stress acting on that plane and ϕ is the angle of sliding friction, and the failure envelope is as shown in Fig. 9.46(*a*). Sliding can occur if the plane is oriented so that the shear stress on it is in excess of that given in Eq. (9.29(*a*)). This occurs if the *normal* to the plane is inclined at an angle between β_1 and β_2 with the axis of maximum principal stress (as shown in Fig. 9.46(*b*)). It can be shown that these angles are given by:

$$\beta_1 = 90° - \tfrac{1}{2}(\cos^{-1} A) + \tfrac{1}{2}(\cos^{-1} BA) \qquad (9.30(a))$$

and

$$\beta_2 = 90° - \tfrac{1}{2}(\cos^{-1} A) - \tfrac{1}{2}(\cos^{-1} BA) \qquad (9.30(b))$$

where

$$A = \tan \phi / (\tan \phi^2 + 1)^{\frac{1}{2}}$$

and

$$B = \sigma_m / \tau_m.$$

Table 9.4.

σ_3/σ_1	$\phi = 18.5°$	
	β_1	β_2
0.20	85.0°	23.5°
0.30	81.0°	27.0°
0.40	75.5°	33.0°
0.5	63.4°	45.0°

If we take $\phi = 18.5°$, a representative value for many clay-rocks, then the angles β_1 and β_2 for different ratios of σ_3/σ_1 are listed in Table 9.4.

It should be noted that the stress ratios given in this table relate to effective stresses. In order to give the equivalent ratio of total stresses it would be necessary to specify the value of the fluid pressure (*p*) and the magnitude of the total vertical stress (*S_z*).

The region in which the normal to the weakness plane must fall, for sliding to be possible, is as shown in Fig. 9.46(*b*). It may be inferred from this figure that a wide range of orientations exists in which later V-S joints may form and be superimposed on an earlier set. As already noted, the sets may be orthogonal. Alternatively, the acute angle between the sets may be as little as 45°. This acute angle relationship would have led some geologists to categorise the two sets of fractures as a system of conjugate shear joints.

External influences

It is emphasised that in the theoretical analysis presented in this chapter, we considered only those fractures which are 'induced' within a basin as the result of downwarp and subsequent uplift. Basins may experience more than one such cycle so that the induced patterns may be correspondingly more complex. In addition, it is unlikely that the various stages of basin development will occur without reactivation of pre-existing fractures in the basement. Such reactivation will give rise to associated fractures in the cover rocks of the basin. These associated or 'inherited' fracture patterns may exhibit trends and movement patterns which are completely different from the internally induced fractures. In addition, the basin may be subjected to concomitant or subsequent extension or compression which can complicate the fracture patterns in the basin even further.

We have considered, briefly, the importance of influences, other than those *internal* factors related to vertical movements, in the text relating to Fig. 9.6. These other influences pertain to *external* factors that relate to lateral compressions and/or extensions induced in flat-lying sediments as the result of *bound-*

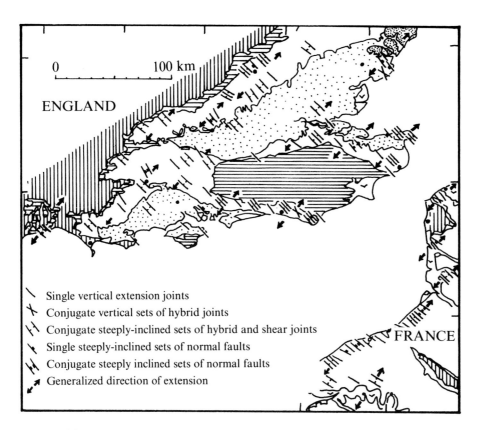

Fig. 9.47. North-west trending mesofracture sets in S. England and N. France. Sets are represented by generalized mesofracture trends, from which generalized directions of the least principal stress were inferred. (After Bevan & Hancock, 1986.)

ary conditions. Such boundary conditions, in this cited example, can be attributed to *plate motions*.

In this example, we considered fracture development in flat-lying sediments in the foreland, adjacent to a fold belt. However, Bevan & Hancock (1986) suggest that sets of fractures in the U. Cretaceous, Tertiary and Quaternary sediments of S. England and N. France can be interpreted as the result of a generalised, or regional extension in the NE–SW direction (Fig. 9.47). Moreover, by comparing the trends of these fracture sets with basement lineaments and orientations of current crustal stresses, inferred from *in situ* measurements or fault plane solutions of active faults (Fig. 9.48), they further suggest that these fracture sets, although they are relatively remote from the fold belt, can be related to the direction of 'Alpine Collision'.

We suggest that it is unlikely that stresses, necessary to cause failure, can be transmitted directly in weak sediments to a distance of 750 km, or more, from the nearest point in the Alpine fold belt. Rather, we suggest, it is reasonable to infer that the fracture patterns are more likely to have formed in response to strains in the lithosphere, which are the result of the mechanisms driving plate motions. These mechanisms are, of course, the subject of fierce debate. However, we would attribute these strains primarily to the gravity-glide mechanism discussed by Price *et al.* (1987). Just how fractures in the upper crust may develop in response to elastic and other strains in the central layers of the lithosphere is only now being

addressed; and it may be some little time before these processes can be established. However, when these processes are understood, it will probably be possible to relate some main fracture trends in sediments situated in intra-cratonic environments to past, or present, plate motions and the mechanisms that drive them.

Commentary

It will be abundantly clear to the reader that the analyses of regional joint pattern in little-deformed sediments are extremely vexing, in that the results of study, even of the same data, are usually confusing and often contradictory. This in part is related to the intrinsic difficulty of the problem of dating and grouping together, or distinguishing between, sets of fractures (or even establishing the relative ages of fractures within one set) which may have formed at different times and possibly as the result of different mechanisms.

For decades, quite heated argument has occurred between the few geologists involved in the analysis of regional joint patterns. Some have advocated that joints are wholly the result of extension in a direction normal to the fracture plane. Others have advocated the possibility that some (possibly a minority) of such joints can result from shear (and these geologists have tacitly, or specifically, assumed that the shear was dominantly strike-slip).

The present authors have demonstrated here

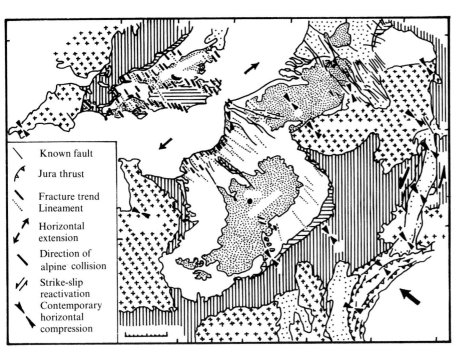

Fig. 9.48. Summary sketch map of N.W. Europe, showing the trend of fracture and lineaments belonging to the north-west system. Also shown is the generalized horizontal extension directions and the directions of horizontal maximum compression based on *in situ* stress measurements (I) and fault plane solutions (F) derived from various sources. (After Bevan & Hancock, 1986.)

and at some length that a third possibility exists. We suggest that many, (possibly the majority) of joints in sediments, and also in metamorphic and some igneous rocks, are the result of vertical-shear induced by differential vertical movements of the rock masses. Moreover, the important influence of plate tectonic effects have yet to be quantified.

Thus, the traditional, published interpretations can often be likened to an attempt at a jig-saw puzzle, from which many important pieces are missing. Consequently, the final presentation is 'full of holes' and the presented picture lacks conviction. We conclude, therefore that the *traditional interpretations of fracture patterns in flat-lying rocks (and also in other geological environments) is suspect. The whole 'data-base' relating to joint patterns is in urgent need of reappraisal.*

References

Bevan, T.G. & Hancock, P.L. (1986). A late Cenozoic regional mesofracture system in S. England and N. France. *J. Geol. Soc. London,* **143**, 355–62.

Brace, W., Paulding, B.W. & Schulz, C. (1966). Dilatancy in the fracture of crystalline rocks. *J. Geophys. Res.,* **77**, 3934–53.

Brown, E. & Hoek, E. (1978). Trends in relationships between measured in situ stress and depth. *Int. J. Rock Mech. Min. Sci. & Geomech. Abstr.,* **15**, 4, 211–15.

Burland, J.B., Longworth, T.I. & Moore, J.F.A. (1977). A study of ground movement and progressive failure caused by deep excavation in Oxford Clay. *Geotechnique,* **27**, 4, 557–91.

Engelder, T. & Geiser, W. (1980). On the use of regional joint sets as trajectories of palaeostress fields during the development of the Appalachian Plateau, New York. *J. Geophys. Res.,* **85**, 6319–41.

Fyfe, W., Price, N.J. & Thompson, A.B. (1978). *Fluids in the Earth's Crust.* Amsterdam: Elsevier.

Hafner, W. (1951). Stress distributions and faulting. *Geol. Soc. Am. Bull.,* **62**, 373–98.

Hancock, P.L. (1969). Fracture patterns in the Cotswold Hills. *Proc. Geol. Assoc.,* **80**, 219–41.

Harper, T.R. & Szymanski, J.S. (1983). Geological processes and the mechanical aspects of rock squeeze. *Tectonophysics,* **91**, 119–35.

Hubbert, M.K. (1951). Mechanical basis for certain familiar geologic structures, *Geol. Soc. Am. Bull.,* **62**, 1355–72.

Jaeger, J.C. & Cook, N.G.W. (1969). *Fundamentals of rock mechanics.* London: Methuen.

Mackenzie, A.S., Mann, D.M. & Quigley, T.M. (1986). Migration of Petroleum Fluids in the sub-surface. *J. Geol. Soc. London* (in press).

Nickelson, P.N. & Hough, V.N.D. (1967). Jointing in the Appalachian Plateau of Pennsylvania. *Geol. Soc. Am. Bull.,* **78**, 609.

Parker, J.M. (1942). Regional systematic joints in slightly deformed sedimentary rocks. *Geol. Soc. Am. Bull.,* **53**, 381.

Price, N.J. (1958). A study of rock properties in conditions of triaxial stress. In *Proc. Conf. on Mech. Prop. Non-metallic Brittle Materials,* ed. W.H. Walton, pp. 106–22. London: Butterworths

(1959). Mechanics of Jointing in Rocks. *Geol. Mag.,* **96**, 149.

(1966). *Fault and joint development in brittle and semi-brittle rock.* Oxford: Pergamon.

(1974). The development of stress systems and fracture patterns in undeformed sediments. In *Advances in rock mechanics.* Proc. 3rd Int. Conf. Soc. Rock Mech. Denver, Colo., 1A, 487.

(1975). Stresses and fractures in granites. *Canadian Geoscience.*

Legend (Fig. 9.48):

Known fault

Jura thrust

Fracture trend Lineament

Horizontal extension

Direction of alpine collision

Strike-slip reactivation Contemporary horizontal compression

(1979). Fracture patterns and stresses in granites. *Canadian Geoscience*.

Price, N.J., Price, G.D. & Price, S.L. (1987). Plate motions and gravity-slide. *J. Geol. Soc. London* (in press).

Ver Steeg, K. (1942). Jointing in the coal beds of Ohio. *Econ. Geology*, **37**, 503.

(1944). Some structural features of Ohio. *J. Geol.*, **52**, 131.

10 Introduction to folding

Introduction

Folds occur on all scales from the microscopic to the regional. When regional folds are viewed on satellite images, oblique aerial photographs or from vantage points in mountain belts they can be awe-inspiring. When, on the mesoscopic scale folds are viewed in profile, their patterns are often spectacular and have great aesthetic appeal, while the almost endless variety of geometric forms they present have stimulated the curiosity and imagination of many geologists. In addition economically interesting concentrations of minerals and hydrocarbons are often found in association with folds. These concentrations may occur either in response to stress gradients that are set up *during* the formation of a fold or, particularly in the case of hydrocrbon concentration, as the result of the entrapment of material *after* the fold has formed. It is little wonder then that these structures have been the subject of extensive study. To understand their development geologists have used three main approaches. These are: (i) a study of the geometry and internal structure of naturally occurring folds, (ii) the theoretical analyses of folding and (iii) the formation of folds experimentally. As a result of these studies, various processes have been proposed to account for geological folding which include *passive folding*, *bending* and *buckling*.

Throughout much of the present and following four chapters we shall be concerned with the mechanisms by which folds form. However for the last hundred years or more structural geologists studying folds have been primarily involved with the description and classification of their geometries. Thus the present chapter opens with the terminology used to describe and classify the geometry of folds. This is followed by a brief discussin of two processes, passive folding and bending, by which folds can form. A vast literature exists in both the geological and material science journals on buckling, the third process of folding. This process is discussed in detail in Chapters 11 to 14. The present chapter continues with the description of two important conceptual models of strain distribution within a folded layer; namely *tangential longitudinal strain* and *flexural slip*.

Because experimental work has played such an important role in advancing our understanding of geological folding, a short discussion on *scale modelling* is given which includes the description of several experiments in which multilayers, composed of rock analogue materials, are folded.

The description and classification of fold geometry given in most texts is restricted to the fold profile section. This is probably because folds are assumed to be *cylindrical* (or *cylindroidal*); i.e. structures whose profile geometry does not change along the fold hinge. However, field observations and experimental work indicate that this is generally not so and we have, therefore, included a section that describes the three-dimensional geometry and spatial distribution of folds.

It is quite common to find that deformed rocks have been affected by more than one phase of deformation. When one episode of folding is superimposed onto another, the resulting fold geometries and outcrop patterns are often complex. However, it is sometimes possible from the outcrop (i.e. interference) pattern of superposed folds to determine the original orientation of the axial planes and axes of the two sets of folds. The interpretation of *interference patterns* is discussed in this chapter as it often plays an extremely important role in the structural analysis of an area of polyphase deformation.

The chapter ends with a brief review of the mechanisms which have been proposed for the formation of fold nappes, large recumbent fold structures that characterise the high grade metamorphic zones of many orogenic belts.

Because the present chapter is the first of six chapters on folding, the reader may find a brief outline of the contents of the remaining five chapters useful. The first (Chap. 11) deals with the buckling of an interface between two unlike materials and also of single layers (Fig. 10.1(*a*) and (*b*)), the second (Chap. 12) considers the buckling of multilayers (Fig. 10.1(*c*)), the third (Chap. 13) examines the buckling of anisotropic materials (Fig. 10.1(*d*)), while the fourth (Chap. 14) deals with the development of large-scale fold structures, i.e. folds that are too large for the effect of gravity to be ignored. In each chapter, relevant field observations, model work and mathematical analyses will be considered. Chapter 15 is entitled 'The life and times of a fold'. In it, the various stages in the development of a fold are discussed and the formation of folds in a complex sedimentary multilayer subjected to layer parallel compression is considered.

Theoreticians have analysed the buckling

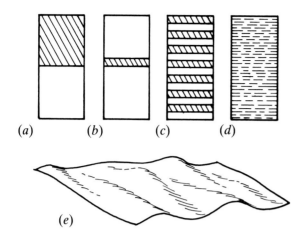

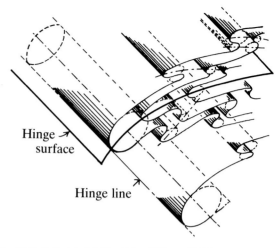

Fig. 10.1. (*a*) An interface between two unlike materials, (*b*) a single layer, (*c*) a multilayer and (*d*) an anisotropic material. The buckling behaviour of these four models is discussed in Chaps. 11, 12 and 13. (*e*) The geometry of folds that form on a surface folded by one episode of deformation.

Fig. 10.2. Cylindrical folds. (After Wilson, 1961.)

behaviour of a large variety of different layered systems, some of which are particularly relevant to the problems of geological folding. It is beyond the scope of this book to present the mathematical arguments used in deriving the various 'buckling equations' and the interested reader is referred to the original works. We do, however, consider the physical implications of the buckling equations and, by careful scrutiny of the assumptions on which the analyses are based, attempt to assess whether the theory may be applied to real materials (rocks) in a natural environment. We attempt to demonstrate to the reader who has not the time or mathematical facility to follow the steps of a mathematical analysis that he or she can still, with a little effort, ascertain the assumptions built into the analyses and thereby determine the relevance of the final buckling equation to the formation of geological structures.

In an attempt to understand the processes involved in geological folding, many experiments using rock analogue materials have been conducted. These experiments have been particularly helpful in increasing our understanding of fold initiation and their growth from low-amplitude folds into finite, large-amplitude structures.

Experimental work on real rocks has also provided invaluable information about the rheological properties which rocks are likely to exhibit during the formation of folds. This enables the geologist to select appropriate rheological parameters to be used in any specific mathematical analysis of folding although, unfortunately, this has not always been done.

Field observations of buckle folds provide the reference frame with which the results of the theoretical and experimental studies must be compared. For example, by considering the similarities of the buckle morphology and the distribution of stress and strain

within a buckle predicted by a specific buckling analysis with the morphology and internal fabric of a natural fold, it is possible to assess the relevance of that analysis to the formation of that geological fold, so let us now consider the elements of fold classification.

Fold description and classification

Folds, one of the most commonly occurring structures to be found in deformed rocks, are formed when planar features such as bedding or cleavage, or linear features such as an alignment of asicular minerals are deflected into curviplanar or curvilinear structures. Folds develop on interfaces, in single layers, in multilayers and in rock fabrics and have a wide variety of geometries and sizes. However, a common feature of these deflections is that they often show a marked periodicity.

Curviplanar surfaces found in nature often have very complex geometries. However, if the rocks have been subjected to only one 'phase of folding' more simple geometries, such as those shown in Fig. 10.1(*e*) may be observed. Some parts of natural folds are approximately cylindroidal (Fig. 10.2) and can be defined by a line moving parallel to itself in space. Such a line, which is known as a rectilinear generator, has no particular location on the folded surface, but is a directional property of the entire surface and is parallel to the *fold axis*. The intersection of the folded surface on a plane perpendicular to the fold axis is known as the *profile* of the fold. Some geometrical features of a fold profile are shown in Fig. 10.3. The *hinge* points are those of maximum curvature and the *inflection points* occur where the rate of change of curvature is zero. The *crest* and *trough* are respectively the highest and lowest point on the fold.

When considering the folded surface as a whole, rather than its profile, the hinge, inflection, crest and

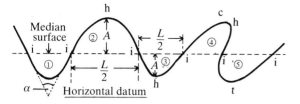

Fig. 10.3. The definition of some of the terms used to describe the geometry of a fold profile: h = hinge, i = inflection point, c = crest, t = trough, α = interlimb angle, L = wavelength, A = amplitude.

Table 10.1.

Interlimb angle	Description of fold
180–120°	Gentle
120–70°	Open
70–30°	Close
30–0°	Tight
0°	Isoclinal
Negative angles	Mushroom or Elastica

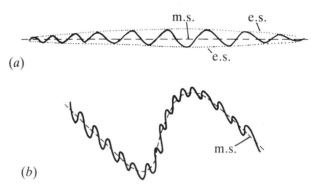

Fig. 10.5. (a) A train of folds shown with their median and enveloping surfaces, m.s. and e.s. respectively. (b) First and second order folds formed during the same episode of folding. Note that the first order folds are defined by the median surface of the second order folds.

trough points become lines. The *hinge line* is parallel to the fold axis, but unlike the fold axis, is at a specific location on the fold. The inflection and hinge points are independent of the orientation of the folded surface in space, and are consequently called invariant features. However, the position of the crest and trough points will probably change if the orientation of the folded surface is changed and are, therefore, described as variant features.

It is generally considered that two adjacent inflection points (i_1, i_2 in Fig. 10.4(a)) mark the limits of an individual fold which can be divided into a *hinge zone* and *fold limbs*. These terms have been defined by Ramsay (1967) who suggested that 'the simplest way of defining the limits of the hinge zone is by comparing the curvature of the fold surface with that (unity) of a circular arc drawn with $i_1 i_2$ as diameter. That part of the fold where the curvature of the fold surface exceeds that of the circular arc can be defined as the hinge zone and the parts of the fold on either side of it where the curvature is less than that of the circular arc are defined as the fold limbs. This definition is useful, but its limitations become apparent when applied to those folds where the curvatures are all equal or less than unity (Fig. 10.4(b)). Using the above definition such folds would not have a hinge zone.

The *interlimb angle* is the minimum angle between the limbs as measured in the profile section (i.e. the angle between the lines tangent to the inflection points on the profile curve, α in Fig. 10.3). Fleuty (1964) has suggested terms that can be used to indicate the *tightness* of a fold which together with the corresponding ranges of interlimb angles are given in Table 10.1.

A number of folds may occur together to form a train of folds (Fig. 10.5(a)). The two surfaces (lines in the profile section) that enclose the folds are known as *enveloping surfaces* and the one joining the inflection lines as the *median surface*. It should be noted that the median surface need not be halfway between the two enveloping surfaces.

During a single phase of deformation folds may develop on different scales (Fig. 10.5(b)). The median surface of the smaller (second order) structures, sometimes called parasitic folds, defines the form of the larger (first order) folds and, as will be discussed in Chap. 18, can be used to determine the position, geometry and orientation of those major flexures which are too large to be seen in outcrop.

The terms *wavelength* and *amplitude* are illustrated in Fig. 10.6(a). The amplitude is the perpendicular distance between the median surface and the fold hinge and the wavelength is the distance between alternate inflection points. However, because an individual fold is only half a wavelength and because natural wavetrains often exhibit large variations in both amplitude and wavelength from fold to fold (Fig. 10.6(b)), it is sometimes convenient to define the wavelength of a fold as twice the distance between adjacent inflection points.

It is emphasised that when referring to the various parameters of a fold such as wavelength, amplitude, interlimb angle etc. it is assumed that the

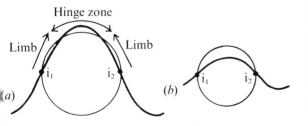

Fig. 10.4. The definition of the hinge zone and limbs of a fold. (See text for detail.)

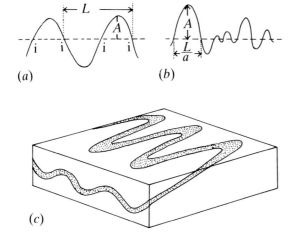

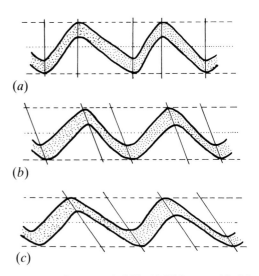

Fig. 10.6. (*a*) and (*b*). The definition of wavelength (*L*) and amplitude (*A*) of a fold. (*c*) Diagram showing the dependence of fold outcrop pattern on the orientation of the plane of erosion.

Fig. 10.8. Types of asymmetric folds: (*a*) With unequal limb lengths and with hinge surface normal to the median surface.(*b*) With equal limb lengths and with the hinge surface not normal to the median surface. (*c*) With unequal limb lengths and with the hinge surface not normal to the median surface.

fold is being viewed in the profile section. It can be seen from Fig. 10.6(*c*) that some of these parameters vary dramatically depending on which section through the fold is taken.

The terms defined so far relate to the geometry of a folded surface. Additional terms are required to describe the geometry of folded *layers* (defined by two adjacent surfaces). For example, a single folded layer will have more than one hinge line and crest line (Fig. 10.7). The planes containing the hinge and crest lines are the hinge and crest surfaces or planes respectively. The intersection of the hinge and crestal surfaces of a fold with the Earth's surface are known as the hinge (or axial) and crestal traces respectively. Although the hinge and axial planes of a fold are parallel, the *hinge plane* is a specific plane related to a specific location (i.e. the hinge) in the fold, whereas the *axial plane* has no specific location.

Examples of a *symmetric fold* (Fold 1) and two *asymmetric folds* (2 and 3) are shown in Fig. 10.7. A symmetric fold is defined as having equal limb lengths and an axial plane which is normal to the median surface. An asymmetric fold is defined as a structure which has limbs of unequal length and/or whose hinge surface is not perpendicular to the median surface (Fig. 10.8). A fold is said to *close* in the direction in which the limbs converge. Folds closing in

opposite directions on a vertical section are shown in Fig. 10.9, those which close upward are called *antiforms* and those that close downwards are termed *synforms*. Sometimes, folded strata contain evidence of the way up of the beds (e.g. certain sedimentary structures such as cross or graded bedding). If it can be shown that the beds are the right-way-up, then the folds ((*c*) and (*d*) in Fig. 10.9) are called *anticlines* and *synclines* respectively. Folds are said to *face* in the direction of the stratigraphically younger rocks along their axial surface, thus fold 'c' of Fig. 10.9 faces upwards and fold 'e' downwards. Vergence (from the German 'vergenz') is the direction in which a structure or family of related structures face. It may be inferred from Fig. 10.9(*g*) and (*h*), which shows a plan view of the anticline and syncline eroded to the level XY shown in Fig. 10.9(*c*) and (*d*), that anticlines have older rocks in their core and that the core of a syncline is occupied by younger rocks. In rock units which have experienced complex deformation the way up of the rocks may be such that antiformal synclines and synformal anticlines may develop (Fig. 10.9(*e*) and (*f*)).

Terms used to describe the orientation of folds

In order to define a fold's orientation, it is necessary to know the orientation of its axial plane and its axis. The axial plane can be defined by a dip and strike value and the fold axis (or hinge), which is a linear feature lying parallel to the axial plane, can be defined either as a *plunge* or a *pitch* (Fig. 10.10(*a*)). The plunge is the angle between the horizontal and the lineation measured in the vertical plane. The lineation orientation can be defined by a plunge angle (θ) and a bearing (N45°E in Fig. 10.10(*a*)) and can be plotted on a stereographic projection (Fig. 10.10(*b*)). The pitch of

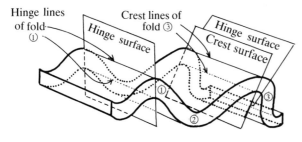

Fig. 10.7. Hinge surfaces and crest surfaces of a folded single layer. Fold 1 is a symmetric fold; Folds 2 and 3 are asymmetric.

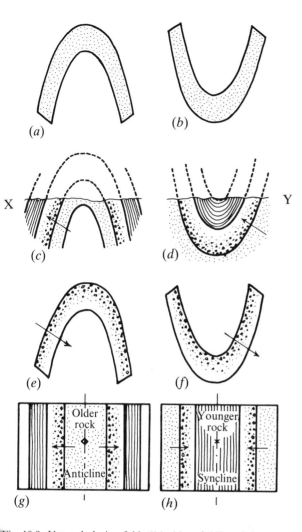

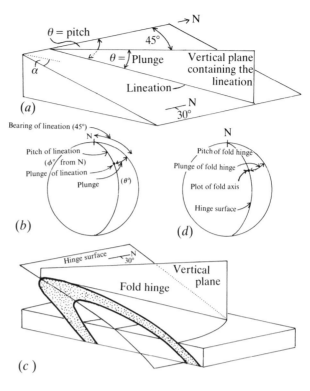

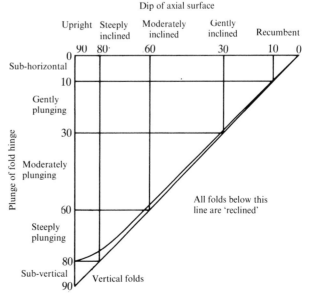

Fig. 10.10. (a) Definition of *plunge* and *pitch* of a lineation. (b) Stereographic representation of the lineation. (c) Hinge surface and hinge line of a fold and (d) their stereographic representation.

Fig. 10.9. Upward closing folds ((a), (c) and (e)) and downward closing folds ((b), (d) and (f)). Folds (a) and (b) are termed antiform and synform respectively. (g) and (h) show plan views of an eroded anticline and syncline respectively. Both folds have vertical axial planes and horizontal hinges. Arrows show direction of younging. See text for details.

a lineation is the acute angle between the lineation and the horizontal, measured in the plane containing the lineation. The lineation orientation can also be defined by the angle of pitch (ϕ) from the north in the example of Fig. 10.10(a) and the orientation of the plane containing the lineation. These data can be plotted stereographically as shown in Fig. 10.10(b).

The relevance of this discussion to the measurement of the orientation of a fold is shown in Fig. 10.10(c) and (d). The fold hinge surface is plotted onto the stereographic projection together with the hinge direction measured as either pitch or plunge. Fleuty (1964) has defined fold attitudes according to the dip of the axial plane and plunge of the fold hinge. The terms he uses are shown in Fig. 10.11.

The classification of folds

Because of the common occurrence of joints approximately at right angles to fold hinges, profile sections of

folds are often to be seen in the field. Various classifications of folds have been proposed, the majority being based on the geometry of the fold in profile section. One of the earliest attempts to classify folds was made by Van Hise (1894), who observed that many geological folds approximate closely either to parallel folds (i.e. folds whose orthogonal thickness remains constant) or similar folds (i.e. folds in which

Fig. 10.11. Classification of fold attitude. (After Fleuty, 1964.)

Introduction to folding   245

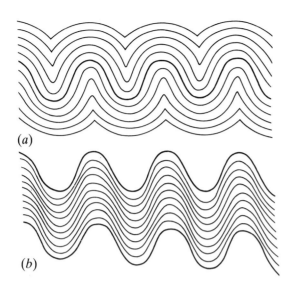

(a)

(b)

Fig. 10.12. (a) Parallel and (b) Similar folds. (After Van Hise, 1894.)

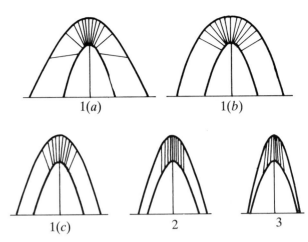

1(a)   1(b)

1(c)   2   3

Fig. 10.14. Classification of fold profiles using dip isogon patterns. 1(a) Strongly convergent, 1(b) parallel, 1(c) weakly convergent, 2 similar, 3 divergent. (After Ramsay, 1967, p. 365.)

the boundaries of the folded layer have identical geometry (Fig. 10.12)). However, it soon became apparent that these two geometries were only two members of a complete spectrum of geometries displayed by natural folds.

An extremely convenient and objective method of classifying fold geometry is the method of dip isogons (lines on the profile section joining points on a fold of equal dip value), proposed by Elliot (1965). To construct dip isogons for a fold profile, a series of tangents are drawn to each folded surface, e.g. surfaces 1 to 3 in Fig. 10.13. Two sets of tangents, one horizontal and the other dipping at θ are shown. The dip isogons are formed by joining points of equal dip on adjacent surfaces. Line abc is the 0° dip isogon and def is the $\theta°$ dip isogon. Folds can be conveniently divided into five classes on the basis of their dip isogon patterns (Fig. 10.14). Class 1(a) (b) and (c) all have dip isogon patterns that converge as they are traced from the outer to the inner arc of the fold. Class 1(a) folds have strongly convergent dip isogons. Class 1(b) folds are the parallel folds of Van Hise (Fig. 10.12(a)) and the dip isogons are perpendicular to boundary surfaces of the folded layer. Class 1(c) folds have weakly convergent dip isogons. Class 2 folds are the similar folds of Van Hise (Fig. 10.12(b)) and the dip isogons

are parallel to each other and to the trace of the axial plane on the profile section. Class 3 folds have dip isogons that diverge when traced from the outer arc to the inner arc of the fold.

This method of fold classification has the advantage that if required, each layer of a multilayer fold can be classified separately. Dip isogon patterns for some natural folds are shown in Fig. 10.15. Although fold geometries can usually be adequately described using the dip isogon classification in conjunction with terms that describe the attitude and inter-limb angle, there are certain fold profile geometries that occur so frequently in nature that they are given specific names. These include similar folds, parallel folds, kink folds (or kink-bands), box folds and chevron folds, the last three of which are shown in Fig. 10.16. Unlike similar folds and parallel folds, the names used to describe the folds of Fig. 10.16 are not restricted to a particular dip isogon pattern. For example, the dip isogon pattern of chevron folds may fall into any of the five classes shown in Fig. 10.14.

Processes of folding other than buckling

A variety of processes other than buckling have been proposed to account for the formation of natural folds. Two of the most important are *passive folding* and *bending*.

Passive folding

Consider the homogeneous deformation of a viscous body of unit dimensions, subject to a uniform and constant compressive stress P (Fig. 10.17(a)). After time t the dimensions of the deformed block are:

$$l_x = e^{-\dot{e}t} \tag{10.1}$$

$$l_y = e^{+\dot{e}t} \tag{10.2}$$

where $\dot{e}$ is the strain-rate caused by P and is given by

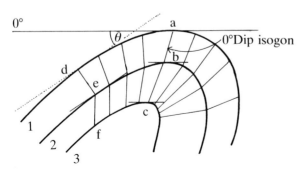

0°

a

θ

0°Dip isogon

b

d

e

c

1

2

f

3

Fig. 10.13. The construction of dip isogons, see text for details.

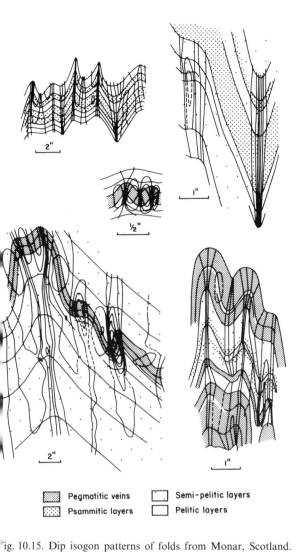

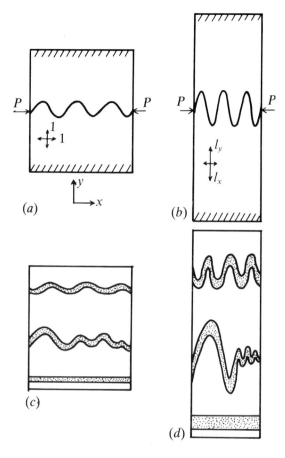

Fig. 10.17. (*a*) and (*b*) show the kinematic (passive) amplification of a passive sinusoidal marker line by pure shear. (After Biot 1965.) (*c*) and (*d*) show the effect of kinematic amplification on three passive marker bands.

▨ Pegmatitic veins	▨ Semi-pelitic layers
▨ Psammitic layers	☐ Pelitic layers

Fig. 10.15. Dip isogon patterns of folds from Monar, Scotland. (After Huddleston, 1973.)

$$\dot{e} = P/4\eta \qquad (10.3)$$

where η is the viscosity of the material.

A sinusoidal line drawn on a block before it is compressed (Fig. 10.17(*a*)) will become deformed, as shown in Fig. 10.17(*b*). The rate of amplification of the folds in this passive marker line increases exponentially with time. Biot (1965) points out that this is purely a *kinematic* effect and is not a true instability in the mechanical sense. Such amplification

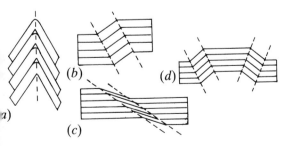

Fig. 10.16. Profile geometry of (*a*) a chevron fold; (*b*) a reverse kink-band; (*c*) a normal kink-band and (*d*) a box fold. The dotted lines are the traces of the hinge surfaces on the profile section.

of a fold is termed kinematic (or *passive*) to distinguish it from the *dynamic* (or *active*) amplification of a fold which occurs in a buckling layer.

The effect of kinematic amplification on three passive marker bands is shown in Fig. 10.17(*d*). If the original band is parallel-sided, the resulting folds will be similar. Passive amplification of geological fold will tend to dominate folding in layered rocks when the contrast in material properties of the different layers is small. In general, the difference in mechanical properties between different rocks becomes less as the metamorphic grade increases. The process of folding by passive amplification will, therefore, usually be more important in high-grade rather than low-grade metamorphic environments.

Bending

Bending is the term used to describe the flexuring of a layer induced by a compression acting at high angle to the layering (Fig. 10.18(*a*)). Geological flexures that are the result of bending are known as *drape folds* and are frequently formed when sediments which cover a more rigid basement, flex in response to components of vertical movement along basement faults (Fig. 10.19(*a*) (*b*) and (*c*)).

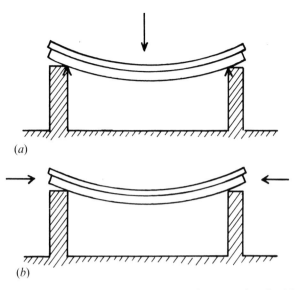

(a)

(b)

Fig. 10.18. The orientation of the principal compression for (a) bending and (b) buckling of the layers.

Examples from the Swiss Alps of drape folds formed in mesozoic cover rocks as the result of basement faulting are shown in Fig. 10.19(d). Such structures are also well developed in the Basin and Range province of west U.S.A. The folding of layered cover rocks as the result of dip-slip movement has been termed *forced folding* by Stearns (1978), who argues that the overall trend and shape of such folds will be dominated by the shape of some forcing member(s) from below.

Despite the relatively early use of model experi-

ments in geology, this approach was not used to study the effect of dip-slip movement of basement faults or cover rocks until 1959 when Sanford addressed this problem using both theoretical analysis and scale modelling. In his experiments the rigid basement was simulated by wooden blocks and the cover rocks by sand. The results of three of his experiments, which involved dip-slip movement on a vertical basement fault, are shown in Fig. 8.30 and 8.31.

The detailed geometry and character of the cover structures (faults and forced folds), are dependent on the mechanism by which the cover accommodates itself to the basement faulting. This in turn depends upon the rheology and physical environment of both basement and cover rocks during deformation and upon the geometry and sense of movement of the basement fault. The effect of normal and reverse movement on a basement fault on a layered cover has been studied experimentally by Ameen (1988). During reverse faulting, horizontal shortening occurred along the cover in the direction normal to the strike of the basement fault and bedding plane slip developed in the vicinity of the fault and died off away from it. Although the strain distribution in the cover was complex, two of the axes of principal strain were always parallel to the layering with one parallel to the strike of the basement fault. The distribution of strain in the forced fold is shown in Fig. 10.20(a).

Normal faulting in the basement caused horizontal extension of the cover in the direction normal to the strike of the basement fault and generated

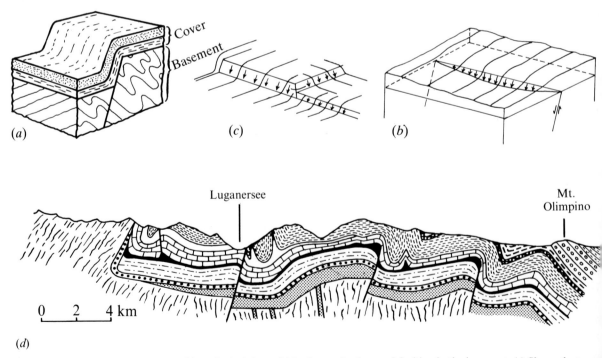

(a)

(c)

(b)

(d)

Fig. 10.19. (a) and (b) Block diagrams of hypothetical drape-folds, the result of normal faulting in the basement. (c) Shows the type of drape-fold geometry which may be associated with block faulting in the basement.(d) Drape-folds, probably over reverse faults in the basement. Swiss Alps. (After Heim, 1922.)

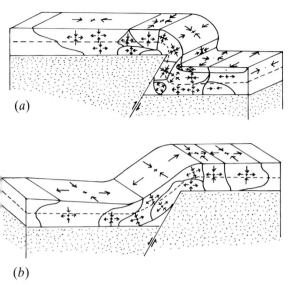

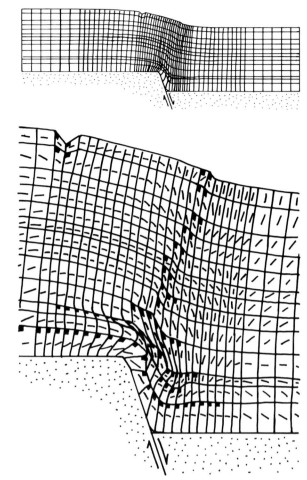

Fig. 10.20. Experiments showing the effects of (*a*) reverse and (*b*) normal faulting in a rigid basement on a layered, ductile cover. (After Ameen 1988.)

relatively simple pattern of strain compared to reverse faulting, Fig. 10.20(*b*). Again, strain in the cover was concentrated in the vicinity of the basement fault. However, no significant bedding plane slip occurred. The cover accommodated itself to movement on the basement fault by thinning and thickening and the general strain pattern was dominated by bedding elongation perpendicular to the strike of the basement fault.

The effect of basement faulting on cover rocks has also been studied using the technique of finite element analysis (a technique discussed briefly in Chap. 11), by Reddy *et al.*, (1982). Their model consisted of a rigid 'basement' and a ductile cover. It was assumed that for stresses up to a particular yield point the material had one viscosity, and for stresses greater than the yield stress the viscosity decreased by a factor of 100 prior to further deformation. Using this model the response of the body to local yielding or fracturing can be modelled. The result of one such 'experiment' in which normal faulting occurred in the basement, is shown in Fig. 10.21. It can be seen that a high angle reverse fault developed in the cover directly above the basement fault together with a near-surface region of normal faulting.

Gentle drape folds of considerable lateral extent may develop as the result of *differential compaction*; for example, where a relatively rigid block of limestone grades laterally into mudstone which originally possessed a high porosity (i.e. 30 per cent or more, of void spaces). As compaction takes place, the void spaces in the mudstone close and the vertical extent of the mudstone decreases. Later sediments are flexured as indicated in Fig. 10.22(*a*).

Nevin (1931) points out that folds which form by

Fig. 10.21. The results of a finite element analysis investigating the effect of faulting of a rigid basement on a viscous cover. Elements with black flags in the upper left hand corner show regions of yielding. The short lines in each element show the directions of least compressive stress. (After Reddy *et al.*, 1982.)

the draping of sediments over the basement fault scarps or by differential compaction, are likely to show a 'thinning of formation upwards, above the crest of the fold'. Such folds (Fig. 10.22(*b*)) have been termed *supratenuous* folds and have the geometry of class 1(*a*) structures (Fig. 10.14).

Models of strain distribution in a buckled layer

Experimental work on folding by Kuenen & de Sitter (1938) brought some interesting features of buckling to the notice of geologists. The effect of layer parallel compression on a slab of unstratified clay, a paraffin wax layer, a pack of paper sheets and a rubber plate are shown in Fig. 10.23(*a*) to (*d*). Layer-normal marker lines were imprinted upon the layers before deformation and this enabled the state of strain that developed in the fold profiles to be determined. It is interesting to note that the profile geometry of the folds which developed in the isotropic rubber plate and the anisotropic pack of paper sheets is the same,

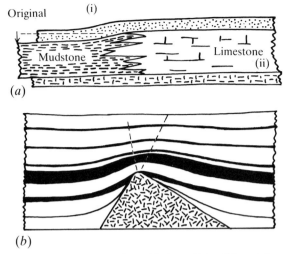

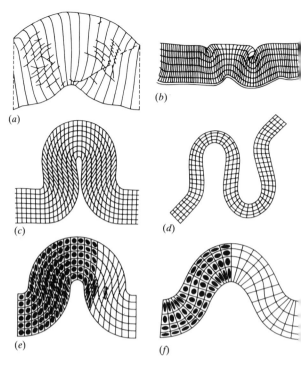

(a)

Fig. 10.22. (a) Drape-fold formed in layer (i) by differential compaction of layer (ii). (b) Supratenuous fold produced in an experiment with sand and clay deposited in successive layers over a prominence. (After Nevin, 1931.)

Fig. 10.23. The effect of layer-parallel compression on (a) a slab of unstratified clay, (b) a paraffin wax layer, (c) a pack of paper sheets and (d) a rubber plate. (After Kuenen & De Sitter, 1938.) Internal strain in a layer deformed, (e) by flexural flow and (f) by tangential longitudinal strain. (After Ramsay, 1967.)

i.e. that of a parallel fold. However, the state of strain within the folds, which very much depends upon the material properties of the layer, is very different (cf. Fig. 10.23(c) and (d)). It is immediately apparent that the strain state within a buckled layer cannot be deduced from its profile geometry. The two strain states are shown more clearly in Fig. 10.23(e) and (f), and folds with these strain states have been termed *flexural flow folds* and *tangential longitudinal strain folds* respectively.

Flexural flow folds

Such structures develop in layers with a high mechanical anisotropy, commonly formed by a layer-parallel mineral fabric or banding. Such structures can be produced by compressing a pack of paper sheets parallel to the layering (Fig. 10.23(c)). As the pack buckles, the individual sheets slide over each other. The maximum amount of slip occurs on the limbs at the inflection points and decreases towards the hinge, where slip is zero. The strain ellipses in Fig. 10.23(e) show clearly that the maximum strain occurs at the inflection points and that no strain occurs at the hinge. A feature of this strain distribution is that one of the lines of no finite longitudinal strain is always parallel to the layering. It follows, therefore, that when the structure develops as a flexural flow fold, there will be no strain on the bedding planes.

If the slip planes are very close together, the layer-parallel shear will be uniformly distributed across the folded layer. Such a structure would be a true *flexural flow fold*. If, however, the slip planes are more widely spaced, as would be the case if the layering was a series of sandstone or limestone units, the layer-parallel shearing would not be uniformly distributed throughout the sequence of layers, but would be concentrated along the bedding planes. Such

structures are referred to as *flexural slip folds* (Fig. 10.24(a)).

Secondary features associated with flexural slip folding

One of the most common features associated with flexural slip folding is the lineations which are frequently and usually incorrectly referred to as slickensides, which forms on the surface of the layers. As has already been mentioned in Chap. 5 and will be discussed in greater detail in Chap. 11, these features

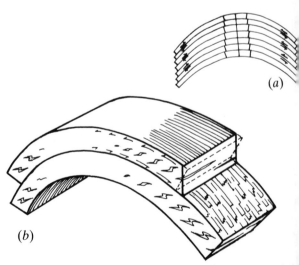

(b)

Fig. 10.24. (a) Flexural slip fold. (b) Crystal fibres and en echelon tension gashes formed by layer-parallel shearing during flexural slip folding.

may be grooves and ridges gouged out as the beds slide past each other. More commonly, however, they are comprised of crystal fibres, often of quartz and/or calcite or pressure solution 'casts' of such fibres. Both slickensides and crystal fibres from parallel to the direction of slip, which is frequently sub-perpendicular to the fold hinge, and generally die out towards the hinge, where the slip between beds is zero (Fig. 10.24(*b*)).

We have noted that there should be no strain on the bedding planes of a flexural slip fold. It is therefore encouraging to find that, in some natural folds, where crystal fibres indicate that a considerable amount of slip occurred between beds during folding, it can be demonstrated, by the occurrence of undeformed fossils which are exposed on the bedding planes, that little or no strain has occurred within the plane of the bedding (Fig. 10.25).

Another structure occasionally found in association with flexural slip folds is an array of en echelon tension gashes, the formation of which is discussed briefly in Chap. 1, 6 and 18 where it is shown that individual gashes tend to form at 45° to the direction of the shear couple that generated them and that with

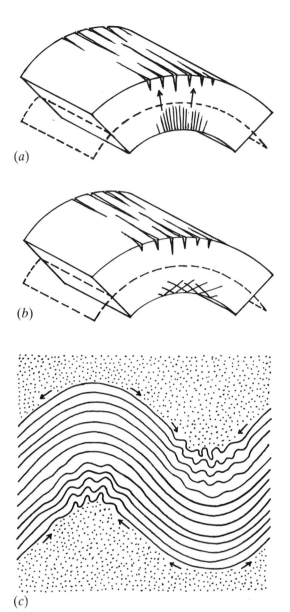

(a)

(b)

(c)

Fig. 10.26. (a) and (b) Secondary features associated with tangential longitudinal strain folding. The dashed plane is the neutral surface. (c) Folded multilayer, showing evidence of compression in the inner arc. (After Ramberg, 1964.)

Fig. 10.25. Undeformed fossils on a bedding plane on the limb of a fold. The crystal fibres indicate that considerable slip occurred on the bedding plane during folding and that folding was probably by flexural slip. (Luarca, N. Spain.)

progressive shear deformation the gashes may become sigmoidal. Because of the decrease in layer-parallel shear from the limbs to the hinge, the tension gashes, should they develop in a fold, tend to become less sigmoidal towards the hinge (Fig. 10.24(*b*)).

Tangential longitudinal strain folds

The strain pattern associated with a tangential longitudinal strain fold (Fig. 10.23(*f*)) develops when a homogeneous, isotropic layer is buckled. In contrast to flexural flow folds in which the maximum strains occur at the inflection point and zero strain at the hinges, in tangential longitudinal strain folds the maximum strain occurs at the hinges and no strain at the inflection point. As its name implies, in this type of folding one of the principal strains is always parallel (tangential) to the layer boundary. The outer arc of the hinge area experiences an extension and the maximum principal extension (major axis of the strain ellipse) is parallel to the layer boundary. Conversely, the inner arc is compressed and the minimum principal extension (minor axis of the strain ellipse) is parallel to the layer boundary. These two states of strain are separated by a line (surface in 3D) along which there is no strain. This surface, which is called the neutral surface (Fig. 10.26), separates an area of layer-parallel compression from one of layer-parallel extension.

Secondary features associated with tangential longitudinal strain folds

The layer-parallel tension generated in the outer arc of a tangential longitudinal strain fold commonly produces *tension gashes* (Fig. 10.26(a) and (b)) which are often infilled with minerals such as quartz or calcite. In the inner arc of the fold, the large layer-parallel compression may give rise to a locally developed *pressure solution cleavage*. The more mobile mineral constituents of the rocks, one of which is commonly quartz, go into solution and migrate across the layer to the low pressure areas in the outer arc where they provide the infilling for the tension gashes. If the response of the rock in the inner arc of the fold to the layer-parallel compression is brittle, conjugate *thrusts* may develop (Fig. 10.26(b)).

One might expect that the two models of strain distribution within a fold discussed in this chapter (tangential longitudinal strain and flexural slip) would develop in perfectly isotropic, and highly anisotropic, layers respectively. However, field observations and experimental work indicate that this is not always the case. For example, although the buckle shown in Fig. 10.23(a) was produced experimentally in a slab of originally unstratified, isotropic clay, it can be seen that folding involved slip along planes parallel to the slab boundary. These planes formed at a very early stage in the folding.

In the Horseshoe Pass quarries north of Llangollen, Wales, field observation of folds in finely laminated shales which seem ideally suited for the formation of flexural flow folds, show that layer-parallel shear is concentrated on discrete, approximately regularly spaced planes and is not uniformly distributed across the layer. In addition, when it is possible to determine the strain distribution in the rock slices between adjacent layer-parallel slip planes, it is often found to show features of both tangential longitudinal strain and flexural slip.

Anticlastic bending

Let us now consider the effect of tangential longitudinal strain folding on the fold geometry in the direction parallel to the fold hinge. The extension parallel to the layer in the outer arc tends to cause a shortening along the fold hinge, and the compression parallel to the layer boundary in the inner arc tends to cause an extension along the fold hinge. The combined effect is to cause a *cross curvature R'* (Fig. 10.27(b)). This effect is called anticlastic bending. The magnitude of the strain at the inner and outer arcs of the fold in the profile section (zx plane of Fig. 10.27) can be determined from the geometry of Fig. 10.27(a). The original length of the inner and outer arc is L ($L = R\theta$) and the change in length $dL(dL = a\theta/2)$. The fibre strains in the x direction at the inner and outer arcs are $+ e_x$ and $- e_x$ respectively, where:

$$e_x = \frac{dL}{L} = \frac{a}{2R}. \qquad (10.4)$$

Because of the effect of Poisson's number (m) (see Chap. 1), these strains in the x direction induce strain (e_y) in the y direction:

$$e_y = e_x/m. \qquad (10.5)$$

Substituting for e_x in Eq. (10.5) from (10.4) we obtain:

$$e_y = a/2mR. \qquad (10.6)$$

From Fig. 10.27(b) it can be seen that the strains in the y direction give rise to a cross curvature of radius R' and that:

$$e_y = a/2R'. \qquad (10.7)$$

Equating Eqs. (10.6) and (10.7) it follows that:

$$R' = mR. \qquad (10.8)$$

That is, the ratio of the radii of curvature is m which, for competent rocks, commonly has a value of 3–5. (The reader can readily demonstrate the validity of anticlastic bending by buckling a rectangular prismatic eraser between his or her fingers). It should, however, be pointed out that the anticlastic bending analysis only holds when the upper and lower surfaces of the layer are free and the sheet is not too extensive. Rock layers in the Earth's crust are confined between other layers which will inhibit anticlastic bending. Nevertheless, the stresses which give rise to the anticlastic bending will still build up parallel to the fold axes. The possible effects of these stresses will be briefly discussed in Chap. 14, where it is argued that they may cause normal faults striking at right angles to the fold hinge.

Multiple folding and interference patterns

So far, we have considered the geometry of folds in rocks that contain only one set of folds which result from one *phase of folding*. The geometries of structures which result from the superposition of two or

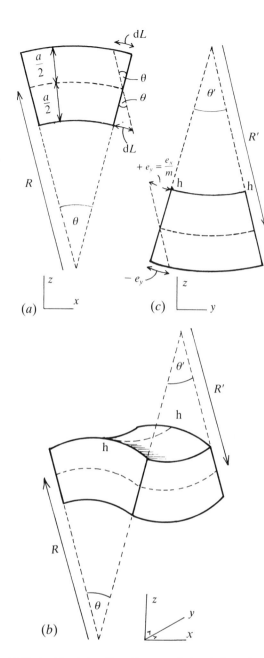

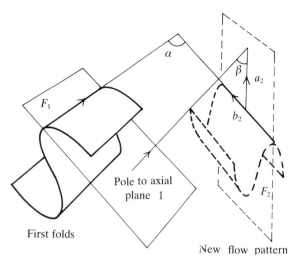

Fig. 10.28. The relationship between two interfering folds. As the angles α and β vary, so different types of interference structures result. (After Ramsay, 1967.)

Fig. 10.27. Anticlastic bending. Curvature induced along the fold hinge as a result of buckling. See text for details. (a) Fold profile, (b) Oblique view of (a) showing curvature induced along fold hinge, (c) Section parallel to the axial plane containing the hinge (hh).

more sets of folds caused by more than one phase of folding are more complex. The unravelling of these geometries was one of the major tasks performed by structural geologists in the two decades following the Second World War. This greater complexity of geometry is reflected in the patterns (known as *interference patterns*) which they form on outcrop surfaces. The geometry of the patterns depends upon the relative orientation of the two sets of interfering folds. This orientation can conveniently be defined using the angle (α) between the two hinges and the angle (β) between the pole of the axial plane of the first folds (F_1) and the normal (a) to the hinge of the second folds

measured in the axial plane of F_2 (Fig. 10.28). Ramsay (1967) points out that there are three main types of interference patterns, Types 1, 2 and 3, with complete gradation between them.

Type 1 Interference
The two sets of folds shown in Fig. 10.29(a) are oriented such that their axial planes (and their axes) are at right angles to each other; i.e. $\alpha = 90°$ and $\beta = 90°$. The superposition of these folds results in a dome and basin structure (Fig. 10.29(b)).

Type 2 Interference
In Fig. 10.30(a), the orientation of the two sets of folds is such that $\alpha = 90°$ and $\beta = 0°$. The superposition of these folds results in the structures shown in Fig. 10.30(b).

Type 3 Interference
The two sets of folds that produce Type 3 inteference are shown in Fig. 10.31(a). They are coaxial, i.e. $\alpha = 0°$ and $\beta = 0°$. The structure that results from this superposition is shown in Fig. 10.31(b).

Geologists normally study these forms of interference on near-planar out-crops which are planes of section through the structures. For interference types 1 and 2, the corresponding interference pattern may have 'dome and basin' and 'mushroom' or crescent geometries respectively (Fig. 12.29(b) and 12.30(b)). A range of interference patterns and their dependence on the values of α and β are shown in Fig. 10.32.

Thiessen & Means (1980) modified and extended Ramsay's classification of interference patterns, using 'angular parameters' which are in part different from those shown in Fig. 10.28. They use the angle between the axis of the first folds and the pole to the axial plane of the second folds (γ) as one parameter and β

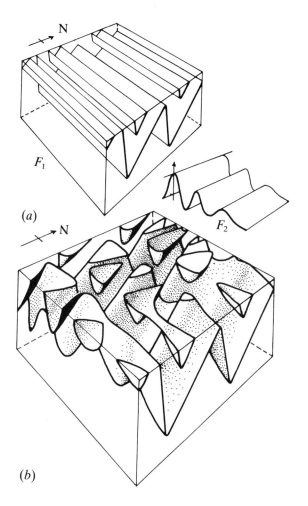

(a)

(b)

Fig. 10.29. (a) The relative orientation of the first (F_1) and second (F_2) folds for Type 1 interference. Erosion of the resulting structure gives rise to a Type 1 interference pattern, (b). (After Ramsay, 1967.)

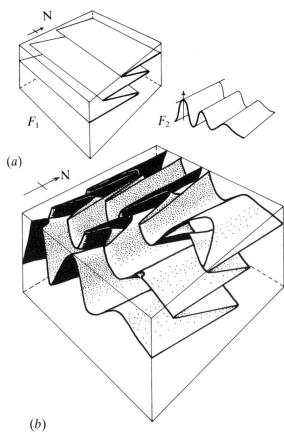

(a)

(b)

Fig. 10.30. (a) The relative orientation of the first (F_1) and second (F_2) folds for Type 2 interference. Erosion of the resulting structure gives rise to a Type 2 interference pattern, (b). (After Ramsay, 1967.)

(Fig. 10.28) as the other and suggest that this provides a more natural basis for classification of three-dimensional patterns, which avoids several ambiguities in Ramsay's scheme. The relationship between α, β and γ is shown in Fig. 10.33. They are the angles between the two sets of axes used to describe the orientation of the two interfering folds. These angles can be plotted along the edges of a cube and any combination of F_1 and F_2 fold orientations will plot as a point on, or inside, the cube. Thiessen & Means demonstrate that not all combinations of α, β and γ are possible and that combinations are restricted to the *orientation volume*, the portion of the cube in Fig. 10.34(a), delineated by full lines. Interference patterns corresponding to various combinations of these angles (points A–P in the orientation volume) are depicted in the surrounding cubes. The first and second axial planes in their final configuration are shown with dotted and dashed lines respectively.

Thiessen & Means also show that a variety of two-dimensional patterns arise from different cuts through any three-dimensional interference geometry.

This is illustrated in Fig. 10.34(b), which shows various sections through a Type 2 three-dimensional structure.

Although the superposition of two separate deformation events may give rise to cross folding and consequently to interference patterns, the occurrence of interference patterns, especially Type 1, does not necessarily imply that two separate deformation phases have taken place. For example, if, as indicated in Chap. 1, Figs. 1.18(b) and 1.21(b), the strain state in the plane of the layer, during a single deformation event, is such that shortening occurs parallel to both the principal strain directions, then the formation of dome and basin structures can result. We have also seen that folding in two directions can form as the result of anticlastic bending (Fig. 10.27). As will become apparent later in this chapter, field evidence and model work indicates that folds, on all scales, often have a periclinal geometry (an elongate dome) particularly in the upper levels of the crust. The erosion of a group of periclines (e.g. Fig. 10.45(c)) may produce a Type 1 outcrop pattern as will the erosion of sheath folds which as we shall see (Fig. 10.60) occur in shear-zones and in association with thrusting.

Interference of folds formed during the same

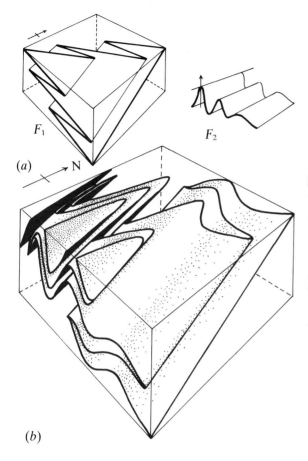

(a)

(b)

(c)

Fig. 10.31. (*a*) The relative orientation of the first (F_1) and second (F_2) folds for Type 3 interference. Erosion of the resulting structure gives rise to a Type 3 interference pattern, (*b*). (After Ramsay, 1967.) (*c*) Example of Type 3 interference pattern, Monar, Scotland.

α = angle between 1ST fold axis and b_2

β = angle between pole to 1ST fold axial surface and a_2

Fig. 10.32. Interference patterns and their dependence on the angles α and β (Fig. 10.28). The patterns are of three types, 1, 2 and 3 with various intermediate forms. (After Ramsay, 1967.)

one of the most illuminating methods of investigating the process of geological folding. The experiments, which often produce folds with geometries remarkably like natural folds, have provided considerable insight into the processes of fold initiation and amplification. Experiments on folding in which rock analogue materials are used can be divided into two groups; i.e. those in which no particular attention is paid to the properties of the model and the question of scale is not considered and those in which the material properties and scale of the model are carefully observed. Both types of experiments may produce folds with geometries encouragingly comparable to natural folds. However, the relevance of the data derived from the experiments to the process of geolo-

deformation may occur when fold trains in two different but adjacent localities propagate and run into each other. This type of interference is discussed in Chap. 15 (see Figs. 15.18 and 15.19).

Modelling of folds using rock analogue material

Experimental work on the buckling of layered models comprised of rock analogue materials has been carried out for over a century and has proved to be

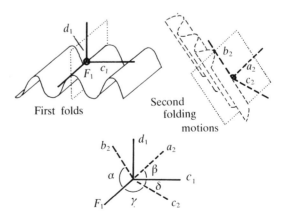

Fig. 10.33. Axes used to describe the relative orientations of the first folds and the second folding movements. Below are shown both sets of axes and the angles α, β, γ and δ. (After Thiessen & Means, 1980).

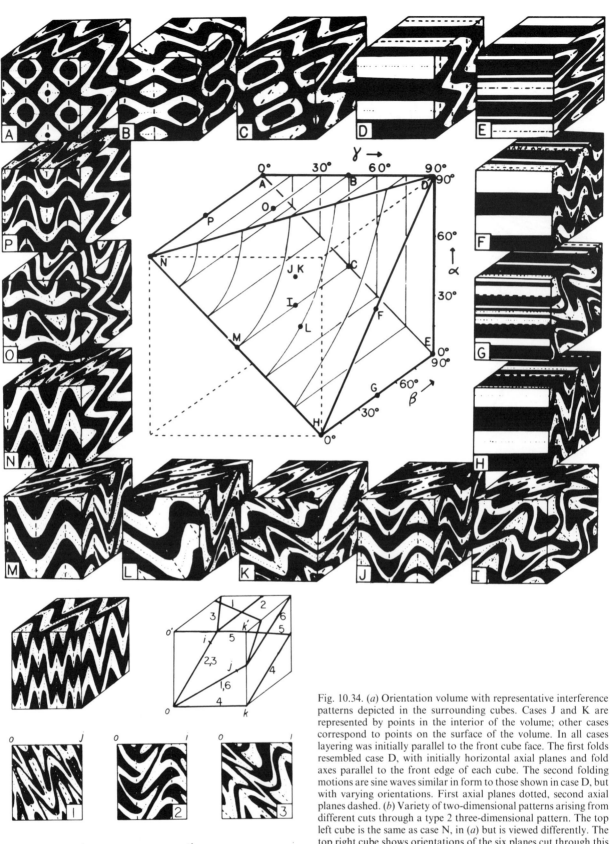

Fig. 10.34. (*a*) Orientation volume with representative interference patterns depicted in the surrounding cubes. Cases J and K are represented by points in the interior of the volume; other cases correspond to points on the surface of the volume. In all cases layering was initially parallel to the front cube face. The first folds resembled case D, with initially horizontal axial planes and fold axes parallel to the front edge of each cube. The second folding motions are sine waves similar in form to those shown in case D, but with varying orientations. First axial planes dotted, second axial planes dashed. (*b*) Variety of two-dimensional patterns arising from different cuts through a type 2 three-dimensional pattern. The top left cube is the same as case N, in (*a*) but is viewed differently. The top right cube shows orientations of the six planes cut through this case that are represented in the six squares below. It depicts the top, front and right side of a cube, as in the left hand cube. the first folds had vertical axial planes, horizontal hinge lines and profile section represented by the right hand face of the cube. The second folds have vertical axial planes parallel to the side of the cube and introduce folds in the first axial planes with a profile shown on the top face of the cube. (After Thiessen & Means, 1980.)

gical folding can only be determined if the model has been correctly *scaled*.

Scale modelling

In order to perform properly scaled experiments, an understanding of scale model theory is necessary (Hubbert, 1937). Although it is not possible to give a detailed account of scale modelling here, it is useful to discuss the concepts on which it is based and to comment on the advantages of scaled models over non-scaled models. Let us now consider the problem of conducting a scale model experiment to study the behaviour of a particular sandstone with a uniaxial strength of 1.0 kbar. If the rock density is 2.5 g/cm³, we know that the sandstone can support a column composed of that sandstone which is 4 km high, before failing at the base under its own weight. If, in a model, 1 cm of model length is to represent 1 km of the prototype being modelled then, if the model is to be correctly scaled in terms of *strength*, the model material should be able to support a column of itself 4 cm high, before failing at the base under its own weight (i.e. it would have the strength of butter at about 25–30 °C).

In scale model studies, it is important to consider the *geometric, kinematic* and *dynamic* similarities between the model and the prototype, so that sets of *model ratios* can be established between corresponding parameters (e.g. strength Z, length L, density ρ, stress S, viscosity η and time t). Two bodies are geometrically similar when all corresponding lengths are proportional and all corresponding angles are equal. If L_p is the length of the original rock and L_m the corresponding model length, then the model ratio for length L_R is given by:

$$L_R = \frac{L_m}{L_p}. \tag{10.9}$$

If two geometrically similar bodies undergo geometrically similar changes of shape or position, or both, then the two bodies are kinematically similar, if the time required for any change in one is proportional to that for the other. This establishes a time model ratio t_R where:

$$t_R = \frac{t_m}{t_p}. \tag{10.10}$$

Because the two bodies are kinematically similar, it follows that the velocities V and accelerations A of corresponding points are proportional, so that:

$$V_R = \frac{V_m}{V_p} = \frac{L_R}{t_R} \tag{10.11}$$

and

$$A_R = \frac{A_m}{A_p} = \frac{L_R}{t_R{}^2}. \tag{10.12}$$

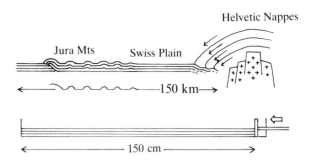

Fig. 10.35. (*a*) Schematic profile through the Helvetic Nappes, the Swiss Plain and the Jura Folds. (*b*) Model used to simulate the formation of the Jura Folds.

Unless the experiment is performed in a centrifuge, the gravitational acceleration (g) in both the model and the rock prototype is the same, i.e. $A_R = 1$. It follows from Eq. (10.12) that:

$$A_R = \frac{L_R}{t_R{}^2} = 1. \tag{10.13}$$

Let us now consider some of the problems that may be encountered when attempting to conduct scale model experiments. Consider a particular experiment in which the formation of the folds of the Jura Mountains is to be simulated using a gelatine model. The original length of the profile section through the Jura fold belt and the Swiss Plain was approximately 150 km, (Fig. 10.35(*a*)). Let the length of the model be 150 cm, (Fig. 10.35(*b*)). The length ratio (L_R) is therefore 10^{-5}. If we take an estimated but reasonable time for the formation of the folds to be 500 000 years, then in order that Eq. (10.13) may be satisfied and the model correctly scaled, the duration of the experiment must be approximately 1500 years. Alternatively, if the model deformation is to take place in the more realistic time of 5 hours, then the time ratio (t_R) would be 10^{-9} and it follows from Eq. (10.13) that, for correct scaling, the model length would have to be 10^{-11} cm! Clearly this means that accurate scale modelling of this example cannot be achieved in a situation where the gravitational acceleration in both the real situation and the model is the same. This problem can be obviated by the use of a centrifuge, for an acceleration ratio (A_R) can be selected that would allow correct scale modelling of the experiment to be achieved. Centrifuges are expensive and the forces involved are such that the total size of the models must be small (i.e. less than 10 cm). Fortunately there is an alternative solution available. It has been argued (Hubbert, 1937) that if it is possible to neglect accelerations and the resultant inertial forces (i.e. if the model deforms sufficiently slowly), we need not satisfy Eq. (10.13). Instead it is merely necessary to choose a time ratio such that it is consistent with the ratios of the viscosities of real and model materials.

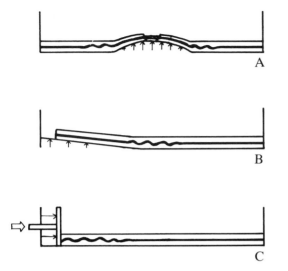

Fig. 10.36. Three types of apparatus used in experiments on folding.

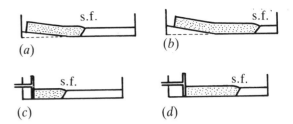

Fig. 10.37. A diagrammatic representation of the propagation of the stress front (s.f.). Examples of the progression of the stress front, (*a*) and (*b*), occurred in the gravity-glide apparatus B of Fig. 10.36. Examples (*c*) and (*d*) occurred in the lateral compression apparatus C of Fig. 10.36.

Model experiments

A considerable amount of model work using rock analogue materials has been performed to investigate the formation of folds in layered materials compressed parallel to the layering. Some of these experiments have confirmed predictions made from theoretical analyses and others have brought to light mechanisms of folding which are completely different from those predicted from theoretical studies. For example, it is implicit in many theories of folding that, prior to buckling, the system is pre-stressed. It is assumed that increments of stress are added to the pre-stressed condition until the system becomes unstable. The process of pre-stressing the material is considered not to play any part in fold initiation. It is also implicitly assumed that once the system becomes unstable, folding will occur pervasively throughout the model. Some experiments by Blay *et al.* (1977) indicate that, although a pervasive wave train of folds with the same wavelength and amplitude does sometimes develop in a pre-stressed multilayer, quite commonly folds develop in a very different manner. The experiments indicate that the process of pre-stressing often plays an important role in fold initiation and that non-synchronous fold initiation and amplification is the rule rather than the exception. Two types of non-synchronous fold development are described. In the first type, adjacent folds initiate and amplify sequentially (*serial folding*) and in the second type, folds amplify randomly at different times and at different points in the multilayer (*stochastic folding*).

The three types of apparatus used in the experiments are illustrated in Fig. 10.36. Each type of apparatus allows a particular set of boundary displacements to be applied to a horizontally layered model. In apparatus A and B a section of the base of the multilayer was slowly raised to simulate basement uplift of a sedimentary pile. Structures developed as a

result of down-slope gravity gliding of the uplifted section of the model on a thin, lubricated base layer. In these experiments, the development of thrusts and folds was generally confined to the flat-lying section of the model. In order to study the development of structures in this region more closely, a third type of apparatus (C in Fig. 10.36) was used, in which lateral compression produced by gravity gliding was simulated by the action of a vertical piston face driven horizontally against the end of a multilayer by a geared electric motor.

The multilayers were composed of identical layers of gelatine separated from each other and the walls of the apparatus by thin layers of the lubricant, liquid paraffin. Gelatin, at the concentrations used in these experiments was a weak, viscoelastic material which was, however, capable of sustaining large elastic strains over the six to eight hours that an experiment lasted. Consequently, the development of structures in the experiments incorporated elastic strains and strain recovery which are an order of magnitude larger than those present in naturally developing, large-scale structures in sedimentary rocks. Despite this limitation, the experiments provide considerable insight into a number of aspects of the development and geometry of natural, large-scale structures. In fact, the importance of the 'wave front' in folding (discussed below), which is so clearly seen in the experiments, might not have been apparent had the elastic properties of the model been correctly scaled.

Before describing the results of individual experiments, some remarks should be made about the initial development of compressive strain along the length of the models.

Initial lateral compression, whether produced by gravity gliding or by piston movement, did not cause an immediate lateral loading of the whole length of the multilayer. During the early stages of an experiment, the model could be divided into a section which had become stressed by the source of lateral compression (stippled region in Fig. 10.37) separated by a stress front (s.f.) from the remainder of the model. This stress front gradually propagated down the

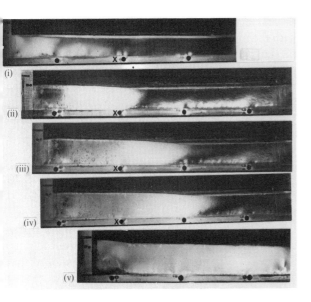

(i)

(ii)

(iii)

(iv)

(v)

Fig. 10.38. Photo-elastic experiment designed to study the effect of base lubrication on the development of stress in an unlaminated gelatine model. The model is separated from the base of the apparatus by liquid paraffin to the left of point X and unlubricated to the right of X.

length of the model until it reached the fixed end wall of the apparatus at which point the whole length of the multilayer had become stressed.

The existence of the stress front is related to the shear stress along the base of the models and the rate of propagation of the front is dependent on the magnitude of the stress and the viscosity of the lubricant at the base of the model. This is illustrated by two photo-elastic experiments, (Fig. 7.37 and 10.38). In these experiments, advantage was taken of the photo-elastic properties of gelatine to study, qualitatively, the build-up of stress caused by lateral compression of an unlaminated block of material. The part of the model which has become stressed by the piston is seen as a zone of relatively high transmitted light intensity, which is related to the isochromatic fringe order produced in the stressed gelatine. The experiments clearly show the division of the models into a relatively highly stressed region adjacent to the piston and a region of lower differential stress. The slight flexure (break in slope) of the surface of the model – the geometric expression of the stress front – coincides approximately with the boundary between the regions of high and low differential stress and is known as the 'wavefront'.

In the experiment shown in Fig. 7.37 the base of the gelatine was lubricated with a film of Vaseline grease, which was the base lubricant used in the majority of the experiments described in this section. The stress front took two hours to propagate through this model of length 60 cm. In the experiment shown in Fig. 10.38 the base of the model was separated into two zones with differential frictional properties. To the left of point X (Fig. 10.38) the base was lubricated

with liquid paraffin. To the right of this point, the gelatine was allowed to set in direct contact with the Perspex base of the apparatus with resulting cohesion between the gelatine and the Perspex. The velocity of the stress front through the zone of paraffin lubrication was seven times faster than through the unlubricated zone, This reflects the low shear stress affecting the base of the model in this well-lubricated zone. After fast propagation of the stress front through this region, the front became arrested at point X. The horizontal differential stress of the model to the left of this point then increased until this stress could overcome the cohesive resistance along the remainder of the base. Once this began to occur, the stress front continued to propagate (although at a much slower rate) and eventually reached the fixed end of the Perspex tank. Propagation in this zone was achieved by a series of stick-slip movements at the base of the model and local sites of high stress concentration and high differential stress gradients were seen to develop at the base. It will be shown that these regions of concentrated stress probably played a significant role in initiating folds in some experiments.

In order to obtain a more quantitative understanding of the stress front, and to estimate the rate at which lateral compressive stress might propagate through a rock pile resting on a decollement zone, a mathematical model described by Bott & Dean (1973) has been adapted. This has already been discussed in Chap. 7, (Fig. 7.36) where it was shown that the time required for the whole length of an elastic layer resting on a viscous substratum to be stressed within 90 per cent of the constant applied end-stress is given by,

$$t = \frac{(\eta/E)L^2}{h_1 h_2} \tag{10.14}$$

(see Eq. (7.16)) where η and h_2 are the viscosity and thickness of the viscous layer and E, h_1 and L the Young's modulus, thickness and length of the overlying elastic layer.

The velocity (v) with which the wavefront moves along the layer is dx/dt and can be obtained by differentiating Eq. (10.14).

$$v = \frac{dx}{dt} = \frac{(Eh_1 h_2)}{2\eta L} \tag{10.15}$$

A series of curves each of which gives the value of horizontal stress within the elastic layer (expressed as a function of the constant applied end-stress P_0) as a function of position along the layer (x) at a particular value of the time parameter, t, is shown in Fig. 10.39(c).

The experiment shown in Fig. 10.40 illustrates the role that the stress front played in certain experiments, in the initiation of folds. The experiment is an example of the development of structures on either side of a zone of progressive uplift. Tensile failure of

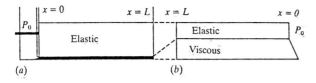

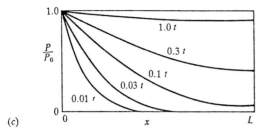

(c)

Fig. 10.39. (*a*) Sketch of an elastic gelatine layer resting on a thin layer of lubricant (black) in the apparatus C shown in Fig. 10.36. (*b*) Model analysed by Bott & Dean (1973) in which an elastic lithosphere rests on a viscous aesthenosphere. (*c*) Graphs showing the stress states in the elastic block at different times expressed as a function of *t*, the time taken for the whole multilayer to become laterally stressed to within 90 per cent of the applied end stress, P_0.

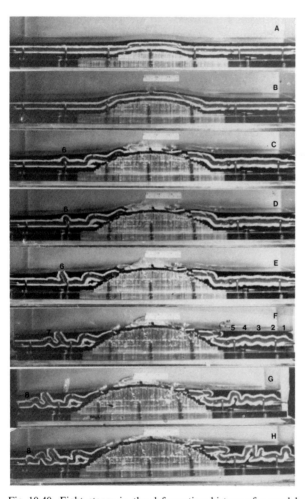

Fig. 10.40. Eight stages in the deformation history of a model multilayer. To the right of the 'basement' uplift, folds develop *synchronously* as a wave-train in a pre-stressed region of the multilayer. To the left of the uplift, the folds are developed *serially* at the wave-front as it propagates by a stick-slip mechanism through the multilayer.

the model occurred at the crest of the uplift together with the initiation of gravity gliding down both flanks of the basement arch (Fig. 10.40(*c*)). Lateral compressive stress, derived from the gravitational potential of the zone of uplift, propagated through the model behind a slow-moving stress front. However, the relationship between fold development and initial stress development to the right of the basement arch was very different from that to the left.

To the right, the stress front propagated through the whole length of the multilayer (Fig. 10.40(*b*) and (*c*)) before there was any significant development of the folds. The folds were then initiated as a uniform train of buckles, (folds 1–5, Fig. 10.40(*e*) and (*f*)). This is the type of folding that is assumed to occur in most theories of folding where it is implicitly assumed that the whole of the layered system becomes pre-stressed before folds are initiated and that once the conditions for folding are achieved, the folds develop uniformly throughout the model.

With continued uplift, three of these buckles were selectively amplified (Fig. 10.40(*g*) and (*h*), folds 4, 1 and 3). In contrast, to the left of the basement arch, the stress front became arrested during its propagation and a fold was initiated and amplified at this point of arrest (fold 6 Fig. 10.40(*c*), (*d*) and (*e*)). The stress built up behind the arrested wave-front until it was of sufficient magnitude to overcome the local resistance to shear on the decollement, probably an area of little, or no, lubrication. Renewed propagation of the stress front then occurred and a repetition of this process generated a series of folds (Fig. 10.40(*e*), (*f*), (*g*) and (*h*), folds 6, 7 and 8), each fold having developed at the temporarily static stress front.

The observations that folding occurred before the whole of the multilayer had been pre-stressed and that individual anticlines become amplified at different times during the course of an experiment was typical of the majority of the experiments. Such non-synchronous (but serial) fold development was not always associated with fold initiation at the stress front and, in a number of experiments, occurred after the whole length of the multilayer had been laterally stressed. A good example of this can be seen in Fig. 10.41, where the stress front reached the fixed end wall of the tank before the folds developed serially from the advancing piston face. The serial development of folds, although a neglected concept, is not a novel idea in geology. For example, it is implicit in Buxtorf's reconstruction of the development of two folds in the Jura Mountains (Fig. 10.42).

It is interesting to contrast the non-synchronous, serial development of folds in the experiment shown in Fig. 10.41 with the stochastic, non-synchronous development of isolated anticlines in the experiment shown in Fig. 10.43. In the latter experiment the

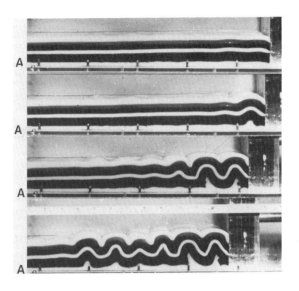

Fig. 10.41. The serial development of folds in a lubricated gelatine multilayer, as the result of the piston moving from right to left, in the lateral compression apparatus C of Fig. 10.36.

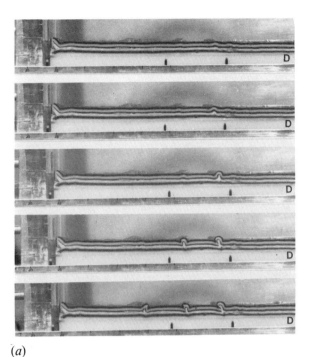

(a)

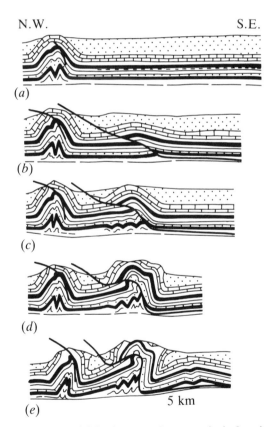

N.W.

S.E.

(a)

(b)

(c)

(d)

(e)

5 km

Fig. 10.42. The serial development of structures in the Jura deduced from field observations. (After Buxtorf, 1916.)

BLOCHMONTKETTE

DELSBERG

(b)

10 km

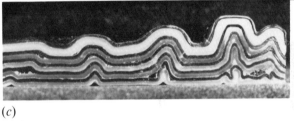

(c)

Fig. 10.43. (a) The stochastic development of isolated folds in a compressed gelatine multilayer, (after Blay *et al.*, 1977) and (b) a cross section through the Jura mountains between Blochmontkette and Delsberg, (after Buxtorf, 1916) which shows close similarity to the model structures. (c) Isolated, straight limbed conjugate folds in a wax multilayer. (After Summers, 1979.)

folds developed randomly within the multilayer after the wave-front had passed right across the model and fold initiation was not caused by the amplification of an adjacent fold, as is the case in serial folding. The cross section through the Jura structures (Fig. 10.43(b)) shows the close similarity between the geometry and spatial distribution of the model and

natural folds. A deformed wax multilayer containing folds whose geometry and spacial distribution are comparable to the folds in Fig. 10.43(a) and (b) is shown in Fig. 10.43(c).

In many of the experiments, the folds were approximately concentric, at least, during the early stages of their development. The basic, geometric features of these folds are shown in Fig. 10.44(b) and (d). Because the folds were concentric, they died out at some depth within the multilayer where the oblique axial surfaces on either side of the fold met. This point usually lay at the decollement between the multilayer and the base unit (Fig. 10.44(a) and (b)) but, in some experiments, lay within the multilayer itself (Fig. 10.44(c) and (d)).

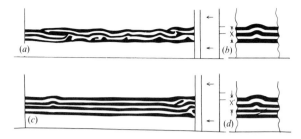

Fig. 10.44. (*a*) and (*c*) are line diagrams of compressed gelatine multilayers. In (*a*) the conjugate, concentric folds die out at the base of the multilayer: in (*c*) they die out within the multilayer. (*b*) and (*d*) are idealised sketches of these two situations.

In a discussion of the geometry of concentric folding, De Sitter (1956) showed how the depth, (X and X' in Fig. 10.44(*b*) and (*d*)) at which concentricity can no longer be sustained is related to the 'span' of the fold in the highest unit of the multilayer. In the experiments, shortening of the multilayer below depth X' was usually achieved by local thrust development. These thrusts, initially confined to the lowest units of the multilayer, often subsequently propagated to the surface of the model and developed into major overthrusts.

The preceding remarks concern observations of the profile section of developing structures. However, the design of the experiments also allowed the study of progressive fold development on the top surface of a model.

Let us now consider the three-dimensional geometry and spatial organisation of folds.

The three-dimensional form and spatial organisation of folds

The terminology and classification of folds described earlier in this chapter relate, in the main, to the profile section and assumes that the fold retains a constant geometry in a direction normal to the profile, for long distances. However, examination of many natural folds and experimentally produced flexures indicates that such buckles are generally not cylindrical but die out, often quite rapidly, when traced along the fold axis. This can be clearly seen in Fig. 10.45(*a*) and (*b*) which shows an oblique and plan view of folds in a gelatine multilayer. This model, which was deformed in the apparatus B shown in Fig. 10.36, clearly illustrates the non-cylindroidal nature of the folds.

The geometry and spatial arrangement of minor non-cylindrical folds on the limbs of a larger 'pod-shaped fold' were described by Campbell (1958) (Fig. 10.46). These minor folds form elongated domes or *periclines* which are arranged in an en echelon array. The ratio of the maximum profile wavelength to the extent of the fold along the fold axis is approximately 1:5. This geometry and spatial arrangement of folds can be found on a wide range of scales. For

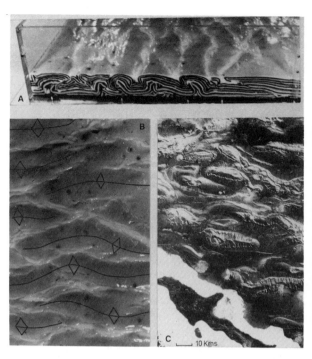

Fig. 10.45. (*a*) An oblique view and (*b*) a plan view of folds produced in the gravity-glide apparatus B of Fig. 10.36. The folds are non-cylindrical. (*c*) Large-scale non-cylindrical folds in the Zagros mountains, Iran (E.R.T.S. image).

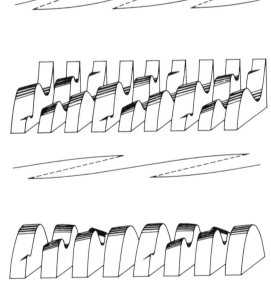

Fig. 10.46. En echelon array of minor 'pod' folds on the limbs of a major pericline. (After Campbell, 1958.)

example minor folds with wavelengths of a few centimetres which exhibit an en echelon arrangement and which die out rapidly along the hinge line are shown in Fig. 10.47. The periclinal folds of the Jura Mountains (Fig. 10.48) have wavelengths which may attain 2 km or even more. In passing, the area provides an excellent example of concordant morphology (i.e. the mountains coincide with the anticlines and the valleys

Fig. 10.47. En echelon periclines in folded metasediments near Luarca, N. Spain.

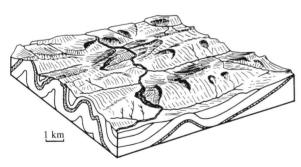

1 km

Fig. 10.48. Large-scale periclinal folds in the Jura Mountains. (After Strahler, 1960).

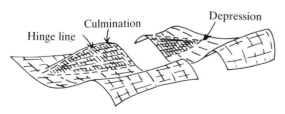

Hinge line | Culmination | Depression

Fig. 10.49. Terms used to describe various parts of a pericline.

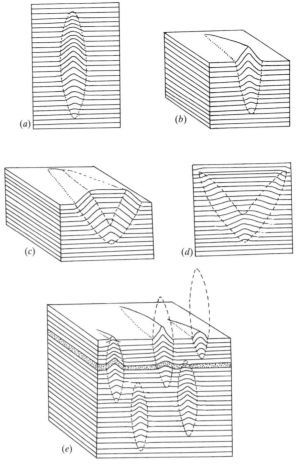

(a)   (b)   (c)   (d)   (e)

Fig. 10.50. The three-dimensional geometry and spatial organisation of some multilayer folds.

with synclines). The folds shown in Fig. 10.45(*c*), which is a satellite view of the Zagros Mountains (Iran), have wavelengths of between 10–20 km. It will be seen that these Zagros folds also display a periclinal geometry. Terms used to describe various parts of such periclines are given in Fig. 10.49.

As well as having a limited extent along their hinges, folds often die out relatively rapidly in profile section. A typical profile section through a fold in a multilayer is shown in Fig. 10.50(*a*). It can be seen that the central part of the fold is more angular than the peripheral parts of the structure. The 'whale back' fold shown in Figs. 10.51 and 10.52 clearly shows that a similar change in profile geometry occurs along the hinge direction. In its central portion the structure has a chevron-like profile, but towards the ends of the pericline the fold has a rounded profile.

Thus, if the plan and profile data are combined, we obtain a general three-dimensional picture of an isolated fold which exhibits the maximum amount of deformation near the centre of an approximately oblate, ellipsoidal space, with the fold progressively 'dying out' away from this central position (Fig. 10.53). The reasons for this geometry are dis-

Fig. 10.51. Minor periclinal fold on the limb of a large fold. Bude, S.W. England.

Fig. 10.52. Profile geometry of fold shown in Fig. 10.51. Note that the fold is chevron-like near the culmination (i.e. central region) of the pericline and becomes progressively more rounded towards the ends of the structure.

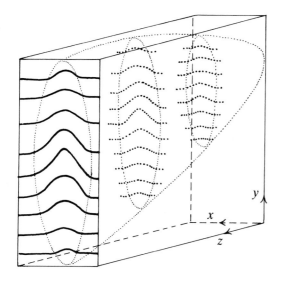

Fig. 10.53. The three-dimensional geometry of an isolated multilayer fold, indicating that the structure occurs within the limits prescribed by an (approximately oblate) ellipsoid.

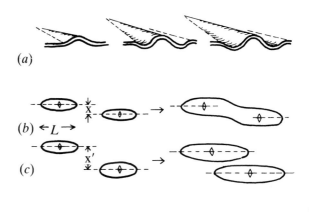

Fig. 10.54. (*a*) The development of a train of folds by the initiation of folds on either side of an amplifying, single buckle. (*b*) The amplification and coalescence of two periclines (shown in plan) separated by a distance X which is somewhat less than half the wavelength of the structure. (*c*) Two periclines separated by X' which is greater than half the wavelength. The structures overlap and lock up, each preventing further propagation of the other.

cussed in Chap. 13. We may, however, anticipate this duscussion and state that generally a buckle fold is initiated at local, often point, irregularities and that the central portion of the structure develops first and the peripheral deformation occurs progressively. In its earliest stages of deformation, the central portion of the structure goes through the geometrical forms that the outer portions of the fold eventually exhibit. That is, the structure exhibits a form of 'recapitulation'.

The random initiation of folds at point irregularities within a multilayer would give rise to the type of fold distribution shown in Fig. 10.50(*e*). It can be seen that although the individual periclines may have a common wavelength and amplitude, this will generally not be apparent from an inspection of a profile section showing any particular layer which runs through several folds. For example, the stippled layer (Fig. 10.50(*e*)), as well as exhibiting unflexured regions, is seen to form relatively rounded folds with small wavelength and low amplitude at some localities and larger wavelength and amplitude chevron folds at others.

As an individual fold amplifies, it may influence the subsequent initiation of other folds on either side and so begin the formation of a wave-train, or wave packet (Fig. 10.54(*a*)). As individual folds or wave-trains grow, they increase in length along their hinge lines and in so doing may interact with other folds. Experimental work and field observations have shown that there are a variety of ways in which interaction can occur. Let us first consider the interaction of two individual folds and then of two wave-trains.

If the hinges of two interfering folds are off-set, but the amount of offset is only a small fraction of the wavelength, the folds link or elide to form a larger fold with a deflection in the hinge line (Fig. 10.54(*b*)). Experimental observations also show that if the

hinges of two folds are separated by more than about half their maximum wavelength, but are still close enough to interact, they lock up, each preventing further propagation of the other, so that they become arranged in an en echelon fashion (Fig. 10.54(*c*)).

As adjacent wave-trains propagate, they too may interact with each other by the process of 'linking' or 'blocking', depending upon the relative wavelengths and positions of the two trains or packets. Some of the different ways in which they can interact are shown in Fig. 10.55. It can be seen that linking of folds from two fold trains can also give rise to structures which, in plan view, have deflections in their hinges (Fig. 10.55(*c*)). Alternatively, folds may bifurcate (Fig. 10.55(*c*) and (*d*)). It should be noted in passing that box folds show hinge bifurcation both in plan and profile section (Fig. 10.50(*c*) and (*d*)). Bifurcation of geological fold hinges are shown in Figs. 10.56 and 10.57 which are plan views of reverse kink-bands and asymmetrical folds respectively. Other natural examples of fold hinges which show some of the geometries and spatial relationships discussed above are illustrated in Figs. 10.58 and 10.59.

Sheath folds

Folds with such a pronounced curvature along their hinge lines that it sweeps through an arc of more than 90° within the hinge surface have been termed *sheath folds* (Figs. 10.60 and 10.61). They occur in many natural shear zones, especially in high-grade metamorphic rocks, and are thought to be the result of passive amplification of initial perturbations in a layer subject to layer-parallel shear (Cobbold & Quinquis, 1980). The effect of simple shear parallel to the layering on an initially symmetrical irregularity in the layering is shown in Fig. 10.61.

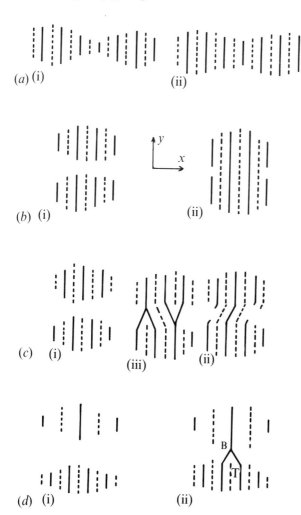

Fig. 10.57. Plan view of asymmetric folds in Devonian meta-volcanics showing bifurcation of the hinge. Boscastle, S.W. England.

Fig. 10.55. Simplified plan view illustration of the interference between two fold complexes of different phase and wavelength. The continuous and boken lines represent anticlinal and synclinal axes respectively. (*a*(i)) Wave-trains in phase approaching along *Z* resulting in (*a*(ii)) direct linking along *Z*. (*b*(i)) Wave-trains in phase approaching along *Y* resulting in (*b*(ii)) direct linking along *Y*. (*c*(i)) Wave-trains out of phase approaching along *Y* resulting in oblique linking (*c*(ii)) and folding bifurcation along *Y* (*c*(iii)). (*d*(i)) Wave-trains of different wavelengths approaching along *Y* resulting in (*d*(ii)) fold bifurcation B and blocking producing steeply plunging terminations T. (After Dubey & Cobbold, 1977.)

We have considered both the geometry of folds in profile and in three dimensions and have noted the large range of scales over which folds occur in nature. Probably the largest of such structures to occur are fold nappes which are thought to form during the culmination of an orogenic event as the result of plate convergence.

Fold nappes

The first person to describe the huge recumbent folds and overthrusts that we now know to be such a common feature of orogenic belts was Escher van der Linth (1841) who described such a structure from the Glarus region of the Swiss Alps. He considered the structures to be *autochthonous* (i.e. the rocks had not moved significantly from their site of deposition) and referred to them as 'nappes de recouvrement', which may be loosely translated as 'overlapping sheets'. This was later abbreviated to *nappes* and also translated into the German language as *Decken*. Schardt (1893) realised that many of the nappe structures were *allochthonous* (i.e. had been transported great distances) but was at a loss to explain how such transport had been achieved. Indeed, this is still one of the more intensely discussed problems of structural geology (see Chap. 7).

Fold Nappes can be subdivided into:

(i) Rooted (Pennine) Fold Nappes. These large recumbent folds are found in the higher grades of metamorphism associated with the deeper regions within an orogenic belt and can be traced to a root zone where the strata are vertical or sub-vertical. The most famous and most closely studied examples are the Pennine Nappes of the Swiss Alps, illustrated in Fig. 10.62(*a*).

(ii) Rootless (Helvetic) Fold Nappes. Such structures occur in less highly metamorphosed regions than rooted nappes. Among the first

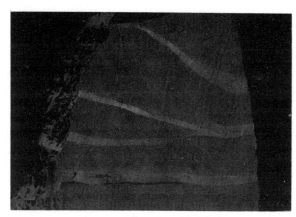

Fig. 10.56. Plan view of reverse kink-bands in slate showing bifurcation of the hinge. Bigbury Bay, S.W. England.

Fig. 10.58. Plan view of reverse kink-bands. Bigbury Bay, S.W. England.

Fig. 10.59. Plan view of crenulated mica schist. Lukmanier Pass, Switzerland.

examples to be studied were the Helvetic Nappes of the Swiss Alps (Fig. 10.62(*a*) and (*b*)).

The mechanisms by which such structures develop and are emplaced are not well understood. Indeed, the most significant contributions on this topic come from model work. For example Bucher (1956) conducted a series of experiments in which he produced large recumbent folds in multilayered models constructed from stitching wax. This material (a resin with plasticiser) has the ability to flow under its own weight. In some of his experiments he heated one end of the multilayer, thereby locally reducing its

strength. Consequently, when the multilayer was subjected to a horizontal compression, folding was concentrated in the heated region. In other experiments he replaced the 'cool', relatively rigid part of the multilayer with a wooden block and produced recumbent anticlines in the layered stitching wax by means of horizontal compression (Fig. 10.63(*a*)). The piston advanced at 2.5 cm/hr, which was several times faster than the wax could creep and spread under the action of its own body-weight. An anticline was formed in the wax next to the advancing piston (Fig. 10.63(*b*)). The advance of the piston was then arrested and the wax in the uplifted anticline, which was laterally confined between the perspex sides of the apparatus, was allowed to spread out under its own weight (i.e. *gravity spreading*, see Chap. 7). The resulting structure (Fig. 10.63(*c*)) was a recumbent nappe with a vertical root zone. Bucher draws attention to the fingering-out of the hinge region in the deeper part of the fold core (Fig. 10.63(*c*) and (*d*)) and points out the similarities between this model structure and the digitations of the Wildhorn Nappe in the S.W. Bernese Alps of Switzerland (Fig. 10.62(*b*)).

It might be argued that the structures produced experimentally by Bucher could be developed in nature if a multilayer sequence contained a metamorphic 'hot spot' and the sequence was then compressed along the layering, or, alternatively, if the sequence

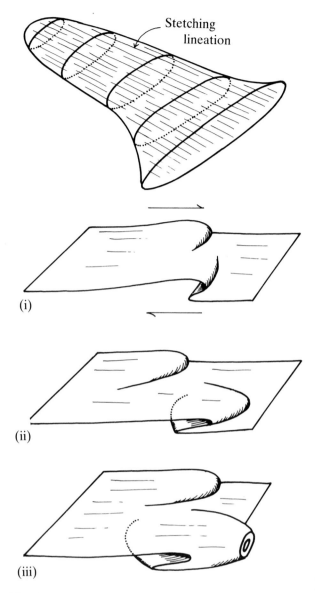

Fig. 10.60. Geometry of sheath folds showing how erosion can produced an 'eye-structure'.

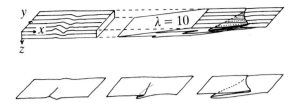

Fig. 10.61. Formation of sheath folds from an initial perturbation. (After Cobbold & Quinquis, 1980.)

contained no hot spot but was compressed against a basement block, as in Fig. 10.63(*a*).

Ramberg (1981) has compared the structures formed in Bucher's experiments with structures he produced in a *centrifuge*, and argues that fold nappes can result from a density inversion instability and diapirism (see Chap. 4). In his experiments, models with inverted density stratification are deformed in a centrifuge; so that the lower density layers rise diapirically through the more dense 'upper' layers to form structures of the type shown in Fig. 10.64.

In some of Ramberg's experiments the nappes were seen to *extrude* at the surface (Fig. 10.64 and 10.65). In nature this could only occur if the material was intrinsically weak and ductile, e.g. salt, or other evaporites and overpressured or uncompacted clays and muds. However, we are here concerned with deep-seated crustal processes. The Pennine nappes, for example, have cores of remobilised basement which are surrounded by an envelope of meta-sediments. Consequently, those of Ramberg's experiments which relate to *nappe intrusion* will have more relevance to the formation of rooted nappes.

When the nappe experiences gravitational spreading (this could apply to an extrusive or an intrusive situation) the advancing front of the nappe compresses the strata of the cover to cause buckles (Fig. 10.65(*a*)) which are comparable with those produced by Bucher (Fig. 10.63(*c*)). As the diapir spreads laterally and overruns these buckles, the folds first become rotated into a recumbent position and may eventually become broken and stacked as imbricate structures (Fig. 10.65(*b*) and (*c*)).

Rootless fold nappes, we have noted, are generally less metamorphosed than the rooted nappes. Some of the possible modes of emplacement of rootless nappes have been discussed in Chap. 7. An experiment by Blay (1974) illustrates how a fold may be generated at a break in the slope down which the material is sliding and which may then develop into a rootless recumbent fold, as the basement block continues to rise and the break in slope becomes a step (Fig. 10.66). Drape folds over fault scarps in the basement of the Swiss Alps are shown in Fig. 10.19(*d*). Gravity gliding of sediments, forming a drape fold, off such fault blocks could result in large recumbent nappes.

Fold nappes have been recognised for almost 150 years. However, despite the experimental and theoretical studies, the actual mechanisms by which they form, whether by horizontal compression as the result of plate collision, or as the result of diapirism, as was suggested in Chap. 4, remain one of the major problems of structural geology.

The stages of fold development

From the consideration of theoretical models and observation of physical models, it can be concluded that there are several stages in the formation of a fold. There is a period of pre-buckle flattening, during which a constant layer-parallel compression induces a homogeneous thickening of the layer. When buckling is initiated, the earlier period of stable deformation

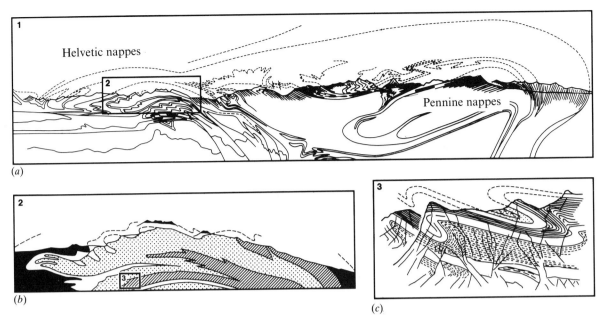

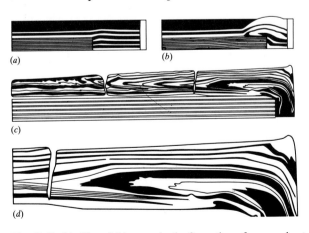

Fig. 10.62. (*a*) Section through the Swiss Alps. The Pennine and Helvetic nappes are examples of rooted and rootless nappes respectively. (After Argand, 1911 from Heim, 1922.) (*b*) Semi-schematic profile through the Helvetic nappes. The digitations at the front of the nappes show remarkable similarity with those produced in the experiment shown in Fig. 10.63. (*c*) Detail of folds in the Diablerets (Helvetic) nappe. (After Heim, 1922.)

gives way to unstable deformation. The resistance to deformation falls, i.e. strain-softening sets in, and the rate of deformation increases. Fold initiation is therefore often followed by an explosive amplification of the buckle, which continues until some geometrical or mechanical constraint results in the on-set of strain hardening and the process of folding slows down and eventually ceases. At this stage, the fold is said to have 'locked up' and further shortening takes place by post-buckle flattening. These various stages in a fold's history are represented in Fig. 10.67, which shows the variation in buckling stress and amplitude of a fold with time. This history can be contrasted with the stress and amplitude development associated with a

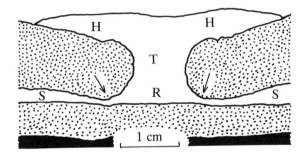

Fig. 10.64. Cross section through a dome of silicone putty (white) pierced through an overburden of painter's putty (stippled) with well-developed marginal sink (arrows), trunk (T), hat (H) and root (R) zone. S is the source layer and the scale bar is 1 cm. (After Ramberg, 1981.)

fold formed by passive amplification of a marker layer (dashed lines, Fig. 10.67). These various stages are discussed in the following chapters and are given special emphasis in Chap. 15.

Geological folds are often the result of more than one of the processes discussed above and, as will become apparent in the following chapters, the combination of active and passive fold amplification commonly occurs. It is sometimes possible to determine the relative importance of these two mechanisms in the formation of particular folds. For example, in the multilayer folds (Fig. 10.68), the intense folding of the quartz veins parallel to the layering shows that a minimum of 90 per cent shortening has occurred parallel to the original orientation of the layering. It is apparent from this, that the amplification of the multilayer folds has been extremely slow compared to that of the quartz veins

Fig. 10.63. (*a*), (*b*) and (*c*) stages in the formation of a recumbent anticline (fold nappe) produced by compressing a short, thick column of stitching wax against a rigid block, beneath an overall cover of the same material. See text for details. (After Bucher, 1956.) (*d*) detail of (*c*) showing digitations in the nose of the fold nappe (c.f. Fig. 10.62(*b*)).

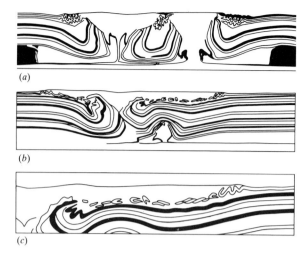

(a)

(b)

(c)

Fig. 10.65. (*a*) Cross section through diapiric model showing the buckling of the layered surface complex as the nappe structures spread out laterally. (*b*) A late diapir (angular, white body) just beginning to rise, gently deforming the surface lobe (nappe) which spread out from an earlier diapir. As the early nappe spread, the early-formed buckles of the type shown in experiment (*a*) were rotated, broken and stacked. (*c*) detail of (*b*). (After Ramberg, 1981.)

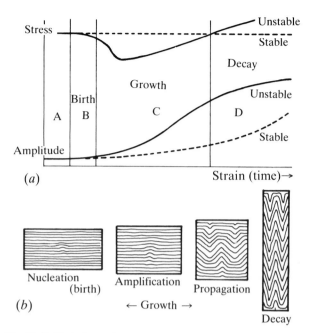

(*a*)

(*b*)

Fig. 10.67. Variations of stress and fold amplitude with strain for stable (kinematic) and unstable (dynamic) folding. (After Cobbold, 1976.)

and that this multilayer folding has been dominated by passive folding.

Commentary

In this chapter the classification and description of folds in profile section has been outlined together with their three-dimensional geometry and spatial organisation. The interference patterns that can sometimes be seen on erosional surfaces in rocks that have experienced more than one episode of folding, have also been classified and two conceptual models of strain distribution within a fold (those of tangential longitudinal strain and flexural flow) have been described.

The theory of scale modelling and the use of rock analogue models in the study of folding has been mentioned and the problem of the formation of fold nappes briefly addressed. The processes of passive folding and bending have been discussed but the formation of folds by buckling has not been considered as this is considered in some detail in the succeeding chapters.

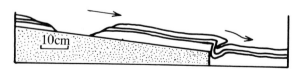

Fig. 10.66. The formation of a recumbent fold in a layered gelatine multilayer by gravity-glide down an uplifted block. (After Blay, 1974.)

References

Ameen, M.S. (1988). Forced folding of layered cover due to dip-slip, basement faulting. Unpublished Ph.D. Thesis, University of London.

Biot, M.A. (1965). Theory of viscous buckling and gravity instability of multilayers with large deformations. *Geol. Soc. Am. Bull.*, **76**, 371–8.

Blay, P.K. (1974). A model study of gravity tectonics. Unpublished Ph.D. Thesis, University of London.

Blay, P.K., Cosgrove, J.W. & Summers, J.M. (1977). An experimental investigation of the development of structures in multilayers under the influence of gravity. *J. Geol. Soc. London*, **133**, 329–42.

Bott, M.P.H. & Dean, D.S. (1973). Stress diffusion from plate boundaries. *Nature, Phys. Sci.*, **243**, 339–41.

Bucher, W.H. (1956). Role of gravity in orogenesis. *Geol. Soc. Am. Bull.*, **76**, 1295–318.

Buxtorf, A. (1916). Prognosen und Befunden beim Hauensteinbasis und Grenchenbergtunnel und die Bedeutung der Letzteren fuer die Geologie des Juragebirges. *Verl. Natur. Ges. Basel.*, **27**, 185–254.

Campbell, J.D. (1958). En echelon folding. *Econ. Geol.*, **53**, 448.

Cobbold, P.R. (1976). Fold shapes as functions of progressive strain. *Phil. Trans. Roy. Soc. London*, **A283**, 129–38.

Cobbold, P.R. & Quinquis, H. (1980). Development of sheath folds in shear regimes. *J. Struct. Geol.*, **2**, 1/2, 119–26.

De Sitter, L.U. (1956). *Structural Geology*, pp. 552. London: McGraw–Hill.

Dubey, A.K. & Cobbold, P.R. (1977). Non-cylindrical, flexural slip folds in nature and experiment. *Tectonophysics*, **38**, 223–39.

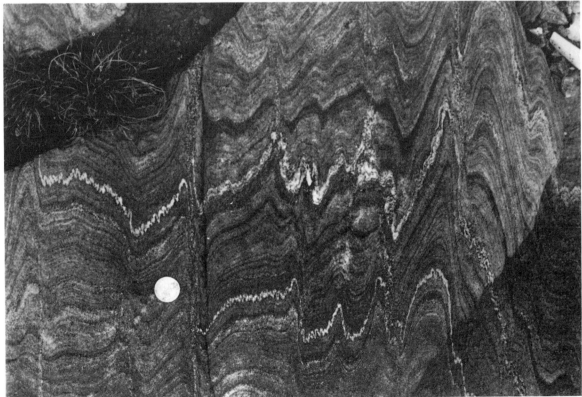

Fig. 10.68. The folded quartz veins show a minimum of 90 per cent layer-parallel shortening. If however the shortening is calculated by measuring the lengths of the other individual layers of the multilayer it can be seen that considerably less buckle shortening has occurred, showing that the folding of the multilayer was dominated by passive (kinematic) rather than active (dynamic) buckling. Monar, Scotland.

Elliot, D. (1965). The quantitative mapping of directional minor structures. *J. Geol.*, **73**, 865–80.

Escher van der Linth, A. (1841). *Ver. Schweizer Natur. Gesell.*, Zurich, vol. 54.

Fleuty, M.J. (1964). The description of folds. *Proc. Geol. Assoc.*, **75**, Pt. 4, 461–89.

Heim, A. (1921/2). *Geologie der Schweiz*. Band II, 2 pts. 1018 pp. Leipzig: Tauschnitz.

Hubbert, M.K. (1937). Theory of scale models as applied to geological structures. *Geol. Soc. Am. Bull.*, **48**, 1459–520.

Huddleston, P.J. (1973). An analysis and interpretation of minor folds in the Moine rocks of Monar, Scotland. *Tectonophysics*, **17**, 89–132.

Keunen, P.H. & de Sitter, L.U. (1938). Experimental investigation into the mechanisms of folding. *Leidsche Geol. Mag.*, **10**, 217–40.

Nevin, C.M. (1931). *Principles of structural geology*, 303 pp. New York: Wiley, London: Chapman & Hall.

Ramberg, H. (1964). Selective buckling of composite layers with contrasted rheological properties; a theory for simultaneous formation of several orders of folds. *Tectonophysics*, **1**, 307–41.

  (1981). *Gravity, deformation and the earth's crust*, 2nd edn., Academic Press.

Ramsay, J.G. (1967). *Folding and fracturing of rocks*, 568 pp. New York: McGraw–Hill.

Reddy, J.N., Stein, R.J. & Wickham, J.S. (1982). Finite-Element modeling of folding and faulting. *Int. Jour. Num. and Anal. Methods.* In *Geomechanics*, **6**, 425–40.

Sanford, A.R. (1959). Analytical and experimental study of simple geological structures. *Geol. Soc. Am. Bull.*, **70**, 19–52.

Schardt, H. (1893). Sur l'origine des Pre-Alpes romandes (zone du Chablais et du Stockhorn). *Eclog. Geol. Helvet.*, **4**, 129–42.

Stearns, D.W. (1978). Faulting and forced folding in the Rocky Mountains foreland. *Geol. Soc. Am. Bull.*, **151**, 1–37.

Strahler, A.N. (1960). *Physical Geography*. New York: Wiley.

Summers, J.M. (1979). An experimental and theoretical investigation of multilayer fold development. Unpublished Ph.D. Thesis, University of London.

Thiessen, R.L. & Means, W.D. (1980). Classification of fold interference patterns: a re-examination. *J. Struct. Geol.*, **2**, 3, 311–16.

Van Hise, C.R. (1894). Principles of North American pre-Cambrian geology. *U.S. Geol. Surv. 16th Ann. Report*, 571–843.

Wilson, G. (1961). The tectonic significance of small-scale structures. *Ann. Soc. Geol. Belg. Bull.*, **84**, 423–548.

11 Surface and single layer buckling

Introduction

Buckling is defined as the flexing or folding of a surface or series of parallel surfaces by a compressive stress directed along that surface or layer. Surfaces, either primary, such as bedding, or tectonically induced, such as cleavage, are common features in many rocks and are often buckled during deformation. In this chapter, the buckling of a single surface and a single layer (defined by two parallel surfaces) will be discussed. In an attempt to understand these processes geologists have used a variety of approaches including field observations, mathematical analyses of buckling, model work and finite element analysis.

The initial approach must be field observations, for these enable the detailed geometry of the buckles, their spatial organisation and strain patterns to be determined. Once these parameters have been clearly established it is possible to test the predictions of various theories of buckling against the field observations.

The majority of buckling theories discussed in this text describe only the first increment of buckling. The advantage of these *first increment* theories is that because the strains involved are infinitesimal, second order terms (i.e. $e_x e_y$ or $(e_x)^2$) will be extremely small and can therefore be ignored. This considerably simplifies the mathematics of the problem. However, these theories should not be used to predict the geometry of the folds as deformation takes them beyond this initial stage in their development. This restriction makes the comparison of the theoretically predicted fold geometry with the geometry of natural buckles difficult unless it is assumed that, as the incremental buckle amplifies into a finite structure its essential geometry is maintained i.e. if it starts as a sinusoidal deflection it will remain sinusoidal as the amplitude of the fold increases.

Although a few theories are now available which can be used to predict deformation beyond the first increment of buckling, geologists have generally turned to model work and finite element analysis for insight into the amplification of buckling instabilities into finite structures.

In the following discussion of the formation of buckles along an interface we will draw on field observations as well as theoretical and finite element analyses.

Buckling of an interface or free surface

Field observations

Small scale buckles at the interface between two rocks are shown in Fig. 11.1(a) and (b) and a larger scale example is illustrated in Fig. 11.2. The larger buckles, which have the geometry of large amplitude *cusp* structures have developed at the junction between a multilayer and a relatively thick quartz-rich horizon. These particular surface instabilities are especially interesting because the intense buckling of the layer-parallel quartz veins in the multilayer clearly indicates that these cusps are associated with a considerable amount (>90 per cent) of interface parallel compression.

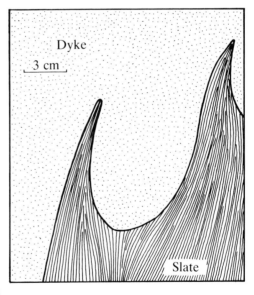

(a)

(b)

(c)

(a)

Fig. 11.1. (*a*) Cusps formed at the interface between a dyke and slate, caused by compression parallel to the dyke–slate junction. Start Bay, Devon, England. (*b*) Cusps at the interface between a quartz-rich (light) and mica-rich layer in the metasediments of Monar, Scotland. (*c*) Flame structures at the interface between a shale and a sandstone. The instability is caused by a density inversion and not by compression parallel to the interface. Bude, Cornwall, England.

It should be noted in passing that interface parallel compression is not the only method by which cusp structures can be formed at the junction between two rocks. For example, the formation of *flame structures* at the interface between a sandstone and an underlying, less dense shale (Fig. 11.1(*c*)) is the result of a density inversion and small-scale diapirism. The maximum principal compression in this situation is *normal* to the interface.

Mathematical analysis
Biot (1964) studied the deformation of a linear, viscous block of material, one boundary of which had an initial, low-amplitude, sinusoidal deflection. Such a model would be obtained if the portion of the model in Fig. 10.17(*a*) above the sinusoidal marker surface, were removed. A free surface of sinusoidal shape would then exist, but the state of homogeneous deformation associated with the complete model would then be disturbed. However, the original deformation will remain undisturbed if surface forces are applied which restore the initial stress field. These forces,

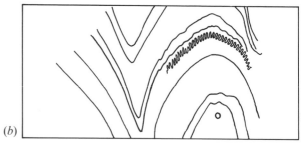

(b)

Fig. 11.2. Cusps at the interface between a multilayer and a quartz/feldspar/mica layer, Monar, Scotland. The two buckled quartz veins, a and b, show that the rock has been shortened at least 93 per cent in a direction parallel to the original interface.

which are shown in Fig. 11.3(*a*), are $P \sin \alpha$, per unit area of surface, where α is the slope of the surface and P the applied compression. The forces shown in Fig. 11.3(*a*) are equivalent to those acting on the surface when the upper portion of the model(Fig. 10.17(*a*)) is present. The effect on the surface of removing the upper portion can be obtained by superposing surface forces which cancel these forces. These, which will be of equal magnitude but of opposite sign, are shown in Fig. 11.3(*b*). They disturb the uniform strain state associated with the model in Fig. 10.17(*a*) thereby causing the concave parts of the interface to be 'pinched' in and the convex parts to remain smooth (Fig. 11.3(*c*)). The original sinusoidal deflection deforms into a series of cusp structures. Similar structures will develop when com-

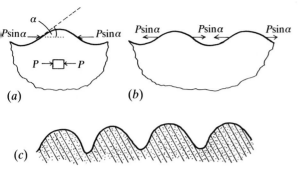

Fig. 11.3. The formation of cusp structures from initial low-amplitude, sinusoidal deflection of an originally horizontal free surface or interface between two unlike materials, by a horizontal compression. See text for details. (After Biot, 1964.)

pression is applied parallel to the interface between two unlike materials. The cusps will point into the more competent (i.e. stronger) material.

It was noted in the introduction that by restricting a mathematical treatment of folding to the first increment of buckling (at which stage in the deformation the strains are infinitely small) it is only necessary to consider first order terms as second order terms will be negligibly small. The buckling instability is often apparent from these first order terms and an expression can be obtained which gives the dominant (i.e. most easily formed) wavelength in terms of the material and geometric properties (i.e. viscosity and layer thickness) of the buckling system.

Fletcher (1982) points out that the development of mullion structures (i.e. lobes and cusps) at an interface between two fluids described by Biot (1964) and shown in Fig. 11.3, is the result of *kinematic* (i.e. passive) amplification of initial surface irregularities, with modification at finite amplitude to give the characteristic lobe and cusp geometry. He agrees with Biot that no first order instability is present and that, consequently, the structure will show no preferred length scale (i.e. dominant wavelength) unless one is present in the initial irregularity. To first order accuracy, the evolution of the interface between two viscous fluids, undergoing shortening parallel to the interface, does not depend upon the viscosity contrast. However, Fletcher demonstrates that when the buckling is considered to second order accuracy, the asymmetric element of the finite amplitude does emerge, showing that these structures do in fact develop by the dynamic (i.e. active) amplification of buckling instabilities. Such second order instabilities are generally not as marked as those of the first order, explaining why field observations (e.g. Fig. 11.2) indicate that for the formation of high amplitude cusps with a marked lobe and cusp geometry, large amounts of shortening are necessary.

In an experimental investigation of the development of lobe and cusp structures by compression

along a sinusoidal interface between two viscous materials of different viscosities, Cobbold (1969) found that large interface irregularities (or presumably large strains, as indicated by the example shown in Fig. 11.2) were necessary before significant cusp structures developed. Moreover, he discovered that there was no marked tendency for a dominant wavelength of cusp to develop; the wavelength of the cusp being controlled by that of the initial sinusoidal disturbances of the interface.

Fletcher further shows that the instability is enhanced if the materials have non-linear properties and that the strength of the instability is proportional to power law exponent of the more competent medium. However, the growth-rate of the instability is the same for both a welded and a perfectly frictionless boundary between the two media. As is discussed later in this chapter, in the section on the theories of single layer buckling, lobe and cusp structures can result from first order mechanical instability by the process of 'resonant folding' when an incompetent layer set in a more competent matrix is compressed along its length (Fig. 11.26). As will be seen, the wavelength of this instability is related to the layer thickness and is very insensitive to the competence contrast between the layer and matrix. However, this mechanism can not account for lobe and cusp development, when only a single interface exists.

Finite element analysis

The development of cusp structures at a free surface has been successfully studied by using the technique of finite element analysis (Dieterich & Onat, 1969). This type of analysis, in which a computer is used to determine the state of stress and strain within a deforming body, is particularly useful when the deformation problem is not amenable to theoretical analysis and for studying the effects of large deformations. The continuous solid body of the 'model' is subdivided into polyhedral (often triangular) elements (Fig. 11.4(b)) within each of which, deformation is approximated by a simple relationship. In this relationship, the displacement function is defined such that the strain within any given element may be determined completely from the displacements of the corners (nodes) of the element. With the aid of the stress–strain relationships specified for each element (and assuming that the stress applied to each element can be related to equivalent forces at nodes) a set of force–displacement relationships for each node can be obtained. To determine the displacement field and then the deformation of the body some form of potential energy minimisation criterion is used and the force–displacement relationships for all the nodes are solved simultaneously for specific boundary displacements which are defined by the problem to be analysed.

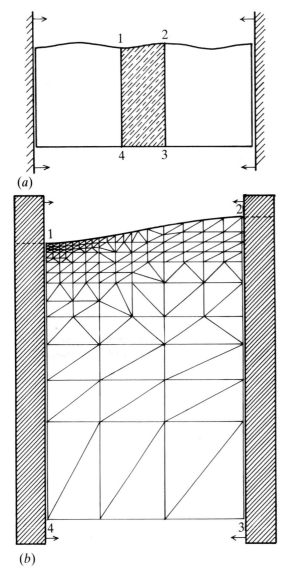

(a)

(b)

Fig. 11.4. (a) The initial configuration of a body with a sinusoidal surface. The basic unit of the body is defined by points 1, 2, 3 and 4. (b) The division of the basic unit into triangular elements. (After Dieterich & Onat, 1969.)

A study of the model of Dieterich & Onat (Fig. 11.4) will make this clear. The model is assumed to be of a linear viscous material and has a low-amplitude, sinusoidal deflection on one of its surfaces (Fig. 11.4(a)). The basic unit of the model, 1234 in this figure, is divided into triangular elements (Fig. 11.4(b)) which are located in space by the coordinates of their three nodes. The material properties (stress–strain relationships) of each element are specified. The model is deformed in increments of 0.01 average strain by displacing one of the vertical boundaries towards the other. The new position of all the nodes and the new stress and strain states within each element that result from the boundary displacement are then calculated. The cumulative effect of 100 increments of 0.01 average strain on the original sinusoidal surface (Fig. 11.4(a)) is shown in

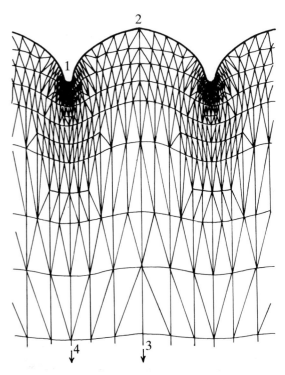

Fig. 11.5. The configuration of Fig. 11.4(a), after 100 increments of 0.01 average strain. (After Dieterich & Onat, 1969.)

Fig. 11.5. By comparing Figs. 11.4(b) and 11.5, it can be seen that a marked concentration of strain occurs around the cusp point. In many geological examples of cusp structures formed at the interface of two unlike lithologies, there is evidence of high strain in the region of the cusp points, for example, the cusps shown in Fig. 11.2. These have formed at the interface between two lithologies, and an intense pressure solution cleavage is locally developed at the cusp points. It was noted earlier that the intensely folded quartz veins clearly indicate that these well-developed cusp structures are associated with a very large compressive strain (≈ 93 per cent) in a direction parallel to the original interface.

Introduction to the buckling of a single layer

Examples of single layer buckles are found in many deformed rocks and have attracted considerable attention from structural geologists. In an attempt to understand their mechanism of formation we shall first consider some field examples and then assess some of the theoretical analyses and model work pertaining to single layer buckling.

Field observations

Careful field observations of natural single layer folds have provided details of fold geometries, including such parameters as wavelength/thickness ratio and amplitude (Sherwin & Chapple, 1968), the strain states in and around the buckled layers as indicated

Fig. 11.6. Single layer buckling of an aplite dyke. Lewis, Outer Hebrides, Scotland.

by rock fabrics (e.g. cleavage patterns) and deformed objects such as ooliths and fossils (Cloos, 1947 and Ramsay, 1967) and the stress state in a buckled layer (Carter & Raleigh, 1969). Such observations provide the basic reference data with which we compare the results of experimental and theoretical studies aimed at understanding the mechanism of single layer buckling in rocks.

Wave-trains found in natural single layers are sometimes composed of folds of approximately the same wavelength (Fig. 11.6). Field observations also indicate that there is a relationship between thickness of the layer and the wavelength that develops. This is well shown by the buckled, tapered quartz veins shown in Fig. 11.7.

Single layers are sometimes found to contain two orders of folds, a smaller set of second order buckles which have been buckled into larger, first order folds (Fig. 10.5). This process is shown diagrammatically in Fig. 11.8, where it can be seen that the wavelength of the second order buckles is determined by the thickness (a_1) of the layer. Once these buckles are fully developed, they no longer provide a mechanism by which shortening along the layer can occur. The buckled layer acts effectively as a thicker layer (a_2) and if subjected to continued 'layer' parallel compression, may buckle again into folds with a correspondingly larger wavelength. A natural example of two orders of folds in a single layer is shown in Fig. 11.9, where the folds formed in a tapered vein and both the first and second order folds show a reduction in wavelength with layer thinning.

Fig. 11.7. Buckled, tapered quartz veins in slate. The fold wavelength is controlled by the thickness of the layer. Porthleven, Cornwall, England.

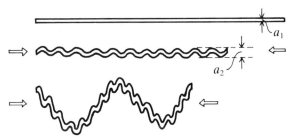

Fig. 11.8. The formation of 2nd and 1st order buckles. The wavelength of the second order folds is determined by the thickness of the layer (a_1) and the wavelength of the first order folds by the effective thickness of the buckled layer (a_2).

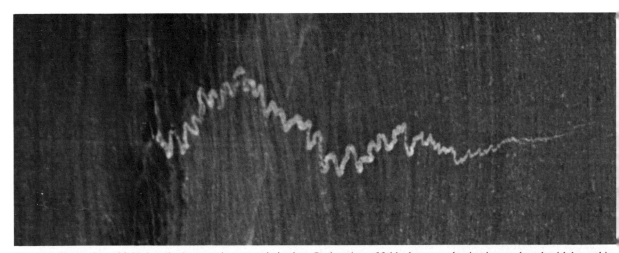

Fig. 11.9. Two orders of folds in a single tapered quartz vein in slate. Both orders of folds shows a reduction in wavelength with layer thinning. From Porthleven, S. Cornwall, England.

Field observations also indicate that different single layers, buckled during the same tectonic event, may show marked differences in their wavelength/thickness ratios (Fig. 11.10(a)). It is clear that there is less shortening by buckling in the layer with the low wavelength/thickness ratio. This can be seen by comparing the length l_1 and l_2 of layers A and B in Fig. 11.10(b), which indicates the lengths which the layers would now have if they were unfolded. It can reasonably be assumed that both layers were deformed by a combination of homogeneous flattening and buckling; however, the deformation of layer A involved more flattening and less buckling than layer B. The reasons for these differences in buckling behaviour in different layers will become apparent when the theories of single layer buckling are considered.

Theories of single layer buckling
The earliest work dealing with the buckling of *elastic* layers was by Euler (1757). Later workers include Smoluchowski (1909), Goldstein (1926), Kienow (1942), Biot (1961) and Ramberg (1961a). Concise and useful reviews of previous work on buckling are given by Smith (1975 and 1977) and Arvid Johnson (1977) in his excellent and clearly written book 'Styles of folding' which focuses primarily on the problems of buckling of single and multilayered elastic materials. However, it was pointed out by both Biot and Ramberg that the behaviour of rocks during the formation and development of folds is dominantly inelastic. Geologists were so influenced by this argument that they have tended to ignore the elastic components of buckling and have almost exclusively considered the buckling behaviour of viscous and elastico-viscous layers.

The model relating to the folding of a single *viscous* layer embedded in a matrix of lower viscosity has been analysed in an influential series of papers by Biot (1957, 1959a and b, 1961) and Biot & Odè (1962) and by Ramberg (1959, 1960, 1961a and b, 1963a, 1970a and b). These authors have considered the relevance of their work to the problem of the folding of layered rocks. Their various papers are restricted to the analysis of fold initiation and the principal results bear upon the relationship between wavelength (arc length)/thickness ratio and the competence contrast between the layer and the matrix.

Inevitably, despite the mathematical sophistication, the theoretical models studies are much less complex than the real geological situations they are meant to represent. Some of the more important simplifying assumptions made in several of these analyses are listed below:

(i)   Body forces are negligible (i.e. it is assumed that the structures are small enough for the effect of their body-weight to be ignored).
(ii)   Both layer and matrix are treated as Newtonian fluids (i.e. they exhibit a linear relationship between the deforming stress and the resulting strain-rate).
(iii) Compression is applied parallel to the layer.
(iv)   The analysis is restricted to plain strain (i.e. no area change accompanying deformation occurs in the profile section of the folds and no strain occurs in the direction normal to this section).
(v)   The amplitude of the folds is very small (i.e. the analyses are restricted to fold initiation and do not consider the amplification of folds into finite structures).
(vi)   The folds are sinusoidal.

In addition, the conclusions of the early analyses of both Biot and Ramberg are based upon assumptions regarding the state of stress within the buckling layer that are only appropriate to folds with wavelengths much larger than the layer thickness. That is, they are presenting what are called 'thin plate'

(a)

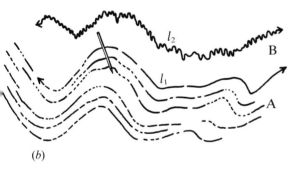

(b)

Fig. 11.10. (a) Marked differences in the wavelength/thickness ratio of the folds in different layers deformed during the same tectonic event. Dalradian metasediments, Ardalanish Bay, Mull, Scotland. (b) is a line drawing of (a).

analyses. In such an analysis, the assumption that the strain, and corresponding stress, distributions in the buckling layer approximate to those of a tangential longitudinal strain fold (Fig. 10.23(*f*)), is reasonable.

Although the majority, usually all, of the assumptions listed above are not valid during the buckling of rock layers in the Earth's crust, the results of these analyses have been used extensively by geologists to explain the formation of natural folds. Some of the limitations of these analyses, when they are applied to the problems of geological folding, will be discussed in the following sections. Although it is not possible (or desirable) in this text to give full details of the various theories of folding that are discussed, we do consider the assumptions built into the analyses and the physical implications of the buckling equa-

tions. Knowing the assumptions on which an analysis is based enables its relevance to the formation of any particular fold or other structure to be more easily assessed.

Theories restricted to the first increment of buckling of linear materials will now be considered and this is followed by a discussion of the theories that deal with the buckling of non-linear materials. Some theories can be applied to folding beyond the first increment of buckling and so can be used to indicate how folds amplify. These theories are discussed later in this chapter.

The earliest treatment of the buckling of an elastic strut (Euler, 1757), dealt with a layer unconfined by a matrix. Euler considered various conditions of pinning at the layer ends. The most pertinent to the problem of geological folding is the one shown in Fig. 11.11, which is known as the end-fixed, direction-fixed condition. The critical buckling force (F_{crit}) for a beam in this condition is given by:

$$F_{\text{crit}} = \frac{4\pi^2 EI}{L^2} \qquad (11.1)$$

where L is the length of the beam, E is Young's modulus and I, the moment of inertia, given by:

$$I = \frac{a^3 b}{12} \qquad (11.2)$$

where b is the width of the beam and a its thickness.

Despite the geologically realistic pinning con-

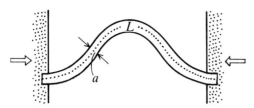

Fig. 11.11. Buckling beam with end-fixed boundary condition.

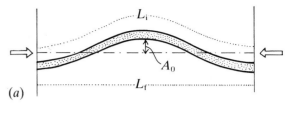

(a)

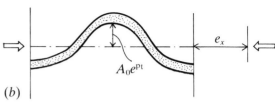

(b)

Fig. 11.12. (a) An initial, sinusoidal deflection amplitude (A_0) and wavelength (L_i). (b) The change in amplitude after time (t). The shortening caused by buckling in the direction of the applied compressive stress is e_x.

ditions (Fig. 11.11), it would not appear that this buckling analysis of an unconfined layer has any direct application to a geological environment in which competent layers are invariably confined in a matrix. So, for reasons which will become apparent, the discussion of this equation and of its relevance to geological folding is held over until the formation of large folds is considered in Chap. 14.

Biot (1961), analysed the buckling behaviour of a linear, elastic layer of thickness a, Young's modulus E and Poisson's ratio v, set in a viscous matrix, and obtained the relationship:

$$L = \pi a \sqrt{\frac{E}{(1 - v^2)P}} \qquad (11.3)$$

where P is the applied buckling stress (i.e. the layer-parallel compression per unit area of layer cross-section) and L the wavelength of the buckles which will develop. It can be seen that this wavelength depends on the material properties of the layer, its thickness and the applied buckling stress. It is, however, independent of the material property of the matrix.

As was noted earlier, because the development of folds in rocks is predominantly inelastic, geologists have concentrated on the buckling analyses of viscous and elastico-viscous layers. Biot (1957) shows that the amplitude (A) of any initial sinusoidal deflection in a viscous layer increases exponentially with time. The equation describing the initial deflection is:

$$y = A_0 \cos wx \qquad (11.4)$$

where A_0 is the original amplitude of the deflection and w is the wave number, given by $w = 2\pi/L_i$, (Fig. 11.12). After time t, the deflection becomes:

$$y = A_0 e^{pt} \cos wx. \qquad (11.5)$$

The parameter e^{pt} is called the amplification factor. For a given buckling stress (P), the exponent (p) is given by:

$$p = P/[(4\eta_2/aw) + (\eta_1 a^2 w^2)/3] \qquad (11.6)$$

where η_1 and η_2 are the viscosities of the layer and matrix respectively. It can be seen from Eq. (11.6) that p and, therefore, the rate of amplification of any initial deflection, will depend on its wave number and therefore its wavelength. This dependence is shown graphically in Fig. 11.13.

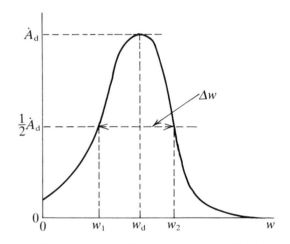

Fig. 11.13. The rate of amplification $\dot{A}$ of a range of initial, sinusoidal layer deflection with wave numbers between 0 and w. There is one wave number, w_d, (the dominant wave number) which amplifies more rapidly than any other. This defines the dominant wavelength, L_d, of the folds ($L_d = 2\pi/w_d$). (After Biot 1961.)

It is apparent from the graph that there is one wavelength that grows faster than any other. This is referred to as the *dominant wavelength* (L_d). After sufficient time, the dominant wavelength will appear in the form of regular, approximately sinusoidal waves.

When a buckling stress is applied to a viscous layer, it experiences homogeneous flattening. If the time required for a significant amplification of an initial irregularity is too long, the flattening of the layer will become so large that the folding phenomenon will be completely masked. In order to clarify this idea let us consider the relationship between time (t) and the amplification factor of the dominant wavelength (A_d). This is:

$$\frac{t}{t_1} = \left(\frac{6\eta_2}{\eta_1}\right)^{\frac{2}{3}} \log A_d \qquad (11.7)$$

where t_1 is the time required to shorten the layer by 25 per cent. Using this equation, the relationship

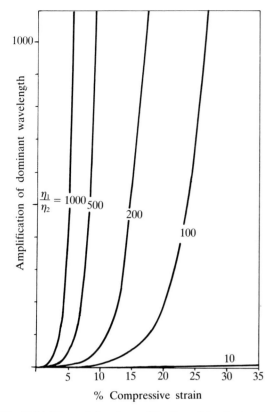

Fig. 11.14. The variation in amplification of the dominant wavelength with compressive strain for various viscosity contrasts (η_1/η_2) between the layer and the matrix. (modified from Ramsay, 1967.)

between the rate of amplification of the dominant wavelength for various viscosity contrasts between the layer and the matrix and time or some time-dependent parameter such as bulk shortening strain, can be determined. These are shown graphically in Fig. 11.14.

One important feature is immediately apparent from this graph. For high values of η_1/η_2 the amplitude of folding will increase at an 'explosive' rate beyond a certain layer parallel bulk shortening. Biot suggests that this occurs when A_d, the amplification factor, is approximately 1000. He argues that when this is so, the resultant fold-train will show a sharply defined *dominant* wavelength, which will occur whatever the distribution of wavelengths of the initial perturbations may have been. It can be seen from the graph of Fig. 11.14 that, as the viscosity contrast between the layer and matrix decreases, the amount of shortening necessary before 'explosive' amplification of the dominant wavelength occurs, increases. When the viscosity contrast is less than 100/1, explosive amplification will not occur until the bulk shortening has exceeded 25 per cent. Biot considers that such a large compressive strain would overshadow folding and concludes that for clear-cut folding (i.e. for a regular wave-train to develop) the viscosity ratio must be greater than 100/1.

Biot (1957) and Ramberg (1961a) both derive the same equation for the dominant wavelength of a buckling single, linear, viscous layer of thickness (*a*) set in a less viscous matrix, Fig. 11.12, which is:

$$L = 2\pi a \sqrt[3]{\frac{\eta_1}{6\eta_2}}. \tag{11.8}$$

η_1 and η_2 are the viscosities of the layer and the matrix respectively. This equation is compatible with the field observations that show a relationship between layer thickness and wavelength (Figs. 11.7 and 11.9). As the arc length (*L*) and thickness (*a*) are the only data that can be directly obtained by field observations, this and other similar equations, which we shall present later, are often expressed in terms of the wavelength/thickness, or *slenderness ratio, L/a*:

$$\frac{L}{a} = 2\pi \sqrt[3]{\frac{\eta_1}{6\eta_2}}. \tag{11.9}$$

Thus, the slenderness ratio depends only upon the ratio of the material properties of the competent layer and the matrix. This accounts, therefore, for the different L/a ratios which developed in layers of different lithologies during the same compressional event (Fig. 11.10).

By comparing Eqs. (11.3) and (11.8), which respectively govern the buckling of an elastic and a viscous layer embedded in a viscous matrix it can be seen that for the elastic layer (Eq. (11.3)) the dominant wavelength is dependent upon the buckling stress and independent of the properties of the matrix, and for the viscous layer, the dominant wavelength is independent of the buckling stress but dependent upon the relative contrast in properties of the layer and surrounding matrix.

In the purely static problem of elastic buckling, where a progressively increasing load is applied parallel to the embedded layer, a mathematically, clearly defined wavelength is associated with the buckling instability. By contrast, for viscous or elasto-viscous buckling, the buckling is less sudden; rather the folding depends upon the imperfections (often referred to as irregularities or perturbations) present in the competent layer.

Let us now scrutinise Ramberg's derivation of the single layer buckling equation (Eq. (11.9)), for this also throws light on the process of buckling. (See Ramberg 1960 and 1961a.) We shall consider the point in his analysis where he has determined the buckling equation for a viscous layer *without* a confining matrix:

$$f_x = \frac{2\pi^2 \eta_1 a^3 \dot{e}_x}{3e_x L_i^2} \tag{11.10}$$

where e_x and $\dot{e}_x$ are the strain and strain-rate parallel to the undeflected layer of thickness (*a*) and viscosity η_1 (Fig. 11.12), and L_i and f_x are the initial wavelength and buckling force respectively. From Eq. (11.10) it follows that:

$$f_x \text{ is proportional to } a^3 \qquad (11.11)$$

$$f_x \text{ is proportional to } \frac{1}{L_i^2} \qquad (11.12)$$

and

$$f_x \text{ is proportional to } \frac{1}{e_x}. \qquad (11.13)$$

Let us consider the implications of Eqs. (11.10) to (11.13) in turn. It can be seen from Eqs. (11.10) and (11.11) that the resistance of a layer to buckling is very sensitive to layer thickness. Thus, if two layers, of identical material, have thickness ratios of 3/1, their buckling resistances will be in the ratio of 27/1. This is a point to which we shall return when we discuss the importance of *control units* in Chap. 12.

It also follows from Eqs. (11.10) and (11.12) that, for a particular layer, the larger the wavelength of a buckle, the smaller the buckling force. The layer will therefore buckle into the largest possible wavelength. The reader can demonstrate this by compressing a playing card between the forefinger and thumb, first along its width and then along its length. The card buckles into a half wavelength, thus forming the largest possible wavelength (Fig. 11.15). The reason for this is that the formation of one large wavelength requires less internal deformation of the layer than the formation of two or more folds of smaller wavelength. However, the single fold that minimises the internal deformation of the layer would maximise the deformation of the matrix if one were present. It is not surprising, therefore, that when a viscous matrix surrounds a more competent viscous layer, the case described by Eq. (11.8), the wavelength of the fold that develops is directly proportional to the ratio of the strength of the layer and matrix, (η_1/η_2), for this indicates that the stronger the layer, the greater the wavelength and the stronger the matrix, the smaller the wavelength.

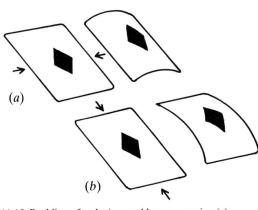

(a)

(b)

Fig. 11.15. Buckling of a playing card by compression (a) across its width and (b) along its length, demonstrating that an unconfined layer will buckle with the largest possible wavelength. See Eq. (11.12).

Perhaps the most interesting conclusion that can be drawn from Eqs. (11.10) and (11.13) follows from the observation that the force necessary for buckling is inversely proportional to the buckling shortening (e_x), i.e. the greater the buckle shortening, the lower the buckling force and the easier it is for the fold to grow. This is because, in order for buckling to occur, the externally applied moments (the layer parallel force multiplied by the amplitude of the buckle) must balance the internal moments generated in the layer as it folds. It follows that the larger the buckle shortening (which of course determines the buckle amplitude), the smaller the layer parallel force required for buckling.

Let us now consider the implications of Eqs. (11.10) and (11.13) regarding the process of fold initiation. At fold initiation e_x (the strain in the x direction resulting from buckling) is zero and it follows from Eq. (11.13) that an infinitely large buckling force will be required before folding can take place. The physical significance of this conclusion is not difficult to find for if, as is assumed in the theory, the layer is perfectly planar and contains no initial irregularities, then the initial amplitude of the 'fold' will be zero and consequently the externally applied moment will be zero regardless of the magnitude of the buckling force. We see, therefore, that buckling will not develop in a perfectly planar layer and that some irregularity is necessary for fold initiation. If an irregularity is present, as for example in Fig. 11.12(a), then the fold has already been initiated. A finite fold amplitude already exists which corresponds to an effective finite value of buckle shortening, e_x $(L_i - L_f)/L_i$. It can be inferred from Eq. (11–13) that because $e_x \neq 0$, fold growth from the irregularity does not require an infinite buckling force.

It is also apparent from Eqs. (11.10) and (11.13) that once folding has been initiated, the force necessary to develop buckles will drop dramatically as e_x increases, i.e. once a fold has been initiated, its development becomes progressively easier. This is compatible with Biot's concept of the *explosive amplification* of the dominant wavelength discussed earlier.

The buckling equation (Eq. (11.9)) can be easily modified to deal with the deformation of a viscous layer in a less viscous matrix when the material properties of the matrix on each side of the competent layer are different (Eq. (11.14)). Then the *slenderness ratio* (L/a) is given by:

$$\frac{L}{a} = 2\pi \sqrt[3]{\frac{\eta_1}{3(\eta_2 + \eta_3)}} \qquad (11.14)$$

where η_1 is the viscosity of the layer and η_2 and η_3 are the viscosities of the matrix above and below the layer respectively.

No mention has yet been made concerning the interface between the layer and the matrix. In Ramberg's derivation of Eq. (11.8) (1961a), he assumes perfect cohesion between the layer and the matrix, whereas Biot (1961), applying fundamentally different techniques, assumes perfect slip. Biot (1959b) also considers the condition of cohesion between the layer and the matrix and shows that there is only a slight difference in predicted wavelength values between the cases of slip and non-slip at the layer–matrix junction. He concludes that cohesion leads to a slightly longer wavelength but that the difference becomes vanishingly small for viscosity ratios greater than 70 $(L/a > 14)$. However, although the effect of slip or non-slip at the layer boundary does not seem to have an important influence on determining the dominant wavelength, it will be shown later in this chapter, in the section dealing with finite element analysis, that the possibility of slip at the layer–matrix interface has a very marked effect on the rate of amplification of folds.

Having considered some of the interesting implications of Biot and Ramberg's analyses for the buckling of a single, linear viscous layer, it is important to consider some of the shortcomings when these analyses are used to account for the formation of geological folds. The graphical expression of Eq. (11.9) is shown in Fig. (11.16), where it can be seen that, even when the viscosity ratio between the layers and the matrix is 1 (i.e. the layer and the matrix are of the same material), folding with a dominant wavelength is still predicted. The explanation of this paradox is found by looking at one of the assumptions built into the analysis. It will be recalled, that Eq. (11.9) is derived from a thin plate analysis, i.e. it is assumed that the wavelength of the buckles is considerably greater than the layer thickness. It is, therefore, unreasonable to expect this equation to provide an accurate description of folding when the slenderness ratio is small, (say $< 10/1$), for in such folds the fibre

stresses associated with the buckling of the layer become prohibitively large in the inner and outer arc areas of the hinge and the tangential longitudinal strain model, (Fig. 10.23(f)), assumed in the analysis, ceases to be appropriate.

A further problem that arises when attempting to use Eq. (11.9) to account for geological folds, occurs when realistic values of viscosity contrasts between layer and matrix are considered. It will be recalled that, if the viscosity contrast is less than 100/1, explosive amplification of the dominant wavelength will not occur until the percentage compression is greater than 25 per cent. If such compressive strains develop, Biot argues that the initial irregularity, or perturbation, will have undergone a degree of passive amplification which exceeds that generated by buckle folding. If it is assumed that, for significant buckle folding, the viscosity contrast between the layers must equal or exceed 100/1, then from Eq. (11.9) it follows that the slenderness ratio should be 16/1 or greater. However, as will be seen in the following section, the slenderness ratios for the majority of single layer folds in rocks is considerably less than this.

Modification of the Biot/Ramberg single buckling analysis

Sherwin & Chapple (1968) measured the slenderness ratio (L/a) of over 800 natural folds and found that the values ranged between 5/1 and 10/1. If we ignore the fact that Eq. (11.19) should not be applied to structures with low values of L/a and use it to determine the viscosity contrasts for these natural folds, values of viscosity contrasts of less than 10/1 are obtained. The occurrence of well-defined folds with L/a ratios that indicate a difference in viscosity contrasts between layer and matrix which was an order of magnitude smaller than is theoretically required, prompted Sherwin and Chapple to modify Biot's analysis.

Biot points out that an embedded layer compressed parallel to the layer can be shortened by thickening or folding or both, although in his derivation of Eq. (11.9), he assumes that the layer experiences negligible thickening. Sherwin and Chapple note that the small slenderness ratios (L/a) of the natural folds they studied correspond to very low viscosity ratios when these are calculated according to Biot's theory, (Eq. (11.8)). They suggest that even with these small viscosity ratios a strong folding instability exists. However, they argue that with such small viscosity ratios it is certain that shortening of the layer during folding can no longer be neglected. They therefore modified Biot's analysis to take into account the effect of significant shortening and thickening of the layer that may accompany folding and derived the following relationship for the dominant wavelength/thickness ratio:

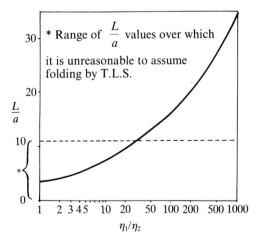

Fig. 11.16. Graphical expression of Eq. (11.9), in which the dominant wavelength/thickness ratio of a buckled viscous layer of viscosity η_1 in a matrix of viscosity η_2, is plotted against the viscosity contrast η_1/η_2. (After Ramsay, 1967.)

$$\frac{L_{\text{d}}}{a} = 2\pi \sqrt[3]{\frac{\eta_1(s-1)}{12\eta_2 s^2}} \qquad (11.15)$$

where $s = (\lambda_1/\lambda_2)^{\frac{1}{2}}$ and λ_1 and λ_2 are the quadratic elongations, $(1 + e)^2$, normal and parallel to the layer respectively, of the homogeneous deformation on which the buckling may be considered superimposed. The difference in rate of thickening on the limb and at the hinge is neglected and it is assumed that the shortening and thickening are those of a flat plate. It can be seen that the only difference between Eqs. (11.8) and (11.15) is the additional factor under the cube root sign.

The modified buckling equation (Eq. (11.15)) indicates that the dominant wavelength/thickness ratio of a single layer buckle depends on both the viscosity contrast between the layer and the matrix *and* on the amount of thickening of the layer. According to this equation, progressive stages of the deformation give rise to a fold which progressively changes its L/a ratio as the structure becomes amplified. By representing the results achieved by Sherwin & Chapple graphically (Fig. 11.17) a comparison can be made with Biot's results (Fig. 11.14). It can be seen that if the effect of layer thickening is taken into account, then explosive amplification of the dominant wavelength for a particular competence contrast between layer and matrix will occur at slightly lower values of bulk shortening than is predicted by Biot's analysis.

The effective of homogeneous shortening during buckling is particularly important when the viscosity contrast between the layer and matrix is small, (Fig. 11.17). The shift of the curves on the fold amplitude/compression plot caused by taking into account the effect of the homogeneous flattening that accompanies buckling, indicates that the minimum

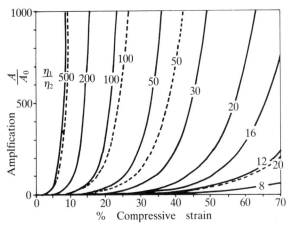

Fig. 11.17. Amplification (expressed as the ratio of the initial A_0 and final amplitude A) of the dominant wavelength as a function of uniform shortening (expressed as compressive strain), for various values of the viscosity ratios. Solid lines after the analysis of Sherwin & Chapple (1968). Dotted lines after Biot's analysis (1961). (After Huddleston, 1973b.)

viscosity contrast of 100/1 suggested by Biot to be necessary before 'clear-cut' folding can occur, is too high. It is interesting to note that Huddleston (1973b) in his experiments on the buckling of a single viscous layer in a viscous matrix, observed 'clear-cut' folding with a marked amplification of the dominant wavelength, with a viscosity contrast between the layer and matrix of only 25/1. (However even this ratio of viscosities when substituted into Eq. (11.9) gives L/a values of more than 10/1, which is greater than the value for the majority of geological single layer folds.)

One of the main problems associated with the application of the buckling theories mentioned so far to the folding of rocks is that they are infinitesimal theories and should be applied only to the first increment of buckling. They should not be used to predict changes of layer thickness or changes in the shape of folds as they develop.

An analysis of single layer buckling which permits small, but finite fold amplification

Fletcher (1977) developed a theory that could be extended beyond the first increment of buckling, to the stage where the amplitude of the fold is small, but finite. In addition, this theory is not based on the *thin plate* assumptions (i.e. it is not assumed that the fold wavelength is significantly greater than the layer thickness). In this *thick plate* analysis, the folding of an isotropic, single, linear viscous layer embedded in a uniformly shortening matrix is treated mathematically by superposing the *mean flow* corresponding to uniform shortening and a *perturbed flow* corresponding to folding. Folding with adherence and folding with slip, between the layer and matrix, are both considered. Fletcher expresses his results in a different form from Biot and Ramberg. He finds that for the case where there is *adherence* between the layer and matrix:

$$q = -2(1 - R)$$
$$\left\{(1 - R^2) - \frac{[(1 + R)^2 e^w - (1 - R)^2 e^{-w}]}{2w}\right\}^{-1}$$
$$(11.16)$$

where R is the ratio of the viscosity of the matrix to the viscosity of the layer (η_2/η_1) and q is a measure of the rate of amplification of a fold with a wavenumber w, where:

$$w = \frac{2\pi a}{L} \qquad (11.17)$$

where, as before, L is the wavelength and a is the layer thickness. The w value of the fold that amplifies at the greater rate, w_{d}, maximises q (Eq. (11.16)) and is related to the dominant wavelength L_{d} by:

$$L_{\text{d}} = \frac{2\pi a}{w_{\text{d}}}. \qquad (11.18)$$

The equation that governs buckling when *slip* occurs between the layer and the matrix is:

$$q = -2\{1 - [(e^w - e^{-w}) + R(e^w + e^{-w} + 2)]/2w\}^{-1}. \quad (11.19)$$

When $w \ll 1$, i.e. the wavelength is much greater than the layer thickness, the exact results (Eqs. (11.16) and (11.19)) approximate to Eqs. (11.20) and (11.21). For (i) adherence and (ii) slip at the layer/matrix interface, the values of q are given by:

$$\text{(i)} \quad q_a \simeq (w^2/12 + R/w + Rw/2)^{-1} \quad (11.20)$$

and

$$\text{(ii)} \quad q_s \simeq (w^2/12 + R/w + Rw/4)^{-1}. \quad (11.21)$$

Here, the first two terms in each equation correspond to the thin plate analysis of Biot and Ramberg. The third term in each is a correction term. Note that in the thin plate approximation, the results are independent of whether the layer adheres or slips at the layer–matrix junction. (It is for this reason that Biot and Ramberg obtained the same equation (Eq. (11.9)) even though Biot assumed a slip at the layer boundary and Ramberg assumed adherence.)

The correction term distinguishes between these two cases. At any wavelength, the instability is greater for the case of the frictionless contacts. If the thin plate approximations are made, the dominant wavenumber given by Eqs. (11.20) and (11.21), is obtained by deleting the correction terms and finding the maximum value of q:

$$w_d = \sqrt[3]{6R} \quad (11.22)$$

which is identical with Eq. (11.9). With the retention of the correction term, the results are:

$$w_d = \sqrt[3]{6R} - R \quad (11.23)$$

and

$$w_d = \sqrt[3]{6R} - \frac{R}{2} \quad (11.24)$$

for welded and frictionless contacts respectively. Note that the corrected dominant wavelengths, which are inversely proportional to the wave-numbers w_d are *increased* in both cases over the thin plate value. The dominant wavelength for the layer with adherence at the layer–matrix junction is larger than that for layers with slip at the junction. It will be recalled that Biot came to the same conclusions. The results of Fletcher's analysis are compared with other thick plate analyses (Biot 1959a and Ramberg, 1960 and 1970b) in Fig. 11.18, in which the dominant wavelength/thickness ratio is plotted against the viscosity contrast between the layer and matrix.

It will be recalled that thick plate analyses have been studied by geologists because of the low L/a ratios of many natural single layer folds (between 5/1

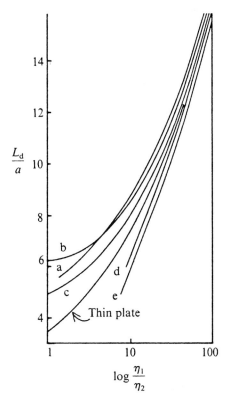

Fig. 11.18. The comparison of the relationship between the dominant wavelength/thickness ratio and the viscosity contrast between the layer and the matrix for five 'thick' layer analyses, curves a to e, and a 'thin' plate analysis. (After Fletcher, 1977.) (*a*) Biot 1959a – Slip (*b*) Fletcher 1977 – Welded (*c*) Fletcher 1977 – Slip (*d*) Ramberg 1970b – Welded (*e*) Ramberg 1962 – Welded.

and 10/1). It can be seen from the curves of Fig. 11.18 that this range of aspect ratios corresponds to low viscosity contrasts (between 1/1 and 50/1) for which it has been argued that deformation would be dominated by homogeneous flattening and that 'clear-cut' folding would not occur. Field geologists who have encountered the numerous natural examples (e.g. Figs. 11.6 and 11.11) of low L/a single layer buckles which occur in regular wave-trains of high amplitude folds, might therefore question the applicability of these curves to geological folding. For this and other reasons that will be discussed later in this chapter in the section on elastic versus viscous behaviour during fold initiation, *the reader is warned against applying, too readily, the results presented in Figs. 11.17 and 11.18 to geological folding.*

An analysis of single layer buckling which permits large, finite fold amplification

Although Fletcher's analysis extends beyond the first increment of buckling, it is only strictly valid for very low amplitude folds. Chapple (1968) extended the linear viscous infinitesimal theory of Biot to a rigorous, mathematical analysis of folds with large finite amplitudes. He assumed that the competent layer was thin (he took 40/1 as the wavelength/layer thickness

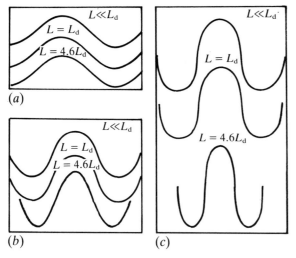

(a)

(b)   (c)

Fig. 11.19. Three stages in the amplification of three different folds in linear, viscous materials; one with the dominant wavelength $(L = L_d)$, one a larger wavelength $(L = 4.6L_d)$ and one with a smaller wavelength $(L \ll L_d)$. (a) Limb dips of approximately 45°: (b) limb dips of approximately 70° and (c) limb dips of 89°. (After Chapple, 1968.)

ratio) and initially folded into a very low amplitude, sinusoidal shape. He calculated progressive shortening across a block containing a single layer in a less viscous matrix. An interesting feature of Chapple's analysis bears on the 'wavelength selection' process proposed by Biot (1961) and discussed earlier in this chapter (Figs. 11.14 and 11.17).

Chapple concludes from his study of the finite growth of folds that when the maximum limb-dip reaches approximately 15° the growth of a given wavelength component of fold shape is no longer independent of the amplitude of the other components. He therefore argues that it seems reasonable to set the limit of the 'wavelength selection' process at this point. Since the wavelength selection process operates only at low amplitudes, the principal factor which determines whether a fold train of the dominant wavelength will develop is the amplitude of the initial shape perturbation. Only if this is small enough will the dominant wavelength assert itself before the finite-amplitude stage (limb dip of 15°) is reached. If appropriate perturbations are present, wavelengths other than the wavelength predicted by Eq. (11.9) may develop. Chapple considered the finite amplification of three different folds, one with the dominant wavelength predicted by Eq. (11.8), one with a larger wavelength and one with a smaller wavelength. He showed that the path of fold development depended only on the ratio of the actual wavelength to the predicted or dominant wavelength L_d and calculated the changes in fold shape for the three folds. Three stages in the growth of these folds are shown in Fig. 11.19, where a progressive departure of fold shape from sinusoidal can be seen, the hinges of the fold becoming more rounded as deformation progresses.

Chapple suggested that the main reason for the lack of correspondence of many natural fold shapes with the shapes developed from his theoretical model was that real rock materials followed non-linear rheological laws. In a later paper (1969) he considered the development of large, finite amplitude folds in a layer with non-linear (viscous-plastic) properties.

Analyses of materials exhibiting non-linear behaviour
A variety of materials with non-linear stress–strain or stress–strain-rate properties have been discussed in Chap. 1. Chapple (1969), in an analysis of the finite growth of folds in a non-linear material, assumed the layer to exhibit viscous-plastic properties (Fig. 11.20(a)). However, his arguments apply equally well to a realistic rheological model, namely the elasto-plastic model (Fig. 11.20(b)). The stresses in the buckling layer are greatest at the layer boundary on the inner and outer arc of the fold hinges and it is at these localities that plastic yielding will first occur. The amplitude of the fold at which the yield stress of the layer is reached at the hinge is given by:

$$\sigma_y < \sigma_b = - F(1 + 6A/a)/a \qquad (11.25)$$

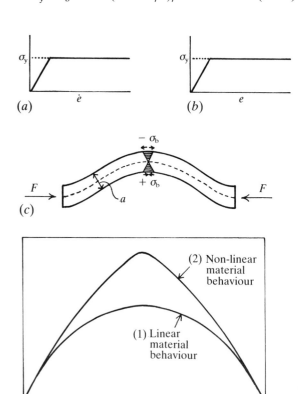

(a)   (b)

(c)

(d)

Fig. 11.20. Stress–strain-rate curve for a viscous-plastic material (a) and stress/strain curve for an elasto-plastic material. (b) σ_y is the yield stress above which plastic deformation (i.e. deformation other than viscous or elastic) takes place. (c) The development of fibre stresses σ_b in the hinge of a fold. When $\sigma_b = \sigma_y$, plastic yielding occurs at the hinge. (d) (1) Profile of a fold in which no plastic yielding occurred at the hinge. (2) Profile of a fold with the same interlimb angle as (1) but in which plastic yielding at the hinge occurred from a limb dip of 4.1° onwards. ((d) After Chapple, 1969.)

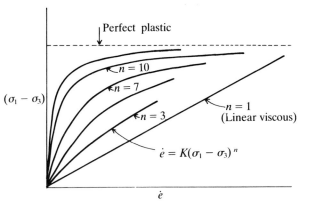

Fig. 11.21. Examples of power law stress–strain-rate relationships. When the stress exponent n is 1 (the material is a Newtonian (i.e. linear) fluid), there is a straight line relationship between stress and strain-rate. For values of n other than 1, the relationship is non-linear.

where σ_y is the yield stress, σ_b is the fibre stress at the boundaries of the layer around the hinge region, F is the buckling force applied along the layer, a is the layer thickness and A is the amplitude of the fold (Fig. 11.20(c)). The shape of the fold will depend primarily on the fraction of the folding history during which plastic yielding occurs. If this fraction is large, the resulting fold will have longer, straighter limbs, a higher amplitude and a higher curvature at the fold hinge than a fold with the same interlimb angle but for which no plastic yielding took place (Fig. 11.20(d)).

Fletcher (1974) and Smith (1977) have both shown how the theory of buckling of an isotropic linear viscous layer can be extended and applied to materials which exhibit non-linear viscous behaviour. In their analyses, they assume that the differential stress $(\sigma_1 - \sigma_3)$ and strain-rate $(\dot{e})$ for such materials are linked by a *Power-law relationship*, so that:

$$\dot{e} = K(\sigma_1 - \sigma_3)^n \qquad (11.26)$$

where K is a constant and n is the power-law exponent. Examples of such relationships for various values of exponent (n) are shown graphically in Fig. 11.21, where it will be noted that for values of $n = 10$, the curves begin to exhibit characteristics of a pseudo-plastic. (Perfect plastic behaviour is indicated by the horizontal dotted line.)

Both the non-linear analyses are restricted to the first increment of buckling and assume that the layer and matrix are incompressible and exhibit the power-law stress–strain-rate behaviour indicated by Eq. (11.26). The analyses take into account the uniform layer shortening that accompanies folding during its early stages and considers the effect of the superposition of a small perturbed flow associated with the buckling instability onto the 'basic state' of uniform compression. One of the most important conclusions of these treatments is that the constitutive

equations (the equations relating stress and strain-rate) that describe the basic state of uniform background flow (Eqs. (11.27)), are different from those that describe the perturbed state (Eqs. (11.28)):

$$\sigma_{11} = 2\eta \dot{e}_{11} - p \qquad (11.27(a))$$

$$\sigma_{33} = 2\eta \dot{e}_{33} - p \qquad (11.27(b))$$

$$\sigma_{13} = 2\eta \dot{e}_{13} - p \qquad (11.27(c))$$

$$\sigma'_{11} = 2\eta'(1/n)\dot{e}'_{11} - p' \qquad (11.28(a))$$

$$\sigma'_{33} = 2\eta'(1/n)\dot{e}'_{33} - p' \qquad (11.28(b))$$

$$\sigma'_{13} = 2\eta'\dot{e}'_{13}. \qquad (11.28(c))$$

Equations (11.27(a) and (b)) relate the normal stresses σ_{11} and σ_{33} to the normal strain-rates $\dot{e}_{11}$ and $\dot{e}_{33}$ and Eq. (11.27(c)) relates the shear stress σ_{13} to the shear strain-rate $\dot{e}_{13}$. The constant of proportionality in these equations (2η) is given by:

$$2\eta = 1/(BJ^{(n-1)/2}) \qquad (11.29)$$

where

$$J = [\tfrac{1}{4}(\sigma_{11} - \sigma_{33})^2] + \sigma_{13}^2. \qquad (11.30)$$

B is a constant, n is the stress exponent of the layer and p is the pressure (mean stress); where mean stress $\bar{\sigma}_m$ is given by:

$$p = \bar{\sigma}_m = (\sigma_{11} + \sigma_{22} + \sigma_{33})/3. \qquad (11.30(a))$$

The deviatoric stresses σ_d which are a measure of how much the stress state deviates from being hydrostatic, are:

$$\sigma_{d11} = \sigma_{11} - \sigma_m \qquad (11.30(b))$$

$$\sigma_{d33} = \sigma_{33} - \sigma_m. \qquad (11.30(c))$$

It will be recalled from Chap. 1 that the sum of the mean and deviatoric stress in a particular direction is the normal stress in that direction.

It can be seen from Eqs. (11.29) and (11.30) that the resistance of deformation, 2η, depends on both the stress state and the value of exponent n of the layer. Comparison of Eqs. (11.27(a)) and (b) with (c) shows that the resistance to shear and resistance to compression for the background flow, are the same (i.e. 2η). Such stress–strain-rate equations characterise the deformation of *isotropic* viscous materials.

In Eqs. (11.28) the primes signify quantities associated with the perturbed flow. Inspection of these equations reveals that the resistance to shear and compression are *not* the same for the perturbed flow. These equations characterise the deformation of an *anisotropic* viscous material whose viscosity in shortening (or extension) parallel to the layer, is less than its viscosity in shear, parallel to the layer, by a factor of $1/n$. It can be seen, therefore, that the incremental stresses associated with the localised, per-

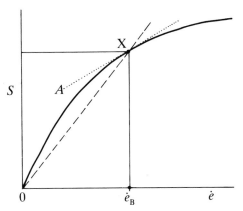

Fig. 11.22. The stress–strain-rate curve for a power-law (non-linear) layer onto which has been superposed a background pure shear (flattening) strain-rate $\dot{e}_B$, parallel to the layer. If an increment of pure shear is added coaxially to this, the modulus relating the increment of stress to the increment of induced strain-rate is given by the slope of the curve at X, whereas when an increment of simple shear is applied parallel to the layer, the modulus relating the increment to that of induced shear strain-rate is given by the slope OX. Because there is a difference between the resistance to compression and shear parallel to the layer, the material is mechanically anisotropic. (After Smith, 1977.)

turbed flow in an initially isotropic power-law material *induce* an anisotropic material response. This anisotropy is caused by a combination of the non-Newtonian behaviour of the material and the basic state of uniform compression that accompanies folding. Although a detailed discussion of this important idea is beyond the scope of this book, the reader may find the following brief discussion useful.

The stress–strain-rate curve for the power-law layer upon which has been superposed a uniform state of pure shear (flattening) represented by $\dot{e}_B$ is shown in Fig. 11.22. Any perturbed flow associated with buckling must be added to this basic state of pure shear. If the increment of perturbed flow is an increment of pure shear, then it must be added to the existing pure shear deformation associated with the basic state. The appropriate modulus that relates the increment of pure shear to the increment of normal stress will, therefore, be the local slope of the stress–strain-rate curve at X (Fig. 11.22). If, however, the increment of perturbed flow is an increment of simple shear parallel to the layer, then, as there is no simple shear parallel to the layer associated with the basic state of background deformation, the basic state will not significantly affect this incremental deformation. The modulus that relates the increment of shear stress to the increment of shear strain it induces is given by the slope of the line OX (Fig. 11.22). The slopes AX and OX are measures of the resistance to deformation when an increment of pure shear and simple shear respectively, associated with the perturbed flow, are applied to the background pure shear. Because these slopes are different, the resistance to increments of pure and simple shear are different. (We have already

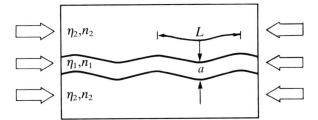

Fig. 11.23. Model for the buckling of a power-law layer in a power-law matrix.

noted that this is so by comparing Eqs. (11.28(*a*) and (*b*)) with Eq. (11.28(*c*)). It follows, therefore, that the material is behaving as an *anisotropic* material in response to increments of the perturbed deformation.

An anisotropy induced by the perturbed flow may either enhance or oppose any *intrinsic* anisotropy that the material may possess in the unstressed state. We shall return to this concept of induced and intrinsic anisotropy in Chap. 13 where we consider the buckling behaviour of anisotropic materials.

Both Fletcher and Smith show that the growth rate (q) of an interface irregularity in an embedded layer will depend on the power-law exponent of each material n_1 and n_2 (for the layer and matrix respectively), the ratio of the wavelength of the disturbance to the layer thickness (*a*) and the ratio (*R*) of the viscosities of the matrix (η'_2) and layer (η'_1) materials in the perturbed state (η' of Eqs. (11.28)), (Fig. 11.23), when:

$$R = \frac{\eta'_2}{\eta'_1}. \tag{11.31}$$

The formula which they developed for the growth rate is:

$$q = 2n_1(1 - R)\{ - (1 - Q^2)$$
$$+ (n_1 - 1)^2 [(1 + Q)^2 e^{wc}$$
$$- (1 - Q)^2 e^{-wc}]/2 \sin Bw\}^{-1} \tag{11.32(a)}$$

or

$$q \simeq n_1 \left(\frac{w^2}{12} + n_1^{\frac{1}{2}} \frac{Q}{w} \right)^{-1} \tag{11.32(b)}$$

where

$$Q = R(n_1/n_2)^{\frac{1}{2}} \tag{11.32(c)}$$

$$w = \frac{2\pi a}{L} \tag{11.32(d)}$$

$$c = (1/n_1)^{\frac{1}{2}} \tag{11.32(e)}$$

and

$$B = (1 - 1/n_1)^{\frac{1}{2}}. \tag{11.32(f)}$$

Using Eq. (11.32(*a*)), Fletcher determined the rate of amplification of a range of wavelength thickness ratios for given values of R, n_1 and n_2. The results are summarised in Fig. 11.24. Each curve on

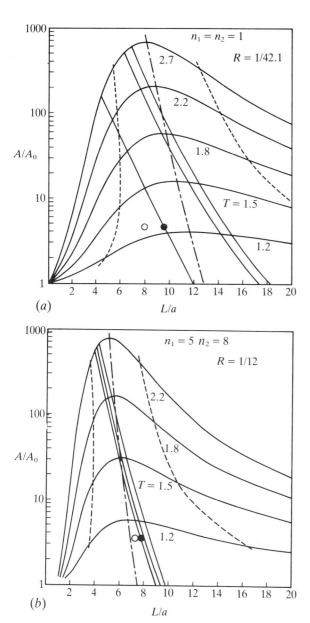

Fig. 11.24. Amplification spectra as a function of layer thickening (*T*), for (*a*) a linear layer and matrix ($n_1 = n_2 = 1$) with a viscosity contrast of 42.1/1; and (*b*) a power-law (non-linear) layer and matrix ($n_L = 5$, $n_M = 8$) with a viscosity contrast of 12/1. Note that as the layer thickness changes (with progressive deformation) the wavelength/thickness ratio of the dominant wavelength also changes. (After Fletcher, 1974.)

the graphs corresponds to a particular amount of layer thickening *T*, (where $T = (a/a_0)^2$, a_0 being the original thickness of the layer and *a* the thickness at the time of interest). The buckling behaviour illustrated by Fig. 11.24(*a*) is for a *linear* viscous layer embedded in a *linear* viscous matrix ($n_1 = n_2 = 1$). The viscosity contrast between the layer and the matrix is 42.1/1 and each curve shows the rate of amplification of a 'spectrum' of fold wavelengths over the range of wavelength/thickness ratios of 1/1 to 1/20. The curves are known as *amplification spectra* and the *L/a* ratio that amplifies most rapidly corresponds to the *dominant wavelength*. It is interesting to

note that the *L/a* ratio that amplifies most rapidly *changes* as the amount of uniform layer shortening (*T*) increases. According to Fletcher, this process of *dominant wavelength selection* effectively ceases when the limb-dips reach approximately 15°. (However, one should note that in order to reach this latter conclusion, it is necessary to ignore the fact that the theory only applies to the first increment of buckling behaviour). At this point, the shortening of the layer by predominantly uniform layer thickening gives way to buckle shortening, and the hinge position of the low amplitude fold configuration that exists at this time is subsequently preserved during finite amplitude folding. Work by Huddleston (1973a) substantiates this conclusion, for he observed that, in folding experiments, layer thickening ceased at the range of limb-dips from 10° to 20°.

The graph in Fig. 11.24(*b*) shows a similar set of amplification spectra for a non-linear viscous layer and matrix. The stress exponents of layer (n_1) and matrix (n_2) are 5 and 8 respectively and their viscosity contrast is 12/1. It can be seen, by comparing the graphs of Fig. 11.24(*a*) and (*b*), that even with a significantly lower viscosity contrast between the layer and matrix, the dominant wavelength exhibits a higher degree of amplification in the non-linear example, relative to that of the linear example for the same degree of layer thickening.

It was noted earlier that when Sherwin & Chapple took into account the flattening that accompanies folding, they found that explosive amplification occurred at values of total compressive strain lower than those predicted by Biot. They therefore concluded that the initial viscosity contrast necessary for rapid amplification of folds would be less than 100/1. We now see that if the layer and matrix are made of non-linear materials, the amplification of folds occurs even more rapidly. This effect would further reduce the viscosity contrast necessary for clear-cut folding to develop. Indeed, Fletcher concludes that folds with an arc wavelength/thickness ratio in the range 4–10 (the range commonly observed in natural folds) can form at small to moderate amounts of uniform layer thickening, provided the layer rheology is strongly non-linear ($n > 3$).

Although the conclusions of Fletcher (1974) and Smith (1977) regarding the buckling behaviour of a single layer with non-linear viscous properties are formally identical, Smith treats a broader class of fluid properties and considers both layer-parallel and layer normal compression. He presents his results in a different form from Fletcher and plots (i) the dominant wavelength/thickness ratio against the competence contrast between the layer and the matrix ($m = 1/R$) and (ii) the growth rate (*q*) of the dominant wavelength against the competence contrast between the layer and the matrix (for various combinations of

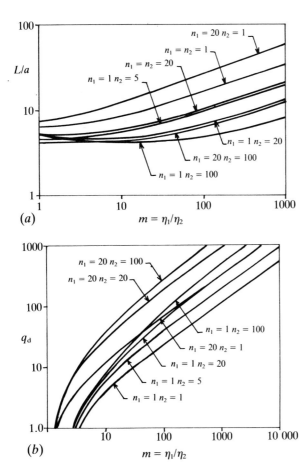

(a)

$m = \eta_1/\eta_2$

(b)

$m = \eta_1/\eta_2$

Fig. 11.25. (a) Plot of dominant wavelength/thickness ratio (L/a) against viscosity contrast (η_1/η_2) between the layer and matrix for various values of power law exponents n_1 and n_2, for the layer and matrix respectively. (b) Normalised, dynamic growth rate (q_d) for folding plotted against viscosity contrast between the layer and matrix for various values of power law exponents n_1 and n_2. (After Smith, 1977.)

non-linearity for both the layer and the matrix). These graphs are shown in Fig. 11.25(a) and (b), and it is immediately apparent from (a) that when both L/a and m (the competence contrast between the layer and the matrix) are high there is a simple relationship between these two parameters. If we apply the thin layer approximations to Eq. (11.32(a)), i.e. assume that the wavelength is considerably greater than the layer thickness (L/a is large), the resulting growth-rate is given by Eq. (11.32(b)) for which the dominant wavelength/thickness ratio is:

$$\frac{L}{a} \approx 2\pi \sqrt[3]{\frac{1}{6R}}. \qquad (11.33)$$

R is the ratio of the resistance to deformation of the matrix and layer. The shear and compressive moduli of viscosity ($\eta^s \; \eta^c$) for a non-linear material are different. For example, for the non-linear material described by Eqs. (11.27) and (11.28) we see that for the perturbed state (Eqs. (11.28)) these viscosities are:

$$\eta^c = \frac{\eta'}{n} \qquad (11.34(a))$$

$$\eta^s = \eta' \qquad (11.34(b)).$$

In order to define R, it is necessary to determine which of these two moduli represents the resistance to deformation in the layer and the matrix. Fletcher selects a value of R for Eq. (11.33) of:

$$R = (\eta_2^c \; \eta_2^s)^{\frac{1}{2}}/\eta_1^c \qquad (11.34(c))$$

and justifies this choice by arguing that because the flow of the matrix during buckling involves about equal shear and shortening parallel to the layering the relevant viscosity of the matrix is the geometric mean of the viscosities of these modes, $(\eta_2^c \; \eta_2^s)^{\frac{1}{2}}$, whereas in the layer, buckling involves predominantly the shortening or extension of fibres lying parallel to the layer. The relevant viscosity for the layer is therefore η_1^c.

The relationship between the normalised dynamic growth rate, q, (i.e. the dynamic amplification of the fold divided by the passive amplification associated with the background pure shear deformation) and the competence contrast for various values of non-linearity of the layer and matrix are shown in Fig. 11.25(b). Smith points out that for large values of η_1/η_2 the relationship approximates to:

$$q = -1.21(n_1 n_2)^{1/3}/R^{2/3}. \qquad (11.35)$$

It can be seen that the growth-rate of a fold is greatly enhanced if either the layer and/or the matrix are non-linear. For example, for a linear layer and matrix with a viscosity ratio of 100/1 the fold growth-rate q is approximately 26. If however the layer and matrix

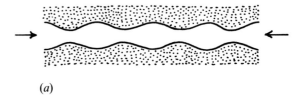

(a)

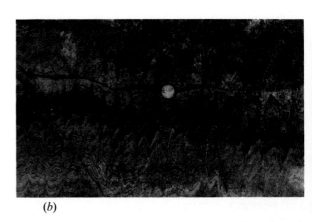

(b)

Fig. 11.26. (a) Mullion structures (discussed in Chap. 16) caused by layer parallel compression of a less competent layer set in a more competent matrix. (After Smith, 1977.) (b) Mullion structures developed at both boundaries of a relatively incompetent (dark) layer, Monar, Scotland. The structure in (b) indicate clearly that the initial sinusoidal geometry becomes cusp-like with progressive deformation.

have the same viscosity ratio but are non-linear materials with stress exponents n, $n_1 = n_2 = 10$, the growth rate q is 126. Smith points out that this difference in growth-rate would make a huge difference to the final amplitude of the fold and that non-linear material behaviour will be especially important for small viscosity ratios ($m < 10$), where it can make the difference between weak and strong growth.

Smith also considers the instability of an *incompetent* layer subjected to layer-parallel compression and embedded in a more competent matrix. He termed the resulting structures *mullions*, Fig. 11.26(*a*). Although the dynamic and kinematic growths co-operate to amplify any disturbance, it is found that the growth-rate of these structures is very slow and Smith concludes that they will develop only in a material in which the rate of deformation increases strongly with increase in strain (i.e. n, the power-law exponent is considerably greater than 1), subjected to large deformations. Inspections of natural examples of these structures shows that the sinusoidal deflections at the layer/matrix interface become cusp-like as the structure amplifies, Fig. 11.26(*b*).

Smith does not restrict his treatment of the deformation of single layers with non-linear material properties to layer-parallel compression. He also considers layer normal compression. However it is convenient to defer discussion of this until Chap. 16.

In a later paper, Smith (1979) used Eq. (11.32(*a*)) to consider the buckling behaviour of a layer with *strongly* non-linear properties. The results of this study are expressed graphically in Fig. 11.27. Two qualitative points are immediately apparent from Fig. 11.27(*a*). These are: (i) If the power exponent of the layer (n_1) is not too large then, for large viscosity contrasts, the dominant wavelength is related to the competence contrast according to the predictions of the Biot–Ramberg Eq. (11.9) which is expressed graphically in Fig. 11.16, and (ii) If the viscosity contrast between the layer and matrix is not too large, the dominant wavelength-thickness ratio lies close to, but above, 4 (usually between 4 and 6).

These two features can also be seen, if the dominant wavelength data of Fig. 11.27(*a*) is replotted as a parameter map (Fig. 11.27(*b*)), which is obtained by plotting the viscosity contrast against n_1, the power exponent of the layer, keeping n_2, the power exponent of the matrix, equal to 1.0. The ratio L_d/a, resulting from any particular combination of viscosity contrast (m) and n_1, is indicated. When m is greater than n_1, (low and to the right of the plot) L_d/a depends on the material properties (in agreement with the classic buckling theory) while, when n_1 is greater than m, L_d/a approaches 4 and is nearly independent of the material properties. Smith has suggested that these two situations correspond to two distinct mechanisms of buckling. The first is classical single layer buckling

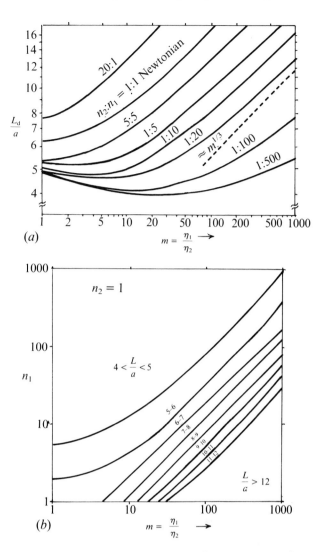

Fig. 11.27. (*a*) Plot of dominant wavelength/thickness ratio (L_d/a) against the viscosity contrast between the layer and the matrix (η_1/η_2) for various values of n_1 and n_2 according to Eq. (11.32(*a*)). At large viscosity contrasts, L_d/a increases as $m^{1/3}$ in agreement with classical buckling theory (Eq. (11.9)). When the layer is very non-linear (i.e. when n_1 is large), the curves cluster between $L_d/a = 4$–6. (*b*) The dominant wavelength data of (*a*) replotted as a parameter map with $n_2 = 1$ (i.e. a linear, viscous matrix). The ratio L/a resulting from any particular combination of viscosity contrast (m) and layer non-linearity (n_1) is indicated. When $m > n_1$ (low and to the right), L_d/a depends on the material properties in agreement with classical buckling theory, while when $n_1 > m$, L/a approaches 4 and is nearly independent of the material properties. (After Smith, 1979.)

as described and analysed by Eq. (11.9), and produced experimentally by Biot *et al.* (1961). The second type of buckling occurs when the non-linearity of the layer is larger than the viscosity contrast between the layer and matrix. This mechanism of folding, in which the material properties (e.g. viscosity contrast) play very little role in determining the dominant wavelength that develops, has been described by Smith as *resonant folding*. In resonant folding the competent layer does not act mechanically as a coherent unit, but instead the irregularities on one interface produce motions that deform the other and vice versa. This

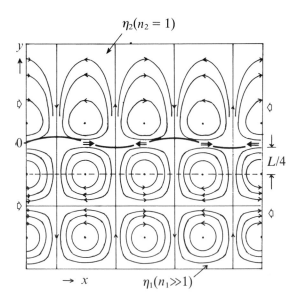

Fig. 11.28. The secondary flow induced by the waviness of the interface between the different materials ($\eta_1 > \eta_2$) undergoing pure shear. The unbalanced normal stresses in the basic, pure shear which drives the secondary flow are shown by the double arrows. For the sake of illustration, the lower material (η_1) is strongly non-Newtonian ($n_1 \gg 1$), whereas the upper material (η_2) is Newtonian ($n_2 = 1$). The relative values of the stream function are given in the diagram. The vertical velocities vanish near the interface ($y = 0$), so the interface waviness will not be amplified by the secondary flow. A passive marker line at $y = L/4$ (shown dashed) will soon become wavy as a result of the strong vertical velocities there. (After Smith, 1977.)

'resonant' growth mechanism leads to a dominant wavelength/thickness ratio between 4 and 6.

Consider the region near the interface between a strongly non-linear, relatively competent layer and a linear, less competent matrix. If the interface is slightly wavy, then a non-uniform motion of the material (i.e. a secondary flow) will be induced by this waviness during layer-parallel pure shear (flattening), Fig. 11.28. It can be seen that the displacements normal to the interface vanish in its vicinity so that its waviness is not amplified. However, a straight, passive marker line, at some distance y from the interface, will soon be deflected into a wavy form. In a strongly non-Newtonian material, the most rapid deflections will occur if the marker-line is located at distance $L/4$ away from the interface. Smith argues that in a single layer (i.e. a two interface) situation, it is largely the secondary flow produced by one interface that deforms the other. We have just seen, in fact, that a single interface is not effective in amplifying its own waviness. Further, it is clear from Fig. 11.28 that in strongly non-Newtonian materials, the deformation of each interface by the other can operate most effectively if the interfaces are at distance $a = L/4$ apart. The selection of the dominant wavelength such that $L_d/a = 4$ is thus seen to be geometrical in nature and insensitive to the exact values of the material parameters (as long as n_1 is large).

To summarise, two changes occur in the buckling behaviour of a single layer as the non-linearity of the layer (n_1) becomes large. First, the viscous resistance to folding occurring in the layer decreases so much that the system no longer has to choose a long wavelength to get optimum growth. Thus, 'classical' theories of the buckling of linear viscous layers are no longer useful. Second, when the non-linearity of the layer is large, the secondary flow in the layer (in a direction normal to the layer) is nearly periodic and decays only slowly away from the driving interface. The condition of optimum distortion of each interface by the other then leads to a wavelength/thickness value equal to, or slightly greater than, 4.

The division between classical single layer buckling described by Eq. (11.9) and short-wave, non-Newtonian folding is, of course, not sharp but continuous and it might be difficult to find a natural fold that was perfectly described by either model. (This would require either extreme contrast between the layer and matrix ($\eta_1/\eta_2 > 100$ and $n_1 \approx 1$) or extreme non-Newtonian properties of the layer (e.g. $n_1 > 20$.)

Smith refers to the average wavelength/thickness values of nine groups of natural folds described in the literature. The average values of these groups, which include over 800 folds, range between 4 and 6.8. He argues that the simplest and most reasonable explanation for the occurrence of folds with $5 < L/a < 6$ is that the folding layer was quite strongly non-Newtonian which in turn caused the resonance mechanism to operate. This not only explains how these short wavelength folds form, but why they are so common. However some folds with L/a ratios between four and six occur at high crustal levels where it is probably unreasonable to assume that competent rocks are viscous; a point to which we shall return.

We have reviewed some of the theoretical work on buckling of single layers and, as we have seen, this relates in the main to fold initiation. Let us now examine two other methods by which the process of single layer buckling has been investigated. These are finite element analysis and model work.

Finite element analysis

The finite element method has been used by various workers on a range of problems associated with single layer buckling. It is particularly useful for examining the finite development of folds from low amplitude buckles. Dieterich & Carter (1969) used this method of analysis to examine the growth of, and to determine the stress and strain distribution in, a linear viscous layer buckling within a linear viscous matrix. The viscosity contrast between the layer and matrix was set at 42/1 and Biot's single layer buckling equation (Eq. (11.9)) was used to determine the dominant wavelength/thickness ratio of the initial, low ampli-

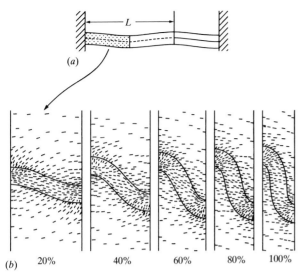

(a)

(b) 20%   40%   60%   80%   100%

Fig. 11.29. (*a*) Initial sinusoidal deflection of a competent layer in a less competent matrix ($\eta_1/\eta_2 = 42$, $L/a = 12$). (*b*) Diagram showing the geometry of the fold in (*a*) and the orientation of maximum, principal compressive stress (σ_1) in the layer and matrix at increments of 20 per cent natural strain. Note that the area of relative tension in the hinge enlarges with increasing strain and that σ_1 rotates to increasingly higher inclinations to the limbs. (After Dieterich & Carter, 1969).

tude, sinusoidal buckle. It was assumed that this initial fold had already grown such that its amplitude was one tenth of the layer thickness Fig. 11.29(*a*). The model was deformed by successive increments of 0.02 compressive, natural strain. The change in geometry of the layer with progressive deformation and the orientations of the principal compressive stresses are shown in Fig. 11.29(*b*). It is found that during the early stages of buckling, the principal, compressive stresses (σ_1) of large magnitude are parallel to the layer throughout. With increasing fold amplification, σ_1 along the limbs decreases in magnitude and rotates making increasingly higher angles with the layer. However, in the hinge region, below the neutral surface, the maximum principal compressive stress is parallel to the layering. Above it, the layer-parallel compression decreases and may become tensile so that locally the minimum principal compression is parallel to the layering.

Microstructures induced by intergranular flow of calcite and quartz have been used to determine the principal stress orientations around natural folds. A review of this method is given in Carter & Raleigh (1969). It is generally found that in specimens from the limbs of open folds, σ_1 is inclined at a low angle to the layer, σ_2 is parallel to the fold axis and σ_3 is nearly perpendicular to the layer. In the hinge area above the neutral surface, although the data is more scattered, the microstructures indicate an effective tensile stress field. Dieterich & Carter conclude that these findings indicate that the stress history during the formation of natural folds is comparable with that predicted by the viscous model studied; and that the fabric data record

cumulative, intragranular flow during folding controlled by the magnitude as well as the orientation of the principal stresses.

The effect of layer/matrix cohesion on the rate of fold amplification

One of the present authors (J.W.C.) has used the finite element analysis technique to determine the effect of adherence and slip at the layer–matrix junction, on the rate of amplification of a single layer fold. He considered a linear elastic layer in a linear elastic matrix

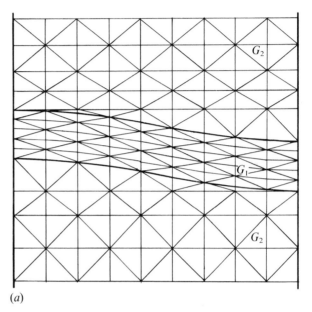

(a)

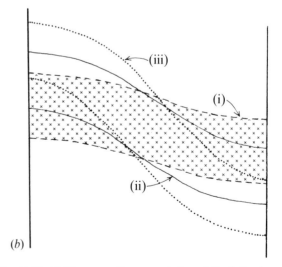

(b)

Fig. 11.30. (*a*) Diagrammatic representation of the finite element model used to investigate the effect of adherence and slip at the layer/matrix junction, on the rate of amplification of a fold. Initial amplitude of fold is 1/10 the layer thickness. Amplitude exaggerated by × 3 for clarity. (*b*) Three fold profiles formed after 20 per cent shortening. The ornamented profile (i) shows the amplification of the initial fold when passive amplification occurs (i.e. when $G_1 = G_2 = G_3$). (ii) Shows the amplification that occurs when adherence exists between the layer and the matrix i.e. when $G_1 > G_2 = G_3$ and (iii) the amplification when slip can occur at the layer/matrix junction ($G_1 > G_2 \gg G_3$). Vertical exaggeration × 3.

and selecting a competence contrast between the layer and matrix of 42/1, used the corresponding elastic moduli in Eq. (11.9) to determine the dominant wavelength/thickness ratio of the initial, low amplitude, sinusoidal buckle, ($L/a = 12.1$). Following Dieterich & Carter, the initial fold was given an amplitude of one tenth the layer thickness. Thin layers (not shown) were placed on each side of the layer, separating it from the matrix (Fig. 11.30(a)). By varying the properties of these two thin layers it is possible to investigate the situation in which no slip occurs between the layer and the matrix during folding (i.e. the strength of the thin layers is made the same as that of the matrix) and the situation when the resistance to slip at the layer–matrix interface is effectively zero (i.e. the shear modulus of the thin layer G_3 was made much smaller than that of the matrix). The amplification of the initial buckle after 20 per cent natural strain when the thin layers were given the same properties as the matrix (simulating cohesion between the layer and matrix) and when the thin layers were given effectively zero strength (corresponding to no resistance to slip at the layer–matrix junction) are shown in Fig. 11.30(b). The amplifications are compared with the amplification of the initial buckle after 20 per cent strain when the properties of the matrix, the layer and the thin layers are all the same. In this last model, because the model is made of only one material, deformation is by pure shear and the initial, sinusoidal buckle amplifies passively. It can be seen from Fig. 11.30(b), that if the resistance to slip at the layer–matrix junction is very low, the rate of amplification of the fold is considerably greater than if there is cohesion at the interface.

Various computer runs were performed in which the strength of the thin layers was varied between that of the matrix and zero. The variation in fold amplification after 20 per cent shortening is shown in Fig. 11.31, where it can be seen that the effect of a weak film at the layer–matrix interface on the amplification of a fold is not very significant unless the strength of the film approaches zero. However, as will be discussed in Chaps. 14 and 15,

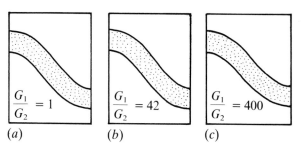

Fig. 11.32. Amplitude of three 'models' after 20 per cent shortening, all with effectively zero cohesive strength between the layer and matrix ($G_3 = 1$). (a) Layer and matrix the same ($G_1 = G_2$); (b) $G_1/G_2 = 42$; (c) $G_1/G_2 = 400$. Vertical exaggeration $\times$ 3.

crystal fibres formed on bedding planes during folding indicate that the resistance to slip at the bedding–matrix junction does become extremely small, possibly zero, during the formation of many natural folds.

The deformation of two other models was considered in which the thin layers, separating the buckling layer from the matrix, were given negligible strength. In the first model, the contrast in competence between the layer and matrix was 1/1 and in the second, 400/1. The results of these 'experiments', after 20 per cent shortening may be compared, in Fig. 11.32, with the model in which the competence contrast was 42/1 and the resistance to slip on the layer–matrix interface was negligible. The final amplitudes of the buckles in all three experiments are approximately the same, indicating that if the cohesion between layer and matrix approaches zero, the fold amplification will be insensitive to competence contrast.

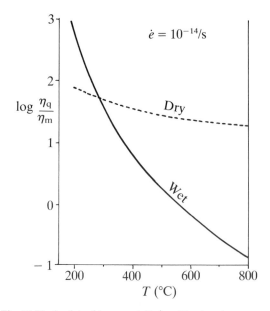

Fig. 11.33. A plot of log quartzite/marble viscosity contrast at a strain rate $\dot{e} = 10^{-14}$/s as a function of temperature. The high temperature-dependence of the wet quartzite/marble viscosity contrast is a direct consequence of the strong temperature-dependence of the viscosity of *wet* quartzite. (After Parrish *et al.*, 1976.)

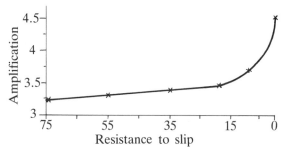

Fig. 11.31. Amplitude of fold after 20 per cent shortening plotted against the cohesive strength of the layer/matrix junction. A significant increase in amplitude occurs as the cohesive strength approaches zero.

In an attempt to get his finite element model closer to the geological prototype, Parrish (1973) took rheological data pertaining to the non-linear behaviour of dry quartzite and marble, and introduced this non-linear behaviour into his analysis. In this way, he simulated the growth of a dry quartzite fold in a matrix of marble. The results were very like those of Dieterich & Carter who used linear material properties for the layer and matrix. As in the models of Dieterich & Carter, the folds remained approximately concentric even after large amounts of amplifications. However, Parrish observed that many geological folds, particularly in rocks with a high metamorphic grade, are approximately similar and not concentric, and in a later paper (Parrish *et al.*, 1976) attempted, by finite element analysis, to form similar folds. He used the same model, of a quartzite layer in a matrix of marble, but performed a series of experiments on quartzites to determine a new flow law for the quartzite layer, one that is appropriate when the quartzite is deformed in the presence of water. The viscosity of wet quartzite is much more strongly temperature dependent than that of marble or dry quartzite (Fig. 11.33). Thus, at a strain-rate of 10^{-14}/s the quartzite/marble viscosity ratio drops from 10 at 375 °C to 1 at 550 °C. (It is interesting to note that if the temperature is raised above 550 °C, the viscosity contrast between the layer and the matrix drops below 1; i.e. the layer becomes less competent than the matrix.) A computer simulation of folding of the wet quartzite layer at 375 °C and a strain-rate of 10^{-14}/s produced concentric folds even after a shortening of 80 per cent. This and the other finite element experiments described so far, all illustrate that the buckling of a relatively competent layer in a less competent matrix generates approximately concentric folds, even after very large shortening. In order to obtain approximately similar folds in these originally competent layers, it is necessary to subject the parallel folds to considerable homogeneous flattening, during which time active buckling is relatively unimportant.

To obtain significant homogeneous flattening of their folds, Parrish *et al.* found it necessary to change the temperature (and, therefore, the viscosity contrast) at some point in the simulated deformation. Fig. 11.34 shows the results of such a computer experiment. The temperature was maintained at 375 °C (viscosity constant at 10/1) until 20 per cent shortening had occurred and then raised to 550 °C (i.e. viscosity contrast dropped to 1/1) for the remainder of the experiment. It can be seen that this 'unnatural' and geologically unrealistic process of single layer buckling gives rise to initially concentric folds and that, during the subsequent homogeneous flattening of these folds, the geometry changes towards that of a similar fold. In the formation of natural folds, no such temperature increase is

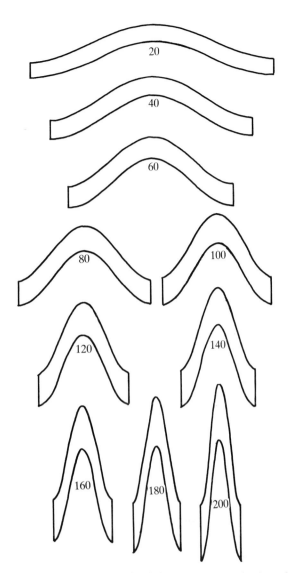

Fig. 11.34. Finite element simulation of wet quartzite layer in marble matrix shortened by 20 per cent at 375 °C (i.e. the competence contrast $\eta_q/\eta_m = 10$, see Fig. 11.33) and then compresses further at 550 °C (competence contrast drops to 1). Stages of total shortening between 40 and 200 per cent total shortening, the folds have flattened to nearly similar geometry. (After Parrish *et al.*, 1976.)

required, and post-buckle flattening of the folds becomes the dominant mechanism of shortening, once the buckles become fully developed and lock-up.

By varying the amount of compression, during which active buckling occurred before flattening was imposed as the dominant mechanism of shortening, and by comparing the resulting fold geometries with natural folds, they concluded that in many natural folds about 40 per cent of the shortening occurred by buckling and the remainder by flattening.

The use of finite element analysis in the study of the development and finite amplification of geological structures has great potential, as can be seen from the examples quoted in this chapter and in Chaps. 12 and 17. The technique involves the use of numerical methods to solve differential equations and the

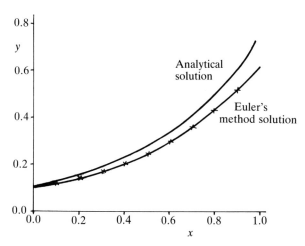

Fig. 11.35. Comparison of the analytic and Euler's method solution of $dy/dx = 2y$ with initial conditions $y = 0.1$, $x = 0.0$. (After Casey, 1976.)

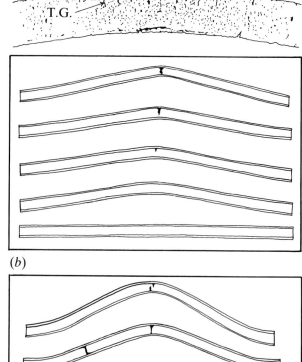

(b)

(c)

Fig. 11.36. (*a*) Tension gashes formed on the outer arc of a folded sandstone beam. (*b*) and (*c*) show experimentally folded beams under 1 kbar confining pressure. All are 0.6 cm thick, 3.2 cm wide and 20 cm long. (*b*) Sandstone beams permanently shortened (from bottom to top) <0.1 to 1.5 per cent. (*c*) Indiana limestone beams permanently shortened (from bottom to top) 0.3, 0.9, 0.6, 4.3, 5.1 and 7.6 per cent. (After Handin *et al.*, 1972.)

simplest method of numerical integration is the Euler method used by Dieterich & Onat (1969) and Dieterich & Carter (1969) in the work discussed above, (Figs. 11.4, 5 and 29). Casey (1976) has determined the errors that accrue in an analysis where this method is used. He points out that in general, equations that describe deformations that are of interest to structural geologists will have an exponential solution as, for example, the exponential growth in amplitude of a single layer buckle, Eq. (11.5), and argues that any method of integration used should be capable of solving such equations accurately. He suggests that the Euler method is not the most suitable and illustrates this point by comparing two solutions to a differential equation, $(dy/dx = 2y)$, (Fig. 11.35). The solution obtained using Euler's method falls progressively behind the analytical solution and thus, when used to trace the development of mechanical instability (as in the finite element analyses discussed above) will always give a solution which errs on the low side and so underestimates the amplification.

Model work on single layer buckling

Experiments on single layer buckling can be divided into experiments in which real rocks are used and experiments using rock analogues.

Relatively few experiments on the buckling of single layers have been performed using real rock and these have been restricted to the range of environmental conditions pertaining to the upper levels in the crust. They include the work of Handin *et al.* (1972) on limestone and sandstone and Gairola & Kern on halite (1982) and marble and limestone (1984).

Handin *et al.* deformed beams of sandstone and limestone at room temperature, under various confining pressures (0.7–2.5 kbar) at a strain-rate of 10^{-4}/s. They found that end-loaded thin beams of rocks with length/thickness ratios of 30/1 or more

were unstable. The beams buckled elastically and their critical buckling stresses were well predicted by the Euler formula for a beam fixed at both ends. This aspect of buckling is discussed at some length in Chap. 14, where the slenderness ratio/buckling stress relationship noted here is shown in Fig. 14.9. Although the sandstone flowed cataclastically and hence had macroscopic ductility, it was locally brittle under the test conditions, and failure first occurred in the hinge region by tensile fracturing, (Fig. 11.36(*a*)). These fractures are consistent with layer-parallel extension in the outer arc and layer-parallel compression in the inner arc associated with the buckling of a homogeneous isotropic layer (Fig. 10.26(*a*) and (*b*)). However, it is interesting to note that the fractures on the inner arc of the fold in Fig. 11.36(*a*) appear to be splitting (i.e. extension) fractures formed by layer-parallel compression rather than the shear

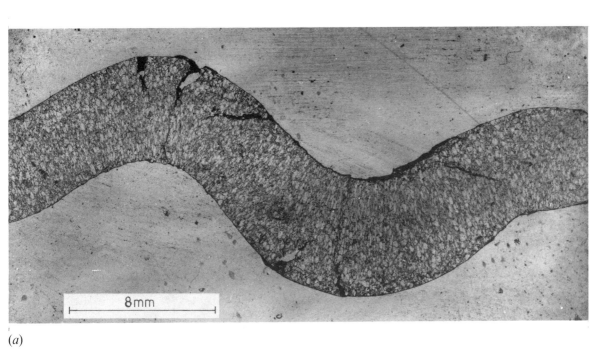

(a)

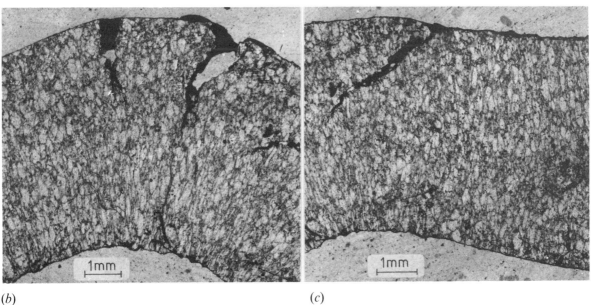

(b)   (c)

Fig. 11.37. (a) Experimentally folded single layer of rock salt in a matrix of Plasticine at 200 °C and confining pressure of 10 bar. (b) and (c) details of (a). (After Gairola & Kern, 1982.)

fractures shown in Fig. 10.26(b). As folding progressed, the limbs rotated rigidly, while the inelastic deformation spread outwards from the hinge region. The final fold shape was that of a chevron, (Fig. 11.36(b)). In contrast, the limestone beams under the conditions of the experiments behaved predominantly as macroscopically ductile materials, their inelastic deformation occurring by intracrystalline twin and translation gliding in the calcite grains. The profile of these folds approximates well to a sinusoidal curve (Fig. 11.36(c)).

Experiments on the buckling of a layer of artificial polycrystalline rock salt embedded in a Plas-

ticine matrix have been conducted by Gairola & Kern (1982). In a layer with a slenderness ratio (L/a) of 13, layer-parallel compression at 200 °C under a confining pressure of 10 bars produced a sinusoidal fold (Fig. 11.37). Folding began after 8 per cent bulk shortening. Layer-parallel shortening and thickening continued with progressive folding, and changes in the arc length, thickness and limb-dip showed that initially the fold growth was slow and then became rapid. The inner arc of the fold was characterised by a strong, preferred dimensional grain orientation, with the longer axis normal to the layer surface. The axial ratio of the grains gradually decreased towards the outer arc, where extension fractures were developed.

The second group of experiments, using rock analogues, can be further divided into those in which no particular attention is paid to the material properties and those in which the material properties and scale of the model have been carefully considered. Experiments of the first type have been used for over a century and have provided insight into the way in which various finite fold geometries develop. However, the amount of detailed information about the processes of folding that can be gleaned from these experiments is considerably less than that which can be obtained from the second group in which the material properties are properly scaled. (The reader is referred to Chap. 10 for a brief outline of scale-model theory.)

The buckling of properly scaled single layer models have been studied by Biot *et al.* (1961) Ramberg (1961a, 1963b and 1964), Ghosh (1966), Huddleston (1973b), and Neurath & Smith (1982). In these various studies, photo-elastic and other techniques have been used to analyse the stress and strain distributions around a fold and to check the validity of buckling equations. Biot *et al.* (1961) conducted experiments to check the validity of the two buckling Eqs. (11.3) and (11.8) which Biot derived for the buckling of an elastic and a viscous layer, respectively, in a viscous matrix. The results for the elastic layer are shown in Fig. 11.38 which, in accord with Eq. (11.3), demonstrates that wavelength is related to the applied load. It can also be inferred from Eq. (11.3) that the wavelength of an elastic layer is independent of the viscosity of the matrix. Experiments in which identical elastic layers are contained and folded in matrices of different viscosities, confirm this prediction. In another set of experiments, in which the competent layer was viscous, it was established, Fig. 11.39, that the wavelength/thickness ratio of the single layer buckle did indeed depend upon the ratios of the viscosities of the layer and matrix, as indicated by Eq. (11.9), see Table 11.1.

Table 11.1. *Dependence of wavelength on viscosity ratio*

η_2 (poise)	η_1/η_2	a (cm)	L/a (calc.)	L/a (experiment)
700	42 800	0.37	121	119–127
13 500	2220	0.35	45.1	35.3–51.4
13 500	2220	0.87	45.1	39.1–47.1
13 500	2220	1.08	45.1	35.2–48.1

Earlier in this chapter it was argued that Eq. (11.8) could be applied to a model of an elastic

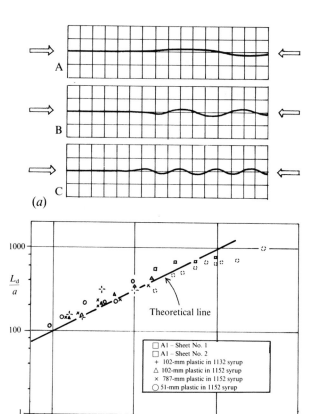

Fig. 11.38. Experimental verification of Eq. (11.3). Buckling of elastic layers in a viscous medium under various loads. (*a*) 1 mm acetate layer in corn syrup (A) 1.6 kg load; (B) 6.6 kg load; (C) 11.6 kg load. All loads applied from right hand side. (*b*) Graph of results of buckling of elastic layers in a viscous medium. L_d is the wavelength; a is the layer thickness; P is the compressive load; E = Young's modulus of the layer; v = Poisson's ratio of layer. (After Biot *et al.*, 1961.)

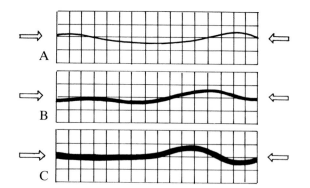

Fig. 11.39. Buckling of viscous layers in a viscous medium details below.

Load (kg)		Time applied (s)	Layer thickness	η_1/η_2
A	1.6	7.5	0.37	40 000
B	11.6	49.0	0.87	2000
C	11.6	19.0	1.08	2000

(After Biot *et al.*, 1961.)

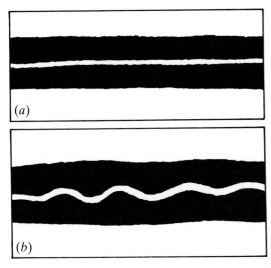

Fig. 11.40. Experiment performed to confirm that Eq. (11.9) holds for an elastic layer in an elastic matrix if the viscosities of the layer and matrix are replaced by the elastic shear moduli *G*. A thin, rubber sheet is embedded between soft rubber (dark) which is sandwiched between two thick, relatively stiff rubber layers. (*a*) Uncompressed, and (*b*) compressed state. (After Ramberg, 1963.)

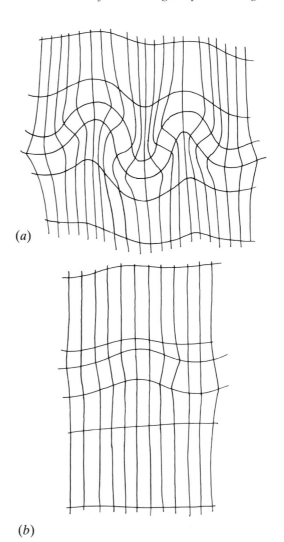

Fig. 11.41. Experimentally produced buckles in viscous materials. (*a*) Viscosity contrast between layer and matrix 24/1 and bulk deformation $S = (\lambda_1/\lambda_2) = 6$, λ is the quadratic elongation. (*b*) Viscosity contrast 5.5/1 and bulk deformation $S = 6.5$. It can be seen that the amount of buckle shortening decreases and the amount of layer-parallel shortening increases as the viscosity contrast decreases. (After Huddleston, 1973b.)

layer in an elastic matrix, if the viscosity coefficients are replaced by the elastic shear moduli (G_1 and G_2). Ramberg (1963b) conducted experiments in which he used thin 'stiff' rubber sheets embedded in soft rubber (Fig. 11.40) and found excellent agreement between the experimentally produced and theoretically predicted wavelength/thickness ratios.

Huddleston (1973b) performed experiments to study the development of folds in viscous materials where the effective ratios of layer/matrix viscosity fell between 5/1 and 100/1. In all instances, folds were observed to form. Analysis of arc length (*L*) and limb dip, with progressive deformation, showed that there was always a stage of initial layer-parallel shortening, during which folds began to develop. This phase then gave way to a stage of development when the arc length of this fold changed only slightly during further deformation. This is the phase of active buckling. The amount of layer-parallel shortening that took place, increased as the ratio of the viscosity contrast decreased: however, the transition from a stage of layer shortening to one of approximately constant arc-length appeared to be independent of the viscosity contrast and took place when the folds developed a limb-dip of 20–30°. No relative thickening of the competent layer in the hinge regions was observed in any of these experiments (Fig. 11.41).

It was pointed out earlier in this chapter that some initial irregularlity is necessary before a fold can be initiated. The growth of folds from irregularities has been studied by Cobbold (1975) who built low amplitude layer deflections into single layers of paraffin wax set in a matrix (competence ratio 10/1). The results of these experiments are summarised in Fig. 11.42. It can be seen that buckling does not occur

simultaneously throughout the whole length of the layer, but propagates from the irregularity. The deformation involves a geometric bending of the layer and a cell-like flow in the matrix. Buckling advances along the layer by the serial formation of new inflection points which delimit these folds. Throughout the deformation, these folds undergo progressive changes in amplitude and arc length, although, in some of the experiments, the arc-length stabilises. This may occur when the arc-length is equal to a dominant wavelength which, according to Eq. (11.8), depends upon the competence contrast and layer thickness. As folding progresses, more and more individual folds are developed and the fold shapes become more sinusoidal. This serial development of folds from an initial deflection in a single layer can be compared with a similar development in multilayers described in

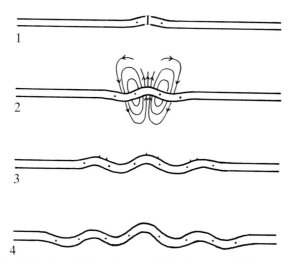

Fig. 11.42. The serial development of folds from an initial irregularity (1) in a single layer of paraffin wax set in a paraffin wax matrix. The competence contrast between layer and matrix is 10/1. (After Cobbold, 1975.)

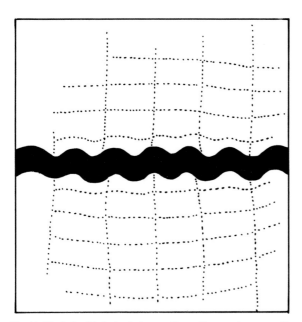

Fig. 11.43. A buckled single layer of non-linear microcrystalline wax ($n = 5$) in a non-linear paraffin wax matrix ($n = 1.8$) after a total strain of 0.5. The competence contrast between the layer and matrix was 28/1. (After Neurath & Smith, 1982.)

Chap. 10 and illustrated in Fig. 10.41 and Fig. 12.20(*d*).

Experiments have also been conducted to investigate the buckling behaviour of non-linear viscous materials by Neurath & Smith (1982), who used two types of wax whose stress–strain-rate relationship had power-law exponents (*n*) of 5 for the layer and 1.8 for the matrix (Fig. 11.43). In their experiments they concentrated on the growth-rate of folds because this exhibits a much greater variation with changing material rheology than does dominant wavelength. For example, over the range of experimental conditions used, theory predicts that domi-

nant wavelengths should only vary by ±20 per cent while the growth rates should vary by ±250 per cent (Fletcher, 1974 and Smith, 1975 (see Fig. 11.25)). Because the waxes have non-linear properties, their viscosities vary with strain-rate. Neurath & Smith describe two folding experiments, one at a relatively low strain-rate (0.29×10^{-4}/s) when the viscosity contrast is 7.3 and one at a higher strain-rate (1.1×10^{-4}/s) when the viscosity contrast is 28. The growth-rate of the folds was found to be considerably faster than could be accounted for by theories assuming linear viscous properties; and could be better explained by the theories based upon non-linear behaviour presented by Fletcher (1974) and Smith (1975, 1977 and 1979) discussed earlier.

Elastic versus viscous behaviour during fold initiation

From the preceding discussion of the theories of buckling of single viscous layers it follows that only when it is assumed that the viscous behaviour is markedly non-linear will folds develop with slenderness ratios (L/a) which are comparable with those that are observed in the field. It is, therefore, apposite to consider the evidence relating to the linear or non-linear viscous behaviour of rocks.

As we saw in Chap. 1 the constitutive equations, or 'equations of state' which describe the steady-state flow of rock takes the form of a power-law relationship, such that strain-rate ($\dot{e}$) is given by:

$$\dot{e} = K(\sigma_1 - \sigma_3)^n \qquad (11.26)$$

where K is a constant and ($\sigma_1 - \sigma_3$) the differential stress. The exponent n is predicted by various theories of rock (or mineral) deformation and depends upon specific dislocation or diffusion mechanisms which are assumed to prevail. In practice, the value of the exponent is determined empirically from rock mechanics experiments, when a variety of mechanisms may contribute to deformation. Such experimentally determined values of n fall in the range from 1 to 100.

It is emphasised that the equations of state relate only to conditions of steady-state flow. These equations are established from experimental data of the type shown in Fig. 1.65. Such curves indicate that there is an initial elastic stress–strain relationship prior to the ductile flow condition so that rock in the conditions represented in Fig. 1.65 behaves as an elastico-viscous, rather than a purely viscous, material.

One may represent an elastico-viscous body by the simple, symbolic diagram shown in Fig. 11.44(*a*), where the spring and dash-pot represent the elastic and viscous behaviour respectively. If such a body is subjected to an instantaneous boundary displacement (Fig. 11.44(*b*)) the whole of the stress incurred by the induced strain is initially taken up by the spring (the

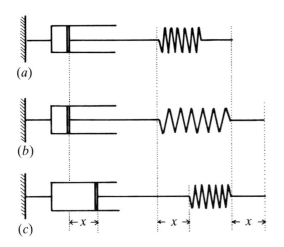

Fig. 11.44. (*a*) Schematic representation of an elastico-viscous body. (*b*) Instantaneous elastic strain as a result of an applied load. (*c*) The development of viscous strain as the elastic strain relaxes.

elastic component of the body). This situation follows from the fact that the development of viscous strain requires a finite time. Should the boundary displacement conditions represented in Fig. 11.44(*b*) be maintained constant, the stress in the spring is gradually able to induce viscous deformation (of the dash-pot). In so doing, the stress in the spring *relaxes* and eventually diminishes to zero (Fig. 11.44(*c*)). The *relaxation time* (T_r) is defined as the time required for stress in the spring to decay to $1/\exp = 37$ per cent of the original magnitude and, for materials with linear properties, is given by:

$$T_r = \frac{\eta}{G} \tag{11.36}$$

where η is the viscosity and G is the elastic shear modulus. When the viscosity is expressed in poise and the shear modulus in dynes/cm², the relaxation time is in seconds.

An inherent element which constitutes a part of the argument presented by Biot, Ramberg and others, that has led to ignoring the contribution of stresses resulting from elastic behaviour, is the commonly held viewpoint that geological deformation is an inherently slow process. Consequently, it would be argued that stress relaxation is to be expected. The shear modulus for many rock types is about 10^{11} dynes/cm². Hence, from Eq. (11.36), it follows that the relaxation time is mainly determined by the value of η. If we take the probable range for η, for competent rock, as 10^{20}–10^{23} poise, then the relaxation time would range from 10^9–10^{12} seconds (i.e. approximately 33 to 3330 years). Hence, it does not at first appear unreasonable that the elastic stresses may be neglected. However, it must be emphasised that the relaxation time is based on the assumption of strain which is held constant through the relaxation period (Fig. 11.44). In an orogenic environment, which will

give rise to folding, layer-parallel compressive strains (e_x) are continually being induced: and these would tend to replenish the reduction in stress attributable to relaxation.

Let us assume that the initial buckling stress (which gives rise instantaneously only to elastic deformation) is 500 bars (5×10^8 dynes/cm²) so that the quantity of stress which decays in time T_r is approximately 300 bars (3×10^8 dynes/cm²). Such a magnitude of stress would be replenished by an increment of compressive strain (de_x), parallel to the layering, given by:

$$de_x = \frac{3 \times 10^8}{G}. \tag{11.37}$$

As already noted, for many competent rocks G will be approximately 10^{11} dynes/cm², so that $de_x = 3 \times 10^{-3}$.

Clearly, it is now necessary to ascertain the probable strain-rates associated with fold initiation. We have noted that folds are frequently initiated and developed sequentially (Chap. 10). Price (1975) used this concept of serial folding to estimate the bulk strain-rates associated with fold development (Fig. 11.45) in which strains are generated by rock transport at an average velocity (V). He argued that strain-rates could vary widely, depending upon the mechanism that gave rise to the rock transport (and so determined the velocity V) and the wavelength of the structure which developed. These arguments were applied to the finite development of a fold (i.e. Phase *C* in Fig. 10.67), but they apply with equal validity to fold initiation (phase *B* in Fig. 10.67).

Let us assume that the average velocity (V) of displacement along the layer (Fig. 11.45) that results in the build up in stress, which in turn causes fold initiation, has the possibly conservative value of 1.0 cm/year. Then the rate of strain that develops in a fold is dependent upon the initial wavelength (L). For values of $L = 10$ cm, 1.0 m and 10.0 m, the strain-rate induced by such an average displacement velocity (V) is 3×10^{-9}, 3×10^{-10} and 3×10^{-11}/s respectively.

It has been noted that the strain (de_x) that needs to be replenished in order to maintain the level of elastic buckling stress is 3×10^{-3}. At strain-rates of 3×10^{-9} to 10^{-11}/s, these stresses would be replenished in a few months, for the small wavelength

Fig. 11.45. Conceptual model of a fold which develops as a consequence of rock transport (e.g. a nappe gliding or plates closing) at an average velocity V. (After Price, 1975.)

structures, and in 33 years for a 10 m wavelength fold. Folds with a wavelength of 100 m and 1.0 km respectively would have their elastic stresses replenished in 330 and 3300 years respectively. Thus, for the examples cited above, for folds with quite large wavelengths, the elastic stress would be maintained at its initial level but relaxation would become progressively more important for structures with very large wavelengths, say greater than one kilometre.

The preceding arguments are based on serial folding, but we have noted that fold initiation may occur by the simultaneous appearance of a wave-train. One may infer from the previously described examples that when dealing with folds with a wavelength of 10 cm; even if the fold in question is one of a hundred identical structures in a wave-train, the elastic stresses would be maintained. Consequently, we argue, the elastic behaviour of competent rocks is likely to prove the dominant influence in the *initiation* of folds of reasonably small wavelengths in most, if not all, crustal environments. At higher levels in the crust, where temperatures are relatively low and diffusion mechanisms are not important, viscosities of competent rocks are high (i.e. $>10^{22}$ poise), the importance of elastic behaviour increases and is pertinent even in the development of large wavelength folds (see Chap. 14).

It is interesting to note that a solution to a viscous problem can readily be adapted to accommodate a comparable elastic situation. Indeed, both Biot and Ramberg point out that their basic equation (11.9) can be expressed in terms of either viscous or elastic moduli. If we make the reasonable assumption that for two specific examples, the ratios of the elastic shear moduli of the layer and the matrix are assumed to be 3/1 and 10/1, the L/a ratios predicted by this equation are 7.45/1 and 5.0/1 respectively. These are very close to the upper and lower limits of the L/a values determined from field measurements.

Commentary

The various analyses by Biot and Ramberg and the workers that follow them are meritorious and have advanced geologists' understanding of fold initiation processes. Nevertheless, in general, their predicted results regarding L/a ratios have not fitted observed natural data. Only by treating rocks as highly non-linear viscous materials has it proved possible to bring theory and field observations into line. This problem has arisen because it has been implicitly assumed, that elastic behaviour of rocks during the initiation of folding can be ignored. Although rock mechanics data are still sparse, sufficient exist to enable certain generalisations to be made. The power-law exponent of rocks (n) is usually only unity under very high pressure and temperature conditions. At the pressures and temperatures associated with intermediate grade metamorphism, n is greater than 1, (say 4–10) whilst during low grade metamorphism in the upper levels of the crust, $n \geqslant 10$ and so the rocks behave in a manner comparable with a Bingham body (i.e. a solid).

In the theoretical treatment of folding, the difference in rock behaviour at fold initiation and during its finite amplification has often been ignored, so that the ductile response of rocks during fold amplification has also been assumed to occur at fold initiation. There have been certain pointers throughout this chapter indicating that elastic behaviour is important during folding. However, the arguments which, in our opinion, indicate that elastic behaviour governs fold initiation in the upper levels in the crust for all scales of folding (and is important for small scale folds at all crustal levels), were left until the end of the chapter. Geologists for the last two decades, have been misleading themselves by ignoring this situation.

References

Biot, M.A. (1957). Folding instability of a layered visco-elastic medium under compression. *Proc. Roy. Soc. London*, **A242**, 444–54.

(1959a). Folding of a layered visco-elastic medium derived from an exact stability theory of a continuum under initial stress. *Quart. Appl. Math.*, **17**, 185–204.

(1959b). On the instability and folding deformation of a layered visco-elastic medium in compression. *Appl. Mech.*, **26**, 393–400.

(1961). Theory of folding of stratified visco-elastic media and its implications in tectonics and orogenesis. *Geol. Soc. Am. Bull.*, **72**, 1595–620.

(1964). Theory of viscous buckling and gravity instability of multilayers with large deformation. *Geol. Soc. Am. Bull.*, **76**, 371–8.

(1965). Theory of similar folding of the first and second kind. *Geol. Soc. Am. Bull.*, **76**, 251–8.

Biot, M.A., Odè, H. & Roever, W.L. (1961). Experimental verification of the folding of stratified, visco-elastic media. *Geol. Soc. Am. Bull.*, **72**, 1621–30.

Biot, M.A. & Odè, H. (1962). On the folding of a visco-elastic medium with adhering layers under compressive, initial stress. *Quart. Appl. Math.*, **19**, 335–51.

Carter, N.L. & Raleigh, C.B. (1969). Principal stress directions from plastic flow in crystals. *Geol. Soc. Am. Bull.*, **80**, 1231–64.

Casey, M. (1976). Application of finite element analysis to some problems of structural geology. Unpublished Ph.D. Thesis, University of London.

Chapple, W.M. (1968). A mathematical theory of finite amplitude rock folding. *Geol. Soc. Am. Bull.*, **79**, 457–66.

(1969). Fold shape and rheology: the folding of an isolated, viscous-plastic layer. *Tectonophysics*, **7**, 2, 97–116.

Cloos, E. (1947). Oolite deformation in South Mountain Fold, Maryland. *Geol. Soc. Am. Bull.*, **58**, 843–918.

Cobbold, P.R. (1969). An experimental study of the formation of lobe and cusp structures by shortening of an initially sinusoidal contact between two materials of different viscosities. Unpublished dissertation, University of London.

(1975). Fold propagation in single embedded layers. *Tectonophysics*, **27**, 333–51.

Dieterich, J.H. & Carter, N.L. (1969). Stress history of folding. *Am. J. Sci.*, **267**, 129–55.

Dieterich, J.H. & Onat, E.T. (1969). Slow, finite deformation of viscous solids. *J. Geophys. Research*, **74**, 2081–8.

Euler (1757). *Sur la force des colonnes*. Mem. de l'acad. de Berlin, t. xiii, 252 pp.

Fletcher, R.C. (1974). Wavelength selection of folding of a single layer with power-law rheology. *Am. J. Sci.*, **274**, 1029–43.

(1977). Folding of a single viscous layer: exact infinitesimal amplitude solution. *Tectonophysics*, **39**, 593–606.

(1982). Analysis of the flow in layered fluids at small, but finite amplitude with application to mullion structures. *Tectonophysics*, **81**, 51–66.

Gairola, V.K. & Kern, H. (1982). Microstructure and texture in experimentally folded single layer rock salt. *Mitt. Geol. Inst. ETH Zurich, N.F.*, **239a**, 106–8.

(1984). Single layer folding in marble and limestone: an experimental study. *Tectonophysics*, **108**, 155–72.

Ghosh, S.K. (1966). Experimental tests of buckling folds in relation to the strain ellipsoid in simple shear deformations. *Tectonophysics*, **3**, 169–85.

Goldstein, S. (1926). The stability of a strut under thrust when buckling is resisted by a force proportional to the displacement. *Cambridge Philos. Soc. Proc.*, **23**, 120–9.

Handin, J.W., Friedman, M., Logan, J.M., Pattison, L.J. & Swolfs, H.S. (1972). Experimental folding of rocks under confining pressure: buckling of single layer rock beams. *Am. Geophys. Union Monograph.*, **16**, 1–28.

Huddleston, P.J. (1973a). Fold morphology and some geometrical implications of theories of fold development. *Tectonophysics*, **16**, 1–46.

(1973b). An analysis of 'single layer' folds developed experimentally in viscous media. *Tectonophysics*, **16**, 189–214.

Johnson, A.M. (1977). *Styles of folding: Mechanics and mechanisms of folding of natural elastic materials*. 406 pp. Elsevier Scientific Pub. Co.

Kienow, S. (1942). Grundzuege einer Theorie der Faltungs- und Schlieferungsvorgaenge. *Fortschr. Geol. Palaentol.*, **14**, 1–129.

Neurath, C. & Smith, R.B. (1982). The effect of material properties on growth-rates of folding and boudinage:

experiments with wax models. *J. Struct. Geol.*, **4**, 2, 215–29.

Parrish, D.K. (1973). A non-linear finite element fold model. *Am. J. Sci.*, **273**, 318–34.

Parrish, D.K., Krivz, A.L. & Carter, N.L. (1976). Finite element folds of similar geometry. *Tectonophysics*, **32**, 183–207.

Price, N.J. (1975). Rates of deformation. *J. Geol. Soc. London*, **131**, 553–75.

Ramberg, H. (1959). Evolution of ptygmatic folding. *Nor. Geol. Tidsskr.*, **39**, 99–151.

(1960). Relationships between lengths of arc and thickness of ptygmatically folded veins. *Am. J. Sci.*, **258**, 36–46.

(1961a). Contact strain and fold instability of a multilayered body under compression. *Geol. Rund.*, **51**, 405–39.

(1961b). Relationship between concentric, longitudinal strain and concentric shearing strain during folding of homogeneous sheets. *Am. J. Sci.*, **259**, 382–90.

(1963a). Strain distribution and geometry of folds. *Bull. Inst. Geol. Univ. Uppsala*, **42**, 1–20.

(1963b). Fluid dynamics of viscous buckling applicable to folding of layered rocks. *Am. Assoc. Petrol. Geol. Bull.*, **47**, 484–515.

(1964). Selective buckling of composite layers with contrasted rheological properties; a theory for simultaneous formation of several orders of folds. *Tectonophysics*, **1**, 307–41.

(1970a). Folding of laterally compressed multilayers in the field of gravity. I. *Phys. Earth Planet. Inter.*, **2**, 203–32.

(1970b). Folding of laterally compressed multilayers in the field of gravity. II. *Phys. Earth Planet. Inter.*, **4**, 83–120.

Ramsay, J.G. (1967). *Folding and fracturing of rocks*. 568 pp. New York: McGraw–Hill.

Sherwin, J.A. & Chapple, W.M. (1968). Wavelengths of single layer folds: a comparison between theory and observation. *Am. J. Sci.*, **266**, 167–79.

Smith, R.B. (1975). A unified theory of the onset of folding, boudinage and mullion structure. *Geol. Soc. Am. Bull.*, **86**, 1601–9.

(1977). Formation of folds, boudinage and mullions in non-Newtonian materials. *Geol. Soc. Am. Bull.*, **88**, 312–20.

(1979). The folding of a strongly non-Newtonian layer. *Am. J. Sci.*, **279**, 272–87.

Smoluchowski, M. (1909). Versuche ueber Faltungserscheinungen schwimmender elastischen Platten. *Anz. Akad. Wiss., Krakau, Math. Phy. Kl.*, 727–34.

12 Multilayer folds and associated structures

Introduction

Geologists encounter multilayers (e.g. sedimentary sequences) more often than single layers (veins and dykes in a massive host). Consequently, there is a considerable body of geological literature which considers the buckling of multilayers. However, before looking at the various analyses of multilayer buckling, it is necessary to define from the viewpoint of the structural geologist what is meant by or constitutes a multilayer. The chapter opens therefore with a brief discussion regarding the definition of a multilayer and continues with a study of some of the mathematical analyses that have been used by geologists to account for the formation of multilayer folds in rocks. These analyses are restricted to the first increment of buckling and the finite amplification of multilayer buckles is investigated using finite element analyses and model work. The chapter ends with a consideration of the second order structures such as accommodation structures, parasitic folds and single layer buckles in control units, which form during the folding of a complex geological multilayer i.e. a multilayer containing numerous layers of different thicknesses and material properties.

The definition of a multilayer

Consider the two folded layer sequences shown in Fig. 12.1. Both examples were originally made up of parallel layers and were, therefore, physically multilayers. In the first example, however, the individual, relatively competent (dark) layers have behaved mechanically as single layers, each layer developing its own wavelength in accord with the single layer buckling equation (Eq. (11.8)). Such folding, where adjacent competent layers in a multilayer develop different wavelengths and different amplitudes, is termed *disharmonic folding*. In the second example, despite differences in thickness and mechanical properties of the individual layers, all the layers have buckled with the same wavelength and amplitude. Such folding is termed *harmonic folding*. Clearly, the units in the second example behaved mechanically as a multilayer, in the sense that we, as structural geologists, shall use hereafter.

These two types of buckling behaviour can be readily explained by considering the strain that develops in the matrix around a single, competent layer as

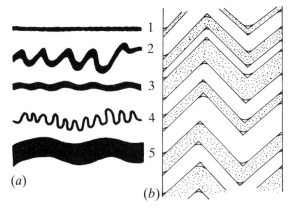

Fig. 12.1. (*a*) A model multilayer that was compressed parallel to the original orientation of the layering. The competent (dark) layers are enclosed in a less competent (light) material and the relative competency of the layers is $4 > 2 > 3 > 5 > 1$. (After Ramberg, 1964.) (*b*) A buckled multilayer; each layer having a different thickness and material properties.

it buckles. The zone of disturbance on each side of the buckling layer is known as the *zone of contact strain*. If the competent layers of a multilayer are sufficiently far apart for there to be no significant overlap of their zones of contact strain, then each layer will buckle as a single layer. If, however, the zones of contact strain of adjacent competent layers do significantly overlap, the layers can no longer buckle totally independently of each other. The zones of contact strain and the associated zones of contact stress of all the competent layers must be compatible and, as a result, all the layers develop the same wavelength, i.e. the multilayer wavelength.

In order to determine how close the competent layers of a multilayer may be before multilayer buckling occurs, it is necessary to know how far the zone of contact strain extends away from the layer into the matrix. Ramberg (1961) addressed this question by making the assumption that (i) the matrix behaved as a Newtonian, viscous fluid, (ii) the wavelength/thickness ratio of the buckling layer is large and (iii) the amplitude of the fold is small. He then calculated the displacements in the matrix both parallel to and normal to the layer. He found that although the disturbance of the matrix caused by the buckling layers theoretically extended an infinite distance away from the layer, the displacements drop to approximately 1 per cent of their maximum value at a distance of one wavelength from the buckling layer (Figs. 12.2(*a*) and

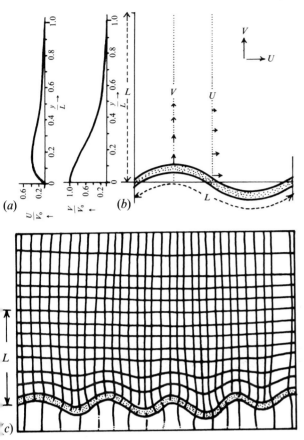

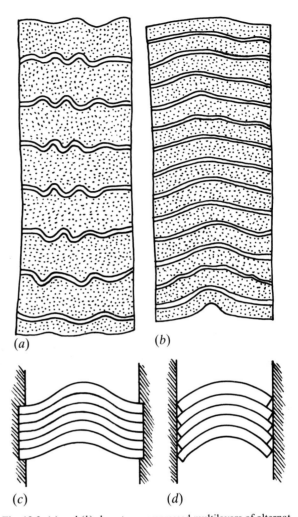

Fig. 12.2. (*a*) Plot of *U* and *V* (displacements parallel and normal to the initial layer orientation respectively), in the matrix as a function of distance (*y*) from the folded layer. V_0 is the rate of change of amplitude of the fold, and *L* is the wavelength. (*b*) Maximum values of *U* occur along lines parallel to *y* and passing through the inflection points, and of *V* along lines parallel to *y*, through the hinges. (*c*) A buckled layer of relatively rigid rubber cemented in a block of softer rubber. The contact strain is shown by the distortion of the marker lines which were originally straight. The matrix deformation becomes vanishingly small at a distance greater than one wavelength from the buckled layer. (After Ramberg, 1961.)

Fig. 12.3. (*a*) and (*b*) show two compressed multilayers of alternating competent (white) and incompetent (stippled) rubber layers. In (*a*), the competent layers are widely spaced and the overlap of the zones of contact strain is low. Some layers buckle harmonically but others buckle disharmonically (*b*) the layers behave mechanically as a multilayer. (After Ramberg, 1963.) (*c*) A buckled multilayer with its ends clamped, and (*d*) a buckled multilayer with its ends supported but not clamped.

b)). Beyond this distance the matrix is effectively undisturbed.

Ramberg tested his theoretical predictions regarding the extent of the matrix disturbance around a buckled layer experimentally. Although he assumed viscous behaviour in his analysis he used elastic models (Fig. 12.2(*c*), 12.3(*a*) and (*b*)) and exploited the 'correspondence' between viscous and elastic behaviour, (see Chap. 11). Ramberg found excellent agreement between theory and experiment and noted that, even after considerable amplification of the buckles, the zone of contact strain is still negligible at distances greater than one wavelength from the buckled layer (Fig. 12.2(*c*)). The extent of the zone of contact strain is also clearly delineated in photo-elastic experiments (Fig. 12.4(*a*)), by the isochromatic fringes set up in a photo-elastic matrix (e.g. gelatine) around a buckling rubber strip. Field evidence regarding the width of the zone of contact strain can be found in the cleavage patterns formed in the matrix

adjacent to buckled rock layers. Close to the layers, the cleavage fans around the outer arcs and converges into the inner arcs of the folds (Fig. 12.5(*a*)), but at distances greater than one wavelength away from the layer the cleavage is undisturbed.

The transition from single layer to mulilayer buckling which occurs as the competent layers become closer together can be seen in the experiments of Fig. 12.4. In experiment (*b*), two identical but widely spaced layers buckle independently as single layers whereas in experiment (*c*) the two layers are closely spaced and behave mechanically as a multilayer, buckling with a larger wavelength. The same type of multilayer buckling is shown in experiment (*d*) which shows five competent layers folding harmonically. If these competent layers are brought even closer together, then the multilayer behaves effectively as a single anisotropic layer (Fig. 12.4(*e*)).

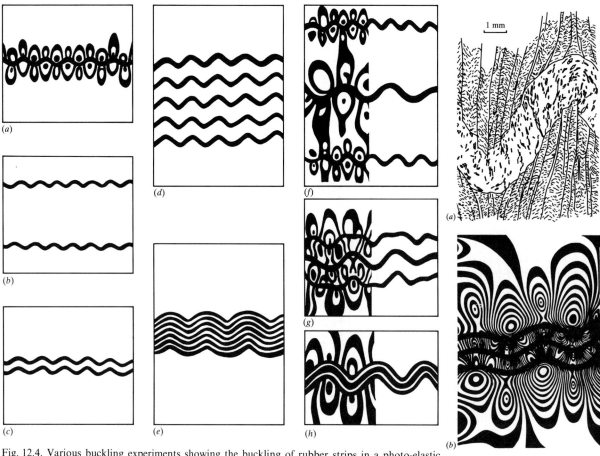

Fig. 12.4. Various buckling experiments showing the buckling of rubber strips in a photo-elastic matrix (gelatine). (*a*) Isochromatic fringes indicate the extent of the zone of contact strain in the matrix. (*b*) to (*h*) Various experiments showing how multilayer buckling is affected by the degree of overlap of the zones of contact strain of adjacent competent layers, see text for details. (After Currie, Patnode & Trump, 1962.)

As will be seen later in this chapter, such a 'layer' can buckle with a much smaller wavelength/thickness ratio (L/a) than an isotropic layer with the same thickness. In the series of experiments (f), (g) and (h), the effect of spacing on the buckling of three competent layers, a central thick layer sandwiched between two thinner layers, is examined. The three layers buckle independently in experiment (f), the thicker layer showing a larger wavelength than the thinner layers in accord with the single layer buckling equation, (Eq. (11.8)). As the spacing between the layers is reduced, the disharmonic folding of experiment (f) gives way to harmonic folding. These can be divided into two types; firstly, multilayer folds with a conjugate (box-like) profile geometry (experiment (g)) and secondly, when the layers are very close to each other, buckles with a more rounded profile geometry, (experiment (h)). In the latter example the multilayer acts as an anisotropic strut.

The overlapping and resultant modification of the zones of contact strain around adjacent competent layers during multilayer buckling is shown by the isochromatic fringes in Fig. 12.5(*b*). The orientation of the strain in the matrix can be calculated from

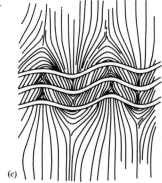

Fig. 12.5. (*a*) Cleavage in a phyllite matrix, fanning around the outer arc and into the inner arc of a quartz microfold. (After Dieterich, 1969.) (*b*) The interference (shown by the isochromatic fringe patterns) of the zones of contact strain of adjacent competent layers in a buckling multilayer. (*c*) The finite strain trajectories (parallel to the principal extension direction) determined from the fringes shown in (*b*). ((*b*) and (*c*) after Roberts & Stromgard, 1972.)

these fringes and the finite strain trajectories (lines parallel to e_3, the maximum principal extension) determined from this experiment, are shown in Fig. 12.5(*c*). The pattern of the trajectories corresponds remarkably well with the cleavage patterns developed in the incompetent matrix around natural folds (Fig. 12.5(*a*)), and we shall discuss this point further in Chap. 17.

The gradual changeover from single layer to

multilayer buckling behaviour as the competent layers of a multilayer become closer together is also clearly illustrated in the natural examples shown in Fig. 12.6. In Fig. 12.6(*a*) and (*b*), the spacing of the competent layers is such that some of them buckle harmonically with adjacent competent layers and others fold disharmonically. In contrast, the spacing of the competent sandstone units in Fig. 12.6(*c*) is such that even though the sandstone layers are of different thicknesses, harmonic multilayer buckling occurs.

Although it is not possible to state exactly how close adjacent competent layers must be before their zones of contact stress and strain overlap sufficiently for the layers to buckle harmonically, an indication of the critical spacing can be obtained by the inspection of model and natural examples of harmonic and disharmonic folding, e.g. Figs. 12.1 to 12.6 and Fig. 15.17. We have already noted that, at a distance of one wavelength from the buckling layer, the disturbance caused by the contact strain has dropped only 1 per cent of its maximum level. However, from these model and field examples we may infer that adjacent layers must be significantly closer than the sum of their dominant wavelengths (perhaps as little as 10 percent of this distance) before they will buckle harmonically.

Theories of multilayer buckling

Sedimentary sequences and layered metamorphic rocks, such as banded gneisses are natural examples of multilayers. These natural examples are often *bilaminates*; that is, sequences comprising only two different types of layers which alternate with each other. Consequently, theoretical work on the development of folds in layered complexes has concentrated on the buckling of such multilayers. For the sake of mathematical simplicity, the treatments were restricted to *regular* bilaminates in which all the layers of one type are of equal thickness. Natural examples of bilaminates rarely meet this extra requirement.

Two different approaches have been applied to the problem of multilayer folding. The first, used by Ramberg (1963 and 1964) treats individual units of a multilayer as separate entities and predicts how the layers interact when the complex is compressed along the layering. A similar approach was used by Johnson (1977) who studied a variety of multilayer problems by considering the resistance to buckling of a 'representative' layer and selecting appropriate interlayer properties. This treatment has produced a highly satisfactory explanation for the formation of kink-bands and is discussed further in Chap. 13. The second approach to the problem of multilayer buckling is that used by Biot (1961, 1963a and 1963b) who considers the *bulk* properties of the multilayer. We will consider these two approaches to the buckling of bilaminates in turn, for they yield different, though not incompatible, results.

Ramberg's analysis

Ramberg's approach to the problem of multilayer buckling is to determine stress and displacement solutions for the individual layers of the multilayer and to match solutions at the interface between adjacent layers, thereby obtaining a solution for the buckling response of the whole multilayer. As we have seen in Chap. 11, he suggests that for slow creep under relatively small differential stresses, rock may, as a first approximation, be treated as a viscous fluid and argues that under such conditions it is possible to apply the theory of fluid dynamics to the problem of the folding of rocks. He considers a simple, unconfined multilayer (Fig. 12.7) made up of Newtonian viscous layers. Two relatively competent layers are separated by a less competent layer. The contacts between the layers are welded to prevent free slip. The analysis ignores the effect of gravity and is restricted to the initial period of deformation, when the fold amplification is infinitesimally small. Ramberg points out that the analysis can be extended to multilayers with any number of layers as long as all the thin, competent layers are mechanically identical and evenly spaced in a uniform incompetent matrix. The expression for the buckling force (F) for any of the competent layers of the multilayer is:

$$F = \left\{ \begin{array}{l} \dfrac{w^2 \eta_1 a_1^3}{6} \\[2ex] + 2\eta_2 \dfrac{(1-b)}{\left[1 + b + 2a_2 w\left(\dfrac{b}{1+b}\right)\right]w} \end{array} \right\} \dfrac{\dot{e}_x}{e_x} \qquad (12.1)$$

where η_1 and η_2 are the viscosities of the competent and incompetent layers respectively, a_1 and a_2 are the thicknesses of the competent and incompetent layers respectively and $\dot{e}_x$ and e_x are the strain-rate and strain in the x direction (Fig. 12.7);

$$b = e^{-wa_2} \qquad (12.2)$$

$$w = \dfrac{2\pi}{L} \qquad (12.3)$$

and L is the fold wavelength.

The total resistance to buckling of the multilayer is obtained by multiplying Eq. (12.1) by the number of competent layers in the multilayer. The buckling force equation is expressed graphically in Fig. 12.8, where a function (f) of the buckling force ($f = Fy_0/\eta_1 a_1 V_0$) is plotted against the wavelength/thickness ratio of the competent layers of the multilayer (where V_0 is the amplitude growth rate and y_0 is the amplitude). In Fig. 12.8(*a*), the viscosity contrast between adjacent layers is held constant at 10/1 and a

(a)

(b)

(*c*)

Fig. 12.6. (*a*) Buckled multilayer in Pre-Cambrian metasediments, Northern Ontario, Canada. Both harmonic and disharmonic folding have occurred. (*b*) Photomicrographs of buckled schist from the Lukmanier Pass, Switzerland. (*c*) Multilayer buckles in Carboniferous turbidites at Millook, N. Cornwall. Note that the sandstone layers have all buckled with the same wavelength and amplitude despite having different thicknesses.

family of buckling force curves plotted for various values of (R_a), the ratio of the thickness of the in-competent layer and the competent layer (a_2/a_1). A similar family of curves for multilayers with a viscosity contrast of 100/1 is shown in Fig. 12.8(*b*).

It can be seen from these graphs that the buckling force curves of the multilayers with high R_a values (competent layers widely spaced) contain well marked minima. Each minimum corresponds to a particular wavelength thickness ratio that is most easily formed and which will, therefore, develop. However, the buckling force curves for multilayers with low R_a values (competent layers closely spaced) show no marked minimum. Instead, the buckling force decreases continuously with increasing wavelength/thickness ratio. From this, one may infer that the larger the value of the wavelength/thickness ratio L/a, the smaller will be the buckling force. The multilayer will therefore buckle either as a single wavelength or half a wavelength depending on whether its ends are clamped or simply supported (Figs. 12.3(*c*) and (*d*)).

It is apparent from the graphs that there will be some R_a value (competent layer spacing) which separates a family of curves with a finite minimum from a family of curves without such minima. Multilayers with a competent layer spacing greater than this limit should buckle to a train of several waves, whereas multilayers with smaller spacing should buckle with a single or half wavelength. These two types of buckling behaviour have been produced experimentally (Fig. 12.3(*a*) and (*b*)).

The condition for the formation of the single fold is:

$$R_a \lesssim 1.5\pi \sqrt[3]{\frac{\eta_1}{6\eta_2}} \qquad (12.4)$$

and the condition for the formation of several multi-layer folds is:

$$R_a \gtrsim 1.5\pi \sqrt[3]{\frac{\eta_1}{6\eta_2}}. \qquad (12.5)$$

Ramberg also considers the effect of enclosing the multilayer (Fig. 12.7) in a viscous matrix. He modifies the buckling force equation to include a term owing to the strain in the matrix generated by the multilayer folds. Because this term is proportional to the fold wavelength, the buckling force will increase with the increasing wavelength. Consequently multilayers, which, when unconfined by a matrix, have no minimum in their buckling force curve (e.g. $R_a < 4$ in Fig. 12.8(*a*) and $R_a < 12$ in

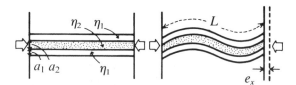

Fig. 12.7. Newtonian, viscous multilayer analysed by Ramberg (1961). Two relatively competent layers sandwich a less competent layer.

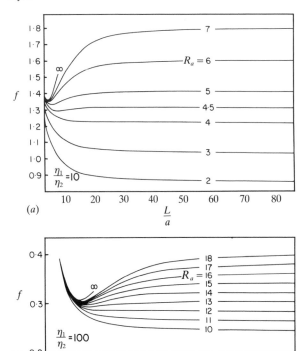

(a)

(b)

Fig. 12.8. A function, *f*, of the buckling force plotted against wavelength/thickness ratio for a variety of multilayers. In (*a*) the viscosity ratio between the competent and incompetent layers is 10/1 and in (*b*) 100/1. Each curve corresponds to a particular multilayer whose R_a value (ratio of thickness of incompetent to competent layers) is indicated. (After Ramberg, 1963.)

Fig. 12.8(*b*)), may well develop such a minimum if surrounded by a matrix.

Biot's analysis

As mentioned earlier in this section, Biot frequently adopted a different approach from Ramberg to the problem of buckling of multilayers. Biot recognised that there are important similarities between the relatively simple problem of the stability of a confined anisotropic plate and the more complex problem of the stability of a confined multilayer plate and describes a method of analysis that can be applied to the buckling behaviour of both systems. An important feature of his approach is the use of an approximation whereby the response to stress of a discretely layered material, such as a multilayer, is analysed in terms of the response of an equivalent homogeneous continuum with anisotropic mechanical properties. This mechanical anisotropy represents the 'bulk' mechanical anisotropy of the discretely layered material.

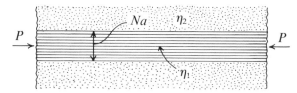

Fig. 12.9. Multilayer consisting of *N* identical, Newtonian viscous layers each of thickness *a* and viscosity (η_1) set in a less viscous matrix of viscosity η_2. Perfect lubrication exists between the layers. (After Biot, 1961.)

In order to determine the buckling behaviour of a multilayer made up of alternating competent and incompetent layers (a model with great geological applicability), Biot (1961) first considered a sequence consisting of *N* identical superposed layers of thickness (*a*) and viscosity η_1, with perfect lubrication between the layers (Fig. 12.9). He argues that this assumption is geologically realistic when (i) very thin incompetent horizons separate the competent beds or (ii) hydraulic lifting in fluid-filled porous rocks occurs along bedding planes (as suggested by Rubey & Hubbert, 1959, in the context of thrusting). The importance of high fluid pressure in the formation of large scale folds is discussed in Chap. 14, where it is argued that, unless very high fluid pressures develop along bedding planes, it is not possible for large folds, with wavelengths in excess of a few hundred metres, to develop.

The multilayer considered by Biot (Fig. 12.0) is embedded in a matrix of lower viscosity (η_2) and is subjected to a horizontal compression (*P*), per unit area. The dominant wavelength L_d that results from this analysis is given by:

$$L_d = 2\pi a \sqrt[3]{\frac{N\eta_1}{6\eta_2}}. \qquad (12.6)$$

The wavelengths of a multilayer (made up of *N* identical, perfectly lubricated layers of thickness *a*) and of a single layer (thickness *Na*) can be compared by using Eqs. (12.6) and (11.8). If it is assumed that the coefficients of viscosity in both examples are identical, then the ratio of the single layer/multilayer wavelengths (L_s/L_m) is:

$$\frac{L_s}{L_m} = N^{2/3}$$

From this relationship it is clear that the wavelength that forms in a multilayer will be significantly smaller than that which forms in a single layer with the same thickness as the multilayer, particularly if *N* is large.

Buckles developed in rock multilayers made up of competent layers separated by very thin incompetent layers, are shown in Figs. 12.10 and 12.11. If friction does occur between the layers of the multilayer represented in Fig. 12.9, then the dominant wavelength will lie somewhere between that of a

Fig. 12.10. Natural multilayer that approximates to the multilayer analysed by Biot (Fig. 12.9). The rock is made up of quartz-rich layers separated by very thin films of mica.

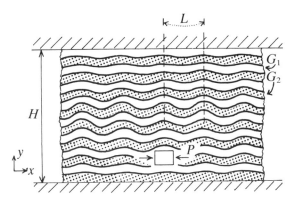

Fig. 12.12. Buckling of a bilaminate under rigid confinement. (After Biot, 1964.)

and α_2 of the total thickness H, the two elastic co-efficients that define the bulk properties of the multilayer are:

$$L' = \frac{G_1 G_2}{\alpha_1 G_2 + \alpha_2 G_1} \qquad (12.7)$$

and

$$M' = G_1 \alpha_1 + G_2 \alpha_2 \qquad (12.8)$$

where M' and L' are respectively measures of resistance to compression and shear parallel to the layering. Such a multilayer behaves as an anisotropic, continuous medium obeying the following constitutive (stress–strain) equations:

$$\sigma_{xx} = 4M' e_{xx} \qquad (12.9)$$

$$\tau = L' \gamma \qquad (12.10)$$

where e_{xx} is the average compressive strain and γ the average shear strain. The materials are assumed to be incompressible and the deformation is assumed to be by plane strain, in the xy plane, (i.e. exhibits no area change in the xy plane and no strain parallel to the z axis). The analysis takes into account the bending resistance of the relatively competent layers and the interstitial flow of material from the limbs towards the hinge that occurs within the incompetent layers. The buckling stress P for this elastic multilayer is:

Fig. 12.11. Photomicrograph of a folded rock multilayer of the type shown in Fig. 12.10.

single layer of thickness Na (Eq. (11.8)) and the wavelength given by Eq. (12.6).

In later papers Biot (1964, 1965a and b) considered the buckling of a multilayer made up of alternating competent and incompetent layers confined between two rigid, straight boundaries (Fig. 12.12), whose vertical separation varies with the amount of horizontal strain. The analysis is applicable to elastic, viscous and visco-elastic materials. If the multilayer in Fig. 12.12 is made up of two incompressible elastic materials of elastic shear moduli (rigidity) G_1 and G_2 and which comprise fractions α_1

$$P = L' + \left(\frac{\pi}{N}\right)^2 \frac{DM'}{c^2} + \frac{16 G_1 \alpha_1{}^3 c^2}{3} \qquad (12.11)$$
$$\quad (1) \qquad\quad (2) \qquad\quad (3)$$

where

$$c = \frac{\pi a}{L}. \qquad (12.12)$$

The thickness of a competent/incompetent layer pair of the multilayer is $2a$ and L is the wavelength. N is the total number of layers and D, which is a parameter related to the interstitial flow is given by:

$$D = \frac{1}{|1 + kc^2|} \qquad (12.13)$$

where

$$k = \frac{16M'\alpha_2^{3}}{3G_2}. \tag{12.14}$$

It is instructive to consider the physical significance of the various elements of Eq. (12.11), as each corresponds to a particular component of resistance offered by the multilayer to buckling. The resistance to sliding of one layer over another is proportional to L' (which is referred to as the slide modulus). The second element (2), represents the combined effect of the resistance to vertical movement, a consequence of the rigid confinement imposed on the model, and the resistance to interstitial flow in the incompetent layer (D). The third element of Eq. (12.11) (3) is a measure of the bending resistance of the component layers. It can be seen from Eq. (12.11) that when the stress (P) is sufficient to overcome all the different components of resistance to deformation offered by a multilayer, buckling will occur. The value of (c) (Eq. (12.12)), for which P is a minimum, determines the dominant wavelength L_d of the multilayer and defines the critical buckling stress.

Having shown that viscous and visco-elastic stress–strain relations may be expressed in a form completely analogous with those of purely elastic media, Biot uses this 'correspondence principle' to adapt Eq. (12.11) for a viscous multilayer made up of alternating layers of viscosity η_1 and η_2. In order to obtain the viscosity coefficients that characterise the deformation of the multilayer, the elastic moduli G_1 and G_2 in Eqs. (12.7) and (12.8) are replaced by the viscous product terms $\eta_1 p$ and $\eta_2 p$ respectively so that the dimensions of the elastic moduli are retained. The parameter p is an amplification factor and gives the rate of amplification of the fold. It is defined as follows. The amplitude A of the viscous buckles is exponentially related to time (t), such that:

$$A = A_0 e^{pt} \tag{12.15}$$

where A_0 is the amplitude of the original, sinusoidal perturbation. The coefficients L'_v and M'_v that describe the bulk rheological behaviour of the viscous multilayer can be obtained by substituting the viscous product terms (ηp) for the rigidities (G) in Eqs. (12.7) and (12.8).

$$L'_v = \eta_1 p = \frac{\eta_1\eta_2 p^2}{\alpha_1\eta_2 p + \alpha_2\eta_1 p}$$

$$= \frac{\eta_1\eta_2 p}{\alpha_1\eta_2 + \alpha_2\eta_1} \tag{12.16}$$

$$M'_v = \eta_n p = \eta_1 p\alpha_1 + \eta_2 p\alpha_2. \tag{12.17}$$

To illustrate the theory further, he simplifies the problem by assuming that the layers are of equal thickness and that the viscosity contrast between the layers is such that $(\eta_1/\eta_2 > 50)$. Eqs. (12.16) and (12.17) then reduce to the following approximations:

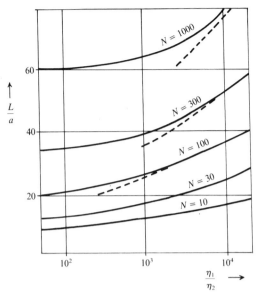

Fig. 12.13. Wavelength/thickness ratio plotted against viscosity contrast for the confined bilaminate shown in Fig. 12.12, when the competent and incompetent layers have the same thickness, a. Each plot corresponds to a multilayer each with the same total thickness H but with a different number of layers N. (After Biot, 1964.)

$$L'_v = \eta_t p \approx 2\eta_2 p \tag{12.18}$$

$$M'_v = \eta_n p \approx 0.5\eta_1 p \tag{12.19}$$

and Eq. (12.11) becomes:

$$\frac{P}{2\eta_1 p} = \frac{\eta_2}{\eta_1} + \left(\frac{\pi}{2N}\right)^2\frac{D}{c^2} + \frac{c^2}{3}. \tag{12.20}$$

The value of (c) (Eq. (12.12)) for which P of Eq. (12.20) is minimum, gives the dominant wavelength L_d. The wavelength/thickness ratio L_d/a is found to be a function of the viscosity contrast η_1/η and the number of layers in the multilayer N. The relationship is shown graphically in Fig. 12.13, where it can be seen that the dominant wavelength is very sensitive to the number of layers but, in contrast to the conclusion drawn from the single layer buckling analysis, is very insensitive to the viscosity contrast particularly for those viscosity contrasts below 1000. The near-horizontal portions of the curves on the left hand side of the diagram are approximately given by the simple relationship:

$$\frac{L_d}{a} = 1.9\sqrt{N} \tag{12.21}$$

which is obtained by neglecting the interstitial flow (i.e. by putting $D = 1$ in Eq. (12.20)). The ascending portion of the curve on the right hand side of the diagram corresponds with relatively large interstitial flow in the incompetent layers which can be represented, to a first approximation, by writing:

$$D = \frac{3\eta_2}{\eta_1 c^2}. \tag{12.22}$$

The simplified formula for the dominant wavelength in this case is:

$$\frac{L_d}{a} = 1.66 N^{1/3} \left(\frac{\eta_1}{\eta_2}\right)^{\frac{1}{6}}. \tag{12.23}$$

Eq. (12.23) is represented by the dashed curves in Fig. 12.13.

Biot (1965a) extended this work on the buckling of multilayers to bilaminates which are not confined between rigid plates but which are set in a less viscous matrix of viscosity η (Fig. 12.14(a)). The buckling condition for this model is given by:

$$\frac{P}{\eta_t p} = \frac{4\eta}{\eta_t w H} + \frac{Bw^2}{(Bw^2 + 1)} \tag{12.24}$$

where P is the compressive stress applied parallel to the layers and p (defined in Eq. (12.15)) is a measure of the rate of amplification of a sinusoidal perturbation wavelength L. The quantity B is given by:

$$B = \frac{1}{3}\frac{\eta_n}{\eta_t} H^2 \tag{12.25(a)}$$

where H is the thickness of the multilayer and w is the wave number defined by Eq. (12.3). The viscosity coefficient η_t relates the resistance of the whole multilayer to layer parallel shear and is given by

$$\eta_t = \frac{\eta_1 \eta_2}{\alpha_1 \eta_2 + \alpha_2 \eta_1} \tag{12.25(b)}$$

where η_1 and η_2 are the viscosities of the competent and incompetent layers respectively and α_1 and α_2 the proportions of the multilayer thickness made up of the competent and incompetent layers respectively. The viscous resistance of the multilayer to layer parallel compression is given by

$$\eta_n = \alpha_1 \eta_1 + \alpha_2 \eta_2. \tag{12.25(c)}$$

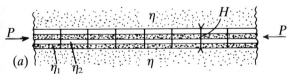

Fig. 12.14. (a) A viscous bilaminate set in a less viscous matrix. (b) The 'first kind' of buckling which occurs when the mechanical anisotropy of the bilaminate is low. (c) The 'second kind' of buckling which occurs when the mechanical anisotropy of the bilaminate is high. (After Biot, 1965a.)

The parameters η_t and η_n are known as bulk moduli and are discussed in more detail in Chap. 13. Eq. (12.24) does not take into account any resistance which the layers may offer to bending during buckling. A discussion of this equation leads to the establishment of two fundamental types of folding which we shall refer to as folding of the first and second kind (Fig. 12.14(b) and (c)). The two kinds of buckling behaviour are attributed by Biot to the influence of the mechanical anisotropy of the model.

For the case when the anisotropy is relatively small (i.e. when the contrast between the viscosity coefficients of the competent and incompetent layers of the multilayer are low) folding of the *first kind* occurs. This folding is characterised by the deformation of the multilayer structure which resembles the pure bending of a homogeneous plate (Fig. 12.14(b)). In such a case, lines which are normal to the layer boundaries in the initial unfolded state, tend to remain normal to the layers. The geometry of the fold is that of a parallel fold (class 1b fold, Fig. 10.14) and the overall deformation has the form called 'tangential longitudinal strain' (see Chap. 10 and Fig. 10.23(f)). This type of folding gives rise to folds with wavelengths which are large in comparison with the total thickness of the multilayer. The equation governing the wavelength is:

$$L = \frac{2\pi}{w} = 2\pi H \sqrt[3]{\frac{\eta_n}{6\eta}} \tag{12.26}$$

which is identical in form with the equation for the classical 'plate buckling' of an isotropic plate embedded in a matrix, Eq. (11.8). The condition for the formation of folds of the first kind is:

$$\frac{\eta}{\eta_t} \ll 0.2 \sqrt{\frac{\eta_t}{\eta_n}}. \tag{12.27}$$

Folding of the *second kind* occurs if the mechanical anisotropy is high; i.e. if the contrast between the viscosity coefficients of the competent and incompetent layers of the multilayer plate is large. If this is so, then the deformation of the multilayer structure involves layer-parallel shear on the fold limbs. In this case, lines in the profile section, originally normal to the layering, tend to remain vertical (Fig. 12.14(c)) and the folds that develop are similar folds (class 2 folds, Fig. 10.14). Their geometry is that which would be produced by heterogeneous simple shear parallel to the fold axial plane. The buckling instability associated with the second kind of folding is related to the mechanical anisotropy of the layered system and is a totally different instability from that associated with the buckling of the first kind. It has been referred to as a *rotational instability* and is discussed in greater detail in Chap. 13 in the section on the amplification of folds in anisotropic materials.

Folding of the second kind is associated with fold wavelengths that are small and of the same order as the total thickness of the multilayer ($L_d/H \approx 1$, see Fig. 12.4(e)). The equation governing the fold wavelength is:

$$L_d = 2\pi a_1 \sqrt[3]{\frac{N_c \eta_1}{6\eta}} \qquad (12.28)$$

where a_1 is the thickness of the competent layers and N_c is the number of competent layers in the multilayer. The condition for the formation of folds of the second kind is:

$$\frac{\eta}{\eta_t} \gg 0.2 \sqrt{\frac{\eta_t}{\eta_n}}. \qquad (12.29)$$

In multilayers whose properties fall between the two conditions (Eqs. (12.27) and (12.29)) one might expect the development of a complete spectrum of flexures with fold geometries and internal deformation patterns between folds of the first kind (i.e. parallel folds with the internal deformation pattern associated with tangential longitudinal strain, which develop when the mechanical anisotropy is not exploited during folding), and folds of the second kind (i.e. similar folds with a strain pattern described by heterogeneous, simple shear parallel to the fold axis, which develops when the mechanical anisotropy totally controls the buckling of the multilayer).

Biot points out that if (η/η_t) is large (i.e. if the competence of the enclosing matrix, η, is large and is, consequently, able to act as a strong confinement) this may prevent the development of folding of the types shown in Fig. 12.14.

A third type of folding in which the boundaries of the multilayer do not fold and in which the folding is restricted to the central, or internal, part of the multilayer may occur. This folding, a specific example of which we have already considered (Fig. 12.12), is called *internal buckling*. The analysis presented by Biot for the deformation of the multilayer shown in Fig. 12.12, is a specific example of a more general analysis in which he considers the buckling behaviour of any mechanically anisotropic material. This general analysis, which can cope admirably with the deformation of both multilayers and rock fabrics, has provided tremendous insight into the deformation of many geological materials, and is discussed in some detail in Chap. 13.

The finite amplification of multilayer buckles

The theories of buckling discussed so far in this chapter apply only to the first increment of buckling; they *should not* be used to predict how the initial low amplitude folds grow into finite structures. Although some attempts have been made to develop theories that can be used to predict the way in which low

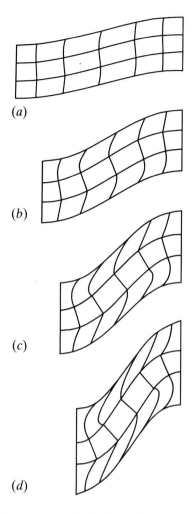

(a)

(b)

(c)

(d)

Fig. 12.15. Four stages in the buckling of a three layer, linear, viscous multilayer in a viscous matrix. The ratio of the viscosity of the central, relatively competent layer, the two adjacent incompetent layers and the matrix is 144/1/12. The marker lines were originally orthogonal to the layering. (After De Bremaecker & Becker, 1978.)

amplitude multilayer folds amplify (e.g. Bayly 1964, 1970, 1971, 1974, Cobbold 1976a and b, Honea & Johnson 1976 and Summers 1979), most of the work on this subject is non-analytic and involves either physical modelling or finite element analysis. The application of these last two techniques to the problem of amplification of multilayer folds is considered in the following section.

Finite element analysis of multilayer buckling

De Bremaecker & Becker (1978) used the technique of finite element analysis to study the finite amplification of a variety of multilayers consisting of either incompressible or compressible and linear or non-linear materials. Four stages in the deformation of a multilayer made up of three Newtonian viscous layers, a central layer of relatively high viscosity sandwiched between two layers of low viscosity are shown in Fig. 12.15. The ratios of the viscosity of the central ayer, adjacent layers and matrix is 144/1/12. Initially, the fold had a low-amplitude, sinusoidal geometry with marker lines normal to the layer boundaries. A constant displacement rate is applied to the left hand side of the model throughout the 'experiment'. De Bremaecker & Becker note that the geometry of the multilayer fold (Fig. 12.15(*d*)) is almost exactly similar (class 2 fold, Fig. 10.14), but are at a loss to explain why this should be so. The strain-rates are largest in the low viscosity part of the model; conversely, the stresses are largest in the high viscosity layer. The experiment shown in Fig. 12.15 was repeated using non-linear viscous materials. A power-law material was chosen, i.e. a material in which the strain-rate is related to the stress through an exponent *n*.

$$\dot{e} \propto \sigma^n. \tag{12.30}$$

An *n* value of 3 was selected and two stages in the deformation of this multilayer are shown in Fig. 12.16. These correspond to the same amount of shortening as that shown in (*b*) and (*c*) of Fig. 12.15. It can be seen that the results are very similar. It is difficult to determine from De Bremaecker & Becker's data whether the non-linear multilayer fold amplifies more rapidly than the linear fold; however, the non-linearity has the effect of lowering the viscosity in the areas of highest strain-rate and of gradually concentrating strain in these regions.

A third model made up of three high viscosity layers alternating with four low viscosity layers in a matrix of medium viscosity is shown in Fig. 12.17. As

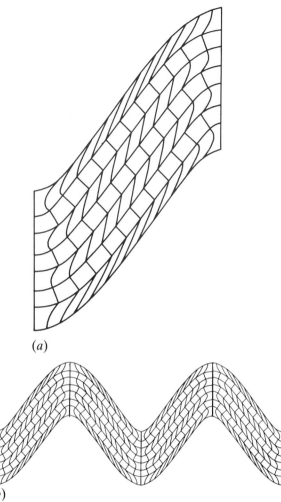

(*a*)

(*b*)

Fig. 12.17. (*a*) The geometry of an initially sinusoidal buckle in a linear, viscous multilayer after 45 per cent shortening. Three relatively competent layers alternate with four incompetent layers. The ratio of viscosity of the competent layers, incompetent layers and matrix is 144/1/12. (*b*) Fold train constructed from the basic unit shown in (*a*). (After De Bremaecker & Becker, 1978.)

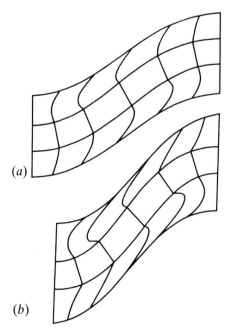

(*a*)

(*b*)

Fig. 12.16. Two stages in the deformation of a model similar to the model shown in Fig. 12.15 but where the layers have non-linear viscous properties. The two stages (*a*) and (*b*) correspond to stages (*b*) and (*c*) respectively of Fig. 12.15. (After De Bremaecker & Becker, 1978.)

before, the fold was initiated from a low-amplitude, sinusoidal buckle but, as can be seen from Fig. 12.17, has the finite geometry of a chevron fold with fairly long straight limbs and quite narrow hinges. It is encouraging to note that this straightening out of the fold limbs during fold amplification is predicted from the theory of finite amplitude folding of anisotropic materials discussed in Chap. 13. (See also Fig. 13.26.)

The tendency for multilayer folds to become chevron-like with progressive deformation is also illustrated by the work of Williams (1980), who also used the technique of finite element analysis to study the amplification of low-amplitude, sinusoidal buckles. His multilayer model is made up of alternating layers of high and low viscosity and he analysed the behaviour of the basic unit of this bilaminate, a pair of layers one relatively competent, the other incompetent. He imposes the constraint that the fold remains similar in geometry throughout the deformation; i.e. although the fold geometry may change as

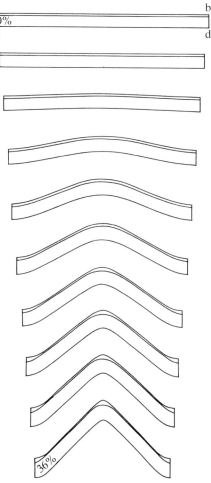

Fig. 12.18. Ten stages ranging from 0 to 36 per cent compression in the development of a chevron fold from an initial sinusoidal fold, as a representative pair of layers from a linear, viscous bilaminate is compressed. The thickness ratio of competent to incompetent layer is 5/1 and the viscosity contrast 653/1. (After Williams, 1980.)

the fold amplifies, at any stage in the deformation, the upper and lower surfaces, ab and cd in Fig. 12.18, will have the same shape.

In order to reduce the amount of layer-parallel shortening (which is neglected in the analysis) a high viscosity contrast (653/1) is chosen and 'experiments' are performed in which the ratio of thicknesses of the competent and incompetent layers are 5/1, 2.5/1 and 1/1. In all three experiments, the initial sinusoidal shape of the fold changes and becomes more chevron-like as the fold amplifies (Fig. 12.18). Initially, the rate of amplification of the folds is exponential, but after the limb-dips reach between 5° and 10°, the limb-dip becomes linearly dependent on the strain. Although the layer-pair maintains a similar geometry throughout the deformation, the competent and incompetent layers making up the pair do not. As in De Bremaecker & Becker's analyses, the competent layer buckles as a parallel fold (class 1B) and the incompetent layer, by thinning in the limb regions and thickening in the hinges, buckles as a class 3 fold (Fig. 12.18).

In natural and model folded bilaminates (e.g. Fig. 12.19(*a*) and (*b*)), the geometry of the competent layers often approximates well to that of a parallel fold (class 1b, Fig. 10.14) and the geometry of the incompetent layers is that of a class 3 fold. The combined geometry of the competent/incompetent layer pair approximates to that of a similar fold. Thus, any number of layer pairs can be stacked together without the generation of any geometric problems. Consequently, the multilayer folds are not restricted to a few layers within the multilayer but can affect many layers and extend for a significant distance in the profile section, in a direction parallel to the axial plane.

Williams argues that the chevron geometry which is seen to develop as multilayer folds amplify from original sinusoidal deflections, can be understood in terms of the symmetry constraints which the fold must satisfy in order to fit into folded multilayers and the energy dissipation during fold amplification which is minimised subject to these constraints. When a single isolated isotropic layer buckles, it forms parallel folds with a sinusoidal shape because this is the configuration that reduces the average curvature, and therefore the strain, to a minimum. The two different layers of the bilaminate would both form parallel folds if free from other constraints. However, layers within a multilayer are not free, and the symmetry requires that the two layers together must form a 'similar' fold. Thus, if the competent layer takes up a sinusoidal shape by buckling as a 'parallel' fold, then the incompetent material must flow in to the hinge regions, so that a similar shape is maintained. This requires considerable flow in the incompetent layer, and hence, great energy dissipation. A shape that requires much less flow is the chevron fold shape, which increases the bending in the hinge regions, but greatly reduces the material flow and is, therefore, the shape which minimises the total energy dissipated. This conclusion is compatible with that reached by Bayly (1974) who, in an interesting series of papers (1964, 1970, 1971 and 1974), considers the formation of chevron folds in some detail.

In either natural or model multilayers, it is remarkably difficult to find examples of folds that are convincing illustrations of structures that have amplified from the sinusoidal buckling instability predicted by the theories of folding discussed in this chapter. As we have already seen in Chap. 10, and as will become apparent in Chap. 13, the folds formed when model multilayers are compressed parallel to the layering exhibit straight limbs and often have axial planes which are markedly oblique to the layering (and hence, to the direction of compression) (Fig. 10.43(*c*)). The reasons for this are discussed in the following chapter, where it is argued that the buckling behaviour of a multilayer is determined by its *bulk* mechanical anisotropy.

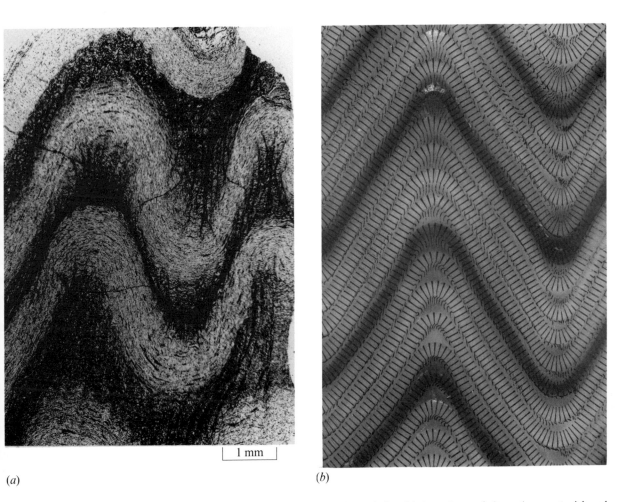

(a)   (b)

Fig. 12.19. (a) Photomicrograph of multilayer buckles in a metamorphic differentiation fabric made up of alternating quartz-rich and mica-rich layers. This fabric approximates well to a regular bilaminate. From the Pre-Cambrian phyllites of Rhoscolyn, Anglesey, N. Wales. (b) Buckled paraffin wax bilaminate. The competent layers have formed parallel folds (class 1b, Fig. 10.14) and the marker bands, that were originally normal to the layering, have remained so. The incompetent layers have formed class 3 folds and on the limbs, the deflection of the marker lines shows that the deformation approximates to simple shear. (Photograph (b) courtesy of J-P. Latham.)

In the simple bilaminate made up of materials with linear properties and which possess welded interlayer surfaces, the competence contrast between adjacent layers gives a direct measure of the anisotropy. However, as we shall see in the next chapter, in a complex multilayer, the anisotropy depends upon several factors. In general, it will be seen that it is low cohesion or lack of cohesion between the layers which is of fundamental importance in determining the anisotropy (and consequently the buckling behaviour) of a rock multilayer. In natural and model multilayers the layer interfaces are only rarely welded. Consequently, it is not surprising to find that structures predicted by the theory of buckling of bilaminates with welded interfaces are rarely found either in nature or in models. One of the most convincing examples, known to the authors, of a rock multilayer buckling in accordance with those theories outlined in this chapter, in which it is assumed that the layers are welded along their contacts, is shown in Fig. 12.19(a). The folds occur in a bilaminate multi-

layer of alternating quartz-rich and mica-rich layers produced by metamorphic differentiation. This process (considered in Chap. 17) produces bilaminates of very regular layer thickness and which exhibit gradational boundaries between layers. Such gradations, of course, inhibit slip along the boundaries during folding. There are other examples which are perhaps somewhat less convincing. These were obtained in the models illustrated in Fig. 12.20(a) and (b).

The change in geometry that accompanies the amplification of multilayer buckles has been studied theoretically by Honea & Johnson (1976). Their work concerns the buckling of an elastic multilayer consisting of alternating competent and incompetent layers. They show that when such a multilayer is compressed parallel to the layering, the initial fold pattern is always sinusoidal. However, depending on the interlayer properties, subsequent growth of the fold can change this pattern. If the contact between the layers cannot resist shear stress, or if the incompetent layers

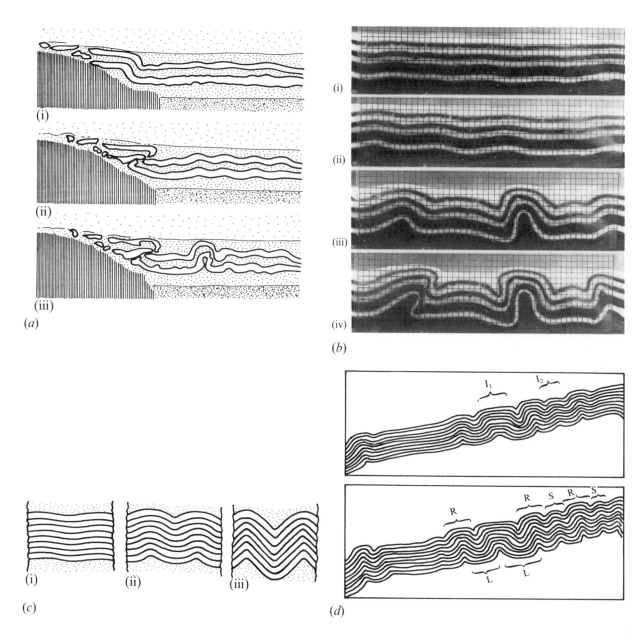

Fig. 12.20. (*a*) and (*b*) show details of approximately sinusoidal folds formed in gelatine multilayers deformed (*a*) in the apparatus *A* and (*b*) in the apparatus *B* shown in Fig. 10.36. The sinusoidal geometry is lost as the folds increase in amplitude. (*c*) Sequence of fold geometries associated with the buckling of an elastic multilayer, (i) sinusoidal, (ii) concentric and (iii) chevron. (After Johnson, 1977.) (*d*) The generation of folds with concentric geometry in a regular multilayer consisting of alternating layers of relatively competent and incompetent modelling clay in a less competent matrix. The layering was originally at 10° to the compression. (After Watkinson, 1976.)

provide uniform resistance to shear between the relatively competent layers, sinusoidal folds (Fig. 12.20(*c*) (i)) with the wavelength given by either Eq. (12.26) or (12.28), grow most rapidly with increased shortening. However, these folds become unstable as they amplify and are transformed into concentric-like folds (Fig. 12.20(*c*) (ii)) and finally into chevron folds. The generation of folds with concentric geometry is shown in Fig. 12.20(*d*). If the multilayer has different interlayer properties from those mentioned above, other structures, notably reverse kink-bands (Fig. 13.19e), may form. These are discussed further in Chap. 13.

Buckling of complex multilayers

The theories discussed in this chapter apply only to those regular bilaminates in which all the competent units are of one thickness (a_1) and all the incompetent layers have a thickness (a_2). However, layers within natural bilaminates, such as turbidites or banded gneisses, are almost invariably of different thicknesses. When such a bilaminate buckles, a variety of structures develop, which are the result of these different thicknesses (particularly those relating to the competent layers). These structures which cannot, therefore, be predicted from the buckling theories of regular bilaminates can be grouped into three main types. These are as follows.

(i) Structures that form when the competent layers are of unequal thickness, but where there is no large difference between the thickness of the various layers. Such structures are called *accommodation structures*.

(ii) Structures that form when one of the competent layers is particularly thick, compared with the mean thickness of the other layers in the sequence. Such a thick unit is called a *control unit*, for, as will be seen, it is of over-riding importance in governing the buckling behaviour of the multilayer.

(iii) Structures that form when one of the competent layers is relatively thin. These structures, which are small wavelength folds, are known as *parasitic folds*.

These three categories of structures are discussed in the following paragraphs.

Accommodation structures

Ramsay (1974) presents a geometric model for the finite development of chevron folds in a multilayer consisting of alternating competent and incompetent layers. The model is shown in Fig. 12.21(a), where θ is the limb-dip, l the limb-length and a_1 and a_2 the thicknesses of the competent and incompetent layers respectively. It follows from the geometry of the model that the shortening, $(1 + e)$, in the direction parallel to the original layering, is related to the competent layer thickness and limb-dip by:

$$(1 + e) = \left(1 - \frac{\theta a_1}{l}\right)\cos\theta + \frac{a_1}{l}\sin\theta. \quad (12.31)$$

This relationship is expressed graphically in Fig. 12.21 for various competent layer thickness/limb-length ratios. From the graph, we can see that if a multilayer contains competent layers of different thicknesses then, according to the model, for any finite percentage shortening, the limb-dips of the different layers will vary (Fig. 12.22). In practice, of course, the limb-dips of the various competent layers of a chevron fold will have the same dip and consequently the relatively thick competent layer of Fig. 12.22 must 'accommodate' itself into the overall limb-dip and amplitude of the multilayer fold.

It can be seen from Figs. 12.21 and 12.22 that a competent layer slightly thicker than the majority of the competent layers of a multilayer will be too long to fit easily into the multilayer fold, and a competent layer slightly thinner than the majority of the competent layers of the multilayer will be too short to fit easily into the multilayer fold. A variety of structures can develop, as layers with anomalous thicknesses accommodate themselves into the multilayer fold.

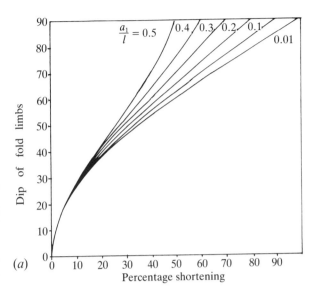

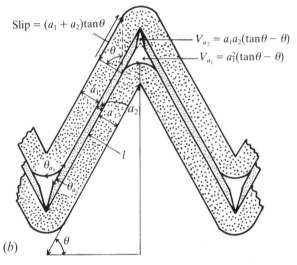

Fig. 12.21. (a) Graphical expression of Eq. 12.31), showing the variations of limb-dip with shortening for various competent layer thickness/limb-length ratios. (b) Geometric model of chevron fold. (After Ramsay, 1974.)

Some of these are shown in Fig. 12.23. Accommodation structures which are the result of a relatively thick layer reducing its effective length are shown in Fig. 12.23(a), (b) and (c). In (a) only ductile deformation is involved, in (b) and (c) both brittle and ductile deformation occur. A relatively thin competent layer increasing its effective length by forming boudinage structures is shown in Fig. 12.23(d).

A variety of accommodation structures which have developed in natural multilayers are shown in Figs. 12.24 to 12.27. If a particularly thick layer, or group of layers, is sandwiched between two thick incompetent layers, then the possibility exists of the thick layer developing a different amplitude from the rest of the layers of the multilayer (Fig. 12.22). A natural example where this has occurred is shown in Fig. 12.28.

Accommodation structures (particularly cari-

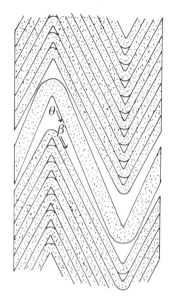

Fig. 12.22. The geometric problem posed when a single, thick, competent bed is incorporated into a succession of folded, thinner layers. For the same total shortening, the dip in the thick layer (θ) is always greater than in the thin layer (β). (After Ramsay, 1974.)

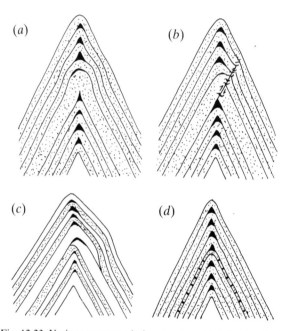

Fig. 12.23. Various accommodation structures that develop during multilayer buckling as layers with anomalous thicknesses adjust to fit into the overall wavelength and amplitude of the multilayer buckles. ((a), (b) and (d) after Ramsay, 1974.)

Fig. 12.24. The accommodations of a relatively thick sandstone layer into a multilayer buckle by the formation of a keel-like (carinate) hinge (cf. Fig. 10.23(a)). Hartland Quay, N. Devon, England.

nate folds, i.e. folds with a keel-like hinge, Fig. 12.23(a) and 12.24) have been formed experimentally in complex multilayers containing competent bands of different thicknesses (Fig. 12.29). However, these accommodation structures have also been formed experimentally in *regular* bilaminates, Fig. 12.30, and in model multilayers made up of identical lubricated layers. They have also been observed in regular natural bilaminates (Fig. 12.31). The occurrence of these structures in the last two types of multilayers can not be explained by the thickness variation arguments given above and illustrated in Fig. 12.22. However, as can be seen in Fig. 12.32, their occurrence is associated with relatively large amounts of slip on certain layer interfaces.

To explain the formation of these accommodation structures, it is necessary to pre-empt the discussion on rotational instability given in Chap. 13 where it is shown that the rate of layer or foliation rotation varies with its orientation with respect to the maximum principal compression. When the principal compression is parallel or sub-parallel to the layering, there is an initial stage of accelerating rotation rate. This is followed by an intermediate stage of steady rotation and a final stage of decaying rotation rate. The amount of shortening after which one stage passes into another is determined by the degree of anisotropy. For a high degree of anisotropy, the initial stage is very short and extremely rapid, rotation rates being high over a small interval of shortening. The reverse is the case for low anisotropies, the initial stage being much longer. The dependence of rate of layer rotation on anisotropy and layer orientation is shown graphically in Fig. 12.33.

Fold initiation (layer rotation) in a multilayer will occur at local sites where rotation is assisted by the presence of heterogeneities which may be local deflections or thickness variations of a layer, or the

Fig. 12.25. Accommodation thrust in the hinge of a fold. The thrust formed as two relatively thick sandstone layers adjusted to the wavelength and amplitude of the multilayer buckle, see Fig. 12.23(*b*). Millook, N. Cornwall.

local occurrence of low frictional contact between adjacent layers. The development of a horizon along which relatively large amounts of slip occur during folding necessitates the production of a volume increase in the hinge areas near the decollement. This may result in the formation of saddle reefs or the migration of relatively incompetent material from the limb to the hinge areas. This can be clearly seen in the experiment shown in Fig. 12.29 in which there are several horizons along which relatively large amounts of slip have occurred and with which thickened hinges are associated, (e.g. C in Fig. 12.29). With the onset of rotation-rate decay (rotation strain hardening), slip between layers is suppressed and most of the deformation is concentrated in the hinge zones. The straight limbs become bent by the production of carinate hinges and hinge collapse. Collapse of inner arcs to form isoclinal fold cores is common.

It seems probable that the formation of accommodation structures in complex, natural multilayers will involve both the effect of having layers of different thicknesses and horizons of relatively easy slip.

An interesting example in which accommodation structures have influenced the economic geology of an area is provided by the famous gold deposits of Bendigo, Australia. The gold occurs in quartz saddle reefs which formed in the hinge areas of multilayer folds formed by the buckling of a complex sedimentary sequence overlying a basement that was undergoing prograde metamorphism during the time of folding. Inspection of Fig. 12.34 shows clearly that numerous accommodation thrusts (Fig. 12.23(*b*) and (*c*)) developed during the folding, thus increasing the permeability of the hinge region. The formation of tension fractures in the outer arcs and/or shear fractures in the inner arcs of the fold hinges of the competent layers (Fig. 10.26(*b*)) may have further increased the permeability of the hinge regions. Dehydration reactions associated with the prograde metamorphism of the basement rocks produced fluids rich in precious metals, and these fluids, together with water driven out of the sediments as they were compressed and folded, were channelled upwards along the hinge surfaces of the folds. Gold and quartz were precipated in the hinge regions as the hot fluids rose into the cooler overlying rocks.

Control units
It will be recalled from Eqs. (11.10) and (11.11) that the resistance to buckling of a layer is proportional to the cube of its thickness. Consequently, in a regular

Fig. 12.26. A carinate fold hinge in one sandstone layer, and a thrust and carinate hinge in another, from a buckled turbidite sequence. Bude, Cornwall, England.

bilaminate, the resistance to buckling of the individual competent layers, which are identical in material properties and thickness, will be the same. However, the existence of an anomalously thick competent layer in a multilayer will dramatically affect its resistance to buckling, for even if the layer is only three times thicker than the other competent layers, it will offer twenty seven times as much resistance to buckling. It is to be expected, therefore, that the buckling of such a relatively thick layer will control the buckling of the multilayer as a whole. The layer will effectively act as a single layer, developing its own wavelength which is given by the single layer buckling equation (Eq. (11.8)), and the multilayer 'matrix' will buckle accordingly. Such relatively thick layers that control the buckling of multilayers and behave essentially as single layers, are known as *control units*. It should be noted, however, that although the initial buckling response of the control unit may be that of a single layer, because the matrix is anisotropic, the finite growth of the fold may be considerably influenced by the tendency of an anisotropic material to form straight limbed folds, (see Chap. 13).

An example of a control unit governing the buckling of a rock multilayer has been reported by Price (1967), who describes folds developed in the Aberystwyth Grits, a turbidite sequence in which the thickness of the individual sandstone beds are very different. Detailed measurements were made of the thickness of the sandstone and shale beds in a 70 metre high section. For ease of presentation of the data, the section was divided into 55 equal parts. The resistance to buckling of each 1.25 m division was taken as the sum of the resistance of every competent band in the division. This is represented graphically in Fig. 12.35(b), where it can immediately be seen that the division that comprises group X (Fig. 12.35) must have profoundly influenced the development of folds and acted as the 'control unit'. It is clear from this example that a control unit may be a group of closely spaced, competent beds instead of an individual, competent bed.

However, it should not always be assumed that a relatively thick layer in a layered sequence will control buckling. An example where it does not occur is shown in Fig. 12.35(c). In this idealised sketch, which is based on several examples of experimentally deformed phyllites, the flexing of the relatively thick quartz layer is controlled by the buckling (kinking) of the anisotropic matrix. Conjugate kink-bands develop in the matrix and the quartz layers are deflected to accommodate these bands. In contrast, in the example shown in Fig. 12.35(d), it seems likely that the buckling of the quartz layer initiated the kink-bands in the matrix.

Fig. 12.27. The formation of second order folds in the hinge region of a multilayer buckle. Trearddur Bay, Anglesey, N. Wales.

Parasitic folds

Folding within a complex multilayer may occur on several different scales during a single phase of compression. The similarity in hinge direction of the different orders of folds is summarised in Pumpelly's rule (1894) which states that ... 'The degree of direction of pitch (i.e. plunge in modern usage) of the (major) fold is indicated by those of the axes of the minor plications on its side'. The variation in geometry of the minor folds associated with a major fold is shown in Fig. 10.5. The minor folds are asymmetric on the limbs of the major fold and symmetric in the hinge areas. Originally, these folds were called *drag folds* (Leith, 1923), indicating a belief that they were caused by slip along bedding planes as the major folds formed by flexural slip (Fig. 12.36(*a*) (i), (ii) and (iii)). However, because the slip between beds decreases from a maximum at the inflection point to zero at the hinge, drag folds ought to die out towards major fold hinges. This does not generally occur (Fig. 12.37). This objection to the mechanism of formation of drag folds proposed above could be overcome by suggesting that the symmetric minor folds in the hinge regions of major folds are the result of post-buckle flattening (Fig. 12.36(*a*) (iv)). However, a more fundamental objection to this mechanism exists. As discussed in Chap. 10 (Fig. 10.23(*e*)), during flexural slip folding, because one of the planes of no finite, longitudinal strain coincides with the bedding, no change in length occurs parallel to the bedding. The

Fig. 12.28. A group of relatively competent layers sandwiched between two thick, incompetent layers. These competent layers are able to develop a different amplitude and wavelength from the rest of the multilayer (cf. Fig. 12.22). New Harbour Series, Holyhead, N. Wales.

suggestion that a shear couple applied parallel to the bedding can cause buckling (i.e. shortening) of the bedding is, therefore, unreasonable (Ramsay, 1967). The term drag fold with its unfortunate genetic implication has been replaced by *parasitic fold* (Shearman in De Sitter, 1958).

Ramberg (1964) has suggested that minor folds in relatively thin, competent layers may form during the early stages of buckling of a multilayer before the development of the major folds (Fig. 12.36(*b*) (i) and (ii)). During the amplification of the major folds, the originally symmetric minor folds become asymmetric on the major fold limbs but remain symmetric in the hinge regions (Fig. 12.36(*b*) (iii)). It can be seen that the mechanism proposed by Ramberg for the formation of asymmetric folds involves the application of a shear couple to originally symmetric folds. The symmetry of the profile sections of the minor folds, *z*, *m* or *s*, Fig. 12.36(*b*) (iii) and Fig. 12.37, indicates their position on the major fold. Thus, careful mapping of minor folds enables the major folds, which are often too large or not well enough exposed

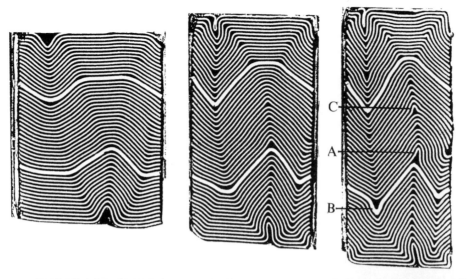

Fig. 12.29. Three stages in the deformation of a complex multilayer made up of Plasticine layers of approximately the same rheological properties. Keel-like hinges form in the relatively thick layer (at B) and in the thinner layers (at A and C). (After Asselin, 1975.)

Fig. 12.30. Keel-like hinges developed experimentally in a regular Plasticine bilaminate.

Fig. 12.31. A natural, regular bilaminate containing folds with carinate hinges. There are no relatively thick layers to account for the formation of these fold hinges and their formation seems to be related to a considerable amount of slip along certain layer boundaries. New Harbour Series, Holyhead, N. Wales.

to be observed directly, to be located precisely. This point is discussed further in Chap. 18.

Parasitic folds are not always single layer buckles and may develop in relatively small multilayer sequences within a larger multilayer as shown in Fig. 12.38.

Minor asymmetric folds

It has been noted that in the limbs of major folds the parasitic structures are asymmetrical. A mechanism which can be invoked to explain the development of asymmetrical buckles has been proposed by Price (1967). He noted that the extant theories relating to single and multilayer buckling assumed that compression acted exactly along the layer. However, from field evidence, it can sometimes be inferred that the axis of maximum compressive stress, prior to the initiation of folding (and also subsequent to fold

Fig. 12.32. Buckled wax bilaminate showing the association of horizons of relatively large amounts of slip and the formation of carinate fold hinges (accommodation structures). (Photograph courtesy of J-P. Latham.)

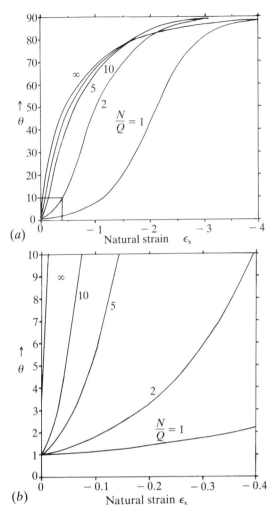

Fig. 12.33. (*a*) Graph showing the dependence of θ, the finite attitude of the layering that occurs as a result of shortening (ϵ_x), on the anisotropy (N/Q) of a material. (*b*) Enlargement of small square in (*a*). (After Cobbold, 1976b.)

development) acted at a significant angle to the layering. An example in which a specific group of structures has developed which permit such stress orientations to be inferred has been described by Skarmeta & Price (1984). (See also Chap. 3.) Price then argued that the layer-parallel shear stresses and the consequent bending moment which results from the application of the maximum compression at an angle to the layering could be invoked to explain the initiation of asymmetric elastic buckles.

The stress situation postulated is indicated in Fig. 12.39(*a*). It is assumed that the horizontal stress σ_H is equivalent to the Euler critical buckling load and so would give rise to a sinusoidal elastic buckle. The deflection of such an elastic buckle is so small that one can assume with small error that the critical buckling force acts in a straight line and that the buckling moment (M_{buc}) will be directly proportional to the displacement y, of the elastic sine curve, (i.e. $M_{buc} \propto P_{crit} y$). Consequently, the magnitude of the buckling moment throughout the wavelength of the elastic buckle will also exhibit the form of a sine curve (Fig. 12.39(*c*)).

It will be noted (Fig. 12.39(*b*)) that because the axis of greatest principal stress is inclined to the layer, a layer-parallel shear stress (τ) is generated. However, it is clear that, for elastic equilibrium, there will be a balancing shear stress which acts normal to the layering (Fig. 12.39(*b*)). For a unit extent in the *z* direction, the vertical shear stresses give rise to a corresponding *shear force* (S_f) acting normal to the surface of the layer, such that:

$$S_f = a\tau \tag{12.32}$$

where *a* is the thickness of the layer. It can be shown (see Salmon, 1952, p. 58) that there is a relationship between a shearing force in a unit and a bending moment (M_{bend}) such that:

$$S_f = \frac{\mathrm{d}M_{bend}}{\mathrm{d}x} \tag{12.33}$$

or

$$M_{bend} = S_f \int_0^x \mathrm{d}x. \tag{12.34}$$

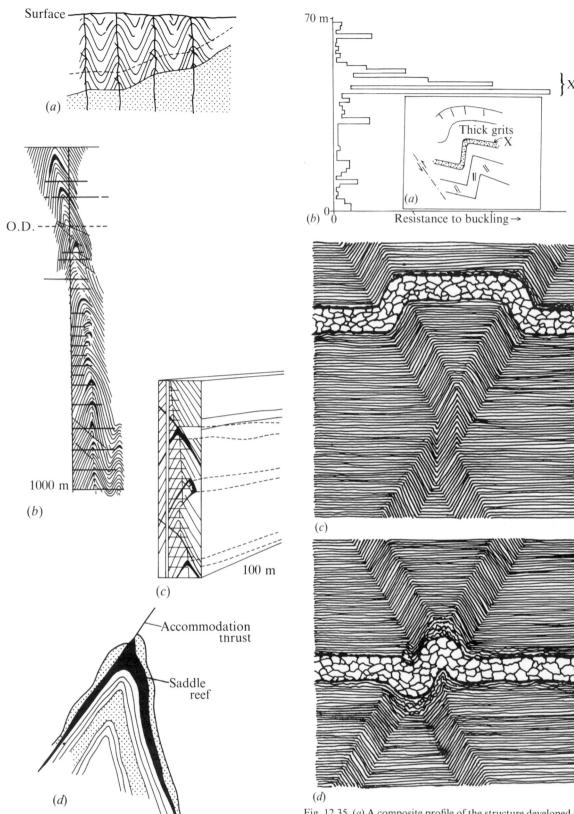

Fig. 12.34. (*a*) A section through the folded sedimentary sequence and basement rocks at Bendigo, Australia. (*b*) A detailed section through the folded sequence of (*a*), showing the persistence of the folds in profile section. The formation of accommodation thrusts in the hinge of the folds is apparent from both (*a*) and (*b*) and the geometry of the thrusts and the saddle reefs is shown in greater detail in (*c*) and (*d*). (After Herman, 1914 and 1923.)

Fig. 12.35. (*a*) A composite profile of the structure developed in the Aberystwyth Grits in the cliff sections near Llanrhwstyd, Dyfed, Wales. (*b*) Graphical representation of the resistance to buckling of the various rock units exposed in the profile shown in (*a*). (After Price, 1967.) (*c*) An idealised sketch of a box, or conjugate fold (in a quartz layer) formed during the experimental deformation of a phyllite. (*d*) Concentric folds in a quartz layer at a kink intersection formed during experimental deformation of the rock. ((*c*) and (*d*) after Paterson & Weiss, 1968.)

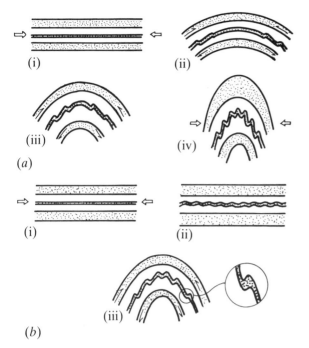

(a)

(i)   (ii)

(iii)

(b)

Fig. 12.36. Diagrammatic representation of two mechanisms that have been proposed for the formation of parasitic folds: (a) as a result of layer-parallel slip during folding followed by post-buckle flattening; and (b) as a result of first order (larger) folds forming after second order (smaller) folds.

(a)

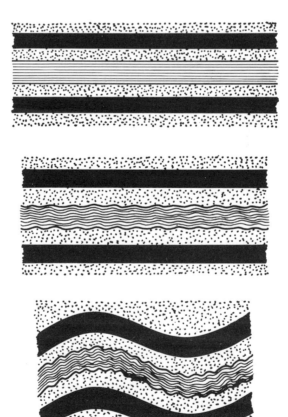

(b)

Fig. 12.37. Parasitic folds in carbonate layers (a) on the limb of a fold and (b) around the hinge of a fold in the Devonian rocks of N. Devon. ((b), after De Sitter, 1958.)

Thus when, as here, S_f is a constant, M_{bend} changes in a linear fashion along the unit. If M_{bend} is arbitrarily put equal to zero at point O in Fig. 12.39(c), the distribution of the bending moment along the length of the fold which results from the shear stress (τ) is represented by the straight line OL'. The relative increase in magnitude of the bending moment along an elastic fold of wavelength L is:

$$M_{bend} = \tau aL. \qquad (12.35)$$

The deflection of the competent layer will be influenced by both the buckling moment and the bending moment. Because the problem at this stage is one of elasticity, one may superimpose the effects of these moments represented by curves OA_2SA_1L and OL' respectively in Fig. 12.39(c). The total moment (M_{tot}) is therefore represented by curve $OA'_2S'A'_1L'$.

It will be noted that the points of maxima on the M_{tot} curve have moved to the left relative to the corresponding points on the M_{buc} curve. The lateral shifts (represented by PQ and P'Q') are the same for both sets of maxima, so that the wavelengths of the curves are the same. However, it can be seen that the minimum value of the M_{tot} curve is to the right of the corresponding point on the M_{buc} curve, by an amount represented by UV.

Price argued that elastic failure (and therefore the establishing of the finite fold hinges) occurs where the radius of curvature is a maximum. Maxima occur on the total moment curve at Q, V and Q' in

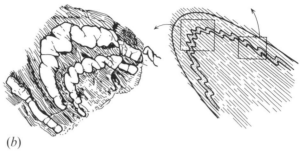

Fig. 12.38. The formation of parasitic folds in a relatively thin multilayer sequence within a larger multilayer. (After Ramberg, 1964.)

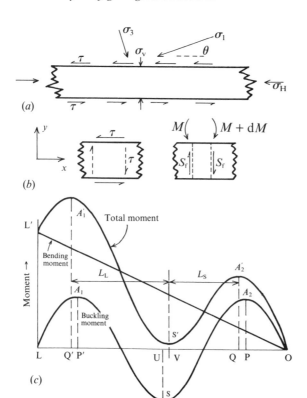

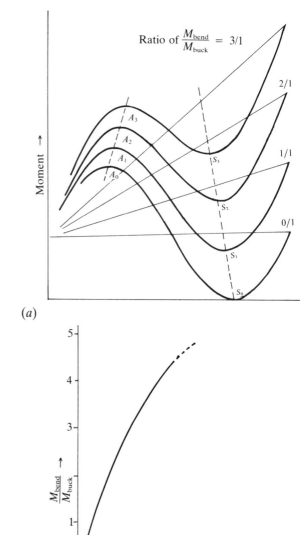

Fig. 12.39. (*a*) The distribution of stresses normal and parallel to a horizontal, competent unit when the axis of greatest principal stress is inclined at an angle θ. (*b*) The relationship between shear stresses (τ), shear force (S_f) and bending moment (M). (*c*) Shows how the bending and buckling moment curves may be combined to give an asymmetric, total moment curve. (After Price, 1967.)

Fig. 12.40. (*a*) Shows the various, total moment curves compiled for ratios of bending to buckling moments from 1/1 to 3/1. (*b*) Shows the relationship between the ratio of moments and the resulting ratio of the limb lengths L_{long}/L_{short}. (After Price, 1967.)

Fig. 12.39(*c*). It will be seen from this figure that the distance VQ is less than VQ', so, given that the antiformal and synformal axes will be initiated and fixed at points Q, Q' and V, the structure that would result from these moment curves would be asymmetric.

It may be inferred from Fig. 12.39(*c*) that the degree of asymmetry that would result from this mechanism will depend upon the slope of the bending moment curve, which is related to the angle between the maximum principal compression direction and the layer. The total moment curves for different slopes of bending moments are shown in Fig. 12.40(*a*). From Fig. 12.40(*b*) it can be seen that for a given ratio of buckling to bending moments between 1/1 and 3/1, the ratios of limb lengths of asymmetric folds will range from 1.3/1 to about 2.5/1. Such ratios of limb lengths are in keeping with those of many small folds observed in the field.

Subsequent to the generation of a major fold, the angle which the axis of compressive stress makes with the layering is likely to be largest in the limbs of the major structure and smallest at the hinge. Hence, one would infer from this proposed mechanism that the degree of asymmetry of any parasitic folds generated at this time would decrease from a maximum in the limbs to a minimum at the hinge line. This is, of course, in keeping with field observations.

In the original paper, Price (1967) uses these arguments to explain the generation of large asymmetric structures. However, as is discussed in Chaps. 14 and 15, it seems likely that high fluid pressures must exist along bedding planes before large folds can develop. Indeed, crystal fibres found on the bedding planes of large-scale folds indicate that these pressures are often sufficiently large to overcome both the overburden and the cohesive strength of the bedding planes and to 'jack' the beds apart during folding. Clearly it would not be possible for significant shear stresses to exist along such a bedding plane, and the mechanism discussed above can *not* be used to explain the formation of large asymmetric folds. It could however be invoked to explain the development of small-scale buckle folds which formed prior to the generation of, or after the locking-up of, a major fold structure.

Commentary

Field and model examples of single layer folds indicate that the finite geometry of such folds can be obtained to a first approximation by the geometric amplification of the initial low amplitude sinusoidal folds that are predicted theoretically. However, finite fold structures developed in natural and model multilayers show a more varied range of geometries, typically straight limbed folds with either chevron or box-like profiles which cannot be obtained by the geometric amplification of the low amplitude sinusoidal folds predicted by the simple theories of multilayer folding. In addition, it is observed that a similar range of finite fold geometries to those encountered in deformed multilayers is found in deformed mineral fabrics whose behaviour approximates to that of a homogeneous, anisotropic material. It is therefore necessary to consider the buckling of such a material and this we do in the following chapter. Having done this the reason for the sparsity of natural examples of large amplitude multilayer folds with a sinusoidal profile geometry becomes apparent.

References

Afdjei, S. (1985). Accommodation structures in folded multilayer complexes. Unpublished M.Sc. Dissertation. University of London.

Asselin, J. (1975). Accommodation structures in Plasticine kink-band models. Unpublished M.Sc. Dissertation. University of London.

Bayly, M.B. (1964). A theory of similar folding of viscous materials. *Am. J. Sci.*, **262**, 753–66.

(1970). Viscosity and anisotropy estimates from measurement on chevron folds. *Tectonophysics*, **9**, 459–74.

(1971). Similar folds, buckling and great circle patterns. *J. Geol.*, **79**, 110–18.

(1974). An energy calculation regarding the roundness of folds. *Tectonophysics*, **24**, 291–316.

Biot, M. A. (1961). Theory of folding of stratified, visco-elastic media and its application in tectonics and orogenesis. *Geol. Soc. Am. Bul.*, **72**, 1595–632.

(1963a). Theory of stability of multilayered continua in finite, anisotropic elasticity. *J. Franklin Inst.*, **276**, 2, 128–53.

(1963b). Stability of multilayer continua including the effect of gravity and visco elasticity. *J. Franklin Inst.*, **276**, 3, 231–52.

(1964). Theory of internal buckling of a confined multi-layered structure. *Geol. Soc. Am. Bul.*, **75**, 563–8.

(1965a). Theory of similar folding of the first and second kind. *Geol. Soc. Am. Bul.*, **76**, 251–8.

(1965b). *Mechanics of incremental deformations.* 504 pp. New York: Wiley.

Cobbold, P.R. (1976a). Fold shapes as functions of progressive strain. *Phil. Trans. Roy. Soc. London*, **A283**, 129–38.

(1976b). Mechanical effect of anisotropy during large finite deformations. *Bull. Soc. Geol. France*, **18**, 1497–510.

Currie, J.B., Patnode, H.W. & Trump, R.P. (1962). Development of folds in sedimentary strata. *Geol. Soc. Am. Bul.*, **73**, 655–74.

De Bremaecker, J.C. & Becker, E.B. (1978). Finite element models of folding. *Tectonophysics*, **50**, 349–67.

De Sitter, L.U. (1958). Boudins and parasitic folds in relation to cleavage and folding. *Geol. en Mijnbouw*, N.S., **20**, 277–86.

Dieterich, J.H. (1969). Origin of cleavage in folded rocks. *Am. J. Sci.*, **267**, 155–65.

Herman, H. (1914). *Economic geology and mineral resources of Victoria.* British Assoc. Handbook to Victoria.

(1923). Structure of Bendigo Goldfield. *Bull. Geol. Survey, Victoria*, **47**.

Honea, E. & Johnson, A.M. (1976). A theory of concentric, kink and sinusoidal folding and of monoclinal flexuring of compressible elastic multilayers, IV. Development of sinusoidal and kink folds in multilayers confined by rigid boundaries. *Tectonophysics*, **30**, 197–239.

Johnson, A.M. (1977). *Styles of folding: Mechanics and mechanisms of natural elastic materials.* 406 pp. Elsevier Scientific Pub. Co.

Leith, C.K. (1923). *Structural geology.* Rev. edn. 390 pp. London: Constable, New York: Henry Holt.

Paterson, M.S. & Weiss, L.E. (1968). Folding and boudinage of quartz-rich layers in experimentally deformed phyllite. *Geol. Soc. Am. Bul.*, **79**, 795–812.

Price, N.J. (1967). The initiation and development of asymmetrical buckle folds in non-metamorphosed, competent sediments. *Tectonophysics*, **4**, 2, 173–201.

Pumpelly, R., Wolff, J.E. & Dale, T.N. (1894). Geology of the Green Mountains. *U.S. Geol. Surv. Mem.*, **23**, 1–157.

Ramberg, H. (1961). Contact strain and folding instability of a multilayered body under compression. *Geol. Rundsch.*, **51**, 405–39.

(1963). Fluid dynamics of viscous buckling applicable to folding of layered rocks. *Am. Assoc. Petrol. Geol. Bull.*, **47**, 485–505.

(1964). Selective buckling of composite layers with contrasted rheological properties, a theory for simultaneous formation of several order of folds. *Tectonophysics*, **1**, 4, 307–41.

Ramsay, J.G. (1967). *Folding and fracturing of rocks.* 568 pp. New York: McGraw–Hill.

(1974). Development of chevron folds. *Geol. Soc. Am. Bul.*, **85**, 1741–54.

Roberts, D. & Stromgard, K-E. (1972). A comparison of natural and experimental strain patterns around fold hinge zones. *Tectonophysics*, **14**, 105–20.

Rubey, W.W. & Hubbert, M.K. (1959). Role of fluid pressure in mechanics of overthrust faulting, II. *Geol. Soc. Am. Bul.*, **70**, 167–206.

Salmon, E.H. (1952). *Materials and Structures.* 638 pp. Oxford University Press.

Skarmeta, J. & Price, N.J. (1984). Deformation of country rock by an intrusion in the Sierra de Moreno,

Northern Chilean Andes. *J. Geol. Soc. London*, **141**, 901–8.

Summers, J.M. (1979). An experimental and theoretical investigation of multilayer fold development. Unpublished Ph.D. Thesis, University of London.

Watkinson, A.J. (1976). Fold propagation and inter-ference in a single multilayer unit. *Tectonophysics*, **34**, T37–T42.

Williams, J.R. (1980). Similar and chevron folds in multi-layers, using finite element and geometric models. *Tectonophysics*, **65**, 323–38.

13 The buckling of anisotropic rocks

Introduction

As we have seen in Chap. 12, two quite different theoretical approaches have been used to analyse the buckling behaviour of geological multilayers. In one, the properties of the individual layers making up the multilayer are considered and the other deals with the 'bulk' mechanical properties of the multilayer. This latter approach, which treats a multilayer as a statistically homogeneous *anisotropic* material, can be applied to materials which range from mineral fabrics to multilayers. It is considered in some detail in this chapter, for it indicates a fundamental relationship between such apparently diverse structures as folds, kink-bands, pinch-and-swell structures, crenulation cleavage, shear-zones and faults, as well as providing considerable insight into the buckling behaviour of many geological multilayers and fabrics.

The first part of this chapter introduces some field observations relating to folds in anisotropic rocks. This is followed by an outline of the theoretical analysis of the buckling of an anisotropic material and a discussion of its geological implications and of the way in which the structures that are initiated in such materials amplify into finite structures.

Fold structures in anisotropic rocks

There are two significant geometric features which characterise many folds developed in geological multilayers and fabrics, that cannot be accounted for by the theories of single layer and multilayer buckling discussed in Chaps. 11 and 12. The first is that the limbs of the folds are very straight and the second is that, even when it may be inferred that the principal compression acted parallel to the layering or fabric, the axial planes are often oblique to the original orientation of the layer or fabric. Folds showing these features can be seen in the deformed phyllites illustrated in Fig. 13.1. Field observations show that folds with axial planes oblique (rather than normal) to the layering may occur as conjugate structures (referred to as box folds or conjugate kink-bands), for example, folds A and B of Fig. 13.2(*a*); or as structures with axial planes in only one direction (referred to as kink-bands) as seen in Figs. 13.1 and 13.3(*b*). These structures occur on many scales, and the individual kink-bands of Figs. 13.3 and 13.4 range in size over at least five orders of magnitude. The conjugate kink-

bands or box folds of the type shown in Figs. 13.4, 13.6, 13.5 and 13.2 range over six orders of magnitude.

It is interesting to note that although it is rarely, if ever, possible to observe the development of folds in nature, it is possible, by studying the geometry of structures arrested at various stages in their development, to reconstruct the way in which a structure amplifies and changes geometry with progressive deformation. For example, it is tempting to suggest from a study of the cross sections of the Jura mountains shown in Fig. 13.2, that structures A, B and C of Fig. 13.2(*b*) are successive stages in the development of a bedding plane thrust from a fold. As will be seen later in this chapter, experimental work confirms that this is so. It is not, however, suggested that all bedding-plane thrusts develop in this way and the various mechanisms that have been proposed for the formation of thrusts are discussed in Chap. 7.

As has been previously noted in Chap. 10, fold structures in anisotropic rocks commonly form in isolation rather than as one of a train of folds. The isolated conjugate folds from the Jura shown in Fig. 13.2 are good examples of this. These observations indicate clearly that folding does not always occur uniformly throughout a layered system once the critical buckling stress has been reached.

It must be emphasised that the axial planes of folds in anisotropic materials are not always oblique to the original orientation of the layering. In fact, a complete spectrum of structures between chevron folds with axial planes normal to the original orientation of the layering (Fig. 13.7(*a*)) and conjugate kink-bands with axial planes at approximately 45° to the layering (Fig. 13.7(*b*)) can be found in nature. Examples of field structures with geometries intermediate between chevron folds and conjugate kink-bands are illustrated in Figs. 13.8(*a*) and (*b*) and 13.9. These structures often have gently diverging axial planes. As we shall see, the factor that determines whether one or a conjugate set of kink-bands develops in a rock is the orientation of the principal compression with respect to the layering. If the compression is parallel, or sub-parallel, to the layering or fabric, conjugate structures tend to develop. If it is markedly inclined, then one of the sets develops preferentially to the possible exclusion of the other.

The few examples cited here merely serve to illustrate some of the general geometrical features of fold structures in anisotropic rocks. In the subsequent

Fig. 13.1. Folds (kink-bands) exhibiting straight limbs and axial planes oblique to the rock layering. Northern Spain.

sections, other examples will be introduced which specifically illustrate some of the predictions of the analyses of the buckling of such materials which will now be discussed.

The theory of buckling of mechanically anisotropic materials

Biot (1964 and 1965) presented an analysis of the buckling of a mechanically anisotropic material confined between two rigid horizons. Although details of the analysis are beyond the scope of this book, the outline of Biot's approach is given, as it affords an excellent example of how a buckling problem is tackled theoretically.

Biot's analysis is concerned with the deformation of a rectangular block. The problem is outlined below by a series of statements which define the material properties (i), (ii) and (iii), the orientation of the applied stress (statement (iv)) and the boundary conditions (statements (v) and (vi)).

(i)   The material is homogeneous and mechanically anisotropic.

(ii)   The material is either linear elastic or Newtonian viscous.

(iii)   The material is incompressible and consequently no volume changes occur during the deformation, (i.e. the deformation is isochoric).

(iv)   The combined symmetry of the material properties and the applied stress field is orthotropic. This occurs when the maximum principal compression is applied either parallel, normal or at 45° to the axes of symmetry of the anisotropy (Fig. 13.10). (These axes of symmetry will be normal and parallel to any mineral fabric or layering in a real geological material.)

(v)   The material deforms by plane strain (i.e. deformation is restricted to one plane, the xy plane in Fig. 13.10 and no strain (change in length) occurs parallel to the z direction).

(vi)   The model is rectangular and no deflection or shear stresses occur at the boundaries

Because of condition (vi) any buckling that occurs is restricted to the central (i.e. internal) region of the model and is called *internal buckling*.

Having defined the buckling problem in terms of verbal statements, these may now be replaced by equivalent mathematical expressions (i.e. equations). For example, the assumptions that the material is incompressible and that deformation occurs by

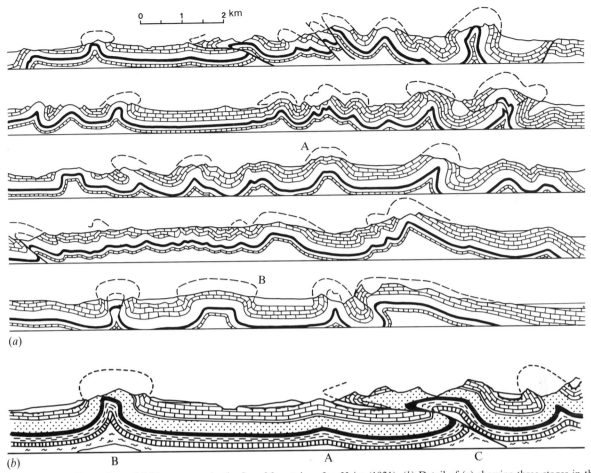

(a)

(b)   B   A   C

Fig. 13.2. (a) Profile sections of fold structures in the Jura Mountains after Heim (1921). (b) Detail of (a) showing three stages in the formation of a thrust from an originally symmetrical fold.

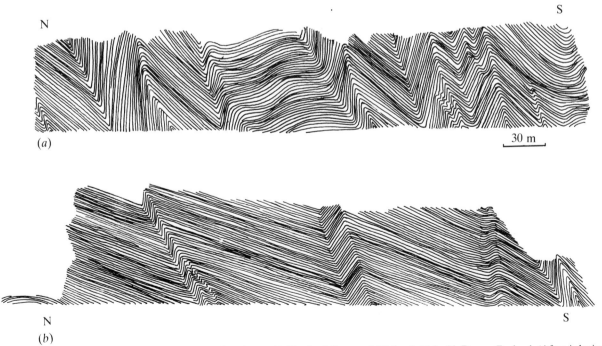

(a)

(b)

Fig. 13.3. Folds developed in turbidites in the cliff section at (a) Hartland Quay and (b) South Hole, N. Devon, England. (After Ashwin, 1957.)

Fig. 13.4. A photomicrograph showing individual and conjugate kink-bands in a Pre-Cambrian phyllite from Rhoscolyn, Anglesey, North Wales.

plane strain (statements (iii) and (v)) imply that:

$$e_x + e_y = 0 \qquad (13.1)$$

and

$$e_z = 0. \qquad (13.2)$$

Similarly, statement (ii) implies certain simple stress–strain or stress–strain-rate relationships. Once equations corresponding to statements (i) and (v) have been obtained, they are combined into a single equation, solutions to which automatically satisfy all the stipulated conditions (i) to (v) of the analysis. For

convenience, the reference coordinate directions x and y are made to coincide with the symmetry axes of the material, (Fig. 13.10).

The material is assumed to be in a state of initial stress, P (where $P = (\sigma_1 - \sigma_3)$), (Fig. 13.10), which induces a uniform state of deformation within the material. We wish to establish the condition which gives rise to the appearance of instability that disrupts this initial uniform state of homogeneous flattening. The equation that describes this instability is derived by combining the individual equations corresponding to statements (i) to (v). That is:

$$\left(Q - \frac{P}{2}\right)\frac{\partial^4 \phi}{\partial x^4} + 2\,(2N - Q)\frac{\partial^4 \phi}{\partial x^2\,\partial y^2}$$
$$+ \left(Q + \frac{P}{2}\right)\frac{\partial^4 \phi}{\partial y^4} = 0 \qquad (13.3)$$

where ϕ is the displacement function that describes the non-uniform deformation of the material and which is related to the displacements u and v in the x and y directions respectively by:

$$u = -\frac{\partial \phi}{\partial y} \qquad (13.4)$$

and

$$v = \frac{\partial \phi}{\partial x} \qquad (13.5)$$

and N and Q are compressive and shear moduli respectively in the direction of the maximum compression.

It can be seen that Eq. (13.3) is an equation in

Fig. 13.5. A box fold (conjugate kink-bands) formed in medial turbidites at Northcott Mouth, Bude, N. Devon, England. Cliffs 50 m high.

Fig. 13.6. An asymmetrical box fold in upper Carboniferous turbidites Bude, Cornwall, England

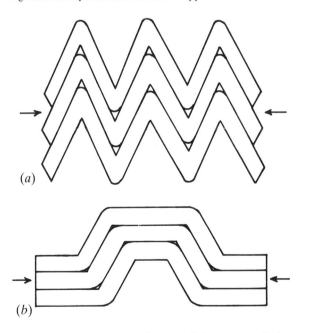

Fig. 13.7. (*a*) Upright chevron folds, axial planes perpendicular to the maximum compression. (*b*) A box fold with axial planes oblique to the maximum compression.

three unknowns which are, (i) the material properties, defined by the moduli N and Q, (ii) the applied stress P and (iii) the deformation (displacements, defined by ϕ) induced in the material by this stress. By defining two of the unknown quantities (usually the material properties and the applied stress), the equation can be solved for the remaining unknown, the displacement function.

Before considering some of the geologically interesting solutions to Eq. (13.3), it is worthwhile considering the moduli N and Q in more detail. Biot makes the important point that the resistance to deformation of a material depends on both the material properties *and* the stresses acting on a body. This concept can be appreciated by considering the resistance to deflection of a string clamped at its ends. If no tension is applied to the string, a sideways deflection can be achieved with relative ease. If, however, a tension is applied, a greater force is needed to cause the same sideways deflection.

This is because, in the first situation, only the material properties of the string resist the deflection, whereas in the second, both the material properties and the imposed stress field resist deflection. N and Q are measures of the resistance to compression and shear in a particular direction when the material is *unstressed*. If the material is under an initial stress (P), then the resistance to deformation will be different and is governed by the following equations:

$$M' = N + \frac{P}{4} \tag{13.6}$$

$$L' = Q + \frac{P}{2}. \tag{13.7}$$

M' and L' give a measure of the resistance to compression and shear in a stressed body and it can be seen that they reduce to N and Q respectively when the stress P is zero. *The ratio M'/L' is a convenient*

(a)   1 mm

(b)

Fig. 13.8. Structures with geometries intermediate between chevron folds and box folds (*a*) photomicrograph and (*b*) a minor fold from the Pre-Cambrian phyllites of Rhoscolyn, Anglesey, N. Wales.

Fig. 13.9. Folds with gently diverging axial planes.

measure of the anisotropy. Materials for which M' and L' are very different have a high degree of mechanical anisotropy and when M' = L' the material is isotropic.

Using Eqs. (13.6) and (13.7), we can replace N and Q in Eq. (13.3) by M' and L':

$$(L' - P) \frac{\partial^4 \phi}{\partial x^4} + 2 (2M' - L') \frac{\partial^4 \phi}{\partial x^2 \, \partial y^2}$$
$$+ L' \frac{\partial^4 \phi}{\partial y^4} = 0. \qquad (13.8)$$

Let us now consider the geological implications of some solutions to Eq. (13.8). The general solution to such a homogeneous partial differential equation is:

$$\phi = f_1(x + \xi_1 y) + f_2(x - \xi_1 y)$$
$$+ f_3(x + \xi_2 y) + f_4(x - \xi_2 y) \qquad (13.9)$$

where ξ_1 and ξ_2 are arbitrary constants and f_1, f_2, f_3 and f_4 are any arbitrary functions. The constants ξ_1 and ξ_2 may be real or imaginary, depending upon the coefficients of Eq. (13.8). If ξ_1 and/or ξ_2 are real, the displacement function ϕ (Eq. 13.9) is a solution of Eq. (13.8). Biot (1965, p. 193) shows that real values of ξ_1 and ξ_2 exist in the following three cases.

Case 1:
$$\frac{M'}{L'} > \tfrac{1}{2}; \; \frac{P}{L'} > 1 \qquad (13.10)$$

Case 2:
$$\frac{M'}{L'} < \tfrac{1}{2}; \; 1 > \frac{P}{L'} > \left(\frac{4M'}{L'}\right)\left(1 - \frac{M'}{L'}\right) \qquad (13.11)$$

Case 3:
$$\frac{M'}{L'} < \tfrac{1}{2}, \; \frac{P}{L'} > 1. \qquad (13.12)$$

Substitution of Eq. (13.9) into (13.8) (see Chap. 12) gives an expression relating the critical stress difference (P) to the moduli M' and L' and the parameters ξ_1 and ξ_2 at the onset of instability:

$$L'\xi^4 + 2(2M' - L')\xi^2 + (L' - P) = 0. \quad (13.13)$$

The conditions represented by Eqs. (13.10) to (13.12) are shown graphically in Fig. (13.11), where the criti-

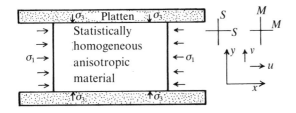

Fig. 13.10. A homogeneous, mechanically anisotropic block of material confined between two rigid plattens. The maximum principal compression is applied such that the combined symmetries of the material properties and the stress field are orthotropic. In a multilayer or mineral fabric which can be regarded as statistically homogeneous, anisotropic material, this occurs when σ_1 is parallel, at 45°, or at 90° to the layering or fabric. S and M are symmetry axes of the stress field and the anisotropic material respectively.

cal stress difference (P) necessary to cause buckling is plotted as a non-dimensional ratio P'/L' against the material properties defined by the ratio M'/L'. Stress states that are theoretically unstable occupy three separate fields representing cases 1, 2 and 3 of Eqs. (13.10) to (13.12). These three cases correspond to three types of instability which we shall refer to as *Types 1, 2 and 3.*

It is at this point in the analysis that the boundary conditions (statement (vi)) must be introduced. These are necessary to define the arbitrary constants and functions of Eq. (13.9) and thus permit a specific rather than a general solution to be obtained.

Type 1 solution

A solution to Eq. (13.8) which satisfies the boundary conditions of no deflections and no shear stresses at the boundaries and which is valid when $P/L' > 1$, in materials where $M'/L' > \frac{1}{2}$, is:

$$\phi = -\frac{C}{2}[\cos w(x - \xi y) + \cos w(x + \xi y)] \qquad (13.14)$$

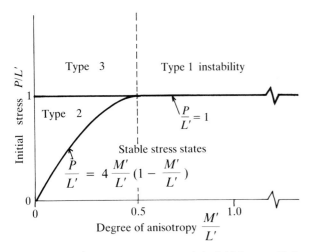

Fig. 13.11. Stable and unstable values of the initial stress (P) in materials with different degrees of anisotropy (expressed by the modulus ratio M'/L'). The thick, black lines represent the critical stresses (P) of Eqs. (13.10) to (13.12).

or

$$\phi = -C\cos wx\cos \xi wy \qquad (13.15)$$

where w, ξ and C are arbitrary constants. The displacement function defines a particular deformation pattern. The displacements u and v in the x and y directions respectively can be obtained from Eq. (13.15), using Eqs. (13.4) and (13.5), so that:

$$u = -Cw\xi\cos wx\sin \xi wy \qquad (13.16)$$

$$v = Cw\sin wx\cos \xi wy \qquad (13.17)$$

and the effect of these displacements on passive marker lines is shown in Figs. 13.12(a) and (b) and 13.13(a) for those originally parallel to the principal compression direction and in Fig. 13.13(c) for those originally normal to the compression direction. Each of the displacement vectors u and v, varies sinusoidally in magnitude, with wavelengths L_x and L_y along the directions x and y respectively, where:

$$L_x = \frac{2\pi}{w} \qquad (13.18)$$

and

$$L_y = \frac{2\pi}{\xi w}. \qquad (13.19)$$

It can be seen from these relationships that the ratio of the two wavelengths is given by ξ:

$$\xi = \frac{L_x}{L_y}. \qquad (13.20)$$

The two wavelengths can be seen in Fig. 13.12. The graphical expression of Eq. (13.13) is shown in Fig. 13.14(a) where the buckling stress is plotted against ξ. The value of ξ associated with the minimum buckling stress is zero. It follows from Eq. (13.20) that the most easily formed wavelength L_x will also be zero.

The reason for this anomalous result is that the material has been analysed solely in terms of its mechanical anisotropy, no account having been taken of any bending resistance the material may have possessed. It will be recalled, that the theory applies to an anisotropic homogeneous material. If the theory is applied to a material which only approximates to such a model and actually contains competent layers or mineral grains that offer resistance to bending, then it is necessary to add a correction term to the analysis (Biot, 1967). If this is done, it is found that the value of ξ, (and hence L_x) associated with the minimum resistance to buckling, is not zero but has some finite value (Fig. 13.14(b)): the larger the resistance to bending, the larger this value, and hence the larger the wavelength (L_x) of the most easily formed fold. This effect is shown diagrammatically in Fig. 13.15(a), which depicts three confined multilayers made up of lubri-

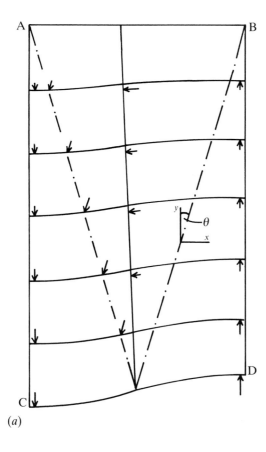

(a)

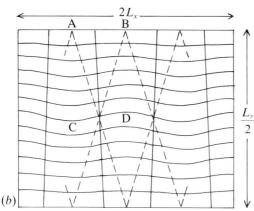

(b)

Fig. 13.12. The effect of the theoretical displacement pattern represented by Eq. (13.15) (Type 1 instability) on an initially rectangular grid of passive marker lines. (*a*) Element ABCD of (*b*) enlarged to show the magnitude of the displacement vectors. (After Cobbold, Cosgrove & Summers, 1971.)

cated, competent layers of the same material. The three multilayers differ only in the thickness (and hence bending resistance) of the layers. The dominant, (i.e. most easily formed) wavelength of the internal buckles increases as the bending resistance increases.

In an extensive study of internal buckles developed in mineral fabrics, one of the authors (J.W.C.) has observed a direct relationship between the wavelengths of the microfolds and the thickness (which governs the bending resistance) of the lepidoblastic

(plate-like) and nematoblastic (needle-like) minerals which make up a fabric.

An arbitrary value value of $\xi = 3$ was chosen when drawing Fig. 13.12 and the oblique, dotted lines shown in this figure are inclined at $\theta°$ to the *y* axis where:

$$\theta = \pm\tan^{-1}\xi. \tag{13.21}$$

These lines, which are referred to as *Characteristics* are fundamental symmetry axes of the deformation field defined by Eqs. (13.16) and (13.17) and have important mathematical properties and physical significance. Nadai (1950, p. 555), for example, points out that local disturbances within a material tend to propagate along characteristic directions. The displacement vectors along the characteristics are all parallel and the deformation along the characteristic approximates to that of heterogeneous simple shear. It can be seen from Fig. 13.14(*a*), that for an ideal homogeneous, anisotropic material which offers no resistance to bending, the orientations of the characteristics (the *characteristic directions*) associated with Type 1 instability will be normal to the applied buckling stress which is acting along the *x* axis ($\xi = 0$ therefore $\theta = 0$). However, if the material does offer a resistance to bending, then the characteristic directions associated with Type 1 instability will be inclined to the direction of principal compression (i.e. $\theta \neq 0$): the greater the bending resistance, the greater the value of ξ and, therefore, θ associated with the minimum resistance to buckling.

It is emphasised that the buckling displacement pattern (Eqs. (13.16) and (13.17)) and Figs. 13.12 and 13.13) is only valid for the first increment of buckling deformation. Nevertheless, in some natural examples of buckled anisotropic materials such as mineral fabrics, it is apparent that the sinusoidal geometry of the initial instability is maintained as the structure amplifies into a finite structure, Fig. 13.15(*b*). However, as will be discussed later in this chapter in the section on the finite growth of folds in anisotropic materials, physical modelling, finite element analysis and mathematical analysis capable of predicting buckling behaviour beyond the first increment of folding, all indicate that there is usually a marked tendency for the fold limbs to straighten out during fold amplification and for the folds to become chevron-like, or for conjugate straight limbed folds (box folds) to grow from low amplitude sinusoidal buckles.

It will be recalled that one of the assumptions made in the buckling analysis of anisotropic materials outlined above is that the combined symmetry of the material properties and the applied stress fields should be orthotropic. This condition is satisfied when the maximum compression is parallel with, at 45° to, or normal to, the fabric or layering. It can be argued,

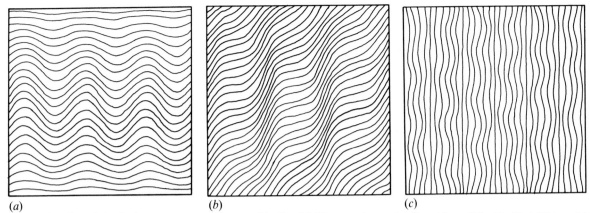

Fig. 13.13. The effect of the displacement pattern represented by Eq. (13.15) on passive marker lines (a) parallel with, (b) at 45° to and (c) normal to the principal compression. (After Cosgrove, 1976.)

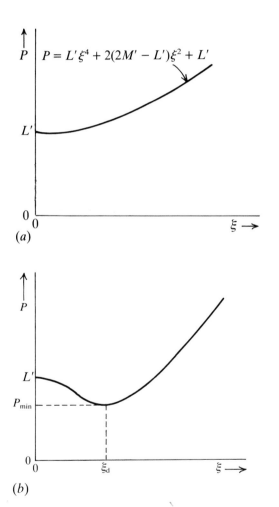

$$P = L'\xi^4 + 2(2M' - L')\xi^2 + L'$$

(a)

(b)

Fig. 13.14. Stress conditions for Type 1 instability (a) in a homogeneous anisotropic material with no resistance to bending and (b) in a similar material with resistance to bending. (Biot, 1965.)

therefore, that the displacement pattern (Figs. 13.12 and 13.13(a)) would also develop if the maximum compression (σ_1) acted at 45° or 90° to the fabric or layering, provided that in the direction of maximum compression, $M'/L' > \frac{1}{2}$ and $P > L'$. The effect of the displacement pattern on passive marker lines in these

two orientations is shown in Fig. 13.13(b) and (c). Asymmetrical folds and interlocking internal pinch-and-swell structures result.

It is interesting to note that experimental work has shown that the development of internal buckling instabilities is not restricted to the condition where the combined symmetries of the material properties and the applied stress field are orthotropic. For example, Fig. 13.16 shows three stages in the buckling of a Plasticine multilayer in which the principal compression was initially at 20° to the layering. The model multilayer is made up of alternating dark and light, mechanically identical Plasticine layers separated by thin films of Vaseline and confined between two rigid plates which, in accordance with the boundary conditions used in the theoretical analysis, prevent deflection from occurring at the boundary.

The asymmetrical buckles that develop invite immediate comparison with those produced by the theoretical displacement pattern (Fig. 13.13(a)), when superposed on passive marker lines at 45° to the direction of maximum compression (Fig. 13.13(b)). We can conclude from this experiment that orthotropic symmetry is not an essential prerequisite for the formation of these buckling instabilities and that the essential predictions of the analysis can probably be extended to situations where the combined symmetry of the material and the stress field are not orthotropic.

Type 2 solutions

For a Type 2 instability to develop, the material must satisfy the condition:

$$M'/L' < \tfrac{1}{2} \tag{13.11(a)}$$

and the stress state P must be such that:

$$1 > P/L' > (4M'/L')(1 - M'/L'). \tag{13.11(b)}$$

As for Type 1 instability, there is a critical value, P_c, of the initial differential stress below which Type 2

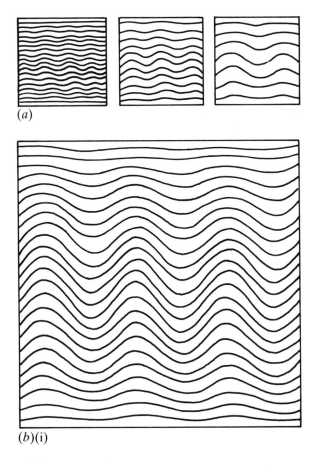

(a)

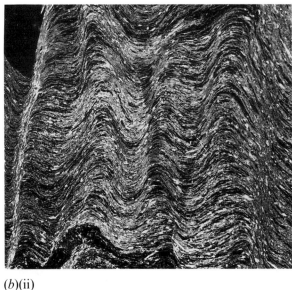

(b)(i)

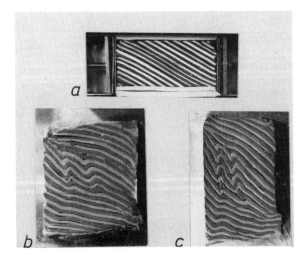

Fig. 13.16. The formation of asymmetrical, internal buckles in a confined Plasticine multilayer. The layering was initially inclined at 20° to the principal compression. (After Cosgrove, 1976.)

(b)(ii)

Fig. 13.15. (*a*) Internal buckles in three confined multilayers made up of lubricated, competent layers of the same material. They differ only in the thickness (*t*) (and hence bending resistance) of the layers. (*b*) (i) The effect of the displacement pattern represented by Eq. (13.15) on passive marker lines originally parallel to the direction of maximum compression. The folds are amplified well beyond the first increment of buckling for which the theory is applicable. (ii) Internal buckles in a mica fabric from the Mesozoic metasediments of the Luckmanier pass Switzerland. ((*b*) after Cosgrove, 1976.)

instability cannot occur. From the relationship (13.11(*b*)), this critical value is:

$$P_c = \left(\frac{4M'}{L'}\right)(L' - M'). \tag{13.22}$$

Substituting this value of P into Eq. (13.13) gives a single value of ξ, namely:

$$\xi = \left(1 - \frac{2M'}{L'}\right)^{\frac{1}{2}}. \tag{13.23}$$

Associated with this value of ξ, there are characteristic directions; the angle $\theta°$ between these directions and the y-axis being given by:

$$\tan \theta = \pm \xi_c. \tag{13.24}$$

Using Eqs. (13.23) and (13.24), we can plot the orientation of the characteristic directions against the material property M'/L' for the range of materials $(0.5 > M'/L' > 0)$ in which Type 2 structures will develop (Fig. 13.17). The solutions to the buckling equation (Eq. (13.8)) for Type 2 instability, unlike those for Type 1 instability, are not analytic, i.e. they do not provide us with an expression that enables the displacements associated with Type 2 instabilities to be plotted in the same way as was done for Type 1 instability (Figs. 13.12 and 13.13).

However, Biot does suggest a possible way in which a Type 2 instability might express itself in a statistically homogeneous, layered material inclined at 45° to the directions of principal compression (Fig. 13.18).

If it is assumed that the resistance to inter-layer slip is negligible, the resistance to compression (M') in the direction of the principal compression will be zero, the resistance to shear (L') in this direction however, will be greater than zero, so that the ratio M'/L' is less than $\frac{1}{2}$, satisfying one of the conditions for Type 2

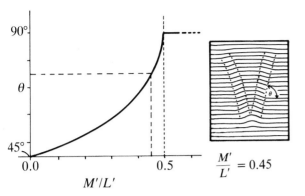

Fig. 13.17. The relationship between the anisotropy (M'/L') of a material and the characteristic direction (θ) for Type 2 internal instability. (After Cosgrove, 1976.)

instability (Eq. (13.11(a))). If $M' = 0$, it follows from Eqs. (13.22), (13.23) and (13.24) that:

$$P_c = 0 \qquad (13.25)$$

and

$$\theta = \pm 45° \qquad (13.26)$$

i.e. the characteristic directions are parallel and normal to the layering and instability will occur after the first increment of stress has been applied. It may be inferred from Fig. 13.18(b) that there is a connection, in this example, between the characteristic direction and the displacement at instability produced by discrete slip between the layers. However, for conditions of no deflection at the boundries, a finite displacement pattern, involving further slip between the layers, will require a complementary shear displacement along the other characteristic direction. This complementary shear may appear as a kink-band (Fig. 13.18(c)). Latham (1983) has criticised this explanation of Type

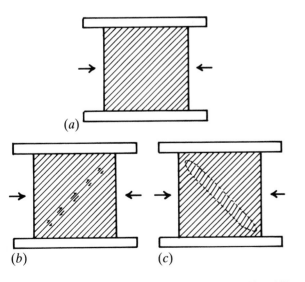

(a)

(b)   (c)

Fig. 13.18. Displacements resulting from Type 2 internal instability of a layered orthotropic material for which $M'/L' < \frac{1}{2}$. (After Cobbold, Cosgrove & Summers, 1971).

2 structures, pointing out that local slip on the layers implies a non-linearity in material response which violates the assumption of material linearity built into the analysis. We shall return to this problem later in this chapter where Biot's analysis of internal buckling is extended to incorporate the effects on non-linear material properties and bending resistance.

Conjugate kink-bands (Fig. 13.7(b)) have been produced experimentally by compressing a multilayer parallel to the layering (Fig. 13.19). Such experiments brought into question the correlation of kink-bands with Type 2 instability, for it was argued that the ratio M'/L' for layer parallel compression in a multilayer will be greater than $\frac{1}{2}$. The theory would, therefore, indicate that Type 1 structure (upright, sinusoidal folds) should develop.

In an attempt to explain this paradox, Cobbold *et al.* (1971) suggested that kink-bands form from small initial or stress-induced disturbances. If the disturbances were absent, the ratio M'/L' in the direction of maximum compression (i.e. x direction, parallel to the layering) would be large in a material with a high anisotropy. However, any irregularity in the layering defines a region in which shortening parallel to the x direction can be achieved by slip between the layers. There is, therefore, in such a region, a significant local reduction in the resistance to compression along the x direction, while the resistance to shearing in the x direction will increase (because in this region of local disturbance the layering is no longer parallel to the x direction). It was argued that this local change in the moduli M' and L' effectively inverts the modulus ratio and that a material with an M'/L' ratio of 100 would, at local perturbations, have an M'/L' value approaching $1/100$. Consequently, instabilities initiated from these irregularities will be of Type 2 and will be generated along the characteristic directions which are oblique to the applied compression. It can be clearly seen that the initiation of the kink-bands in the experiment shown in Fig. 13.19 occurs at a local irregularity in the layering, developed at the boundary of the multilayer during the early stages of layer-parallel compression.

Based on this idea of moduli inversion, Cobbold *et al.* concluded that kink-bands will develop in materials with a high mechanical anisotropy ($M'/L' < \frac{1}{2}$ or > 2 which could invert to $< \frac{1}{2}$ at a perturbation) and that upright, sinusoidal internal buckles will develop in materials with a low anisotropy ($2 > M'/L' > \frac{1}{2}$).

Honea & Johnson (1976) presented an analysis of kink-band formation which is in complete contrast to the continuum approach of Biot's analysis used by Cobbold *et al.* They consider an isolated, representative layer from a uniform multilayer sequence and examine the elastic buckling profiles which can be produced when various shear stress conditions are

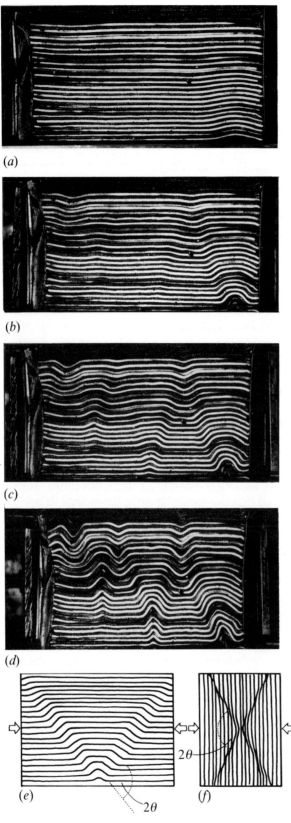

Fig. 13.19. (*a*)–(*d*) Stages in the development of Type 2 internal structures in a model composed of layers of Plasticine lubricated with graphite powder. The compression direction was parallel to the layering. (After Cobbold, Cosgrove & Summers, 1971.) (*e*) Conjugate reverse kink-bands and (*f*) conjugate normal kink-bands formed when the maximum principal compression is parallel to and normal to the layering respectively.

applied to the surface of the layer. They suggest that the local loss of interlayer cohesion in the limbs of low amplitude buckles can lead to the development of conjugate kink-bands. In contrast to Cobbold *et al.*, Honea & Johnson conclude that low mechanical anisotropy favours kink-band formation and high mechanical anisotropy favours sinusoidal folding. As will be seen later in this chapter, by incorporating non-linear material properties into Biot's analysis of internal buckling, these apparently conflicting ideas on kink-band formation can be resolved.

If, as is highly probable, we are justified in correlating kink-bands and characteristic directions, then it can be seen from Fig. 13.17, that the angle θ between the kink-bands and the layer normal will depend upon the material properties. When the mechanical anisotropy is large (i.e. when M'/L' is considerably less than 1), the angle between the layer normal and kink-band axis will approach 45°. As the anisotropy becomes less marked ($M'/L' \to \frac{1}{2}$), this angle gradually decreases until it reaches 0° when $M'/L' = \frac{1}{2}$; i.e. the conjugate kink-bands will be indistinguishable from an upright fold with its axial plane normal to the principal compression. The inequalities (Eqs. (13.10) and (13.11)) show that Type 1 structures develop in materials when $M'/L' > \frac{1}{2}$, and Type 2 structures in materials when $M'/L' < \frac{1}{2}$. It is interesting to note that for materials in which $M'/L' = \frac{1}{2}$, it is irrelevant whether a Type 1 or Type 2 solution is used to predict the structure that will result, as the same structure is predicted by both solutions. It follows that there is a complete gradation between conjugate kink-bands inclined at 45° to the layering or fabric and upright, symmetrical folds. Examples of structures with geometries between conjugate kink-bands and upright sinusoidal folds are shown in Figs. 13.8 and 13.9.

It can be seen from Figs. 13.13 and 13.17, that the manifestation of 'buckling' instabilities in anisotropic materials depends on both the M'/L' ratio and the orientation of the fabric or layering with respect to the direction of maximum compression. Some of the possible modes of expression of internal instabilities in materials with different M'/L' values and at different angles to the direction of maximum compression are shown in Fig. 13.20. A transition between upright, symmetrical folds and conjugate, reverse kink-bands (Fig. 13.20(*a*), (*b*) and (*c*)) occurs with the increase in M'/L' values when the principal compression acts parallel to the fabric or layering, and a transition between internal pinch-and-swell structures (Fig. 13.20(*g*)) and conjugate, normal kinks (Figs. 13.19(*f*) and 13.20(*i*)), which form at an angle of $\geqslant 45°$ to the direction of compression, occurs with an increase in M'/L' values when the principal compression is normal to the layering or fabric.

The close relationship between Type 1 structures

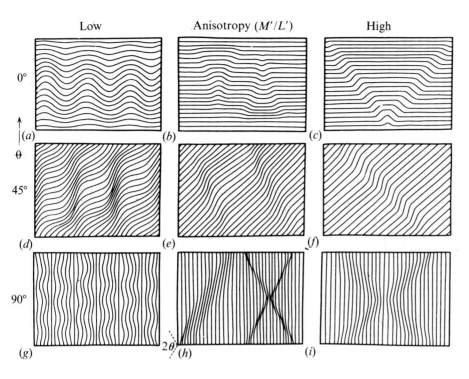

Fig. 13.20. Some possible modes of expression of internal instability in materials with different anisotropy (M'/L') with the fabric elements inclined at different angles (ϕ) to the direction of maximum compression. See also Figs. 13.44 and 17.9. (After Cosgrove, 1976.)

(e.g. sinusoidal multilayer folds and internal pinch-and-swell structures) and Type 2 structures (e.g. conjugate normal and reverse kink-bands) also becomes apparent when the displacements necessary to form the two structures are considered. The displacements associated with Type 1 instability result from the *superposition* of two simple shears of opposite sense parallel to the characteristic directions and those for Type 2 instability, from the *juxtaposition* of two opposing simple shears parallel to these directions (e.g. Fig. 13.20(*a*) and (*c*) respectively).

Type 3 solution

It follows from Eq. (13.12) that for the formation of Type 3 structures, the material must satisfy the condition $M'/L' < \frac{1}{2}$: and the stress state P must be such that $P'/L' > 1$. However, the range of materials in which Type 3 structures appear to have the potential to develop is the same as that for which Type 2 structures will develop (i.e. $M'/L' < \frac{1}{2}$). It can be seen from the relationship $P'/L' > 1$ (which relates to Type 3 structures) that P must be greater than L before the third type of structures can develop. However, from Eq. (13.11(*b*)), it follows that Type 2 structures will occur when P is *less* that L'. Hence, because in a real situation it is to be expected that the value of P will gradually increase until it results in the development of one or other of the structures, it follows that Type 2 structures will develop in a material before the conditions conducive to the development of Type 3 structures can be reached. It is interesting to note that Type 3 instabilities are identical to those of Type 1. We see then, that sinusoidal buckling is predicted for all anisotropic materials when the compressive stress

$P > L'$. However, in the range of anisotropy $0 < M'/L' < \frac{1}{2}$ another type of instability (Type 2) will occur at a lower magnitude of stress (i.e. when $P < L'$). For this range of material, the development of sinusoidal buckling will be pre-empted by the occurrence of Type 2 instabilities. Therefore, we will never see Type 3 structures in real materials.

The analysis for the buckling of linear anisotropic materials, outlined above, applies only to the first increment of buckling. Let us now consider some of the work that has been carried out to investigate the finite development of structures in anisotropic materials, for this will indicate whether or not the geometry of the initial instabilities will be maintained as they amplify into finite structures.

Finite development of folds in mechanically anisotropic materials

It has already been noted at the beginning of this chapter that the finite development of folds can sometimes be deduced from field observations of folds arrested at various stages of amplification (Fig. 13.2(*b*)). In addition to such field observations, three other approaches have been adopted to study the finite development of geological structures. These are (i) theoretical work, (ii) finite element analysis and (iii) physical modelling, which we shall now consider in turn.

Theoretical approach

Cobbold (1976) and Summers (1979) both considered the effects of mechanical anisotropy on the finite development of folds and represented the bulk

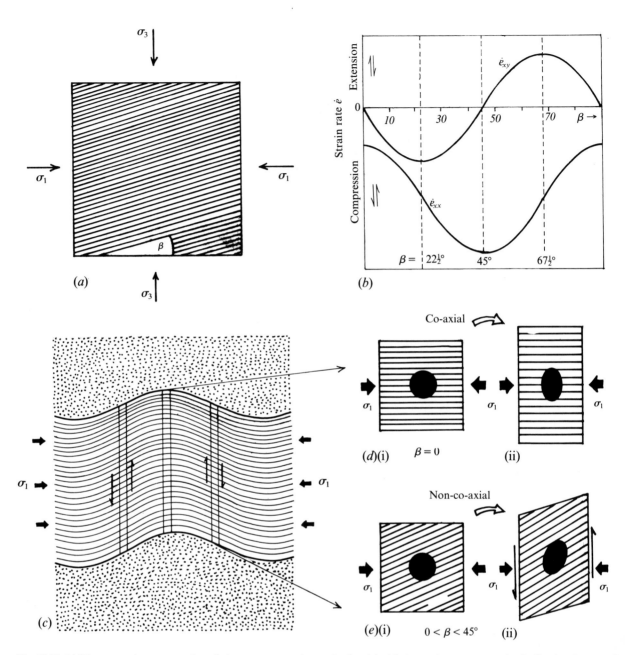

Fig. 13.21. (*a*) Diagrammatic representation of a homogeneous anisotropic material with the maximum compression inclined at β to one of the axes of material symmetry. (*b*) Graphical expression of Eqs. (13.27) and (13.28) which are the stress–strain-rate equations for material (*a*). (*c*)–(*e*) The relationship between the phenomenon of non-coaxial deformation in an anisotropic material and the development of heterogeneous, simple shear parallel to the axial plane of a fold. See text. (From Summers, 1979; after Cobbold, 1976.)

mechanical properties of a uniform layered or foliated material in terms of the properties of an equivalent anisotropic, homogeneous continuum. In order to do this, Cobbold derived expressions for the relationships between incremental stress and strain (or strain-rate), making assumptions that deformation occurred at constant volume under conditions of plane strain and that the combined symmetry of the material properties and applied stress was originally orthotropic. Resistance to bending was ignored. These relationships can be expressed as:

$$\dot{e}_{xx} = \frac{\sigma'_{xx}}{4N} \left\{ -\left[\left(\frac{N}{Q} + 1\right) \right.\right.$$
$$\left.\left. - \left(\frac{N}{Q} - 1\right) \cos 4\beta \right] \right\} \quad (13.27)$$

$$\dot{e}_{xy} = \frac{\sigma'_{xx}}{4N} \left[-\left(\frac{N}{Q} - 1\right) \sin 4\beta \right] \quad (13.28)$$

where $\dot{e}_{xx}$ is the incremental strain-rate in the x direction, $\dot{e}_{xy}$ is the incremental shear strain-rate in the x and y directions and σ'_{xx} is the maximum principal deviatoric stress $[(\sigma_1 - \sigma_3)/2]$. Further, (N/Q) is the

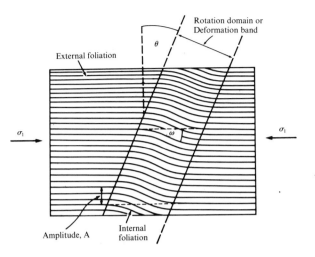

Fig. 13.22. A kink-band forming the initial perturbation, the amplification of which is determined by the mechanism of rotational instability. See text for details. (After Summers, 1979.)

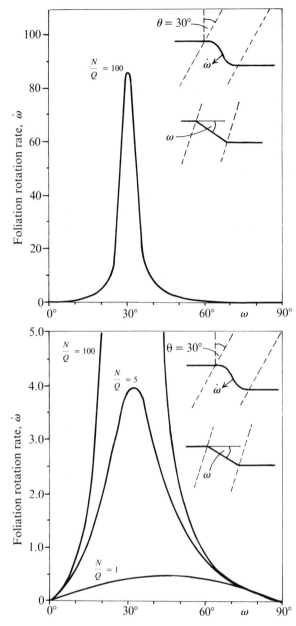

Fig. 13.23. The relationship between the orientation of the internal foliation of the kink-band in Fig. 13.22, and the rate of rotation ($\dot\omega$) of the foliation that occurs in response to an increment of compressive stress (a) for an anisotropic material for which $N/Q = 100$. The equivalent curves for materials with $N/Q = 5$ and $N/Q = 1$ (i.e. an isotropic material) are shown in (b). (After Summers, 1979.)

ratio of the bulk compressional stiffness coefficient to the bulk shear stiffness measured *along the plane of the layering* and β is the angle between the direction of principal compression and the material symmetry axes. In a real foliated or layered material, these axes will be normal and parallel to the foliation or layering.

The variations in strain-rate and shear strain-rate with orientation, described by Eqs. (13.27) and (13.28) are shown graphically in Fig. 13.21(b) together with a diagrammatic representation of the material under consideration (Fig. 13.21(a)).

As will be seen from the following illustration, an important mechanism of fold amplification, which is related to the mechanical anisotropy of the material, is implied by these equations. It is the same mechanism associated with the formation of folds of the 'second kind', shown in Fig. 12.14(c), and the formation of internal buckling (Figs. 12.12 and 13.13) and is a completely different mechanism from that of isotropic layer buckling Fig. 12.14(b).

Consider the implications of Eqs. (13.27) and (13.28) on the amplification of a low amplitude, sinusoidal fold in the multilayer shown in Fig. 13.21(c). The multilayer, which is made up of many thin layers, has a high mechanical anisotropy. In the hinge region of the fold, β (the angle between the maximum principal compressive stress and the layering) equals 0. Consequently, from Eq. (13.28), $\dot e_{xy} = 0$ and the incremental deformation will be co-axial with the stress (Fig. 13.21(d)). However, in the limb regions, the principal compression is inclined at a small angle to the layering, and an increment of stress produces both a shortening e_{xx} (associated with a shortening strain-rate $\dot e_{xx}$) along its line of action and a shear e_{xy} (associated with a shear strain-rate $\dot e_{xy}$) parallel to the axial plane of the fold, both of which are achieved by layer-parallel shearing and rigid layer

rotation. This *rotation* causes the incremental deformation, represented by the strain ellipse in Fig. 13.21(e) (ii) to be non-co-axial with the applied stress. The sense of shear (the half arrows in Figs. 13.21(c) and (e)) will be opposite on either side of the fold axial plane, and the systematic variation in the amount of shear is such that it can be described as heterogeneous simple shear. This shearing instability is related to the non-co-axiality of incremental stress and strain, or strain-rate, which occurs in mechanically anisotropic materials. Cobbold shows that once

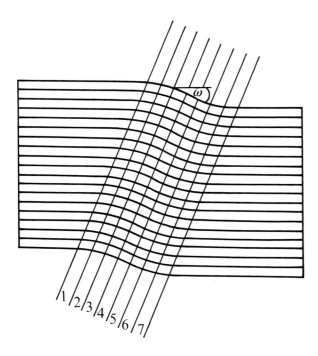

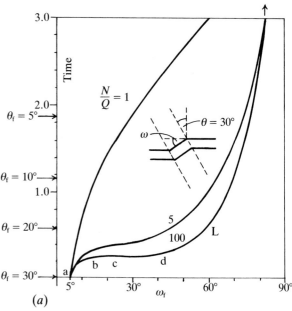

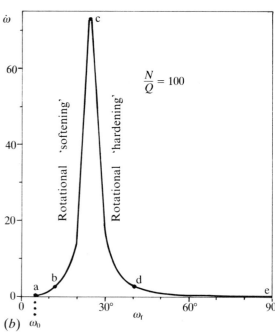

Fig. 13.24. The initial sinusoidal kink-band deflection of Fig. 13.22, divided into elements.

layer rotations have been initiated, this instability produces significant rates of layer rotation and mechanically active fold amplification, during the initial stages of the growth of a symmetrical, similar fold in materials with a high mechanical anisotropy. Summers (1979) refers to this mechanism of fold amplification as a *rotational instability* and extends Cobbold's work, to consider the amplification of perturbations of the type illustrated in Fig. 13.22, which shows a band of sinusoidal flexures inclined at an angle θ to the normal to the undeflected 'layering'. He assumes that the boundaries of the perturbed region, which he refers to as a rotation domain, remain parallel throughout the deformation and that the geometry of the fold remains that of a similar fold. It is assumed that the minimum principal compression acts normal to the undeflected foliation and that the maximum principal compression acts parallel to the foliation and normal to the rotation axis. External and internal foliation are defined in Fig. 13.22. Summers develops a relationship between the orientation of the internal foliation ω and the rate of rotation $\dot{\omega}$ of the foliation that occurs in response to an increment of compressive stress. The relationship is mathematically complex but can be clearly seen when it is expressed graphically (Fig. 13.23). The rate of rotation is found to depend primarily on the mechanical anisotropy of the material (N/Q), the orientation ω of the layering inside the rotation domain (Fig. 13.22) and, to a lesser extent, the orientation θ of the deformation band (Fig. 13.22) with respect to the external layering. It can be seen from the graphs of Fig. 13.23 that the rate of rotation is very sensitive to the orientation of the

Fig. 13.25. (*a*) Graphs showing the relationship between finite change in the orientation (ω_f) of the internal foliation with time for an isotropic ($N/Q = 1$) and two anisotropic materials ($N/Q = 5$ and 100). The initial orientation of the foliation was 5°. θ_0 is the original orientation of the rotation domain boundary and θ_f its orientation after time T. (*b*) Change in rotation rate ($\dot{\omega}$) with change in finite foliation rotation (ω_f) corresponding to the finite foliation rotation curve for $N/Q = 100$ in (*a*). (After Summers, 1979.)

internal foliation. It is maximum when $\omega = \theta$, that is, when the internal foliation is normal to the rotation domain boundary.

The division of the original sinusoidal deflection into elements is shown in Fig. 13.24. It will be noted that the orientation of the foliation in adjacent elements is different. It follows from the graphs of Fig. 13.23 that for any increment of stress, the foliation within adjacent elements will rotate at different rates and, therefore, by different amounts, depending upon the orientation ω. Consequently, the sinusoidal geometry of the initial perturbation changes as the fold is amplified.

It should be noted that the graphs of Fig. 13.23 represent only the variation in the incremental foliation rotation as a function of internal foliation attitude within a rotation domain at a *specific angle* (θ) to the plane of external foliation. Let us now consider the variation in incremental foliation rotation rate with time. In order to do this, it is necessary to take into account the fact that not only will the attitude of the local foliation within a rotation domain change with finite deformation, but also that the boundaries of the rotation domain will themselves undergo a finite rotation.

Summers derives a relationship between the finite change in the orientation of the internal foliation with time. This is shown graphically in Fig. 13.25(a), for an element in which the internal foliation was initially at 5° ($\omega_0 = 5°$) to the compression direction, in a rotation domain inclined at 60° to the external foliation. Each curve corresponds to a different anisotropy. It can be seen that, with progressive deformation, the rate of rotation of the foliation builds up to a peak value and then drops off. The greater the anisotropy, the more pronounced is the increase (and subsequent decrease) of the rotation rate. Using the graph of Fig. 13.25(a), the rotation rate of the internal foliation ($\dot{\omega}$) can be determined for any value of ω. The relationship between these two parameters is shown graphically in Fig. 13.25(b).

Let us now consider the effect of this difference in rotation rates of the internal foliation of the various elements in Fig. 13.24 on the geometry of the fold as it amplifies. Initially, the sinusoidal deflection is amplified as the foliation in the central element 4 (Fig. 13.24) rotates faster than that in the adjacent elements 3 and 5 which rotate faster than that in elements 2 and 6. When the condition $\omega = \theta$ is reached in element 4, it can be seen from the graphs in Figs. 13.23 and 13.25 that the rate of foliation rotation will begin to slow down. However the rotation rates in the adjacent elements 3 and 5, where the condition $\omega = \theta$ has not yet been reached, is still increasing and consequently, the foliation rotating in these two elements begins to 'catch up' with that in element 4. This process continues and, as the rotation rates in the central elements slow down as ω equals and exceeds θ, the foliation in the peripheral elements progressively 'catch up', with the result that the original sinusoidal deflection straightens out.

The rate of limb straightening is determined by the anisotropy; the higher the anisotropy, the faster the limbs straighten out. It seems, therefore, that if fold amplification is governed by the mechanism of

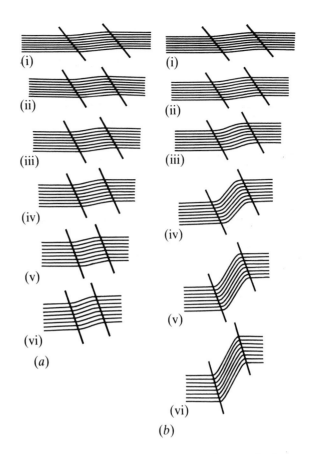

Fig. 13.26. Computer simulation of the amplification of oblique, originally sinusoidal deflections by the mechanism of 'rotational instability' discussed in the text; (a) in a homogeneous, isotropic material, and (b) in a homogeneous, anisotropic material ($N/Q = 5$). Stages (i) to (vi) represent 0, 10, 18, 26, 33 and 39 per cent shortening respectively. (After Summers, 1979.)

'rotational instability', the fold will be characterised by being angular and having straight limbs. Summers, using the equations represented graphically in Fig. 13.25, and a computer, determined the change in geometry of a variety of low amplitude, sinusoidal bands oriented at various angles θ in various materials (N/Q) with progressive deformation. The results of two such 'experiments' are shown in Fig. 13.26, which depicts various stages in the amplification of an identical perturbation in two different materials. In the model shown in Fig. 13.26(a), which is *isotropic*, the initial sinusoidal deflection and the marker lines are amplified only by the mechanism of passive (i.e. kinematic) amplification (see Chap. 10) and the sinusoidal geometry is maintained. In the model in Fig. 13.26(b), which is *anisotropic*, amplification occurs both kinematically and by the mechanism of 'rotational instability'. The limbs of the folds straighten out during amplification and the original sinusoidal geometry is lost.

Summers points out that it can be inferred from the graphs of Fig. 13.25 that there are several clearly defined stages in the amplification of folds in anisotropic materials.

Stage 1 is defined by the curve segments a–b in Fig. 13.25(a) and (b). During this initial phase of foliation rotation, there is a gradual increase in the foliation rotation-rate ($\dot\omega$) within the rotation domain. However, $\dot\omega$ over the range a–b is relatively slow and of the same order as the rate of rotation of the rotation–domain boundaries. This is expressed in Fig. 13.25(a) by the development of a significant rotation of the rotation–domain boundaries during the initial stages of internal foliation rotations.

Stage 2 is known as the *rotational softening* or *rotational instability* stage and is defined by the curve segments b–c in Fig. 13.25(a) and (b). The subhorizontal section of the finite foliation rotation curve (Fig. 13.25(a)) corresponds to the development of relatively rapid rates of foliation rotation. This stage of maximum rotational-instability occurs as the internal foliation approaches and passes through the domain normal foliation attitude $\omega_f = \theta_f$ and corresponds to the peaked section of the rotation-rate curve plotted in Fig. 13.25(b). During this stage of rotational-instability, the rate of rotation of the internal foliation in highly anisotropic materials is significantly greater than that of the domain boundaries ($\dot\omega \gg \dot\theta$). Thus the domain orientation, θ_f, remains relatively stable and deformation is concentrated in the rotation domain (sometimes called a deformation band) as the internal foliation undergoes a large finite rotation. The point c of the finite foliation rotation curve corresponds to the maximum incremental rate of internal foliation rotation and marks the inflection point in the curve.

Stage 3, known as the *rotational hardening* or *rotational stability* stage is defined by the curve sections c–d in Fig. 13.25(a) and (b) and is associated with a dramatic reduction in the rotation rate of the internal foliation. During the later stages of finite foliation rotation, beyond the point d, the incremental rate of foliation rotation is very low in comparison with the peak rate developed during the phase of maximum rotational instability. The magnitude of $\dot\omega$ is reduced to a level which is (as in Stage 1) of the same order as the rate of rotation of the domain boundaries, $\dot\theta$. Indeed beyond the point L shown on the curve corresponding to $N/Q = 100$ in Fig. 13.25(a), the magnitude of $\dot\omega$ is *less* than the rate of rotation which would be achieved for a corresponding value of foliation attitude, ω_f, if the foliation was mechanically passive ($N/Q = 1$). Thus, the *effect of mechanical anisotropy is to stabilise the attitude of the internal foliation planes during the later stages of finite foliation rotation*. This is a particularly important prediction of the analysis. Casey (1976) and Cobbold (1976) have demonstrated that this effect also occurs in symmetrical folds developed in anisotropic materials. The development of *rotational stability* or *locking* during the later stages of finite foliation rotation is a con-

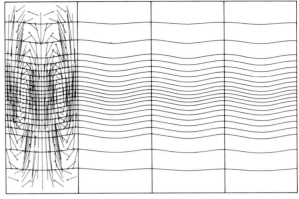

(a)

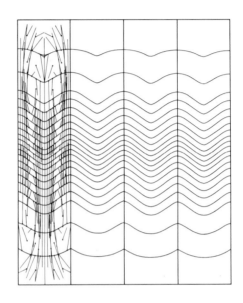

(b)

Fig. 13.27. Two stages in the amplification of an originally sinusoidal train of low amplitude deflections in a homogeneous, anisotropic material. As the folds amplify, the limbs in the central portion of the model straighten out and the folds become progressively more chevron-like. The anisotropy of the model (M'/L') is 2.22. (After Casey, 1976.)

sequence of the effects of mechanical anisotropy alone. These stages in fold amplification have been briefly mentioned in Chap. 10 and are discussed further in Chap. 15.

Having looked at some of the theoretical work on the finite growth of folds in anisotropic materials, let us now consider some finite element analyses that have been used to study the problem.

Finite element analyses

Casey (1976) used the technique of finite element analysis to study the finite development of internal buckles in a homogeneous, anisotropic continuum. The boundary conditions he chose are those used by Biot (1964) which require that the model be confined between rigid, frictional boundaries. As has been discussed previously, because the resistance of the material to bending is ignored, the wavelength predicted for the internal buckles is infinitely small.

Although Casey also neglects bending resistance in his analysis, he arbitrarily selects a finite value of wavelength ($\xi = \frac{1}{2}$, see Eq. (13.20)) for the geometry of the initial perturbation. Two stages in the finite amplification of folds from such perturbations are shown in Fig. 13.27. During the initial stages of amplification of the perturbation (up to limb-dips between 10° and 15°), the rate of fold amplification is found to be an exponential function of time, as predicted by Biot (see Eqs. (11.5) and (11.6)):

$$A = A_0 e^{pt} \tag{13.29}$$

where A_0 is the amplitude of the initial perturbation and A its amplitude after time t. Beyond limb-dips of 15°, the fold amplification rate is found to become approximately a linear function of time.

It follows from these observations that, if a series of folds is generated from a particular perturbation, the amplitude of the initial fold during the early stages of development is relatively large compared with that of the adjacent folds. However, this means that the reduction in fold growth rate which occurs when the limb dip exceeds 15° first affects the initial fold whose amplification rate then begins to slow down. The amplitude of adjacent, younger folds, therefore, 'catch up' with the early folds and the result is the development of a zone of approximately constant fold amplitude across a substantial portion of the confined multilayer. Casey's 'experiments' show this to be so and in addition illustrate that once the amplification of the fold becomes stabilised, i.e. after the initial exponential growth, the folds become increasingly chevron-like as the deformation continues (Fig. 13.27(b)). This straightening of the limbs is the result of the 'rotational instability' discussed above.

Modelling of folds in anisotropic materials using rocks and rock analogues

A considerable amount of experimental work using rock and rock analogue materials has been performed to investigate the formation of folds in anisotropic materials (usually multilayers). Some of these experiments have confirmed predictions made from the analysis of the buckling behaviour of anisotropic materials regarding the geometry of folds. Other experiments, as discussed in Chap. 10, have brought to light mechanisms of folding which are completely different from those predicted from theoretical studies.

Two important indications that the buckling behaviour of a material has been controlled by its bulk mechanical anisotropy are the formation of straight-limbed folds (Fig. 13.28(a)) and the formation of kink-bands and related structures such as box folds (Fig. 13.1 and 13.2) which have axial planes oblique to any layering or fabric and to the principal compress-

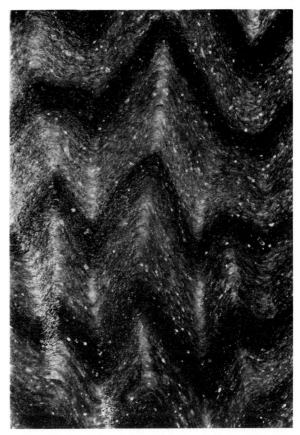

(a)

(b)

Fig. 13.28. (a) A micaceous fabric set in a paraffin wax matrix deformed by compression parallel to the mica flakes. (Photograph courtesy of J-P. Latham.) (b) A Plasticine multilayer with graphite as an interlayer lubricant. Straight-limbed, conjugate structures typical of folds in anisotropic materials have been formed by compression parallel to the layering.

ion. Examples of the latter structures are the straight-limbed, conjugate kink-bands in the layered Plasticine model shown in Fig. 13.28(b), which were produced by compression parallel to the layers.

Cobbold (1976) approached the problem of finite development of folds in anisotropic materials by using a combination of field observations and model

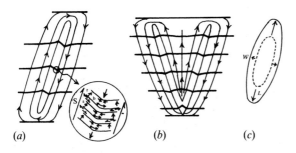

Fig. 13.29. Flow cells in a buckling multilayer. (*a*) Single flow cell. Flow lines define a shear (S, insert), with small components of rotation (r), flexural flow or slip (s) and bending (b). (*b*) Conjugate flow cells. (*c*) Propagation of flow cell by lengthening (*L*) and widening (*w*). (After Cobbold, 1976.)

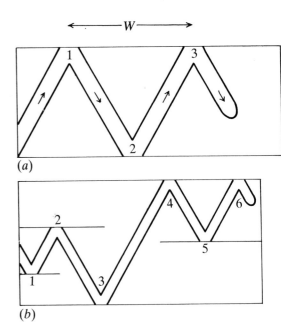

Fig. 13.30. Reflection of flow cells in a multilayer, (*a*) between constantly spaced, parallel boundaries which would generate a periodic complex of folds and (*b*) between parallel boundaries with different separations which would give rise to a less periodic complex of folds (c.f. Fig. 13.19). (After Cobbold, 1976.)

work. He observed that heterogeneities within a multilayer often nucleate folding and, like single layer folds (Fig. 11.42), those developing in anisotropic materials have associated flow cells (Fig. 13.29).

To a first approximation, the flow pattern within one cell is a shear parallel to the axial plane of the folds (kink-bands in Fig. 13.29). In his experiments, Cobbold found that the flow cells changed in shape and increased in size during deformation. As the flow cells propagate, the zone of folding spreads into surrounding areas which were previously unfolded. Propagation can occur in all directions simultaneously but is generally fastest in one 'characteristic direction', approximately parallel to the long axis of the flow cell. The long axis of the flow cell is generally oblique to the layering and to the direction of principal compression and represents a direction of 'easy shear' in the multilayer. As a multilayer fold propagates, it may encounter a change in its average properties. This might occur for example, at an internal or external boundary. If the boundary offers a resistance to transverse flow, either because it is relatively rigid or because it facilitates tangential displacements (for example a decollement horizon), the fold may cease to propagate. If there are conjugate directions of maximum propagation rate, the fold may reflect from one direction to another (Fig. 13.30). The reflection of a flow cell can be observed in the experiment shown in Fig. 13.19, in which conjugate kink-bands, initiated by a stress-induced deflection at the lower boundary of the multilayer, propagate across the model and are reflected off the opposite external boundary. The flow cell then proceeds to ricochet between these two reflecting boundaries. Cobbold points out that if the boundaries are parallel and planar, the successive reflections result in a complex of adjoining folds (Fig. 13.30(*a*)), with a 'wavelength' determined by the angle of incidence, the angle of reflection and the distance between the reflecting surfaces. If some internal boundaries do not persist laterally, reflection may be staggered and the resulting fold complex will not have a regular wavelength (Fig. 13.30(*b*)). During the

early stages of the experiment shown in Fig. 13.19, the kink-bands were reflected between the external boundaries of the multilayer. However, the fourth reflection occurred at a 'boundary' *within* the model.

The models described so far in this section have been made up of Plasticine layers. However, conjugate, straight-limbed structures develop in a wide variety of layered materials. For example, Paterson & Weiss (1966) have produced conjugate, reverse kink-bands in a phyllite under a confining pressure of 5.0 kbar and at room temperature, by compression parallel to the mineral fabric (Fig. 13.31). They performed a series of experiments with various amounts of shortening and found that the conjugate kink-bands frequently intersected in the more deformed specimens. A chevron fold forms in the area of intersection. Paterson & Weiss infer that the kink-

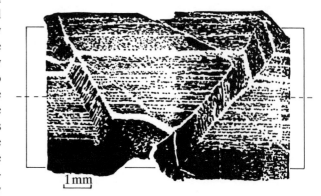

Fig. 13.31. Photomicrograph of a deformed phyllite after compression parallel to the mineral fabric. (After Paterson & Weiss, 1966.)

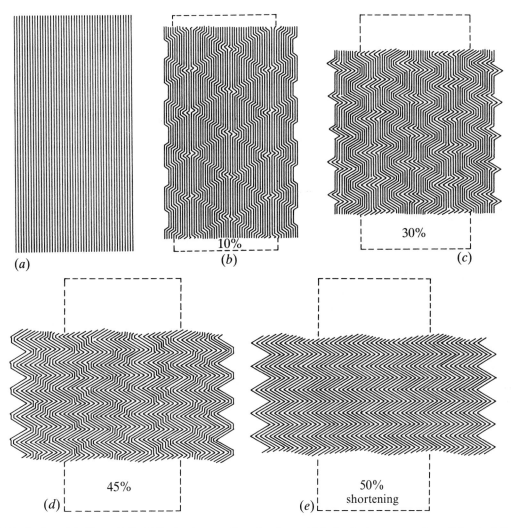

Fig. 13.32. Idealised model of flexural slip folding by the spreading of conjugate kinks in a body compressed parallel to the foliation; initial boundaries shown by broken line. (After Paterson & Weiss, 1966.)

bands spread laterally during deformation, so that the chevron fold also increases in size. An idealised model of chevron fold development by this process is shown in Fig. 13.32. Chevron folds have also been formed by this process in experiments using 'card decks' (Fig. 13.33) and Plasticine multilayers (Fig. 13.34). This particular Plasticine model was compressed at 10° to the layering and the transition from conjugate kink-bands to chevron folds as the bands widened and coalesced can be clearly seen.

Field observations also indicate a relationship between chevron folds and kink-bands. For example, the 75 m high cliff section sketched in Fig. 13.35(a) shows a series of chevron folds with axial planes normal to the original orientation of the layering. These folds are separated from a relatively unfolded area by an axial plane which is oblique to the layering. This geometrical configuration could be produced by the mechanism of spreading conjugate kink-bands, as indicated in Fig. 13.35(b). However, one of the objections to this mode of formation of chevron folds is that parts of the layers would have to be folded and

subsequently unfolded as the kink-band boundaries migrated. If hinge migration does occur, then the second order structures sometimes found in the hinge areas, such as tension gashes in the outer arc and pressure solution cleavage in the inner arc, should not be localised around the final positions of the hinges, but should also exist along the parts of the layers through which the hinge has migrated.

An alternative method of forming the structures shown in Fig. 13.35(a) is illustrated diagrammatically in Fig. 13.36. If a train of low amplitude folds developed as the result of localised buckling above a decollement horizon of rather limited extent, or if a localised train of folds formed by the buckling of a relatively thick competent layer, or group of layers, then, as the limb-dips increase, inversion of the modulus ratio would occur (see page 341) which would facilitate the formation of kink-bands. In the centre of the wave-train, the established geometry of the folds would not allow the kink-bands to propagate and these central folds would continue to amplify and become chevrons. However, the kink-bands are not

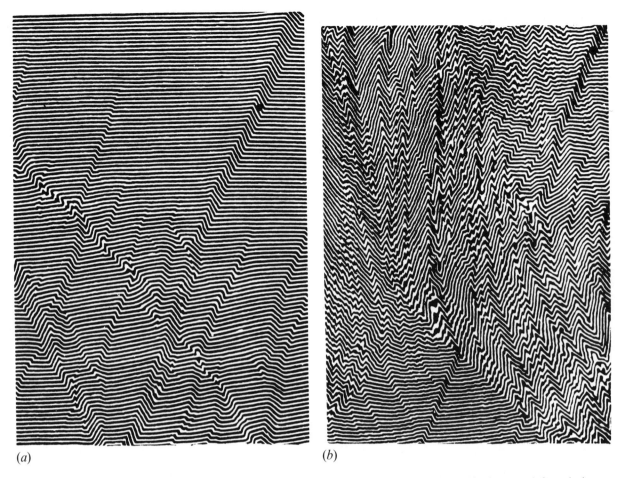

(a)   (b)

Fig. 13.33. Two stages in the formation of multilayer folds (b) from initially conjugate kink-bands (a). The dot on each figure is the same point in the model. (After Weiss, 1968.)

prevented from propagating into the relatively unde-flected layering, outside the wave train, from the bounding limbs A and B. The most important difference between the two mechanisms proposed for the formation of the structures in Fig. 13.35(a) is that in the first (Fig. 13.35(b)) hinge migration is necessary and in the second (Fig. 13.36) it is not.

Inspection of the sandstone units of the folds shown in Fig. 13.35(a) shows that the secondary features mentioned above (i.e. veins and pressure solution cleavage) are restricted to the present position of the hinges, indicating that for this specific example, it is probably the second mechanism which was operative.

Study of fold amplification in gelatine multilayers

Various stages in the development of folds in a gelatine multilayer are shown in Fig. 13.37. The experiment was performed in the apparatus B illustrated in Fig. 10.36. The base of the model to the left of the layering shown in Fig. 13.37 was tilted and the uplifted gelatine slid down the imposed slope under the influence of gravity, applying a layer-parallel compression to the gelatine layers shown in Fig. 13.37. A rectangular grid was placed in front of

the model so that the change in geometry of the folds that accompany fold amplification could be monitored more precisely. Fold initiation and early amplification produced an approximately uniform sinusoidal fold train (Fig. 13.37(a) and (b)). In the light of the theoretical work on the development of folds in anisotropic materials discussed earlier, it is not surprising to find that kink-bands developed on the limbs of these low-amplitude, sinusoidal folds. If kink-bands grew from both limbs of a fold, an upright box fold resulted (central fold, Fig. 13.37(d)). If a kink-band grew from only one limb, and asymmetrical fold resulted (the left and right hand folds of Fig. 13.37(d)). It is of interest to note that the two asymmetrical folds, formed in this experiment, face in opposite directions. These structures were developed in a gravity-glide experiment in which the 'tectonic transport direction' was from left to right. Thus, the asymmetrical fold on the far right of Fig. 13.37 is facing in the opposite direction from the tectonic transport. *It is apparent, therefore, that considerable caution should be exercised when attempting to determine the direction of tectonic transport from the direction in which folds face.*

It was commonly observed in the experiments

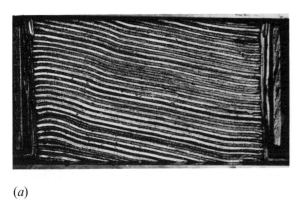

(a)

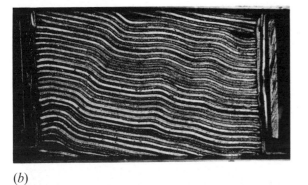

(b)

(c)

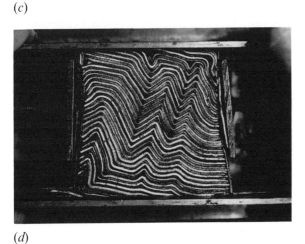

(d)

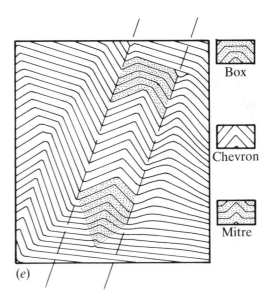

(e)

Fig. 13.34. (a)-(d) Four stages in the formation of chevron folds in a Plasticine multilayer compressed at 10° to the layering. The interlayer lubricant was graphite and the multilayer was confined between rigid plates. (e) Line drawing of (d), showing the division of the structures into box, chevron and mitre folds. (After Cobbold, Cosgrove & Summers, 1971.)

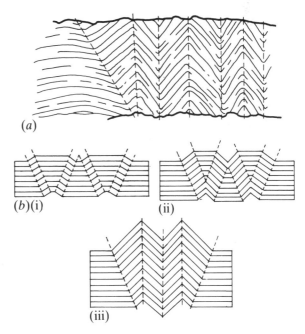

(a)

(b)(i) (ii)

(iii)

Fig. 13.35. (a) Chevron folds in the cliffs at Hartland Quay, N. Devon, England. (b) Folds with the geometry of those in (a) produced by the spreading of conjugate kink-bands.

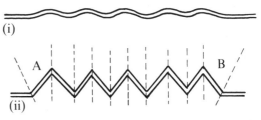

(i)

A   B

(ii)

Fig. 13.36. The amplification of a localised train of low amplitude folds giving rise to folds with the geometry shown in Fig. 13.35(a). See text for discussion.

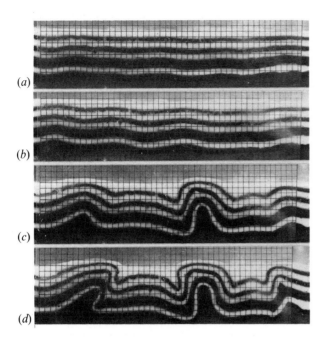

(a)

(b)

(c)

(d)

Fig. 13.37. (a) to (d) show four stages in the formation of symmetric and asymmetric box folds from low amplitude sinusoidal buckles in a gelatine multilayer. (After Blay, Cosgrove and Summers, 1977.)

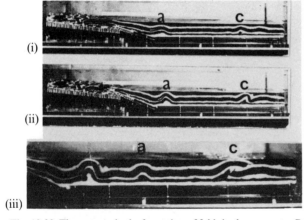

(i)

(ii)

(iii)

Fig. 13.38. Three stages in the formation of folds in the apparatus B shown in Fig. 10.36, showing the progression from sinusoidal deflection (i), through symmetric box fold, asymmetric box fold (ii) to thrust (iii).

using gelatine multilayers that, with progressive deformation, initial, sinusoidal deflections developed first into an upright, symmetrical box fold, then into asymmetrical folds and finally into an overthrust. This sequence of events is shown in Fig. 13.38.

It is interesting to note that the development of folds in many of the experiments is limited in one direction by a decollement horizon. For example, Fig. 13.37 shows the formation of folds above a decollement horizon at the base of a gelatine multilayer. The fold geometries and spatial organisation compare well with those shown in the cross sections through the Jura Mountains (Fig. 13.2) which are

known to have formed above a decollement horizon of Triassic evaporites.

In a complex sedimentary pile, it is reasonable to expect that several potential decollement horizons exist. Each of these may become active at different times during any deformation. The activation of more than one decollement horizon during a single deformation is shown in the experiment (Fig. 13.39(a)), and represented diagrammatically in Fig. 13.39(b). Tensile failure initially occurred only in the uppermost layers of the model, which began to slide off the zone of uplift, using the lubricated interface between themselves and the layer below as a glide plane. As further uplift occurred, tensile failure developed in successively lower units of the multi-layer, allowing gravity-gliding to occur on a lower decollement plane. The wave train of four folds seen in Fig. 13.39(a), developed on two decollement planes within the multilayer. Folds 1 and 2 rest on an upper decollement plane, folds 3 and 4 on a lower decollement plane.

Having considered the finite development of buckles in an anisotropic material, it is apparent that the marked differences in geometry between Type 1 and 2 structures are often maintained and preserved as they grow into finite structures; however, the development of both structural types gives rise to a progressive straightening of the limbs

Buckling of anisotropic materials with non-linear properties

In the theory of buckling of anisotropic material discussed earlier in this chapter, it is assumed that the bending resistance of the material is negligible and that the material has linear stress–strain (or stress–strain-rate) properties. Both these assumptions are geologically unrealistic. Latham (1983, 1985a and b) incorporated the effect of non-linear material properties and bending resistance into Biot's theory of internal instability and, by doing so, has added substantially to our understanding of the geological implications of the theory, as well as resolving some of the problems regarding the mechanical significance of kink-bands. He considered the initial unstable stage of deformation of non-linear, elastic or viscous bilaminates with alternating competent and incompetent layers of thickness h_1 and h_2, respectively. The form of non-linearity considered is that of a power-law material i.e. a material for which the stress–strain or stress–strain-rate equations in a given direction take the form:

$$e = K\sigma^n \qquad (13.30)$$

$$\dot{e} = K\sigma^n \qquad (13.31)$$

where K is a constant of proportionality and n is a material constant known as the power-law exponent.

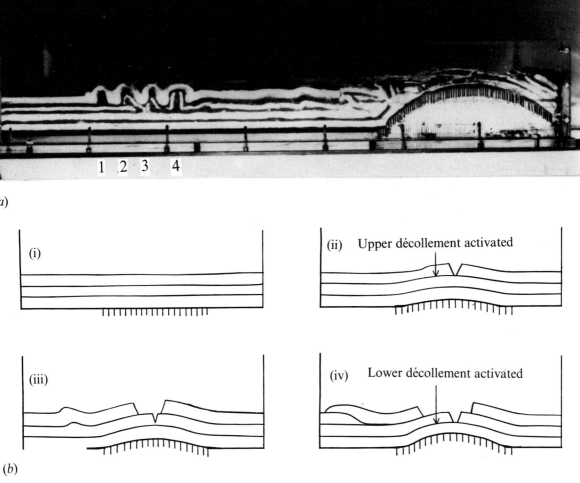

(a)

(b)

Fig. 13.39. (a) A gravity-glide experiment in which two planes of decollement developed within a multilayer. Folds 1 and 2 rest on an upper decollement plane; folds 3 and 4 rest on a lower decollement plane. (b) Diagrammatic representation of the sequential activation of two decollement horizons illustrated in (a). (After Blay, Cosgrove & Summers, 1977.)

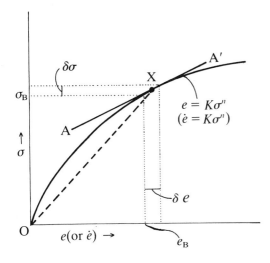

Fig. 13.40. The stress–strain (or strain-rate) curve for a power-law material. The modulus relating the bulk stress σ_B to the bulk strain e_B is given by the slope of the line OX and is known as the secant modulus. If an increment of stress ($\delta\sigma$) is added to the bulk stress, an increment of strain (δe) results. The modulus relating the incremental stress and strain is given by the slope of the line AA' and is known as the tangent modulus.

Before discussing the results of Latham's work, it is instructive to consider briefly the properties of a power-law elastic material represented by the graph in Fig. 13.40. At a particular strain (point X on the curve) it is possible to define two moduli, the tangent modulus, given by the slope of the curve at point X and the secant modulus given by the slope of line OX (these moduli have already been briefly discussed in Chap. 11 with reference to power-law viscous materials). In the unstrained state, i.e. at the origin, these two moduli have the same value but, with progressive deformation, their values change so that the differences between them progressively increases. In a linear material, the two moduli are the same at any state of strain.

The secant modulus is that which relates the stress to the total strain, e.g. a bulk strain e_B (Fig. 13.40) occurs when a stress σ_B is applied and the relationship between σ_B and e_B is given by the secant modulus. The tangent modulus relates incremental stresses and strains, e.g. if an increment of stress $\delta\sigma$ is added to the stress state (σ_B) it induces an increment of strain (δe) (Fig. 13.40). The effect of the two moduli being different during the deformation of a non-linear material (such as a power law material) is to *induce* an anisotropy into the material. The greater the difference between the two moduli, the greater is the *induced anisotropy*.

Clearly, such anisotropy will not be induced in linear materials because the two moduli are always the same. It is interesting to note, however, that there are relatively few real materials which exhibit a perfectly linear stress–strain or stress–strain-rate relationship, especially for strains in excess of 5 per cent. It follows, therefore, that many so-called 'isotropic' rocks (e.g. some granites and sandstones) will exhibit non-linear material properties during deformation and that the incremental response of these rocks to deformation will become increasingly anisotropic.

If the geometry of the deformation (e.g. pure shear, simple shear etc.) and the type of non-linear behaviour is known, it is possible to determine the value of the induced anisotropy after a given amount of deformation. For example, for the case of pure shear and a power-law *elastic* material, Latham (1985a) shows that the induced anisotropy, can be given approximately by the ratio of the secant and tangent moduli. For power-law materials this ratio is the stress exponent n, which is constant. We have already seen from the discussion of the deformation of a power-law *viscous* material (Chap. 11) that the induced anisotropic response in this material is also determined by the power-law exponent.

From this discussion, it follows that the *total* anisotropy of a material will be made up of the *intrinsic* anisotropy that the material possesses prior to deformation as the result of any layering or fabric, plus a component of *induced* anisotropy introduced during deformation as a consequence of any non-linear material properties.

Let us now consider the modification of Biot's analysis of internal buckling to incorporate non-linear (power-law) materials. Although the treatment of the non-linear elastic problem (Latham 1985a and b) is more rigorous than that for the equivalent viscous problem (Latham 1983), the results of the latter are simpler to illustrate and are consequently considered below.

Following Biot, the multilayer is analysed in terms of the response of an equivalent homogeneous continuum with an intrinsic anisotropy. This is done by using effective moduli $\bar{M}$ and $\bar{L}$ which have the dimensions of the corresponding effective elastic modulus and which depend upon the thickness and material properties of the individual layers. For a bilaminate made up of such power-law viscous materials, the effective moduli $\bar{M}$ and $\bar{L}$ are:

$$\bar{M} = \frac{p}{n}(\alpha_1 \eta_1 + \alpha_2 \eta_2) = \eta'_n p \qquad (13.32)$$

and

$$\bar{L} = p\left(\frac{\eta_1 \eta_2}{\alpha_1 \eta_2 + \alpha_2 \eta_1}\right) = \eta_t p \qquad (13.33)$$

(cf. Eqs. (12.16) and (12.17) where η_1 and η_2 are the viscosities of the competent and incompetent layers making up the bilaminate, n is the stress exponent (Eq. (13.31)) which, for convenience, is assumed to be the same for both layers and α_1 and α_2 are the proportional thicknesses of the layers, where:

$$\alpha_1 = \frac{h_1}{h_1 + h_2} \qquad (13.34)$$

and

$$\alpha_2 = \frac{h_2}{h_1 + h_2}. \qquad (13.35)$$

h_1 and h_2 are the thickness of the competent and incompetent layers respectively and p is a measure of the amplification (A) of an instability after time, t, where

$$A = A_0 e^{pt} \qquad (13.36)$$

and A_0 is the amplitude of the initial irregularity. η'_n and η_t are the average viscosity coefficients for the bilaminate while it is undergoing *perturbed* normal straining and shear straining respectively, i.e. they are the effective viscosities operating during the initial growth of the instability. They are analogous to the tangent and secant moduli described in connection with the power-law material represented in Fig. 13.40. It can be seen from Eq. (13.32) that the viscosity coefficient for perturbed normal straining ($\bar{M}$) involves the power-law exponent n. This is in contrast to the viscosity coefficient (η_n) associated with a uniform compressive stress P which produces a *steady* background normal flow at a rate p_0

$$\eta_n = \alpha_1 \eta_1 + \alpha_2 \eta_2. \qquad (13.37(a))$$

For a steady, uniform pure shear deformation of the material:

$$P = 4\eta_n p_0. \qquad (13.37(b))$$

We have seen (Eq. (13.13)) that the expression relating the critical buckling stress (P) to the material properties M' and L' for a linear, elastic bilaminate is:

$$L'\xi^4 + 2(2M' - L')\xi^2 + (L' - P) = 0. \qquad (13.13)$$

Latham (1983) shows that the equivalent expression for a power-law viscous bilaminate is:

$$4\eta_n p_0 = \eta_t p + 2(2\eta'_n - \eta_t)p\xi^2 + \eta_t p\xi^4 \qquad (13.38)$$

which is obtained by substituting $\bar{M}$, $\bar{L}$ and p of Eqs. (13.32), (13.33) and (13.37(b)) into Eq. (13.13). p and p_0 are, respectively, measures of the rate of amplification of the instability (Eq. (13.36)) and of a

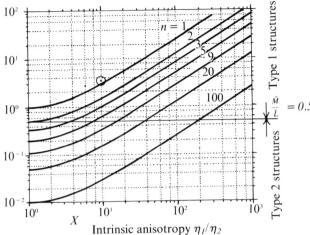

Fig. 13.41. The variation of total anisotropy $\bar{M}/\bar{L}$ of a power-law viscous bilaminate with stress exponent n (= n_1 = n_2) and viscosity contrast η_1/η_2. The thickness of the incompetent and competent layers is the same. $\bar{M}/\bar{L} = 0.5$ separates the fields of Type 1 and Type 2 instability. (After Latham, 1983.)

passive marker line with the same geometry as the instability (i.e. kinematic amplification). η'_n is given by:

$$\eta'_n = \frac{\eta_n}{n} \qquad (13.38(a))$$

where n is the stress exponent (Eq. (13.31)) and is a measure of the non-linearity of the material. ξ, as we have already seen (Eq. (13.21)), is related to the characteristic direction θ by:

$$\tan^{-1}\xi = \pm\theta \qquad (13.21)$$

The dependence of the *total anisotropy* of this bilaminate on the *intrinsic anisotropy* (which depends upon the competence contrast between the layers, η_1/η_2) and the *induced anisotropy* determined by the stress exponent n of the layers, is shown in Fig. 13.41. The line separating the materials in which Type 1 and 2 structures will form occurs at $\bar{M}/\bar{L} = 0.5$. The graph also shows the variation in total anisotropy $\bar{M}/\bar{L}$ as the power-law exponent n increases from the linear case when $n = 1$ to $n = 100$. If we consider a multilayer with an intrinsic anisotropy (η_1/η_2 value) of X (Fig. 13.41), it is clear from the graph that the total anisotropy is very sensitive to the power-law exponent n of the layers. If $n = 1$, the total anisotropy will be well above 0.5 and the formation of Type 1 structures would be expected. If $n = 20$, the total anisotropy is well below 0.5 and the formation of Type 2 structures would result. It can be seen from this graph that a high intrinsic anisotropy favours the formation of Type 1 structures and a high induced anisotropy the formation of Type 2 structures.

Latham presents this as an alternative explanation to that of modulus inversion discussed earlier, to account for the formation of kink-bands in mater-

ials whose intrinsic anisotropy is greater than 0.5. He argues that if the materials are markedly non-linear, the induced anisotropy will generally cause the total anisotropy to fall below 0.5 and thus facilitate the formation of kink-bands (i.e. Type 2 structures).

Latham, having considered the effect of non-linearity on the deformation behaviour of a bilaminate, then proceeded to introduce the effect of the 'bending resistance' of the individual layers making up the multilayer into his analysis. The equation which describes the condition of instability for a confined non-linear, viscous bilaminate with bending resistance is given by:

$$\frac{4\eta_n p_0}{\eta_t p} =$$

$$1 + \frac{\bar{a}}{\xi^2} + 2\left(\frac{2\eta'_n - \eta_t}{\eta_t}\right)\xi^2 + \xi^4 \qquad (13.39)$$

where $\bar{a}$ is a dimensionless quantity that expresses the influence of bending resistance on internal buckling. If one of the layers of the bilaminate is much more competent than the other (i.e. $\bar{M}_1 \gg \bar{M}_2$ and $\bar{L}_1 \gg \bar{L}_2$) then the value of $\bar{a}$ approximates to:

$$\bar{a} = \frac{\pi^2 h_1^2 \bar{M}_1}{3H^2 \bar{L}} \qquad (13.40)$$

where $\bar{L}$ is defined by Eq. (13.33) and H is the total number of layers in the bilaminate. After the discussion presented in Chap. 11, it should come as no surprise to find that the resistance to bending depends on both the thickness of the competent layer h_1 and its material properties that resist bending, $\bar{M}_1$. One effect of introducing bending resistance into the analysis is that the condition separating Type 1 and Type 2 structures ($\bar{M}/\bar{L} = 0.5$) changes to:

$$\frac{\bar{M}}{\bar{L}} = 0.5 + 0.75\bar{a}^{\frac{1}{3}} \qquad (13.41)$$

A convenient way of understanding the physical significance of Eq. (13.39) and of seeing the way in which varying the geometric and material properties of the layers of the model will affect the total anisotropy and, therefore, the type of structure that will develop, is to express the equation graphically. For a specific bilaminate, the only two variables in Eq. (13.39) are p/p_0 (the normalised amplification rate which is a measure of the rate of amplification of an instability) and ξ which is associated with a direction θ (Eq. (13.24)) along which the instability will tend to propagate, where θ is the angle between this direction and the normal to the layering. Such a plot is termed an 'amplitude spectrum' and it indicates clearly the orientation of the most rapidly amplifying instability. This orientation is the dominant characteristic direction i.e. the direction along which heterogeneous simple shear will occur most easily. If this direction

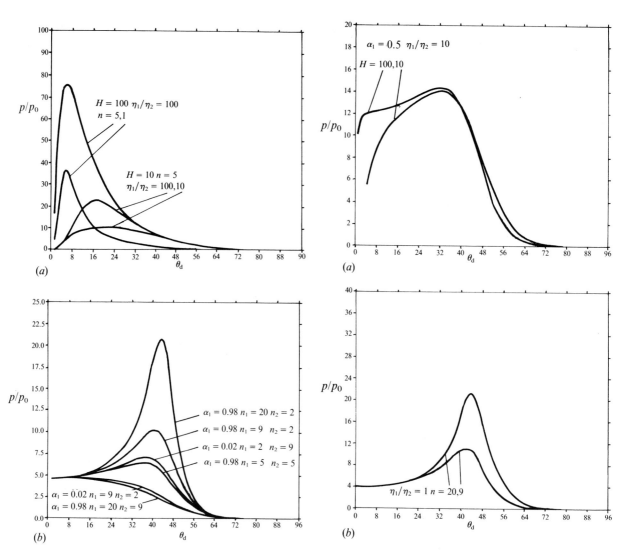

Fig. 13.42. (*a*) The amplification spectra for four power-law viscous multilayers whose geometries (*H*) and physical parameters (*n*, η_1, η_2) are such that the multilayers have a high *intrinsic* anisotropy ($\bar{M}/\bar{L} > 0.5$). (*p/p_0 is a measure of the rate of amplification of the structure.*) (*b*) The amplification spectra for six power-law, viscous multilayers with physical (n_1, n_2) and geometric (α_1, α_2) parameters which favour a high *induced* anisotropy compared with intrinsic anisotropy, ($\bar{M}/\bar{L} < 0.5$). For each multilayer, the viscosity contrast η_1/η_2 at the applied strain-rate is 10, see text for discussion. (After Latham, 1983.)

Fig. 13.43. (*a*) The amplification spectra for two power-law, viscous multilayers with geometric and physical parameters which give them a significant *intrinsic* and *induced* anisotropy. M'/L' just > 0.5. The stress exponent (*n*) of the layers is 9. (*b*) The amplification spectra for two power-law, viscous multilayers with geometric and physical parameters which give them *no* intrinsic anisotropy. The only anisotropy in the materials is an *induced* anisotropy. (After Latham, 1983.)

falls within the range $\theta = 0°$ to $15°$, then the characteristic directions are at a high angle ($75°–90°$) to the direction of principal compression. Consequently, the structures which form will be Type 1. This is the case for the multilayers represented by the graphs of Fig. 13.42(*a*). If, however, the maximum value of θ falls in the range $\theta = 35°$ to $45°$, then the characteristic direction will be at $55°$ to $45°$ to the layering so that the structures that form will be Type 2. This is the case for the multilayers represented by the graphs shown in Fig. 13.42(*b*).

A multilayer may have a combination of geometric and mechanical properties such that a sharp peak does not develop in the amplification spectrum.

The multilayers represented in Fig. 13.43(*a*) are of this type. It can be seen that there is little difference in the amplification rates of instabilities propagating over a wide range of orientations. It follows that such a multilayer will show no preference for the formation of either Type 1 and Type 2 structures, and in such multilayers it is likely that the type of structure which develops will be very sensitive to the geometry and orientation of the perturbation from which the instability grows.

The interesting amplification spectra for two materials with no intrinsic anisotropy are shown in Fig. 13.43(*b*). The only anisotropy which these materials will possess at the onset of the instability is induced anisotropy. It can be seen from their amplification spectra that the dominant characteristic direction is approximately 45° and that the type of structures to develop will be Type 2. From the various amplification spectra shown in Figs. 13.42 and 13.43, we can see that a high intrinsic anisotropy favours the formation of Type 1 structures and a high induced anisotropy favours the formation of Type 2 structures. Latham recognises Type 1 and Type 2 structures as being two end members of a complete spectrum of structures. Type 1 structures will develop typically as diffuse (i.e. pervasive) internal buckles in materials with a high intrinsic anisotropy and Type 2 structures as localised shears, oblique to the bulk compression direction in materials with a high induced anisotropy.

It will be immediately apparent that a material such as granite, which often exhibits no intrinsic anisotropy, may develop an induced anisotropy during deformation which will allow internal instabilities to develop. These will be expressed as shear zones, cutting an apparently homogeneous and isotropic rock. Latham considers this to be a Type 2, end member structure.

He illustrates four distinct types of geological structures (I–IV of Fig. 13.44) that may develop in rocks, depending on the amount and type of anisotropy that exists in the rock at the time it becomes unstable. If the anisotropy is predominantly intrinsic, internal buckles form (Fig. 13.44 II). For the situation in which the anisotropy is predominantly induced, shear zones or shear fractures are developed (Fig. 13.44 III). In materials in which there are significant amounts of both intrinsic and induced anisotropy, kink-bands form (Fig. 13.44 IV) and in materials in which neither type of anisotropy is well developed, passive (kinematic) amplification of any initial irregularity occurs (Fig. 13.44 I). There will, of course, be a complete spectrum of structures between these four end members. The diagram also shows the effect of a further, finite pure shear deformation on the initial structures. It is emphasised that this theory only applies to the first increments of instability and that the finite deformations sketched in Fig. 13.44 I to IV are based on experimental evidence and finite deformation theories.

We have restricted the discussion of the deformation of power-law bilaminates with bending resistance to viscous bilaminates. However, as was pointed out earlier in this chapter, the viscous analysis is based on a more rigorous, elastic analysis. Consequently, the conclusions reached regarding the deformation behaviour of a viscous power-law material also pertain to elastic power-law materials.

The two models A and B shown in Figs. 13.45 and 13.46 are made up of 1 mm thick layers of identical 53.5 °C melting point paraffin wax lubricated by liquid soap. However, the experiments were performed at different temperatures (and confining pressures) which resulted in the models having differ-

ent rheological properties. During the deformation of the model in Fig. 13.45, the confining pressure was low and the temperature was 33 °C; a temperature at which the layers offer a relatively high resistance to deformation. The model, therefore, had a relatively high intrinsic anisotropy and the structures which developed reflect this in that their geometries lie between the geometries of structures IIb and IVb of Fig. 13.44. The model shown in Fig. 13.46 was deformed under a relatively high confining pressure at a temperature of 39 °C, at which the layers offer a much lower resistance to deformation. This reduction in the intrinsic anisotropy of the model compared to the model shown in Fig. 13.45 is reflected in the structures which develop. The dominant characteristic directions associated with these structures lie between 2–5° compared to the 15–20° range of the model shown in Fig. 13.45 and their geometries lie between the geometries of structures Ib and IIb of Fig. 13.44.

In the light of this theoretical and experimental work on the deformation of non-linear bilaminates, let us reappraise the conclusions of Cobbold *et al.* (1971) who considered some of the geological implications of Biot's theory of internal buckling in linear materials. It will be recalled that they considered that kink-bands were associated with Type 2 instablity and developed in materials with a high anisotropy. In contrast, Johnson & Honea (1975) argued that kink-bands develop in materials with low anisotropy. From Latham's work on the deformation of non-linear materials, discussed above, the reason for this apparent conflict becomes clear. As he points out, the formation of kink-bands is favoured by a high induced anisotropy. The multilayer analysed by Johnson & Honea had a small intrinsic anisotropy, but developed a large induced anisotropy as a result of local loss of interlayer cohesion and the associated non-linear behaviour.

It follows from this work on non-linear materials that it is more appropriate to consider:

(i) internal buckles as the end member Type 1 structures, and

(ii) shear zones (rather than kink-bands, as implied by Cobbold *et al.*) as the end member Type 2 structures.

Instabilities of the second kind (Type 2) can express themselves as a variety of structures. If the deforming material had a marked intrinsic anisotropy, such as a layering or fabric, then the instability may manifest itself as either a reverse, or normal, kink-band, depending upon whether the maximum compression acted parallel or perpendicular to the layering or fabric. If the material does not possess a significant intrinsic anisotropy (as for example a homogeneous sandstone or granite), then deforma-

INTRINSIC ANISOTROPY *(e.g. ratio of power law constants)*

(e.g. average power law stress exponent) INDUCED ANISOTROPY

Viscous→Irregular, passive folds
(Elastic : Faulting)

Ib

Ia

Viscous→Similar folds
(Elastic→Regular similar folds)

IIb

IIa

Viscous→Shear zones
(Elastic→Faulting)

IIIb

IIIa

Viscous→Folds, kinks
(Elastic→Kinking)

IVb

IVa

Fig. 13.44. Summary diagram to illustrate the genetic relationships between four types of instability and the kinds of geological materials and conditions in which the incremental anisotropy will favour the production of these structures. The idealised confined multilayer consists of 100 layers and has a relatively low resistance to bending. The square sections represent the forms of the first expression of instability. The probable effect of a further finite pure shear is shown in the rectangular sections which are based on experimental evidence and finite deformation theories. In a viscous bilaminate k_1/k_2 is the ratio of the viscosities of the two layers η_1/η_2 and in an elastic bilaminate the ratio of the elasticities is G_1/G_2. See text for discussion. (After Latham, 1985b.)

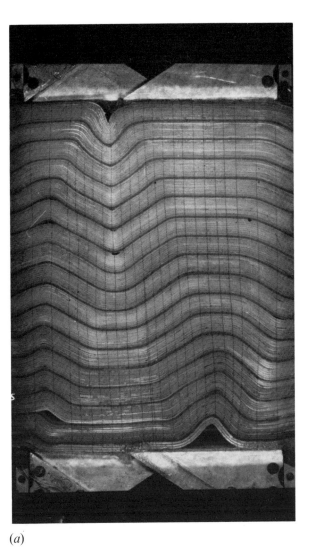

(*a*)

(*b*)

Fig. 13.45. Stages in the deformation of a model A after a bulk shortening of (*a*) 0.081 and (*b*) 0.225 logarithmic strain. The model is made up of identical 53.55 °C melting point wax layers and deformed at 33 °C. (After Latham, 1983.)

tion may induce an anisotropy which may fall within the range of values compatible with the formation of Type 2 instabilities. If this latter situation occurs, the instability may express itself as a shear zone. Depending upon a variety of parameters, including rock type, pressure and temperature conditions and strain-rate, the shear zone can range in geometry from the 'classical' ductile shear zone (Fig. 13.47) where continuity of the rock is maintained, to discrete planes where continuity and cohesion are lost. These latter structures would, of course, be described as faults. It must be emphasised that the orientation and movement sense of conjugate shears which result from this mechanism cannot be explained in terms of the criteria of brittle failure (see Chaps. 1, 5 and 8).

We have seen that two basic approaches have been used to analyse the buckling behaviour of multi-layers; i.e. that of considering (i) the individual layers as separate entities and (ii) the average properties of the multilayer. Theories that adopt the first approach predict the buckling behaviour in terms of the properties of the individual layers making up the multi-layer (e.g. the viscosities η_1, η_2 and thicknesses h_1 and h_2 of a viscous bilaminate). Theories which use the second approach predict the buckling behaviour in terms of the anisotropy of the material (M'/L'). In order that both types of theory can be applied to a particular multilayer, it is necessary to be able to express the bulk anisotropy of a specific multilayer in terms of the geometrical and material properties of the

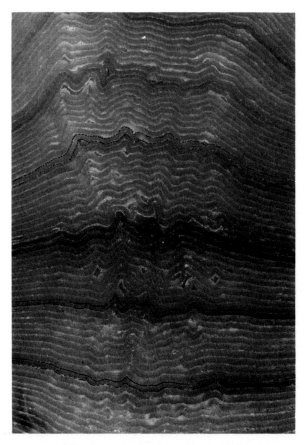

Fig. 13.46. Model B, a deformed wax multilayer identical to that in Fig. 13.45 but deformed at 39 °C and at a higher confining pressure. See text for details. (After Latham, 1983.)

individual layers. It will be recalled (Eqs. (12.16) and (12.17)) that the relationships between the moduli (M' and L') and the properties of the competent and incompetent layers (η_1 and η_2) with fractional thicknesses α_1 and α_2 (where $\alpha_1 + \alpha_2 = 1.0$) are:

$$M' = (\eta_1 \alpha_1 + \eta_2 \alpha_2)p \tag{13.42}$$

$$L' = \frac{\eta_1 \eta_2 p}{\eta_1 \alpha_2 + \eta_2 \alpha_1} \tag{13.43}$$

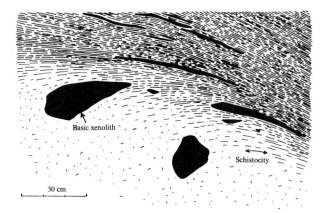

Fig. 13.47. Shear zone formed in a relatively homogeneous and isotropic granite. (After Ramsay and Graham, 1970.)

or

$$\frac{1}{L'} = \frac{\alpha_1}{\eta_1 p} + \frac{\alpha_2}{\eta_2 p}. \tag{13.44}$$

It follows that the M'/L' ratio of this bilaminate is:

$$\frac{M'}{L'} = \frac{(\eta_1 \alpha_1 + \eta_2 \alpha_2)(\eta_1 \alpha_2 + \eta_2 \alpha_1)}{\eta_1 \eta_2} \tag{13.45(a)}$$

$$\frac{M'}{L'} = \frac{\alpha_1 \alpha_2(\eta_1^2 + \eta_2^2) + \eta_1 \eta_2(\alpha_1^2 + \alpha_2^2)}{\eta_1 \eta_2}. \tag{13.45(b)}$$

It can be seen that Eqs. (13.42) and (13.44) are examples of the more general equations:

$$M' = \sum_1^N \eta_i \alpha_i p \tag{13.46}$$

$$\frac{1}{L'} = \sum_1^N \frac{\alpha_i}{\eta_i p} \tag{13.47}$$

where N is the number of layers making up the repeating unit of the multilayer (4 in Fig. 13.48). These expressions enable us to use the theory of internal buckling to analyse the buckling behaviour of the multilayer made up of linear materials. For non-linear multilayers the equations will be modified. For example, if the multilayers are made of power-law materials, Eqs. (13.42) and (13.43) should be replaced by Eqs. (13.32) and (13.33).

The relationship between the various structures that develop in anisotropic materials

It has been noted in this and the preceding chapter, that the geometry of structures which form in anisotropic materials when they are deformed can be very varied (e.g. Fig. 13.20). This is because the geometry depends on several factors including the degree and type of anisotropy (intrinsic or induced), the amount of amplification of the structure and the orientation of the maximum principal compression with respect to any layering or fabric. However, despite the diversity

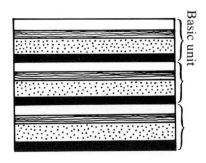

Fig. 13.48. A multilayer made up of a repeating unit containing four layers.

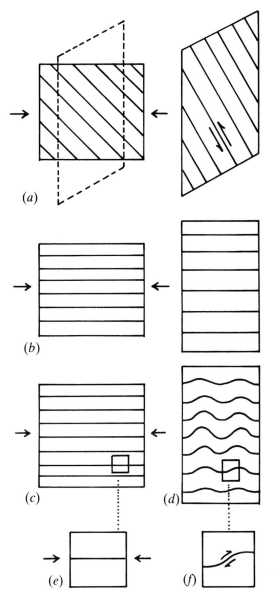

Fig. 13.49. The deformation of a layered material by (*a*) slip between the layers (*b*) compression of the layers (*c*) compression followed by (*d*) slip between the layers. The change over is facilitated by a flexing (buckling) of the layer ((*e*) and (*f*)).

and requires less energy (Fig. 13.49(*a*)). However, when the maximum compression is either parallel or normal to the layering, the lower modulus cannot be utilised. Consequently, deformation does not involve slip between the layers and occurs only by layer-parallel (or normal) compression (Fig. 13.49(*b*)). If however, rearrangement of the layering can occur so that, in some localities within the multilayer, the layering is oblique to the maximum principal compression direction, then shortening in the compression direction can occur at these localities by slip between the layers. This then, is the reason why a state of homogeneous flattening (Fig. 13.49(*c*)) gives way to one of buckling (Fig. 13.49(*d*)). On the limbs of the folds, shortening in the compression direction occurs by slip between the layers. In the example of the multilayer under consideration, it is apparent that, before the changeover from flattening to buckling can take place, the applied stress must be of sufficient magnitude to overcome the internal resistance of the layering to flexing. This stress is known as the *critical buckling stress*. The essential element of the instability is a flexing of the layering (Fig. 13.49(*e*) to (*f*)), and these instabilities can be arranged in a variety of ways to achieve the required bulk shape changes.

For example, when the principal compression is parallel to the layering, the instabilities are arranged either into buckles (Fig. 13.50(*a*)) or, if conditions are such that the deformation develops more locally, into reverse kink-bands (Fig. 13.50(*c*)). When the compression is normal to the layering, the bulk shape changes are achieved by arranging the instabilities to form pinch-and-swell structures (Fig. 13.50(*b*)) normal kink-bands or shear zones (Fig. 13.50(*d*)). It can be seen, therefore, that the spatial organisation of this basic instability and consequently the geometry of the structure that develops, is determined primarily by the orientation of the applied stress with respect to any mechanical anisotropy (intrinsic or induced) of the material.

of geometric forms, these structures are all the result of a common instability. To identify this instability we must briefly consider why a layered (and, therefore, anisotropic) material becomes unstable when compressed parallel to the layering and changes from an initial state of homogeneous flattening to one of buckling. The material is anisotropic because its resistance to compression and shear are not the same. Let us consider a multilayer, made up of identical layers, with very low interlayer cohesion, so that resistance to layer-parallel compression is much greater than layer-parallel shear. Such a material can deform either by slip between the layering or by deformation of the layers themselves. Whenever possible, deformation will occur by slip between the layers, for this exploits the lower of the two moduli

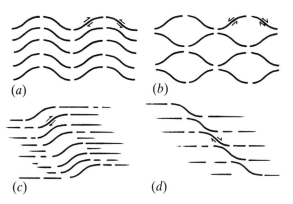

Fig. 13.50. Various arrangements of the basic instability (Fig. 13.49(*f*)) to produce (*a*) folds (*b*) internal pinch-and-swell structures and reverse and normal kink-bands (*c*) and (*d*).

It was noted earlier in this chapter (see Fig. 13.20(*a*)–(*c*)) that there is a complete range of structures between Type 1 (e.g. buckles and internal pinch-and-swell structures) and Type 2 structures (e.g. normal and reverse kink-bands) and that the displacement field associated with Type 1 structures (Eq. (13–14) and Fig. 13.12), which tend to be pervasive, is equivalent to the *superposition* of two simple shear deformations parallel to the two *characteristic directions*: whereas Type 2 structures (conjugate kink-bands and shear zones) which are the result of a more localised deformation, are equivalent to the *juxtaposition* of two simple shear deformations parallel to the two characteristic directions. It is now clear that there is a further link between these two types of structures in that they are both the result of the same basic instability.

Commentary

In this chapter Biot's theory relating to the deformation of anisotropic materials has been used to explain the formation of a variety of structures including folds, kink-bands, shear zones and faults, all of which are important field structures.

It has been argued that it is the bulk mechanical anisotropy of a rock that is of over-riding importance in determining the type of structure that forms as a result of deformation. This bulk anisotropy is made up of an *intrinsic* anisotropy i.e. that possessed by the rock prior to deformation as the result of any layering or fabric, and an anisotropy which is *induced* when a rock with non-linear material properties is deformed.

In addition, the theory has been used to explain the development of some types of listric faults (Chap. 8) and is also applied to explain certain aspects of crenulation cleavage and the development of some pinch-and-swell structures (Chaps. 16 and 17).

It is apparent from field and laboratory observations that the distribution of deformation in a deformed rock lies between two extremes; that of *brittle failure* (in which the rock consists of planes of extremely high deformation and blocks of almost undeformed material) and *homogeneous flattening* (ductile failure) in which there is a uniform distribution of strain throughout the rock. Between these two extremes lies a vast range of possible arrangements of instabilities which divide the deformed body into areas of relatively high and low strain.

We have noted (Figs. 13.20, 13.49 and 13.50) that a large range of geological structures (including folds, kink-bands, shear zones and faults) are fundamentally related, in that they represent different spatial arrangements of a common instability.

The various concepts discussed in this chapter relating to the deformation of anisotropic rock masses, are extremely versatile and are some of the most useful available to geologists; for they can be used in the interpretation of a wide range of forms and scales of geological structures which may have developed in many varied geological environments and rock types.

It is emphasised that the theoretical analysis of folding discussed here and earlier in Chaps. 11 and 12 have ignored the influence of gravity. Strictly, the conclusions so far reached should therefore be restricted to relatively small folds, with wavelengths which do not exceed a few tens of metres. We have, however, quoted natural field examples from the Jura and Zagros Mountains with wavelengths well in excess of this limit. It would appear that we have made unjustified correlations. In the following chapter, we shall consider the effect of gravity and the attendant body-weight problem on folding and shall demonstrate that the theories expressed in this chapter can be used to cover these much larger wavelength structures.

References

Ashwin, D.P. (1957). The structure and sedimentation of the Culm sediments between Boscastle and Bideford, N. Devon. Unpublished Ph.D. Thesis, University of London.

Biot, M.A. (1964). Theory of internal buckling of a confined multilayer structure. *Geol. Soc. Am. Bull.*, **75**, 563–8.

(1965). *Mechanics of incremental deformations*. New York: Wiley.

(1967). Rheological stability with couple stresses and its implication to geological folding. *Proc. Roy. Soc. London*, **A2298**, 402–23.

Blay, P., Cosgrove, J.W. & Summers, J.M. (1977). An experimental investigation of the development of structures in multilayers under the influence of gravity. *J. Geol. Soc. London*, **133**, 329–42.

Casey, M. (1976). Application of finite element analysis to some problems in structural geology. Unpublished Ph.D. Thesis, University of London.

Cobbold, P.R. (1976). Mechanical effects of anisotropy during large finite deformations. *Bull. de la Société Géologique de la France*, **18**, 1497–510.

Cobbold, P.R., Cosgrove, J.W. & Summers, J.M. (1971). Development of internal structures in deformed anisotropic rocks. *Tectonophysics*, **12**, 223–53.

Cosgrove, J.W. (1976). The formation of crenulation cleavage. *J. Geol. Soc. London*, **132**, 155–78.

Heim, A. (1921). *Geologie der Schweiz*. Leipzig: Tauchnitz.

Honea, E. & Johnson, A.M. (1976). A theory of concentric, kink and sinusoidal folding and of monoclinal flexuring of compressible, elastic multilayers. IV. Development of sinusoidal and kink folds in multilayers confined by rigid boundaries. *Tectonophysics*, **30**, 197–239.

Johnson, A.M., & Honea, E. (1975). A theory of concentric, kink and sinusoidal folding and of monoclinal flexuring of compressible, elastic multilayers. III. Transition from sinusoidal to concentric-like to chevron folds. *Tectonophysics*, **27**, 1–38.

Latham, J-P. (1983). The influence of mechanical anisotropy on the development of geological structures. Unpublished Ph.D. Thesis, University of London.

(1985a). The influence of non-linear material properties and resistance to bending on the development of internal structures. *J. Struct. Geol.*, **7**, 2, 225–36.

(1985b). A numerical investigation and geological discussion of the relationship between folding, kinking and faulting. *J. Struct. Geol.*, **7**, 2, 237–49.

Nadai, A. (1950). *Theory of flow and fracture of solids*. New York: McGraw–Hill.

Paterson, M.S. & Weiss, L.E. (1966). Experimental deformation and folding in phyllite. *Geol. Soc. Am. Bull.*, **77**, 343–74.

Ramsay, J.G. & Graham, R.H. (1970). Strain variation in shear belts. *Can. J. Earth. Sc.*, **7**, 786–813.

Summers, J.M. (1979). An experimental and theoretical investigation of multilayer fold development. Unpublished Ph.D. Thesis, University of London.

Weiss, L.E. (1968). Flexural slip folding of foliated model materials. In *Proc. Conference on Research in Tectonics*, ed. A.J. Baer & D.K. Norris. Canada Geol. Survey, Ottawa. Paper 68-52.

14 Initiation of large buckle folds and fracture–fold relationships

Introduction

In the previous chapters on folding, the influence of gravity (i.e. the body-weight of the beds or units) on the process of folding has been largely neglected, so that the theories of buckling presented there should only be applied to folds with wavelengths that do not exceed a few tens of metres. The mechanical effects of fluid pressure in the rock mass has received no quantitative evaluation. Moreover, the theoretical analyses already discussed place a disproportionate emphasis upon viscous models. As will have become apparent to the reader, this emphasis merely reflects the dominance of such analyses in the literature. It is the authors' current conviction that they do not correctly model rock behaviour during fold initiation. Other aspects relating to buckling tend to be ignored or neglected, especially for those folds which develop in the upper levels of the crust. For example, major folds are ramified by fractures, some of which certainly developed while the fold was developing and may supply evidence of how finite folds develop. In an attempt to rectify these various 'sins of omission and commission' we accordingly divide this chapter into two parts.

The first part is given to the discussion of the initiation of elastic buckling and flexural slip folding in the upper levels of the crust, with special emphasis on the magnitude of the fluid pressure that obtained during folding and its mechanical importance in counteracting the effects of body-weight. The second part deals with the relationships of fracture types and patterns to folds, and what one may deduce regarding the mechanics of fold development from these patterns.

Development of large scale, flexural slip folds

In the first part of this chapter we shall discuss the initiation of large scale folds; i.e. structures which eventually exhibit wavelengths between several hundred metres and several kilometres. It is assumed that the environment in which the folds develop is in the uppermost levels of the crust, where the initial deformation of competent beds will certainly be elastic. That is, we consider the buckling of such folds to take place in an environment where the wavelength of the fold is comparable with, or even exceeds, the depth of burial of the units to be folded. It is not then necessary to take into account layer-normal resistance to buckling, other than the body-weight effects, for the topographic surface, which we assume initially to be horizontal and parallel to the undeformed beds, will be free to move as the folds are initiated and develop. Structures which satisfy these conditions occur in fold belts such as those seen in the Jura or the Zagros.

In this conceptual model, we shall consider one single unit which may either:

(i)   exist within a sequence of units with identical properties and dimensions, subjected to identical horizontal compressive stresses, (so that if the conditions are met for fold initiation in one unit, they are also met in the other units), or

(ii) be sufficiently thick and competent to be considered as a control unit which determines the dimensions and features of the fold.

It is further assumed that either the units (*a*) possess smooth non-cohering bedding planes, or else (*b*) are separated by thin layers of weak clay-rocks. It is considered that the folds formed by flexural slip (see Chap. 10), and finally, it is assumed that the rocks are saturated but exhibit low permeability.

It is perhaps unlikely that geological conditions which exactly satisfy these conditions are to be found: a criticism which applied with equal validity to all the multilayer models proposed by other authors. However, geological conditions which approximate sufficiently closely to these models do exist, so that conclusions regarding the behaviour of real systems can be drawn from an analysis of the idealised models.

From the analyses and comments made in previous chapters, we infer that the total resistance to buckling of the competent layers (which together initially exhibit relatively low intrinsic anisotropy) is made up of a number of component elements. These include:

(i)   the internal stresses and strains incurred by flexing of the competent units

(ii) the body-weight of the units and

(iii) the bedding-parallel shear stresses which are induced by the flexural slip model.

In the conceptual model dealt with here, we shall assume that the bedding-parallel shear stresses are induced by frictional sliding (rather than by viscous

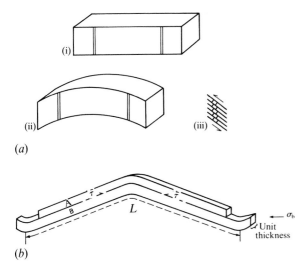

Fig. 14.1. (a) (i) Undeformed sequence of layers with vertical passive markers. (ii) Folded unit involving flexural slip. (iii) Relative movement of individual layers. (b) Shear stress (τ) which resists buckling by flexural slip mechanism induced by buckling stress (σ_b). (After Price, 1975a.)

drag) of one unit past another: and it is the importance of these stresses that will be considered first.

Layer-parallel stresses and slip between layers

The shear stresses which develop along the bedding planes which bound the competent layer resist the development of buckling. These shear stresses may develop because of strain which results from the cohesion between layers, and also as the result of frictional resistance to slip between layers, as represented in Fig. 14.1(a), which is inherent in flexural slip folding.

Let us consider the maximum role which such shear stresses play in the process of buckle folding. This may be done by completely neglecting (for the time being) all other factors which influence fold initiation and development. (The interrelationships of the various factors will be discussed later). Then, from Fig. 14.1(b) one may infer (if one considers only a unit distance parallel to the fold axis) that the buckling force (F) is given by:

$$F = \sigma_b a \qquad (14.1)$$

where a is the thickness of the layer, and σ_b is the buckling stress. The shearing force (F_s) which resists movement of bed A over bed B is given by:

$$F_s = \tau L \qquad (14.2)$$

where L is the arc length of the bed which will determine the wavelength of the fold and τ is the average shear stress acting parallel to the bedding (here we can ignore the sign of the shear stresses, which are opposite on either side of the fold axis). If the layer-parallel, frictional sliding, shear stress is, for the time being, assumed to be the only factor causing resistance to buckling then, in folding:

$$F = F_s$$

therefore,

$$\sigma_b a = \tau L$$

or

$$\frac{L}{a} = \frac{\sigma_b}{\tau}. \qquad (14.3)$$

Clearly, the ratio of the buckling to shear stresses is directly proportional to the slenderness ratio (L/a) of the folded bed. For natural, single-layer buckles the L/a ratio can be determined directly. Unfortunately, when dealing with real multilayer folds, one cannot determine the L/a ratio so readily, for a in this context may not refer to the thickness of a single bed. Several beds may cohere so that bedding-plane slip may occur only at specific, favoured horizons. However, in a multilayer fold, with an initial wavelength of a kilometre or so, it is to be expected that the L/a ratio of individual competent layers will, in general, be at least 100/1.

Because the buckling stress cannot exceed that at which shear failure occurs, it is reasonable to set the upper limit of σ_b at about 1 kbar, (for those conditions in which the confining pressure is small) for at this stress level it is to be expected that the majority of competent rocks would fail in shear. This limit is somewhat arbitrary and, on occasion, it may be apposite to use a somewhat higher limit of 1.5–2.0 kbar. However, using this value, one can use Eq. (14.3) to calculate the relationship between τ and L/a shown in Fig. 14.2. It will be seen that for values of L/a greater than 100/1, τ is equal to, or less than, 10 bar.

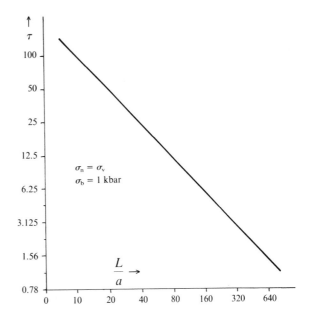

Fig. 14.2. Relationship between τ and L/a when $\sigma_b = 1.0$ kbar.

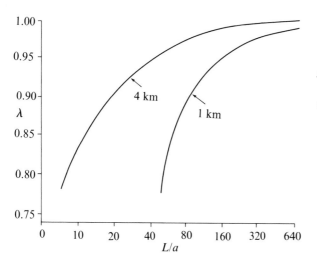

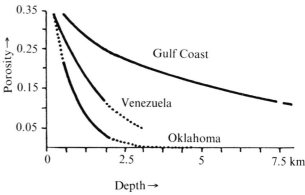

Fig. 14.4. Relationship between porosity and depth of cover in mud-rocks in three areas. (After Hubbert & Rubey, 1959.)

Fig. 14.3. Relationship between λ and L/a for depths of cover of 1 and 4 km at fold initiation for the data and relationship given in Fig. 14.2.

For slip on a non-cohesive boundary:

$$\tau = \mu\sigma_n \qquad (14.4)$$

where μ is the coefficient of sliding friction, which we initially assume to have a representative value of 0.5 (This value for the coefficient of friction will apply if competent units abut one another, without an intervening thin layer of mud or clay-rock.) Consequently, for $L/a = 100/1$, and $\tau = 10$ bar, it follows from Eq. (14.4) that the effective vertical stress ($\sigma_z = \sigma_n$) is 20 bar. If one assumes that this hypothetical fold begins to develop at a depth of say 4 km, and also that the average density of the cover rocks is 2.5 g cm^{-3}, the total vertical stress (S_z) will be 1000 bar. If the total stress is 1000 bar and the effective stress is 20 bar, it follows that a fluid pressure (p) of 980 bar must exist, i.e. $\lambda = p/S_z = 0.98$. (We shall consider later the situation in which clay-rocks, with a coefficient of friction of less than 0.5, are interbedded with the competent units.)

The interrelationship between L/a and the value of λ which must exist at depths of 1 and 4 km before a buckle stress can overcome the layer-parallel shear stress and initiate a buckle is shown in Fig. 14.3. It will be seen that the greater the depth of cover which exists over a layer about to fold, the higher is the value of λ which must obtain. Moreover, if a flexural slip fold is to be initiated under a cover of 4 km, or more, the value of λ must exceed 0.95 even for modest values of L/a.

At this point, we must emphasise once again that, because the effects of all the other factors on buckling are neglected here, the value of buckling stress used to calculate the various values given in Fig. 14.3 is a high one. Consequently, we must conclude that, even though the values of λ indicated in this figure may seem surprisingly high, they are, in fact

conservative, for the postulated model. When one includes the effects of the other factors, the value of λ, for any specific depth and L/a ratio, must be in excess of the value indicated in Fig. 14.3. *We have no doubt that many geologists may instinctively baulk at this conclusion, so it is necessary to adduce supportive arguments and evidence.*

The fundamental and classic companion papers by Hubbert & Rubey (1959) were, as we saw in Chapter 7, mainly concerned with the overthrusting problem. However, they also presented a considerable amount of information regarding fluid pressures and compaction rates in sedimentary sequences. For example, they presented a diagram (Fig. 14.4) which shows the relationships between porosity in mud-rocks and their depths of burial for three localities, namely, the U.S. Gulf Coast, Venezuela and Oklahoma. It is probable that the Oklahoma mud-rocks are fully compacted, or nearly so; hence, the Gulf Coast sediments must be undercompacted, for they have a porosity of about 12 per cent, even at a depth of about 7.5 km. Moreover, these latter sediments often exhibit a value of λ in excess of 0.7, so that they are not only undercompacted, they are also overpressured.

This overpressuring can be attributed to a number of factors (Smith, 1971) which include increase of ambient temperature with depth of burial and dehydration reactions. However, it is thought that a significant proportion of the overpressuring is the result of the rate of sedimentation. It is suggested that the increase in fluid pressure develops when the accumulation of sediments occurs at a faster rate than the interstitial fluids can escape from the sedimentary pile. Consequently, the interstitial fluids are forced to bear a significant proportion of the gravitational load.

The average rate at which sediments have accumulated in the Gulf Coast, according to Hubbert & Rubey, is approximately 100 m/10^6 yrs. Price (1975b) used this information and the curve shown in Fig. 14.4 to calculate the sedimentary strain-rate. He indicated that 5 per cent change in porosity took place as the result of burying the sediments from a depth of 4 km to a depth of 6 km. At a burial-rate of 100 m/10^6 yrs, this change in depth of cover took place in approximately 20 million years and caused 5 per cent vertical strain (e). Hence, the average sedimentary strain-rate ($\dot{e}$), at this depth range, for these sediments is:

$$\dot{e} = e/t = (5 \times 10^{-2})/(2 \times 10^7) \text{ yr}$$
$$= (5 \times 10^{-2})/(6 \times 10^{14}) \text{ s}$$

so that the average sedimentary strain-rate was about 10^{-16}/s. (At shallower or greater depths, the sedimentary strain-rate will be slightly faster and slower respectively.) This is a slow geological strain-rate, nevertheless it is still capable of sustaining a degree of fluid overpressure. Moreover, this overpressure may decay quite slowly. For example, Smith (1971) calculated that a 1 km thick layer of mud-rock, at a depth of 1.3 km, which initially exhibited a value of $\lambda = 0.8$ would have a value of $\lambda = 0.7$ even after a period of 100 million years. This preamble may appear to have little to do with the problem of folding. However, as we shall shortly see, it is extremely pertinent.

A question which we need to answer is, how long does it take for a fold to develop? This seemingly academic question has considerable practical significance. When this question is posed to a group of geology students regarding a specific fold with a wavelength of 50 m, the answers usually range from 5×10^4 to 5×10^6 years. This answer is predicated in part on the tacit assumption that geological processes are slow and also that theoretical analyses appear to indicate that folds in extensive fold trains develop simultaneously and concomitantly. As we have seen from model work (Chap. 10), folds may often develop serially or sequentially. Because hundreds, or even thousands, of folds may occur in a section through a fold belt which formed in a finite time measured in terms of a few millions of years, it follows that an individual fold must have developed in a significantly shorter period.

It was argued by Price (1975b) that fold development can be presented by the simple model shown in Fig. 14.5 in which an originally horizontal unit is caused to shorten into a fold. This conceptual model is

based on observations of 'gravity-glide' and 'horizontal' push experiments, which simulated natural gravity tectonics and the compression resulting from plate collision. It was assumed that the shortening of the model took place at a constant velocity ($\bar{v}$). Price took the original length of bed (L) to be 1000 m which then shortened to form a fold with a wavelength of 500 m (i.e. the fold represents a bulk strain of 50 per cent (5×10^{-1})). He further assumed that this shortening took place at 10 cm/year. The time required for the fold to form would be that taken for reference point A to move 500 m to A' (at 10 cm/year). That is, the fold would form in about 5000 years.

The bulk strain-rate ($\dot{e}$) in the development would be given by:

$$\dot{e} = e/t = (5 \times 10^{-1})/(5000 \times 3 \times 10^7) \text{ s}$$
$$= 3.3 \times 10^{-12}/\text{s}$$

This figure regarding strain-rate is, of course, specific to the size of fold and the average velocity of compression. If the velocity is reduced by a third (to a value somewhat more representative of plate velocities) the strain-rate for the same size fold would be 10^{-13}/s, while a fold an order of magnitude larger (i.e. 5 km wavelength) would form, at the lower velocity, at a strain-rate of 10^{-14}/s.

The important conclusion is that the strain-rates associated with the unstable phase of development of large-scale folds (0.5–5.0 km wavelength) are a hundred to a thousand times faster than that induced by sedimentation. However, it has been noted that even these sedimentary strain-rates are able to contribute to the development of overpressuring in shales. Therefore, provided that the folded sequence contains a significant proportion of low permeability clay-rock, then at strain-rates 100 to 1000 times faster than sedimentary strain-rates, it is reasonable to conclude that these immensely faster tectonic strain-rates will result in the development of fluid overpressure so that $\lambda = 1.0$. It is emphasised that this is a transient effect that, depending upon the size of the fold, need only last for a period of 10^4 to 10^6 yrs. This represents a significant range of times, of course, but is negligible compared with the period of 10^8 yrs noted earlier which was the time for λ to decay from 0.8 to 0.7. *Thus, on purely theoretical grounds, we argue that it is to be expected that high fluid pressures will obtain during folding.*

Let us now turn to some field evidence. High fluid pressure is not only important in the development of buckle folds. The flexural slip mechanism may also operate during the development of drape folds. Indeed, it has been noted (Stearns, 1964) that during the generation of major drape folds in the Rocky Mountains, fluid pressures became so high that the sediments themselves 'fluidise'.

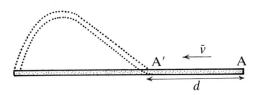

Fig. 14.5. Conceptual model used to estimate the bulk strain-rate associated with the finite development of unstable buckle folds. (After Price, 1975b.)

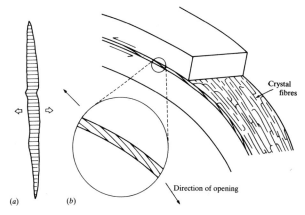

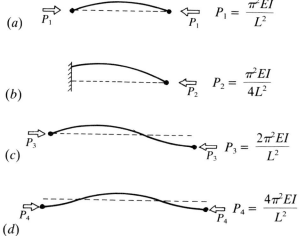

Fig. 14.6. (*a*) Crystal fibre in vein material, indicating direction of opening. (*b*) Diagram representing crystal fibres on a bedding plane indicating opening of the bedding plane as well as differential bedding slip.

Fig. 14.7. (*a*)–(*d*). Four different end conditions considered by Euler and the critical buckling force equations pertaining to each set of end conditions. *E* is the Young's modulus and *I* is a geometrical quantity.

Striations on bedding planes, often called slickensides, have long been recognised as a feature which demonstrates that movement along bedding planes occurs during folding. In Chap. 11, we emphasised that many examples of 'slickensides' are actually layers of crystal fibres (generally quartz or calcite) or their imprints; and are not, in fact, scratch marks formed by beds in physical contact, sliding past each other. It can be seen from the mineral fibres that develop on bedding planes (Fig. 14.6(*a*) and (*b*)) that there is a significant component of movement normal to the beds. Thus, not only do the beds move over each other during folding but they also separate slightly. For such separation, one is forced to the conclusion that $\lambda = p/S_z = 1$ and the cohesion along the bedding plane is lost. It will be recalled that we demonstrated in Chap. 11 the dramatic effect which resistance to slip between layers has upon the rate of fold amplification. Clearly then, it is at this point in the deformational history, when the sedimentary pile contains a significant number of bedding planes with no cohesion and virtually no shear resistance to frictional sliding, that unstable buckling can most readily occur.

Euler buckling equation, wavelength and body-weight
The earliest and perhaps the most well-known buckling equations are those derived by Euler (1757). These equations were designed to enable engineers to forecast the forces at which 'elastic struts' buckled. Euler developed equations which related to four sets of 'end conditions' of the struts (Fig. 14.7(*a*)–(*d*)).

A variety of geological situations can be envisaged (Fig. 14.8), in which one or other of all four different sets of end conditions will obtain. For example, condition P_1 (Fig. 14.7(*a*)) may be satisfied by geological conditions in which a layer, situated in a graben between two near-vertical fractures, is subjected to compression and where frictional constraints on the vertical fractures prevent either end of the bed

from moving (Fig. 14.8(*a*)). If one end of the compressed unit is bounded by a vertical fracture and the other by a lubricated vertical, or an incline fracture, so that slip can occur, (Fig. 14.8(*b*)), the condition represented by Fig. 14.7(*b*) may obtain. The third condition (Fig. 14.7(*c*)) may result if the layer is terminated at one end by a fracture, but this end is frictionally constrained, while the other end is a continuous horizontal layer which is constrained both as to position and direction (Fig. 14.8(*c*)). The end condition which has the more general geological significance, however, is that in which the ends are fixed in position and direction, as indicated in Fig. 14.7(*d*), for these are the end conditions which apply during the development of a continuous fold-train (Fig. 14.8(*d*)).

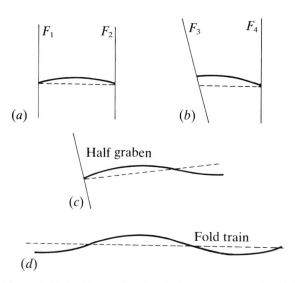

Fig. 14.8. (*a*)–(*d*). Four different geological environments in which the four sets of end conditions represented in Fig. 14.7 may be expected to obtain.

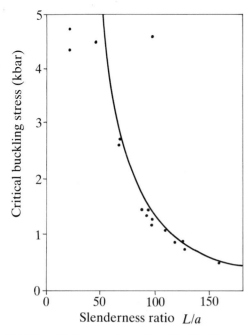

Fig. 14.9. Theoretical curve relating the critical buckling stress and slenderness ratio. The points relate to experimental data. (After Handin *et al.*, 1972.)

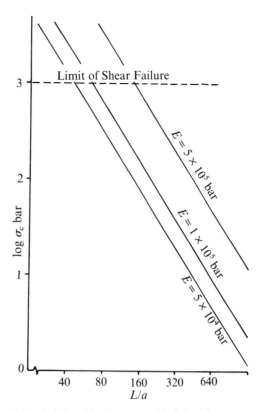

Fig. 14.10. Relationship between critical buckling stress and slenderness ratio on a log-log scale for different values of Young's modulus (E) in bars.

The various Euler equations (the symbols in which are defined below) given in Fig. 14.7 are based purely upon the internal resistance to flexure of the struts. For the end-fixed, direction-fixed condition (Fig. 14.7(*d*)) the solution which is of greatest interest to us here, the critical buckling force (F_{crit}) for a strut length L, is given by:

$$F_{crit} = \frac{4\pi^2 EI}{L^2} \tag{14.5}$$

where E is the Young's modulus of the material and I is a geometrical quantity – the moment of inertia of the strut. The resulting flexure follows a sine curve.

By expressing the buckling equation in terms of I rather than the dimensions of the strut, the equation gains in mathematical elegance, but it results in the obfuscation of a very important relationship. For a strut of rectangular cross-section, $I = a^3 b/12$, where a and b are the thickness and width of the strut, respectively. It follows, therefore, that the critical buckling force is related to the cube of the thickness of the strut. This is a point which has already been discussed, in Chap. 12, with regard to the importance of control units.

By substituting for I in the Euler equations they can be rewritten in terms of the critical buckling stress (σ_{crit}). For example, Eq. (14.5) becomes:

$$\sigma_{crit} = \frac{4\pi^2 E}{12(L/a)^2}. \tag{14.6}$$

Thus, the critical buckling stress is determined only by the Young's modulus and the slenderness ratio (L/a). The graphical expression of Eq. (14.6) is the hyper-

bola shown in Fig. 14.9. Handin *et al.* (1972) conducted a series of experiments to check the validity of this relationship and showed that for large slenderness ratios in excess of 150/1, the experimental data are in excellent agreement with the theoretical curve. Only for relatively low slenderness ratios, when there is considerable doubt about the experimental end conditions, do the data differ from the theoretical predictions.

The Euler equations are, however, more easily used if they are plotted on a log–log basis. As can be seen in Fig. 14.10, Eq. (14.6) then exhibits a straight line relationship with a negative slope. The lines shown in Fig. 14.10 correspond to the $\sigma_{crit}/(L/a)$ ratios for the three different values of Young's modulus.

A limit is placed on the validity of the Euler equations for relatively small slenderness ratios, for then the required critical buckling stress becomes so large that it exceeds the fracturing strength of the material. As already noted, for most strong, water-saturated sedimentary rocks, we set this limit at about 1 kbar.

One of the paradoxical conclusions one would draw from the Euler equations is that the larger folds, i.e. those with a high L/a ratio, form at low buckling stresses. This conclusion is at variance with a commonsense assessment of the situation. Thus, if a large

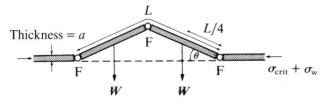

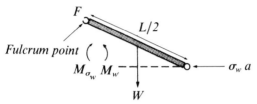

Fig. 14.11. (*a*) Conceptual model used; and (*b*) body-weight analysis.

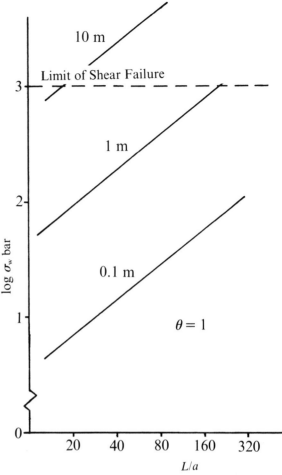

Fig. 14.12. Relationship between stress (σ_w) needed to counteract body-weight of rock units of different thickness (*a*) and L/a ratios, for the analysis indicated in Fig. 14.11 when $\theta = 1.0°$.

sheet of paper is placed on a table top so that about 5 mm of the sheet extends beyond the edge of the table: then by restraining the other edge of the paper, the sheet can be caused to buckle by pushing the paper at the edge which extends beyond the table. The buckle which results will not usually extend throughout the length of the sheet of paper. The fold frequently extends for only about half the length of the sheet. The remainder of the paper remains flat on the table top. Thus, contrary to the prediction of the Euler Theory, long wavelength folds with high slenderness ratios do not develop. This behaviour of the sheet of paper derives from the fact that Euler ignored the effects of body-weight. The agreement between the experimental results and the theoretical predictions at high slenderness ratios (shown in Fig. 14.9) is only possible because the test specimens are small (approximately 20 cm long) and, therefore, their body-weight is insignificant relative to the large compressive stress necessary to induce buckling.

Let us now consider the stresses necessary to overcome the body-weight of a horizontal strut which is caused to buckle. This problem has been addressed by Biot (1965), Ramberg (1970) and Johnson (1977). Here however, we shall present a simpler approach. In order to simplify the problem, we shall take the model represented in Fig. 14.11(*a*), in which we consider the buckle to consist of two straight, rigid limbs, of thickness *a* and unit width, hinged at their ends. The hinges are regarded as having frictional properties, so that they require the 'critical buckling stress' (σ_{crit}) to cause them to move. They therefore accommodate the component of compressive stress due to the Euler buckling concept. To this horizontal compression, one must add a second component σ_w which will

provide the clockwise-acting moment M_{σ_w}, shown in Fig. 14.11(*b*), which will fit the right-hand limb. (There will, of course, be a comparable moment acting on the left-hand limb.) This clockwise-acting moment will be balanced by the anticlockwise moment due to the weight of the limb. From Fig. 14.11(*b*) it can be seen that:

$$\sigma_w = \frac{Ra\rho \cdot \cot \theta}{4} \tag{14.7}$$

where *R* is the ratio L/a and ρ is the weight per unit volume. The stress required to overcome the body-weight effect is clearly related to the angle of dip of the limb. When $\theta = 0$, σ_w is infinitely large. This again emphasises the point made earlier that perfectly flat planes cannot be buckled. However, if some perturbation exists and folding begins, it will be apparent that σ_w decreases very rapidly as θ increases. If we use the simple model shown in Fig. 14.11 and take $\theta = 1.0°$ (cot $\theta = 57.3$) and $\rho = 2.5$ g/cm^3, then:

$$\sigma_w = 36.0\, aR \text{ g/cm}^2.$$

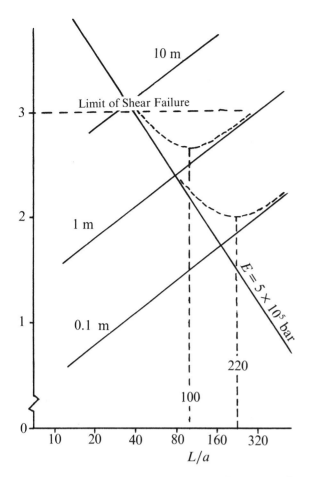

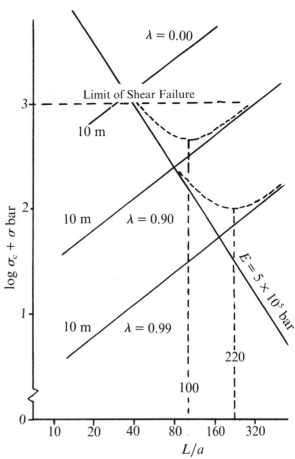

Fig. 14.13. Combined plot of critical buckling stress for $E = 5 \times 10^5$ bar and the body-weight stress (σ_w) for 10.0 m, 1.0 m and 0.1 m thick units. The two curves are obtained by adding the critical buckling and body-weight data. The minima in the curves enables one to define at what combined stress conditions folds of a specific L/a ratio will develop.

Fig. 14.14. Plot of the type shown in Fig. 14.13, but here the unit effective thickness, which takes into account the influence of specific values of λ is taken as 10.0 m.

The relationship between σ_w and L/a, (for remember that $R = L/a$) for various thicknesses of bed (a) are shown in Fig. 14.12.

Because both the critical buckling stress (σ_{crit}, see Eq. (14.6)) and the body-weight stress (σ_w, see Eq. (14.7)) conditions have to be satisfied before folding can take place, we must combine these factors, as shown in Fig. 14.13. The combined stress which satisfies both these factors is required for buckling, and is obtained by adding σ_{crit} and σ_w data. It will be seen that the combined data form a series of curves. If, for the moment, we consider the influence of these two factors only, folding will be initiated when the layer-parallel stress builds up to the minimum value given by the appropriate, combined stress curve. These minima on the various curves, for a given value of bed thickness, indicate that a specific wavelength will be initiated.

Clearly, these conclusions are not compatible with certain field observations, so the analysis needs to be modified. For example, taking the elastic modulus of $E = 5 \times 10^5$ bar, it appears from

Fig. 14.13 that it would not be possible to fold a bed with a thickness of 10.0 m, or more, for it would fail in shear at a lower stress than that required to initiate buckling. Moreover, the L/a ratio of a 1.0 metre thick bed would be fixed at about 100/1. These conclusions are obviously untrue, for from field observations, we know that large folds contain individual sedimentary units which greatly exceed a thickness of 10.0 m and that 1.0 m thick beds are not restricted to folds where they exhibit an L/a ratio of 100/1.

The conclusions, based on Fig. 14.13 are non-sensical, because an important factor has been omitted from the argument; namely fluid pressure. It has already been demonstrated that before flexural slip folds may develop, the fluid pressure must be very high: indeed λ almost certainly exceeds 0.90. The physical significance of a high fluid pressure in this context is that it reduces the effect of the body-weight of beds. If the value of $\lambda = 0.9$ holds everywhere within a sequence, this fluid pressure reduces the *effective* body-weight of a bed by 90 per cent. Thus, a 10 m thick bed would have an effective body-weight equivalent of a 1.0 m bed not subject to a fluid pressure. Similarly, if the value of λ attained a value of

0.99, then, as far as the body-weight is concerned, the 10 m bed would be equivalent to a 10 cm bed of dry rock.

The data shown in Fig. 14.13 can now be replotted in terms of the critical buckling stress and the effective body-weight for a 10 m thick unit for various values of λ (Fig. 14.14). The objections raised earlier are obviated. A 10 m thick unit can be folded, provided the fluid pressure exceeds a value of about $\lambda = 0.9$. Alternatively, the curve for a 1.0 m thick unit (Fig. 14.13) represents the effective body-weight of a 100 m thick unit, if $\lambda = 0.99$.

It will be noted, that at $\lambda = 0.9$, a 10 metre thick 'control' unit would have an initial wavelength of about 1000 m. If $\lambda = 0.99$, the initial wavelength would be about 2200 m. Clearly, the wavelength of a major fold is determined by the value of the elastic modulus, total thickness of the control unit and the precise value of the fluid pressure which obtains when the elastic instability is initiated. The nearer λ approaches the value of unity, the larger is the wavelength of the fold which develops.

The Euler/body-weight analysis presented in this section completely neglects the effect of the layer-parallel shear considered earlier, where, it will be recalled, it was assumed that the average, layer-parallel shear stress was generated by an applied horizontal stress of 1000 bar; i.e. the limiting stress set by the shear failure condition. It will be noted from Fig. 14.14 that the combined critical buckling stress (σ_{crit}) and body-weight stress (σ_w) required to cause flexure of a 10 metre thick bed, when $\lambda = 0.99$, is approximately 100 bar. Clearly then, provided an additional component of compressive stress necessary to overcome resistance to layer-parallel shear is less than 900 bar, folding of the 10 m bed can take place, for then the total compressive stress will be less than 1000 bar, i.e. the total stress is below the limit set by the shear failure condition. For the postulated conditions of E and λ, it transpires that $L/a \approx 220/1$. It follows from Eq. (14.3), that if the compressive stress component is 900 bar, the average layer-parallel shear stress (τ) is 4.09 bar.

It will be recalled that, in order that this shear stress may permit the initiation of flexural slip folding, (τ) must (by Amonton's Law) also be compatible with the effective normal stress (σ_n) acting on the layer, according to the relationship:

$$\tau = \mu\sigma_z.$$

(N.B. The cohesive strength between layers is assumed to be zero.) If, as before, $\mu = 0.5$, then the maximum value of σ_z that can exist, if folding is to occur, is 8.2 bar. Since $\lambda = 0.99$, the corresponding maximum total vertical stress is $S_z = 820$ bar. This is the vertical pressure which would develop at a depth of about 3.5 km.

However, if, instead of competent rock sliding on competent rock, we consider the situation in which the competent units are separated by thin layers of clay-rock, the frictional resistance to flexural slip may be greatly reduced. Thus, some clay-rocks exhibit a coefficient of friction of about 0.27. Then, taking the value of $\tau = 4.09$ bar, the maximum value of effective normal stress on the bedding plane, if folding is to occur, is $\sigma_z = 15.2$ bar, which, if $\lambda = 0.99$, corresponds to a maximum total vertical stress $S_z = 1520$ bar. This is a vertical stress which would develop at a depth of about 6–7 km.

Alternatively, if the unit is 100 m thick and $\lambda = 0.99$, then the minimum on the appropriate curve in Fig. 14.14 is 435 bar, which leaves a potential of 565 bar to overcome frictional sliding on the beds. For these conditions, it will be seen that the L/a ratio is 100/1. From these data it can be inferred that the maximum shear stress that can be generated if folding is to occur is $\tau = 5.65$ bar. If we assume that the competent beds are separated by clay-rock layers, with values of $\mu = 0.27$, then $\sigma_z = 21.1$ bar and folds could occur with overburden stresses up to $S_z = 2110$ bar. Such structures, with an initial (arc-length) wavelength of 10 km could, therefore, be initiated at depths of over 8 km.

Once again, it is emphasised that the limit of 1.0 kbar set on the compressive stress, although realistic, is an arbitrary one. For some very competent rock types (such as dolomites and low porosity quartzites) the limit may more reasonably be set at 1.5 or even 2.0 kbar. If the other elements in the calculations presented earlier remain the same, but the stress limit is set at 1.5 kbar, then such folds could be initiated at depths of more than 10 km.

Clearly, because we are neglecting the layer-normal stress (other than body-weight stress) which resists fold initiation, this mechanism can only be applied to the generation of folds in near-surface environments where the folds may express themselves in surface topography; provided of course that an abnormally high, if transient, fluid pressure of $\lambda > 0.95$ obtains. It may be inferred that such wavelength and depth conditions etc. are compatible with the situation which obtained during the initiation and finite development of the Jura and Zagros Mountains and similar folds belts.

The critical buckling stress, which we have considered at some length in the preceding paragraphs, is that for direction- and position-fixed conditions at both ends of the strut. This is the situation which most frequently applies during the development of a fold-train. If one of the other combination of end conditions shown in Fig. 14.7 obtains, folds with a smaller L/a ratio will be initiated at smaller buckling stresses and lower values of λ than have been cited for the end- and position-fixed conditions considered

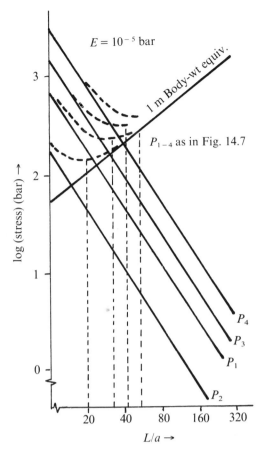

Fig. 14.15. Plot of critical buckling conditions for the four sets of end conditions treated by Euler combined with an 'effective' thickness of 1.0 m and the resulting L/a ratios and combined stresses.

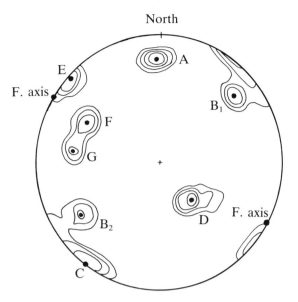

Fig. 14.16. Stereogram showing the maxima for each set of fractures (Max. > 7.0 per cent), 625 poles. (After Ladeira, 1978.)

above. The buckling conditions for all four sets of end conditions are indicated in Fig. 14.15, together with the curve representing the effective body-weight of a '1.0 m' thick unit. It will be seen that the resulting L/a ratios for the different end conditions becomes progressively smaller as the end conditions become progressively more 'free' and unconstrained. It may also be inferred that the value of λ required to permit flexural slip for these more free end conditions will be commensurately smaller.

As with the other theories discussed in earlier chapters, the one presented above relates only to elastic fold initiation. Before we proceed, in the next chapter, to discuss the general finite development of major folds, let us look at the fracture patterns that occur within folds, to see what can be inferred from them regarding the mechanism of finite fold development.

The relationship between folds and fractures

As we have seen, folds are associated with many of the major thrusts systems (Chap. 7). They form an important component of cover rock deformation when

strike-slip or dip-slip fault movement occurs in the basement (Chaps. 6, 8 and 10). Such interrelationships which are noted elsewhere in this book do not concern us here, for we will now restrict our attention to those fractures that are either (i) the same order of magnitude of size as the mesoscopic folds (which can be seen in outcrops such as quarry and cliff sections) or (ii) are considerably smaller than the larger scale folds (that usually can only be defined by mapping, or remote sensing). In particular, we wish to use the relationship of certain of the observed fracture patterns, with respect to folds, to gain insight into the mechanics of fold development. However, before we can do this it is clearly necessary to establish which fractures developed synchronously with folding and in what geological environments such fractures are likely to develop. This task, we suggest, can best be attempted by a process of elimination.

Let us first consider high grade metamorphic rocks of amphibolite or granulite facies. As has already been noted in Chap. 10, such rocks have usually experienced several phases of deformation and exhibit interference patterns that are testimony to repeated ductile deformation. When seen in outcrop, these folds and patterns are cut by systematic, clean, barren, brittle fractures. The ductile deformation occurred when the rocks were at high temperature (perhaps 600–700 °C) whereas the fractures would have developed when the rocks were at a much lower temperature (probably in the range 0–150 °C).

Examples of fracture sets in the high grade metamorphic rocks of the Scourie area of N.W. Scotland, where the late stage, fold axes are approximately horizontal are shown on a stereogram (Fig. 14.16). It will be seen that one fracture set (E) is oriented approximately normal to the fold axis whilst

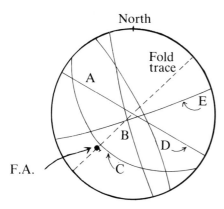

Fig. 14.17. Stereogram of the fracture patterns in the area around Monar Lodge. (After Ramsay, 1958.)

another set (C) is approximately orthogonal to set (E) and so is vertical and roughly parallel with the fold axis. Fracture sets B_1 and B_2 may represent conjugate shear fractures with the orientation of normal faults, which may be related to stress modification during uplift (Chap. 9). The diagram also shows that the rock mass is cut by other sets of fractures (e.g. D F and G), but these cannot readily be related to the final phase of folding.

If the fold axes of the final deformation phase are steeply plunging, the relationship between folds and fractures may be more obscure, as can be inferred from the stereograms of the fracture patterns and the plunge of the fold axes shown in Fig. 14.17. Even the fractures which are most closely related to the fold axis cannot convincingly be interpreted in terms of the stresses related to folding mechanisms. Fracture sets E and B appear to be completely unrelated to this final phase of folding and may be the result of the differential uplift and tilting mechanism described in Chap. 9.

Thus, in the two examples cited above, only a minority of the total number of fractures observed exhibit an orientation that can readily be related to fold geometry. The majority of barren fractures must be attributed to subsequent geological events. Indeed, we suggest that many of the fractures must be related to the phase of uplift and exhumation which these rocks must have experienced before they became exposed at the surface (Chaps. 4 and 9). The orientation of those relatively few fractures which can be related to the final phase of ductile deformation are possibly controlled by fabric elements induced during this phase and/or to residual and remanent stresses. It is reasonable to expect that the differential stress that gives rise to ductile deformation in such high grade metamorphic rocks will be relatively small; with the result that the subsequent differential remanent stresses related to this folding would also tend to be small, so that such stresses may be radically altered both as to magnitude and orientation by subsequent events and the resulting fracture pattern would bear little relationship to the fold structures.

Shear zones and 'ductile fractures' may have developed during any of the phases of deformation. If such structures developed in the earlier phases they themselves would be deformed by the later phases and would not, in any case, be confused with the brittle fractures which quite clearly post-date the final phase of folding.

Layering in these high-grade rocks tends to reflect changes in colour rather than any profound difference in the physical response of the various layers to deformation. At higher levels in the crust, however, in the temperature and pressure environment which gives rise to medium grade (green schist) metamorphism, the disparity between the physical response of layers of different composition (e.g. psammatic or pelitic) to deformation is more marked. The differential stress required for deformation of such rocks in this environment is usually greater than that required in high-grade metamorphic environments. As a result, shear joints are also commonly observed, as well as the extension fractures which may be associated with folds that develop in higher grade deformation.

Multiple phases of deformation are common in these intermediate grades of metamorphism. Fractures may, of course, develop in any of the deformation phases. Such fractures will, however, tend to be ductile or semi-ductile or will be infilled with vein minerals such as quartz or calcite. These fractures and veins which formed in the earlier phases of deformation will be deformed by the later or last phase of folding. The veins are usually easily seen, and the way in which structural geologists tackle the important task of identifying the various phases of deformation from a study of these deformed veins and other structural features is discussed in the final chapter of this book.

The fracture patterns which developed in the Cambrian Marbles about the Estremoz-Vila Vicoca anticlinorium in Portugal have been established by Ladeira (1978) (Fig. 14.18). Not all the sets can be found in each locality. Moreover, the main anticlinal axis is slightly arcuate, so that the trend of individual sets of fractures varies somewhat throughout the structure. It will be seen that fractures of the set marked B in the various stereographic plots are either normal, or make a slight angle, to the anticlinal axis. It is thought that most of the fractures in this set are extension fractures, though they may include hybrid 'shear and extension fractures'. Fractures of Set A are almost certainly 'shear fractures', though they rarely exhibit obvious signs of strike-slip movements. The interpretation of Set C is uncertain: they may be shear fractures in some portions of the structure, but, in general, their trend is usually not much different from that of the fold axis, so that they could be considered as extension features. Alternatively, these fractures

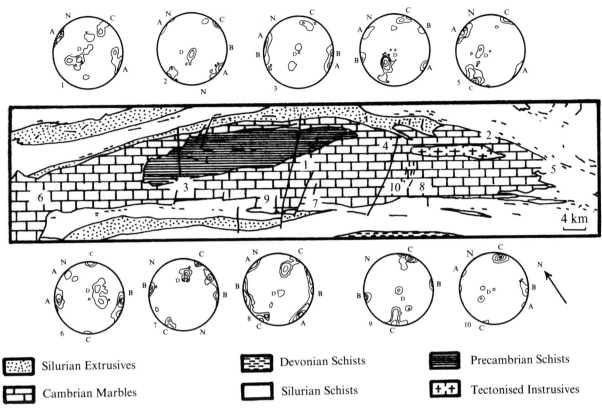

Silurian Extrusives   Devonian Schists   Precambrian Schists

Cambrian Marbles   Silurian Schists   Tectonised Intrusives

Fig. 14.18. Simplified map of the Estremoz-Vila Vicoca anticlinorium (Portugal) with stereograms of the fracture patterns at different localities. (After Ladeira, 1978.)

may have been induced by regional differential uplift, which itself has been influenced by the fold structure. The fractures of Set D are usually sub-horizontal and are exfoliation features. As we have seen in Chap. 4, such exfoliation features develop close to the topographic surface. The fractures of Sets A, B and C are also clear-cut, 'brittle' fractures, so we may reasonably conclude that the majority of fractures which were observed and plotted in the various stereograms of Fig. 14.18 developed during the uplift and exhumation of the marbles. This conclusion will also apply to most other situations in which the rocks have experienced deformation in middle grade metamorphic environments.

By elimination, we conclude, therefore, that for evidence relating to fractures, and how they provide information regarding folding mechanisms, it is best to study folds which have developed at higher levels in the crust, where the rocks exhibit one phase of folding and have undergone little or no metamorphism. However, as we shall see, it is helpful if some of the fractures are infilled by vein material. Hence, if the infilling mineral is quartz, we shall be concerned with an environment in which the rock temperature was probably in excess of 150°C, so that the rocks may have experienced low grade metamorphism.

One of the earliest of the few papers written on this aspect of structural geology, dealing with the interrelationship between fractures and folding mechanisms, was by Hans Cloos (1948). Unfortunately, because this paper is written in German, it has made little impact upon English-speaking (and reading) geologists. Indeed, as far as we aware, reference is given to this paper in only one text-book on Structural Geology, that by De Sitter (1956). (Even this lonely citation was withdrawn from the 2nd Edition).

This paper by Cloos deals with a fold which is exposed near Schuld, in a series of cliffs and hill slopes about an incised meander of the River Ahr, which is a tributary of the Rhine, entering it between Koblenz and Bonn in Germany. An excellent field sketch of this fold and associated fractures is shown in Fig. 14.19. As will be seen from the profile, the fold is markedly asymmetrical. Cloos, somewhat whimsically likens the hinge zones to a human 'knee' and 'ankle' with the straight upper and lower limbs being the 'thigh' and 'foot' respectively, while the steep relatively straight limb is, of course the 'shin'.

As will be seen from the field sketch the fold is cut by sequences of extension fractures, normal faults and thrusts. The stepped quartz veins appear to have been off-set by bedding plane slip, but as we saw in Chap. 3, such appearances can be deceptive. Before we consider how fractures can throw light on the mechanism of folding, let us indicate the various types of orientations of fracture patterns which, from field observations, are commonly seen to develop in meso-

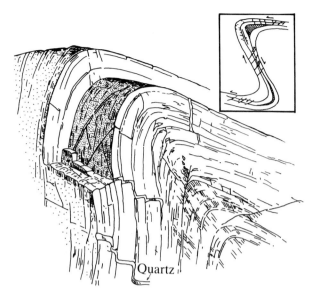

Fig. 14.19. Field sketch showing detail of the Schuld structure (inset). (After Cloos, 1948.)

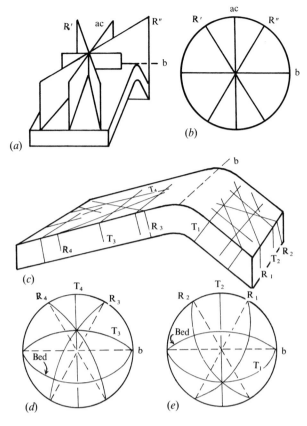

Fig. 14.20. (*a*) Ideal relationship of master joints to a small fold. (*b*) Stereographic plot of fractures shown in (*a*). (*c*) Trends of minor fractures in a folded competent unit. (*d*) and (*e*) Stereographic plots of fractures in the two limbs. R and T are shear and extension fractures respectively. (All after Price, 1966.)

scopic and larger folds that comprise a large number of layers, with various degrees of competence.

Fracture patterns within a fold

As we have seen, fractures can be divided into three groups (Chap. 2) namely: (i) dilation fractures, (ii) shear fractures and (iii) the hybrid, shear-dilation fractures. The third group is gradational between types (i) and (ii) and can provide important information for the structural geologist, regarding the fluid pressure and the differential stresses which obtained when fracturing was induced. However, for the sake of simplicity, we shall initially group (ii) and (iii) together and consider fractures only as *dilational* or *shear* structures.

There are few studies which are specifically designed to establish the spatial relationship between relatively minor fractures and major folds. One such study is that conducted by McQuillan (1973) on the large buckle folds in the Zagros. As noted in Chap. 2, he collected data which were extremely important as regards the relationship between fracture spacing and bed thickness. Unfortunately, in this study, he only noted the orientation of the various *fracture traces* on the bedding. It is not surprising, therefore, that he found it impossible to establish a meaningful relationship between fracture trace orientation and fold geometry.

Price (1966) published a series of diagrams illustrating some of the typical interrelationships between minor fractures and folds. These diagrams (Fig. 14.20) were the result of a compilation of field observations and information culled from the literature. Moreover, they relate fracture patterns and orientations only to cylindrical fold forms. We noted

in Chap. 10 that folds, especially in the upper levels of the crust, are usually markedly non-cylindrical.

As far as we are aware the first and most comprehensive study relating fractures to a pericline, or domal anticline was that conducted by Stearns (1964) in which he established the macrofracture patterns on the Teton Anticline, in northwestern Montana. The present authors and their students have subsequently substantiated the validity of the patterns first established by Stearns (and have been envious of the wonderful exposures he must have enjoyed working with).

The reader must be warned that the various fracture patterns which will be presented in the following sections, although valid, are often only a part of the complex of fractures frequently observed in folded beds, many of which resist all efforts at interpretation. As an illustration of such difficulties encountered in interpretation, the notice of the interested reader is directed to the paper by Phillips (1964) in which he attempts to use the observed fractures to interpret the history of development of the fold structure between Durdle Dor and Lulworth Cove, in Dorset, S. England. He was particularly hampered in this study by the fact that the fold axis and the

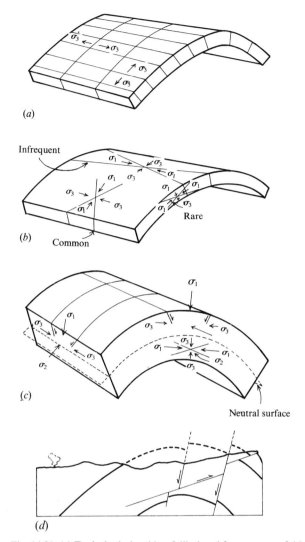

(a)

(b)

(c)

(d)

Fig. 14.21. (a) Typical relationship of dilational fractures to a fold. The orientation of the least principal stress with the sets (of different ages) is also shown. (b) Typical orientation of shear fractures in a thin, bedded layer, with associated stress systems. (c) Typical orientation of normal faults and thrusts which may develop in a thick, flexured unit. (See Fig. 14.22 for relationship between shear fractures and stresses.) (d) Interaction of thrusts and normal faults.

dipping features. The orthogonal set which trends parallel to the axes is also perpendicular to the bedding, so these fractures will vary in dip through an arc which is determined by the tightness of the fold structure. When the crest is well rounded such fractures form a fan. Alternatively, if the fold exhibits a chevron profile, these fractures will form two distinct sets. These relationships will be discussed later.

Such extension fractures are sometimes infilled with calcite or quartz. As they are 'hydraulic fractures', one may indicate that, at the instant of development, the least principal stress acted at right angles to the fracture plane (unless fibre-growth indicates otherwise). From these statements, it can be inferred that (i) in general, the extension fractures perpendicular and parallel to the fold axes did not develop at the same time and (ii) the orientation of the least principal stress (S_3) was mainly parallel to the bedding.

The relationships which shear fractures exhibit relative to the fold geometry are much more complex. The more commonly observed orientations of shear fractures relative to the limbs and axis of a fold are shown in Figs. 14.21(b) and (c), where it will be seen that these shears include *normal, thrust, strike-slip* and *oblique-slip faults*. It will be noted that the orientation and direction of slip on the oblique-slip fractures is determined by the orientation of the bedding. Genetically, they are wrench faults and are called oblique-slip only because it is usual to have the horizontal rather than a bedding plane as one's term of coordinate reference.

As with the dilational fracture patterns, it is possible to infer the orientation and relative magnitudes of the principal stresses which were associated with the initiation of the various fracture systems. For the interpretation of the shear fractures in terms of stress orientation, the Navier–Coulomb criterion of failure is used (Fig. 14.22), where the orientation of

southern limb of the structure had long since disappeared below the waves of the English Channel.

With such warnings borne in mind, we shall now proceed to establish the ideal, but representative fracture patterns which are commonly associated with folds. As has been noted, some of this information has been culled from the literature (here we would wish to express our particular indebtedness to Cloos and Stearns), but mainly the information presented will be related to the authors' own field observations combined with those made by their students.

The relationship between dilational fracture patterns and fold geometry is usually very simple and is indicated in Fig. 14.21(a). It will be seen that the major dilational fractures which occur usually cut the fold axes at 90° and are also perpendicular to the bedding, so that they are usually vertical or steeply

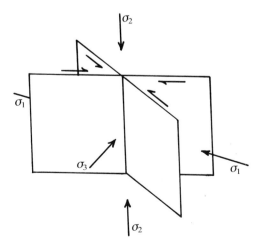

Fig. 14.22. Relationship between conjugate shear fractures and axes of principal stress at fracture initiation, according to the Navier–Coulomb theory of failure.

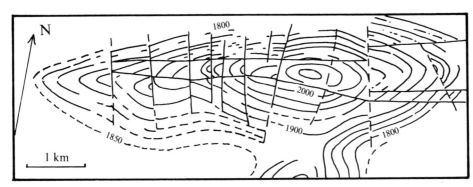

Fig. 14.23. Normal faults trending approximately normal and parallel to the fold axis of the Big Muddy Anticline. (After Wisser, 1960.)

the stress systems which gave rise to the various types of fractures is indicated. As with the dilational fractures, one can conclude that the fractures which develop around a fold are not of the same age and that the orientation of the bedding planes profoundly influenced the orientation of the axes of principal stress. These points will be discussed later.

Normal faults tend to be aligned parallel, or perpendicular, to the fold axis, as indicated in Fig. 14.23. Thrusts, as has been noted, may be well developed in the leading limb of an asymmetrical anticline (Fig. 13.38). They may also occur as accommodation structures within a fold (Fig. 12.25). Less frequently, but more surprisingly, normal faults may be cut and displaced by thrusts and vice versa (Figs. 14.21(*d*) and 14.24). However, the most frequently developed shear fractures are strike or oblique-slip faults, many of which exhibit very little movement and have often been termed 'joints'.

It must be emphasised that not all individual fracture sets shown in Figs. 14.21 are likely to develop in one fold. Moreover, the idealised orientation of the fractures relative to a fold axis shown in these figures and in many texts holds only in the central portions of a pericline, or pod-fold. Field studies indicate that, away from the central part of the flexure, the fracture patterns tend to be related to the slip direction along the bedding, which can be inferred from the flexural-slip model and from the crystal fibres on the bedding planes Fig. 14.24.

Mechanics of development of fracture patterns within a fold

What may at first seem confusing as regards the fracture patterns shown in Figs. 14.21 and 14.23 is that many of the dilational fractures and the various types of shear fractures, i.e. normal, wrench and thrust faults, occur in close proximity to each other and appear to give conflicting information regarding the stress systems which obtain during folding. It should always be borne in mind, however, that the development of even geometrically simple folds is, in fact, a complex process in which rock stresses and fluid pressures are in a continuous state of flux. Even

the behaviour of the rock units may change throughout the evolutionary history of the structure. Some of the stress states leave fossil testimony of their existence by giving rise to sets, or systems, of fractures. Indeed, it is the study of these very fractures which helps one to unravel the mode of development of major structures.

Which of the types of fracture develops depends upon the magnitude of the differential stress and the fluid pressure which obtains in the rock. For the development of dilational fractures, the fluid pressure (*p*) must be high, so that:

$$p > S_3 + T$$

where T is the tensile strength of the rock and S_3 is the least principal total stress. However, it will be recalled that there is a second factor which must be satisfied, namely that the differential stress is relatively low, so that:

$$S_1 - S_3 < 4T.$$

This condition is inherent in the Griffith theory of fracture (Chap. 1). Conversely, for the development of shear fractures, the differential stress must be higher, so that:

$$S_1 - S_3 > 4T$$

and the fluid pressure must be lower than for hydraulic fracturing, so that:

$$p < S_3 + T.$$

With these conditions in mind, let us consider how the principal stresses must be oriented during folding and how the magnitude of the differential stress and the fluid pressure may vary from place to place and from time to time in the fold.

One may infer from the theoretical treatments given in earlier sections of this chapter that the magnitude of the principal stress (σ_1) necessary to initiate a buckle will often be high (near the value necessary to induce shear failure). Not all beds or units will possess the same elastic moduli. Consequently, if a sequence of beds is subjected to a uniform compressive strain (as indicated in Fig. 14.25) then those beds with a high value of elastic

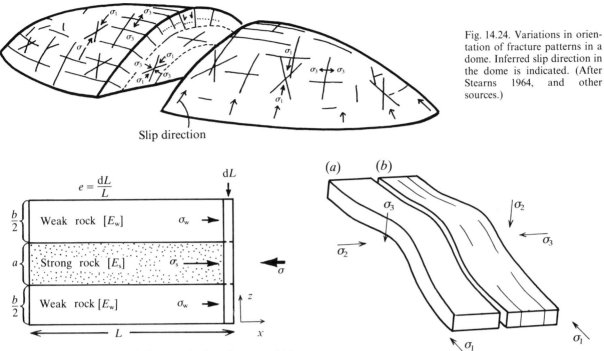

Fig. 14.24. Variations in orientation of fracture patterns in a dome. Inferred slip direction in the dome is indicated. (After Stearns 1964, and other sources.)

Fig. 14.25. Representation of the magnitudes of stress which develop in layers of different Young's moduli, subjected to an identical strain (e).

Fig. 14.26. Initial stress states considered in text, with (a) σ_2 and (b) σ_3 parallel to the fold axis respectively.

modulus (E) will have a higher magnitude of stress (σ_x) than those with a lower value of E. It has been established (Price, 1974) that there is a general empirical relationship between Young's modulus and the strength of a rock. *Strong rocks, of comparable lithology, exhibit high values of elastic modulus.* A three-fold increase in rock strength may be accompanied by a ten-fold increase in value of the elastic modulus. For this reason it is the strong beds (but not, of course, the control units) in which the differential stress first becomes sufficiently high that they fail by fracturing. If the intermediate principal stress (σ_2) is parallel to the incipient fold axis (Fig. 14.26(a)), minor thrusts may develop. A field example in which such structures formed prior to the development of buckling has been discussed in Chap. 3, (Figs. 3.18 to 3.21). If the strong beds are sufficiently thin they may buckle to form minor folds, or where the stress conditions are such that the shear failure criterion and the buckling conditions are simultaneously satisfied, the beds may fold and fracture (as indicated in Fig. 14.27). (It may be noted in passing that for these small folds, body-weight is of little significance and so they may develop when $\lambda \ll 1.0$.) If it is the least principal stress (σ_3) that is aligned parallel to the fold axis (Fig. 14.21(b), then the resultant shear fractures would be strike-slip fractures.

It is emphasised that shear fractures may not necessarily form at this stage in the development of a fold, or if they form, may be restricted to a few of the stronger beds with high elastic moduli. Moreover, as

the fold develops in amplitude, the stress necessary to overcome the body-weight and also to induce internal strains within a bed dramatically decrease in magnitude (Fig. 14.11), so that even if some of the stronger beds do develop fracture patterns, the average stress magnitude, soon after fold initiation, falls to a level such that shear failure is usually no longer possible.

However, we have argued that as compressive strain develops in the sedimentary unit, the pressure of the interstitial fluid increases. If the vertical stress (S_z) is also the least principal stress (S_3) then, as the fluid pressure approaches, or just exceeds, the vertical pressure, the conditions conducive to bedding plane slip and the development of non-stable folding are induced or enhanced. If, however, the vertical stress (S_z) equals the intermediate principal stress (S_2) (Fig. 14.26(b)), then, as the fluid pressure (p)

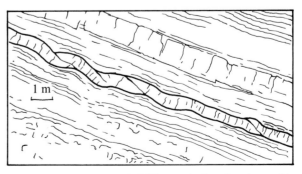

Fig. 14.27. Association of folding and thrusting in a thin, competent sandstone in the Aberystwyth grits.

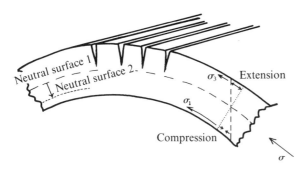

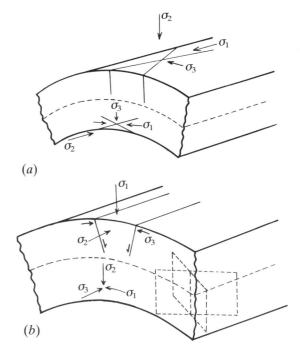

Fig. 14.28. Tangential, longitudinal strain in a thick unit so reduces the stress above the neutral surface that the erstwhile stress σ_1 becomes σ_3 and dilation fractures develop. The fractures reduce the effective thickness of the unit so that the neutral surface (and the fractures) migrate downwards.

approaches S_z (or S_2), clearly:

$$\frac{p}{S_3} = \lambda > 1.0.$$

Moreover, if the difference in magnitude of S_2 and S_3 equals, or exceeds, the tensile strength of the stronger rocks in the sequence, so that:

$$p = S_2 > S_3 + T$$

then hydraulic fracturing will take place with the dilation fractures oriented perpendicular to the fold axis. (See Fig. 14.26(*b*).)

As the amplitude of the non-stable buckle develops with a concomitant increase in the curvature of the beds, then, assuming that the strain distribution in the buckling layer approximates to that of a tangential longitudinal strain fold (Figs. 10.23 and 14.28), considerable reduction of the compressive stress will take place above the 'neutral surface' of a unit, so that locally σ_3 will act parallel to the main compressive direction. Hydraulic fracturing may then occur at the outer fibres of the flexured unit. These fractures will be parallel or sub-parallel with the fold axis and will change the effective thickness of the unit and cause the neutral surface to migrate downwards, as indicated. The extent of these fractures parallel to the fold trend may be considerable. Their extent perpendicular to the bedding, however, is usually much more limited. They are often restricted to a single bed, or portion of a bed.

If the fluid pressure is not quite high enough to cause hydraulic fracturing and if, at the same time, the stress parallel to the fold axis is σ_1, and has a magnitude sufficiently high (so that $(\sigma_1 - \sigma_3) > 4T$), then strike-slip faults will develop, as in Fig. 14.29(*a*). If the stress acting parallel to the fold axis is σ_2 then, as indicated in Fig. 14.29(*b*), normal faults which trend parallel to the fold axis may develop.

It is emphasised that the object of fault development, indeed of any deformation, is to reduce the energy of the system and, more specifically, to reduce

Fig. 14.29. (*a*) and (*b*) Types and orientation of faults which can develop above and below the neutral surface for stress conditions during flexure represented in Fig. 14.26(*a*) and (*b*).

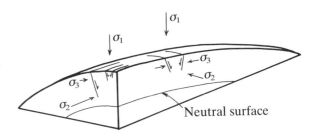

Fig. 14.30. Change of stress magnitude parallel to the fold axis (related to fold axis curvature) and the development of normal faults parallel and perpendicular to that axis.

the differential stress which brings about that deformation. Thus, in the situation represented in Fig. 14.29(*b*), once normal faults are formed, the value of σ_3 will increase in magnitude and could exceed the erstwhile value of σ_2 (Chap. 8).

As has been noted earlier, anticlines are almost invariably periclinal and exhibit 'cross-curvature'. This curvature of the fold axis will, of course, also influence the stress distribution in a thick, competent unit (Fig. 14.30). From this diagram, it will be apparent that if one or more normal faults first develop parallel to the fold axis (with a concomitant local increase in the stress perpendicular to the fold axis) then, as folding progresses, and if the tendency to increased cross-curvature is suppressed, this can result in the stress parallel to the fold axis becoming σ_3. In these conditions, normal faults perpendicular to the fold axis can develop. As a fold amplifies, it extends along its axis, with the result that the cross-curvature

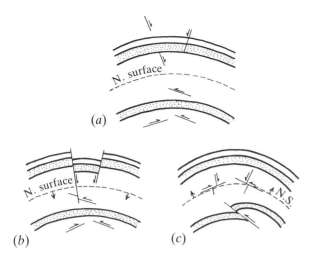

Fig. 14.31. (*a*) Neutral surface situated between two control units, showing 'early' normal faults and thrusts. (*b*) Fracture of upper control unit, with concomitant downward migration of neutral surface, so that normal faults may cut thrusts. (*c*) Fracture of lower control unit with the concomitant upward migration of neutral surface, so that thrusts cut normal faults, (see also Fig. 14.21(*d*)).

is reduced in the central part of the fold. However, the cross-curvature reaches hitherto unaffected areas, and further normal faulting, both parallel and perpendicular to the fold axes, may develop.

The modification of fault development as a result of fold growth occurs not only along the fold axis but also in profile. For example, as may be inferred from Fig. 14.29(*a*), accommodation structures in the form of thrusts may occur below the neutral surface, when the horizontal stress parallel to the fold axis is greater than the vertical stress. Such a situation is also indicated in Fig. 14.31(*a*), but here the neutral surface is determined by two distinct control units. If the upper unit is cut by normal faults, so that it no longer acts as a control (Fig. 14.31(*b*)), the neutral surface migrates down to the lower control unit and the normal faults can also migrate lower and cut thrusts which developed when the neutral surface was in its original position. Similarly, if the resistance of the lower control unit is destroyed by thrusting (Fig. 14.31(*c*)), the neutral surface migrates upwards so that thrusts develop higher in the sedimentary sequence and may cut earlier formed faults, (Fig. 14.21(*d*)).

Commentary

It has been demonstrated that the simple, classical theory of elastic buckling propounded by Euler in the 18th century is capable of being applied to explain the initiation of large-scale buckle folds in the upper levels of the crust, provided the effective body-weight of the rock units is taken into account. It transpires that the wavelength of the elastic buckle is determined by the ratio of the fluid pressure to the total vertical stress,

and that this must often approach unity in order that large-scale flexural slip buckle folds may develop. It follows, therefore, that if a high fluid pressure is assumed to exist, the theories of folding expounded in Chaps. 11 to 13, which ignore the influence of gravity can also be applied to explain the initiation of large folds.

The interrelation of fold geometry and fracture styles and patterns, coupled with the application of simple mechanical principles, provide an important insight into the complex interplay of factors and parameters involved in the finite development of folds. Although some of the aspects relating to finite fold development have been considered in this chapter, it is apposite that a more complete synthesis of the various parameters and factors involved be given special treatment. Accordingly this is presented in the following chapter.

References

Biot, M.A. (1965). *Mechanics of incremental deformation.* New York: Wiley.

Cloos, H. (1948). Gang und Gehwerk einer Falte. *Zeitschr. deutsch. Geol. Ges.*, **100**, 290–303.

De Sitter, L.U. (1956). *Structural geology.* (2nd Impression), New York: McGraw–Hill.

Euler (1757). Sur la force des collonnes. *Mem de l'Acad. de Berlin*, **13**, 252.

Handin, J., Friedman, M., Logan, J.M., Pattison, L.J. & Swolfs, H.S. (1972). Experimental folding of rocks under confining pressure: buckling of single-layer rock beams. *Am. Geophys. Union Monograph*, **16**, 11–28.

Hubbert, M.K. & Rubey, W.W. (1959). Role of fluid pressure in mechanics of overthrusting. *Geol. Soc. Am. Bull.* Pt. i, **70**, 115–66. Pt. II, **70**, 167–205.

Johnson, A.M. (1977). *Styles of folding. Mechanics and mechanisms of natural elastic materials.* Elsevier Scientific Publ. Co.

Ladeira, F.L. (1978). Relationship of fractures to other geological structures in various crustal environments. Unpublished Ph.D. Thesis, University of London.

McQuillan, H. (1973). Small-scale fracture density in the Asmari Formation of S.W. Iran and its relation to bed thickness and structural setting. *Am. Assoc. Petrol. Geol. Bull.*, **57**, 12, 2367–85.

Phillips, J.W. (1964). The structures in the Jurassic and Cretaceous rocks of the Dorset coast between White Nothe and Mupe Bay. *Proc. Geol. Assoc.*, **75**, 373–407.

Price, N.J. (1966). *Fault and Joint development in brittle and semi-brittle rocks.* Oxford: Pergamon.

(1974). The development of stress systems and fracture patterns in undeformed sediments. In *Proc. 3rd Int. Conf. Soc. Rock Mech.*, **1A**, pp. 487–98. Denver, Colo.

(1975a). Fluids in the crust of the Earth. *Science Progress*, **62**, 59–87.

(1975b). Rates of deformation. *J. Geol. Soc. London*, **131**, 553–75.

Ramberg, H. (1970). Folding of laterally compressed multi-layers in the field of gravity. *J. Phys. Earth Planet. Interiors*, **2**, 203–32.

Ramsay, J.G. (1958). Superimposed folding at Loch Monar, Inverness-shire and Ross-shire. *Q. J. Geol. Soc. London*, **113**, 271–308.

Smith, J.E. (1971). The dynamics of shale compaction and evolution of pore fluid pressures. *Math. Geol.*, **3**, 239.

Stearns, D.W. (1964). Macrofracture patterns on Teton Anticline N.W. Montana (abstract) (Eos). *Trans. A.G.U.*, **45**, 107.

Wisser, E.H. (1960). Relation of ore deposits to doming in the North American cordillera. *Geol. Soc. Am. Mem.*, **77**.

15 'The life and times of a buckle fold'

Introduction

In this chapter we shall briefly review the various events which occur during the formation of folds, from initiation through to finite fold development. To this end, we shall make use of the conclusions and concepts outlined in previous chapters (Chaps. 10–14) and attempt to present a unified philosophy regarding the way in which folds develop. Particular emphasis will be given to the 'life and times' of folds which form in the upper levels of the crust, where the rheological contrasts between the layers tend to be large, for it is in situations such as these that the mechanisms involved and the features that develop are most diverse. As we have noted, insight into the process of fold initiation and development has been derived from a number of sources, which include theoretical and model studies and field observations. An important evaluation of the process of folding using field evidence, by Hans Cloos (1948), has been mentioned in Chap. 14. Indeed, the title of this chapter is an extremely loose translation of Cloos' paper entitled 'Gang und Gehwerk einer Falte'. However, the treatment given here far exceeds in scope that outlined by Cloos.

This chapter inevitably contains some degree of repetition of ideas and conclusions presented in earlier chapters. However, we feel that this is justified and hope that the reader will find useful the general and possibly, to some extent, contentious discussion presented here.

From the treatments of the mechanisms involved in folding, given in the previous five chapters, we can list the various factors which should be taken into account when analysing the buckling behaviour of any rock containing a layering or fabric. These include:

(i) the rheological properties of the deforming material,
(ii) the mechanical anisotropy of the rock mass,
(iii) the stress field acting on the rock, including
(iv) the body-weight stresses,
(v) the influence of any inhomogeneity, and
(vi) the boundary conditions.

Even if all this information were available it would be unrealistic to develop a theory of buckling for each and every possible layered system encountered in nature. However, in specific situations, the buckling

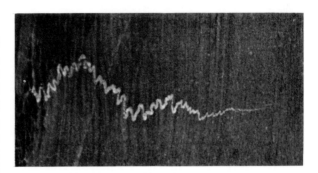

Fig. 15.1. Buckled quartz veins in a pelitic matrix: an example of single layer buckling on two scales.

behaviour of a system can often be accounted for by using one, or a combination, of the theories of buckling of an interface, a single layer, a multilayer and an anisotropic material outlined in the preceding chapters. For example, the buckling of a quartz vein in a pelitic matrix (Fig. 15.1) can be adequately explained using the theories of single layer buckling. Similarly, the microfolding of a pervasive mineral fabric, such as a slaty cleavage, can be accounted for satisfactorily using the theories of the buckling of anisotropic materials.

In the theories of folding discussed in Chaps. 10 to 13, the effect of gravity was ignored. Consequently, these theories ought only to be applied to small-scale folds where body-weight is small compared with the stresses required to overcome the internal resistance of the layer(s) to buckling. However, it was demonstrated in Chap. 14 that the effects of body-weight can be largely, or even completely, nullified by high fluid pressures. In general terms, therefore, by assuming that such fluid pressures exist during folding, the various theories can be used quantitatively to explain the initiation of large structures, even though they ignore the effect of gravity. Field evidence such as crystal fibres on bedding planes is sometimes available to demonstrate that high fluid pressures and low interlayer cohesion existed during folding. However, we would not wish the reader to have the impression that fluids are important only in those instances where fibres are present. The passage of fluids through fractures and along bedding planes in rock most frequently leaves little or no evidence of that passage.

In a complex sedimentary sequence whose mechanical properties can be represented by an average mechanical anisotropy but which contains a

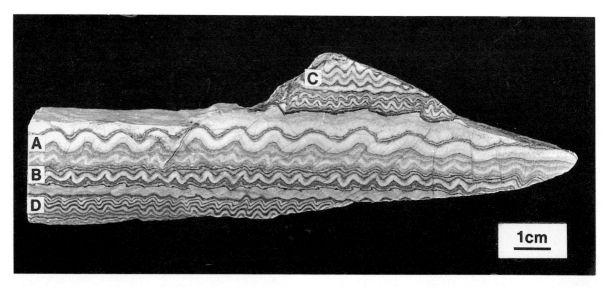

Fig. 15.2. Harmonic and disharmonic folding in a specimen of the Castile and Todillo evaporites, New Mexico, U.S.A. (cf. layers A and B). Some of the folding (A and B) can be described by the theory of single layer buckling, some (C) by the theory of multilayer buckling and some (D) by the theory of buckling of an anisotropic material.

large variety of different layers, each with its own thickness and rheology, features associated with the buckling of single layer, multilayer and anisotropic materials can often be found. A beautiful example of the juxtaposition of these three types of behaviour within a hand specimen of a buckled multilayer is shown in Fig. 15.2.

In this chapter, we shall discuss the deformation behaviour and the various structures that can form in a complex multilayer when it is subjected to layer-parallel compression. The deformation history can be conveniently divided into four stages:

(i)   pre-buckle shortening
(ii)   fold initiation
(iii)   the finite development of folds, and
(iv)   post-buckle flattening.

Associated with these stages are four distinct stress levels each of which may result in the formation of various minor structures. These four stages, together with their stress states and minor structures, will be discussed in turn.

Pre-buckle shortening

When layers of rock or rock fabrics are compressed parallel or sub-parallel to the layering or fabric, they generally experience a period of homogeneous shortening before some less uniform and less pervasive mechanism of shortening, such as buckling or faulting, is initiated. The way in which this pre-buckle shortening takes place depends upon the types of rock involved, the environmental conditions (pressure, temperature, fluid pressures etc.) and the compressive strain-rate. Let us consider what might happen when a rock multilayer, made up of many beds, each with its own thickness and rheology, is subjected to layer-parallel compression.

It was argued in Chap. 11 that the initiation of buckles in the upper levels of the crust is governed primarily by the elastic properties of the rocks. However, if the compressive strain-rate is slow (say less than 10^{-14}/s) the elastic element of the deformation will be small because the associated stresses will be dissipated by the process of relaxation (Chap. 11). Even during the early stages of compression the straining of the rock may be dominated by ductile deformation. In such a situation, buckling may never be initiated and the formation of any folds which might develop will be as a result of the passive amplification of geometrical irregularities in the layering which was, initially, only approximately planar (Chap. 10). The amount of pre-buckle strain that can develop is governed only by the length of time for which the slow strain-rate operates. For example, if a rock mass is subjected to a strain-rate of 10^{-14}/s for 10^5 years (i.e. 3×10^{12} s), the resulting strain will be 3 per cent; while if the period is extended to 10^6 years, the strain would become 30 per cent. In high and intermediate grade metamorphic environments, such strain will tend to develop pervasively throughout the layers and may result in the formation of a uniform mineral fabric (e.g. slaty cleavage or schistosity). However, in low-grade environments, the strains may not be developed evenly. In limestones, for example, the shortening may be accommodated by pressure solution at localised and sometimes periodically spaced sites, which results in the development of stylolites (Fig. 15.3). The reader is referred to Guzzetta (1984) for an interesting discussion on the development of stylolites.

The transition from homogeneous pre-buckle

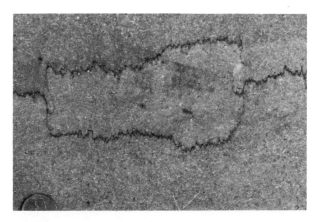

(a)

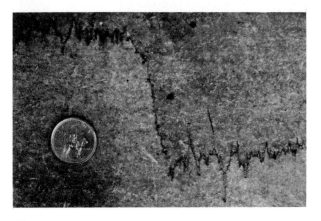

(b)

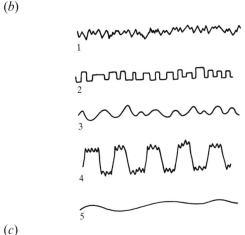

(c)

Fig. 15.3. (a) and (b) Stylolites in Niagara limestone, polished slabs, Grand Union station, Toronto, Canada. (c) Classification of stylolite profiles (after Guzzeta, 1984). Types: (1) sharp-peak, (2) rectangular, (3) wave-like, (4) composite and (5) smooth.

flattening to multilayer folding is unlikely to be abrupt. It will, more probably, be characterised by the progressive localisation of deformation within the minor (i.e. relatively thin) competent beds resulting in a variety of structures, such as minor folds and faults which are usually of secondary importance to the multilayer buckles which eventually develop.

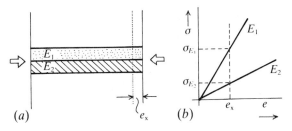

Fig. 15.4. (a) Two linear elastic layers with Young's moduli E_1 and E_2 respectively, compressed parallel to the layering. The same strain (e_x) is indicated in both layers, but the stress level in the two layers is different (see b). The more competent layer (E_1) supports a larger stress than the less competent layer (E_2), i.e. $\sigma_{E_1} > \sigma_{E_2}$.

Fold initiation

As mentioned earlier (in Chap. 11), some irregularity or perturbation is necessary for fold initiation, regardless of whether buckling occurs as an elastic or viscous instability. Such a perturbation from which a fold starts can be of several types. Commonly, *geometric irregularities* such as a local layer thickening or thinning act as fold initiators. Geometric irregularities may be an intrinsic property of the layered system or may be induced by the local failure (e.g. faulting or folding) of a minor competent unit. Alternatively, local variations in the rheological properties of a layer, or a local reduction in the cohesive strength along a layer interface, can cause an area of heterogeneous stress to develop in an otherwise homogeneous stress field. Such *stress irregularities* may be sufficient to cause fold initiation.

Let us now consider how folding might be initiated in a sedimentary multilayer under layer-parallel compression. It is likely that single layer buckling will be started in isolated, relatively thin, but extremely competent layers, before the multilayer as a whole begins to buckle. It is therefore appropriate to consider first the initiation of single layer folds, when the compressive stress within the multilayer as a whole has not yet attained the critical buckling stress of the multilayer, or any single control unit it may contain.

The initiation of single layer folding

During the initial period of homogeneous, pre-buckle flattening, relatively large differential stresses begin to develop in the competent layers if the compressive strain-rate is fast enough. The reason for this is discussed in detail in Chaps. 14 and 17 and the principle can be easily demonstrated by considering two linear, elastic layers with Young's moduli E_1 and E_2 compressed parallel to the layering (Fig. 15.4(a)). If the two layers are shortened by the same amount (e_x), then it follows from the graph in Fig. 15.4(b) that the stress in the two layers will be different. The stronger layer (i.e. the one with the larger Young's modulus E_1) will support a larger stress. The response of a competent layer to the increase in compressive stress will

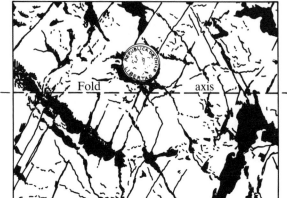

Fig. 15.5. Shear joints in Jurassic sediments of the Sierra de San Moreno, N. Chile.

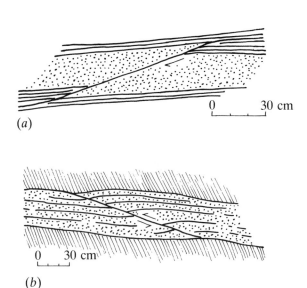

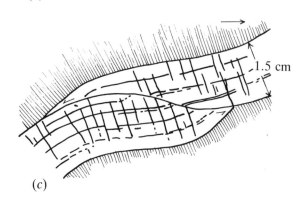

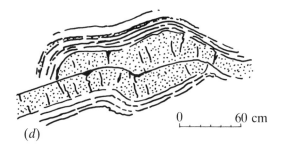

Fig. 15.6. Examples of pre-buckle thrusts developed in relatively competent layers (*a*) in a sandstone layer, Val de Fir, S.W. Geneva, Switzerland (*b*) and (*c*) in a sandstone in the Bloomsburg formation, West Virginia, U.S.A. (*d*) Folded thrust in a sandstone layer. (After Cloos, 1961.)

depend upon the relative magnitude of the compressive strength of the layer and its critical buckling stress, and on the proximity and thickness of other relatively competent layers. Let us first consider the structures that form when *the compressive brittle strength is less than the critical buckling stress*. The differential stress in the competent beds is relatively high and consequently shear, rather than tensile failure, will probably occur. The orientation of the shear failure planes will depend upon the orientation of the principal stresses. We shall consider two possible orientations, (i) when the intermediate principal stress (σ_2) acts normal to the layering and (ii) when the minimum principal stress (σ_3) is normal to the layering.

Shear failure when σ_2 is normal to the layering

When σ_2 acts normal to the layering, the resulting fractures will have the orientation of strike-slip faults. However, because of interlayer traction, large displacements on such fractures will be inhibited and consequently, a large number of miniature strike-slip faults or 'shear joints' will form (Fig. 15.5). The orientation of the maximum principal stress is, of course, likely to be approximately normal to the direction which will eventually be the trend of the multilayer fold axis.

One cannot be absolutely sure that the fractures illustrated in Fig. 15.5 were actually formed prior to folding. However, because fractures which develop before, or early in the history of, a fold become paths of fluid migration (and so may subsequently be infilled with vein material) we suggest that the fractures shown in this figure are probably early features. We interpret these fractures as conjugate shear sets formed at a relatively low fluid pressure. (The acute angle between the shear sets is greater than 45° and we can conclude from the theory of brittle failure (Chap. 1) that there was a compressive normal stress across the fracture during its formation which would

Fig. 15.7. Examples of conjugate thrusts in a 75 cm thick sandstone bed from Widemouth Bay, Cornwall, England.

tend to inhibit it from opening and forming a vein.) The opening and infilling of vein material happened later when the fluid pressure was significantly higher. As we have seen, high fluid pressures are to be expected when large buckle folds develop.

Shear failure when σ_3 is normal to the layering

Should the axis of least principal stress (σ_3) act perpendicular to the competent layer, then one or more thrusts will develop (Fig. 15.6). Movement on the thrust plane may cause the wedges of competent

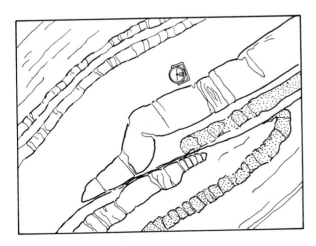

Fig. 15.8. Buckling of the tip of a thrust wedge, Northcott Mouth, Cornwall, England.

material to be driven into the adjacent, less competent layers Fig. 15.6(*a*) or alternatively the thrust tips may be driven along the bedding plane between the competent and incompetent beds, Fig. 15.6(*b*) and (*c*). The latter alternative requires the thrust planes to be bent. Occasionally, both the thrust planes of a conjugate set develop at a locality, Fig. 15.7. The formation of thrusts either singly or as a conjugate pair, produces a local thickening of the layer, Figs. 15.6(*c*) and 15.7. If the brittle strength of the competent layers is considerably lower than their critical buckling stress, then there will be no tendency for the layers to buckle except, perhaps at the thrust tips where the layer-parallel compressive stresses may become locally very high, Fig. 15.8. However, if the brittle strength and critical buckling stress of the layer are of approximately the same magnitude, then both thrusting and folding will occur, Fig. 15.6(*d*) and 15.9(*b*) and (*c*).

The association of thrusting and folding can be very varied and a complete spectrum of structures, from undeflected thrusts (Fig. 15.9(*a*)) through regularly spaced thrusts with some flexing of the thrust blocks (Fig. 15.9(*b*)) to asymmetrical folds with thrusts through their short limbs (Fig. 15.9(*c*)), exists in nature.

Another association between thrusts and folds has already been mentioned. It was noted that the formation of a single thrust in a relatively thick competent unit may lead to a considerable local

(a)

(b)

(c)

Fig. 15.9. Three examples of thrusts in sandstone layers; (a) Thrusts with no element of buckling. Northcott Mouth, Cornwall, England. (b) Thrusts with incipient buckling. Combe Martin, Devon, England. (c) Thrusts and associated, well developed buckles, Monks Cave, south of Aberystwyth, Wales.

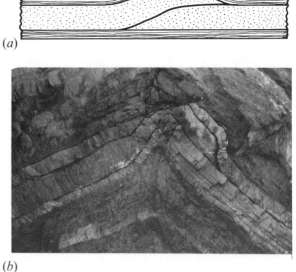

(a)

(b)

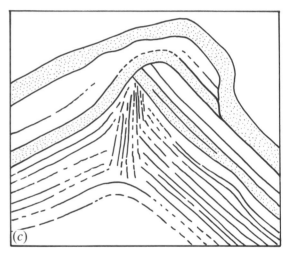

(c)

Fig. 15.10. (a) A pre-folding thrust in a competent sandstone unit. This acted as a perturbation from which fold (b) was initiated. (c) Line drawing of (b).

thickening of the layer (Fig. 15.10(a)). This local irregularity in layer thickness may become a perturbation from which a major multilayer buckle subsequently grows. An example where this has occurred is shown in Fig. 15.10(b), where the thrust is too extensive to have been formed as an accommodation structure (Chap. 12). Such thrusts do not always act as sites of initiation of major folds and the pre-buckle thrust shown in Fig. 15.11 is situated on the *limb* of a major fold. Displacement on this thrust has caused a minor fold to develop in the overlying beds, Fig. 15.11(b). The localisation and initiation of folds by faults has also been described by Laubscher (1977), who has convincingly demonstrated that several of the large-scale folds in the Jura were generated from relatively small irregularities in the decollement horizon, caused by movement on underlying reverse faults or thrusts (Fig. 15.12; see also Chap. 10).

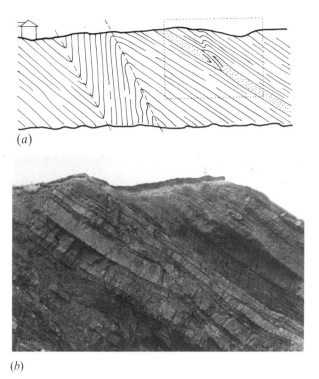

(a)

(b)

Fig. 15.11. (a) Pre-buckle thrust in the limb of a fold. (b) Detail of (a) (dotted rectangle) showing the thrust and associated minor folding. Northcott Mouth, Cornwall, England.

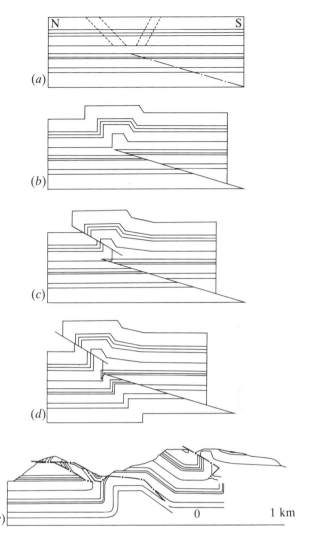

Fig. 15.12. (a)–(d) stages in the development of the Goumois-S anticline (e) which is exposed in the Doubs gorge near Goumois in the Jura Mountains. The folding was initiated by movement on an underlying thrust. (After Laubscher, 1977.)

Structures that form when the compressive brittle strength and critical buckling stress of a competent layer are approximately equal

In the examples (Figs. 15.6 to 15.12) of the association of shear failure and folding, the shear failure generally pre-dates the folding. However, examples where folds and thrusts develop synchronously (see the discussion on accommodation structures, Chap. 12, Figs. 12.23 to 12.34) and where folding pre-dates and initiates thrusts (see example of bedding plane thrusts initiated from conjugate folds in the model of Fig. 13.39) are probably more common (e.g. Fig. 15.9(c)). A model made up of alternating layers of hard (black) and medium (grey) Plasticine in which shear failure and multilayer buckling were seen to occur together is shown in Fig. 15.13. It is probable that flexures often slightly predate fractures (even in this 'synchronous' mode) because the compressive fibre stresses which develop below the neutral surface will increase the likelihood of shear failure (Fig. 10.26(b)).

Structures that form when the compressive failure strength of a competent layer is greater than the critical buckling stress

When the compressive failure strength of a competent layer is greater than its critical buckling stress, folds may be initiated at suitable irregularities. The spacing of competent layers within a multilayer is critical in determining whether harmonic or disharmonic folding of these layers will occur. It was shown in Chap. 12 (e.g. Fig. 12.6(a)), that if no other competent layer of comparable, or greater, thickness lies closer to the competent layer of interest than approximately 0.25L (where L is the fold wavelength), the buckling will be governed by the theory of single layer buckling. Such symmetrical, single layer buckles in a relatively isolated, competent layer may eventually become asymmetrical parasitic folds on the limbs of larger folds that develop when the multilayer as a whole begins to buckle (Fig. 12.36(b)). In situations where the competent layer is relatively thick and other thinner, competent layers fall within its effective zone of contact strain, then, as the thick layer, which acts as a local control unit, buckles, the thinner layers buckle harmonically with it.

Thus, as the multilayer is compressed, the strongest, relatively thin layers (i.e. too thin to act as control units) fracture and/or form small-scale folds and experience a stress drop. Their role of resisting deformation passes to somewhat weaker, but thicker,

Fig. 15.13. Deformed Plasticine multilayer in which shear failure and multilayer buckling occurred synchronously.

competent units which may either become control units or experience a similar mode of failure (fracturing or small scale-folding) and stress drop. This process may occur several times until the 'identity' of the eventual control unit, or group of units, is established. As deformation proceeds beyond this point, the multilayer as a whole begins to buckle. The type of buckling that occurs depends principally on the presence or absence of important control units. If they are present, the buckling of the multilayer will be governed by the buckling of the control unit and the problem is essentially one of single layer buckling in an anisotropic matrix. If a control unit is not present, then the buckling will be governed by the bulk properties of the multilayer and its buckling behaviour will best be described by the theories of multilayer buckling discussed in Chaps. 12 and 13.

While this intricate process of layer-parallel compression of a complex multilayer (i.e. one containing layers of markedly different thicknesses and/or competence) is going on, the incompetent rocks which, in the environment under consideration, are often impermeable clays or mud rocks, are experiencing compaction and a commensurate build-up in fluid pressures. This deformation may remain wholly within the elastic range but, more frequently, will include permanent strain.

The amount of pre-buckle, elastic or elastic-plastic deformation which may take place will largely be controlled by the degree of induration and cementation experienced by the rocks prior to layer-parallel compression, for this will largely determine their porosity, and hence the amount of strain that can be induced by compaction. In soft sediment deformation, this pre-buckle strain could approach 50 per cent.

The stress history in the competent units will depend upon their rheological properties. If they behave as perfect elastic-plastic materials, there will be an initial, relatively rapid increase in the lateral compressive stress, followed by a period in which the stress remains constant. In general, however, it is likely that these rocks will exhibit a degree of strain hardening, so that during the plastic phase of their deformation, the stresses will continue to rise, though at a slower rate. Consequently, although a few of the more competent units may fail and so experience a stress drop, the overall level of lateral compressive stress in the sedimentary sequence increases progressively. When the compressive stress in the multilayer reaches some critical level, the multilayer as a whole begins to buckle.

The initiation of multilayer folding

Let us assume that no control unit is present in the multilayer and that buckling will occur when the compressive stress reaches the critical buckling stress of the multilayer as a whole. Buckling will be governed by the theories of multilayer buckling of the type described in Chaps. 12 and 13 and this will lead either to the formation of upright folds with axial planes at right angles to the layering or, more commonly, to conjugate kink-bands (box folds), the axial planes of which are oblique to the directions of the layering and principal compression.

Experimental work on multilayers (Chaps. 10 and 13) throws considerable light on the problem of fold initiation. It demonstrates convincingly that when models are subjected to compression from one end, they do not immediately experience a uniform stress throughout the multilayer. Such a state is only achieved after a slow-moving, elastic 'wavefront' has traversed the layers. It was observed in these experiments, that fold initiation sometimes occurs at the wavefront as it passes through the model and that the serial development of folds could occur in this way (Fig. 10.40, folds 6, 7 and 8). Alternatively, the wavefront may move through the model without starting folds, so that extensive lengths of the system become subjected to an essentially uniform stress condition. Two different types of fold initiation were observed in these pre-stressed regions. The first was the random initiation of fold, *sequentially* at various points within the multilayer, presumably at various perturbations (Fig. 10.43(*a*)). The second was the initiation of a train of elastic buckles which initially amplified *synchronously* and at the same rate (Fig. 10.40 folds 1 to 5)..

It will be apparent from this brief discussion, that folds formed at a wavefront as it sweeps through the sequence will be 'older' than those formed after the multilayer has been totally pre-stressed.

From these experiments, we see that multilayer folds may be localised and occur at random sites throughout the layered sequence, or they may be developed uniformly throughout part, or all, of the layering as a wave-train. Let us consider briefly why some natural multilayer folds are locally and sequentially initiated.

One of the effects of pre-buckle flattening is to compact the sediments and drive interstitial fluids out of the beds and along bedding planes. As compression continues, the fluid pressure along the bedding planes increases until it is sufficient to overcome the weight of the overlying sediments and the cohesive strength of the bedding plane. Hydraulic fracturing may then take place locally along the bedding at sites of relatively low cohesion (provided that σ_3 is normal to the bedding) and it is at these places, where the interlayer cohesion is reduced to zero and where the effect of the body-weight of the overburden is exactly countered by the fluid pressure, that the folds are most likely to be initiated. This process would lead to the random, sequential initiation of folds within the multilayer.

It should be remembered that the development of high fluid pressure along bedding planes is not only associated with the diagenetic or tectonic compaction of relatively unindurated, water-laden sediments. They may also be generated by the dehydration reactions that accompany prograde metamorphism. This becomes particularly important at depth but, as we saw in Chap. 3, may also occur at depths of only a few hundreds of metres, when triggered by contact metamorphism. Even without this effect, high fluid pressure may be generated by the gypsum/anhydrite reaction which may be induced at a depth of less than 1.5 km.

The finite development of folds

After folds have been initiated, either by the buckling of a control unit or the buckling of the multilayer as a whole, they rapidly amplify into finite structures. During this amplification, a variety of second order structures may develop and the geometry of the fold profile may change dramatically.

Let us first consider the amplification of folds in a relatively competent control unit within a multilayer and then the amplification of multilayer folds when such control units are absent.

The amplification of multilayer folds with a control unit

We have argued that most buckle folds developed in the middle and higher levels of the crust begin as elastic flexures. It has been noted in Chap. 8, that the fibre stresses (σ_f) induced in a layer as a result of buckling are:

$$\sigma_f = \frac{Ea}{2R} \qquad (15.1)$$

where E and a are the Young's modulus and thickness of the layer respectively and R is the radius of curvature of the fold hinge. If the compressive and tensile strengths of the rock are known, it is possible to determine the amplitude at which the fibre stresses exceed the strength (i.e. the elastic limit) of the layer. This is the amplitude at which the buckling ceases to be totally elastic and at which permanent deformation begins.

It was shown by Price (1967) that the maximum elastic deflection that can occur before the onset of permanent deformation, depends upon the wavelength/thickness (L/a) ratio. In his estimates, he used a value of $E = 7 \times 10^4$ bar and for an L/a ratio of 15/1, a total horizontal stress of 2.5 kbar. The maximum, elastic deflections he obtained for L/a ratios of 20/1, 15/1 and 10/1 were $0.75a$, $0.5a$ and $0.375a$ respectively. However, these estimates are based on high, initial horizontal stress and a relatively low value of Young's modulus. In addition, it was assumed, for mathematical convenience, that the fold followed arcs of circles rather than a sine curve. Consequently, the estimates of the amplitude at which elastic buckling breaks down, are probably a factor of 2–5 too high. We shall assume, therefore, that the maximum, elastic deflection that can be experienced by a buckling layer with an L/a ratio of 20/1 is $0.2a$.

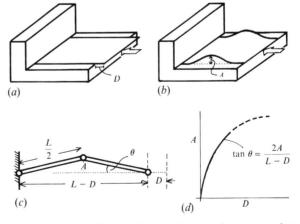

Fig. 15.14. (*a*) and (*b*) A simple experiment to demonstrate that during the early stages of folding a small displacement (*D*) produces a much larger fold amplitude (*A*). (*d*) The relationship between displacement (*D*) and fold amplitude (*A*) for the straight-limbed fold (*c*).

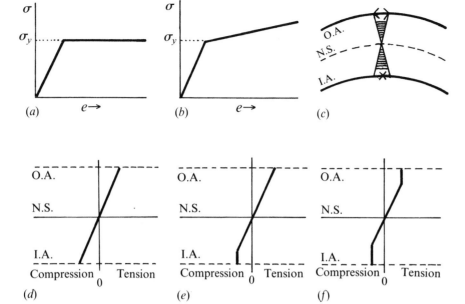

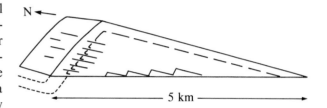

Fig. 15.15. (*a*) Stress–strain curve for an elastic-plastic material. The behaviour changes from elastic to plastic at the yield stress σ_y. (*b*) An elastic-plastic material displaying strain hardening. (*c*) Linear variation in layer-parallel, normal stress (fibre stress) across the fold hinge. (*d*) Normal stress distribution across a fold hinge when no plastic yielding has occurred, (*e*) when yielding has only occurred at the inner arc of the fold and (*f*) when yielding has occurred at both the inner and outer arc. *O.A.* and *I.A.* are the inner and outer arcs of the fold hinge respectively and *N.S.* = Neutral Surface.

The displacement D, which gives rise to an elastic buckle (amplitude 0.2*a*), is, in fact, a very small fraction of the amplitude. The reader may demonstrate this by buckling a sheet of paper in the manner indicated in Fig. 15.14(*a*) and (*b*). The layer is deflected in a sine curve, however, for small deflections, the difference in length between the sine curve and a straight line joining adjacent hinges, will be negligibly small. Hence, as a close approximation, we can take the ratio of amplitude/displacement (A/D) from the geometry of the 'straight limbed' fold indicated in Fig. 15.14(*c*) and its associated graph, Fig. 15.14(*d*).

It was noted that the elastic limit of a control unit with a L/a ratio of 20/1 would be reached for a deflection of as little as 0.2*a*. From the geometry of Fig. 15.14, it can be seen that, at this point in the fold amplification, the A/D ratio would be about 50/1. For example, if a was taken to be 20 m, L, the wavelength, would be 400 m and the displacement required to generate a buckle that would reach the elastic limit is only 8 cm. It can, therefore, be concluded that this phase of elastic behaviour will require a relatively brief period in relation to the finite, non-elastic development of the fold.

Once the elastic limit has been reached, individual buckling layers may fail in the hinge region by the development of tensile fractures above the neutral surface, possibly in conjunction with thrusts below (Fig. 10.26(*a*) and (*b*)). Alternatively, the layers may fail by ductile flow. If it is assumed that there is a linear variation in the layer-parallel normal stress across the layer and that the layer behaves as an elastic-plastic material, then the stress/strain behaviour and the possible stress conditions that can exist at the fold hinge are as shown in Fig. 15.15. If the units behave as ideal elastic-plastic materials (Fig. 15.15(*a*)), the zone of plastic deformation will

Fig. 15.16. A block diagram representing the type of major structure formed in the Aberystwyth Grits. Smaller, disharmonic, asymmetric chevron folds form beneath the main structural canopy, which has resulted from the buckling of a sequence of thick sandstone beds. (After Price, 1962.)

remain very localised and the fold will develop an angular hinge. However, if the material experiences strain hardening during the 'plastic' deformation (Fig. 15.15(*b*)), the zone of permanent deformation will spread and the folds will tend to become more rounded.

A complex sedimentary sequence may contain a number of control units each of which will have its associated zone of contact strain (Chap. 12). Although this zone extends, in theory, an infinite distance from the buckling layer, the displacements induced in the matrix by the buckling layer have already fallen to about 1 per cent of their maximum layer at a distance of one wavelength from the layer (Chap. 12). We noted earlier in this chapter that from field observations, (e.g. Fig. 12.6(*a*)) we can infer that other competent layers are not significantly affected by the zone of contact strain unless they are closer to the first control unit than 20–25 per cent of the wavelength of the first control unit. For example, a control unit with a wavelength of 4 km would have an effective zone of contact strain of approximately 1 km above and below it. Within this zone, thinner com-

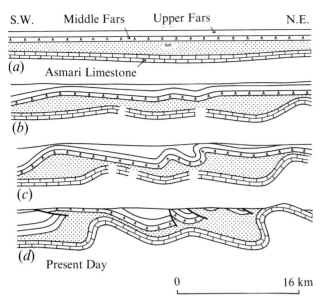

S.W. Middle Fars Upper Fars N.E.

(a) Asmari Limestone

(b)

(c)

(d) Present Day

0   16 km

Fig. 15.17. The evolution of folds in the Haft Kel area of the Zagros mountains, Iran. The competent Asmari limestone and middle Fars beds are separated by a thick salt layer (lower Fars, stage 1) and buckle disharmonically. (After O'Brien, 1957.)

petent beds will fold harmonically with the control unit. Beyond this limit, the thinner beds can form folds with a shorter wavelength. Such a disposition of structures has been reported from the Aberystwyth grits (Fig. 15.16). These turbidites contain a sequence of relatively thick sandstone beds, which act as a control 'unit', overlying more distal (i.e. thinner) units. Large wavelength folds form in the thick sandstone control unit, and smaller folds in the distal units that fall outside its zone of contact strain. A well known example of two control unit (the middle Fars beds and the Asmari limestone) separated by incompetent material (salt), which buckle disharmonically is shown in Fig. 15.17.

The amplification of folds in multilayers which contain no control unit

The finite development of folds in multilayers which do not contain control units has been extensively studied experimentally by means of a variety of models (Chaps. 10 and 12). It was argued in Chaps. 12 and 13 that the deformation behaviour of a multilayer is governed by its bulk mechanical anisotropy, which is determined by the contrast in rheological properties of the layers and by the amount of interlayer cohesion. When cohesion is maintained along the layer boundaries, the bulk intrinsic anisotropy is determined only by the contrast in rheological properties between the various layers. However, in multilayers with low or zero interlayer cohesion, the effect of the layer properties is often only of secondary importance, and the intrinisic anisotropy is governed predominantly by the interlayer properties.

In all the multilayer experiments described in

Chaps. 10 and 12, lubricant was placed between the layers and consequently interlayer cohesion was low. As the mechanical behaviour of these multilayers was governed by the interlayer properties, it is not surprising that structures with the same geometry, i.e. that of kink-bands or box folds, both of which have axial planes oblique to the layering and are characterised by straight limbs, were formed in all the experiments regardless of the type of layer material (e.g. gelatine, wax, Plasticine, rubber and cardboard). Although high amplitude, upright symmetrical folds were seldom formed, it was apparent in several experiments that kink-bands and box folds are often initiated from low amplitude, sinusoidal buckles (Fig. 13.37).

This phenomenon has been discussed in Chap. 13, where it was argued that the deflection of the layering by the low amplitude buckles causes a local change in the resistance to deformation (i.e. a modulus inversion) which enables kink-bands to form. The regular spacing of the deflections of the layering (i.e. the limb regions of the low amplitude folds) may result in a regular spacing of kink-bands, and the alternating sense of layer rotation of adjacent deflections ensures that adjacent kink-bands will be conjugate to each other. In this way, a train of low amplitude, sinusoidal buckles can grow into a series of box folds (Fig. 13.37). The stage in the amplification of the low amplitude folds when the kink-bands are initiated from the limbs is probably related to the stage when cohesion between the layers is totally lost, and a significant amount of slip between the layers begins to occur. Honea & Johnson (1976) have argued (Chap. 13) that it is this local, markedly non-linear behaviour that is responsible for the formation of kink-bands. Evidence of local loss of cohesion on bedding planes during the folding of rocks is provided by crystal fibres (so-called slickensides) which are to be found on the limbs of many natural folds (Figs. 10.24 and 10.25).

The interference of multilayer folds

We have seen from both field observations and from some of the experiments discussed in Chaps. 10 and 13, that folds within a multilayer are frequently initiated in isolation. As they amplify, they may initiate folds in the immediately adjacent layers by a similar process to that operating to produce the fold train in the single layer shown in Fig. 11.42. These new folds will, in turn, initiate other folds and in this way the folding spreads into the surrounding undeflected layering to form a packet, set or train of folds. As two sets of folds spread they may interfere in the manner illustrated in Fig. 15.18, which shows four stages in the experimental deformation of a multilayer (Watkinson, 1976). The multilayer, which consists of lubricated layers of equal thickness and competence, is set in a less competent matrix and compressed at 10° to

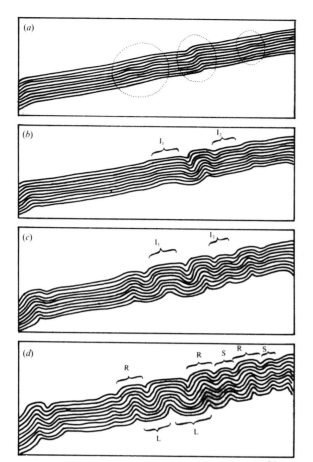

Fig. 15.18. Stages in the deformation of a multilayer unit, (*a*) three areas of initiating folds. (*b*) Two interference zones (I₁ and I₂) become apparent. (*c*) Amplification of the initiating folds and the converging of cusps in the interference zone I₁. (*d*) Convergence to a single cusp point in I₁. R = regular, S = small and L = large span folds. (After Watkinson, 1976.)

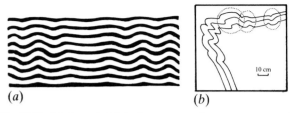

Fig. 15.19. (*a*) A central interference zone between two sets of regular folds in a modelling clay multilayer. (*b*) Line drawing of a fold from Bovisands, S. Devon showing the discrete nature of the parasitic folds in the upper limb (small rings) and the box fold geometry of a possible interference zone (large ring). (After Watkinson, 1976.)

propagation from the end boundary folds, creating new zones of interference with the centre folds'. Other examples of interference zones in a multilayer and a phyllite are shown in Fig. 15.19.

Stress variations accompanying multilayer folding

It will be recalled from the discussion of fold development in anisotropic materials (Chap. 13) that various distinct stages in the history of fold growth in such materials can be recognised. These include a period of strain softening, during which there is a rapid increase in the rate of foliation rotation followed by a period of strain hardening, during which the rate of foliation rotation decreases dramatically and the fold locks up. These stages reflect the various stages of 'rotational' instability which can develop in anisotropic rocks, and have been represented diagrammatically in Fig. 15.20.

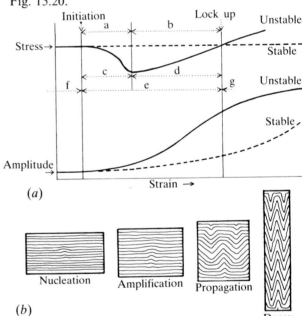

Fig. 15.20. (*a*) Stress versus strain and amplitude versus strain for stable (dashed lines) and unstable (solid lines) deformation. Letters a–g represent periods of a) strain softening b) strain hardening c) increase in rate of fold amplification d) decreasing rate of fold amplification e) rotational instability f) pre-buckle flattening g) post-buckle flattening of rotational stability. (*b*) Stages of fold development. (Slightly modified from Cobbold, 1976.)

the layering. Three folds were initiated in isolation (Fig. 15.18(*a*)). They were not in phase and so, during continued deformation, two interference zones between the folds became apparent, (I₁ and I₂ Fig. 15.18(*b*)). Watkinson points out that ' ... the first interference zone, I₁, is a "dead" zone which develops to a double cusp/box fold geometry (Fig. 15.18(*c*)). In the other interference zone a cusp/chevron/concentric fold is developed between the two dominant folds but it has a significantly smaller span than the dominant folds. With further deformation the initial folds, which now have almost identical span lengths, continue to amplify (Fig. 15.18(*c*)). The double cusp structure is however unstable and starts to converge to form a single cusp point creating two anomolously large span folds (*L*) on either side of the now single cusp point. Note that the convergence of this double cusp structure occurs at finite fold amplitudes, showing that in this particular model some fold spans adjust even at high fold amplitudes and limb dips (see also Hara *et al.*, 1975, p. 129). Concomitantly with the development of these interference patterns, new folds appear by lateral

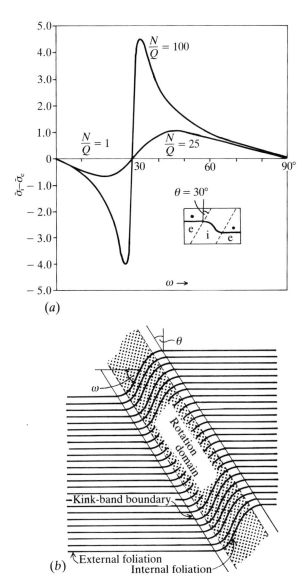

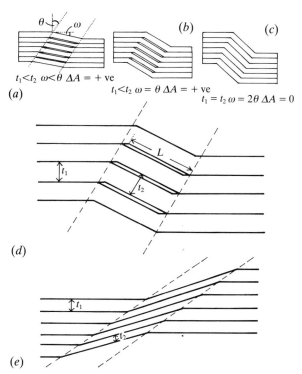

Fig. 15.22. (*a*)–(*c*) Variation in volume in a reverse kink-band as the layering rotates during kink-band amplification in a card-deck multilayer. t_1 is the layer thickness outside the layer; t_2 is the thickness of the space available to accommodate the layer in the kink bands. (*d*) Detail of (*b*). (*e*) A normal kink-band. The layering inside the kink-band is always thinner than that outside and must continue thinning as the structure amplifies.

Fig. 15.21. (*a*) Variation in mean stress ($\bar{\sigma}_i$) within a kink-band (*b*) with ω, the orientation of the internal foliation. The curves are for two anisotropic materials, $N/Q = 5$ and 100. (After Summers, 1979.)

It can be demonstrated (Summers, 1979) that the magnitude of stress in the rotational domain (a fold limb or kink-band), like foliation rotation rate, also varies as the fold amplifies. This can be conveniently shown graphically (Fig. 15.21) by plotting the difference in mean stress between the internal and external foliation ($\bar{\sigma}_i - \bar{\sigma}_e$) against the orientation of the internal foliation ω, using the relationship:

$$\bar{\sigma}_i - \bar{\sigma}_e = \tau_{e(max)} \left[\frac{\left(\dfrac{N}{Q} - 1\right)(\cos 2\theta - \cos 2(2\varpi - \theta)}{\left(\dfrac{N}{Q} + 1\right) - \left(\dfrac{N}{Q} - 1\right)\cos 4(\theta - \varpi)} \right] \quad (15.2)$$

where $\tau_{e(max)}$ is the maximum shear stress in the external foliation, N/Q is a measure of the anisotropy

of the material and θ and ω are as defined in Fig. 15.21(*b*). The relationship (Eq. (15.2)) is plotted for two anisotropic materials, one for which $N/Q = 5$ and the other for which $N/Q = 100$. In both materials the kink-band is oriented at 60° to the external foliation (i.e. $\theta = 30°$). Inspection of this graph shows that, during the early stages of kink-band amplification, a stress gradient exists which tends to drive fluids out of the surrounding rock and into the kink-band. With continued amplification, however, this gradient, and consequently the direction of fluid migration, is reversed.

In deriving Eq. (15.2) it was assumed that no volume changes occurred within the kink-band during its amplification. Let us now consider a simple geometric model in which volume changes are allowed to occur during folding (Fig. 15.22). The material shown in this figure has the laminar properties of a 'card-deck' and it is assumed that the laminae within the kink-band (i.e. the rotation domain) rotate as rigid plates which are effectively hinged at the domain boundaries. The kinked laminae retain both constant length (L) and constant orthogonal thickness during rotation and the kink-band boundaries remain fixed.

It is a simple matter to demonstrate that, under the conditions noted above, the kinked laminae must separate during initial rotation (Fig. 15.22(*a*)) up to the domain normal position ($\omega = \theta$) when the dilation within the rotation domain is maximum (Fig. 15.22(*b*)). With continued rotation, the laminae will close and return to a position of contact when the angle of rotation $\omega = 2\theta$ (Fig. 15.22(*c*)). Further rotation requires a reduction in the thickness of the laminae. This initial expansion and subsequent contraction in the volume of the rotation domain can be related to the contrast in internal and external mean stress which develops in the constant volume model (Fig. 15.21).

Summers demonstrates that the local expansion and subsequent contraction in the width of the kink-band (Fig. 15.22) would enhance the rate of foliation rotation and therefore that of the overall instability. However, once the laminae have returned to a position of contact (Fig. 15.22(*c*)) continued volumetric strain within the rotation domain requires a finite decrease in the volume of the rotation domain, i.e. a thinning of the layers. In the card-deck model, this can be achieved only by a process that involves the indirect removal of material from this zone by pressure solution or by diffusion. In rocks, such processes are likely to occur at rates which are insignificant in comparison with the rates of volumetric expansion and contraction achieved during the initial phase of delamination. The structures will therefore become effectively locked up when $\omega = 2\theta$ (Fig. 15.22(*c*)) as has been suggested on geometric grounds by Donath (1968).

During the formation of a fold or kink-band in an anisotropic rock one therefore expects an initial reduction in mean stress within the fold limb (Fig. 15.21), and/or an increase in volume (Fig. 15.22). This will cause a reduction in fluid pressure within the kink-band, setting up a hydraulic gradient with the fluids external to the kink-band. As a result, fluids migrate from the surrounding rock mass into the kink-zone. Eventually, as the layering or foliation within the kink-band continues to rotate, the volume of the kink-band begins to decrease. Because of the preceding phase of fluid 'immigration', the band, at this stage, contains relatively large quantities of fluids. Hence, as the volume of the band decreases with continued deformation, high fluid pressures will develop which can induce new fractures to form within the kink-band and enforce the 'emigration' of these fluids from the kink-band.

We have discussed the migration of fluids into and out of a kink-band or fold during its amplification. However, stress gradients, and therefore migration of fluids, will also occur *within* the structure. During the early stages of amplification, fluids will tend to migrate from the edge towards the centre of a kink-band. This tendency will be maintained until the foliation in the central element(s) (Fig. 13.24) has rotated so that $\omega = \theta$ (Fig. 15.22(*b*)). Further amplification of the kink-band results in a volume reduction of the central elements and the migration trend is reversed.

When the layering within the kink-band has rotated to the stage shown in Fig. 15.22(*c*), the volume within the kink-band is reduced to the original value; the excess fluids have been expelled and further rotation requires a volume decrease. The loss of fluids makes slip between the layers progressively more difficult and this, combined with the geometric constraints on further rotation, causes the fold to 'lock up'.

From these arguments, we may infer that fluid pressures change dramatically from time to time and place to place within a fold, during its formation. Moreover, from field evidence, we can conclude that these changes in fluid pressure are not necessarily achieved smoothly and continuously. Thus, when crystal fibres are seen on bedding surfaces, they are often found to occur as composite layers, each layer representing a separate episode of bedding plane slip. The direction of slip for each layer can be inferred from the orientation of the crystal fibres; and it is commonly found that the slip directions are far from constant from layer to layer (Fig. 15.23). It seems probable, therefore, that fluid flow is intermittent and stochastic and that quantities of fluid are transferred rapidly from one part of a fold to another. (In this respect, these fluids are comparable with the flow of magma which gives rise to the emplacement of dykes and sills.) Alternatively, fluids may be ejected from the fold along appropriately oriented fractures into unfolded regions of the multilayer, where they may cause hydraulic fracturing along bedding planes and the initiation of new folds. In the intervening periods, the rock stresses may change their orientations, so that later slip directions are different from the earlier ones. Because fluid pressures play such an important role in the finite development of major folds, it follows that these folds also develop in an intermittent stick-slip manner. Thus, the fold acts as a 'pump' which, if the external energy is sufficient, continues to extrude fluids from itself until it is almost dry. Alternatively, if the external energy is limited and becomes balanced by the internal energy of the fold, the structure may cease to act as a pump and, indeed, cease to develop further, long before it becomes dry.

It was argued earlier in this chapter, in the section on fold initiation, that, during the compression of a sedimentary multilayer, hydraulic fracturing along the bedding planes at sites of low cohesion may cause the *initiation* of folds. We now see that local high fluid pressures are equally important during fold *amplification*.

Stress variation and fracture formation accompanying the amplification of multilayer folds

Let us now consider the stress states within an amplifying kink-band with regard to fractures that may be

Fig. 15.23. Several layers of crystal fibres on a folded bedding plane in Wenlock shales, Pant Glas quarry, Llangollen, N. Wales. Each layer records a pulse of bedding plane slip and the direction of slip is indicated by the fibres.

generated during folding and examine the fractures around some natural folds. We have seen how the variation of mean stress within a kink-band during its amplification (Fig. 15.21) may cause the influx and expulsion of fluids and how the level of compressive stress may vary during the formation of a fold (Fig. 15.20). A similar variation in stress during fold amplification to that shown in Fig. 15.20 was proposed by Price & Hancock (1972) who argued that before a sedimentary multilayer is subjected to a layer-parallel, tectonic compression, it is reasonable to expect that the total vertical stress (S_V) will be greater than the horizontal (S_H) (Chap. 3) and that a fluid pressure (p) exists which cannot be much greater than S_H (or else hydraulic fracturing and fluid drainage would be in progress). These conditions are indicated in Phase 0 of Fig. 15.24. With the onset of layer-parallel tectonic compression, the total lateral stress will increase, as shown in Phase I, until it exceeds the total vertical stress. Because the strain-rates involved in the process are generally 2–3 orders of magnitude higher than those associated with sedimentation, the fluid pressure (p) will increase, perhaps until it almost equals the total vertical pressure (S_V). During this phase of pre-buckle shortening, the over-

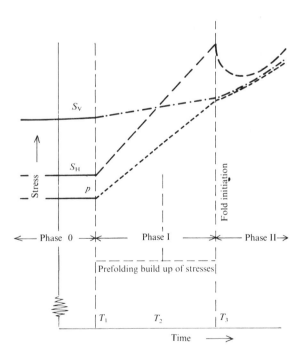

Fig. 15.24. Variation in the magnitude of stress with time throughout the development of a fold. S_V and S_H are the vertical and horizontal stresses respectively and p is the fluid pressure. See text for discussion. (After Price & Hancock, 1972.)

burden and the total vertical pressure may increase slowly because of the vertical strain induced.

When the horizontal pressure reaches a critical value (which, in part, is determined by how closely p approximates to S_V) elastic buckling of any control unit(s) will be initiated (Phase II, Fig. 15.24). We have seen that individual, thin but very strong beds may fold or fracture before this point is reached. The stress curves represented in this figure are, therefore, smoothed generalised representations of the stress conditions within any control unit(s) or, if none exist, within the multilayer as a whole.

Once Phase II has been entered and the buckling strains exceed the elastic limit, the stress required to overcome bending and, consequently, the resistance to buckling, decreases. What happens to the stress levels within the competent units during finite buckling, however, depends upon the external (i.e. boundary) conditions. If the deformation is 'catastrophic' as, for instance, would be the result of the stress-shock wave associated with a meteoritic impact, then the internal stress level would remain essentially constant. Accelerated deformation would take place within the fold with a commensurate increase in internal strain rates. However, in the majority of geological situations, folds result either from horizontal movements or from gravity tectonics. Although gravity-glide may occasionally give rise to catastrophic failure, both these conditions of deformation will generally be associated with a regional constant velocity of rock transport.

If the rock mass yields in a ductile manner, such a constant velocity of transport may be reflected locally in a constant strain-rate though, of course, this may vary from locality to locality. For example, the ductile strain-rate in a fault zone is likely to vary systematically from zero to a maximum and back to zero in a traverse at right angles to the fault trace. If a portion of the fault becomes locked, then the stresses may build up until seismic faulting takes place at very high strain-rates. Such an event is followed by a stress drop.

If a fold belt is generated by the closure of plates at a near-constant, relatively low velocity of 1–5 cm/yr then, although the bulk strain-rates are also likely to change from place to place throughout the area undergoing deformation, the overall slow rate at which plate displacement occurs, prevents the catastrophic yielding of a single fold. Thus the stress levels in the vicinity of a fold would drop. In a real situation, the boundary conditions will be neither those of constant strain-rate nor of constant stress, but some intermediate condition. Thus, it follows that, because the resistance to buckling of the rock units decreases once a fold has been initiated, there is a resulting decrease in the magnitude of S_1 which is the total horizontal stress (Fig. 15.24). The vertical stress

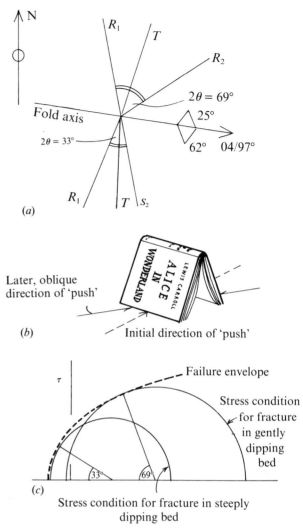

Fig. 15.25. (*a*) Conjugate shear fractures S_1 and S_2 on the limb of an asymmetrical anticline in the turbidites south of Bude, N. Cornwall. The interfracture angle 2θ is different on the two limbs and the fractures are not symmetrical with respect to the fold axis. Tension fractures T also occur on both limbs. (*b*) Experiment using a book to illustrate that once a fold hinge is established, it may be maintained even if the direction of compression changes. (*c*) Stress condition during fracture formation on the two limbs of the anticline shown in (*a*).

(which is now S_2 or S_3) is likely to increase, as the fold either causes changes in the surface topography or encounters progressively greater resistance to continued amplification by some rigid, horizontal boundary. Consequently, after a period in which S_1 decreases in magnitude, it begins to increase once more, so that the reduced magnitude of the differential stress ($S_1 - S_3$) may be maintained, or even augmented, and finite fold development may continue. Eventually, as the fold locks up, the value of S_1 will reach, or even exceed, the value which it had attained at fold initiation (Fig. 15.24). At this point in the deformation history, rather than causing this fold to amplify further, the high compressive stress will give rise to the initiation and generation of folds in other localities in the multilayer. This variation in stress associated with folding is further discussed in Chap. 17, where it is argued that the stress-drop during the formation of one fold may facilitate the formation of a fracture cleavage in an adjacent fold.

The stress-drop associated with the onset of finite fold development will be most prominent in the folding of strong rocks in near-surface environments, where they are able to sustain high differential stresses prior to failure. At deeper levels in the crust, at relatively high temperatures, where diffusion-controlled deformation mechanisms are important, the differential stresses which the rocks can sustain before they fail are smaller, so that the magnitude of the stress-drop is correspondingly less.

Using the concept of stress variation during the growth of a fold (Fig. 15.24) and the variation in fluid pressure within a kink-band, which was discussed earlier in the chapter, we can analyse the fractures on the limbs of the asymmetrical fold shown in Fig. 15.25(a). This fold is representative of many of the folds in the turbidites exposed on the coast south of Bude, North Cornwall. The shear fractures R_1 and R_2 on the two limbs of the anticline are not symmetrical with respect to the fold axis, nor are the interfracture angles 2θ the same for the two limbs. From this we may infer (i) that these fractures develop in different physical conditions and (ii) that the maximum principal compression need not act perpendicular to the fold axis. With reference to this latter point, once the position of the fold hinge is established then, because the limbs will exhibit considerable rigidity, further folding will be controlled by the orientation of the fold hinge and limbs even if the compressive stress generating the fold changes its orientation. This can be illustrated using a book (Fig. 15.25(b)) which can be 'folded' by pushing slightly oblique to the 'fold axis'.

The difference in angles between the shear fractures R_1 and R_2 on either limb can be attributed to the differences in the physical conditions of fluid pressure and differential stress. On the more gently dipping limb, conjugate shear fractures form with an acute angle greater than 45° ($2\theta = 69°$), while on the steeply dipping limb, conjugate hybrid shear/extension fractures have developed with an acute angle of less than 45°. We may now infer that the fractures which developed on the steep limb formed during the folding while the beds were rotating and water was being 'pumped' out of the limb. We conclude that, because of the initial stress drop associated with finite fold development (Fig. 15.24), the differential stress was not sufficient to generate shear fractures and the fluid pressure was insufficient to induce hydraulic fractures, with the result that hybrid shear/extension fractures were formed (Chap. 1). The sets of shear fractures which formed in the more gently dipping limb formed later, when the differential stress became

sufficiently large. (It will be noted that these fractures do not exist in the steeply dipping limb, which means that they could not have developed prior to finite fold development.)

The orientation of these sets of fractures (A and B) indicates that the maximum and least principal stress acted almost parallel to the bedding. This is completely compatible with the conclusions reached in Chap. 14 and earlier in this chapter; namely, that the mechanism of flexural slip is commonly important in the finite folding phase and for this to take place, shear resistance along all, or selected, bedding planes must be exceedingly low.

We have demonstrated that such low shear resistances are the result of high fluid pressures. By definition, if the shear stress on a plane is zero, then the stress normal to that plane is a principal stress and, because axes of principal stress are orthogonal, two of the principal stresses will act parallel to that plane. At this stage in finite fold development, the bedding plane shear stresses are so small that the axes of principal stress will be within a few degrees of being either perpendicular or parallel to the bedding planes. This situation of high fluid pressures during late finite fold development is clearly represented in Fig. 15.24. The possible importance of such high fluid pressures in the development of 'fracture cleavage' is discussed in Chap. 17.

Locking up of folds

Eventually, the slip between layers, which is the essential element of folding associated with 'rotational instability', ceases as the folds lock up, by a process that may be controlled by either geometrical or stress conditions. Thus, we have noted that, before and during this phase of fold development, fractures form which will permit fluids to drain progressively and, when fluid pressures are high, relatively rapidly from the fold. If the quantity of water initially contained by the rocks is low, drainage of relatively small quantities of fluid may result in a fall in the fluid pressure which would then give rise to an increase in the shear resistance to flexural slip. Hence, the fold could lock up after only a slight, or moderate, degree of fold amplification. However, if the folded rocks were highly porous and contained large quantities of fluids and also contained many impervious layers, the high fluid pressures may be maintained until the fold is forced to lock up because of geometrical constraints (Fig. 15.22). It will be recalled from this figure that reverse kink-bands tend to 'lock up' when $\omega = 2\theta$. This is because further amplification would require a thinning of the layering within the band. De Sitter (1958) and Ramsay (1967) have considered the amplification and locking up of chevron folds, using the model shown in Fig. 15.26. It can be shown that the variation in shearing strain increment $(d\gamma/de)$ with variation in total shortening $(1 + e)$ is:

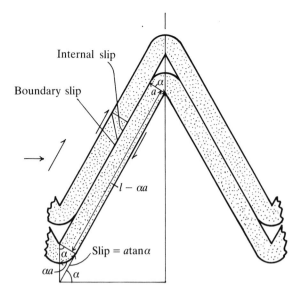

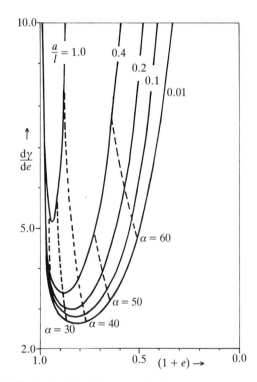

Fig. 15.26. (*a*) Model of chevron fold. (*b*) The variation in the shearing-strain increments $d\gamma/de$ with variation in total shortening and L/a ratio. The dashed lines show values of limb dip α. (After Ramsay, 1967.)

$$d\gamma/de = \frac{\sec^2 \alpha \csc \alpha}{\dfrac{\alpha a}{l} - 1} \qquad (15.3)$$

where a/l is the ratio of layer thickness to limb length. This expression (Eq. (15.3)) is represented graphically in Fig. 15.26(*b*) for various values of a/l. It can be seen from this graph that, at fold initiation, the increments of shearing strain are infinitely high. This re-emphasises the problem of fold initiation discussed in Chap. 11. The increments decrease rapidly to a

minimum as the fold amplifies and then rise almost equally rapidly to very high values. Because of this rapid increase in the increment of shearing strain, it is argued by these authors that further growth of the fold is inhibited. De Sitter suggests that this is the reason why many natural chevron folds have an inter-limb angle of about 60°.

A variety of second order structures can be found associated with multilayer buckles, particularly those formed in complex multilayers made up of various layers each with its own thickness and mechanical properties. Some of the 'minor' structures, such as the single layer buckles, small thrusts and strike-slip faults, discussed earlier in this chapter, develop before the initiation of the multilayer folds. Others, such as the accommodation structures (Chap. 12, Figs. 12.23 to 12.34), develop as the multilayer buckles amplify and the various layers adjust to the wavelength and amplitude of the multilayer buckles. Still other structures form after the folds have locked up during the period of post buckle flattening, and some of these are discussed in the following section.

Post-buckle flattening

In rocks which are deforming at depth in the crust in a high temperature environment, the post-buckle deformation will be controlled by diffusion mechanisms which will give rise to the development of fold flattening with the production of schistosity, or cleavage, coupled perhaps with the generation of boudinage or pinch-and-swell structures on the fold limbs. These structures will be discussed later in Chaps. 16 and 17, so we shall not deal with them here. Instead, we shall concentrate on those folds which develop in lower temperature environments where diffusion mechanisms do not play an important role; where the competent contrast between layers is often considerable and post-buckle flattening may involve physical migration of the relatively incompetent material.

As we have already seen, during finite fold development, because the resistance to layer- or bedding-slip is so small, the axes of principal stress act either parallel or perpendicular to the layering. When lock up occurs, however, the resistance to shear along the erstwhile planes of slip becomes significant and the maximum principal compression is no longer restricted to being parallel to the layering. This can be clearly seen in the photo-elastic experiments of Currie *et al.* (1962).

If the folds are markedly asymmetrical (e.g. a kink-band), it is often possible to infer from the relative thickness of the same incompetent layers in the long and short limb and from the structures which form in the short limb only that, during the late stages of fold development, the stress axes were not symme-

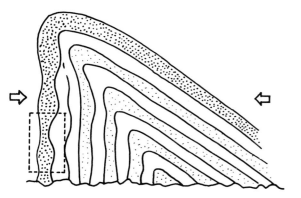

Fig. 15.27. Detail of the fold shown in Fig. 15.11(*a*), showing a thinning of the incompetent layers and the formation of pinch-and-swell structures in some of the competent layers on the steep limb. The dotted rectangle is the area shown in Fig. 16.5.

trically disposed about the fold. For example, in the fold shown in Fig. 15.27, individual incompetent layers which can be traced from the short to long limb, are found to be significantly thinner in the short limb. In addition, pinch-and-swell structures or boudinage may be developed locally in the competent layers of the short limb (Fig. 15.27) or these layers may be cut by minor thrusts (Fig. 15.28). All these features may best be explained if the axis of maximum principal compression acts at a high angle (or even normal) to the layering in the short limb.

It may be noted that the incompetent layers in the short limb may have thinned, purely as the result of the closing of the pores and void spaces. In addition, or alternatively, they may have thinned by the migration of material from the limbs to the hinge region in response to stress gradients, which tend to become greater during the last stages of fold amplification. They are analogous to the gradient established during micro-folding that cause the migration of mineral species between the limb and the hinge, and result in the formation of crenulation cleavage (Chap. 17).

The formation of thrusts in the leading limbs of small scale folds (Fig. 15.28), may also occur in larger folds. Indeed, one or more of these fractures may become so important, that it becomes a dominant

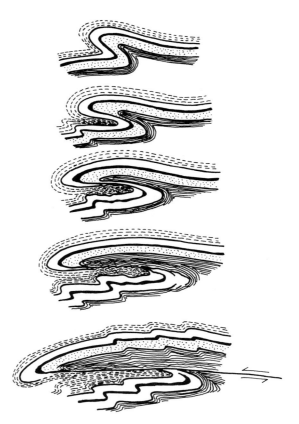

Fig. 15.29. Various stages in the formation of a major thrust from an asymmetrical fold. (After Heim, 1921.)

thrust feature, which is so large that the folds may be only second order features at the later extremes of the thrust (cf. the reconstruction by Heim, 1921, Fig. 15.29).

If the fold is subsequently uplifted and exposed by erosion, it experiences the changes in strain described in Chap. 9. Near the surface, the residual and remanent stresses, inherited from the period of active folding, give rise to the generation of barren fractures. If the rocks exhibit marked differences in competence, the anisotropy will ensure that the remanent, principal stresses tend to act approximately parallel, or near perpendicular, to the layering. Hence, barren fractures induced at this late stage are often similar in orientation to sets of fractures which formed during active fold development.

Commentary

In this chapter we have discussed the formation of folds in a complex, sedimentary multilayer. It will be clear to the reader that the range of parameters controlling fold development is very large, and that no single theory of buckling exists which can completely account for buckling behaviour. We have tried to draw from the various theories, concepts and conclusions set out in the previous five chapters, and present the possible sequence of events and stages which are likely to occur during multilayer folding in nature.

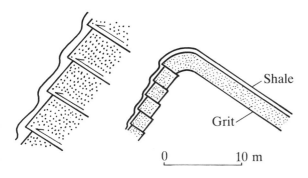

Fig. 15.28. Minor thrusts in an asymmetric fold in the Aberystwyth grits between Borth and Aberystwyth, Wales. (After Price, 1953.)

To do this, it is necessary to select those various, theoretical aspects which are most pertinent to the specific fold(s) that are of particular interest in a given situation. Only in those special geological conditions in which a theoretical model accurately simulates nature may a single theory prove sufficient. Even then, it must be borne in mind that most theories relate only to fold initiation and provide little, if any, insight into final fold development. Moreover, it is vital that the correct rheological model be applied, and this can only be done by quantifying the various parameters involved in any theory of buckling. Fold initiation and finite development may occur relatively quickly. Because of this, it is possible, even necessary, to postulate conditions which may at first seem extremely surprising.

To understand how folds develop, geologists must have at their finger tips the essential, mechanical elements outlined in these last six chapters and be aware of their strengths and their limitations. They should then be applied to natural examples with care, understanding and controlled imagination – potentially the most powerful tool available to any geologist.

References

Cloos, E. (1961). Bedding slips, wedges and folding in layered sequences. *Bull. Commn. geol. Finl.*, **196**, 105–22.

Cloos, H. (1948). Gang und Gehwerk einer Falte. *Zeitschr. deutch. Geol. Ges.*, **100**, 290–303.

Cobbold, P.R. (1976). Fold shapes as a function of progressive strain. *Phil. Trans. Roy. Soc. London*, **A283**, 129–38.

Currie, J.B., Patnode, H.W. & Trump, R.P. (1962). Development of folds in sedimentary strata. *Geol. Soc. Am. Bull.*, **73**, 655–74.

De Sitter, L.M. (1958). Boudins and parasitic folds in relation to cleavage and folding. *Geol. Mijnbouw.*, **20**, 277–86.

Donath, F.A. (1968). Experimental study of kink-band development in Martinsburg slate. In *Proc. conf. on research in tectonics*, ed. A.J. Baer & D.K. Norris, pp. 255–93. Geol. Surv. Canada paper 68–52.

Guzzeta, G. (1984). Kinematics of stylolite formation and physics of pressure solution process. *Tectonophysics*, **101**, 383–94.

Hara, I., Yokojama, S., Tsukuda, E. & Shiota, T. (1975). Three-dimensional size analysis of folds of quartz veins in the psammitic schists of the Oboke district, Shikoku. *J. Sci. Hiroshima Univ. C.*, **7**, 3, 125–32.

Heim, A. (1921). *Geologie der Schweiz*. Leipzig: C.H. Tauchnitz.

Honea, E. & Johnson, A.M. (1976). A theory of concentric, kink and sinusoidal folding and of monoclinal flexuring of compressible elastic multilayers. IV. Development of sinusoidal and kink folds in multilayers confined by rigid boundaries. *Tectonophysics*, **30**, 197–239.

Laubscher, H.P. (1977). Fold development in the Jura. *Tectonophysics*, **37**, 337–62.

O'Brien, C.A.E. (1957). Salt diapirism in South Persia (Iran). *Geol. Mijnbouw*, **19**, 357.

Price, N.J. (1953). Structural and Tectonic development of the Aberystwyth Grits. Unpublished Ph.D. Thesis, U.C.Wales.

(1962). The tectonics of the Aberystwyth Grits. *Geol. Mag.*, **99**, 542–57.

(1967). The development of asymmetric buckle folds in non-metamorphosed sediments. *Tectonophysics*, **4**, 173–201.

Price, N.J. & Hancock, P.L. (1972). The development of fracture cleavage and kindred structures. *Proc. Int. Geol. Cong. (Canada)*, Session 24, Section 3, pp. 584–92.

Ramsay, J.G. (1967). *Folding and fracturing in rocks*. New York: McGraw-Hill.

Summers, J.M. (1979). An experimental and theoretical investigation of multilayer fold development. Unpublished Ph.D. Thesis, University of London.

Watkinson, A.J. (1976). Fold propagation and interference in a single multilayer unit. *Tectonophysics*, **34**, T37–T42.

16 Boudinage and pinch-and-swell structures

Introduction

Relatively little attention has been given to geological structures that form when layered sequences, or rocks with a fabric, are compressed at a high angle to, and/or extended in, the plane of the layering or fabric. The reason for this is not clear, for 'extension' structures are much more common in nature than one would infer from the paucity of literature concerning them. In this chapter two of the main structures which result from such deformation, namely boudins and pinch-and-swell structures, are discussed.

The term 'boudin' was first used by Lohest in 1909 to describe structures which he observed in the metamorphosed Devonian sediments at Bastogne in the Ardennes, Belgium. These structures were barrel-shaped in profile, each barrel separated from the next by a quartz vein (Fig. 16.1). In three-dimensions the structures appear as large cylinders or sausages (boudins) laid side by side. Boudins had been described previously, e.g. by Ramsay (1866) and Harker (1889), but Lohest was the first to consider that the structure warranted a name. The term has subsequently been misused and Ramberg (1955) noted that 'there has been a tendency to apply the term "boudin" to any isolated body which has been held to be formed by the tectonic disruption of any original, more or less, extensive layer. Such a wide use appears to be undesirable and, in fact, obscures the different modes of origin of such bodies'. Wilson (1961) also complains that 'now one finds that the term (boudin) is being misused. It is not uncommon for any detached free-floating rock inclusion in a metamorphic complex to be referred to as a "boudin" regardless of its shape and origin.' Ramberg suggests that such detached bodies be called by the general term of 'tectonic inclusion'.

The terms used to describe boudins are given in Fig. 16.2. The geometry of the type boudins (Fig. 16.1) is only one of a complete spectrum of geometries (Fig. 16.3(a), (b) and (c)) which exists between isolated rectangular blocks and a regular thickening and thinning of the layer known as *pinch-and-swell structure*, all of which are found in nature. (Terms used to describe pinch-and-swell structures are also given in Fig. 16.2.)

The range of structures shown in Fig. 16.3 reflects a range of rheological behaviour of the relatively competent layer from brittle (when tensile

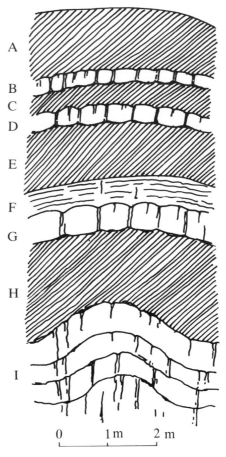

Fig. 16.1. The original boudins described by Lohest in 1909. The boudins are formed in the sandstone beds (unornamented) by layer-normal compression and are separated from each other by quartz veins. (The sandstones have subsequently been folded by layer-parallel compression.) The shale beds, A, C, E and H are cleaved but the cleavage does not seem to be related to either the layer-normal or layer-parallel compression.

failure occurs by the formation of discrete extension fractures, Fig. 16.3(a)), to ductile (when 'failure' occurs by localised necking, Fig. 16.3(c)). The formation of the type boudins, Fig. 16.1 and 16.3(b), involves a combination of both brittle and ductile behaviour.

The boudins of Figs. 16.1 and 16.3(a), (b) and (c), are formed by failure in extension. However, in nature, boudins are sometimes formed by shear failure. It will be recalled from Chap. 1 that the factor which determines whether tensile failure or shear failure occurs, is the magnitude of the differential stress $(\sigma_1 - \sigma_3)$. When the conditions for brittle fracture are satisfied, if the differential stress is less than

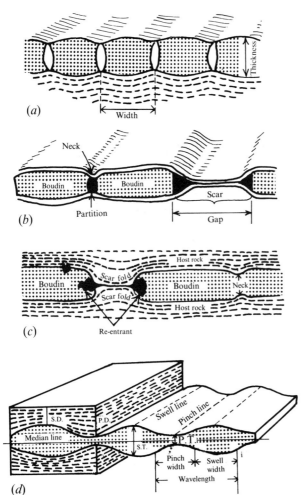

(a)

(b)

(c)

(d)

Fig. 16.2. (*a*)–(*c*) Terminology used to describe boudins. (After Wegmann, 1932; Jones, 1959; and Wilson, 1961.) (*d*) Terminology used to describe pinch-and-swell structures. S.T. and S.D. are swell thickness and swell disturbance and P.T. and P.D. pinch thickness and pinch disturbance. i denotes an inflection point. (After Penge, 1976.)

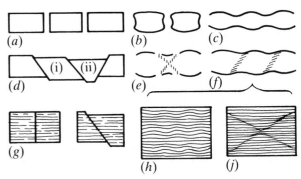

(a) (b) (c)

(d) (i) (ii) (e) (f)

(g) (h) (j)

Fig. 16.3. A variety of structures formed when a layer ((*a*)–(*f*)) or rock fabric ((*g*)–(*j*)) is compressed normal to the layering. The response may be brittle or ductile and involve tensile or shear failure.

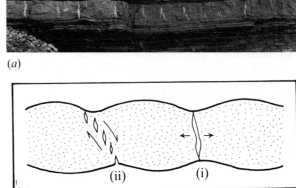

(a)

(b)

(ii) (i)

Fig. 16.4. (*a*) Incipient boudins in a sandstone layer from Millook, N. Cornwall. Some of the boudins are separated from each other by a single tension vein ((i) in (*b*)), others by an en echelon array of tension veins ((ii) in (*b*)).

four times the tensile strength of the rock, tensile failure prevails; if it is greater, shear failure occurs (Eq. (1.69(*b*))). The geometry of the boudins formed as the result of shear will depend on the rheology of the layer. If the layer behaves in a brittle manner, shear failure occurs by the formation of discrete failure planes inclined at approximately 30° to the maximum principal compression direction. These may occur as a single set of fractures which divide the layer into rhomboidal blocks (Fig. 16.3(*d*) (i)), or as conjugate sets which divide the layer into blocks with a trapezohedral profile (Fig. 16.3(*d*) (ii)). If the shear failure involves both ductile and brittle behaviour, then the boudins are separated from each other by shear zones, often containing en echelon tension gashes (Fig. 16.3(*e*)). Natural examples, showing boudins separated by shear zones, are shown in Figs. 16.4 and 16.5. Because the geometry of boudin profiles depends on both the rheological behaviour of the layer and the type of failure (tensile or shear) that occurs, a large

variety of profile geometries can develop (Fig. 16.3(*a*)–(*f*)).

The three-dimensional geometry of the type boudins is cylindrical (Fig. 16.2), and natural examples of boudins that are approximately cylindrical are shown in Figs. 16.5 and 16.6. However, many natural and experimentally produced boudins are found not to exhibit this simple cylindrical geometry. Wegmann (1932) describes an example where two perpendicular sets of boudin necks occur and produce what he termed 'chocolate tablet structures' (Fig. 16.7). Non-cylindrical boudins, produced experimentally, are shown in Fig. 16.8.

Although pinch-and-swell structures in a layer may have a constant wavelength, and boudins may have a constant width and separation, it should be noted that the distribution of these structures within a single layer may not be regular. The development of boudins and pinch-and-swell structures in localised 'packets' is analogous to the localised development of fold trains observed in single layers and multilayers. This is a point to which we shall return in the commentary to this chapter, where some of the simi-

Fig. 16.5. Cylindrical boudins separated from each other by en echelon arrays of tension gashes. Note that both the conjugate arrays of tension gashes are formed and that the beds are on the steep limb of the fold shown in Figs. 15.11(*a*) and 15.27. Northcott Mouth, N. Cornwall, England.

Fig. 16.6. Plan view of cylindrical boudins, Rillage Point, N. Devon, England.

larities between pinch-and-swell structures and folds are discussed.

Let us now consider the various experimental and theoretical studies which relate to the formation of boudins and pinch-and-swell structures.

Experimental work

The experimental work on boudins and pinch-and-swell structures can be divided into four groups; namely, experiments on: (i) rocks, (ii) rock analogues, (iii) photo-elastic rock analogues and (iv) experiments using the finite element method.

The finite development of these structures cannot usually be determined from mathematical analyses of the problem for, as with most analyses of folding, these are generally restricted to the first increment of boudin development. Experimental work is therefore particularly relevant to the study of their finite development. The experiments are also useful for studying the initiation of boudins and pinch-and-swell structures in an initially continuous layer.

Experiments on rocks

In an attempt to produce boudins experimentally in rocks, Griggs & Handin (1960) extended concentric cylinders of rock under high confining pressure. In these experiments, the inner core of dolomite was less ductile than the enclosing limestone. As may be seen from Fig. 16.9, the inner core was ruptured and individual pieces were engulfed by the influx of the ductile matrix into the cracks. Sometimes, the inner core fractured into several approximately equant blocks. The bounding fractures were either shears or, less commonly, tension fractures which terminated against the outer cylinder. A natural example of boudins probably formed by shear failure is shown in Fig. 16.9(*b*).

Paterson & Weiss (1968) have also produced boudins experimentally in quartz veins by compressing a phyllite containing thin quartz veins under a confining pressure of 5 kbar at room temperature. However, the 'boudins' produced are extremely small and it is difficult to determine the mechanism by which they formed.

Experiments on rock analogues

Because experiments on rocks have not been (and probably never will be) performed under the correct combination of conditions of pressure, temperature and strain-rate necessary to endow the rocks with the rheology they possess during the formation of natural boudins and pinch-and-swell, it is necessary to study the formation of these structures using rock analogues. Moreover, by using material that is ductile at room temperature and under zero confining pressure, it is possible to observe the detailed development of the resulting structures.

Ramberg (1955) was one of the first to investigate the formation of boudins experimentally. He used Plasticine and putty to simulate the competent and incompetent rock materials respectively. Layers of these materials were compressed between rigid

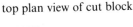

top plan view of cut block

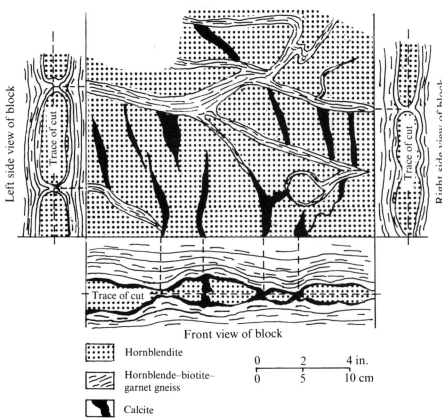

Fig. 16.7. Boudinage of a hornblendite layer (stippled). The mineral lineation in the host rock is parallel to the front face of the block. (After Jones, 1959.)

Front view of block

Hornblendite

Hornblende–biotite–
garnet gneiss

Calcite

0   2   4 in.
0   5   10 cm

plates and allowed to expand in either one or two directions. The competent layers were either ruptured to form boudins or were locally necked down to form pinch-and-swell structures. The most competent layers formed relatively sharp-edged, rectangular boudins, (Fig. 16.10(*a*)). Slightly less competent layers formed barrel-shaped boudins (Fig. 16.10(*b*)), and the least competent of the competent layers formed lenticular boudins and pinch-and-swell structures (Fig. 16.10(*c*)).

This work on single layer boudinage has been complemented by the experiments of Woldekidan (1982) on the formation of boudins and related structures in wax or Plasticine multilayers, in which the layer interfaces are lubricated by a thin film of silicone grease or Vaseline respectively. These experiments show the development of various structures that form in a multilayer compressed at a high angle to the layering. In addition, they show how these structures may be related in both time and space. A summary of the structures that formed in these experiments is given in Fig. 16.11. The rectangular boudins shown in Fig. 16.11(*a*) were developed in experiments on paraffin wax multilayers containing relatively thick and highly competent layers (dark) separated by several thinner, less competent layers (light).

It can be seen in Fig. 16.12, that the competent layers of wax failed by tension and that the spacing of

the failure planes is related to the layer thickness. This experiment can be compared directly with the natural example of the rectangular boudins shown in Fig. 16.13.

The individual competent layers of the multilayer of Fig. 16.12 have behaved mechanically as single layers. However, when the contrast in rheological properties between the various layers of the multilayers is smaller, and/or if the competent layers are more closely spaced than those in the multilayer of Fig. 16.12, its mechanical behaviour may be dominated by its bulk mechanical anisotropy rather than by the properties of any individual competent layer. As a result, either internal pinch-and-swell structures (Fig. 16.3(*h*)) or normal kink-bands (Figs. 16.3(*j*), 16.11(*c*) and 16.14(*c*)), or both, may develop.

The deformed multilayer shown in Fig. 16.14 is made up of different coloured layers of wax with identical properties, separated from each other by a thin film of lubrication. It was found, in this and similar experiments, that internal pinch-and-swell structures and normal kink-bands formed and that these were localised in zones adjacent to the compressing plattens, presumably initiated at, or near, the model-platten interface at local stress or layer orientation irregularities. It appears that strain softening accompanies the initiation of these structures so that, once they have been formed, it is easier to continue the

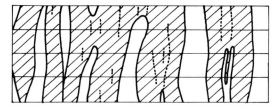

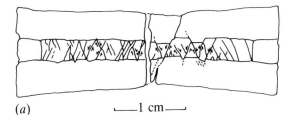

Fig. 16.8. Plan view of experimentally produced boudins in a competent, Plasticine layer (shaded) in a less competent Plasticine matrix (blank). Dotted lines are neck regions. (After Sadrabadi, 1978.)

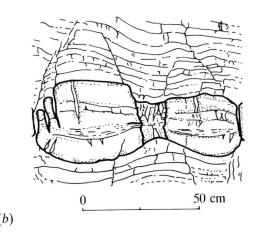

(a)   └─1 cm─┘

(b)

Fig. 16.9. (a) Section through concentric cylinders of dolomite and limestone that have been extended along their length. The inner, less ductile dolomite fails by shear failure. (After Griggs & Handin, 1960.) (b) Natural example of boudins probably formed by shear failure. The shear fractures can be seen extending into the layered matrix. (After Wunderlich, 1962.)

bulk deformation of the model by the amplification of these structures rather than to induce the formation of new structures in another part of the model. Thus, an initial 'homogeneous', anisotropic multilayer becomes one in which a relatively undeformed, 'competent' block is sandwiched between two relatively 'incompetent' horizons of deformation (X in Fig. 16.14(b)) in which strain softening has occurred. The central, competent, anisotropic horizon, (Y in Fig. 16.14(b)), may subsequently develop boudins or pinch-and-swell structures, in a manner comparable with that in which such structures form in competent, isotropic layers set in a less competent matrix (Fig. 16.11(a)).

In a series of experiments of the type illustrated in Fig. 16.12, Woldekidan (1982) investigated the relationship between boudin width and separation during progressive deformation, and observed that

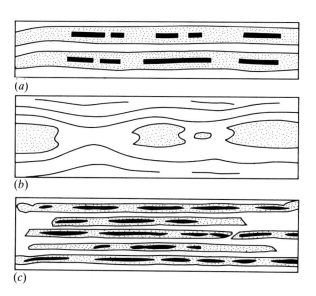

(a)

(b)

(c)

Fig. 16.10. Sections of compressed cakes of putty simulating incompetent rocks and Plasticine or cheese simulating competent rocks. In (a) the boudin layer is cheese and in (b) and (c) Plasticine. (After Ramberg, 1955.)

during the initial stages of the experiments, the first-formed tension fractures break the competent layer into rectangular boudins, with relatively large widths. As deformation continues, these wide boudins are broken down by the development of new extension fractures into narrower structures. The process continues until an optimum aspect ratio (i.e. width to thickness) is obtained. Histograms of the width of boudins in 1 mm thick, relatively competent, wax layers, at various stages in the progressive deformation of a multilayer, are shown in Fig. 16.15. It is interesting to note that the 'dominant' aspect ratio, which is eventually established, becomes apparent even after only a relatively small amount of deformation. For example, in the 1 mm thick layers represented in Fig. 16.15, the optimum aspect ratio is approximately 4:1; and this can be seen from the histogram peak which relates to 16 per cent layer-normal shortening.

It was also noted in these experiments, that, when a relatively wide boudin divides during progressive deformation, the resulting two boudins do not always have the same width. This can be seen in Fig. 16.16 which shows a histogram of the ratio of widths of the two resulting boudins (wide/narrow).

The processes observed in the experiments described above lead to large variations, within a particular layer of boudin width and separation, particularly during the low strain conditions obtaining during the early part of the experiments. (See also the natural example of Fig. 16.13.) However, these variations become less apparent at the higher strains which obtained later in the experiments.

A considerable amount of literature exists on the deformation of metals in both compression and

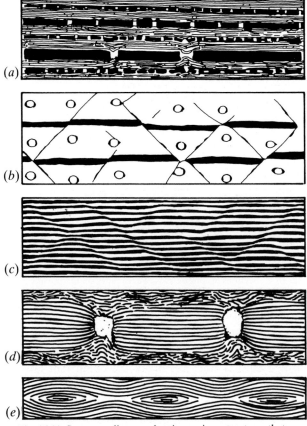

(a)

(b)

(c)

(d)

(e)

Fig. 16.11. Summary diagram showing various structures that were formed when different multilayers were compressed at right angles to the layering. (After Woldekidan, 1982.)

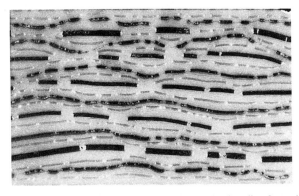

Fig. 16.12. Experimentally produced rectangular boudins formed in competent (black and dark grey) wax layers. Note that the spacing of the tensile failure planes is related to the thickness of the layers. (After Woldekidan, 1982.)

Fig. 16.13. Rectangular boudins in basic (dark) layers in a metamorphosed limestone succession at Bancroft, Ontario, Canada.

tension. A few comments on this work are included here, although its relevance to rock deformation has not yet been fully assessed. Nadai (1950) describes the deformation of a flat, polished, mild steel bar tested in tension. At the instant of the drop in load at the yield point, fine, dull lines appear on the polished surface of the bar at an angle to the axis of tension (Fig. 16.17(a)). These lines are known as Luders lines or bands, slip lines or deformation bands. A similar phenomenon is observed in specimens subjected to compression (Fig. 16.17(b)). The Luders lines are the expression of planar zones of high deformation on the surface of the metal, and the angle α they make with the axis of the test piece in the case of tension is usually a little greater (47°) than 45°, and in the case of compression a little less. Luders bands are laminar zones of plastic distortion in which the macroscopic strain seems to be simple shear. The remainder of the specimen only undergoes elastic (i.e. recoverable) deformation. The conjugate sets of Luders lines shown in Fig. 16.17(a) and (b) may intersect (Fig. 16.17(c) and (d)) and either one or several bands may develop (Fig. 16.17(e)). Nadai shows that the two systems of orthogonal lines of slip that develop during the plane strain of an ideally plastic substance are the *characteristics* of the differential equations of plane plastic strain. The characteristics are straight lines under a state of homogeneous strain, but under heterogeneous states of plane strain in an ideal plastic substance they are, in general, two systems of orthogonal curves. A detailed discussion of plasticity theory and the localisation of deformation bands is beyond the scope of this book, and the interested reader is referred to Nadai (1950), Hill (1950) and Rudnicki & Rice (1975).

Two examples of necking in metals are shown in Fig. 16.18. A copper rod tested in tension is shown in Fig. 16.18(a). It was observed that after necking had occurred brittle failure was initiated as a crack which formed at the centre of the minimum section and subsequently spread out. The profile and plan of a flat steel bar tested in tension are shown in Fig. 16.18(b) and (c), which illustrates just one of a series of tests on flat steel bars, whose thickness had been reduced by 20 per cent by cold rolling; a process which induces an anisotropy into the metal. It was found that if the bars have a width/thickness ratio of less than about 6 or 7:1 they neck down symmetrically around a section *normal* to the bar axis when tested in tension along their length. If the width/thickness ratio is greater than about 6 or 7:1, necking occurs along a section *oblique* to the bar axis as shown in Fig. 16.18(b) and (c). Continued extension of the bar results in failure along the necked region which is inclined at between 55° and 60° to the axis of maximum extension.

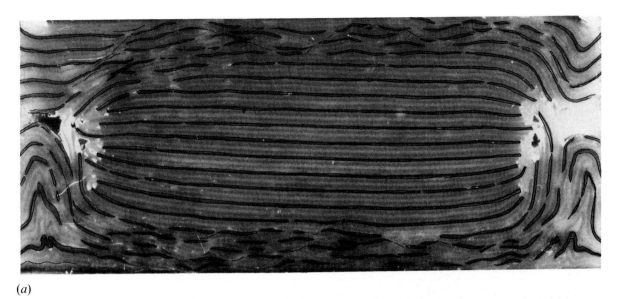

(a)

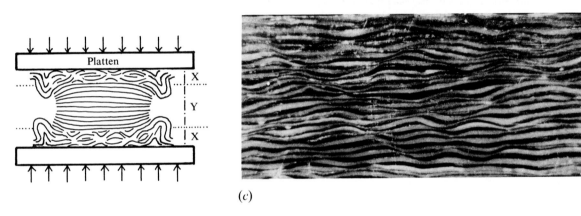

(b)   (c)

Fig. 16.14. (a) A wax multilayer compressed normal to the layering by two rigid plattens. (b) Line drawing of (a). (c) Zones of intense deformation develop adjacent to the plattens. The central portion of the multilayer (Y) remains relatively undeformed. (After Woldekidan, 1982.)

Experiments using photo-elastic rock analogues

Photo-elastic experiments are particularly useful for investigating the initiation of boudins and pinch-and-swell structures, for they show how, as the principal compression is increased, the initial uniform stress field becomes disturbed. They also indicate the way in which the instability governs the spacing and initiation of these structures. Consider for example, some photo-elastic experiments by Sowers (1973) which involve a model made up of a thin, competent layer of agar (a gelatinous carbohydrate) embedded between two thicker, less competent layers of gelatine. At low stress levels, the entire model took on a nearly uniform 'colour', showing a homogeneous distribution of stress. However, at a critical applied load, the stress patterns became non-uniform. A sketch of the isochromatic fringe patterns that develop when the elastic instability occurred, is shown in Fig. 16.19. The regularly spaced circles of fringes indicate local high shear stress concentrations which appeared just before fractures, which subsequently developed along the 'dashed lines'. In experiments where the contact

between the layer and matrix was well lubricated, Sowers noted that few, if any, fractures developed; dramatically illustrating the influence of interfacial shear on the formation of the instability.

Stromgard (1973) conducted similar photo-elastic experiments, in which he studied the change in stress distribution caused by the development of

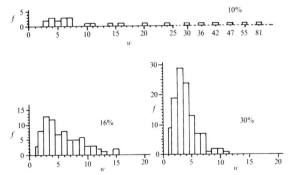

Fig. 16.15. Histograms showing the frequency of widths (w) of boudins formed in 1 mm thick, competent layers at 10 per cent, 16 per cent and 30 per cent layer-normal shortening of a wax multilayer compressed normal to the layering. (After Woldekidan, 1982.)

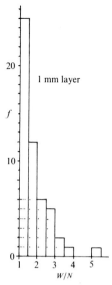

Fig. 16.16. Histogram showing the frequency of different ratios of the widths (wide/narrow) of the two boudins that form as relatively wide boudins in 1 mm thick, competent, wax layers subdivided during the progressive deformation of a wax multilayer compressed normal to the layering. (After Woldekidan, 1982.)

tension fractures in the component layers when there was cohesion between the competent layer and the matrix. He was able to demonstrate that the fracture reduced the tension in the competent layer only over a limited area, and that the length of the boudins so formed falls within the range 2 to 4 times the thickness of the competent layer; the precise ratio being controlled by the amount of reduction in tension near the crack, (Chap. 2, Fig. 2.19).

Experiments using the finite element method

Stephenson & Berner (1971) used the finite element method to investigate the process of boudinage. They attempted to determine the stress distribution and finite strains in a boudin and the enclosing rock, but did not consider the problem of boudin initiation. Their model consisted of two rectangular boudins in a less competent matrix, with the principal compression acting as indicated in Fig. 16.20(*a*). The deformation is by plane strain and the material is assumed to be elastic. The effect of shortening normal to the boudin width is shown in Fig. 16.20(*b*). The boudins separate slightly to become barrel-shaped with concave ends. The mean stress distribution around them is shown in Fig. 16.20(*c*).

As can be seen from this latter figure, the areas at the ends of the boudins are at relatively low pressure. Thus, stress gradients are established during the development of boudins which may lead to the migration of the more mobile mineral constituents through, and from, the surroundings towards the low pressure area. This would account for the concentration of relatively mobile minerals such as quartz and calcite, frequently found in the neck regions between natural boudins.

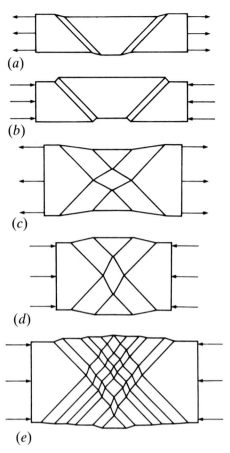

Fig. 16.17. Conjugate sets of Luders bands developed in a flat, mild steel bar (*a*) under axial tension and (*b*) under axial compression. Conjugate Luders bands may intersect (*c*) and (*d*), and frequently several bands may form (*e*). (After Nadai, 1950.)

Selkman (1978) also investigated the process of boudinage using the finite element method. He assumed elastic behaviour, and based the geometry of his model on a natural example, which occurs in the Udden ore body in the Skelleftea field of the Pre-Cambrian supracrustal rocks of Sweden. An amphibolite, chlorite layer within the ore body has formed boudins. The mineral and grain size distribution around a particular boudin neck is shown in Fig. 16.21(*a*) and (*b*) respectively. The finite element model used to determine the stress distribution around this natural model is shown in Fig. 16.21(*c*). It is a more complex model than the one shown in Fig. 16.20, in that one of the boudins is rotated, and the matrix and neck regions contain different kinds of material. The contours of maximum principal stress predicted by the analysis are shown in Fig. 16.21(*d*) and correspond well with the contours of grain size (Fig. 16.21(*b*)). As an increase in grain size lowers the free energy, one would expect relatively large crystals in areas of low stress and relatively small crystals in areas of high stress. This is exactly what is observed to have occurred.

Lloyd & Ferguson (1981) used the technique of finite element analysis to study the finite development

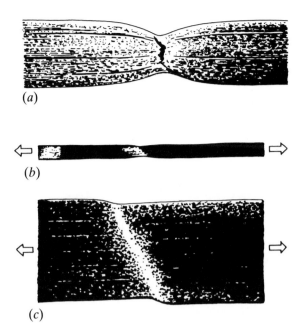

Fig. 16.18. (*a*) A circular section copper bar tested in tension. (*b*) Section and (*c*) plan, of a flat steel bar tested in tension, showing necking along a direction oblique to the maximum principal extension. (After Nadai, 1950.)

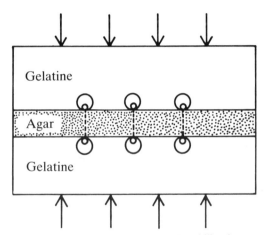

Fig. 16.19. Early indication of tensile instability in a central, competent agar layer. The small circles are isochromatic fringes that appear in the gelatine adjacent to the agar layer. These indicators of high stress concentration appear before the fractures occur. Fractures will appear along the dashed lines. (After Sowers, 1973.)

of boudins in materials with non-linear properties. They assumed elastic–plastic material properties which exhibit strain hardening for both the matrix and the boudins (Fig. 16.22). The materials behave elastically until their yield strengths ($\sigma_{y(B)}$, $\sigma_{y(M)}$) are reached, at which point irreversible plastic deformation begins. All models comprise an initial isolated block and, therefore, only simulate post-fracture behaviour of originally rectangular boudins. The simulations show that the *boudin competence*, or hardness, (defined by the local slope of the stress–strain curve) and the amount of deformation determine boudin *shape*, while the *matrix competence*, or hardness, and amount of deformation determine the

boudin *separation*. As with previously described finite element analyses, maximum stresses develop at the corners of the rectangular boudins. This tends to cause barrelling, and the amount of barrelling depends upon whether the deformation is predominantly elastic or whether the material behaves plastically and yields. If the deformation is predominantly elastic, only weak barrelling occurs (Fig. 16.23(*a*)). If, however, the deformation is predominantly plastic, the barrelling is more pronounced (Fig. 16.23(*b*)). The lower the amount of work-hardening, the more barrel-shaped the boudin becomes.

Tvergaard *et al.* (1981) examined necking instabilities in plastic and non-linear elastic materials, again using the techniques of finite element analysis. The initial irregularity in the 'layer' is a thickness inhomogeneity defined by two out of phase sinusoidal deflections of the layer boundaries (Fig. 16.24(*a*)). As the layer is extended, the initial layer thinning grows into a neck and subsequently, at a sufficiently high local strain level, bands of intense shear deformation form in the necking region. Various stages in the

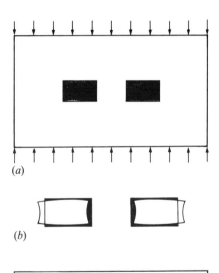

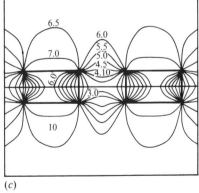

Fig. 16.20. Deformation of two rectangular, elastic bodies (*a*) in a less competent matrix using the technique of finite element analysis. The 'boudins' separate and become barrel-shaped (*b*). The stress distribution around the boudins is given in (*c*) which shows contours of mean stress in units of 10^8 dynes/cm^2. (After Stephenson & Berner, 1971.)

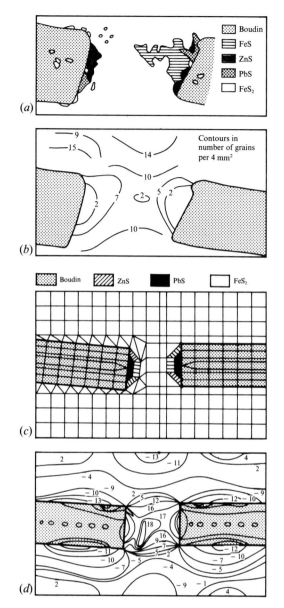

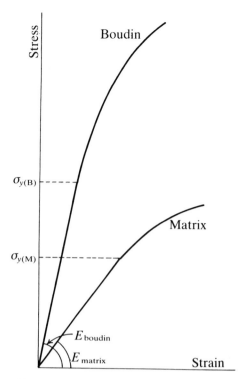

Fig. 16.22. Generalised, elastic–plastic stress–strain curves for boudins and matrix, used in the finite element model shown in Fig. 16.21. $\sigma_{y(B)}$ and $\sigma_{y(M)}$ are the yield stresses for the boudins and matrix respectively. (After Lloyd & Ferguson, 1981.)

Fig. 16.21. Distribution of (*a*) minerals and (*b*) grain size in the neck region between two natural boudins from the Udden mine, Sweden. (*c*) Finite element model of the natural boudins of (*a*) and (*b*). (*d*) The stress distribution around the boudins determined from the finite element analysis. Contours of maximum principal stress values are given, negative is compression, positive tension. (After Selkman, 1978.)

extension of such a non-linear elastic layer are shown in Fig. 16.24.

Summary of experimental work

The experiments discussed above confirm that structures such as normal kink-bands, boudins (whether separated by tension, or shear fractures) and pinch-and-swell structures are the result of compression at a high angle to the layering. They show that when a single competent layer in a matrix, or a multilayer made up of layers with different properties, is compressed normal to the layering, local stresses are set up at the layer interfaces and that these stresses may cause 'necking' instabilities which determine the spacing of the neck regions in pinch-and-swell structures, or the spacing of fractures which separate boudins. The experiments also demonstrate that the stronger the competent layer is, the more rectangular the boudin will be. In less competent layers, progressive deformation gives rise to the development of barrel-shaped and eventually lensoid-shaped boudins as the matrix flows into the neck regions between the separating rectangular boudins, (Figs. 16.10 and 16.23(*b*)). This is exactly the deformation history deduced by Wegmann (1932) from his field observations, (Fig. 16.25).

From the experiments and observations noted above, it can be concluded that the spacing of fractures in the competent layer depends upon the layer thickness and layer boundary properties. Similarly, the (quantitative) stress distribution around boudins, determined from finite element analyses, is comparable with the distribution of (qualitative) stresses that may be inferred from grain size and mineral distribution found around some natural boudins.

Theoretical analyses

Theoretical analyses relating to the formation of boudins and pinch-and-swell structures have been developed by Goguel (1948), Ramberg (1955), Biot (1965), Voight (1965), Stromgard (1973), Smith (1975, 1977 and 1979), Lloyd & Ferguson (1981) and Lloyd

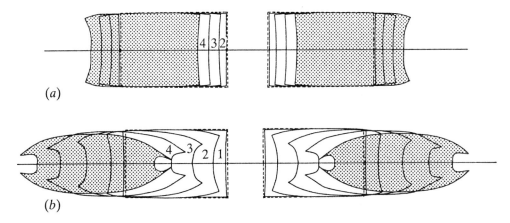

Fig. 16.23. Variations in boudin shape and separation during progressive deformation (increments 1–5). In (a) only elastic deformation of the boudins is permitted and in (b) the stresses in the boudins exceed the yield stress $\sigma_{y(B)}$ and plastic deformation occurs. (After Lloyd & Ferguson, 1981.)

et al. (1982). These studies can be grouped into three i.e. those considering (i) single layers, (ii) multilayers and (iii) anisotropic materials.

Theories of boudin formation in single layers by brittle failure

From geological observations, Goguel (*op. cit.*) concluded that the ductile flow of relatively incompetent members of a geological sequence controlled the deformation of the whole sequence, including that of the more competent layers. He used Prandtl's (1925) solution for plastic flow between two rigid plattens moving towards each other, to obtain the displacement and flow 'velocity' fields that might develop in the relatively incompetent layers. ('Velocity' is placed in inverted commas as time does not enter into plasticity theory). He postulated that the relatively competent layers were fractured to form large rectangular boudins by longitudinal extension induced by 'flow' of the ductile matrix material which could then flow into the voids created by the fractures. These large, rectangular boudins then assumed the role of the rigid plattens of the Prandtl model.

Goguel determined the interfacial forces at the junction between the competent and incompetent layers and, by computing the strain energy, was able to estimate the width of the boudins that would form in the competent layer.

Ramberg (*op. cit.*) considered a model in which a competent elastic layer, of length $2l$ and thickness t_1, was sandwiched between relatively incompetent viscous layers of viscosity η and thickness t_2, as indicated in Fig. 16.26(a). The incompetent layers were assumed either to flow out of the ends of the unrestricted model, or into internal sinks created by the separation of blocks of the disrupted competent layer, and in so doing exerted shear stresses on the interface between the competent and incompetent layers. The magnitudes of these stresses are found to depend upon the thickness and viscosity of the in-

competent layers. The integration of these stresses provides the tensile force that is available to form fractures in the competent layer. Ramberg derived an expression for the tensile stress σ_x in the x direction, generated in the competent layer at a distance x from the middle of the layer:

$$\sigma_x = \frac{-6\eta\dot{e}_z(l^2 - x^2)}{t_1 t_2^2} \qquad (16.1)$$

where $\dot{e}_z$ is the strain-rate normal to the layer (i.e. the rate of compression). It can be seen from this equation, that the stress along the layer increases from zero, at the layer ends, to a maximum at the centre, where failure will eventually take place, when the tensile stress equals the tensile strength of the layer. It follows from Eq. (16.1), that the stress at the centre of the layer ($x = 0$) is:

$$\sigma_0 = \frac{-6\eta\dot{e}_z l^2}{t_1 t_2^2}. \qquad (16.2)$$

We can consider the competent layer of Fig. 16.26 as a wide boudin that will continue to divide into smaller boudins by successive 'mid-point' fracturing, until the tensile stress at the centre of the layer cannot attain the tensile strength of the layer (S_L), i.e. when

$$\sigma_0 < S_L. \qquad (16.3)$$

By substituting S_L for σ_0 in Eq. (16.2) and rearranging, we can determine the width of boudin ($2l$) that will eventually develop:

$$2l = t_2 \sqrt{\frac{S_L t_1}{1.5\eta\dot{e}_z}}. \qquad (16.4)$$

It can be seen that the boudin width is proportional to the thickness and tensile strength of the competent layer and inversely proportional to the compressive strain-rate.

Ramberg presents a summary diagram showing the successive stages in the formation of boudins (Fig. 16.27). This conceptual model, made up of a

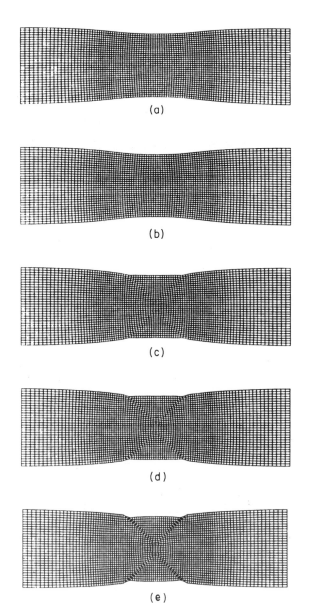

(a)

(b)

(c)

(d)

(e)

Fig. 16.24. Deformation of a quadrilateral mesh during the progressive extension of a non-linear elastic layer. Uniform necking occurs at the site of the original layer thinning, followed by the formation of more localised shear bands. (After Tvergaard *et al.* 1981.)

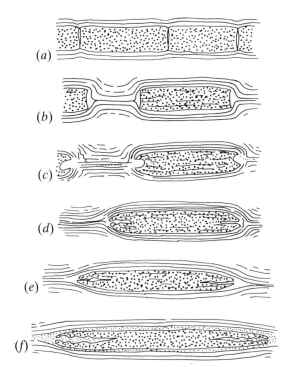

(a)

(b)

(c)

(d)

(e)

(f)

Fig. 16.25. Various stages in the development of lens-shaped boudins from originally rectangular boudins formed by tensile failure of a relatively competent layer in a less competent matrix. (Compiled from field observations by Wegmann, 1932.)

central competent layer (black) between two incompetent layers, is sandwiched between too rigid boundary layers, through which the layer-normal compression is applied.

However, Sowers (1973) is critical of Ramberg's assumption that the incompetent matrix is viscous. He argues that the ductile flow in the physical experiments (conducted by Ramberg (1955) and Sowers (*op. cit.*) to authenticate their theoretical analyses), is plastic rather than viscous. He accepts the quantitative relationships developed by Ramberg, but considers that the simple viscous model leads to large errors when calculating layer boundary stresses.

Voight (*op. cit.*) considered a slightly more complex model than that dealt with by Ramberg

(Fig. 16.26(*a*)). In Voight's model a competent elastic layer, thickness t_1, is sandwiched between two relatively incompetent layers of different thicknesses (t_2 and t_3) but with the same viscosity η (Fig. 16.26(*b*)). He derived an equation for the width of boudin (L) that would form in the competent layer:

$$L = \sqrt{\frac{S_L t_1 t_2 t_3}{c \eta \dot{e}_z (t_2 + t_3)}} \qquad (16.5)$$

where c is a constant and S_L is the tensile strength of the competent layer.

It can be seen from Eq. (16.5) that Voight's analysis, like Ramberg's, indicates that the boudin width is proportional to the tensile strength of the competent layer and inversely proportional to the strain-rate in the matrix.

We should note in passing, that in both Ramberg's and Voight's analyses it is assumed that no initial compression exists along the layers. In addition, any normal stress that may be exerted on the ends of the boudins by the incompetent material in the neck regions is ignored.

The analyses of Goguel, Ramberg and Voight discussed above all indicate that the formation of boudinage by the development of tensile failure is controlled by the transfer of stress from the matrix to the more competent layer. These stress-transfer theories predict that the tensile stress in a layer segment rises from a minimum at the end of a segment to a maximum at the centre (Eq. (16.1)). Thus, boudinage

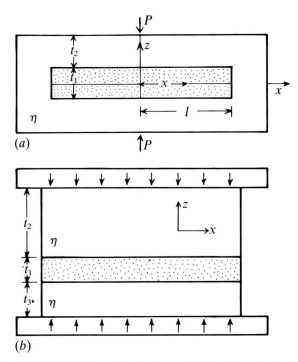

(b)

Fig. 16.26. (a) Model used by Ramberg (1955) in his analysis of boudin formation. A competent, elastic layer, thickness t_1 and length $2l$ is enclosed in a viscous matrix of viscosity η. The model is compressed normal to the layer. (b) Model analysed by Voight (1965) consisting of a central, relatively competent layer sandwiched between less competent viscous layers of thicknesses t_2 and t_3. The model is compressed normal to the layering by rigid plattens.

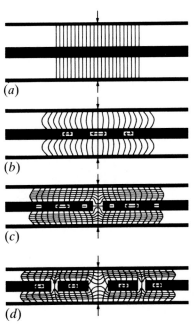

Fig. 16.27. Successive steps during formation of boudinage structures: (a) state prior to compression: (b) compression has started: the plastic flowage in the incompetent layers is indicated by the distortion of the originally vertical lines. Arrows in the competent black layer indicate tensile stress: (c) a more advanced step than (b). Competent layer is now ruptured in the middle where tension is greatest. The network in the incompetent layer indicates pattern of flowage. This network is not a further evolution of the deformed network in (b), but rather the evolution of an imaginary rectangular network not shown in (b). Horizontal arrows show the tensile stresses in the competent layer: (d) a stage more advanced than (c). Competent layer has ruptured at two new places. The pattern of the network indicates plastic flowage in the incompetent layers during evolution from (c) to (d). Again, the deformed network in (d) is supposed to have developed from a rectangular, imaginary network not shown in (c). (After Ramberg, 1955.)

develops by successive 'mid-point' fracturing of segments, until the layer is reduced to segments (boudins) all of which are shorter than some critical length (i.e. the length for which the tensile fracture strength of the layer is equal to, or greater than, the tensile stress at the mid point (Eq. (16.4))). Lloyd *et al.* (1982) point out that this mechanism implies that boudin-defining fractures occur sequentially, so that the spacing of boudins in a particular layer will be unequal. This can be clearly seen in the wax model in Fig. 16.12 and in the natural example shown in Fig. 16.13.

Theories of pinch-and-swell formation in single layers

Let us now consider the non-brittle instability associated with the formation of pinch-and-swell structures. Smith's (1975, 1977, 1979) study of the deformation of a single layer discussed in Chap. 11 is not restricted to layer-parallel compression. He considers four linear-viscous models (Fig. 16.28(a)–(d)). In models (a) and (b) the single layer is more competent than the matrix and in (c) and (d) less competent. The maximum principal compression is parallel to the layer in (a) and (d) and normal to it in (b) and (c). It is argued that the structures shown in this figure are all the result of the same general type of instability (see resonance folding, Chap. 11). The dominant wavelength of the folds, pinch-and-swell structures, mullions and inverse folds that will develop in a particular layer, are equal. However, their growth rates are very different.

When the principal compression is parallel to the layer, the pervasive background pure shear deformation, which causes a kinematic amplification of any layer irregularity (see the discussion of passive folding, Chap. 10), and the localised secondary deformation around the irregularity which causes a dynamic amplification, cooperate to form mullions or regular folds. When the principal compression is normal to the layering, these two effects compete, i.e. the background pure shear tends to deamplify any layer irregularity. This is one reason why boudins and pinch-and-swell structures are less common in nature than folds, a point we shall return to later in this chapter, when the problems of initiation of layer-normal compression structures are considered.

Smith points out that the structures (Fig. 16.28) will only grow if the total growth rate (kinematic plus dynamic) is positive. For Newtonian (i.e. linear) materials this only occurs for fold and mullion formation (models (a) and (d) Fig. 16.28). Pinch-and-swell structures will *not* amplify and develop into finite structures. However, if certain types of non-Newtonian layers are considered, for example a strain-rate softening layer, it is found that the dynamic

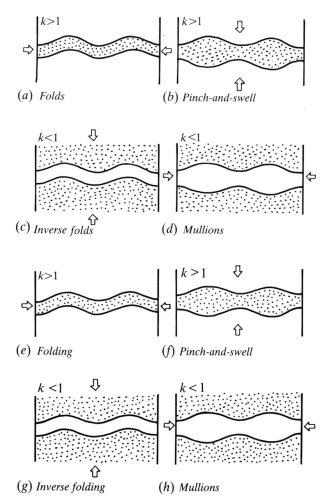

(a) *Folds*

(b) *Pinch-and-swell*

(c) *Inverse folds*

(d) *Mullions*

(e) *Folding*

(f) *Pinch-and-swell*

(g) *Inverse folding*

(h) *Mullions*

Fig. 16.28. (a)–(d) Four cases of dynamic instability in a single linear-viscous layer. The arrows indicate the direction of maximum, principal compression. The materials are linear-*viscous* and the stippled areas have a higher viscosity than the undecorated areas. k is the ratio of the viscosity of the layer and matrix. Only the 'fold' and 'mullion' instabilities have growth rates large enough to produce finite structures. (After Smith, 1975.) (e)–(h) Diagram summarising the four single layer instabilities when the layer and matrix are *non-linear* (specifically strain-rate softening) viscous materials. The growth rates of all four instabilities are significantly higher than the rates for linear materials ((a)–(d)). (After Smith, 1977.)

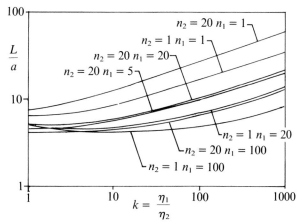

Fig. 16.29. Plot of dominant wavelength/thickness ratio (L/a) for folds and pinch-and-swell structures against viscosity ratio (k) between layer and matrix for various non-linear materials. If the materials are power-law materials, then n_1 and n_2 are the stress exponents of the layer and matrix respectively. Only strain-rate softening materials ($n > 1$) are shown. (After Smith, 1977.)

growth rates are increased (Smith, 1977 and 1979). Hence, folds and mullions will form more readily in certain non-linear materials. In addition, the total growth rate can become positive for pinch-and-swell structures (model (f) in Fig. 16.28) and so instabilities can develop into finite structures. However, Smith concludes that in order for pinch-and-swell structures to form, the layer must be strongly non-linear, otherwise the rate of amplification of the structure will be too slow to avoid being swamped by the background flattening.

The relationships between the wavelength/thickness ratio (L/a) of pinch-and-swell structures (and folds) and the viscosity contrast (k) between layer and matrix are shown graphically in Fig. 16.29,

for a variety of different combinations of non-linear properties of layer and matrix. Smith's analysis is for a general class of non-Newtonian materials and is not restricted to power-law materials. However, if the materials are power-law materials, n_1 and n_2 in Fig. 16.29 are the exponents for the layer and matrix respectively. As was discussed in Chap. 11, the curves can be divided into two parts. The portion of the curves in the upper right of the graph correspond to high viscosity contrast between the layer and matrix and low values of material non-linearity. The relationship between L/a and k in this region of the graph is adequately described by the classical single layer buckling equation:

$$L = 2\pi a \sqrt[3]{\frac{\eta_1}{6\eta_2}}. \tag{16.6}$$

We see therefore, that this equation also relates the wavelength of the pinch-and-swell structure to the layer thickness and viscosity contrast.. However, the portion of the curves in the lower left of the graph correspond to low viscosity contrasts and high values of non-linearity of the layer. The curves in this region are such that the L/a value is relatively insensitive to the competence contrast, and falls within the range 4–6. Deformation of layers that plot in this region of the graph is dominated by the 'resonance' mechanism described in Chap. 11.

We have just seen that a layer must have highly non-linear material properties before pinch-and-swell structures can form, and that high non-linearity favours the formation of these structures by the mechanism of 'resonance folding'. If this is so their L/a values should fall between 4 and 6. Inspection of the natural examples shown in Figs. 16.30 and 16.31 support this.

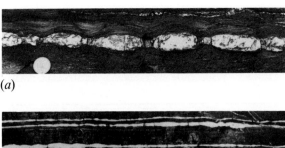

(a)

(b)

(c)

Fig. 16.30. (a)–(c) Pinch-and-swell structures in quartz veins, (Combe Martin, N. Devon, England). (a) Conjugate, normal, kink-bands have been initiated in the matrix above the layer in the neck region to the right of the coin.

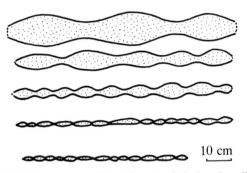

10 cm

Fig. 16.31. Comparison of the L/a ratio of pinch-and-swell structures in aplite and pegmatite dykes which exhibit a range of thicknesses. From the Haygby area, 50 km NNW of Stockholm, Sweden. (After Ekström, 1975.)

The effect of non-linear material properties on the growth rate of folds and pinch-and-swell structures was investigated experimentally by Neurath & Smith (1982). One of the reasons that they focused attention on the growth rate rather than wavelength is that the former exhibits a much greater variation with change in material rheology than does dominant wavelength. Growth rates are, therefore, more characteristic of a particular rheological combination. For example, over the range of experimental conditions used, the theory of deformation of non-linear viscous layers (Smith, 1977; see Chap. 11) predicts that dominant wavelengths should vary only by about ±20 per cent while the growth rates should vary by ±250 per cent for folding and ±300 per cent for the formation of pinch-and-swell structures. Experiments on single layers of microcrystalline wax (a power-law material with a stress exponent, $n = 5$) embedded in a matrix of paraffin wax ($n = 1.8$), show that, in agreement with

the theoretical predictions of Fletcher (1974) and Smith (1975), the growth rates of folds and pinch-and-swell structures are much higher than can be accounted for by theories assuming Newtonian material properties. However, although the growth rates of the *folds* formed in these experiments are reasonably close to those predicted by the non-Newtonian theory, the *pinch-and-swell* structures grew almost three times faster than is predicted by the non-Newtonian theory. Neurath & Smith suggest that this is because the waxes do not exhibit steady-state creep as assumed in the theoretical analysis. They extend the theory to include strain softening of the layer and show that this can lead to greatly increased growth rates for pinch-and-swell structures, while having little influence on the growth of folds.

The experiments also show that with progressive deformation the geometry of the pinch-and-swell structures deviates from the sinusoidal shape assumed in Smith's (1975 and 1977) theory. The pinches develop into cusps which point into the layer, and which eventually produce quite sharp separations between adjacent 'boudins'.

Theories of boudin formation in multilayers

Stromgard (1973) analysed the behaviour of a multilayer made up of alternate, incompressible layers of competent and incompetent linear viscous layers, (Fig. 16.32(a)). The analysis is not restricted to the condition where the maximum principal compression is normal to the layering, and can be used to determine the behaviour of the multilayer when the principal compression is inclined to the layering. In this analysis, it is assumed that the competent layer fails in a brittle manner by tensile failure and the possibility of shear failure is not considered. The conditions under which tensile failure can occur in the competent material was shown to be governed by (i) the viscosity contrast between the layers ($k = \eta_1/\eta_2$), (ii) the ratio of the thickness of the incompetent and competent layers (a/b) and (iii) the ratio of the stresses acting parallel and normal to the layers (σ_x/σ_y).

The relationship between these three parameters is presented graphically in Fig. 16.32(c), which shows the stress conditions necessary for the formation of boudins in the competent layers of a multilayer in which the viscosity contrast (k) is 2 and the principal compression acts normal to the layering, (i.e. $\tau_{xy} = 0$). If the stress state and thickness ratio of the multilayer are such that the multilayer plots to the left of the instability envelope, tensile failure and therefore boudinage of the competent layers can occur. If the plot is to the right of the envelope, tensile failure will not occur and deformation will be by viscous flow alone. This does not, however, preclude the formation of pinch-and-swell structures. Instability envelopes for a range of viscosity contrasts are shown in Fig. 16.32(c).

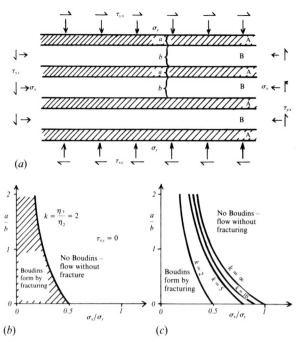

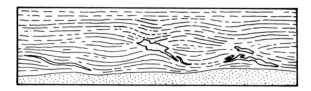

Fig. 16.33. Internal pinch-and-swell structures in a shale (now slate) horizon in Devonian turbidites, Combe Martin, N. Devon, England.

Fig. 16.32. (*a*) A multilayer made up of alternate layers of competent (A) and incompetent (B) linear, viscous materials. The maximum principal compressive stress (σ_1) is at a high angle to the layering. (*b*) The conditions for the formation of boudins by tensile failure of the competent layers when σ_1 is normal to the layering and the viscosity contrast (*k*) between the competent and incompetent layers is 2. (*c*) Instability envelopes for a variety of competence contrasts. (After Stromgard, 1973.)

We will return to this analysis later in the chapter when the formation of rhomboidal boudins is considered.

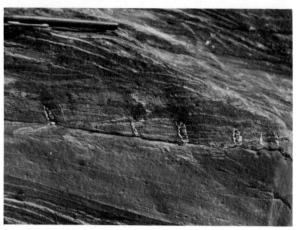

Fig. 16.34. An array of parallel tension gashes in a Devonian slate, Porthleven, Cornwall, England.

Theories of boudin formation in anisotropic materials

We have already considered the deformation of an anisotropic material, such as a sedimentary pile or a rock fabric, when the maximum principal compression acts at a high angle or normal to the layering or fabric (Chap. 13). It will be recalled, that if the anisotropy is primarily intrinsic (i.e. the result of a pre-deformation layering or fabric), then internal pinch-and-swell will form (Figs. 16.3(*h*) and 16.33), whereas if the anisotropy is mainly induced during the deformation, normal kink-bands or shear zones will form (Fig. 16.3(*j*)). The regularly spaced necking, associated with internal pinch-and-swell structures, prompted Sowers (1973) to use Biot's theory of deformation of anisotropic materials to account for the formation of regularly spaced fractures and boudins in such materials. From a general consideration of the incremental deformation of such a medium, Sowers noted that several types of stress-induced, elastic instabilities have been defined. These are (i) internal, (ii) interfacial and (iii) layer instabilities, each of which has an associated periodic stress concentration which might serve to localise fractures. He suggested that these stress concentrations, under appropriate

conditions, may give rise to the regularly spaced veins sometimes seen in anisotropic material (e.g. Fig. 16.34). Similar stress concentrations, formed when a single layer in a less competent matrix is compressed normal to the layer, are shown in Fig. 16.19.

If deformation continues after the development of the regularly spaced veins, then they may act as rigid struts so that the surrounding rock deforms in the manner shown in Fig. 16.35(*a*). Such quartz veins, which are initially aligned parallel to the compression direction, may act as stress concentrators and will either tend to undergo bulk rotation or buckle (Fig. 16.35(*a*) and (*b*)).

It is apparent that the formation of the veins in Fig. 16.34 occurred without any significant development of pinch-and-swell in the rock fabric. However in other examples of deformed rock fabrics (Figs. 16.35(*c*) and 16.36), it is equally apparent that the quartz veins did not form until pinch-and-swell structures had become well developed. These veins are relatively late structures, filling gaps between separating swells.

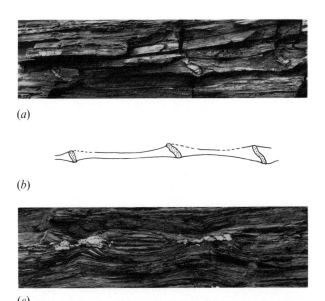

(a)

(b)

(c)

Fig. 16.35. (a) Quartz veins (formed by compression normal to the rock fabric) acting as rigid struts which become either rotated or buckled as the compression continues. (b) A line diagram of (a). (c) Internal pinch-and-swell structures formed in a phyllite compressed normal to the rock fabric. Concentrations of quartz form in the low pressure areas between separating swells. In (a), the matrix adjacent to the quartz dips away from the veins while in (c) it dips towards the quartz concentration. Trebarwith Strand, N. Cornwall, England.

Zones of contact strain around boudins

One of the dominant factors that will determine whether a competent layer in a complex geological multilayer behaves mechanically as a single layer, or as part of the multilayer whose deformation behaviour is governed by its bulk mechanical anisotropy is the spacing of the competent layers. It was noted (Chap. 12) that, if competent layers in a buckling multilayer are sufficiently far apart, so that there is no significant overlap of the zones of contact strain, the individual layers behave mechanically as single layers and develop their own dominant wavelength. The folding is disharmonic. Similarly, when a multilayer is compressed normal to the layering, if the competent units are sufficiently far apart, so that the zones of contact strain do not overlap, each layer will develop its own 'wavelength' (i.e. boudin width or pinch-and-swell wavelength). However, if the zones of contact strain of adjacent competent layers do overlap, then the deformation of the multilayer can best be accounted for by using the theories that treat the material as a mechanical multilayer or mechanically anisotropic material. The extent of the zone of contact strain around boudinage and pinch-and-swell structures can often be clearly seen if the matrix is layered or possesses a fabric (Figs. 16.30(a) and 16.37). No discernible deflection of the matrix occurs at distances greater than approximately half the wavelength of the pinch-and-swell structure.

Often, in deformed complex multilayers (and

most geological multilayers are complex) both single layer and multilayer behaviour can be observed. An example of such behaviour can be seen in Fig. 16.38(a) and (b), where the development of single layer instability in the relatively competent layers has initiated normal kink-bands in the adjacent anisotropic matrix. Conjugate normal kink-bands/shear zones can also be seen in the matrix near the neck regions of the natural pinch-and-swell structures (Fig. 16.30(a)) and has been illustrated by Halbich (1978) in the gneisses of the Okiep copper district of South Africa, (Fig. 16.38(c) and (d)). The deflection of the layering of the matrix as it flows into the neck region between two separating swells or rectangular boudins, produces an ideal irregularity from which normal kink-bands can be initiated.

Rhomboidal boudins

Some natural boudins exhibit rhomboidal cross-sections (Fig. 16.39), (see, for example Cloos, 1947; Beloussov, 1952; Rast, 1956 and Uemura, 1965). Several different mechanisms have been proposed for

Fig. 16.36. Detail of an internal pinch-and-swell structure developed in a schist. Quartz veins have developed in the low pressure area between two swells as they begin to separate.

Fig. 16.37. (a) A boudinaged Tertiary dyke in the Durness Limestone, Loch Slapin, Skye, Scotland. The deflection of the fabric and layering of the metasediments in the neck region of the boudin shows clearly the extent of the zone of contact strain.

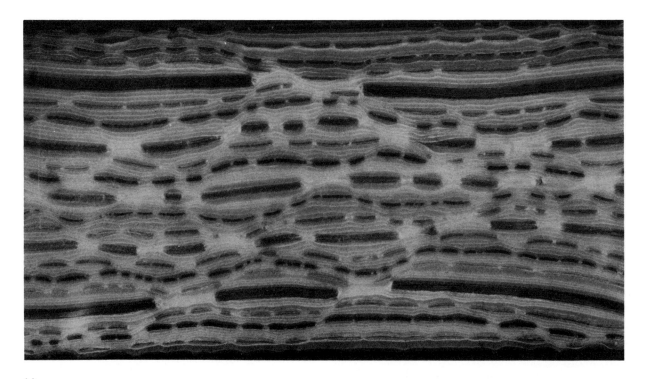

(a)

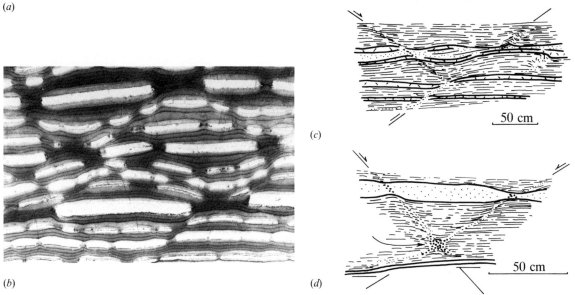

(b)

(c)

(d)

Fig. 16.38. (a) A paraffin wax multilayer compressed normal to the layering. The competent (dark) layers are divided into rectangular boudins by tensile failure. As these separate, the anisotropic 'matrix' is deflected into the neck regions and normal kink-bands are initiated. (b) Detail of (a) but the competent layers are light. (c) and (d) Examples of necking and brittle failure of a competent layer initiating normal kink-bands in the matrix. (After Halbich, 1978.)

their formation. One proposal assumes failure of the competent layer by shear fracturing along planes oblique to the layering (Fig. 16.3(d)). Another proposal assumes that boudins, which were originally rectangular, become rhomboidal during progressive deformation because of shear parallel to the boudinaging layer while undergoing extension. This idea is clearly illustrated by two experiments conducted by Hossein (1970) (Fig. 16.40), which shows two examples of the separation of boudins with progressive deformation. In the first example (Fig. 16.40(a)),

the layer undergoing boudinage was normal to the principal compression and in the second example (Fig. 16.40(b)), it was oblique. In these experiments, the competent layer of wood, which is cut into rectangular blocks (so that the spacing of the boudins is pre-determined) is set in an incompetent matrix of Plasticine. During the experiments, the wooden blocks behave as rigid bodies and do not change shape. However, if the blocks were to behave in a more ductile manner, then, in the experiment with the compression oblique to the layer, we contend that the

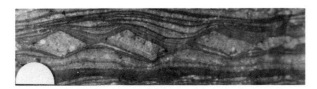

Fig. 16.39. Rhomboidal boudins in metasediments at Castanera Beach, N. Spain.

shear couple generated parallel to the competent layer could give rise to rhomboidal shaped boudins. It is interesting to note that the rate of rotation of the 'layer' is considerably greater than the rate of rotation of the individual boudins. As a result, the individual boudins become arranged en echelon (Fig. 16.40(*b*)). A natural example of differential rotation of individual boudins and the boudinaged layer is shown in Fig. 16.41. The dependence of the rate of rotation of a planar element on its length/width ratio is further discussed in Chap. 17 in the section of fabric development in slaty cleavage.

Another mechanism by which rhomboidal blocks can be formed may be inferred from field observations. It is found that pre-existing planes of weakness, such as cleavage, may become planes of separation for the blocks and thus control their shape. If the cleavage is not normal to the layering, the resulting 'boudins' will be rhomboidal. This idea is shown diagrammatically in Fig. 16.42. As the fold amplifies and the fold limbs begin to extend along their length, the blocks may separate and 'boudins' may form. A natural example of a spaced cleavage breaking up a layer into rhomboidal blocks is shown in Fig. 16.43. Even if the blocks were initially rectangular, as the result of either a cleavage being normal to the layering or the blocks being formed by extensional failure of an uncleaved fold limb, the shear stress generated parallel to the boundaries of the layer during limb rotation may cause the blocks to become rhomboidal. The effect of simple shear deformation on originally rectangular 'boudins' has been investigated experimentally by Ghosh & Ramberg (1976). They demonstrate that as well as becoming rhomboidal the 'boudins' may also develop barrelling.

We have seen (Fig. 16.3(*d*)), how rhomboidal boudins can be produced by shear failure of a layer, Stromgard (1973) has suggested that they can be produced by extensile failure as well. He argues that if the principal compressive stress is oblique to the layering, extensile failure will occur oblique to the layering and the fracture blocks will be rhomboidal. The refraction of the principal stresses through a multilayer, and the differential stress at which extensile failure of the relatively competent layers will occur is found to be very dependent on the relative thickness of the competent and incompetent layers.

Four models, in which the orientation of the

applied stress and the competence contrast between the layers are the same, but in which the relative thicknesses of the competent and incompetent layers are different, are shown in Fig. 16.44. The maximum principal compressive stress trajectories (parallel to which tensile failure will occur) are shown. It can be

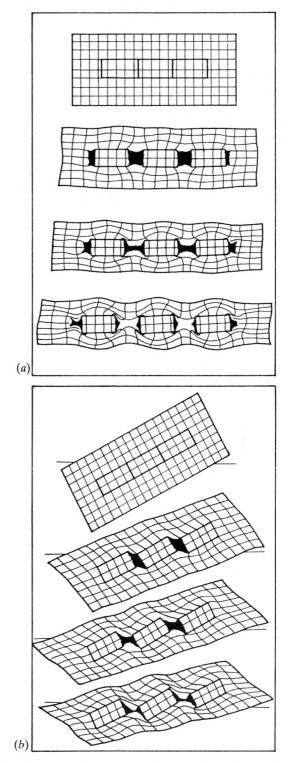

Fig. 16.40. Stages in the progressive separation of rectangular, wooden blocks (*a*) when compressed normal to the 'layer' and (*b*) when compressed oblique. (After Hossein, 1970.)

Fig. 16.41. Rotated boudins from the Moine series rocks. Ross of Mull, Scotland. (Photograph courtesy of G. Wilson.)

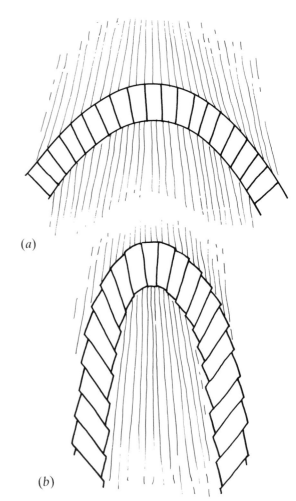

Fig. 16.42. Rhomboidal 'boudins' formed during the stretching of a cleaved, competent layer on the limb of a developing fold.

Fig. 16.43. Spaced pressure solution cleavage in a quartzite layer breaking the limb of a minor fold into rhomboidal blocks. Rhoscolyn, Anglesey, N. Wales.

seen that for widely spaced, competent layers, boudins will form which are very rectangular (Fig. 16.44(*a*)), whereas, for closely spaced competent layers, rhomboidal boudins may form (Fig. 16.44(*c*) and (*d*)). In his analyses, Stromgard determined the critical ratio of the principal stresses (σ_3/σ_1) necessary to cause tensile failure in the competent layer. He found that this ratio was significantly different for the models shown in Fig. 16.44 (e.g. less than 1.0 for (*a*), 0.48 for (*b*), 0.35 for (*c*) and less than 0.18 for (*d*)), and that a relatively high differential stress is required before tensile failure can occur in model (*d*). Although he did not actually quantify the magnitude of the stresses, he suggests that 'such high differential stress would not be expected at great depths' and that this is the reason

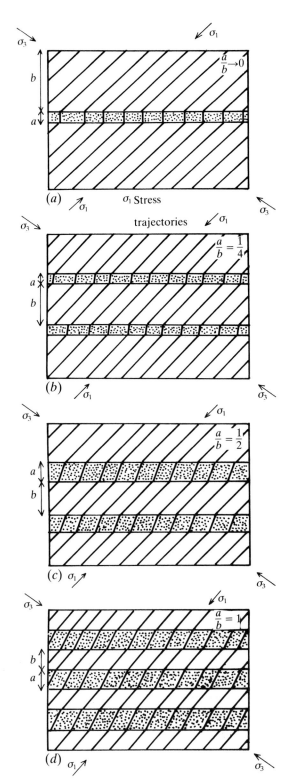

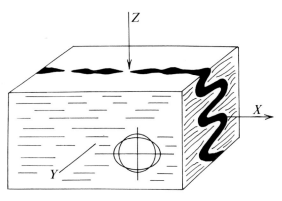

Fig. 16.45. Folds and boudins (pinch-and-swell structures) produced simultaneously in a relatively competent layer in a less competent block of material deformed by pure shear. X = maximum extension; Y = zero strain; Z = maximum contraction. The axes of principal stress coincide with those of principal strain. (After Ramberg, 1959.)

why, in regional metamorphic terrains, the separation planes between boudins are generally normal to the layering. However, it should be noted that care must be taken when attempting to deduce the magnitude of the differential stress ($\sigma_1 - \sigma_3$) from the ratio σ_3/σ_1 for if the least, principal stress is close, or equal, to zero, the ratio goes to very small values, even when the differential stress is small.

The association of folds and boudins

There are at least four different ways in which folds and boudins can be related in the field. These are:

(i) Folds and boudins may form synchronously as the result of a compression in one direction and an extension in the other.

(ii) The structures may form in sequence during a single deformation.

(iii) Boudins and pinch-and-swell structures formed during one deformation may be folded during a later deformation and vice versa.

(iv) Depending upon the orientation of the maximum compression with respect to the layering or fabric, a complete range of structures exists between symmetrical folds, asymmetrical folds, asymmetrical pinch-and-swell structures and symmetrical pinch-and-swell structures.

These four associations are discussed in turn.

1. The synchronous formation of folds and boudins

Ramberg (1959) illustrated how folds and boudins can develop together (Fig. 16.45), where a relatively competent layer, enclosed in a less competent matrix, is deformed by pure shear in the plane of the layer (the XZ plane). The layer will adjust to shortening in the Z direction by folding about an axis parallel to X, while, in an attempt to adjust to the extension parallel to the X direction, the layer may develop either pinch-and-

Fig. 16.44. The orientation of the maximum principal compressive stress in a single layer (*a*) and three multilayers (*b*), (*c*) and (*d*) with the same competence contrast ($\eta_a/\eta_b = \infty$ (i.e. infinity)) but with different thickness ratios between competent and incompetent beds. In (*a*) tensile failure would result in approximately rectangular boudins and in (*b*), (*c*) and (*d*) it would result in rhomboidal boudins. The multilayer is compressed at 55° to the layering. (After Stromgard, 1973.)

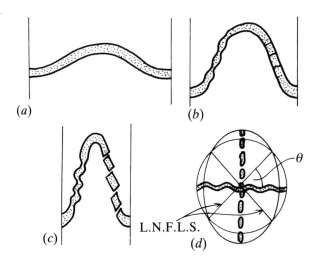

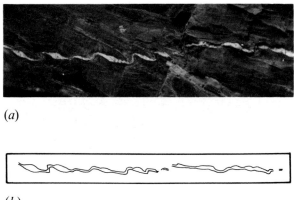

(a)

(b)

Fig. 16.47. (a) Folded pinch-and-swell structures in a quartz vein. West of Luarca, N. Spain. (b) Line drawing of (a).

Fig. 16.46. (a)–(c) Various stages in the amplification of a fold and the formation of boudins or pinch-and-swell structures on the limbs as they rotate into the finite extension field. (d) Circle representing the undeformed state and the ellipse represents the strain ellipse. L.N.F.L.S. = lines of no finite longitudinal strain.

swell structures or boudins with necks parallel to the Z direction.

As discussed in Chap. 1 (Figs. 1.18(*c*) and 1.21), the synchronous formation of folds and boudins can best be understood by considering the strain state in the plane of the layering in which the structures are formed.

The field of possible strain states can be divided into three sub-fields; 1, 2, and 3 (Fig. 1.18(*c*)). In Field 1, extension occurs in all directions within the plane of the layer. The boudin necks will have no uniform orientation, although most will probably be sub-normal to the principal extension. This is the strain field associated with the formation of 'chocolate tablet' structure (Fig. 16.7). In Field 2, an extension occurs parallel to one of the principal strain axes and a contraction parallel with the other. The layer will, therefore, tend to fold in one direction and form boudinage in the other (Fig. 1.18(*c*)), so that the fold hinges will be at right angles to the boudin necks. It is interesting to note that cylindrical boudins (i.e. boudins with no significant contraction or extension along their length) form only when the strain state in the plane of the layer plots in the narrow strain field around the boundary between Fields 1 and 2. It is therefore somewhat surprising to note that a large proportion of natural boudins are approximately cylindrical.

2. The sequential formation of folds and boudins

When the boudins and folds are formed together in the manner described above, the boudin necks are *normal* to the fold hinges. However, boudins are frequently found on the limbs of folds with their necks *parallel* to the fold hinge, (Figs. 18.4(*a*) and 16.41). Such a relationship can be associated with the sequen-

tial development of folds and boudins during a single deformation.

Various stages in the amplification of a fold are shown in Fig. 16.46(*a*)–(*d*), together with the strain ellipse representing the strain in the fold profile section and a circle representing the undeformed state. The two lines of no finite longitudinal strain (lines which are the same length before and after deformation), which make an angle of θ with the direction of maximum compression, are also shown.

It can be seen by inspection, that any line (plane in three-dimensions) at an angle less than θ to the direction of maximum compression will undergo a shortening, parallel to its length during deformation, and any line at an angle greater than θ will undergo an extension. The lines of no finite longitudinal strain separate regions of extension from regions of contraction. During the early stages of the folding, all parts of the folding layer are inclined at an angle less than θ to the maximum compressive strain direction and, consequently, only contraction will occur. This will continue until the limb dip becomes parallel to one of the lines of no finite, longitudinal strain. As folding continues beyond this point, first the limb at the inflection point and then, gradually, more of the limb, moves into the region of extension. The possibility now exists for pinch-and-swell structures or boudins to develop on the limb. The length of the limb along which boudinage can occur increases as the fold tightens. This is shown schematically in Fig. 16.46.

3. (i) The folding of pinch-and-swell structures

The folds in natural examples of folded pinch-and-swell structures (e.g. Figs. 16.47 and 16.48) are almost invariably found to have triclinic symmetry. It can be inferred from these figures, that the pinch-and-swell structure governs the wavelength of the folds that develop. Generally, one pinch-and-swell structure is associated with one fold although, occasionally, two pinch-and-swells are incorporated into a single fold.

The effect of the existence of pinch-and-swell

(a)

(b)

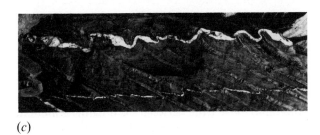

(c)

Fig. 16.48. Folded pinch-and-swell structures, (a) in Devonian limestone (now dolomitised) Hele Bay, N. Devon, England, (b) from the Dent du Morcle area, Switzerland, and (c) west of Luarca, N. Spain.

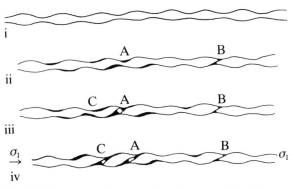

Fig. 16.49. Four stages in an experiment in which a Plasticine layer with pinch-and-swell structures was compressed along its length. At necks A and C buckling occurs initially. Later in the deformation these necks are each cut by two thrusts. At neck B, a single thrust formed which was not preceded by buckling. (After Penge, 1976.)

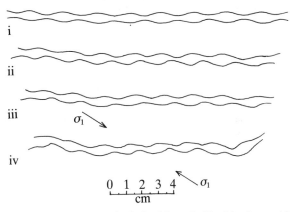

Fig. 16.50. Various stages in the buckling of a Plasticine layer with pinch-and-swell structures. The compression was applied at an acute angle to the layering. (After Penge, 1976.)

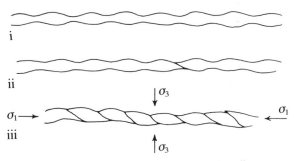

Fig. 16.51. Plasticine layer containing pinch-and-swell structures compressed parallel to the layer. Buckling does not occur and thrusts form at each neck region. (After Penge, 1976.)

structures on the buckling behaviour of a single layer has been investigated experimentally by Penge (1976), who found that when a layer was compressed parallel, or at an acute angle, to its length, the deformation was concentrated in the pinch zones. Two types of deformation of the pinch zones were observed, and these two types often developed sequentially. The first was the buckling of the pinch zones (Fig. 16.49(ii)). The folds that formed were always found to have triclinic symmetry regardless of the original symmetry of the pinch-and-swell structure, or whether the principal compression was parallel, or at an acute angle, to the layer (Figs. 16.49 and 16.50, respectively). The second type of deformation of the pinch zones was by thrusting (Fig. 16.51). In experiments where thrusting, rather than folding, occurred, only one of the possible conjugate sets of thrusts generally develops, even though the principal compression was parallel to the layer. Buckles and thrusts often occur in sequence during progressive deformation (Fig. 16.49 fold A) and either one (Figs. 16.51 and 16.49(B)) or two

(Fig. 16.49(A)) parallel thrusts develop at each pinch zone. The deformation in the pinch zone is such that, regardless of whether buckling or thrusting occurs, the swells are rotated so that they have in imbricate or en echelon arrangement. Folded pinch-and-swell structures have sometimes been referred to as 'fish-hook' folds. Such a geometrical allusion is particularly obvious when intense pressure solution results in the removal of the thin limb (Fig. 16.52). The folding shown in Figs. 16.47 to 16.52 has been

Fig. 16.52. Folded pinch-and-swell structures in the Devonian metasediments of Combe Martin, N. Devon. The limbs containing the pinch regions have almost been removed by pressure solution.

controlled by pre-existing pinch-and-swell structures. An example of such structures having very little effect on subsequent folding is shown in Fig. 16.53.

3. (ii) The folding of layers of rectangular boudins

We have noted in the previous section, that the folding of a continuous layer containing pinch-and-swell structures is, as far as the authors are aware, generally controlled by the wavelength of the individual pinch-and-swell. However, field observations (Fig. 16.54) and experimental work by Sengupta (1983) (Fig. 16.55) on the buckling of a discontinuous layer comprised of a series of approximately rectangular boudins, indicates that folding is not generally controlled by the width of the individual boudins. The experimental work shows that if the competence contrast between the boudins and matrix is relatively large during the folding, then the individual boudins retain their rectangular shape and may be thrust and 'stacked' over each other (Fig. 16.55(a)). Alternatively, if the competence contrast is rather small during the folding, the shape of the boudins may be modified into trapezoidal forms, often with flame-shaped projections (Fig. 16.55(b) and (c)). The boudin width does not seem to play an important role in determining the fold wavelength.

Although no mention has been made of the effect of the boudin separation on the subsequent folding of the disjointed layer it is to be expected that,

when such separation is small, the ability of the material to move from between the boudins is limited. If it is assumed that this material is linear-viscous, then it follows (Chap. 4) that the velocity (V) of migration is related to the fourth power of the distance (d) between the boudins. Hence, when d is small, folding occurs more rapidly that 'extrusion', and the disjointed units behave as a continuous layer.

We have considered the buckling of single layers that contain boudins or pinch-and-swell structures, and have noted that rectangular boudins have very little effect on the wavelength of folds that subsequently develop in the layer, whereas pinch-and-swell structures often determine the wavelength of the folds that form. However, field observations of pinch-and-swell structures in the multilayers in which the layers containing these structures are close enough together for their zones of contact strain to overlap sufficiently for multilayer rather than single layer folding to occur (Chap. 12), indicate that the pinch-and-swell structures do *not* control the wavelength of the subsequent folds (Fig. 16.56).

4. The range of forms from folds to pinch-and-swell structures

It was noted in Chap. 13 that one of the factors which determines the geometry of the structures that develop in an anisotropic rock, such as a rock fabric or well-bedded, sedimentary sequence when it is

Fig. 16.53. Folded pinch-and-swell structures from the Ross of Mull, Scotland. The wavelength of the folds seems unrelated to the wavelength of the pinch-and-swell structure.

deformed, is the orientation of the principal compression with respect to the layering or fabric. We see therefore that during the deformation of such material, structures ranging from symmetrical folds through asymmetrical folds and pinch-and-swell structures to symmetrical pinch-and-swell structures may develop depending upon whether the compression is parallel, oblique or normal to the layering (Fig. 13.20(*a*), (*d*) and (*g*)).

Boudin neck regions

It was noted earlier in the chapter, how the geometry and arrangement of the tension gashes which form in the neck region of boudins (Figs. 16.4 and 16.5) can be used to obtain information regarding the state of differential stress that existed during the development of the boudins. It has also become apparent from finite element work (Fig. 16.21), that the redistribution of minerals and variation in grain size around a boudin neck reflect the variation of stress, and indicate the direction of stress gradients that pertain during the formation of boudins. In addition, the geometry of the neck region can be used to obtain

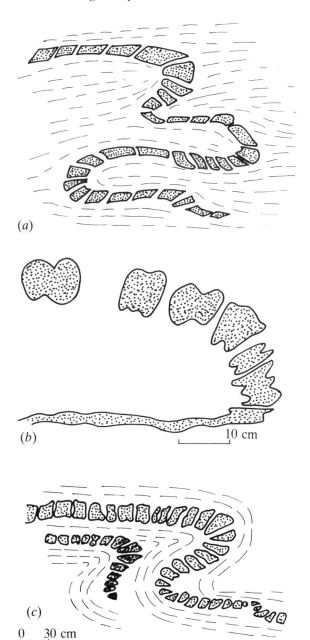

(*a*)

(*b*)   10 cm

(*c*)

0   30 cm

Fig. 16.54. Natural examples of folded boudinaged layers. (*a*) amphibolite bands in a quartzo-feldsparthic gneiss. (After Ramberg, 1952.) (*b*) Amphibolite layer in quartz-gneiss from the Pre-Cambrian rocks of Jashidih, E. India. (After Sengupta, 1983.) (*c*) Quartzite band in a semi-pelite. (After Gindy, 1952.)

information regarding the state of strain that existed during boudin formation. It has already been noted, Fig. 1.18, that the state of strain in the plane of a boudinaged layer is indicated by the plan geometry of the boudins. Let us now consider the state of strain in the profile section.

The space created between two boudins as they move apart may be filled either by the matrix flowing into the neck region (Fig. 16.57(*a*)) or by the neck region becoming filled with mobile minerals (Fig. 16.57(*b*)) commonly quartz or calcite. The factor which determines which of these types of neck develops is the strain-state in the profile section (Ramsay,

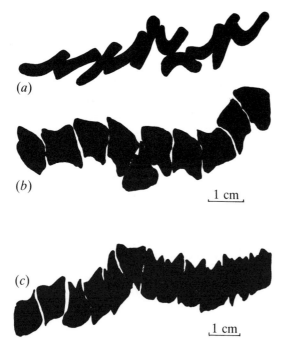

Fig. 16.55. Three rows of originally rectangular boudins which have been compressed parallel to the layering (the original aspect ratio of the boudins for (*a*), (*b*) and (*c*) was 5, 3 and 2.8 respectively). The amount of compression for (*a*) and (*b*) is the same; layer (*a*) is modelling clay and layers (*b*) and (*c*) are a less competent mixture of modelling clay and putty. The matrix is putty. The boudins now have flame-like projections at their edges. (After Sengupta, 1983.)

1967). The strain ellipses thought to be associated with the formation of both types of boudin necks are shown in Fig. 16.57, together with the circles representing the undeformed states. In both examples, considerable extension has occurred along the layer. However, in the example where the neck region is infilled with quartz, or some other mobile mineral, the contraction normal to the layer is relatively small: whereas, in the example in which the matrix flowed into the neck region, the contraction normal to the layering is relatively large. Depending upon the rheological properties of the boudins, the shear stresses generated at and near the 'corners' of the rectangular boudins by the flow of the matrix in to the neck regions, may cause the boudins to become barrel-shaped (Fig. 16.23(*b*)).

The initiation of layer-normal compression structures

Although structures formed by *layer-normal* compression, such as boudins, are probably more common than is realised, they are still far outnumbered by structures formed by *layer-parallel* compression. One possible reason for this may be related to the problem of the initiation of structures. The effect of layer-parallel compression on a low amplitude, initial irregularity in the layer from which a fold might develop is to amplify the irregularity, so that even if it originally had a very small amplitude, it will

Fig. 16.56. Folded multilayer containing pinch-and-swell structures. The zones of contact strain of adjacent competent layers overlaps sufficiently for multilayer buckling to occur and folding is independent of the wavelength of the pinch and swells in the individual layers. Burela, N. Spain.

eventually be of sufficient magnitude to initiate a fold. If, however, we consider the effect of layer-normal compression on this same irregularity, it will be suppressed rather than amplified and the chances of boudin or pinch-and-swell structures being initiated will be reduced. In addition, if slip at the layer interface is governed by the law of frictional sliding (Eq. (1.58), Chap. 1), layer-normal compression will increase the normal stress on the interface, rendering layer-parallel slip more difficult and further reducing the possibility of instability.

These factors may, in part, be responsible for the relatively low number of boudins and related structures found in nature. Because the irregularities from which many layer-normal compression structures are initiated have to be relatively large, it is not surprising that such structures can often be traced to their source perturbation (e.g. Figs. 16.30(*a*) and 16.38).

Commentary

Following the method adopted in the chapters on folding, the discussion of boudins, and other struc-

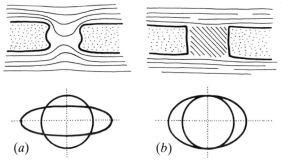

Fig. 16.57. Strain states in the profile section of a boudinaged layer; (a) when the matrix flows into the neck region; (b) when the matrix does not flow into the neck region, which is infilled with some mobile mineral such as quartz or calcite.

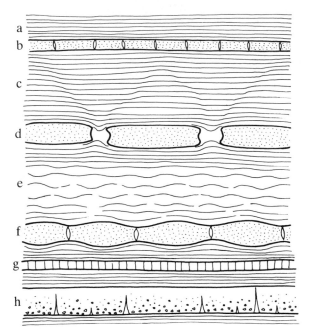

Fig. 16.58. A sketch showing a variety of structures all of which may form when the maximum principal compression is normal, or at a high angle, to the rock layering or fabric. (a) Homogeneous flattening (b) extension veins (c) normal kink-bands (d) boudins (e) internal pinch-and-swell structures (f) pinch-and-swell structures (g) extension fractures (h) tapered veins in graded bed.

tures formed by compression at a high angle to the layering, has been divided into sections on single layer boudins, multilayer boudins and the formation of boudins in anisotropic materials. Boudins, like folds, are surrounded by zones of contact strain and within a multilayer several 'orders' of boudins may develop.

The response of a complex geological multi-layer, which will have a bulk mechanical anisotropy as well as a variety of different, relatively competent layers, to compression at a high angle to the layering, can be very varied (Fig. 16.58). A variety of structures including 'joints', veins, boudins, pinch-and-swell structures and normal kink-bands may form. In addition, the profile geometry and spacing of boudins may also vary, depending on several parameters including the differential stress, Fig. 16.3(*a*) and (*d*), the rheo-

logy of the boudinaging layer, Fig. 16.3(*a*)–(*c*) and (*d*)–(*f*), the rheological contrast between the layer and the matrix, the amount of deformation, Fig. 16.25, the strain-state, Fig. 16.57 and the strain-rate. The range of structures between boudins and pinch-and-swell structures has already been discussed, Fig. 16.3, and it will be apparent from Fig. 16.59 that a gradation between boudins through veins to extension joints also exists. However, although the orientation of the principal compression and the rock layering are the same for the formation of boudins and extension joints normal to the layering, the geological environments in which they form are generally quite different. Boudins form either during diagenesis or during tectonism and are associated with an increase in layer-normal compression. Joints generally form by the release of residual stresses during exhumation when layer-normal stresses are being reduced.

References

Beloussov, V.V. (1952). Spacing of fractures in rocks. *Alcad. Nank. S.S.S.R. Trad. Geofiz. Inst.*, **17**, 144.

Biot, M.A. (1965). *Mechanics of incremental deformation*. New York: Wiley.

Cloos, E. (1947). Boudinage. *Trans. Am. Geophys. Union*, **28**, 626–32.

Ekström, T.K. (1975). Pinch-and-swell structures from a Swedish locality. *Geol. Foren. Stockholm. Forh.*, **97**, 180–7.

Fletcher, R.C. (1974). Wavelength selection in the folding of a single layer with power-law rheology. *Am. J. Sci.*, **274**, 1029–43.

Ghosh, S.K. & Ramberg, H. (1976). Reorientation of inclusions by combination of pure shear and simple shear. *Tectonophysics*, **34**, 1–76.

Gindy, A.R. (1952). The plutonic history of the district around Traweagh Bay, Co. Donegal. *Q. J. Geol. Soc. London*, **108**, 377–411.

Goguel, J. (1948). *Introduction a l'étude méchanique des déformations de l'écorce terèstre*, 2nd edn., Chapter 15, Paris: France Service Carte Geol. Mem.

Griggs, D. & Handin, J.W. (1960). Rock deformation (a symposium). *Geol. Soc. Am. Mem.*, **79**, 1–382.

Halbich, I.W. (1978). *Minor structures in gneiss and the origin of steep structures in the Okiep Copper District.* Geol. Soc. South Africa, special publication 4, No. 18.

Harker, A. (1889). On the local thickening of dykes and beds by folding. *Geol. Mag.*, New Ser., **6**, 69–70.

Hill, R. (1950). *The mathematical theory of plasticity*. Oxford: Clarendon Press.

Hossein, K.M. (1970). Experimental work on boudinage. Unpublished M.Sc. Dissertion, Imperial College, University of London.

Jones, A.G. (1959). Vernon map-area British Columbia. *Mem. Geol. Surv. Brch. Can.*, **296**, 1–186.

Lloyd, G.E. & Ferguson, C.C. (1981). Boudinage structure: some interpretations based on elastic-plastic finite element simulations. *J. Struct. Geol.*, **3**, 117–28.

Lloyd, G.E., Ferguson, C.C. & Reading, K. (1982). A stress transfer model for the development of extension

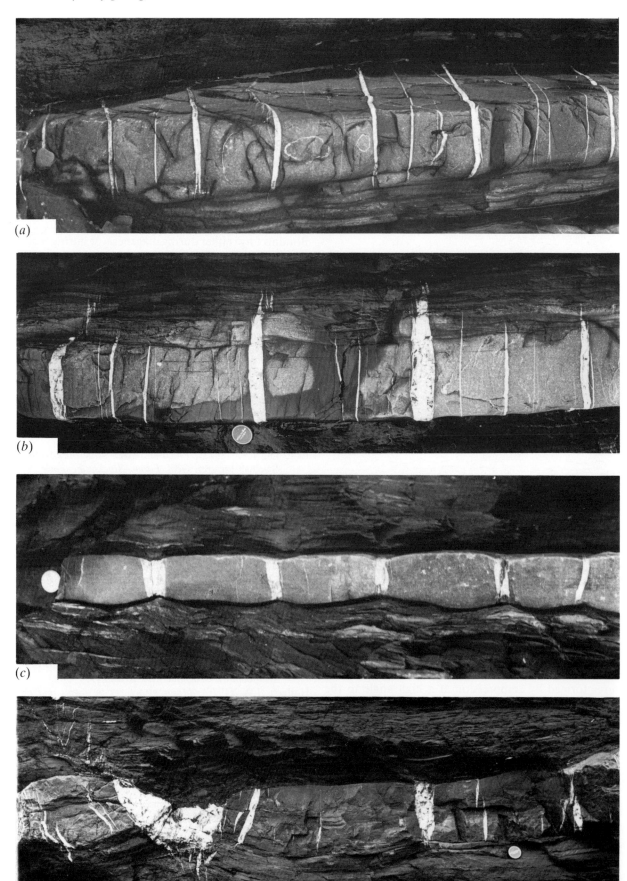

Fig. 16.59. A series of sandstone beds showing the gradation of structures between joints, veins and boudins. Crackington Haven, N. Cornwall, England.

fracture boudinage. *J. Struct. Geol.*, **4**, 355–72.

Lohest, M. (1909). De l'origine des veines et des géodes des terrains primaries de Belgique. Troisième note. *Annls. Soc. Géol. Belg.*, **36** (Bull for 1908–9), 275–82.

Nadai, A. (1950). *Theory of Flow and Fracture of Solids*. Vol. 1, New York: McGraw–Hill.

Neurath, C. & Smith, R.B. (1982). The effect of material properties on the growth rates of folding and boudinage: experiments with wax models. *J. Struct. Geol.*, **3**, 215–29.

Paterson, M.S. & Weiss, L.E. (1968). Folding and boudinage in quartz-rich layers in experimentally deformed phyllite. *Geol. Soc. Am. Bull.*, **79**, 795–812.

Penge, J. (1976). Experimental deformation of pinch-and-swell structures. Unpublished M.Sc. Dissertation, Imperial College, University of London.

Prandtl, L. (1925). Spannungsverkeilung in plastischen Koerpern. *Proc. 1st Intern. Congr. Applied Mechanics*. Delf, Holland.

Ramberg, H. (1952). *The origin of metamorphic and metasomatic rocks*. Chicago: University of Chicago Press.

(1955). Natural and experimental boudinage and pinch-and-swell structures. *J. Geol.*, **63**, 512–26.

(1959). Evolution of ptygmatic folding. *Norsk. Geol. Tidsskr.*, **39**, 99–152.

Ramsay, A.C. (1866). The geology of North Wales. *Mem. Geol. Surv. Gt. Britain*, **111**, 1–381.

Ramsay, J.G. (1967). *Folding and fracturing of rocks*. New York: McGraw–Hill.

Rast, N. (1956). The origin and significance of boudinage. *Geol. Mag.*, **93**, 401–8.

Rudnicki, J.W. & Rice, J.R. (1975). Conditions of localisation of deformation in pressure-sensitive dilatant materials. *J. Mech. Phys. Solids*, **23**, 371–94.

Sadrabadi, H.E. (1978). Experimental formation of single layer boudinage and pinch-and-swell structures. Unpublished M.Sc. Dissertation, University of London.

Selkman, S. (1978). Stress and displacement analysis of boudinage by finite-element method. *Tectonophysics*, **44**, 115–39.

Sengupta, S. (1983). Folding of boudinaged layers. *J. Struct. Geol.*, **5**, 197–210.

Smith, R.B. (1975). Unified theory of the onset of folding, boudinage and mullion structures. *Geol. Soc. Am. Bull.*, **86**, 1601–9.

(1977). Formation of folds, boudinage and mullions in non-Newtonian materials. *Geol. Soc. Am. Bull.*, **88**, 2, 312–20.

(1979). The folding of strongly non-Newtonian layers. *Am. J. Sci.*, **279**, 272–87.

Sowers, G.M. (1973). *Theory of spacing of extension fractures*. Geol. Soc. Am. Bull. Engineering case history, number 9.

Stephenson, O. & Berner, H. (1971). The finite element method in tectonic processes. *Phys. Earth Planet Inter.*, **4**, 301–21.

Stromgard, K.E. (1973). Stress distribution during formation of boudinage and pressure shadows. *Tectonophysics*, **16**, 215–48.

Tvergaard, V., Needleman, A. & Lo, K.K. (1981). Flow localisation in the plane strain tensile test. *J. Mech. Phys. Solids*, **29**, 2, 115–42.

Uemura, T. (1965). Tectonic analysis of the boudin structure in the Muro Group, Kii Peninsula, S.W. Japan. *J. Earth. Sci. Nagoya Univ.*, **13**, 99–114.

Voight, B. (1965). Plane flow of viscous matrix with an interned layer compressed between two rectangular, parallel rigid plates. A geological application and a potential viscometer in distorted rocks. *Geol. Soc. Am. Bull.*, **52**, 1355–418.

Wegmann, C.E. (1932). Note sur le boudinage. *Bull. Soc. Geol. France* (5th ser.), **2**, 477–91.

Wilson, G. (1961). The tectonic significance of small-scale structures, and their importance to the geologist in the field. *Annls. Soc. Geol. Belg.*, **4**, 423–548.

Woldekidan, T. (1982). Deformation of multilayers compressed normal to layering. Unpublished Ph.D. Thesis, University of London.

Wunderlich, H.G. (1962). Falten stereometrie und Gesteinsverformung. *Geol. Rdsch.*, **52**, 417–26.

17 Rock cleavage and other tectonic fabrics

Introduction

A variety of planar features form in rocks, and these can be conveniently divided into *primary* and *secondary* planes. Primary planes are formed when the rocks are deposited, extruded or intruded, and include bedding and flow banding. Secondary planes are produced as the result of tectonic processes and include joints, faults, mineral fabrics and metamorphic banding. A variety of terms such as schistosity, foliation and cleavage have been used to describe secondary planar features of the type that will concern us here. Wilson (1961) discusses the definition of these terms and is careful to distinguish between schistosity and foliation. He points out that *schistosity* is mainly used to describe the prominent planar structures which develop in schists and phyllites, as a result of the alignment of platy minerals. Foliation has a definite significance in Britain that has unfortunately been lost in America. British geologists use the word foliation to describe banded structures in metamorphic rocks in which the laminae are discrete. Darwin (1846) defines *foliation* as 'a more or less pronounced aggregation of particular constituent minerals of a metamorphic rock into lenticles or streaks or inconsistent bands, often very rich in some one mineral, contrasting with contiguous lenticles or streaks rich in other minerals'. Foliation is therefore a particularly appropriate term for the alternating quartz/felspathic-rich and amphibolite layers so common in many gneisses. Thus, a foliated rock is not necessarily schistose, nor is a schistose rock necessarily foliated.

It will become apparent from the following discussion that some of the various structures which are grouped together as cleavage may correspond to a schistosity and/or foliation. Accordingly, Wilson defines *rock cleavage* (and schistosity) as a planar structure, usually distinct from stratification, which permits rocks to be fractured or cleaved into thin slices. Dennis (1977), defines cleavage as a set of closely spaced secondary, planar, parallel fabric elements that impart mechanical anisotropy to a rock without apparent loss of cohesion. This latter definition appears more precise; however, as we shall see, it presents many difficulties in dealing with those forms of cleavage which are mainly defined by foliation or by fractures.

The chapter opens with a discussion of the classification of cleavage and the relationship between cleavage and folds. This is followed by a section on the mechanisms involved in the formation of the various types of rock cleavage, particular attention being paid to slaty cleavage, crenulation cleavage with the associated processes which lead to metamorphic differentiation fabrics and fracture cleavage. Refraction of cleavage and the relationship between cleavage and finite strain is then considered and the chapter ends with a discussion on the environments in which cleavage forms.

Classification of cleavage

It is possible for a mineral fabric to be uniformly developed and totally pervasive throughout a rock (Fig. 17.1). However, it has become apparent, largely as a result of electron microscopy, that the majority of cleaved rocks have a domainal structure, i.e. zones of strongly preferred mineral orientations, known as cleavage domains, separated by zones, known as microlithons, of differently and commonly less well oriented minerals. Powell (1979) has proposed a morphological classification of cleavage based on:

(i)   the spacing of cleavage domains
(ii) the shape of the cleavage domains
(iii) the microlithon fabric, and
(iv) the proportion of rock occupied by the cleavage domain.

1 mm

Fig. 17.1. Photomicrograph of a coarse, slaty cleavage showing a good crystallographic and shape alignment of the mica flakes.

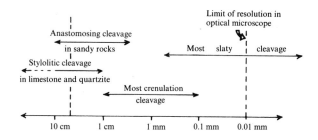

Fig. 17.2. Average spacing of cleavage domains for the various types of cleavage. (After Bayly *et al.*, 1977.)

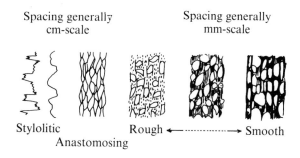

Fig. 17.3. Variation of cleavage domain shape in spaced cleavage in rocks with no pre-existing planar anisotropy. (After Powell, 1979.)

Spacing of cleavage domains

In rock cleavages such as that shown in Fig. 17.1, the cleavage is the result of an alignment of mineral grains and is a totally pervasive property of the rock down to the scale of the individual mineral. Such a pervasive or *continuous* cleavage can be subdivided on the basis of grain size, i.e. whether it is fine, as in some slates, or coarse as in schists and gneisses. A continuous cleavage represents the end member of a series of cleavages in which the cleavage domains range from very thin discontinuities, which occupy less than 1 per cent of the rock, to domains which occupy 100 per cent of the rock (i.e. a pervasive or continuous cleavage). Cleavages in which the cleavage domain occupies noticeably less than 100 per cent of the rock can be termed *spaced cleavage*. The range of spacing of cleavage domains of several types of cleavage are indicated in Fig. 17.2. An arbitrary upper limit of 5 cm has been suggested by Price & Hancock (1972), coupled with a further restriction that the distance between the cleavage planes should be less than 5 per cent of the bed thickness.

Shape of cleavage domain

Spaced cleavage can be subdivided on the basis of the planarity of the cleavage domains. From observations in the field and in thin sections, it is known that a large variety of cleavage domain shapes exists in nature ranging from irregular stylolitic seams (described by Alvarez *et al.*, 1978) through anastomosing networks of cleavage surfaces (Crook, 1964) and short discontinuous cleavage domains which envelope detrital grains (Powell, 1969) to planar discontinuous domains free from irregularities or unevenness. These cleavages, which are respectively termed stylolitic, anastomosing, rough and smooth, are shown in Fig. 17.3.

Microlithon fabric

Another subdivision of spaced cleavage can be made on the basis of whether or not there is a preferred orientation of minerals, which predates the cleavage. In rocks with a pre-existing mineral fabric (or some other form of anistropy such as fine banding) the cleavage that develops is generally one of the varieties

of *crenulation cleavage*. Spaced cleavage that develops in rocks containing no preferred mineral orientation or other form of anisotropy may be termed *disjunctive* cleavage.

Proportion of rock formed by cleavage domains

Spaced cleavages can be further subdivided on the width of the cleavage domain relative to that of the microlithon into *discrete*, when the cleavage domain occupies a small percentage of the rock, and *zonal* when the cleavage domains and microlithons occupy comparable percentages of the rock (Fig. 17.4).

Spatial relationship between folds and cleavage

The geometrical relationship between folds and cleavage was first noted by Sedgwick (1835) who commented on the general parallelism of cleavage strike and the trend of the fold axes. Darwin (1846) and Rogers (1856) noted that there is usually a general parallelism of cleavage and the axial surface of folds (Fig. 17.5(*a*)). The term *axial plane cleavage* sprang from such observations. It should be noted that this

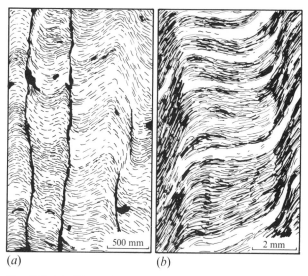

(*a*)   (*b*)

Fig. 17.4. (*a*) 'Discrete' cleavage, pressure solution seams in a siltstone. (*b*) 'Zonal' cleavage, a crenulation cleavage in a mica-rich rock.

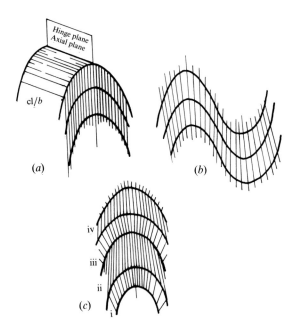

Fig. 17.5. (*a*) The relationship of axial plane cleavage to a fold. (*b*) The fanning of cleavage about a fold. (*c*) The 'refraction' of cleavage as it passes from one bed to another. Note that no refraction occurs at the hinge and that the amount of refraction increases away from the hinge.

term is not restricted to any particular form of cleavage, but simply indicates the geometrical relationship between the two structures.

One of the questions considered by the early workers on cleavage was whether axial plane cleavage predates, is synchronous with, or postdates the folding. Some, including Sedgwick (*op. cit.*) suggested that the cleavage was superimposed upon the folds; while Tyndall (1856) concluded that the formation of the fold and the cleavage was more or less synchronous. Harker (1886), accepting comments by Fischer (1884), concluded that flow cleavage and folding are both manifestations of the same compression and suggested that they are approximately concomitant, though not exactly synchronous structures. He concluded that in the main, cleavage formation is initiated later than folding. There is still some uncertainty regarding the timing of events; nevertheless, it is generally accepted that when a fold exhibits axial plane cleavage, this implies that cleavage and folding occurred during the same tectonic event. Conversely, when there is a marked difference between the orientation of the cleavage and the axial plane, this is usually adduced as evidence to indicate that the cleavage and folding are the result of completely different tectonic events.

Except in the hinge region of a fold, 'axial plane cleavage' is not, in fact, generally parallel to the axial plane. The cleavage may fan about the fold (Fig. 17.5(*b*)) or refract as it passes from one bed to another (Fig. 17.5(*c*)). The fanning and refraction of cleavage were described by Sorby (1853), and the

reasons for these and other variations in cleavage orientations are discussed later in this chapter.

Mechanisms of cleavage development

Having defined rock cleavage and discussed its classification let us now consider the different mechanisms by which the various types of cleavage are thought to form. Accordingly, we shall now discuss five main mechanisms, which individually or in combination, give rise to different cleavage forms. These mechanisms are:

(i)   The mechanical rotation of platy or needle-like minerals.
(ii) Syntectonic recrystallisation of such minerals.
(iii) Microfolding.
(iv) Pressure solution.
(v)   Hydraulic fracturing.

Mechanical rotation

Sorby (1853) suggested that a continuous cleavage, commonly termed flow or slaty cleavage (Fig. 17.1), develops by the rotation of platy silicate minerals during the process of strain development. Such a rotation has been studied using three different models, which are:

(i)   the March Model (March, 1932 and Owens, 1973) which considers the rotation of passive markers during compression,
(ii) the rotation of rigid bodies in a viscous fluid (Jeffrey, 1922), and
(iii) the rotation of viscous bodies in a viscous fluid (Gay, 1968).

The theory developed by March treats the modification of an initial, uniform angular distribution of linear or platy elements, by homogeneous strain (Fig. 17.6). He assumed that the lines or planes respond passively to strain, and deform with the medium in which they are embedded. He found that with increase in strain, the initial random distribution of elements is lost and the elements become progressively more and more aligned. If the elements have mechanical properties which are different from those of the matrix, as in the models discussed by Jeffrey and Gay, then both passive and active rotation of the elements occur and a preferred orientation is produced more rapidly. The passive and active rotation of the elements is analogous to the passive and active amplification of folds (Chaps. 10 and 11). The amount of active rotation depends upon several factors, including the aspect ratio of the particle (a/b in Fig. 17.6(*a*)). This can be seen in the experiment shown in Fig. 16.40(*b*), where the rotation of the individual rectangular boudins which have a relatively

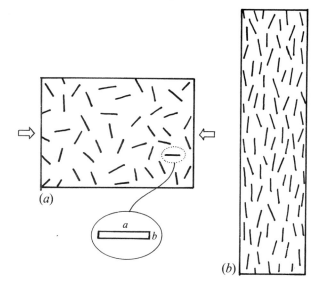

Fig. 17.6. Passive rotation of originally randomly oriented passive markers (*a*) by homogeneous flattening to produce a fabric (*b*).

small aspect ratio, is slower than the bulk rotation of the layer.

On the basis of his experimental work Tullis (1971 and 1976) concluded that, for mineral elements with the aspect ratios of most layer silicates, the model assuming passive rotation of the platy minerals predicts essentially the same degree of preferred orientation for a given strain as the model assuming active rotation. In a series of experiments, it was shown that if a mica aggregate is deformed, at temperatures too low to permit recrystallisation, the degree of preferred orientation predicted by the models is only attained when the mica flakes are loosely compacted. At higher strains, when there is greater interference between grains, the degree of preferred orientation that developed is less than that which is theoretically predicted. However, in high temperature experiments, when recrystallisation is possible, the observed degree of preferred orientation continues to fit the theoretical predictions; even when the mica flakes are in close contact. Thus, the effects of grain interference are cancelled out in conditions in which grain boundary sliding is facilitated and grains can change shape by diffusion-dependent processes.

An active rotation of platy minerals within a rock requires that the minerals displace the matrix. It

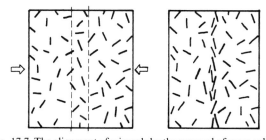

Fig. 17.7. The alignment of minerals by the removal of more soluble minerals from a planar zone by pressure solution.

has been suggested that this may occur in sediments which are not completely consolidated at the time of cleavage formation (Maxwell, 1962). Rotation in lithified sediments can occur if the matrix minerals are susceptible to deformation by pressure solution. This mechanism, which is discussed later in the chapter, is shown diagrammatically in Fig. 17.7. The preferred orientation of platy minerals results from their rotation as the soluble minerals of the 'matrix' (e.g. quartz) are removed. This process may occur uniformly throughout a rock unit (e.g. at quartz/ mica junctions) to produce a uniformly distributed fabric, or along localised seems to produce spaced cleavage.

Syntectonic recrystallisation

It was suggested by Harker (1886) that systectonic recrystallisation and the growth of platy minerals in a preferred orientation in response to compressive stress is a mechanism which can give rise to the formation of slaty cleavage. Kamb (1959) pointed out that anisotropic minerals in a stress field are thermodynamically more stable in some orientations than in others. Thus, in an aggregate of grains of different orientations, some grains will tend to dissolve and others will tend to grow. In the case of micas, for example, new grains will grow with their basal planes (001) perpendicular to the axis of maximum compressive stress (σ_1).

It is difficult to determine the relative importance of the various orientation mechanisms mentioned above in the formation of any particular example of flow cleavage, for all these mechanisms may play a role. The mechanism which dominates will depend upon the metamorphic condition (pressure, temperature etc.), and the strain-rate existing in the rock mass at the time of cleavage development.

Microfolding

The association of microfolding with the formation of a spaced cleavage, now commonly known as crenulation cleavage, has been recognised for over a hundred years. This cleavage develops in rocks which already possess a mineral fabric or fine layering, and the early fabric which cross-cuts the microlithons which separate the cleavage planes, is folded. This type of cleavage has been variously termed *strain-slip cleavage* (Bonney, 1886), *transposition cleavage* (Weiss, 1949) *herringbone cleavage* (Mead, 1940) and *shear cleavage* (Wilson, 1946). However, in recent years, the term *crenulation cleavage*, coined by Rickard and used by Knill (1960) and Rickard (1961), has been widely adopted. This variety of names reflects the variability of the geometry of the microstructures encompassed by this form of cleavage. Two main hypotheses regarding its origin have been proposed. The first of these suggests that the cleavage is the result of microfolding; and the second, that it is the result of microfaulting. The two natural examples

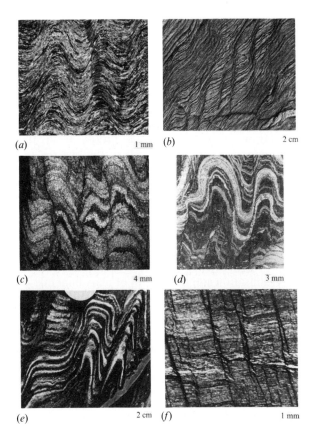

(a)   1 mm   (b)   2 cm

(c)   4 mm   (d)   3 mm

(e)   2 cm   (f)   1 mm

Fig. 17.8. (*a*) Crenulations and an embryonic crenulation cleavage in a quartz muscovite schist from the Lukmanier Pass, Switzerland. (*b*) A buckled, quartz-rich multilayer. Trearddur Bay, Angelsey, N. Wales. (*c*) Crenulation cleavage in a buckled multilayer from the Pre-Cambrian phyllites of Rhoscolyn, Anglesey, N. Wales. (*d*) Pressure solution planes associated with the buckling of quartz-rich layers in the Quartenschiefer, Brigels, Switzerland. (*e*) Crenulation cleavage developed in association with multilayer buckles. Rhoscolyn, Anglesey, N. Wales. (*f*) Crenulation cleavage in the Devonian slates of the Mosel Valley, Berncastle, Germany. (*a*) and (*b*) should be compared with Fig. 17.9 (*a*) and (*d*).

shown in Fig. 17.8(*a*) and (*f*) indicate why these two opposing hypotheses arose.

Many of the mineral fabrics and micro-multilayers in which crenulation cleavage forms, approximate closely to the statistical homogeneous and anisotropic material, the buckling behaviour of which has been discussed at length in Chap. 13. Some of the structures that may develop in such materials when they are deformed are shown in Fig. 17.9. It will be recalled from Chap. 13, that the geometry of the structures depends upon the amount and type of anisotropy, and the orientation of the fabric with respect to the maximum principal stress direction, (Fig. 13.20). The relationship of these various structures to each other through their common 'instability' is further discussed in Chap. 13 and illustrated in Figs. 13.49 and 13.50. With this understanding of the deformation behaviour of anisotropic materials, the reason for the large variety of geometries of micro-structures associates with crenulation cleavage becomes clear. In addition, the dilemma regarding

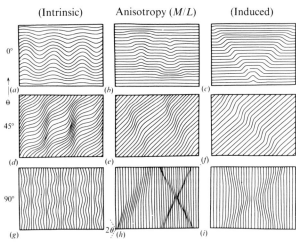

Fig. 17.9. Some possible variations in the geometry of structures that may form in anisotropic materials depending on the type of anisotropy (intrinsic or induced) and the orientation (*θ*) of the axes of symmetry of the anisotropy (in a material with an intrinsic anisotropy these axes are the axes of symmetry of the fabric elements which give rise to the anisotropy) with respect to the maximum compression direction. (Modified from Cosgrove, 1976.)

whether the cleavage types shown in Fig. 17.8(*a*) and (*f*) are the result of folding or fracturing is resolved (cf. Fig. 17.8(*a*) and (*f*) with 17.9(*a*) and (*h*) respectively).

Conjugate crenulation cleavage

This type of cleavage was first recorded by Muff (1909) in the Dalradian phyllites of Craignish, Argyllshire. He concluded that the conjugate cleavage planes have the same significance as conjugate faults.

The relationship between the conjugate crenulation cleavage and the slaty cleavage in the Craignish area is summarised in Fig. 17.10. It can be seen that the crenulation cleavages are symmetric about an axial plane slaty cleavage, and were assumed to be the result of the same stress that caused the slaty cleavage. If the crenulation cleavages were the result of 'brittle' shear failure, then the angle between the conjugate cleavage planes (2*θ*, Fig. 17.11) should be less than 90°. In almost all the examples measured, the angle 2*θ* exceeded 90°. Knill (1957 and 1960) suggested that this angle was originally less than 90° and that subsequent flattening had increased the angle to more than 90°. However, there is an alternative interpretation of this conjugate crenulation cleavage. It was pointed out in Chap. 13, that when a layering or fabric is compressed along the layering, conjugate reverse kink-bands may

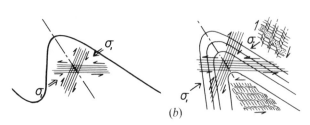

Fig. 17.10. (*a*) The formation of conjugate crenulation cleavage. (After Muff, 1909.) (*b*) A modification of (*a*) by Wilson (1961).

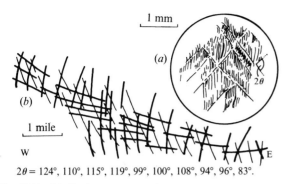

$2\theta = 124°, 110°, 115°, 119°, 99°, 100°, 108°, 94°, 96°, 83°.$

Fig. 17.11. (*a*) Conjugate crenulation cleavage from Craignish, Argyllshire, Scotland. (*b*) The relationship between conjugate crenulation cleavage planes (heavy lines) and slaty cleavage (light lines). (After Knill, 1957.)

form (Fig. 17.12(*a*)) and when compressed normal to the fabric, conjugate normal kink-bands may form (Fig. 17.12(*b*)). In both situations the maximum principal compressive stress bisects the *obtuse* angle between the kink-bands. It is suggested that the conjugate crenulation cleavage of the Craignish area is the result of the formation of 'normal' kink-like structures (Fig. 17.12(*b*)) and not of the subsequent flattening of conjugate faults, which formed by satisfying the criteria for brittle shear failure (Chap. 1).

During the formation of slaty cleavage, the rock fabric of which it is composed becomes more and more pronounced and the anisotropy of the slaty cleavage fabric increases. At some stage in its development, the slaty cleavage fabric may become unstable with respect to the maximum principal compressive stress. As discussed in Chap. 13, the types of structures that develop will depend upon the type and amount of anisotropy and may be internal pinch-and-swell structures or conjugate normal kink-bands (Fig. 17.9(*g*) and (*h*) and (*i*) respectively).

The recognition of this process enables yet another conflict in view-points to be resolved. Thus, as a slaty cleavage develops, instabilities are more and more likely to form. This tends to inhibit the development of a perfectly uniform slaty cleavage fabric. The

development of the incipient shear instabilities, which express themselves as internal pinch-and-swell structures and/or normal kink-bands, which form at a high angle to the compression, may explain why early workers, after examination of thin sections, found considerable difficulty in determining whether slaty cleavage was the result of either a mineral fabric which developed at right angles to the maximum compression direction as a result of grain rotation and recrystallisation, or of failure along conjugate shear planes, which, owing to subsequent flattening, were rotated into orientations approximately normal to the principal compression.

Conjugate crenulation cleavage may also occur in association with conjugate reverse kink-bands (Fig. 17.12(*a*)) which develop in a mineral fabric or finely layered material, when the maximum principal compression acts parallel to the fabric. It can be seen from the geometry of Fig. 15.22, that if the length of foliation (L) inside the reverse kink-band is kept constant, then, during the early stages of its development, there is a volume increase within the kink-band until the internal foliation is normal to the kink-band boundaries ($\omega = \theta$). As the band amplifies beyond this point, the volume begins to decrease and when the internal foliation has rotated so that ($\omega = 2\theta$), no net change in volume has occurred. Further amplification requires a reduction in volume. If, under the conditions of deformation no mechanism is available to achieve this reduction, then the kink-band must lock. If, however, volume reduction (i.e. layer thinning) can be achieved (for example, by the process of pressure solution), then amplification can continue. If the kink-band is significantly depleted in soluble minerals (such as quartz) there is a concentration of the insoluble minerals (such as mica). These residual insoluble minerals are aligned within the kink-band, causing it to be a planar zone of relative weakness. When the original foliation is a mineral fabric, the resulting kink-bands are often narrow and the planes of weakness produced by the process described above constitute a crenulation cleavage (Fig. 17.13(*b*)).

In the development of a normal kink-band, there is no early stage of volume increase and the layering, or fabric, within the kink-band is required to thin at *all* stages of kink-band amplification (Fig. 15.22(*e*)). This difference between reverse and normal kink-bands may help to explain why the latter occur less frequently than the former in nature. Thinning of the layering or fabric within the normal kink-band can be achieved by extension parallel to the layering or fabric within the kink-band and/or by the removal of material. An example of a crenulation cleavage which is the result of the formation of normal kink-bands in a mica fabric is shown in Fig. 17.8(*f*).

From the above arguments we see that the formation of crenulation cleavage can generally be

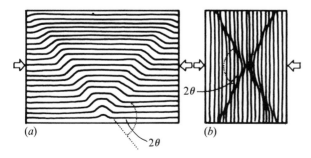

Fig. 17.12. (*a*) Conjugate reverse kink-bands formed by compression parallel to the fabric. (*b*) Conjugate normal kink-bands formed by compression normal to the fabric. In both cases the principal compressive stress bisects the *obtuse* angle between the kink-bands.

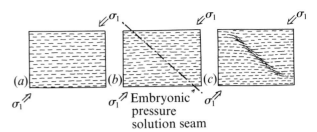

Fig. 17.15. Removal of material from a locality by pressure solution, giving rise to a structure resembling a crenulation cleavage.

Fig. 17.13. Photomicrographs of crenulation cleavage in (a) the Devonian slates of the Mosel Valley, Germany, (b) the Pre-Cambrian phyllites of Rhoscolyn, Anglesey, N. Wales, (c) the metasediments of the Lukmanier Pass, Switzerland and (d) a slate from Tor Cross, S. Devon, England.(e) Cleavage exposed in an outcrop of Devonian slate, Tor Cross, S. Devon, England.

divided into two stages. The first is the development of the stress-induced instability in an anisotropic rock which gives rise to the crenulations, and the second is the onset of mineral redistribution which is often associated with the development of the discrete planes or zones of weakness which constitute the cleavage.

It is possible, however, for a crenulation cleavage to form simply as a result of microfolding and without the occurrence of mineral redistribution. Consider the development of microfolds in a rock fabric made up predominantly of platy or acicular minerals (Fig. 17.14(a)). The orientation of the mineral flakes is approximately constant along any line drawn parallel to the axial trace of the crenulations. As the folds develop and become tighter, the mineral flakes on the limbs become aligned so that they all lie approximately on one plane, producing planes of weakness (Fig. 17.14(b) and (c)), which occur at regular intervals (i.e. on each limb) and constitute a crenulation cleavage which is the result of the mechanical rotation of minerals during folding.

Similarly, structures looking remarkably like crenulation cleavage can result from the development of pressure solution planes in a mineral fabric without the development of buckling instabilities. Durney (1971) has described and discussed the generation of

pressure solution planes in a variety of rock types including limestones and shales. If planar seams which are the result of intense pressure solution are formed in a rock with a quartz/mica fabric which is being compressed oblique to the fabric (Fig. 17.15(a)), then local removal of quartz along the seam causes a collapse of the mica fabric and the formation of a structure that is indistinguishable from some varieties of crenulation cleavage (Fig. 17.8(f)).

However, in general, the formation of crenulation cleavage involves both microfolding and mineral redistribution (Fig. 17.8). The mineral redistribution is often the result of pressure solution, mineral migration and redeposition, and these processes have been grouped together under the heading of *metamorphic differentiation*. The cleavages in Fig. 17.8 can be divided into two groups. The first, which includes the crenulation cleavages shown in Fig. 17.8(a) and (f), is the result of metamorphic differentiation associated with the microfolding of a mineral fabric. The second group, Fig. 17.8(b)–(e) results from metamorphic differentiation associated with the microfolding of a multilayer. In the following section, the process of metamorphic differentiation and its association with the buckling of mineral fabrics and fine multilayers is examined.

Metamorphic differentiation

The three processes involved in metamorphic differentiation in low-grade, metamorphic environments (i.e. pressure solution, mineral migration and redeposition) are governed by the laws of thermodynamics, which indicate the direction in which a reaction or process will tend to proceed under a particular set of physical conditions. We wish to know the direction of these processes under the various pressure and temperature conditions that exist around a developing fold. The thermodynamic potential which determines the direction of a reaction at a particular pressure (P) and temperature (T) is the Gibbs free energy, G. For a reaction occurring at constant pressure and temperature to be spontaneous, $dG < 0$.

The chemical potential μ_i of a component i gives a measure of the stability of that component under

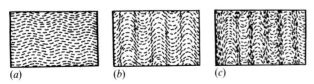

Fig. 17.14. (a), (b) and (c) Various stages in the buckling of a micaceous fabric to produce regularly spaced planes of weakness as the result of the alignment of the mica flakes on the limbs.

particular pressure and temperature conditions and is defined by:

$$\mu_i = \left(\frac{dG}{dn_i}\right)_{PTn} \quad (17.1)$$

where n_i is the number of moles of component i. The larger the chemical potential, the less stable the component and the greater its tendency to react chemically or undergo pressure solution.

If a phase contains only one component and if the temperature remains constant, it follows from the Gibbs–Duhem relationship (Turner & Verhoogen, 1960) that the change in chemical potential ($d\mu$) induced by a change in pressure dP is:

$$d\mu = VdP \quad (17.2)$$

where V is the volume. This relationship indicates that an increase in hydrostatic stress (pressure) will increase the chemical potential of a mineral.

However, in most geological situations, particularly those in which pressure solution and mineral migration are likely to occur, the stress will be non-hydrostatic. The application of non-hydrostatic thermodynamics to geological processes including pressure solution, has been considered by Kamb (1961), McLellan (1966) and Paterson (1973). The condition of equilibrium between a stressed solid and its solution is found to be:

$$\mu_1 = \mu_s - TS_s + \sigma_n V_s \quad (17.3)$$

where μ_s, S_s and V_s are the chemical potential, entropy and volume of the solid respectively. This equation shows that the chemical potential of a solid in a surrounding solution μ_1 is directly related to the normal stress σ_n acting on its surface.

As can be seen from Eqs. (17.2) and (17.3), the chemical potential of a mineral, and therefore its tendency to undergo pressure solution, will be directly related to the magnitude of the applied stress.

In considering pressure solution and mineral migration around microstructures, it will be assumed that the temperature remains constant and that pressure solution and mineral migration occur only in response to stress gradients. For convenience, we shall assume that a 'phase' exists along grain boundries and that mineral species can leave their parent crystal, enter the grain boundary phase, migrate through it and precipitate out onto some other crystal or some other part of the parent crystal.

Because the chemical potential of a mineral depends on the state of stress (Eqs. (17.2) and (17.3)), there will be a gradient in the chemical potential parallel to any stress gradient. The chemical potential at any point in a crystal can be considered as a measure of the tendency of that part of the mineral to leave the parent mineral and move into the grain boundary 'phase'. It follows, therefore, that there will

be a corresponding concentration gradient established in the grain boundary phase which runs parallel with the chemical potential gradient. Mineral species will migrate through the grain boundary phase as a consequence of this concentration gradient, moving from areas of relatively high to relatively low concentration.

During the deformation of rock, stress gradients may be established on various scales, for example around individual grains or, on a larger scale, between the hinge and limb areas of a fold. In order to explain the development of stress gradients around the folds in a buckling mineral fabric, it is useful to consider the stress distribution within a stressed multilayer made up of alternating layers a and b, which may be either elastic or viscous (Fig. 15.4). It was argued in Chap. 15 that, if such a multilayer is compressed parallel to the layering, the layers with the higher valued modulus support a larger proportion of the compressive force than the layers with the smaller modulus. The less 'competent' layers are 'protected' by the more 'competent' layers. This effect can be used to explain how stress gradients develop in certain minerals and layers during folding. Let us now use these concepts to explain the process of metamorphic differentiation and the formation of crenulation cleavage.

Stress gradients and metamorphic differentiation associated with the microfolding of a mineral fabric
In this section, we shall consider the occurrence of metamorphic differentiation in two mineral fabrics; (i) quartz and muscovite and (ii) chlorite and muscovite. Both these examples have been discussed by Cosgrove (1976). In example (i), shown schematically in Fig. 17.16, the limbs of the folds are almost 100 per cent mica and the hinges consist predominantly of quartz. Using the buckling theory for anisotropic materials (Chap. 13) and the arguments regarding stress distribution in a multilayer, presented above, it is possible to account for this differentiation.

Metamorphic differentiation in a quartz and muscovite fabric
Because of the ease with which quartz deforms by pressure solution in low-grade metamorphic environ-

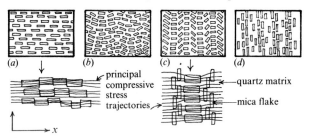

Fig. 17.16. A schematic representation of microfolding and metamorphic differentiation in a quartz/mica fabric. (See text for discussion.)

ments (Cosgrove, 1972, Kerrich, 1975), it can be assumed that during the folding of a quartz/mica fabric, in this environment, the mica offers a larger resistance to compression than the quartz. It follows from the discussion of the stress distribution within a multilayer (Fig. 15.4) that σ_1 in the mica flakes would be greater than σ_1 in the quartz. We may also assume that the stress difference $\sigma_1 - \sigma_3$ increased to some critical value, so that the mineral fabric, being anisotropic, became unstable and developed buckling instabilities (e.g. Type 1 buckling instabilities, Fig. 17.9(a)). Various stages in the process of metamorphic differentiation are shown schematically in Fig. 17.16. Initially, the principal compressive stress is concentrated in the micas (Fig. 17.16(a)), and even after the onset of buckling, when the micas begin to rotate away from parallelism to the *x* direction, the axis of maximum principal stress continues to act along the mica flakes. (The orientation of stress trajectories during folding is well illustrated in the finite-element work of Dieterich & Carter, 1969). However, as the folds develop and the mica flakes on the limbs continue to rotate away from the *x* direction, the trajectories begin to cut 'across' the fabric (Fig. 17.16(c)), and act once more approximately parallel to the *x* direction. As folding proceeds beyond this stage, the quartz in the limb areas becomes subjected to a progressively larger compressive stress as it is not longer 'protected' by the mica framework (cf. Fig. 17.16(a) and (c)). The magnitude of the stress on the quartz in the limb will increase with the dip of the limb, reaching a maximum when the folds become isoclinal. Because the micas in the hinge area have not rotated, the quartz in the hinge area is still protected. It follows that the stress acting on the quartz in the limbs is greater than the stress acting on the quartz in the hinge. A stress gradient will therefore exist between the quartz in the hinge and the limb and the magnitude of this gradient will increase as the folds develop. This stress gradient will tend to cause migration of the quartz *from the limb towards the hinge*.

However, the stress gradient established in the micas during the development of the folds shown in Fig. 17.16, will act in the opposite direction to the gradient established in the quartz. The mica in the hinge areas supports most of the compression and is under a relatively large compressive stress compared with the mica in the limbs (Fig. 17.16(c)). Pressure solution of mica will therefore occur more readily in the hinge areas and there will be a tendency for it to migrate *from the hinge areas towards the limbs*.

Metamorphic differentiation in a chlorite and muscovite fabric

Microfolds in an originally uniform chlorite/muscovite fabric, from the Pre-Cambrian of Rhoscolyn, Anglesey, are shown in Fig. 17.17(a). The limbs of these folds have been completely recrystallised within very well-defined zones. X-ray diffraction analysis of the limb and hinge areas confirms that the rock is made up almost entirely of muscovite and chlorite. Observations in areas where the mineral fabric is unfolded show that, originally, the proportion of muscovite to chlorite was approximately constant. However, the relative proportions of these two minerals in the limb and hinge areas of folds are very different (Fig. 17.17(b) and (c)). The ratio of muscovite to chlorite is approximately 4/1 and 1/2.5 in the limbs and hinge respectively.

During the conversation of chlorite to muscovite, water is released. Because this water can migrate away from the locality of its liberation (cf. Gresens, 1966, on migration of water towards fold hinges) and because chlorite has a larger unit cell volume than muscovite, the conversion of chlorite into muscovite at a locality results in a local volume reduction. Such a reaction will be facilitated by an increase in compressive stress, whereas the breakdown of muscovite to chlorite, which involves a volume increase, will not. The chlorite will therefore be less capable of supporting stress than the mica and it is suggested that the system chlorite/muscovite is analogous to the quartz/muscovite system (Fig. 17.16), where the muscovite acts as the more competent material relative to chlorite. Following the arguments used in the discussion of the quartz/muscovite folds of Fig. 17.16, there will exist a stress gradient in the chlorite which will tend to cause migration of this mineral from the limbs to the hinges. However, because of the complexity of the chlorite structure, it is unlikely that 'unit chlorite' molecules go into solution. It is more likely that the chlorite undergoes incongruent solution, i.e. parts of the molecule enter solution, and that these migrate and are deposited at various sites around the parent crystal.

Because the conversion from chlorite to muscovite and water can achieve local volume reduction and requires little alteration of the basic silicate framework, it is probable that as the chlorite begins to recrystallise in response to the relatively high stress it experiences on the limb, it is converted to muscovite by absorbing the K^+ ions and rejecting the Mg^{++} and Fe^{++} ions. This will result in a concentration gradient in Mg^{++} and Fe^{++} ions which will parallel the stress gradient in the chlorite. Consequently, these mafic species will migrate towards the hinge. At the hinge zone, muscovite is under a relatively high stress compared with the muscovite on the limbs (cf. Fig. 17.16(c)) and will therefore tend to recrystallise. In doing so, it is probable that the free Mg^{++} and Fe^{++} ions and the water, concentrated at the hinge, will be incorporated into the new lattice, resulting in the formation of chlorite. This will result in the build up of K^+ ions in the hinge area, since the K^+ ions

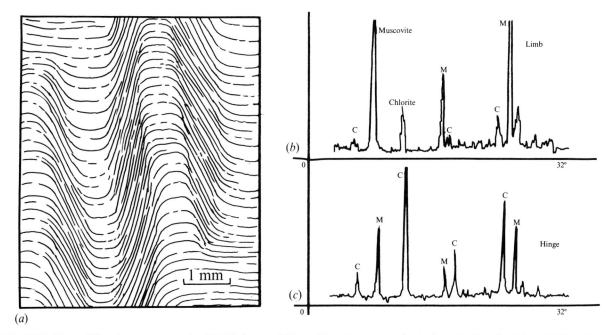

Fig. 17.17. X-ray diffraction pattern peaks (CuK$_A$) from a folded chlorite/muscovite fabric from the Pre-Cambrian of Rhoscolyn, Anglesey, N. Wales (*a*). (*b*) from the limb and (*c*) from the hinge. M = muscovite, C = chlorite. (After Cosgrove, 1976.)

from the muscovite lattice are not incorporated into the chlorite lattice. The build up of K$^+$ ions at the hinge, as a result of the formation of chlorite from muscovite, and the absorption of K$^+$ ions in the limbs as muscovite is formed from chlorite, will result in a concentration gradient in K$^+$ ions which will tend to cause migration from the hinge to the limb. In this way, migration of a few simple species, Mg^{++}, Fe^{++}, K$^+$ and H$_2$O, in response to stress gradients and the resulting concentration gradients, can account for the 'redistribution' of muscovite and chlorite during the development of microfolds.

Although the limb areas are totally recrystallised during microfolding, not all the chlorite has been converted to muscovite. It can be inferred from this that the migration of K$^+$ ions to the limb areas is the rate-determining step. As the K$^+$ ions are provided by the conversion of muscovite to chlorite, a process involving a volume increase in an essentially compressive environment, this is likely to be the rate-determining process.

Metamorphic differentiation, in originally uniform fabrics, described in the preceding paragraphs, will produce a very fine metamorphic banding, the width of which is governed by the wavelength of the microfolds and is generally a few millimetres wide (Figs. 17.16 and 17.17). Let us now consider a process which gives rise to a coarser banding, namely the buckling of a fine multilayer.

Metamorphic differentiation associated with the folding of a multilayer

The banding made up of alternating quartz and mica-rich layers, each with a thickness of approxi-

mately 2 mm (Fig. 17.16(*d*)), produced by metamorphic differentiation of an original mineral fabric, is itself susceptible to folding during subsequent deformation. When this occurs, crenulation cleavage may develop on two quite different scales; on a smaller scale, as the result of the buckling of the mica-rich layers and on a larger scale, as the result of the buckling of the multilayer as a whole (Fig. 17.18).

Using the arguments presented above to explain the metamorphic differentiation that occurs in a quartz/muscovite fabric during folding (Fig. 17.16), it is suggested that the quartz-rich layers in the limb areas within the folded multilayer (Fig. 17.18(*d*)) are under a relatively high compressive stress compared to the quartz-rich layers in the hinge areas which are protected by the layers which are rich in mica. The quartz-rich layers on the limbs will therefore tend to undergo pressure solution and migrate towards the hinge areas. This is shown schematically in Fig. 17.18. The quartz-rich layers in the limbs may be completely removed, leaving isolated hinges, in which the multilayer fabric is still discernible, separated by mica-rich bands (Figs. 17.18(*e*), 17.19, 17.20 and 17.21)).

Some of the folds that develop in such multilayers have slightly divergent axial planes (Type 2 structures, Fig. 17.8(*e*) and 17.9(*b*)). In such folds, the limb areas where intense pressure solution of the quartz-rich layers has occurred are particularly well-defined. They are separated from the hinge areas by a narrow zone where the change in dip of the layering takes place very rapidly (cf. also Fig. 17.21).

It is apparent from Fig. 17.18, that the wavelength of the folds determines the thickness of the layering produced by metamorphic differentiation. It

Fig. 17.18. Various stages in the transposition of a multilayer made up of alternating quartz-rich (q) and mica-rich (m) layers. (After Cosgrove, 1976.)

is also clear from this figure that, as one layering is transposed by metamorphic differentiation into another, the thickness of the resultant layers is increased (cf. Fig. 17.18(*a*) and (*f*)). Metamorphic transposition from one fabric to another, by this mechanism, appears to occur with remarkable ease. For example, in the Pre-Cambrian phyllites of Holy Isle, Anglesey, N. Wales, evidence for at least four such transpositions can be found. Three of these can clearly be seen in Fig. 17.20. These fabrics are only locally developed, and it is often possible to trace a particular fabric, in the field, from its 'undeformed'

Fig. 17.19. An undeflected crenulation cleavage with the fold closures still discernible in the microlithons. From the New Harbour Group, Holy Isle, Anglesey, N. Wales.

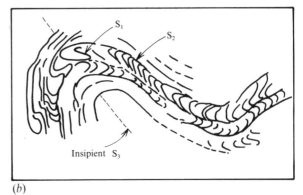

(b)

Fig. 17.20. (*a*) A gently folded crenulation cleavage from the New Harbour Group, Holy Isle, Anglesey, N. Wales. (*b*) Line drawing showing three transposition structures.

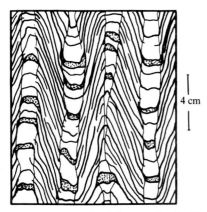

Fig. 17.21. The transposition of bedding associated with folding in the Cambrian Cabitza slates of the Sa Duchessa area, S.W. Sardinia.

state (Fig. 17.19) into an adjacent area where it begins to fold (Fig. 17.20) and is eventually entirely transposed into a new fabric (Fig. 17.18(f)). Thus, various stages in the process of transposition are preserved so that, providing sufficient field exposure and evidence are available, the 'total deformation history' can be reconstructed (Fig. 17.18).

In gneisses, metamorphic banding and layering on a much larger scale can be seen. Here, alternating layers of quartzo/feldspathic and amphibolitic material, sometimes several metres in thickness, occur. It is tempting to infer that a process of metamorphic differentiation comparable with that already discussed gave rise to this coarse banding.

Crenulation cleavage may also develop in a multilayer consisting of alternating quartz- and mica-rich layers in a manner slightly different from that described above and illustrated in Fig. 17.18. A study of thin sections (e.g. Fig. 17.8(d)) enables one to argue that, at some stage in the folding of a multilayer, the quartz-rich layers physically 'collide' so that further buckling is inhibited. The multilayer 'locks up' and stress concentrations develop at the point contacts, resulting in the interpenetration of the limbs (Fig. 17.22(d)). (The situation is analogous to the interpenetration of pebbles by pressure solution at point contacts.) This process leads to the development of intense pressure solution on the limbs of the multilayer fold and extreme thinning, or complete removal, of the quartz-rich layers in these areas. Many of the crenulation cleavages developed in fine multilayers containing layers of different competencies are produced in this way.

However, it is not necessary for the limbs of buckles to collide for intense pressure solution to occur on the limbs, and single layer buckles have been observed in the Devonian slates of Tor Cross, S. Devon, with well developed pressure solution planes at the limb/matrix junctions. These pressure solution planes often extend well into the surrounding matrix.

In order to understand the occurrence of such pressure solution planes at the fold limb/matrix junction and the propagation of these planes into the matrix, it is useful to consider the stress distribution around a relatively rigid particle set in a less rigid matrix. The principal stress trajectories and the variation in mean stress and principal compressive stress around such a rigid inclusion are shown in Fig. 17.23 (i) and (ii). Both the mean and greatest principal stresses are maximum where the particle/matrix interface is normal to the bulk σ_1 direction, and lowest where the interface is parallel to this direction. If pressure solution, mineral migration and redeposition occur as a consequence of stress gradients, as suggested earlier, it is to be expected that pressure solution will occur in regions 'a' (dotted) and deposition in regions 'b' (black) in Fig. 17.23(ii). Mobile minerals will migrate from areas 'a' and it is argued that, because of loss of material, these areas become analogous to *less* rigid particles (or even holes) set in a more rigid matrix. The stress distribution around an elliptical hole is shown in Fig. 17.23(iii). Marked stress concentrations occur at the ends (d and e); hence these will be the sites of most active pressure solution. Pressure solution planes will propagate from these regions in a direction normal to the maximum principal compressive stress, i.e. parallel to the σ_3 stress trajectories (Fig. 17.23(ii)).

The initiation of pressure solution planes at fold limb/matrix contacts and the propagation of these planes into the matrix (point X in Fig. 17.8(d)) can be explained, if we consider a buckled layer to be analogous to a series of relatively rigid particles set in a less rigid matrix (Fig. 17.23(iv)).

Development of cleavage in a complex anisotropic material

The type of cleavage developed in the 'slates' of the Tor Cross area S. Devon depends primarily on lithology. The slates are predominantly pelitic but contain numerous siltstone and sandstone layers of various thicknesses. The pelitic horizons exhibit a finely spaced crenulation cleavage, the result of the buckling of a bedding fabric in the original shale, on which a much coarser, rather irregularly spaced system of pressure solution planes, parallel or sub-parallel to the finer cleavage, is superposed. The silt and sandstone layers are generally buckled and cut by well-developed pressure solution planes, which are often too closely spaced for all of them to be accounted for by pressure solution at the limb/matrix junction. However, despite their close spacing there seems little doubt that large number of these solution planes are formed by the mechanism illustrated in Fig. 17.23(iv).

If the siltstone and sandstone layers are sufficiently far apart for their zones of contact strain not to

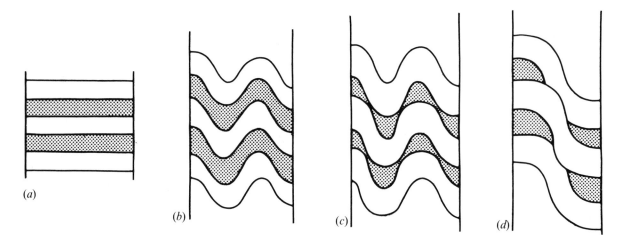

Fig. 17.22. Various stages in the development of a crenulation cleavage in a multilayer, made up of quartz-rich (light) and mica-rich (dark) layers from the Quartenschiefer of Brigels, Switzerland.

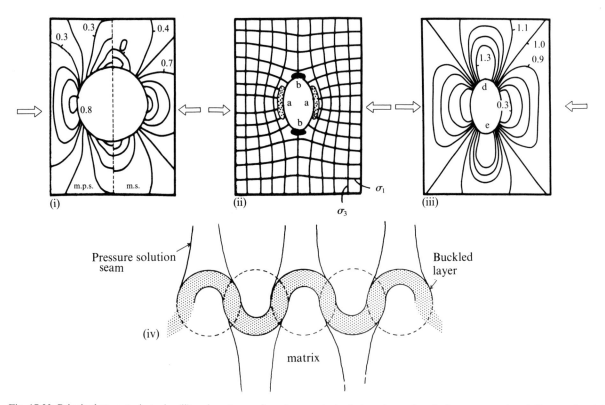

Fig. 17.23. Principal stress trajectories (ii) and contours of maximum principal stress (m.p.s.) and of mean stress (m.s.) (i) around a rigid inclusion set in a less rigid matrix. (iii) shows lines of equal maximum principal compression around an elliptical hole under a compression parallel to the minor axis. (After Savin, 1961.) Maximum compression occurs at d and e. (iv) A buckled layer acting as a series of rigid particles; pressure solution planes are initiated at the limb–matrix contact and propagate into the matrix, parallel to the minimum principal stress trajectories. (After Cosgrove, 1976.)

overlap, each layer will develop its own characteristic wavelength (L) depending on its thickness (a), and its competence (η_1) relative to the competence of the matrix (η_2), according to the single layer buckling equation (11.8). Pressure solution will be initiated at the limb/matrix contact and these planes will propagate into the matrix (Fig. 17.24(b)). They may eventually cut other quartz-rich layers (Fig. 17.24(c)). The end result of this process will be a closely spaced system of pressure solution planes cutting both the matrix and the buckled layers. Some of these pressure solution planes will be obviously associated with particular buckles (e.g. 'a' and 'b' of Fig. 17.24(c)), and others will not (e.g. 'd' and 'e'). The amount of pressure solution that has occurred along a pressure solution plane is found to vary, e.g. two solution planes may coalesce and accentuate each other (x and y of Fig. 17.24(c)).

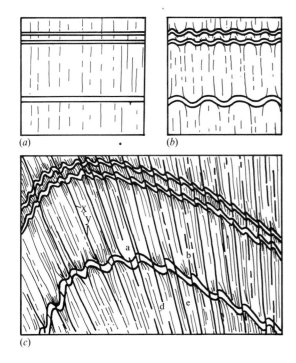

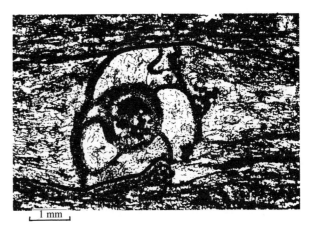

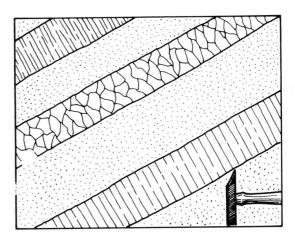

Fig. 17.24. Possible stages in the development of pressure solution planes in the slates of Tor Cross, S. Devon, England.

The development of a disjunctive or spaced cleavage defined by pressure solution planes is not, of course, always found in association with folds. Pressure solution seams can be initiated at any relatively rigid particle, e.g. a fossil (Fig. 17.25), or fossil fragment, an oolite or mineral grain in a host rock, and extend into the rock along the σ_3 stress trajectories.

Fracture cleavage: the formation of a spaced cleavage by hydraulic fracturing

This is a type of spaced cleavage about which there is considerable debate. The morphology of this cleavage varies widely and a variety of terms have been suggested, such as 'false cleavage' or 'close-joint cleavage' to describe the divers structures which are here grouped together under the heading of fracture cleavage. This cleavage may be superimposed on a pre-existing fabric, or pre-existing planes of weakness such as a crenulation cleavage, and it is probable that the fractures in both these instances developed during the process of uplift and exhumation (cf. Chap. 9). Such fractures have the same significance as joints but their orientation is controlled by a pre-existing cleavage. Here, we shall discuss only those fracture cleavages which develop in rocks which are devoid of an obvious fabric. As both the morphology and the orientation of fracture cleavage appear to be related to the rock type it cuts, we shall consider those forms which are exhibited in (i) incompetent mud-rocks and (ii) competent limestones and sandstones.

Fig. 17.25. Pressure solution seams cutting a fossil.

Fracture cleavage in incompetent rocks

Although the structures are often superficially similar, fracture cleavage differs from slaty cleavage in a number of important respects. For example, when a rock exhibits true slaty cleavage (a continuous cleavage) it is possible to split the rock parallel to its planar fabric into progressively thinner and thinner sheets. This is not possible when the rock is cut by fracture cleavage (a spaced cleavage), for the microlithons between the fractures exhibits little or no weakness parallel to the cleavage direction. Microscopic study of thin sections of the microlithons shows little sign of recrystallisation. When it is well-developed, fracture cleavage in incompetent, mud-rocks consists of closely spaced parallel, planar fractures, as shown in Fig. 17.26. It will be noted that the spacing between fractures differs from unit to unit, so that, in this example, the intensity of cleavage development is probably related to minor variations in lithology.

However, it can be demonstrated in certain areas, such as the Silurian turbidites of Mid Wales and the Culm of the north coast of Devon (England), that gradation exists between fracture cleavage and true slaty, or flow cleavage. Moreover, there are, in addi-

Fig. 17.26. An example of fracture cleavage in incompetent units from the Culm of N. Devon, England.

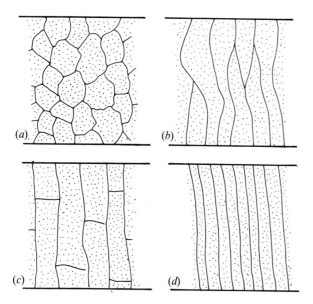

Fig. 17.27. Representation of morphological types of fracture cleavage, ranging from (*a*) the most rudimentary type to (*d*) 'classical' fracture cleavage.

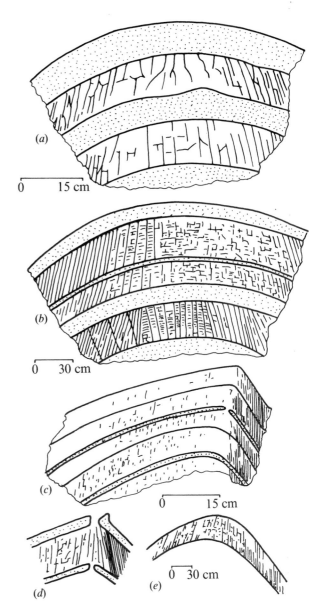

Fig. 17.28. Field sketches showing variations in fracture cleavage development in incompetent units in the Aberystwyth Grits. (After Price, 1953.)

tion, rudimentary or degenerate forms of fracture cleavage. For example, in the Aberystwyth Grits (turbidites of Llandovery age), of mid Wales, the morphology of fracture cleavage and related forms, in incompetent units, varies considerably to form a transitional sequence. In some instances, the fractures are almost randomly oriented and enclose irregularly shaped 'oblate spheroids' which may be several centimetres across (Fig. 17.27(*a*)). In other localities, the fractures are more regularly oriented, the discrete fractures are relatively widely spaced and are sometimes non-planar, with minor cross fractures connecting the main cleavage planes (Fig. 17.27(*b*)). When the main fractures and the cross fractures are equally well-developed, the sections are then approximately square or rhomboid. Such forms have sometimes been termed *pencil cleavage*, or *rab* (Keeping, 1878). As the cleavage form further improves, the cross fractures become less closely spaced (Fig. 17.27(*c*)) and eventually the ideal fracture cleavage form is attained (Fig. 17.27(*d*)).

Because of its transitional nature, it is likely that all the forms shown in Fig. 17.27 are the result of one basic mechanism. It is for this reason that the example of rab and other 'degenerates' or 'rudimentary' forms have been included in the description and discussion of fracture cleavage.

It has been shown by Price (1953), that variations in cleavage development are not necessarily restricted to lithology. Field sketches showing how fracture cleavage forms may vary in a single sedimentary unit over a distance of a metre or less are given in Fig. 17.28. Fracture cleavage in the limbs of folds in the Aberystwyth Grits commonly exhibit a better form than that which developed near the hinge.

Also, fracture cleavage in the short limb of an asymmetrical fold usually exhibits a better form than that which develops in the longer limb (Fig. 17.28(*c*)). In addition, for those examples in the Aberystwyth Grits where it proved possible to establish the position of the neutral surface within a fold, it was found that the cleavage formed below the neutral surface was better than that which developed above it (Price, 1967).

From such observations, it can be inferred that the degree of development of fracture cleavage is a sensitive indicator of lithology and, when lithology is constant, it is a measure of the tectonic strain. Unfortunately, because of the absence of strain markers in the argillaceous turbidites, it is not possible in the examples cited above to indicate the quantitative amount of strain which must be induced in a rock of

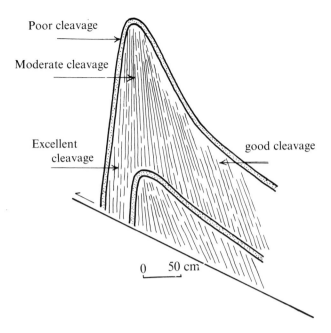

Fig. 17.29. Field sketch showing cleavage fan in incompetent units of the Aberystwyth Grits. (After Price, 1953.)

specific lithology before fracture cleavage will develop.

The relationship of the orientation of fracture cleavage planes relative to the fold in which they develop is comparable with that exhibited by slaty cleavage. That is, the fracture cleavage planes, in the incompetent units, are either parallel to the axial plane of the fold or else arranged in a fan about the hinge surface, which diverges when traced from the outer towards the inner arc of the folds (Fig. 17.29 and layers (ii) and (iv) of Fig. 17.5(*c*)).

Fracture cleavage in competent rocks
Descriptions are given by Price & Hancock (1972) of closely spaced fractures (including fracture cleavage) developed in competent beds with occur in continental Devonian rocks in S.W. Wales and in Palaeogene rocks in the Aragon synclinorium of the S. Pyrenees. When seen in profile sections, these cleavage planes make angles which range from 60–90° with the bedding plane; and the cleavage/bedding intersection lineation is parallel to the fold axis (Fig. 17.30). Thus, as will be seen in this figure, the cleavage planes form a fan which converges when traced towards the inner arc of the fold. This is the opposite relationship from that exhibited by fracture cleavage planes in incompetent rocks. Different lithologies often exhibit different angles of fracture, so that the cleavage 'refracts' as it passes from one bed to another.

In both regions, the principal types of fractures belonging to the group being discussed are joints, fracture cleavage, small fissures and, less commonly, minor faults and small quartz- or calcite-filled veins. The morphology of the fracture cleavage and kindred

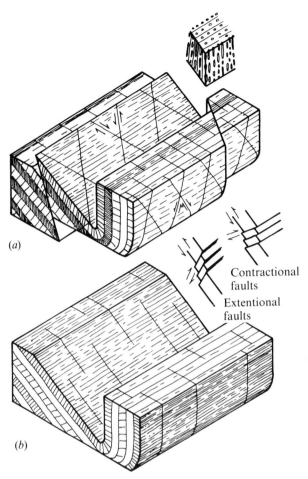

Contractional faults

Extentional faults

Fig. 17.30. Schematic diagrams showing the relationships between folds and hydraulic fractures (cleavage) and some of the fractures which cut the cleavage. (*a*) Dyfed (South Pembrokeshire), Wales: the small diagram depicts the arrangement of the carbonate secretions. (*b*) Aragon: the small diagrams show the arrangement of the rarely developed minor faults. (After Price & Hancock, 1972.)

structures (which some geologists may term 'close-joints') is illustrated in Fig. 17.31. Most fractures are restricted to individual beds and they are slightly irregular surfaces which form a sub-parallel, sometimes anastomosing, array.

In both areas, there is every gradation from widely-spaced fractures to recognisable fracture cleavage. This gradation led to an arbitrary definition; the structure is termed fracture cleavage if the distance between fractures is less than 5 per cent of the thickness of the bed. For beds which are more than 1 m thick a second arbitrary limit is necessary; namely, that the fracture separation should not exceed 5 cm.

Shear movement along cleavage planes is extremely rare in both Dyfed and Aragon. The sense of slip on the few planes which exhibit shear is antithetic to the sense of shear which results from folding (Fig. 17.30(*b*)). Cleavage planes are not deflected by drag along bedding, nor do they show any other evidence of deformation during folding. The

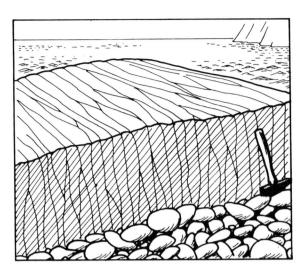

Fig. 17.31. Anastomosing fracture cleavage in Devonian limestone, Dyfed, Wales.

inference is, that fracture cleavage and kindred structures were formed after, or towards the close of, the development of the fold in which they occur.

As the fractures seldom exhibit any evidence of shear, it is probable that the cleavage results from extension of the rock mass in a direction perpendicular to the fractures. Indeed, the slightly non-planar morphology exhibited by many fracture cleavage planes in these areas would preclude shear as a general mechanism of formation. It is significant, therefore, that cleavage planes are cut and distorted by shear zones (represented by shear fractures in Fig. 17.30(*a*)) and thrusts in Dyfed (Hancock, 1964), and by smooth tension joints in Aragon which appear to have developed in response to the same system of stresses which gave rise to the fold (Fig. 17.30(*b*)). This apparently paradoxical situation requires explanation, for this conclusion seems to be at variance with the widely held concept of fold development, namely that folds form in a period of continuous compression.

We have noted in Chap. 15 that during folding the stress history is complex, with some parts of the fold experiencing high levels of compression while others undergo extension. As we have seen, the degree of cleavage development may vary from place to place in a fold. However, in general, it is a well-distributed feature that is not restricted to particular parts of a fold. Hence, we cannot always invoke local extension to explain the development of fracture cleavage throughout the whole fold. It is, therefore, necessary to indicate some mechanism whereby a period in which the effective stresses approximately perpendicular to the axial plane change from compression to tension, and then return to compression.

The only hypothesis known to the authors which attempts to explain how such a hiatus in the compression phase could arise is that proposed by Price & Hancock (1972), who point out that the fracture cleavage planes often have the characteristics exhibited by those induced by hydraulic fracturing. They emphasise the fact that high fluid pressures will exist during the amplification of a fold and that folds develop serially or sequentially. (These are concepts which have been considered at length in Chaps. 10, 11, 13, and 15). Price & Hancock also argue that the transition from stable (elastic) buckling to unstable (inelastic) buckling (the initiation of fold A, Fig. 17.32) is accompanied by a stress drop (cf. also Chap. 15). They suggest that a fold which has locked up and is 'transmitting' the compressive stress to an adjacent extremely low amplitude elastic buckle will also experience a stress drop when the second structure (fold B, Fig. 17.32) amplifies and moves into the inelastic buckle phase. It is at this time (Fig. 17.32), they propose, that hydraulic fracturing induces the formation of fracture cleavage in the first fold. This process is repeated as fold succeeds fold in their initiation and development.

Relationship between cleavage and finite strain

The relationship between slaty (flow) cleavage and the finite strain ellipsoid has been considered for well over a century. In his lucid review of flow cleavage, covering the period from 1815 to 1972, Siddans (1972) points out that a remarkable understanding regarding the mechanisms of formation and the mechanical significance of slaty cleavage existed as early as 1886. Based on numerous studies of deformed objects (e.g. Heim, 1878, Ramsay, 1881, Harker, 1886 and Wettstein, 1886), slaty cleavage has been recognised as a flattening phenomenon, forming normal to the short axis of the finite strain ellipsoid. In addition, the important mechanisms of rotation and deformation of the constituent grains of the rock, pressure solution transfer and recrystallisation, had all been demonstrated and the common geometrical features of slaty cleavage, refraction and fanning, had been described and accounted for in terms of strain theory. Nevertheless, not all geologists were convinced that slaty cleavage and the finite strain ellipsoid were related. Laugel (1855) for example, suggested that slaty cleavage formed along planes of high shearing strain. Becker (1893) supported this view, citing as evidence an example of crenulation cleavage illustrated by Heim (1878) (Fig. 17.33), as indicating that shearing along cleavage planes does occur. Such evidence is, at best, controversial for, as we have seen, the apparent displacement along the limbs of the microfolds can be accounted for by the combination of microfolding and mineral redistribution (as discussed in the section on crenulation cleavage) Fig. 17.34. Similarly, pressure solution along discrete, spaced planes can provide the appearance of shear displacement along the cleavage planes (Fig. 17.15(*c*)) but, as can be seen

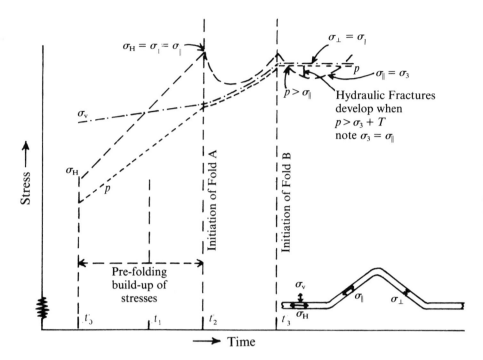

Fig. 17.32. Variations in magnitude of stress with time throughout most of the development of a single fold (A). The axes of the diagram are not to scale. The inset diagram indicates the nomenclature used in describing the orientation of axes of principal stress. (After Price & Hancock, 1972.)

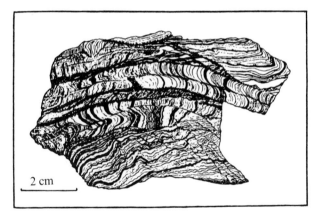

Fig. 17.34. Buckled quartz vein showing various stages in the removal of a limb by pressure solution. (Photomicrograph.)

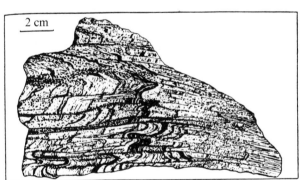

Fig. 17.33. Crenulation cleavage of the type illustrated by Heim (1878) and cited by Becker (1893), as evidence for shearing along a cleavage plane.

from Fig. 17.35, such displacements are 'apparent' and not real.

Alternatively, movement on cleavage planes may occur either at a late stage in the deformation which generated the cleavage or during a much later,

unrelated deformation. For example, vertical shear on discrete planes may subsequently develop during uplift (cf. Chap. 9). The diagnostic feature that will enable the real displacements associated with uplift to be differentiated from the apparent displacements associated with pressure solution, will be the sense of displacement on the cleavage planes cutting the limbs of folds. The displacement sense will reverse when passing from one limb to another if displacement is caused by pressure solution whereas the sense of displacement would be constant across a fold if the displacement were caused by uplift.

Since Laugel and Becker suggested that slaty cleavage formed along planes of high shearing strain, a number of studies of deformed objects have been carried out. All these studies show that slaty cleavage forms approximately normal to the short axis of the

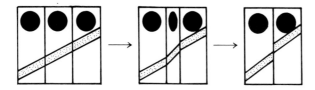

Fig. 17.35. Apparent shear displacement of a marker band by the volume reduction in a local domain; e.g. by the processes of pressure solution and mineral migration.

finite strain ellipsoid that the shearing played no role in cleavage development. One of these classic studies was by Cloos (1947) who determined the strain distribution around the South Mountain Fold in Maryland (U.S.A.) using deformed ooliths as strain markers. He was able to map out the regional variation of strain orientation and intensity, and relate them to the fold geometry and cleavage development (Fig. 17.36) and showed that the cleavage contains the maximum and intermediate axes of the ellipsoidal ooliths (which correspond to the finite strain ellipsoid, (Fig. 17.37).

This conclusion is given further support by the finite element work on the folding of a single layer in a matrix (Dieterich & Carter, 1969, and Dieterich, 1969). This work enabled the magnitude and orientation of stress and strain around and within a growing buckle to be determined. By comparing the stress and strain fields in the model with the cleavage patterns developed around and within natural single layer folds, Dieterich concludes that axial plane foliation forms in response to the *total strain* and develops normal to the direction of maximum total shortening (Fig. 17.38).

Despite the weight of field evidence that exists in the form of deformed objects indicating that slaty cleavage corresponds to the XY plane of the strain ellipsoid (i.e. the plane containing the intermediate

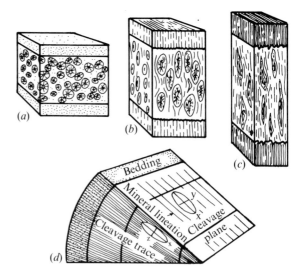

Fig. 17.37. Block diagrams of (a) an undeformed oolitic limestone sandwiched between beds of different lithologies; (b) appearance of (a) after 50 per cent deformation and (c) after 100 per cent deformation. (d) The relationship between the deformed ooliths and the cleavage. The major and intermediate axes of the oolith lie in the cleavage plane and the major axis is parallel to the mineral lineation. (After Cloos, 1947.)

and major axes of finite strain Fig. 17.37), a number of interesting questions regarding its formation (and that of cleavage in general) still remain. Indeed, Siddans (*op. cit.*) concluded that a central problem which still remains unanswered is how and why the various *incremental* stress-controlled mechanisms (e.g. recrystallisation and pressure solution) and strain-controlled mechanisms (e.g. mechanical rotation of platy minerals) combine with mimetic changes to produce a final rock fabric geometrically related to the *finite* strain ellipsoid, even when the deformation increments are likely to have been superposed non-coaxially.

In a review of the relationship between axial plane foliations and strains, Williams (1976) addresses this problem and attempts to determine whether cleavages develop parallel to a principal plane of finite strain and stay parallel to it throughout their development, or whether cleavage forms in some other orientation (e.g. parallel to the principal plane of the stress or incremental strain ellipsoid) and rotates towards parallelism under the influence of strain.

He considers two types of cleavage, one defined by a planar discontinuity such as a fracture cleavage plane, or several varieties of crenulation cleavage, and the other defined by the preferred dimensional orientation of mineral grains.

If it is assumed that no material migrates across the discontinuity and in or out of the system, then cleavages defined by planar discontinuities can only parallel the XY plane of the finite strain ellipse in special cases and can only 'track' the XY plane (i.e. be parallel to it at all times during the deformation) if the strain history is coaxial. If the discontinuity is origi-

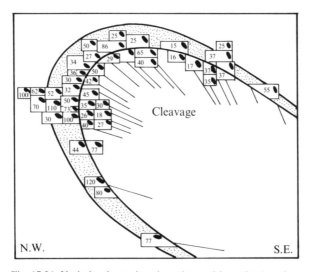

Fig. 17.36. Variation in strain orientation and intensity (numbers represent per cent strain) around a major fold (the South Mountain Fold) determined from measurement of deformed ooliths. (After Cloos, 1947.)

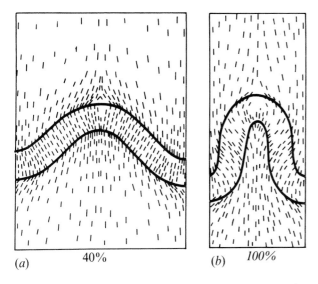

(a) 40% (b) 100%

Fig. 17.38. Two stages (40 per cent and 100 per cent compressive strain) in the finite element modelling of a buckling single layer. The competence contrast between the layer and matrix is 42/1 and the short lines are parallel to the long axis of the finite strain ellipse. (After Dieterich, 1969.)

nally not parallel to the *XY* plane and/or if the strain history is non-coaxial, the discontinuity will not 'track' the *XY* plane.

Cleavages defined by a preferred dimensional orientation of minerals or other bodies differ from those defined by discontinuities in that they *are* capable of tracking the *XY* plane. This can be seen in Fig. 17.39, which shows an example of both coaxial and non-coaxial deformation of an initial, random distribution of passive planar markers. The orientation of the resultant cleavage is defined by that of the

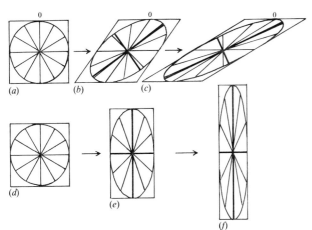

(a) (b) (c)

(d)

(e)

(f)

Fig. 17.39. The development of a preferred orientation by the rotation of passive markers during progressive strain: (*a*)–(*c*) non-coaxial strain history; (*d*)–(*f*) coaxial strain history. In both cases, the preferred orientation is symmetrical about the principal strain axes for all strains. In coaxial deformation, the same markers define the preferred orientation maximum at all stages. In non-coaxial deformation, different markers define the preferred orientation at different stages. The plane of the diagram is the *XZ* plane and the long axes of the strain ellipses represent the trace of the *XY* plane. (After Williams, 1976.)

maximuim concentration of markers and is parallel to the *XY* plane of the finite strain ellipsoid for both examples.

If the markers have a preferred orientation prior to deformation then, in general, the deformation fabric will not coincide with the *XY* plane. In addition, the deformation fabrics predicted by the March model (Fig. 17.39) are modified when the fabric elements do not behave passively and when they are close enough to each other to interfere. These modifications have already been briefly discussed earlier in this chapter in the section on the mechanical rotation of platy minerals.

If the rock contained initially spherical, passive markers, such as calcite oolites in a limestone matrix or lapilli in a matrix of identical tuffacious material, the markers deform into ellipsoids during homogeneous strain which are a direct measure of the finite strain ellipsoids, and their *XY* planes are parallel to any foliation that is formed (Cloos, 1947, 1971; Badoux, 1970) irrespective of whether the strain is coaxial or non-coaxial. However, if the strain markers do not act as passive markers (i.e. have different mechanical properties from the host rock) then, if the strain history is non-coaxial, the resulting foliation defined by the *XY* planes of the deformed markers, may not be truly parallel to the *XY* plane of the finite strain ellipsoid. Further, if the strain markers were originally ellipsoidal, the *XY* planes of the deformed ellipsoids will not generally be parallel to that of the finite strain ellipsoid.

From this brief discussion, we would expect that cleavage would not always be parallel to the *XY* plane of the finite strain ellipsoid. However, from field observations it may be inferred that the two features *are* parallel, any difference between the two is small, and is within the limits of accuracy that can be expected from the measurement and processing of field data.

Refraction of cleavage

The refraction of cleavage as it passes from one lithology to another (Figs. 17.5(*c*) and 17.40) was described by Sorby (1853) and Harker (1886). Both observed that in passing from slate to grit, the cleavage makes a higher angle with bedding in the grits than in the slates and that cleavage refraction is greatest when it cuts the bedding at moderate angles. Sorby attributed this to greater deformation and dilation in the slate than in the grit, but concluded that, since the deformation of both lithologies must be the same in the plane of junction, the direction of greatest compression must be less inclined to the bedding in the slate than in the grit. Harker verified this idea theoretically and numerically and the concept is shown diagrammatically in Fig. 17.41. The

Fig. 17.40. Photograph of refracted cleavage. Combe Martin, N. Devon, England.

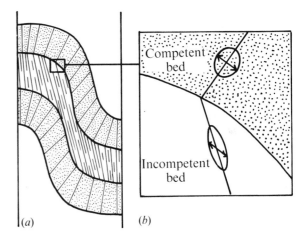

Fig. 17.41. (*a*) Refraction of cleavage between beds of different competences. The magnitude and orientation of the strain is represented by the strain ellipses shown in (*b*). For strain compatibility at the layer boundary, the state of strain (diameter of the strain ellipse) parallel to the boundary must be the same in both layers.

state of strain (represented by the strain ellipses) in the different layers, which have different mechanical properties and, therefore, different resistances to deformation, will be different. If it is assumed that cohesion exists along the junction between layers then, as Sorby argued, the state of strain parallel to the junction must be the same in both layers. For this condition of strain compatibility to be satisfied, it will generally be necessary for the principal extension axes of the strain ellipsoids in the two layers to be inclined at different angles to the layer boundary. Thus, if the cleavage forms parallel, or approximately, parallel to the *XY* plane of the strain ellipsoid, it will refract as it passes from one bed to the next.

Environment in which cleavage develops

Fourmarier (1951) has dealt at length with the role which depth of burial plays in influencing the development of cleavage. From a study of cleavage development in many areas, which include the Canadian Rockies, the Appalachians, the Pyrenees and the Alps, he reached the conclusion that several zones may be defined. From the surface downwards, they are successively:

I   a zone devoid of cleavage, limited at its base by the 'Upper Cleavage Front'
II a zone of fracture cleavage which grades into
III a zone of slaty (flow) cleavage, limited at its base by the 'Lower Cleavage Front'.

According to Fourmarier, the depths at which these limits are reached cannot be accurately defined, for they depend upon the lithology of the rocks and the amount of compressive strain. Thus, where the strata are only gently flexured, the upper cleavage front occurs at a lower level than that where folding is intense. Fourmarier, nevertheless, estimates that the depth at which the upper cleavage front normally occurs is at 5–6 km.

However, it has been suggested that rock cleavage may develop at much shallower levels in the crust. Thus, Maxwell (1962) noted a parallelism between the small sedimentary dykes and the cleavage in the Ordovician shales of New Jersey (Fig. 17.42), and proposed that both cleavage and dykes developed

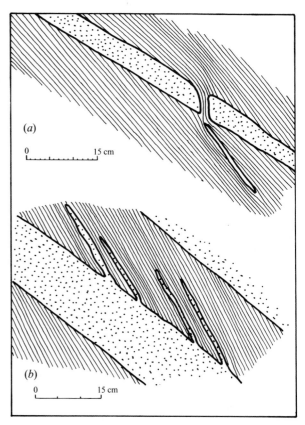

Fig. 17.42. Sandstone dykes parallel to slaty cleavage: (*a*) in the Martinsburg formation, south of Columbia, New Jersey, U.S.A. (*b*) in the Hudson River Group, west of Poughkeepsie, New York, U.S.A. (After Maxwell, 1962.)

Fig. 17.43. Recumbent slump folds in Carboniferous sediments of County Clare, Eire. A rudimentary axial plane fabric is developed.

during dewatering induced by a weak tectonic event. Sedimentary dykes form perpendicular to the least principal stress. Hence, if they are parallel to and contemporaneous with the cleavage it would be necessary to conclude that the cleavage is an extension feature. More recently, the relationship between the cleavage and dewatering structures has been carefully studied and both Gregg (1979) and Boulter (1983) conclude that the dykes *pre-date* the cleavage. They observe that the dykes sometimes form part of a polygonal network of dykes which is probably the result of dessication. These dessication cracks were originally approximately perpendicular to bedding but have subsequently been rotated and deformed under the action of the forces which produced the cleavage. The strain induced was sufficiently large for the dykes to be reoriented parallel, or sub-parallel, to the cleavage. Thus, this parallelism of structures cannot be used to indicate that the cleavage developed in soft sediments in a near-surface environment.

Another example, for which it is claimed that cleavage developed under shallow cover occurs in the Carboniferous delta deposits exposed in the cliffs south of Galway Bay, Eire. This cleavage is sub-horizontal and consequently axial planar to recumbent isoclinal folds, Fig. 17.43. These folds are thought to be the result of sedimentary slumping, and the 'axial plane cleavage' to be the result of diagenesis rather than tectonism. Compaction of sediments during burial will enhance any bedding plane fabric generated during deposition, and, as the pressure and temperature increase, so the possibility of mimetic growth of minerals parallel to the bedding (and thus to the axial plane of the slump folds) becomes more likely.

It has been estimated from the stratigraphy of the area that these rocks never experienced a cover greater than a few hundred metres (at most 1 km). However, the sediments are now well indurated and contain quartz veins. Samples of quartz from a vug were collected by the present authors, and the homogenisation temperature was obtained using the technique of bubble geothermometry. The maximum, uncorrected bubble temperature obtained from the completely undeformed quartz specimens was 140 °C. To this temperature must be added a 'pressure correction factor'. In order that the magnitude of this correction may be established it is, of course, necessary to know the depth at which crystallistion took place, or the geothermal gradient that obtained at the time. Neither of these is known. However, even if the geothermal gradient had approached 100 °C/km, the depth of burial would have been around 2 km; while if the geothermal gradient was nearer the average for sediments at approximately 30 °C/km, the depth of burial could have exceeded 5 km. This latter figure is more in keeping with the degree of induration exhibited by these sediments. The estimated figure of greater that 5 km is in general agreement with the depth limit set by Fourmarier for the upper cleavage front.

We suggest that the main environmental parameter determining whether or not cleavage will develop (for a given strain) is temperature; though, of course, total pressure also plays a role. Temperature controls the dehydration reactions associated with zeolite and greenschist facies metamorphism and determines whether or not diffusion and pressure-solution processes are important.

There is an important aspect of the hydraulic fracture mechanism which lends support to Fourmarier's contention that fracture cleavage develops only below a certain depth. This aspect has to do with the law of effective stress, specific characteristics of the rocks and whether the fluid source and pressure which induce fracture are internal or external to the rock system under consideration.

The fluid source and pressure may be external to the system. Such an external source operates for the emplacement of dykes and sills or veins and during the formation of some barren fractures. However, fracture cleavage is so well distributed throughout folds and the fracture planes are so closely spaced, that it is reasonable to conclude that the fluid source and pressure were internal to the system.

As we saw in Chap. 1, (Eq. 1.70) 'internal' hydraulic fracturing can take place only if the effective stress in the rock mass is governed by the general relationship:

$$\sigma = S - p(1 - a)$$

where a has a small value (0.05–0.15). Rocks which obey this general relationship are those having relatively small porosities and include cemented or well compacted sediments.

As noted in Chap. 1, rocks for which $a = 0$ disaggregate when fluid pressures are suitably large, whereas rocks with $a > 0$ can fail by developing closely-spaced, hydraulic fractures. Unfortunately, rock mechanics data regarding the degree of compaction, and therefore depth of burial, required to produce a suitably large value of a for the formation of closely spaced hydraulic fractures is not known. However, from basic principles, it is clear that Fourmarier's observations regarding the existence of an upper cleavage front is correct. Moreover, his estimated depth to that front of about 5–6 km is also probably correct.

The temperatures at which pressure-solution takes place depends upon the mineral type and chemistry of any ambient fluids. For quartz in contact with fluids at the usual pH values obtaining in the crust, the solubility at temperatures below 150 °C is exceedingly small. Hence, significant recrystallisation in quartzose rock is only likely to occur at temperatures probably in excess of 200 °C. For average geothermal gradients of 30 °C/km, this would result in slaty cleavage beginning to become important at depths in the crust which are consistent with those cited by Fourmarier.

Pressure solution in carbonate rocks is inti-mately related to the ambient fluid chemistry, a topic outside the scope of this book. However, provided the chemical conditions are conducive, it is probable that pressure solution cleavage may develop in some carbonate rocks at very shallow depths.

Commentary

Five main mechanisms have been proposed to explain the development of the various forms of cleavage. However, it will have become apparent to the reader that, in most instances, more than one mechanism was involved in the development of a specific example of cleavage.

Dominance of a specific mechanism may change from point to point within a small domain. For example, White & Knipe (1978) have shown that cleavage development in the vicinity of the crest of a small fold involves crenulation, recrystallisation, body rotation and microfracture. Indeed, in microcosm, cleavage development has imitated several of the various modes of deformation that give rise to larger-scale structures that characterise an orogenic fold belt. The theory of buckling of anistropic rocks has been presented in Chap. 13 where it was applied to the formation of macroscopic folds. This work has also been of fundamental importance in the understanding of crenulation cleavage. In addition it is now realised that after an initial phase of 'homogeneous' flattening, the process of microfolding often plays an important role in the development of other cleavages including slaty cleavage.

The mechanisms by which rock fabrics are formed has intrigued geologists for almost two hundred years. The relationship between slate, schist and gneiss was recognised as early as 1815 by Bakewell, who realised that these three metamorphic rocks may represent the same parent rock at different grades of metamorphism. He states that the . . . 'mica slate (phyllite or schist) is frequently incumbent on gneiss or granite and covered by common slate; it passes by gradation into both these rocks, the coarser grained resembling gneiss and the finer kind by insensible transition becoming clay slates'. Although our understanding of the formation (and deformation) of tectonic fabrics has increased enormously since 1815, the ability of the subject to attract and intrigue structural geologists shows no sign of abating.

References

Alvarez, W., Engelder, T. & Geiser, P.A. (1978). Classification of solution cleavage in pelagic limestones. *Geology*, **6**, 263–6.
Badoux, H. (1970). Les oolites déformées du Vidar (Massif de Morches). *Eclogue Géol. Helv.*, **63**, 539–48.
Bakewell, R. (1815). *An Introduction to Geology*, 2nd edn. 429 pp. London: Harding.

Bayly, B.M., Borradaile, G.J. and Powell, C.McA. (eds.) (1977). Atlas of rock cleavage. Provisional edition. Hobart: Univ. of Tasmania.

Becker, G.F. (1893). Finite homogeneous strain, flow and rupture of rocks. *Geol. Soc. Am. Bull.*, **4**, 13–90.

Bonney, T.G. (1886). Anniversary address of the President. *Q. J. Geol. Soc. London*, 38–115.

Boulter, C.A. (1983). Post-lithification deformation of sandstone dykes: implication for tectonic de-watering. *Am. J. Sci.*, **283**, 876.

Cloos, E. (1947). Oolite deformation in the South Mountain fold, Maryland. *Geol. Soc. Am. Bull.*, **58**, 843–918.

(1971). *Microtectonics*. John Hopkins University Studies in Geology, No. 20, Baltimore: John Hopkins Press.

Cosgrove, J.W. (1972). The development and interrelationship of microfolds and crenulation cleavage. Unpublished Ph.D. Thesis, University of London.

(1976). The formation of crenulation cleavage. *J. Geol. Soc. London*, **262**, 153–76.

Crook, K.A.W. (1964). Cleavage in weakly deformed mudstones. *Am. J. Sci.*, **262**, 523–31.

Darwin, C. (1846). *Geological observations in South America*. London: Smith–Elder.

Dennis, J.G. (1977). In *Atlas of rock cleavage*, eds. B.M. Bayly, G.J. Borradaile & C.McA. Powell. Provisional edition. Hobart: University of Tasmania.

Dieterich, J.H. (1969). Origin of cleavage in folded rocks. *Am. J. Sci.*, **267**, 155–65.

Dieterich, J.H. & Carter, N.L. (1969). Stress history of folding. *Am. J. Sci.*, **267**, 129–55.

Durney, D. (1971). Deformation history of the Western Helvetic Nappes, Valais, Switzerland. Unpublished Ph.D. Thesis, University of London.

Fischer, O. (1884). On cleavage and distortion. *Geol. Mag. New Ser.*, **111**, 1, 396–406.

Fourmarier, P. (1951). Schistosité, foliation et microplissement. *Arch. Sci. Phys. Nat. Genève*, **4**, 5–23.

Gay, N.C. (1968). Pure shear and simple shear deformation in inhomogeneous, viscous fluids. 1. Theory. *Tectonophysics*, **5**, 3, 211–34.

Gregg, W.J. (1979). The redistribution of pre-cleavage clastic dykes by folding. *J. Geol.*, **87**, 99–104.

Gresens, R.L. (1966). The effect of structurally produced pressure gradients on diffusion in rocks. *J. Geol.*, **74**, 307–21.

Hancock, P.L. (1964). The relations between folds and late-formed joints in South Pembrokeshire. *Geol. Mag.*, **101**, 174–84.

Harker, A. (1886). On slaty cleavage and allied rock structures, with special references to the mechanical theories of their origin. *Rep. Br. Assoc. Adv. Sci.*, 1885 (55th meet.), 813–52.

Heim, A. (1878). *Untersuchungen über den Mechanismus der Gebirgsbildung*, 2 vols. Basel: Schwarbe.

Jeffrey, G.B. (1922). The motion of ellipsoid particles immersed in a viscous fluid. *Proc. Roy. Soc. London*, **A102**, 161–79.

Kamb, W.B. (1959). Theory of preferred crystal orientation developed by crystallisation under stress. *J. Geol.*, **67**, 153–70.

(1961). The thermodynamic theory of non-hydrostatically stressed solids. *J. Geophys. Res.*, **66**, 259–71.

Keeping, W. (1878). Notes on the geology of the neighbourhood of Aberystwyth. *Geol. Mag.*, **5**, 532–47.

Kerrich, R. (1975). Aspects of pressure solution as a deformation mechanism. Unpublished Ph.D. Thesis, University of London.

Knill, J.L. (1957). The pre-Tertiary geology of the Craignish–Kilmelfort district, Argyllshire. Unpublished Ph.D. Thesis, University of London.

(1960). A classification of cleavages, with special reference to the Craignish district of the Scottish Highlands. *Inst. Geol. Congr. XXI, Copenhagen*, **18**, 317–25.

Laugel, A. (1855). Du clivage des roches. *C. R. Acad. Sci.*, **40**, 182–5, 978–80.

March, A. (1932). Matematische Theorie der Regelung nach der Korngestalt bei Affiner Deformation. *Z. Krist.*, **81**, 285–97.

Maxwell, J.C. (1962). Origin of slaty and fracture cleavage in the Delaware Water Gap area, New Jersey and Pennsylvania. *B.G.S.A., Petrologic studies, a volume to honour A.F. Buddington*, pp. 281–311.

McLellan, A.G. (1966). A thermodynamic theory of systems under non-hydrostatic stresses. *J. Geophys. Res.*, **71**, 4341–7.

Mead, W.J. (1940). Folding, rock flowage and foliate structures. *J. Geol.*, **48**, 1007–21.

Muff, H.B. (1909). In *The geology of the seaboard of mid Argyll, including the islands of Luing, Scarba and Garvellachs and the Lesser Isles, together with the northern part of Jura and a small portion of Mull.* (Explanation of sheet 36), ed. B.N. Peach, H. Kynaston & H.B. Muff, pp. 1–121. Mem. Geol. Surv., U.K.

Owens, W.H. (1973). Strain modification of angular density distributions. *Tectonophysics*, **16**, 249–62.

Paterson, M.S. (1973). Non-hydrostatic thermodynamics and its geological applications. *Rev. Geophys. Space Physics*, **2**, 355–89.

Powell, C.McA. (1969). Intrusive sandstone dykes in the Siamo slate near Negaunee, Michigan. *Geol. Soc. Am. Bull.*, **8**, 2585–94.

(1979). A morphological classification of rock cleavage. *Tectonophysics*, **58**, 21–34.

Price, N.J. (1953). Structural and Tectonic development of the Aberystwyth Grits. Ph.D. Thesis, University of Wales.

(1962). Tectonics of the Aberystwyth Grits. *Geol. Mag.*, **99**, 542.

(1967). The development of asymmetrical buckles in non-metamorphosed sediments. *Tectonophysics*, **4**, 173–201.

Price, N.J. & Hancock, P. (1972). Development of Fracture Cleavage. *Proc. Int. Geol. Cong.* (Canada), Session 24, Section 3, pp. 584–92.

Ramsay, A.C. (1881). *The geology of North Wales*. Geol. Surv. G.B., 2nd edn., Mem. 3, pp. 1–623.

Rickard, M.J. (1961). A note on crenulated rocks. *Geol. Mag.*, **98**, 324–32.

Rogers, H.D. (1856). On the laws of structure of the more disturbed zones of the Earth's crust. *Trans. Roy. Soc. Edinburgh*, **21**, 431–72.

Savin, G.N. (1961). *Stress concentration around holes*. Oxford: Pergamon Press.

Sedgwick, A. (1835). Remarks on the structure of large mineral masses, and especially on the chemical changes produced in the aggregation of stratified rocks during different periods after their deposition. *Trans. Geol. Soc. London*, 2nd Ser., **3**, 461–86.

Siddans, A.W.B. (1972). Slaty cleavage – a review of research since 1815. *Earth Sci. Rev.*, **8**, 205–12.

Sorby, H.C. (1853). On the origin of slaty cleavage. *New Philo. J.* (Edinburgh), **55**, 137–48.

Tullis, T.E. (1971). Experimental development of preferred orientation of mica during recrystallisation. Unpublished Ph.D. Thesis, University of California at Los Angeles.

(1976). Experiments on the origin of slaty cleavage and schistosity. *Geol. Soc. Am. Bull.*, **87**, 745–53.

Turner, F.J. & Verhoogen, J. (1960). *Igneous and metamorphic petrology.* New York: McGraw–Hill.

Tyndall, J. (1856). Observations on 'The theory of the origin of slaty cleavage' by H.C. Sorby. *Philos. Mag.*, **12**, 129–35.

Weiss, J. (1949). Wissahickon schist at Philadelphia, Pennsylvania. *Geol. Soc. Am. Bull.*, **60**, 1689–726.

Wettstein, A. (1886). Über die Fischfauna des Tertiaren Glarner Schiefers. *Schweiz. Paläontol. Ges. Abh.*, **13**, 1–101.

White, S.H. & Knipe, R.J. (1978). Microstructures and cleavage development in selected slates. *Contrib. Mineral. Petrol.*, **66**, 165–74.

Williams, P.F. (1976). Relationship between axial plane foliation and strain. *Tectonophysics*, **30**, 181–96.

Wilson, G. (1946). The relationship of slaty cleavage and kindred structures to tectonics. *Proc. Geol. Assoc.*, **57**, 263–302.

(1961). The tectonic significance of small-scale structures and their importance to the geologist in the field. *Annls. Soc. Geol. Belg.*, **84**, 423–548.

18 Structural analysis

Introduction

Throughout this book, we have been concerned with various aspects of analysing geological structures and, in particular, we have concentrated on the mechanics and mechanisms which give rise to the various features. However, to many geologists, 'structural analysis' has a different and more restricted meaning which applies to determining the structural history of an area and the geometry and orientation of the major structures, from a study of the types and relationships of minor structures. The techniques of such an analysis are outlined in this chapter.

The definition of minor structures (sometimes called small-scale structures) and the use of these structures to the geologist has been concisely stated by Wilson, (1961, 1982): 'small-scale structures are those structures of tectonic origin that can be observed with the naked eye in the field. Their scale varies broadly between that of the hand-specimen to that of the exposure or even mountain side . . . Recognition of these minor structures and the appreciation of their origin and significance assist the field geologist to elucidate the larger-scale geological structure of his area. Commonly, some can be used in deciphering the order of stratigraphical succession in regions of strongly folded unfossiliferous beds and, in ground which has suffered superimposed tectonic movements, the minor structures may provide evidence of successive phases or events in the tectonic history.'

The importance of small-scale structures had already been recognised at the end of the last century by Pumpelly et al. (1894), who noted the 'general parallelism which exists between the minute and general structure'. This point was taken up by van Hise (1896) who quotes 'Pumpelly's Rule' that 'the degree and direction of the "pitch" (N.B. in modern terminology this would be "plunge") of the fold are indicated by those of the axes of the minor plications on its side'. Since then, the relationship between minor and major structures has been examined in detail, and the study of minor structures has become a powerful tool used by the structural geologist to determine the geometry and orientation of major structures and to unravel the sequence of tectonic events that has affected the area being studied.

The chapter begins with a brief discussion of the geometry of the more important minor structures such as minor folds, kink-bands, cleavage, joints, boudins and pinch-and-swell structures, mullions, rodding, cusps, extension fractures (veins), shear fractures, shear zones and lineations, and their relationships to major structures and to the stress and strain ellipsoids. Mechanisms of formation are only given brief mention unless they have not been covered elsewhere in the book. The text continues with a section on sedimentary structures that can be used to indicate the 'way up' of beds and a consideration of the criteria that enable tectonic and sedimentary structures to be differentiated. This is followed by a discussion of the techniques of stereographic analysis of areas that have experienced either single or multiple deformations. These techniques are then applied to two field areas that have been affected by multiple deformations.

Minor folds

The association of minor folds with major folds was discussed in Chap. 12, where it was shown how the geometry of the minor folds varied around a major fold (Figs. 12.36 and 12.37). It can be seen from these figures, that the minor folds have an 'S', 'M' or 'Z' profile geometry depending on which part of the major fold they occur. Careful mapping of the S, M or Z geometry of minor folds in the field will often enable the hinge and limb regions of a major fold to be delineated (Fig. 18.1). When recording the geometry of minor folds in the field, the folds should be viewed consistently looking down the plunge.

Minor folds which accord with the major fold structure obey Pumpelly's rule so that the axes and axial planes of the minor folds are parallel to those of the major fold (Fig. 18.2).

Cleavage

Cleavage is discussed in detail in Chap. 17, where it is concluded that, except for a variety of crenulation cleavage, cleavage lies parallel to, or within a few degrees of, the XY plane of the finite strain ellipsoid. Field observations show that cleavage is often parallel, or sub-parallel, to the axial plane of the folds formed at the same time as, or in association with, the cleavage. Such fabrics have therefore been termed 'axial plane' cleavage, a term which has no genetic significance. Indeed, one often finds slaty and crenulation cleavages as axial plane fabrics within different lithologies of the same fold. Because cleavage often

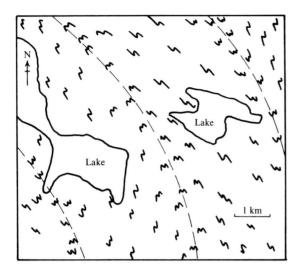

Fig. 18.1. Map showing the profile geometry of minor folds which delineate the limb and hinge regions of major folds. Dashed lines are the traces of the hinge surfaces.

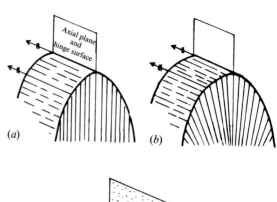

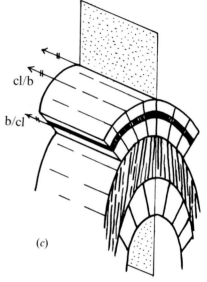

Fig. 18.3. (*a*) Ideal, axial plane cleavage. (*b*) Cleavage fanning around a fold hinge. (*c*) Refraction of cleavage. In all three examples, the cleavage/bedding intersection lineation (cl/b) and the bedding/cleavage intersection lineation (b-cl) are parallel to the fold axis.

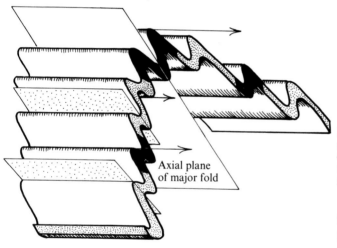

Fig. 18.2. Geometric relationship between major and minor folds. The axes and axial planes of the minor folds are parallel to those of the major fold.

fans about a fold hinge and refracts as it passes between beds of different lithologies (Chap. 17), cleavage is rarely exactly axial planar, except at the fold hinge. However, as can be seen from Fig. 18.3, regardless of whether the cleavage is parallel to the axial plane, or fans or refracts, its intersection with the bedding is usually parallel to the axis of the major fold with which it is associated.

Boudinage and pinch-and-swell structures

The origin of boudins and pinch-and-swell structures has been discussed in Chap. 16, where it was shown that both boudins and folds may be formed during a single deformation. If the boudins are formed because

the limbs of the fold rotate into the extension field as the folds amplify, the boudin lengths will be parallel to the fold axis (Fig. 18.4(*a*)). However, as discussed in Chap. 1 (Figs. 1.18 and 1.21) boudins may also form with their lengths normal to the fold hinges when the strain associated with folding is such that a significant extension occurs along the fold hinge (Fig. 18.4(*b*)).

Mullions and rodding structures

Mullions and rodding structures are both forms of a coarse lineation developed in rocks which have been strongly deformed. They are generally parallel to local fold hinges. Wilson (1953, 1961) distinguishes the two structures, confining the term, mullion, to structures which are formed from the country rock (Fig. 18.5). Rods are monomineralic, stick-like bodies, generally quartz, enclosed in metamorphosed country rock (Fig. 18.6). The terms, mullion and rodding, have no genetic significance. Mullions may be formed in a variety of ways and they have been classified into three groups: (i) Fold-mullions or bedding-mullions, (ii)

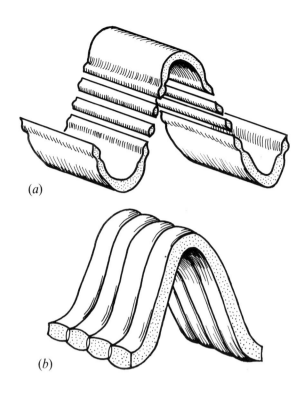

(a)

(b)

Fig. 18.4. Possible associations of folds and boudins formed during a single deformation. See text for discussion.

(a)

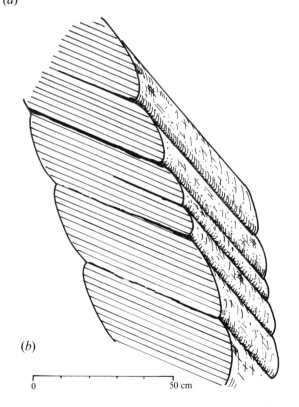

(b)

Fig. 18.5. (*a*) Mullion structures from Dedenborn, Ardennes. (Photograph courtesy of G. Wilson.) (*b*) Diagrammatic representation of same. (After Bruehl, 1969.)

Cleavage mullions and (iii) Irregular mullions. Many are a combination of two of these types.

Fold-mullions possess regular, curved, cylindrical surfaces which correspond to original bedding or to pre-existing planes of foliation. Commonly, they are largely composed of detached or 'strangled' hinges of parasitic folds and the bedding lamination within the mullions accords with the external surface.

Bedding mullions are undulations of the bedding plane surfaces, which have been smoothed, polished or striated. Locally, they may be formed by pinch-and-swell structures developing in a single bed, elsewhere they may be gentle flexures or large corrugations.

Cleavage mullions are long rock prisms which may be more or less angular, or partially rounded in cross section. The prism surfaces are dominantly cleavage surfaces. Several of the sharp edges of cleavage slices may be worn away, so that one is left with a cylinder having an approximately oval cross-section. Other varieties may show curved surfaces on one side while the other surfaces are relatively flat. All, however, are characteristically polished, mica covered or striated.

Irregular mullions are the most common variety. They are long, cylindrical structures, but in cross section they are very irregular and interlock like pieces of a jig-saw puzzle. The cylindrical surfaces are grooved, like worn cog wheels, and striated or covered with a micaceous veneer. The internal structure of the mullion may show contorted bedding laminae; these

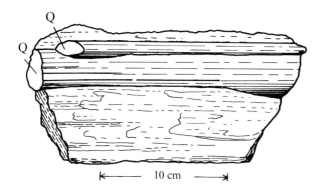

Fig. 18.6. Quartz rods (Q) in a siliceous schist from the Moine Series, Arnisdale, Inverness-shire, Scotland. (From Wilson, 1961 after Ramsay, 1960).

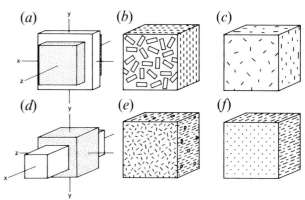

Fig. 18.7. The effect of pure flattening deformation represented by (a) (shaded area is the undeformed cube; unshaded the deformed cube) on an originally random distribution of (b) platy minerals and (c) acicular minerals. Note that both (b) and (c) are planar fabrics (i.e. S-tectonites). (d) represents a constrictional deformation and its effect on an originally random distribution of platy and acicular minerals are shown in (e) and (f) respectively, both of which are linear fabrics (i.e. L-tectonites).

may locally accord with the external surface but, for the most part, are truncated by it.

All varieties of mullions may occur together in any one area and they are parallel to each other and the major fold axes. Although the work of Smith (1975, 1977 and 1979 and Fletcher, 1982 cf. Chaps. 11, 13 and 16) has thrown considerable light on the possible mechanism by which some mullions may form, the mode of origin of many mullions is still unclear. They probably form by either a single constrictional deformation or by two approximately coaxial deformations.

Lineations

The major types of lineation encountered in deformed rocks are:

(i)   mineral lineations (sometimes on a cleavage or schistosity plane),
(ii) bedding/cleavage intersection lineations – either on a cleavage plane or bedding plane,
(iii) relatively coarse lineation defined by the hinges of minor folds and the axes of boudins, mullions and rods,
(iv) crenulation lineation which is a finer feature than that described in (iii) and is defined by the hinges of microfolds, and
(v)   slickensides, or crystal fibres which form lineations on fault planes and bedding planes.

(i) The mechanisms by which minerals are realigned in a rock, during deformation, to produce mineral fabrics, have been briefly discussed in Chap. 17, where it was noted that the alignment may be the result of the growth (or recrystallisation) of a mineral into a preferred orientation in response to an applied non-hydrostatic stress. This tendency is particularly marked in minerals such as micas, where there is a significant mechanical anisotropy. In addition, if, like micas, the minerals are inequant, then they may also be aligned by either passive or active

mechanical rotation. A detailed discussion of rock fabrics, (petrofabrics), is outside the scope of this book and the interested reader is referred to treatments of the subject by Paterson & Weiss (1961), Hobbs *et al.* (1976) and Poirier (1985).

Rocks that possess a tectonically induced fabric are known as *tectonites*, and the elements making up the fabric are termed *fabric elements*. These elements may be individual minerals, or aggregates of mineral grains and often have either a planar or linear morphology. Mineral fabrics that form in a deformed rock reflect the shape of the minerals (equant, platy or needle-like), the mineral's mechanical anisotropy and the type of strain field (flattening or constrictional cf. Fig. 1.21) in which the fabric developed.

States of three-dimensional strain can be conveniently represented on a Flinn Diagram (Chap. 1, Fig. 1.21) which can be divided into a stretching and a flattening field. Extreme examples of these are represented by the prolate (constrictional) and oblate (flattening) strain ellipsoids respectively. Let us consider the mineral fabrics that will form in association with these two types of strain.

As can be seen, in a pure oblate (flattening) strain field (Fig. 18.7(a)) a planar fabric (an *S-tectonite*) will be produced regardless of whether the fabric elements are planar or linear (Fig. 18.7(b) and (c)). Conversely, in a pure prolate (constrictional) strain field (Fig. 18.7(d)) both these fabric elements are reoriented to form linear fabrics (*L-tectonites*) as shown in Fig. 18.7(e) and (f).

There is a complete spectrum between these two extreme strain states, with a corresponding spectrum of rock fabrics. These intermediate fabrics have both planar and linear characteristics and are consequently called *L–S-tectonites*. In his classification of fabrics

Table 18.1. *Classification of fabrics*

1. Planar structures (S-tectonites)
   (a) Planar parallelism of planar fabric elements (Fig. 18.7(*b*))
   (b) Planar parallelism of linear fabric elements (Fig. 18.7(*c*))
2. Linear structures (L-tectonites)
   (a) Linear parallelism of linear fabric elements (Fig. 18.7(*f*))
   (b) Linear parallelism of planar fabric elements (Fig. 18.7(*e*))
3. Composite structures (L–S-tectonites)
   (a) Combined structures: two or more planar and/or linear structures in combination.
   (b) Complex structures: two fabrics formed from either linear or planar fabrics only.
      (i) Linear + planar fabrics formed from linear elements (Fig. 18.8(*a*))
      (ii) Linear + planar fabrics formed from planar elements (Fig. 18.8(*b*))

(After Den Tex (1954) and Ragan (1985).)

(Table 18.1) Den Tex (1954) has termed such tectonites, 'composite' fabrics.

Many slaty cleavages are composite fabrics. In addition to the parallel alignment of mica flakes, the long axes of the micas are aligned in the cleavage plane (Fig. 18.8(*b*)) producing a lineation. When mineral lineations on a cleavage or schistosity plane develop in rocks containing fossils, or other strain markers, the mineral elongation is seen to be parallel to the maximum extension (Fig. 18.9). Consequently, they are called stretching lineations. In addition, the mineral lineation or 'grain' of a slate is often normal to the fold hinge (Fig. 18.9). Experimental work on the folding of layered media shows that even when the model is not confined in the direction normal to the fold hinges (Fig. 18.10), there is very little tendency for extension parallel to the fold axis to occur *during* folding. The fold hinge generally forms parallel to the intermediate axis of strain and the maximum extension occurs within the axial plane, normal to the fold hinge.

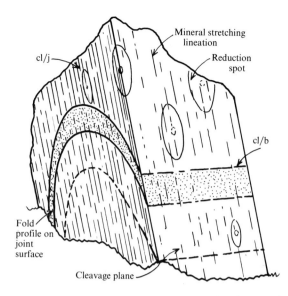

Fig. 18.9. The relationship between the mineral lineation on a slaty cleavage plane and the maximum extension direction (as indicated by deformed reduction spots). Mineral 'stretching' lineations are often normal to the fold hinge.

However, mineral lineations are sometimes found to lie parallel to the fold hinge. This may be the result of extension parallel to the fold axis which can occur when the folds lock up and buckling ceases to be an effective mechanism of shortening. Alternatively, extension parallel to the fold axis can occur during the growth of a fold when the boundary conditions are such that extension in the direction normal to the layering is considerably more difficult than extension within the plane of the layering. This has been observed in experiments by Watkinson (1975).

An interesting alternative explanation for the occurrence of mineral lineations parallel to the fold axis has been presented by Cobbold & Watkinson (1981). They consider the problem of the folding of a layer with bending anisotropy, which is defined as the property of the layer causing it to bend more easily in one specific direction rather than any other. A rock with a strong linear fabric would exhibit such a bending anisotropy. In such rocks, folds will be con-

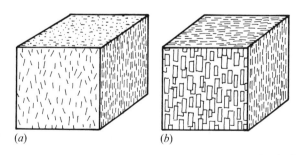

(*a*)   (*b*)

Fig. 18.8. Fabrics with both linear and planar properties (L–S-tectonites) formed from (*a*) linear fabric elements and (*b*) planar fabric elements. See Table 18.1.

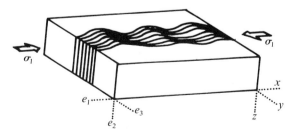

Fig. 18.10. Block diagram of an experimentally deformed multilayer embedded in a less competent matrix. Even when there is no confinement normal to the fold hinge, it is observed that there is very little tendency for extension parallel to the fold hinge to occur. e_1, e_2 and e_3 are the minimum, intermediate and maximum extension directions respectively.

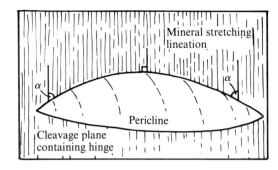

Fig. 18.11. Idealised relationship between a mineral stretching lineation on an axial plane cleavage, and the hinge of a pericline.

strained to form with their axes sub-parallel to the linear fabrics. They illustrate this effect using a sheet of foam rubber, stiffened in one direction by the insertion of thin steel rods. This 'fibre' reinforced sheet bends easily about axes nearly parallel to the rods, but with difficulty about any other axis. This bending anisotropy is also evident during buckling. If the sheet is compressed at an angle between 45° and 90° to the rods, buckling is easily induced, but the fold axes are always within 20° of the rod lineation.

The layer used in these experiments has an exceptionally strong linear fabric. Unfortunately the amount of bending anisotropy of geological L- and L–S-tectonites is not known. However, Cobbold & Watkinson argue that a phyllite containing quartz rods may approximate to the reinforced layer used in their experiment, provided that the quartz rods are much stronger than the phyllite matrix.

It must be concluded, therefore, that in materials with strong bending anisotropy, the orientation of late folds is strongly dependent upon the orientation of a linear fabric, and less so on the attitude of the principal stresses. Consequently, they argue, 'correlation of fold attitudes with the stress field is extremely hazardous, unless the rheology of the rocks is well known. Hence the geologist should beware of attempting such correlations. Also, if early structures vary in orientation from place to place, then later ones are likely to do so as well, even if the whole region is subject to relatively simple boundary conditions. Thus, folds (developed in rocks with linear fabrics) at different sites cannot simply be diagnosed as synchronous (i.e. associated with the same major phase of compression) if they have similar orientations; nor as representing separate phases of folding merely because their orientations differ'.

Mineral lineations in the cleavage plane are also found oblique to the fold hinge. One possible reason for this is illustrated in Fig. 18.11. The majority of geological folds are periclines, and it will be apparent that only at the culmination of the pericline will the stretching lineation be normal to the fold hinge.

(ii) Provided that cleavage and folding have occurred in response to the same tectonic event, i.e.

provided the cleavage has not been superposed on a fold formed by an earlier deformation, then the lineation caused by the intersection of bedding and cleavage will be parallel to the major fold axis. In the field, this lineation can be measured on either the cleavage planes (b/cl) or bedding plane (cl/b), as shown in Fig. 18.3(c).

(iii) The lineations that result from minor fold hinges, mullions and rods (Figs. 18.2, 18.5 and 18.6) will generally be parallel to the hinge of the major fold with which they are associated. However, when boudins and folds form during the same deformation, the boudins may form either parallel or normal to the major fold hinge (Fig. 18.4).

(iv) Microfolds, formed by the buckling of a mineral fabric such as slaty cleavage, are often found in association with crenulation cleavage and the hinges of the microfolds form a fine lineation on the slaty cleavage. Although microfolding can develop in primary fabrics, such as bedding fissility in shales, it more frequently occurs in tectonically produced fabrics and its presence, therefore, often indicates that more than one deformation event has affected an area. Unless the two deformations were coaxial, the orientation of the microfold hinge lineation on the rock fabric will have no specific geometric relationship to the major folds associated with that fabric.

(v) When one surface of a rock is in contact with and slides over another surface, the two surfaces may develop a polish with linear grooves and ridges parallel to the direction of movement. This is termed *slickensiding* and the grooves and ridges, *slickensides*. If one runs one's fingers over the surface parallel to the striations in the direction of movement it will feel smooth, whereas against the direction of movement it will feel rough. It is now realised that many 'slickensides' are in fact crystal fibres (cf. discussion and implications Chaps. 11, 14 and 15) and not the result of mechanical attrition and polishing. However, these fibres have the same orientation as the 'classical' slickensides and can be used in the same way to indicate the direction (and sometimes amount) of movement on a discontinuity such as a fault or bedding plane. If folds form by the process of flexural slip, then crystal fibre lineations may develop on the bedding planes as the beds slide past each other. In an ideal cylindrical, flexural slip fold, these 'slickensides' would be at right angles to the fold axes and decrease in intensity from the inflection line to the fold hinge where, because no slip occurs between the beds in this region, no lineation would develop (Fig. 18.12(a)). Commonly, however, slickensides are found to pass over the fold hinges. One explanation for this is that the folds are minor folds which formed after the development of a major fold. Consequently, the slickensides associated with the formation of the major fold will be folded round the hinges of the minor fold

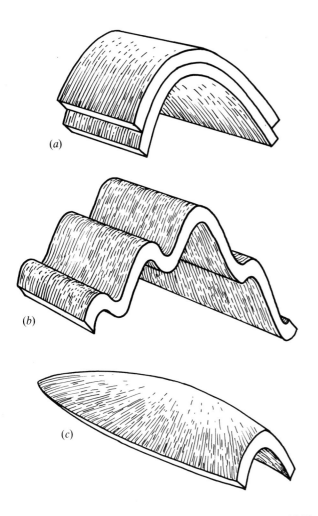

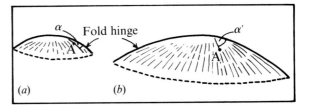

Fig. 18.13. The change in orientation of slickensides at a particular point A on a pericline as it grows. (*a*) An early stage and (*b*) a late stage in the formation of a fold. α and α′ are the angles between the slickenside and fold axis at the two stages of fold amplification shown.

fold, so that the angle between slickenside and fold hinge will begin to approximate to 90° (α′ Fig. 18.13(*b*)). Hence, ideally, if each layer of a composite layer of crystal fibres formed on a bedding plane represents an increment of fold growth, as suggested in Chap. 15, then the orientation of the crystal fibres should change systematically when traced from one layer to the next in the composite band. In some situation, however, the difference in orientation of the crystal fibre in the composite band is too great to be accounted for in the manner just described. These large variations in fibre orientation may reflect the superposition of slickensides associated with a major pericline and those of a secondary, parasitic pericline. It can be seen from Fig. 18.12(*c*), that almost any angular relation between crystal fibres of adjacent layers could occur in this way.

In other instances it is to be expected that differences in fibre orientation may reflect a more general, perhaps regional, change in the stress orientations during the development of the fold.

Shear zones

The distribution of deformation within a rock may be extremely localised, as for example in *brittle* deformation, or evenly distributed, as for example the *ductile* deformation associated with homogeneous flattening. The distribution of strain within rocks frequently lies between these two extremes. Commonly deformed rock can be divided into areas of low deformation cut by planar zones of shear, which exhibit relatively high deformation. These *shear zones*, which may be arranged in networks, are found on all scales, (Figs. 18.14, 18.15 and 18.16).

The geometry of these planar zones of high deformation and the development of fabrics and minor structures within them have been extensively described in the literature (e.g. Wilson, 1961 and Ramsay & Graham, 1970) and the interested reader is referred to a collection of papers edited by Carreras *et al.* (1980).

A generalised section through a major shear zone (Fig. 18.17) has been presented by Sorensen (1983), which is based on his detailed mapping of the

Fig. 18.12. (*a*) Classical relationship between slickensides and fold. The slickensides are at right angles to the fold hinge and decrease in intensity from the line of inflection to the hinge. (*b*) Slickensides occurring over minor fold hinges. See text for discussion. (*c*) Idealised orientation of slickensides around a pericline.

(Fig. 18.12(*b*)). Slickensides are also commonly found oblique to the fold hinge. This is often the result of the folds being non-cylindrical.

The direction of slip associated with a pericline is shown in Fig. 18.12(*c*), where it can be seen that the slip is only at right angles to the hinge near the central portion of the structure. Near the extremities, the slip direction is actually parallel to the hinge. The development of several layers of crystal fibres on a bedding plane has been described in Chap. 10, where it was suggested that each layer is associated with an increment of fold amplification. The orientation of the crystal fibres in each layer is often different and there are several reasons why this might be so. For example let us consider the formation of slickensides at point A on a pericline (Fig. 18.13(*a*)). Initially point A is close to the end of the pericline, in the early stage of its development, and the crystal fibres will be oriented at an angle α which may be considerably less than 90° to the fold hinge. As the pericline grows, position A effectively moves closer to the central portion of the

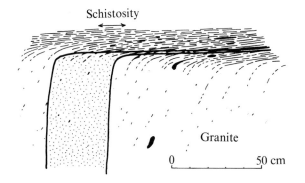

Fig. 18.14. Deformation of an aplite dyke (stippled) by a shear zone. (After Ramsay & Graham, 1970.)

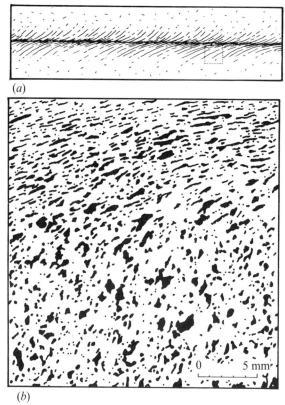

Fig. 18.15. (*a*) The dotted box shows the location of Fig. 18.14 with respect to the shear zone and (*b*) shows the development of schistosity in the deformed granite at the edge of the shear zone. The aggregates of mafic minerals are shown in black. (After Ramsay & Graham, 1970.)

Norde Stromfjord shear zone (Fig. 18.16). It can be seen clearly that an increase in strain occurs from the surroundings into the zone. The shear zone consists of converging and diverging belts of sub-parallel lithologies enclosing less, or even un-, deformed augen: 1, 2, 3 and 4.

For mathematical simplicity it is usual to assume that the deformation in these zones is that of *heterogeneous simple shear*, (see Chap. 1) implying that no volume changes occurred. However, it can quite frequently be demonstrated that material entered or left the shear zone during its formation.

Conjugate shear zones in rock can be con-

veniently divided into two groups. In the first group (Fig. 18.18), the inward-moving wedge occurs in the acute angles between the shear zones. Such 'acute' shear zones can range from brittle to ductile (Fig. 18.18(*a*)–(*d*)). In the second group (Fig. 18.19), it is the obtuse wedges that moved inward. Such 'obtuse' shear zones are often arranged to form a closely linked network of structures (Fig. 18.19(*b*)).

Two mechanisms whereby 'shear zones' may be formed have been dealt with. The first of these we consider to be a ductile, or semi-ductile equivalent of brittle failure (Chaps. 1 and 2; Fig. 18.18(*c*) and (*d*)). This mechanism would give rise to the 'acute' shears described above. The second mechanism of shear zone formation (Chaps. 6 and 13) we suggest relates to formation of planar shear zones (Biot's Type 2 instability) which develop in anisotropic materials; particularly when the anisotropy is induced by deformation rather than being an intrinsic property of the material. Consequently, although they are akin to normal kink-bands, they can develop in materials such as granite, which may exhibit little or no intrinsic anisotropy. This type of deformation, which will produce the 'obtuse' shears described above, is therefore a Type 2 instability (cf. Chap. 13) which becomes manifest as a localised simple shear deformation parallel to one or other of the conjugate characteristic directions (Fig. 18.19).

The critical stress conditions for the formation of the two types of shear zones described above, which conform to the Navier–Coulomb criteria and the Biot Type 2 instability, can be expressed respectively as:

$$P = (\sigma_1 - \sigma_3)$$
$$= \sin \phi \, [(\sigma_1 + \sigma_3) - 2C_0/\tan \phi] \qquad (18.1)$$

and

$$P = (\sigma_1 - \sigma_3) = \frac{4M'}{L'} (L' - M') \qquad (18.2)$$

where C_0 is the cohesive strength of the rock, ϕ is the angle of internal friction and M' and L' are measures of the resistance to compression and shear in the direction of the maximum principal stress σ_1. Which of these two modes of shearing develops will depend, among other things, upon which of the two conditions is first satisfied.

The geometric expression of both modes of shear instability will be sensitive to rock type, the deformational environment and the strain-rate. For example, it has been demonstrated by Sibson (1977) (see also Chap. 7) that a discrete brittle thrust plane, near the Earth's surface can broaden into a ductile shear zone at depth (Fig. 18.20(*a*)). Similarly, near-surface, discrete wrench faults may broaden into wider shear zones at depth (Sorensen, 1983) (Fig. 18.20(*b*)).

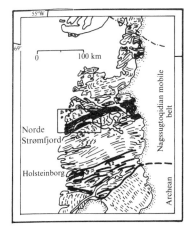

Fig. 18.16. Foliation trend depicting the Norde Strømfjord shear zone (shown as closely spaced lines) within the Nagssugtoqidian of W. Greenland. (After Sorensen, 1983.)

Minor structures associated with shear zones

Homogeneous and heterogeneous simple shear has been discussed in Chap. 1 where it was shown that, during such deformation, the maximum principal compressive stress acts at 45° to the shear zone margins (Fig. 18.21(*a*)). A variety of minor structures commonly found in association with shear zones form in response to this stress. One of the most common are

mineral filled tension veins. These form parallel to the maximum principal compression and are often arranged 'en echelon' within the shear zone (Fig. 18.21(*b*)). With progressive deformation the veins may rotate but, because the orientations of the principal stresses are fixed with respect to the shear zone margin, the veins will continue to grow at 45° to the margin. This results in the formation of curved or sigmoidal tension veins (Fig. 18.21(*c*) and (*d*) and Fig. 18.22(*a*)). The amount of rotation of the veins

Fig. 18.17. A generalised model of the shear zone structure of Fig. 18.16. An increase in strain from the surroundings into the zone occurs across two marginal zones (m.z.). The inner part of the zone consists of converging and diverging belts of subparallel lithologies enclosing less deformed, or even undeformed, augen; 1, 2, 3 and 4. Some augen contain granitic or charnokitic rocks (1): some are characterised by significant changes of strike (2 and 3), often in the form of fold structures (4). (After Sorensen, 1983.)

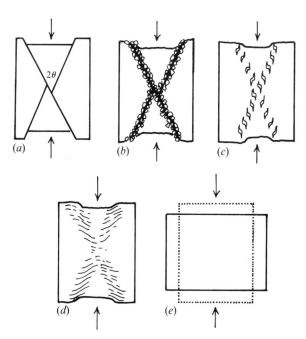

Fig. 18.18. (*a*) Brittle, (*b*) and (*c*) semi-brittle and (*d*) ductile expression of Andersonian shear failure, $2\theta < 90°$; (*e*) homogeneous flattening.

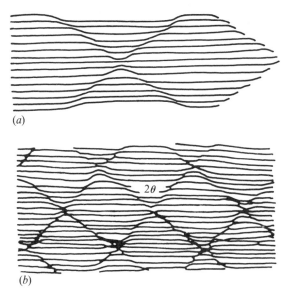

(a)

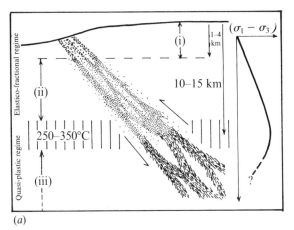

(b)

Fig. 18.19. (a) Conjugate ductile shears, the expression of Biot's Type 2 instability. (See Chap. 13). (b) Conjugate shears forming a network of high deformation enclosing lozenges of less deformed rock, $2\theta > 90°$.

(a)

(b)

Fig. 18.20. (a) Conceptual model of major thrust. (i) Incohesive gouge and breccia (psuedotachylyte if dry). (ii) Cohesive random-fabric crush breccias, rocks of the cataclasite series, pseudotachylyte if dry. (iii) Cohesive, foliated rocks of the mylonite series and blastomylonites. (After Sibson, 1977.) (b) A schematic vertical section through the shear zone of Fig. 18.16, showing a widening of the shear zone with depth and a 'telescoping' of the amphibolite to granulite facies transition downwards in the crust. Black: pyroxene-bearing quartzo-felsparthic gneisses. White: pyroxene-free gneisses. (After Sorensen, 1983.)

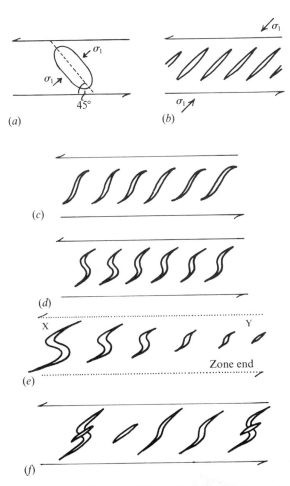

Fig. 18.21. (a) The orientation of the maximum principal compressive stress with respect to a shear zone and the strain ellipse associated with the first increment of shear deformation. The eccentricity of the ellipse is exaggerated for clarity. (b)–(f) The formation of en echelon tension gashes within a shear zone in response to this stress. See text for discussion.

reflects the amount of shear. This will vary along the length of the shear zone, being maximum in the central region, X, and minimum at the shear zone ends, Y (Fig. 18.21(e)). More complex arrays of veins may occur if they develop at different times during the formation of the shear zone; early veins at a particular locality being more sigmoidal than later ones (Fig. 18.21(f)).

If the rock inside the shear zone responds in a ductile rather than a brittle manner, then tension veins do not form and, instead, the rock may develop a fabric at right angles to the maximum principal compression (Fig. 18.23(b)). As deformation continues, this fabric will rotate but, because the orientations of the principal stresses are fixed with respect to the shear zone boundary, new fabric is always formed at 45° to the margin. Thus, the shear zone fabric is commonly curved, making an angle of 45° to the boundary near the margin and becoming approximately parallel to the shear zone near the centre (Figs. 18.23(c), 18.14, 18.15 and 18.16). The shear zone fabric may be a

(*a*)

(*b*)

Fig. 18.22. Natural example of conjugate shear zones containing (*a*) sigmoidal en echelon tension gashes and (*b*) a shear zone fabric. Marloes Sands, Pembrokeshire, S. Wales.

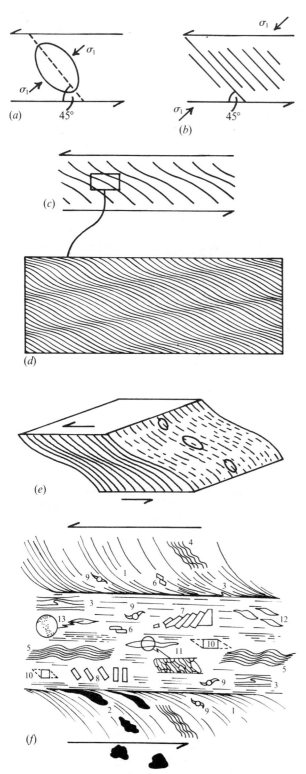

pervasive mineral fabric or closely spaced pressure solution seams. Mineral fabrics formed in shear zones are generally S–L-tectonites and the orientation of the stretching direction is shown in Fig. 18.23(e). A conjugate pair of shear zones, in which a shear zone fabric is developed but which contains no en echelon tension gashes, is shown in Fig. 18.22(b).

A number of planar features are also found which cut the shear zone fabric. It has been suggested that some of these may correspond to the Riedel shears discussed in Chap. 6 and shown in Fig. 6.14. One however, a type of crenulation cleavage, can be readily explained by considering the deformation history of the shear zone fabric. Initially, the fabric formed at right angles to the principal compression (Fig. 18.23(a) and (b)). However, with progressive deformation, it rotates towards parallelism with the shear zone margin and increases in intensity (Fig. 18.23(c)). The fabric may eventually become unstable with respect to the principal compression and the expression of this instability is often the formation of normal kink-bands. Because the principal compression is not acting at right angles to the rotated fabric, only one of the possible conjugate sets of kink-bands develops. With progressive deformation, these crenulation cleavage planes will rotate towards parallelism with the shear zone fabric and may be cut by later cleavage planes which have been generated in the same way as themselves. Some of the minor structures found associated with shear zone fabrics are shown in Fig. 18.23(d) and 18.23 (f).

Sigmoidal tension veins and shear zone fabrics are sometimes formed together (Fig. 18.24). In these examples the fabric is a series of pressure solution seams. The quartz dissolved from these seams has been deposited in the tension veins.

Sheath folds are commonly found in large shear zones and their formation has been discussed in Chap. 10, (see Figs. 10.60 fand 10.61).

In many shear zones continuity across the shear zone is maintained. This, combined with the formation of a shear fabric and/or en echelon tension gashes, enables the structure to be easily recognised. However, with progressive deformation, continuity

(1) rotation of a pre-existing or generated foliation:
(2) rotation of deformed markers:
(3) asymmetry of intrafolial folds;
(4) normal kink-bands (microshears) in the margin or
(5) central fabric of the shear zone:
(6) sheared porphyroclasts;
(7) rotation of fragments owing to shear fractures;
(8) rotation of fragments owing to tensile fractures;
(9) asymmetry of trails growing around rotating clasts;
(10) asymmetry of trails growing around non-rotating clasts;
(11) asymmetry of elongate recrystallised quartz grains;
(12) asymmetry of dragged-out mica porphyroclasts;
(13) asymmetry of quartz c-axia fabrics.

((f) after White *et al.*, 1986).

Fig. 18.23. (a) The orientation of the maximum principal compressive stress associated with the first increment of shear deformation. The eccentricity of the strain ellipse is exaggerated for clarity. (b) The tectonic fabric induced by this compressive stress. (c) Rotation of the shear zone fabric during shearing. New fabric continues to grow at 45° to the shear zone margin. (d) Crenulation cleavage formed by compression of the shear zone fabric. (e) Block diagram showing the fabric as an S–L-tectonite. The stretching lineation is parallel to the long axis of the ellipses. Progressive deformation may result in the development of a mylonite in the central part of the shear zone and kinematic indicators of the movement sense in such mylonite zones are shown in (f). These include:

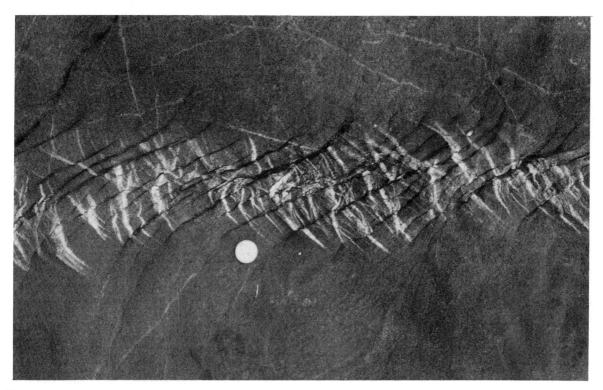

(a)

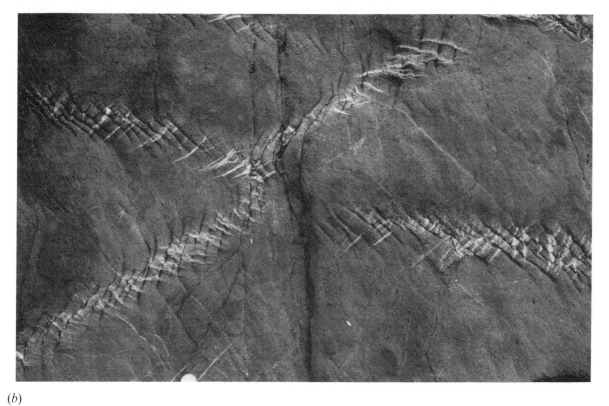

(b)

Fig. 18.24. Sigmoidal tension gashes and a pressure solution shear zone fabric developed synchronously in (a) a single shear zone and (b) a conjugate shear zone. Marloes Sands, Pembrokeshire, S. Wales.

across the shear zone may be lost and the structure may become indistinguishable from a fault.

It can be seen from this brief discussion that shear zones (and therefore faults) can develop at any angle to the maximum principal compression direction (Figs. 18.18 and 18.19). Great care is therefore needed when attempting to determine the orientation of the principal stresses from these structures when they are not present as a conjugate set.

Use of minor structures

Minor structures can be used to determine the geometry and orientation of major structures and the tectonic history of an area. They can also be used in certain situations to determine the stratigraphical succession. In order to ascertain whether an antiform is an anticline or a syncline, it is necessary to know the 'way up' of the beds. The small scale structures that may be used in determining the stratigraphical succession can be divided into two groups, namely sedimentary and tectonic, and these are sometimes referred to as primary and secondary structures respectively.

Sedimentary structures are those which form as a direct result of the sedimentary process. Many of these structures can be used to determine the top and bottom of the individual beds. Such structures include graded bedding, ripple marks, truncated cross bedding, worm burrows, dessication cracks, pillow lavas (Fig. 18.25(*a*)–(*f*)), flame structures (Fig. 18.26), sole marks (Fig. 18.27) and truncated slump structures (Fig. 18.28).

Certain *tectonic structures* and the relationship of these structures to bedding can also be used to indicate whether the beds have been inverted or not during a particular phase of deformation. Wilson (1961), points out that . . . 'In regions where folding and cleavage have resulted from the same deformation . . . one can, by observing the relationship between the cleavage and the stratification, deduce not only the geometrical form of the folds but also the stratigraphical succession'. Thus, in an area of overturned folds, the dip of the normal limbs is less steep than the dip of the fold axial plane; and the dip of the inverted limb is steeper than that of the axial planes (Fig. 18.29). Therefore, if the cleavage is parallel to the axial planes of the folds, we can say that: 'where the cleavage and beds dip in the same direction, if the cleavage is steeper than the stratification, the succession is in the correct order; if the cleavage dips less steeply than the bedding, then the beds are probably inverted'. These general rules do not hold for those areas where the beds were not in their correct order of succession before the folding, with which the axial plane cleavage is associated, took place. In areas of multiple deformation and in areas where sedimentary

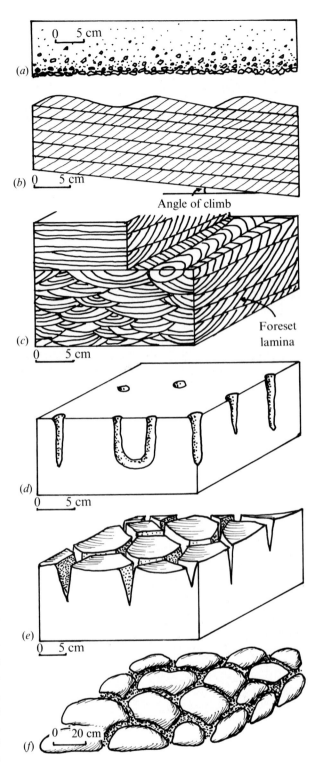

Fig. 18.25. Various sedimentary structures that can be used to determine the way-up of beds. (*a*) Graded bedding. (*b*) Ripple marks. (*c*) Truncated cross bedding. (*d*) Burrows. (*e*) Dessication cracks. (*f*) Pillow lavas. ((*b*), (*c*) and (*d*), after Collinson & Thompson, 1982.)

structures are absent or not clearly preserved, it is often not possible to demonstrate that the strata were the right way up prior to folding. In such situations, the rules can still be used to determine whether the beds have been 'overturned' as the results of the

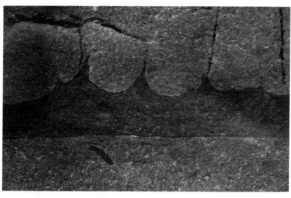

10 cm

Fig. 18.26. Shale, flame structures at the base of a sandstone layer in the Carboniferous turbidites, Bude, Cornwall, England.

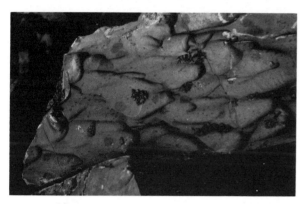

20 cm

Fig. 18.27. Sole marks on the base of a turbiditic sandstone.

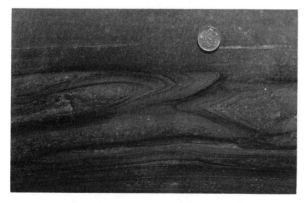

Fig. 18.28. Truncated slump fold indicating way-up in the Dalradian metasediments, Ballachulish, Scotland.

folding with which the cleavage is associated. However, this does not necessarily imply that the beds are now stratigraphically upside down.

The variation in rheological properties across a graded bed, which may be predominantly sandstone at the base and clay rock at the top, gives rise to a commensurate variation in response to deformation. For example, compression normal to such a unit may result in extensional, brittle failure of the sandstone part and homogeneous, ductile flattening of the clay

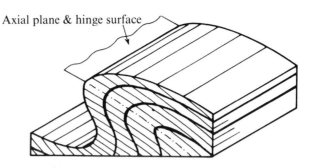

Axial plane & hinge surface

Fig. 18.29. The relationship between axial plane cleavage and bedding in an overturned fold. (After Wilson, 1961.)

rock portion. The result is the formation of tapered veins, which point in the direction of 'younging' (Fig. 18.30(*a*)). If cleavage forms in a graded bed, then this same change in rheological properties will result in a gradual refraction of cleavage (cf. Chap. 17), so that the cleavage forms at a greater angle to bedding in the sandstone (bottom) part of the unit than in the clay rock (top), (Fig. 18.30(*b*)).

The differentiation of sedimentary and tectonic structures

As can be seen from Fig. 18.30(*c*)–(*f*), there is considerable similarity between some tectonic and sedimentary structures and the task of differentiating between them in the field can be extremely perplexing, particularly when the rocks have been metamorphosed. Sometimes the structures look alike because although they are formed in very different environments they are formed by the same mechanism. For example slump folds, which develop in unconsolidated sediments, can look like and can form by the same buckling mechanisms as folds which form in the rock when they are completely lithified. Such mechanisms can operate over a wide range of crustal conditions. It might therefore be argued that because there is a gradation between structures that are definitely sedimentary, through those formed during diagenesis, to those formed when the rock is completely compacted and indurate, the problem of distinguishing between the two types of structure is really one of semantics. However, sometimes the mechanisms involved in producing a structure in a soft sediment environment are very different from those which produce structures with the same geometry in tectonic regimes. Two examples of this are shown in Fig. 18.30(*c*)–(*f*).

Because of the different mechanisms that operate in the two environments (intergranular flow involving no ductile deformation of grains tending to be associated with sedimentary environments and intragranular flow with grain shape changes and recrystallisation associated with many tectonic environments), it might be expected that the rock

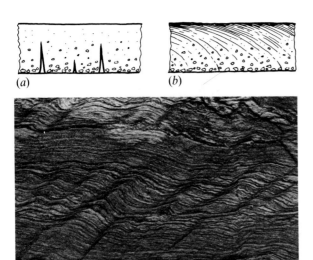

(a)   (b)

Fig. 18.30. Tectonic way-up structures developed in a graded bed. (a) Tension gashes pointing in the direction of younging. (b) Curved cleavage. See text. (c) Asymmetric crenulations, Rhoscolyn, Anglesey, N. Wales (d) Ripple drift a sedimentary structure with a geometry similar to the crenulations in (c) but produced by a completely different mechanism. (Photograph courtesy R. Walker.) (e) Tectonically formed cusp structures at an interface produced by compression parallel to the original orientation of the interface, Monar, Scotland. (f) Sedimentary flame structures formed during sedimentation by the foundering of the more dense (light) sandstone into the underlying shale which begins to rise through the sand diapirically. Bude, Cornwall, England.

(c)

textures could be used to differentiate the two types of structure. However, once a rock has been deformed and metamorphosed the original rock fabric is generally lost. It follows therefore that because the problem of differentiation only occurs in deformed and metamorphosed rocks, the structures cannot generally be separated on textural evidence.

Despite the lack of textural evidence it is sometimes possible to determine, with reasonable certainty, whether a structure is sedimentary or tectonic. For example, the occurrence of a planar rock fabric parallel to the axial plane of a fold, generally indicates that the fold is tectonic, although recumbent slump folds may have a diagenetically induced planar compaction fabric parallel to their axial planes. In addition to an axial plane cleavage folds may contain certain second order features such as mineral filled tension gashes in the outer arc which indicate convincingly that the fold is tectonic. Some structures deform older, tectonic structures or actually require the existence of other features such as a cleavage before they can form. These can also be confidently described as tectonic.

Some sedimentary structures, especially in relatively competent beds, can be particularly tenacious and may survive several episodes of deformation. If these structures are not recognised as sedimentary and are included in the structural analysis of an area, they may considerably distort the structural data and render the stereographic projection of the data more difficult to analyses. It has become apparent that in many areas that have experienced multiple deformation the earliest discernible folds were originally flat lying and isoclinal. Such fold forms are also typically found in slumped sediments. These slump folds may, of course, have experienced significant flattening as the result of burial, and their geometries and orientations been altered during subsequent deformations. It is interesting to speculate on how frequently the F_1 folds in metamorphic areas are of sedimentary rather than tectonic origin.

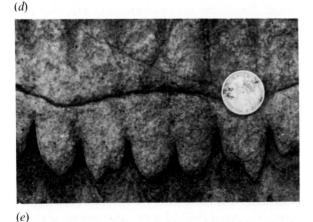

(d)

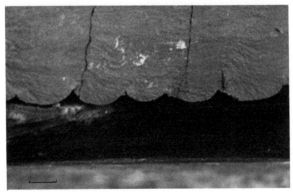

(e)

Structural analysis of an area that has experienced a single episode of deformation

Structural data collected in the field will include the dip and strike of bedding planes and cleavage planes, fold profile geometry (S, M, Z), the angle of plunge

(f)

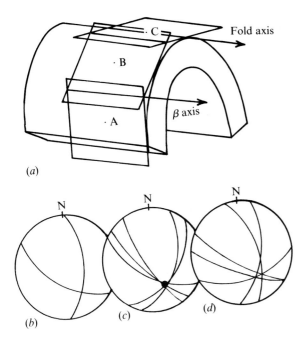

Fig. 18.31. (*a*) Bedding orientation taken around a fold at A, B and C. The line of intersection of any two bedding planes is parallel to the fold axis. (*b*) Stereographic projection of two bedding plane readings: the intersection shows the orientation of the fold axis. (*c*) β diagram for a perfectly cylindrical fold. (*d*) Typical β diagram.

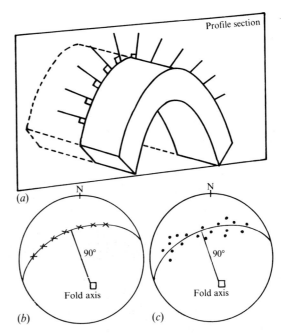

Fig. 18.32. (*a*) Poles to bedding on a cylindrical fold lie parallel to the profile section. (*b*) Fold axis declared by the pole to the plane (great circle) containing the poles to bedding. (*c*) Scatter of poles to bedding about a great circle.

and plunge-bearing of the fold axes, axial plane orientations of minor folds, the orientation and type of fractures and veins, the plunge of pitch of lineations; e.g. intersection lineations (bedding/cleavage), mineral stretching lineation and slickensides. The axes of boudins, mullions, rods, minor folds and cusp structures often form a coarse lineation which should be recorded, as well as data that enable one to infer the 'way-up' of the beds.

The structural data listed above are recorded as the orientation of either planes or lines, and the most convenient methods of studying the angular relationships between the various features is to plot them on a stereographic projection. The reader unfamiliar with the technique for plotting planes and lines on a stereographic projection is directed to the standard reference on this topic by Phillips (1971) and also to Ragan (1985) and Priest (1985). We will now consider the analysis of minor structure data when they are plotted on such a projection.

Bedding planes

When a bedding measurement is taken at any point A, B or C on a fold, the measurement taken is the orientation of the tangent to the fold at that point (Fig. 18.31(*a*)). It can be seen from this figure that the intersection of any two tangents (called a beta axis) is parallel to the fold axis. Consequently, if bedding data are plotted on a stereographic projection as great circles, the orientation of the fold axis (bearing and plunge) can be determined (Fig. 18.31(*b*)). The resulting diagram is known as a beta diagram. All great

circles which represent bedding planes on a cylindrical fold will pass through a common point which defines the axis (Fig. 18.31(*c*)). As most folds are noncylindrical, the bedding plane great circles do not generally intersect at a single point. Instead, the intersections cluster around an area (Fig. 18.31(*d*)), the centre of which can be taken to be the orientation of the fold axis. Because there is often a large number of bedding measurements to plot, the construction of a beta diagram can be very time consuming and may result in a very cluttered projection. A more convenient method of representing a large number of planar features on a stereographic plot is to plot the normals (poles) to the planes.

The poles to bedding in a cylindrical fold are all parallel to one plane, i.e. the fold profile section (Fig. 18.32(*a*)), and will all therefore lie on the same great circle when plotted on a stereographic projection (Fig. 18.32(*b*)). From the geometry of Fig. 18.32(*a*), it can be seen that the normal (pole) to the great circle containing the poles to bedding, is parallel to the *fold axis*. Because natural folds are generally non-cylindrical, the stereographic projection of poles to bedding usually shows a scatter of poles about a greater circle (Fig. 18.32(*c*)).

The *interlimb angle* and the *style* of the major fold can be determined from the distribution of poles to bedding along the great circle. The more open the fold, the less spread out the poles will be (Fig. 18.33(*a*), (*b*) and (*c*)). The interlimb angle (α) can be determined by rotating the stereographic projection until the best fit great circle of the poles to

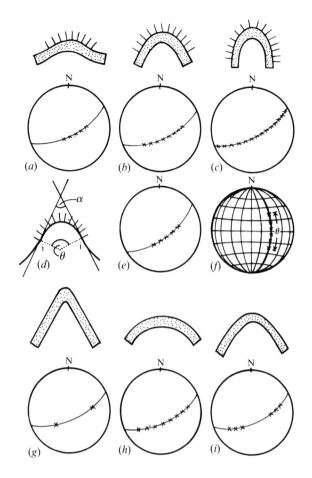

Fig. 18.33. The interlimb angle and style of a fold can be determined from the distribution of poles to bedding. See text for discussion.

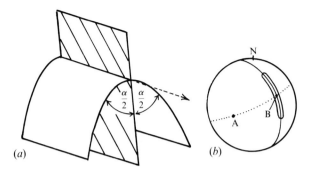

Fig. 18.34. Determination of fold axial plane orientation from a stereographic plot of poles to bedding. See text.

bedding are not spread along a great circle but, because the fold has a chevron geometry, are clustered around two points, then there will be some ambiguity when determining the inter-limb angle. For example, the two chevron folds shown in Fig. 18.35(a) have identical pole to bedding distributions. In order to distinguish between these two possibilities, additional data are required e.g. the orientation of the axial plane which may be declared by either an axial plane cleavage or by direct observation of a minor fold.

To summarise, it can be seen that even though major folds may be too large and/or too poorly exposed to be observed directly in the field, by plotting poles to bedding on a stereographic projection, it is possible to determine the orientation of the fold's axis and axial plane as well as indicating the style of the fold and its interlimb angle. However, it is not possible to distinguish between an antiform and a synform from a plot of poles to bedding. For example, the two

bedding lies over one of the great circles of the stereographic net (Fig. 18.33(e) and (f)). The small circles of the net intersect the great circle at two degree intervals. The angular spread of the poles, θ, can be read off (Fig. 18.33(f)). As can be seen from Fig. 18.33(d), the interlimb angle α is 180° minus θ. The distribution of poles to bedding on the great circle is also effected by the style, or angularity of the fold. For example, if the fold is a chevron fold, the poles will cluster about two points on the great circle (Fig. 18.33(g)). If the fold is rounded, the poles will be evenly distributed along the great circle (Fig. 18.33(h)). Fold geometries intermediate between these two will have pole distributions of the type shown in (Fig. 18.33(i)). Finally, the orientation of the axial plane of the major folds can be determined from the plot of poles to bedding. The axial plane of a fold is parallel to the plane which bisects the interlimb angle and which contains the fold hinge (Fig. 18.34(a)). The great circle representing the axial plane must, therefore, contain the pole to the great circle of poles to bedding (i.e. the fold axis A in Fig. 18.34(b)) and the mid-point of the spread of poles to bedding (B, Fig. 18.34(b)).

It should be pointed out that if the poles to

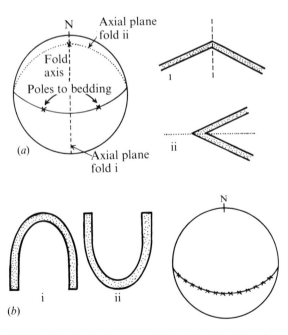

Fig. 18.35. Ambiguities in the interpretation of bedding data when plotted on a stereographic projection. The data from folds (i) and (ii) in (a) will be identical, as will those for (i) and (ii) in (b).

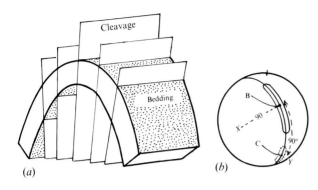

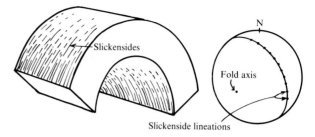

(a)   (b)

Fig. 18.36. (a) The geometric relationship between bedding and cleavage. (b) The relationship between poles to bedding and poles to cleavage when plotted on a stereographic projection.

Fig. 18.37. Spread of slickenside lineations formed during flexural slip folding. If these form normal to the fold hinge they will lie on the same great circle as the poles to bedding and cleavage.

folds shown in Fig. 18.35(b) have identical geometries and will have identical pole to bedding plots. This ambiguity can generally be easily resolved by looking at the spatial distribution of the data on the field map.

Cleavage

Because cleavage which results from the same tectonic event is generally statistically parallel to the axial plane of a fold, poles to cleavage will lie on the same great circle as the poles to the bedding and, ideally, will plot 90° from the mid-point (B) of the bedding pole scatter (point C of Fig. 18.36). Because cleavage refracts as it passes through layers of differing competences and because it generally fans about the hinge plane, the poles to cleavage will not generally plot at a single point but will be scattered along the great circle about a point (C, Fig. 18.36(b)). It should be noted, that if the cleavage refracts or fans, the cleavage planes from different parts of the fold, or from different lithologies, will have different dips. When these cleavage planes are plotted as planes on a stereographic projection, they will intersect, and this intersection will declare the orientation of the fold axis.

Lineations

Most lineations generated during a particular deformation will cluster about a point when plotted on a stereographic projection. For example, the lineations formed by the intersection of bedding and cleavage (Fig. 18.3(c)), minor fold axes (Fig. 18.2), and some boudin axes (Fig. 18.4(a)) will all cluster around the axis of the major fold. Mineral lineations may also be parallel to the fold axis. Even if they are oblique or at right angles to the fold axis they will still cluster about some point on the projection. If the lineations mentioned above do not plot around a point but are spread across the projection, this generally indicates that more than one episode of deformation has affected the area. The lineations belong to an early deformation and have been deformed by a later deformation.

Slickenside lineations formed on bedding planes during folding do *not* cluster around a point but are spread, often along the great circle containing poles to bedding, if the slickensides are at right angles to the fold hinge (Fig. 18.37). Therefore a spread of slickenside lineations does not necessarily indicate that more than one deformation has occurred.

The detailed mapping and analysis of minor structures in an area enables one to elucidate the *geometry* and *symmetry* of the tectonic fabric(s) (either planar, linear or mixed fabrics) and the major structures of that area. Structures (and fabrics) may have orthorhombic, monoclinic or triclinic symmetry (Fig. 18.38) and structural symmetry commonly varies in different parts of a major structure. For example, small monoclinic, parasitic folds may occur on the limbs of a major orthorhombic fold, or orthorhombic parasitic folds may form in the hinge zone of a larger monoclinic fold. Clearly, the symmetry shown by minor structures in isolated exposures does not always reflect the symmetry of the major structure.

Having determined the geometry and symmetry of the tectonic structure and the fabric of an area, the question arises as to what can be deduced about the tectonic processes and stress states that gave rise to them. The major contribution of Bruno Sander (1930, 1948–50) to structural geology was his attempt to determine the kinematics (displacements, including strain) and dynamics (stresses) associated with the formation of a structure or fabric, from its symmetry. Siddans (1972) points out that the essence of Sander's method for study of tectonite fabrics (i.e. fabrics induced by tectonism) is the analysis of geometric relations of all measurable elements in a rock; for example, crystallographic orientations of individual grains, grain-shapes and the arrangement of particular kinds of grains into some geometric configuration. From the geometric analyses, a kinematic interpretation is made in terms of translation, rotation and strains, i.e. Sander's *Bewegungsbild* or 'movement picture'. Dynamic interpretation is then made in terms of body forces and surface forces. However, it should be noted that these steps are made with increasing uncertainty. In deriving the kinematic from the geometric analysis, the key factor advocated by Sander is

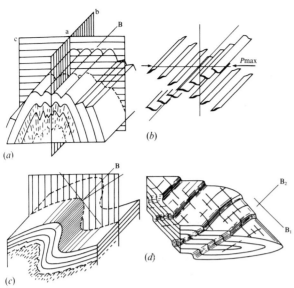

Fig. 18.38. Examples illustrating the various forms of structural symmetry. (*a*), (*b*) Orthorhombic; (*c*) Monoclinic; (*d*) Triclinic symmetry. The orthogonal axes a, b and c are those used by Sander & Schmidegg (1926) to describe the geometry of symmetrical cylindrical folds. b coincides with the hinge line of the folds (sometimes referred to as *B*), c is at right-angles to the axial plane and a is at right-angles to b and c. (After Wilson, 1961.)

symmetry; the concept being that the symmetry of the 'movement picture' is reflected in the symmetry revealed in the geometric analysis. The dynamic analysis follows from assumed constitutive laws (stress–strain relationships).

Sander used symmetry classes and orthogonal *fabric axes a, b and c*, to emphasise the dominant structural features of the rocks. Thus, 'spherical fabrics' have randomly orientated fabric elements; a 'planar fabric' is the ab-plane, with any conspicuous linear element in the plane defining the b axis. Planar fabrics may be of four types: 'axial', 'orthorhombic', 'monoclinic' or 'triclinic', according to the relationship of the various fabric–element sub-fabrics (Fig. 18.39). Sander also used orthogonal *kinematic axes, a, b, c* which are only meaningful in simple shear (Fig. 18.40) and *orthogonal axes a, b, c* to describe fold geometry, which are only meaningful for cylindrical folds (Fig. 18.41). As Siddans points out . . . 'the presentation of these sets of orthogonal axes, identically labelled yet quite independent in their meaning, provided a situation fraught with confusion and with considerable scope for abuse. Kinematic axes were equated with finite strain ellipsoid axes, previously denoted by a, b, c (Heim, 1921), a totally invalid practice in most geological deformation (see Fig. 6.45 in Ramsay, 1967); and ill-conceived references to 'movement', 'flow' and 'tectonic transport' based on fabric and fold axes, and thought to define the kinematic axes in situations where they are quite indefinable, proliferated in the literature'.

It is apparent from the above discussion, that

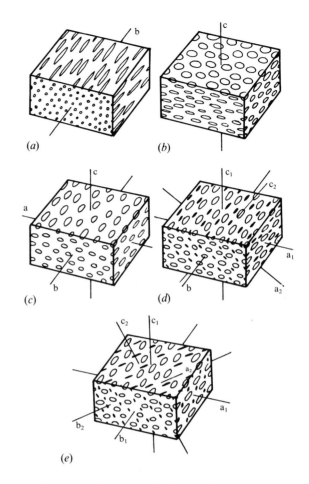

Fig. 18.39. Relationship between *fabric axes* a, b and c and symmetry. (*a*) and (*b*) Axial fabrics. (*c*) Orthorhombic fabric. (*d*) Monoclinic fabric. (*e*) Triclinic fabric. (After Turner & Weiss, 1963.)

the steps leading from a geometric analysis to a kinematic interpretation and then to a dynamic interpretation should be taken with great care. It is generally incorrect to assume that Sander's fabric (i.e. geometric) axes a, b, c coincide with the maximum, intermediate and minimum extension directions of the strain ellipsoid and that these coincide in orientation with the minimum, intermediate and maximum com-

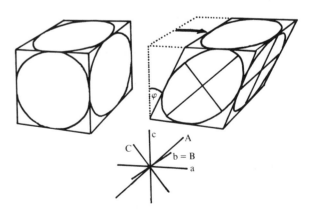

Fig. 18.40. Simple shear deformation showing the *kinematic* a, b and c *axes* and the axes of the finite strain ellipsoid. (After Siddans, 1972.)

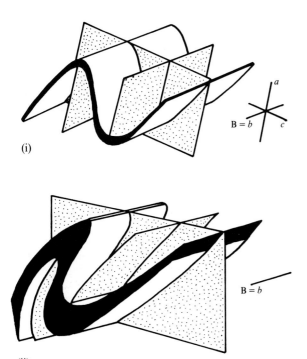

(i)

(ii)

Fig. 18.41. (i) Cylindrical folds referred to orthogonal axes, a, b and c. The b axis is parallel to the fold axis. The ac plane is the symmetry plane normal to the fold axis. (After Siddans, 1972.) (ii) The a and c axes are only uniquely defined when the axial surface of the fold is planar. (After Whitten, 1966.)

pression directions of the stress ellipsoid. Nevertheless, it is not surprising that with the advent of Sander's ideas, many geologists enthusiastically attempted to deduce the 'movement picture' associated with the structures they were studying.

Other workers were more circumspect. Wilson (1951), for example, studied the minor structures in two areas (the A'Mhoine region in north Sutherland, Scotland and the region round Tintagel, Cornwall) where the tectonic 'movement picture' was known to be controlled by a major thrust. He was able to demonstrate that the majority of the minor structures were related to the major structures and that the symmetry of the structures were compatible with the 'movements' that were thought to have occurred. Idealised diagrams showing the grouping of the different varieties of structures observed in the two areas are shown in Fig. 18.42. The structural synthesis of the Tintagel area shown in this figure is, however, oversimplified for it has been subsequently recognised that the area has been subjected to multiple phases of deformation.

Multiple (superimposed) deformation

So far, it has been assumed that the rocks of the region being studied have been subjected to only one episode of deformation. Let us now consider the structural geometries that result from the superposition of two

(or more) deformations which may be approximately coeval, or result from deformations widely spaced in time. Work in the 1950s and 1960s in the Scottish Highlands, where more than one episode of deformation has occurred, focused attention on the difficulty of structural analysis in such rocks, and resulted in the recognition of features that could be used to determine that an area had been subjected to more than one phase of deformation. Three main ways of recognising and analysing multiple deformation are:

(i)   the observation of deformed minor structures in the field,
(ii) the recognition of *interference patterns* caused by an early set of folds being refolded by a later set, and
(iii) the study of the stereographic projections of the structural data.

These will be discussed in turn.

1. The analysis of deformed minor structures

Generally, the first indication of multiple deformation comes from the direct observation of superposed minor structures; for example, a late cleavage cutting across earlier structures or the folding of a cleavage, lineation or minor fold (Fig. 17.20).

Folded lineations

In areas of multiple deformation, early lineations are often deformed by later folding (Fig. 18.43). *In theory*, it is possible to determine whether this later fold formed by tangential, longitudinal strain, flexural flow or heterogeneous simple shear by measuring the orientation of the folded lineation at various points around the fold (points a–g on Fig. 18.43(*c*)) and plotting the data on a stereographic projection (Fig. 18.43(*d*)). (See, for example, Wilson 1961, Ramsay 1967.) The lineations folded around a flexural flow fold, for instance, will plot on a small circle (Fig. 18.43(*d*)). This is because there is no strain on the bedding plane in such folds (see Chap. 10). Consequently, the original angle (θ) between the fold hinge and the lineation does not change during folding. In order to determine whether the plot of a deformed lineation lies on a small circle it is necessary to rotate the fold axis (and with it the lineation data) until it is horizontal. When this has been done, the loci of the folded lineation data can be compared directly with the small circles on the stereographic net, by rotating the fold axis until it lies N–S.

If the lineation is folded round a tangential longitudinal strain fold, then unless the bedding plane on which the lineation is measured coincides with the neutral surface, (on which there is no strain), the deformed lineation will *not* plot on a small circle. If the bedding plane lies above the neutral surface, extension parallel to the layer boundary (in the direct-

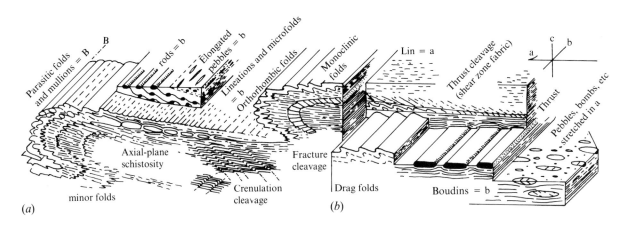

Fig. 18.42. The relationships of minor structures to the large-scale structures. (*a*) Recumbent fold structures, based on the general structure found at A'Mhoine, N. Sutherland, Scotland. (*b*) An area of flat thrust tectonics, based on the Tintagel area, Cornwall, England. (After Wilson, 1961.)

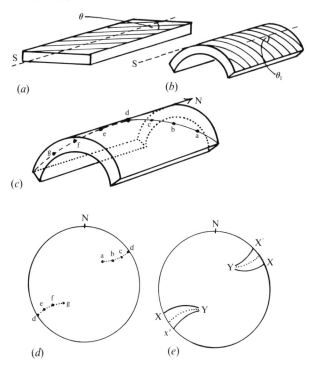

Fig. 18.43. (*a*) An unfolded and (*b*) a folded lineation. The orientation of the lineation at various points around the fold (*c*) depends upon the mechanism of folding. The stereographic projection of a lineation folded around a flexural flow fold and a tangential longitudinal strain fold are shown in (*d*) and (*e*) respectively.

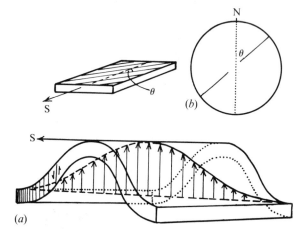

Fig. 18.44. The stereographic projection of a lineation folded by heterogeneous simple shear. The shear is parallel to the axial plane of the fold. See text.

Fig. 18.43(*e*). If the lineation fell on the neutral surface, it would plot along the small circle shown dotted in this figure.

It will be recalled from Chap. 10 that the profile geometry of tangential, longitudinal strain and flexural flow folds is that of a *parallel* fold. If folding took place by heterogeneous simple shear (Chap. 1) the resulting folds will be *similar* and the deformed lineation will lie in a plane (Fig. 18.44(*a*)) and plot on a great circle (solid lines Fig. 18.44(*b*)).

In practice, the 'ideal' plots of lineations around tangential, longitudinal strain and flexural flow folds shown in Fig. 18.43(*d*) and (*e*) are generally modified by post-buckle flattening. This changes the profile geometry of the folds from that of a *parallel* towards that of a *similar* fold; and a corresponding change in the deformed lineation plots from those shown in Fig. 18.43(*d*) and (*e*) to that shown in Fig. 18.44(*b*) occurs. Consequently, *in practice*, such studies are unlikely to provide definitive information regarding the types of deformation the rocks have experienced.

ion normal to the fold hinge) will have occurred, and the angle between the lineation and the fold axis will increase. This increase is maximum at the hinge and reduces to zero at the inflexion point. Consequently a deformed lineation on such a bedding plane will plot along lines XY in Fig. 18.43(*e*). If the lineation lies on a bedding plane below the neutral surface, then contraction parallel to the layer boundary (normal to the fold hinge) will occur and the angle between the lineation and the fold axis will be reduced. Again, the reduction will be a maximum at the fold hinge and will reduce to zero at the inflexion point. The deformed lineation will therefore plot along lines X'Y in

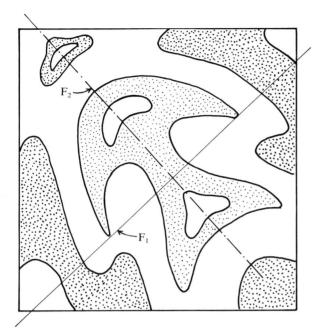

Fig. 18.45. F$_1$ and F$_2$ fold axis orientation determined from a Type 2 interference pattern. See text.

2. Superimposed folding

As discussed in Chap. 10 (Figs. 10.29 to 10.34), the occurrence of two superposed folds can often be recognised from the outcrop pattern (interference pattern) that appears on an erosion surface cutting through these folds. The geometry of the pattern depends upon the relative orientation of the axial planes and axes of the two sets of folds and upon the orientation of the erosion surface (Figs. 10.28 and 10.32). For example it can be seen from Fig. 10.30 that the interference pattern shown in Fig. 18.45 is the result of an early, almost recumbent fold being refolded by an upright fold whose axis is at 90° to that of the early fold. In addition, it is possible to locate the axial traces of both the early and late folds (Fig. 18.45). The tips of the crescent, or 'mushroom', shapes are F$_1$ fold closures; and a line joining the tips has the same bearing as the F$_1$ fold axis. The axis of bilateral symmetry of the interference pattern corresponds to the trace of the F$_2$ hinge surface.

Interference patterns may develop on all scales from the very small, observable only in thin section, to patterns that cover several hundred square kilometres.

3. Stereographic projections

Multiple deformation of a region can also be recognised by the study of the stereographic projection of structural data that has been gathered from the region. Features such as minor fold axes, boudin lengths, intersection and stretching lineations etc., which, in areas affected by only one deformation, plot around a point on a stereographic projection, are

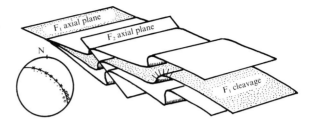

Fig. 18.46. Coaxial deformation causing the poles to the early cleavage to spread along the great circle containing poles to bedding. (Bedding, marked by crosses; first cleavage, solid dots; second cleavage, open dots.)

generally found to be spread over a wider area of the projection if they have been affected by a second deformation. Similarly, features such as poles to bedding, which, in areas affected by only one deformation, plot along a great circle, are generally found to be spread over a larger area when a second deformation has occurred. Any deviation from the single deformation data distribution on the stereographic projection mentioned above, and shown in Figs. 18.31 to 18.37, indicates that more than one deformation has occurred. The exception to this arises when the two deformations are coaxial (Type 3 interference pattern, Fig. 10.31). Because both early and late folds have the same axis, the poles to bedding will still plot on a great circle even after the second deformation (Fig. 18.46). Although coaxial folding, particularly of an early isoclinal fold, is more difficult to detect on the stereographic plot than other types of superposed folding, some features of the data distribution do differ from that associated with a single deformation. For example, any cleavage formed during the early folding will be folded by the second folding. Consequently, poles to this cleavage will not cluster around a point but will be spread along a great circle (the same great circle that contains the poles to bedding) (Fig. 18.46). Similarly, any lineation not parallel to the fold axis, and which was formed during the first deformation, will be spread by the second folding and will not plot around a point.

When two deformations have affected a region, it is sometimes possible to differentiate between structures associated with the first and second deformations from the distribution of data on a stereographic projection. For example, the plot of minor fold axes from an area may show a concentration of points around a point maxima in one part of the projection and a spread of points in another. The minor folds associated with the point maxima were formed during the second deformation, and those associated with the spread of points formed during the early deformation. It should be noted, however, that minor fold axes associated with a second deformation do not always cluster around a point maxima (Fig. 18.47). In the example of superposed folding shown in this figure,

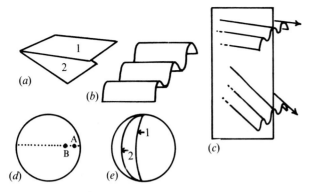

Fig. 18.47. The affect of an early fold (*a*) on the orientation of minor folds formed during a later deformation represented by the second folds (*b*). See text. (*c*) Second folds on limbs A and B of the first fold (*a*). (*d*) shows the F_2 axial plane (dotted). A and B are the F_2 minor fold axes from limbs 1 and 2 respectively (see (*a*)). Great circles containing poles to bedding from limbs 1 and 2 (after the second deformation) are shown in (*e*).

the first fold is an almost recumbent chevron fold (Fig. 18.47(*a*)) and the second folds are upright folds with axes at right angles to the first fold axis (Fig. 18.47(*b*)). The effect of superposing these folds is shown in Fig. 18.47(*c*). Although all the second folds have a common axial plane, their plunge will vary, depending upon which limb of the first fold they develop. Consequently, the axes of the F_2 minor folds will cluster around two points, A and B in Fig. 18.47(*d*), and the great circle containing these points represents the axial plane of the F_2 folds. Similarly, if the poles to bedding of the refolded fold of Fig. 18.47(*c*) are plotted on a stereographic projection, the data from one limb of the early fold will lie on one great circle and those from the other limb on another great circle (Fig. 18.47(*e*)). If the first fold had been rounded with the same interlimb angle as the chevron fold (Fig. 18.47(*a*)), then the poles to bedding after the formation of the second fold would be scattered between the two great circles of Fig. 18.47(*e*). Similarly, the axes of the second folds would be spread between the two points of Fig. 18.47(*d*).

Examples of structural analyses of areas of multiple deformation

The superposition of minor structures can occur either by the sequential development of structures with progressive deformation during a single major phase of deformation, or as the result of two or more separate tectonic events.

Major phases of deformation are sometimes associated with a 'group' of minor structures, some of which may be deformed by structures which develop slightly later, but which are related to the same deformation phase. If such 'groups' of structures can be recognised in the field, the major phases of tectonism that have affected an area can be discussed

without recourse to the listing of countless minor deformation episodes that characterise and confuse many structural descriptions of areas of multiple deformation.

Each group of structures can be considered separately, and a detailed picture of the major deformation phases deduced. In addition, the relative importance of each major phase can be determined by using criteria such as the regional extent and intensity of the deformation.

The recognition of groups of minor structures associated with a particular deformation is not always easy. Sometimes, folds of one specific deformation will have a particular style and orientation. For example, F_1 folds may be recumbent, rounded and isoclinal and F_2 folds upright, open and chevron. Unfortunately, however, as discussed in this chapter in the section on mineral lineations, the orientation and style of folds are not always reliable criteria for the grouping of folds into separate episodes of deformation. It is only by combining the detailed mapping of minor structures, interference patterns and the analysis of stereographic projections of the structural data, that the unravelling and elucidation of the phases of multiple deformation in an area can be successfully achieved.

We shall now illustrate how minor structures and deformed minor structures have been used to determine the major structures and tectonic history of two areas; one the south-west corner of Holy Isle around Rhoscolyn in North Wales and, the other the area around Bude on the north coast of Cornwall, S.W. England.

1. Analysis of the multiple deformation of the Rhoscolyn area, Holy Isle, North Wales

The geographical location and geological sketch map of the Rhoscolyn area are shown in Fig. 18.48. Four rock units have been recognised in the metasediments of the area, which are, in stratigraphic succession, the South Stack series, the Holyhead Quartzite formation, the Rhoscolyn formation and the New Harbour series. The South Stack formation is a mixed sequence of schistose grits, greywackes and sandstones with interbedded shales. Individual lithological units are commonly about one metre thick. The Holyhead Quartzite formation is a pure white, recrystallised sandstone made up of massive sandstone beds with some pelitic intercalations. The Rhoscolyn formation is a sequence of schistose greywackes, which are essentialy similar to the South Stack beds but contain fewer thick pelitic horizons and the New Harbour series is a thick sequence of pelitic rocks. Sedimentary structures indicate that the beds are the right way up. A simplified geological cross section through the south-western corner of Holy Isle is shown in Fig. 18.49. The section is dominated by the

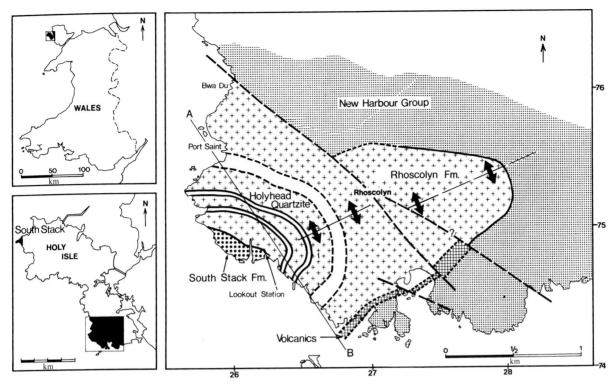

Fig. 18.48. Geographical location and geological sketch map of the Rhoscolyn area, Holy Isle, N. Wales.

overturned Rhoscolyn anticline, the steeper limb of which faces to the south-east.

General discussion of the major and minor structures

Although the rocks of Rhoscolyn have been subjected to more than one phase of deformation, the rapid alternations of what were originally sandstones and shales can still be clearly seen, especially when the rocks are viewed from a distance of more than ten metres. Moreover, in the 'sandstones' (now quartzites), sedimentary structures (e.g. graded bedding and dewatering structures) are still preserved. However, in the pelitic horizons almost all trace of bedding has been lost and the planar fabrics in these horizons are transposition fabrics (cf. Chap. 17). At various localities evidence can be found for two, three or even more transpositions.

From the preceding paragraph, it may be inferred that in order to determine the tectonic history of an area, it is useful to consider, first the structures in the more massive, competent units which only responded to the major tectonic events and, secondly, to consider the more complex array of structures that is formed in the less competent, pelitic horizons, which developed structures in response to the slightest tectonic 'hiccup'.

Some of the minor structures found in the area of the Rhoscolyn anticline are shown in Fig. 18.49. These include minor folds, kink-bands, pressure solution cleavage, pinch-and-swell structures, intersection lineations, quartz veins and cusp structures. The

figure represents a summary of field observations from which the tectonic history of the area is to be determined.

Before discussing these structures in detail and considering their implications regarding the tectonics of the area, the reader may find it of interest to consider the logic used in arranging the structures chronologically and in sub-dividing them into three groups associated with three separate deformation phases.

There is no unique method of analysing the information presented by the minor structures of Fig. 18.49, or of synthesising the tectonic history of the area from these data: each geologist will have his own method. The arguments set out briefly below are those used by Cosgrove (1980).

The Holyhead Quartzite which is folded around the Rhoscolyn anticline contains three main cleavages. One cleavage is statistically parallel to the axial plane of the anticline; an earlier cleavage is folded around the anticline and a later cleavage cuts across the other two. These cleavages are all deflected by kink-bands which are developed sporadically in the area and represent a very weak, late deformation, D_4. From these observations, it can be concluded that at least three important phases of deformation D_1, D_2 and D_3 affected the area. The first cleavage, S_1, associated with the earliest deformation, D_1, is folded around the Rhoscolyn anticline. The second cleavage, S_2, associated with the second deformation, D_2, is axial planar to the Rhoscolyn anticline itself a D_2

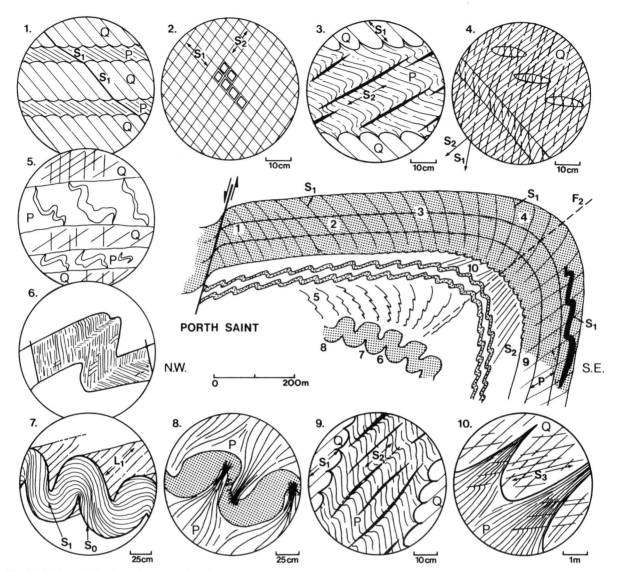

Fig. 18.49. A partially schematic cross section through the southwestern part of Holy Isle (section A–B, Fig. 18.48). Minor structures found at localities 1–10 are shown in the inserts 1–10 respectively (Q = quartzite; P = pelite). (After Cosgrove, 1980.)

structure, and the third deformation, D_3, deforms both these cleavages and produces a locally developed cleavage dipping very gently to the N.W., Fig. 18.49 (10).

Having established that at least three phases of deformation affected the area, it is useful to attempt to group the various minor structures with the appropriate deformation. One simple method that can be used to help sub-divide these structures into D_1, D_2 and D_3 groups is that of determining whether or not they have been deformed.

The easiest group to identify are the structures (folds and kink-bands) associated with the third deformation. These are undeformed and are only sporadically developed, and it is clear that the D_3 deformation was much less intense than the D_1 and D_2 events both of which produced more pervasive deformations. Because the D_3 deformation is only sporadically developed, many D_2 structures are also undeformed.

However, the direction of principal compression during the D_3 deformation was sub-vertical, whereas the direction of principal compression during the D_2 deformation was sub-horizontal. Consequently, the orientation of axial planes of F_2 and F_3 folds are generally sufficiently different to enable F_3 folds and undeformed F_2 folds to be easily distinguished in the field.

There are numerous minor folds associated with the Rhoscolyn anticline, and their geometry 'S', 'M' or 'Z' depends upon their position on the anticline (Fig. 18.49). Some of these small-scale D_2 folds in the quartzite layers of the South Stack beds fold pre-existing pinch-and-swell structures, lineations and a cleavage. These pinch-and-swell structures, lineations and the cleavage must, therefore, be associated with the D_1 deformation.

The above is a brief outline of the method used to identify the separate deformation episodes affect-

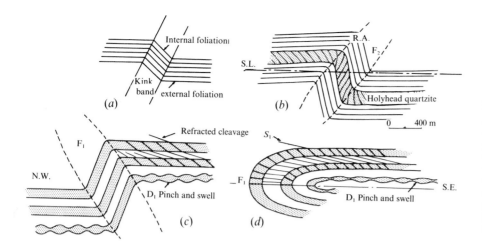

Fig. 18.50. (*a*) Sketch of a kink-band defining terms used in the text. (*b*) The Rhoscolyn anticline shown as part of a large kink structure. (*d*) A recumbent fold (a D₁ structure) whose geometry is compatible with field observations of the D₁ cleavage and pinch-and-swell structures. At present, there are insufficient field data available to enable the scale of the F₁ fold to be determined or to ascertain whether or not the structure is a recumbent fold, as indicated in (*d*), or a kink structure with the kink-band boundary dipping to the south-east (*c*).

ing the area and to associate each minor structure with the appropriate episode.

Once all the minor structures have been assigned to one of the major deformation phases, the groups of structures related to each deformation phase can be considered in turn, in order to obtain a more detailed understanding of the different phases.

Structures associated with the first deformation phase D_1

One of the most pervasive minor structures on Holy Isle is the early cleavage S_1 which is folded around the D_2 Rhoscolyn anticline (Fig. 18.49). This cleavage is refracted as it passes through the boundaries between the different types of sediments (Fig. 18.49 (1)), and the morphology and mode of formation of this cleavage depend upon the lithology of the rock in which it developed.

In the massive Holyhead Quartzite, the S_1 fabric is a closely spaced, pressure solution cleavage, whereas in pelitic horizons of the South Stack Formation and the Rhoscolyn Formation it is developed as a crenulation cleavage. The change in orientation of the S_1 cleavage around the Rhoscolyn anticline is shown in Fig. 18.49.

In order to determine the original orientation of the S_1 cleavage which will be statistically parallel to the original orientation of any F_1 fold axial planes, the anticline must be 'unfolded'. To do this, it is necessary to know by what mechanism of folding the Rhoscolyn anticline formed. Although this is not known with certainty, it is possible, as is indicated below, to argue on the basis of field observation and experimental evidence that the anticline may be a large kink structure (Fig. 18.50(*b*)). Experimental and theoretical work on kink-bands (see Chaps. 10 and 13) show that they often develop in materials with a high mechanical anisotropy (such as the metasediments of the Rhoscolyn area), and that layer rotation is confined primarily to the kink-band, with very little rotation of the external foliation (Fig. 18.50(*a*)). It can therefore be argued that the gently dipping limb of the anticline

has probably undergone very little rotation during the D_2 folding, whereas the sub-vertical limb has undergone considerable rotation.

Although the S_1 cleavage in the pelitic horizons within the Holyhead quartzite and between the quartzite layers of both the Rhoscolyn Formation and the South Stack Formation has been mainly transposed into an S_2 crenulation cleavage, it can still be demonstrated in the gently dipping limb of the Rhoscolyn anticline that the original orientation of the cleavage was either subhorizontal or very gently dipping to the south-east.

The quartzite layers in the South Stack beds underlying the main quartzite are folded by D_2 folds. However, it can still be clearly seen that the S_1 cleavage in these quartzite layers originally dipped gently to the south east (Fig. 18.49 (7)).

Many of the thin quartzite layers in these beds developed pinch-and-swell structures during the D_1 deformation. If folds formed during the D_1 deformation, then the pinch-and-swell structures may have developed at a late stage in the fold amplification when the limbs were at high angles to the direction of maximum compression (Fig. 16.46). (An example of these pinch-and-swell structures which have subsequently been folded is shown in Fig. 18.49 (8).) The influence of these pinch-and-swell structures on the refolding of the quartzite beds during the second deformation will be discussed later.

Before the occurrence of the D_2 deformation, a set of quartz veins developed parallel, or sub-parallel, to the early cleavage, S_1 in the pelitic horizons (Fig. 18.49 (5)). These veins are extremely useful strain markers and often enable the D_1 and D_2 deformations to be easily separated.

These field data regarding the orientation of the early cleavage in the unrotated limb of the Rhoscolyn anticline indicate that any early folds were flat-lying and faced towards the north-west. It is interesting to note that, although the S_1 cleavage is so well developed, F_1 folds are nevertheless remarkably elusive and difficult to identify, and the scale of these early folds is

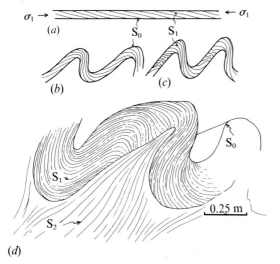

Fig. 18.51. (*a*), (*b*) and (*c*). The buckling of two mechanically active foliations, the bedding S_0 and the early cleavage S_1 during the second deformation. The two foliations do not buckle synchronously: S_0 buckles first. (*d*) A line diagram of Fig. 18.52 showing the folding of two fabrics S_0 and S_1.

not known. Several interpretations are possible. These include (i) a large recumbent nappe (Fig. 18.50(*d*)) of the type postulated by Greenly (1919 and 1930), (ii) a series of small recumbent folds, or (iii) a series of reverse kink-like structures (Fig. 18.50(*c*)) with axial planes dipping to the south-east.

Structures associated with the second deformation D_2

Folds The area under consideration is dominated by structures formed during the D_2 deformation. F_2 folds are developed on several scales, ranging from the Rhoscolyn anticline with a wavelength of 2 km controlled by the buckling or kinking of the Holyhead quartzite, to minor folds with wavelengths between 5 mm and 5 cm formed by the buckling of the early cleavage, S_1.

The quartzite layers in the South Stack Formation in the core of the Rhoscolyn anticline developed buckles during the D_2 deformation with wavelengths ranging between 0.5 and 5 m (Fig. 18.49 (7 and 8)). Many of these quartzite layers had, as already mentioned, developed pinch-and-swell structures during the D_1 deformation. These structures are commonly found to govern the wavelength of the F_2 folds. (A discussion of the folding of pinch-and-swell structures is given in Chap. 16).

Prior to the onset of the second deformation, the quartzite layers of the South Stack and Rhoscolyn Formations contained two mechanically active fabrics, i.e. the bedding S_0 and the cleavage S_1 (Fig. 18.51(*a*)). Both these fabrics buckled during the second deformation but not synchronously. Initially, the quartzite layers buckled to develop folds with a low wavelength/thickness ratio ($L/a \approx 7$) (Fig. 18.51(*b*)); the wavelength sometimes being con-

trolled by pre-existing pinch-and-swell structures. As these folds amplified, the efficiency of the buckling of S_0 as a mechanism of layer shortening decreased and the folds eventually 'locked up'. At this point in the deformation, the early cleavage began to fold wherever it was suitably oriented with respect to the maximum principal compression. Suitable orientations of the cleavage occurred on alternate limbs, and the type of folds that formed were 'internal buckles' (Fig. 18.51(*c*)) (see Chaps. 12 and 13). A line drawing of the photograph shown in Fig. 18.52 can be seen in Fig. 18.51(*d*). In the limbs in which the cleavage does not buckle, the cleavage is at a high angle (often 90°) to the direction of principal compression associated with the second deformation. In these limbs, the intensity of the early cleavage is increased by the process of pressure solution.

Cleavages The intensity of the cleavage associated with the second deformation is variable. Like the early cleavage (S_1), the type of cleavage developed during the second deformation depends upon the lithology of the rock in which it forms.

In the Holyhead Quartzite and the quartzite members of the Rhoscolyn and South Stack Formations, both the S_1 and S_2 cleavages are generally pressure solution cleavages. (Locally, however, where the S_1 pressure solution cleavage is very well developed, the S_2 cleavage may develop as a crenulation cleavage.) The development of the second pressure solution cleavage does not usually destroy the first. However, on the overturned limb of the Rhoscolyn anticline, the S_1 and S_2 pressure solution cleavages are parallel. Consequently, on this limb, the formation of the S_2 cleavage simply intensifies the S_1 cleavage and only one cleavage is apparent. Where S_1 and S_2 are not parallel, e.g. around the hinge and on the gently dipping limb of the Rhoscolyn anticline (Fig. 18.49), both cleavages can be clearly seen. They slice the Holyhead Quartzite into lozenge-shaped rods and constitute a pencil cleavage (Fig. 18.49 (2)).

The cleavage that develops in the pelitic horizons during the second deformation is a crenulation cleavage, the development of which is restricted to rocks that possess a good mechanically active fabric or layering.

An example of a transposition fabric (see Chap. 17) from the New Harbour beds which overlie the Rhoscolyn formation is shown in Fig. 17.19. The remnants of fold hinges can still be seen in some of the quartz-rich layers. At this locality the fabric is not folded; at others it is gently folded (Fig. 17.20) and at others it is completely transposed into a new fabric.

The S_2 cleavage in the pelitic horizons within and around the Holyhead Quartzite is a crenulation cleavage (Fig. 18.53(*a*)) formed by the microfolding of the S_1 cleavage, which is commonly itself a crenulation

Fig. 18.52. Folds in a quartzite layer in the South Stack formation at Rhoscolyn, formed during the second deformation D_2. Both the bedding (S_0) and the early cleavage (S_1) are buckled. A line diagram of this figure is shown in Fig. 18.51(d).

cleavage. It can be seen, that both the fabric being folded to form the S_2 cleavage in this figure and the fabric that was folded to form the transposition fabric shown in Fig. 17.19, are made up of quartz-rich and mica-rich layers, i.e. are themselves transposition fabrics. It is interesting to speculate as to what 'fabric' was folded to produce these fabrics. They could have been formed either by the microfolding of a sedimentary fabric in the shales or by the microfolding of a pre-S_1 tectonic fabric.

Various stages in the formation of the S_2 crenulation cleavage can be found (Figs. 17.19, 17.20 and 18.53) and these throw considerable light on the origins of some of the lineations which occur throughout the area.

Lineations A labelled line diagram of Fig. 18.53(b) is given in Fig. 18.54, and shows the relationship between the S_1 crenulation cleavage, the S_2 crenulation cleavage and the lineation L_2 on the S_2 cleavage plane. The lineation is caused by the intersection of the S_1 crenulation cleavage (a transposition fabric of alternating mica-rich and quartz-rich layers) with the S_2 crenulation cleavage, i.e. a cleavage/cleavage intersection lineation. This lineation appears as numerous long stripes of quartz separated by thin stripes of mica and often looks deceptively like the quartz crystal fibre slickensides that are often found on fold limbs.

The L_1 intersection lineations can be conveniently divided into two types, those formed on the bedding by S_1 intersecting S_0 (Fig. 18.55) and those formed on the early cleavage by S_0 intersecting S_1; this latter lineation being formed in a similar way to the lineation on the S_2 crenulation cleavage planes shown in Figs. 18.53(b) and 18.54. The former lineation (i.e. the early cleavage/bedding intersection lineation) indicates the orientation of the first fold axes. The angle between this lineation and the F_2 fold axes is, therefore, the angle between the F_1 and F_2 fold axes (approximately 15–20°). The D_1 and D_2 deformations seem approximately coaxial. It must be remembered that, unless the F_2 folds in the quartzite layers are formed by flexural flow folding and have not been affected by post buckle flattening, this angle would have changed during the F_2 folding. If the folding was by tangential, longitudinal strain, the angle (α) between the F_2 fold hinge and the cleavage bedding intersection on the outer arc of the fold (Fig. 18.55) would increase, (see Fig. 18.43 and discussion). Conversely, post-buckle flattening would cause this angle to decrease. However, the folds (Fig. 18.55) have approximately parallel geometries, indicating that very little post-buckle flattening occurred. It can therefore be argued that the angle between the F_1 and F_2 axes never exceeded 15–20° (i.e. the deformation was approximately coaxial).

(a)

(b)

Fig. 18.53. Folds in the Rhoscolyn Formation photographed normal to the fold profile (a) and oblique to the profile (b). An S_2 crenulation cleavage has been formed by the buckling of an S_1 crenulation cleavage. Note on (b) the lineation caused by the intersection of the S_1 crenulation cleavage and the S_2 crenulation cleavage. Fig. 18.54 is a labelled diagram of Fig. 18.53(b). The diameter of the coin is 2.3 cm.

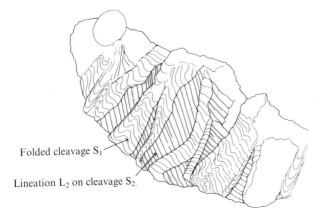

Folded cleavage S_1

Lineation L_2 on cleavage S_2.

Fig. 18.54. Line diagram of Fig. 1853(b), showing a lineation (L_2) caused by the intersection of the early crenulation cleavage S_1, on the later crenulation cleavage S_2.

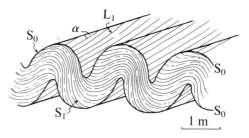

Fig. 18.55. An intersection lineation (L_1) caused by the intersection of an S_1 pressure solution cleavage with the bedding S_0 deformed by the F_2 buckles in the South Stack Formation

The effect of the second deformation D_2 on the quartz veins developed in the pelitic horizons during the early deformation D_1 (Fig. 18.49 (5)) can be used to demonstrate that the formation of the F_2 multilayer folds in the South Stack Formation and the Rhoscolyn Formation was by the process of flexural slip. This, of course, does not preclude the possibility that the individual quartzite layers may have buckled by a process that approximates to tangential, longitudinal strain folding. The veins are parallel to or sub-parallel to the early cleavage and were deformed during the formation of the F_2 folds. The effect of the D_2 deformation is found to depend upon which limb of an F_2 fold the veins occur. On the north-west dipping limbs, a dextral shear couple was generated in any pelitic layer sandwiched between quartzite layers. Consequently, the veins in the pelitic horizons are folded. On the south-east dipping limb, the shear couple is sinistral and the veins are generally not folded (Fig. 18.56). This difference of behaviour is shown schematically in Fig. 18.57.

Structures associated with the deformation phase D_3

Evidence for a deformation phase D_3 is provided by minor folds which affect both the S_1 and S_2 cleavages. In the pelitic horizons, F_3 folds can be found folding the S_2 cleavage (Fig. 18.49 (10)). They are approximately co-axial with the F_1 and F_2 folds but have sub-horizontal axial planes which dip gently towards the north-west. The maximum principal compression during the D_3 deformation was, therefore, steeply inclined. Occasionally, a crenulation cleavage is found in association with these F_3 folds which is restricted to just one limb (Fig. 18.58(c)). This could be interpreted as being the result of a later deformation, but more probably indicates that the D_3 deformation was non-coaxial (i.e. rotational, see Chap. 1, Fig. 1.15), which resulted in the relative rotation of the stress field with respect to the folds and pre-existing cleavages. This is shown schematically in Fig. 18.58(a) to (c). It can be seen that, during the folding of the S_2 cleavage, one limb rotates into a position approximately parallel to the principal compression and the other into a position approximately normal to it. The S_2 cleavage on the former limb is appropriately oriented for the

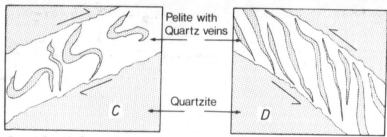

Fig. 18.56. F_2 folds in the South Stack Formation at South Stack. During the development of these folds, quartz veins in the pelitic layers were folded in the north-west dipping limbs (*C*) and underwent a body rotation in the south-east dipping limbs (*D*). See also Fig. 18.57.

formation of a crenulation cleavage by the microfolding of the S_2 cleavage, whilst the same cleavage on the later limb will be at a high angle to the principal compression and will not be folded. Instead, this limb will be flattened and the S_2 cleavage intensified. The result is the production of a coarse banding, in which bands or layers with a well developed, cross cutting crenulation cleavage alternate with layers which have a well developed planar fabric parallel to the layering, but not cross cutting crenulation cleavage. This banding is a form of metamorphic differentiation or transposition (see Chap. 17), each layer corresponding to one limb of an F_3 fold (Fig. 18.58).

Having considered the effect of the D_3 deformation on the S_2 cleavage in the pelitic horizons, let us now consider its effect on the psammites. It was noted earlier that, during the D_2 deformation, the rocks possessed two mechanically active fabrics, the

bedding and the S_1 cleavage (Figs. 18.51 and 18.52). In this example, it is apparent that of the two the bedding was initially the dominant mechanically active fabric during D_2. During D_3, the two dominant rock fabrics are bedding and the S_2 cleavage. However, during this deformation, probably because of the relative orientation of the bedding and cleavage with respect to the maximum principal compression, it is the S_2 cleavage that buckles and the bedding which behaves passively. This can be clearly seen in the examples shown in Fig. 18.58(*d*) and (*e*).

A weak D_4 deformation produced localised kinking of the various cleavages. These kink-bands commonly occur in conjugate sets and, occasionally, one of the conjugate bands may develop as an array of en echelon bands (Fig. 18.49 (4)).

It will be recalled that the quartz veins, associated with the deformation D_1, developed in the pelitic

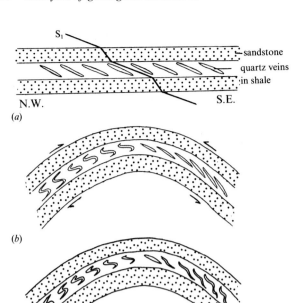

(a)

(b)

(c)

Fig. 18.57. Various stages in the formation of the folds shown in Fig. 18.56. (*a*) Horizontal beds after the deformation (*D*₁). The quartz veins are restricted mainly to the pelitic horizons and are sub-parallel to the cleavage (*S*₁). (*b*) Beds folded into F₂ folds. (*c*) The only visible effect of the D₃ deformation on these F₂ folds is the low amplitude buckling of quartz veins in the S.E. dipping limb. See also Fig. 18.56, limb D.

horizons and were parallel or sub-parallel to the S_1 cleavage which originally dipped very gently to the south-east. As a consequence of the D_2 folding, some of these quartz veins were folded (Figs. 18.56, limb C and 18.57(*b*)) and others (i.e. those in the south-east dipping limbs of the F_2 folds (Figs. 18.56, limb D, and 18.57(*b*)) underwent a body rotation which resulted in their dip being increased. The vertical principal compression associated with D_3 tended to buckle these steeply dipping veins. However, the F_3 folds in the veins are generally only poorly developed (e.g. Figs. 18.56, limb D, and 18.57(*c*)).

2. Analysis of multiple deformation of an area on the North coast of Devon and Cornwall

Let us now consider briefly an example where Pumpelly's Rule, that minor structures reflect major structures, has proved invaluable in determining both the geometry *and* possible mode of formation of major structures. The rocks in which the structures are exposed are to be seen in the magnificent cliffs, between Hartland Quay and Tintagel on the north coast of the south-western peninsular of England (Fig. 18.59), which provide a cross section through the Carboniferous and Devonian sedimentary and volcanic rocks which were deformed during the Variscan orogeny. From the outcrop pattern of the major rock units, it can be inferred that the present distribution of lithologies is the result of a large synclinorium, trending E–W, its axis passing through Bude.

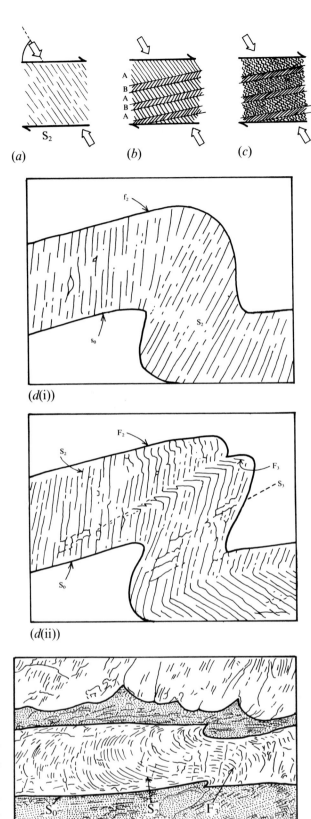

(a)   (b)   (c)

(*d*(i))

(*d*(ii))

(*e*)   0   15 cm

Fig. 18.58. (*a*), (*b*) and (*c*) show the effect of a rotational D₃ deformation (represented by the shear couple) on the S₂ cleavage. Chevron folds form which subsequently develop a crenulation cleavage on only one of their limbs (A). (*d*) and (*e*) show folding of a second cleavage during the D₃ deformation, where the bedding behaves passively.

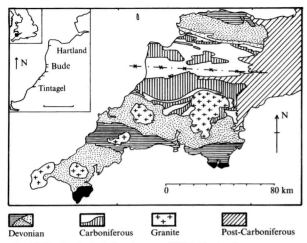

Fig. 18.59. Geological sketch map of S.W. England.

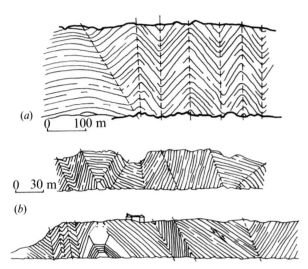

Fig. 18.60. Chevron folds, kink-bands and box folds in the Carboniferous turbidites between Hartland Quay and Bude, Cornwall, England. (*a*) Hartland Quay (*b*) Northcott Mouth, 3 km north of Bude.

The mesoscopic folds developed in the Carboniferous turbidites between Harland Point and Bude are upright chevron folds, box folds and kink-bands (Figs. 18.60 and 13.3) formed as a result of compression, sub-parallel to bedding. An abrupt change in the orientation of the folds occurs just north of Millook Haven, where the folds are recumbent (Fig. 18.61).

A possible explanation for this abrupt change in fold orientation can be inferred from the minor structures developed in the finely bedded, volcanoclastic sediments at Boscastle. In this locality, the folds have the same geometries as those developed in the turbidites, further north, but, because of the finer bedding, are developed on a much smaller scale. Consequently, their temporal and spatial relationships are more easily determined.

The earliest recognisable deformation at Boscastle produced a series of recumbent, isoclinal folds with a well-developed axial plane cleavage. This marked, horizontal fabric, made up of both bedding and cleavage, was deformed by a later deformation into minor box folds and kink-bands (Figs. 18.62 and 18.63(*e*)), the width of which ranges between a few centimetres and a few metres. The kink-band width is largely governed by the thickness of local arenaceous bands.

It is clear that the folds of this second phase of deformation did not all form at the same time, as can be illustrated by considering the folds shown in Fig. 18.63(*e*), which is a line diagram of Fig. 18.62. These figures show a relatively large kink-band within which there are smaller chevron folds and an axial plane crenulation cleavage. The crenulation cleavage and the axial planes of the kink-bands and the chevron folds are all parallel. The kink-band is one of a set developed at this locality but, significantly, the conjugate to this set is not present. We can establish the relative age of the kink-band and the chevron folds within it by considering experimental work on the formation of kink-bands. It will be recalled (Chap. 13), that when the principal compression acts oblique to the layering or fabric only one set of kink-bands develops (Fig. 18.63(*b*)); the conjugate set is suppressed. From the geometry of the chevron folds and crenulation cleavage within the kink-band (Figs. 18.62 and 18.63(*e*)) we can infer that the principal compression responsible for their formation was sub-parallel to the layering within the kink-band. It can be seen from Fig. 18.63(*b*), that the principal compression that generated the kink-band is, in fact, conveniently oriented in this direction. We see, therefore, that the kink-band predates the chevron folds, although both are formed by the same compression.

Outside the kink-band, a variety of folds (box folds and upright chevron folds), with geometries similar to the folds formed at Bude and further north, have developed in the horizontal layering (Fig. 18.63(*e*)). The symmetry of these structures indicates that the compression responsible for their formation acted sub-parallel to the layering. They may predate or postdate the formation of the kink-band and chevron folds, discussed above or, alternately, may be associated with the same deformation if, during the deformation, the principal compression rotated.

The chevron folds formed within the kink-band of Fig. 18.62 may be considered as analogous to the far larger recumbent folds exposed in the cliffs at Millook (Fig. 18.61). If one continues the analogy, we can suggest that the folds at Millook also developed after a major kink-band formed in the turbidites. We need not, however, suggest that the principal compression was inclined to the bedding. It could be argued that, because of the larger scale of the kink-band at Millook, during or after its formation, the body-weight of the sub-vertical beds within it would be enough to cause the development of recumbent chevron folds.

Fig. 18.61. Recumbent chevron folds in turbidites at Millook, England.

We use this example of structural analysis as an illustration of Pumpelly's rule which indicates that the geometry of minor structures often reflects that of the major structure with which they are associated. In addition we have argued that the stages in the development of the minor structures at Boscastle (Fig. 18.63), deduced from field observations, probably apply to the formation of the larger scale recumbent folds at Millook (Fig. 18.61).

In the area of study, observations of structures that can be seen in their entirety, (i.e. structures ranging in size from hand specimen to cliff face), indicate that the deformation was dominated by folding. However it was noted earlier (Chap. 15 and Figs. 15.9, 15.10 and 15.11) that pre-folding thrusts are common in the relatively competent beds. In addition it is possible to recognise larger thrusts which are not restricted to one or two competent layers but which climb through numerous beds. These thrusts are often folded and are less apparent than the folds. Nevertheless they probably played an important role in the horizontal shortening of the multilayer. It has been suggested that on a regional scale the area is dominated by major thrusts (see e.g. Coward & Ries (eds.) 1986). We see therefore that the conclusion drawn from structures directly observable in the field, regarding the dominant mechanism of horizontal shortening in the area, may be misleading and that, compared to thrusting, the spectacular arrays of folds (Figs. 13.3 and 15.11) may in fact be of only secondary importance.

This brief discussion illustrates how minor structures, in addition to indicating the geometry and orientation of major structures which are too large to be visible in outcrop, can also be used to indicate the way in which major structures may have formed.

In addition to the relationship between major and minor structures, discussed in this chapter, there is also a close relationship between many of the minor structures, including kink-bands, symmetrical and asymmetrical folds, boudins, pinch-and-swell structures, shear zones and faults. This interrelationship can be most clearly seen by considering the *fundamental instability* associated with the formation of these structures and how this instability may be organised and arranged within a material to produce the various minor structures under consideration. (Chap. 13, Figs. 13.49 and 13.50).

Commentary

This final chapter has mainly been concerned with the traditional methods of structural analysis, i.e. it has dealt with the methods of determining the geometries

Fig. 18.62. Kink-band containing smaller-scale chevron folds in the Devonian metasediments of Boscastle, Cornwall, England.

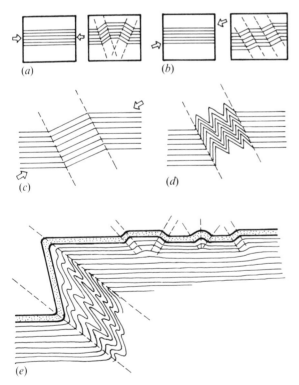

Fig. 18.63. Possible mechanism for the formation of the kink-band and chevron folds shown in Fig. 18.62. (*a*) The formation of conjugate, reverse kink-bands in a layered material compressed parallel to the layering. (*b*) Compression oblique to the layering resulting in the formation of only one of the conjugate set of kink-bands. (*c*) Continued action of the oblique compression causes folding of the foliation within the kink-band (*d*). (*e*) Line diagram of Fig. 18.62.

of major structures and of ascertaining their spatial and temporal relationships. For decades, this has been the primary interest of structural geologists mapping orogenic areas. This traditional form of structural analysis, dealing with the geometry of structures and the changes of these geometries with time, is of great importance. However, such an analysis should never be considered an end in itself. Rather it should be considered as the basis on which new and exciting analyses, based on mechanical principles outlined and exemplified at length in the rest of this text, are made and which eventually will lead us to greater understanding of the processes of deformation and the structural and tectonic evolution of the earth's crust.

References

Bruehl, H. (1969). Boudinage in den Ardennen und in der Nordeifel als Ergebnis der inneren Deformation. *Geol. Mitt.*, **8**, 263–308.

Carreras, J., Cobbold, P.R., Ramsay, J.G. and White, S.H. (eds.), (1980). Shear zones in rocks *J. Struct. Geol.* (special issue), **2**, 1/2.

Cobbold, P.R. & Watkinson, A.J. (1981). Bending anisotropy: a mechanical constaint on the orientation of fold axes in an anisotropic medium. *Tectonophysics*, **72**, T1–T10.

Collinson, J.D. & Thompson, D.B. (1982). *Sedimentary structures.* pp. 1–194. George Allen & Unwin.

Cosgrove, J.W. (1980). The tectonic implications of some small-scale structures in the Mona Complex of Holy Isle, North Wales. *J. Struct. Geol.*, **2**, 4, 383–96.

Coward, M.P. & Ries, A.C. (eds.), (1986). Collision tectonics. *Geol. Soc. Lond.*, Special Publ. No. 19. Blackwell Scientific Publications.

Den Tex, E. (1954). Stereographic distinction of linear and planar structures from apparent lineations in random exposure planes. *J. Geol. Soc. Australia*, **1**, 55–66.

Fletcher, R.C. (1982). Analysis of the flow in layered fluids, at small, but finite, amplitude with application to mullion structures. *Tectonophysics*, **81**, 51–66.

Greenly, E. (1919). *The geology of Anglesey*. Mem. Geol. Surv. U.K. (2 vols.).

(1930). Foliation and its relation to folding in the Mona Complex at Rhoscolyn. *Q. J. Geol. Soc. Lond.*, **86**, 169–90.

Heim, A. (1921). *Geologie der Schweiz*, vol. 2. Leipzig: Tauschnitz. 476 pp.

Hobbs, B.E., Means, W.D. & Williams, P.F. (1976). *An outline of structural geology*. Int. edn, Wiley.

Paterson, M.S. & Weiss, L.E. (1961). Symmetry concepts in the structural analysis of deformed rocks. *Geol. Soc. Am. Bull.*, **72**, 841–82.

Phillips, F.C. (1971). *The use of stereographic projection in structural geology*. Third edn, Edward Arnold.

Poirier, J.P. (1985). *Creep of crystals*. Cambridge University Press.

Priest, S. (1985). *Hemispherical projection methods in rock mechanics*. London: George Allen & Unwin.

Pumpelly, R., Wolff, J.E. & Dale, T.N. (1894). Geology of the Green Mountains. *U.S. Geol. Surv. Mem.*, **23**, 1–157.

Ragan, D.M. (1985). *Structural geology. An introduction to geometric techniques*. 3rd edn, 393 pp. Wiley.

Ramsay, J.G. (1960). The deformation of early linear structures in areas of repeated folding. *J. Geol.*, **68**, 75–93.

(1967). *Folding and fracturing of rocks*. 568 pp. New York: McGraw–Hill.

Ramsay, J.G. & Graham, R.H. (1970). Strain variations in shear belts. *Can. J. Earth Sci.*, **7**, 786–813.

Sander, B. (1930).*Gefugekunde der Gesteine*. Vienna: Springer. 352 pp.

(1948–50). *Einfuhrung in die Gefugekunde der geologischen Koerper*. 1. 1948; 215 pp. 2. 1950, 409 pp. Vienna: Springer.

Sander, B. & Schmidegg, O. (1926). Zur petrographisch-tektonischen Analyse. *III Jahrb. Geol. Bundesanst. Wien*, **76**, 323–8.

Sibson, R.H. (1977). Fault rock and fault mechanism. *J. Geol. Soc.*, **133**, 191–213.

Siddans, A.W.B. (1972). Slaty cleavage – a review of research since 1815. *Earth Sci. Rev.*, **205–32**.

Smith, R.B. (1975). Unified theory of the onset of folding, boudinage and mullion structure. *Geol. Soc. Am. Bull.*, **86**, 1601–9.

(1977). Formation of folds, boudinage and mullions in non-Newtonian materials. *Geol. Soc. Am. Bull.*, **88**, 312–20.

(1979). The folding of a strongly non-Newtonian layer. *Am. J. Sci.*, **279**, 272–87.

Sorensen, K. (1983). Growth and dynamics of the Nordre Stromfjord shear zone. *J. Geophys. Res.*, **88**, B4, 1419–37.

Turner, F.J. & Weiss, L.E. (1963). *Structural analysis of metamorphic tectonites*. New York: McGraw–Hill.

van Hise, C.R. (1896). *Principles of North American Pre-Cambrian Geology*. pp. 1–630. 16th Ann. Rep. U.S.G.S.

Watkinson, A.J. (1975). Multilayer folds initiated in bulk plane strain, with the axis of no change perpendicular to the layering. *Tectonophysics*, **28**, T7–T11.

White, S.H., Bretan, P.G. & Rutter, E.H. (1986). Fault-zone reactivation: kinematics and mechanics. *Phil. Trans. Roy. Soc. Lond.*, **A317**, 81–97.

Whitten, E.H.T. (1966). Structural geology of folded rocks. Chicago, Ill: Rand McNally.

Wilson, G. (1951). The tectonics of the Tintagel area, N. Cornwall. *Q. J. Geol. Soc. Lond.*, **106**, 393–432.

(1953). Mullion and rodding structures in the Moine Series of Scotland. *Proc. Geol. Assoc. Lond.*, **64**, 118–51.

(1961). The tectonic significance of small-scale structures and their importance to the geologist in the field. *A. Soc. Geol. Belg.*, **84**, 423–548.

(1982). *Introduction to small-scale geological structures*. George Allen & Unwin.

Index